# Rotifera IX

Developments in Hydrobiology 153

*Series editor*
H. J. Dumont

# Rotifera IX

Proceedings of the IXth International Rotifer Symposium, held in Khon Kaen, Thailand, 16–23 January 2000

*Edited by*

L. Sanoamuang,[1] H. Segers,[2] R.J. Shiel,[3] & R.D. Gulati[4]

[1] *Khon Kaen University, Thailand*
[2] *Ghent University, Belgium*
[3] *Murray–Darling Freshwater Research Centre, Albury, Australia*
[4] *NIOO/Centre of Limnology, Nieuwersluis, The Netherlands*

*Reprinted from Hydrobiologia, volumes 446/447 (2001)*

Springer-Science+Business Media, B.V.

**Library of Congress Cataloging-in-Publication Data**

Rotifera IX / edited by L. Sanoamuang ... [et al.].
p. cm. -- (Developments in hydrobiology ; 153)
Proceedings of the IXth International Rotifer Symposium held at Khon Kaen, Thailand, during 16-23 January 2000.
ISBN 978-94-010-3820-1 ISBN 978-94-010-0756-6 (eBook)
DOI 10.1007/978-94-010-0756-6
1. Rotifera--Congresses. I. Title: Rotifera nine. II. International Rotifer Symposium (9th : 2000 : Khon Kaen, Thailand) III. Sanoamuang, L. (La-orsri) IV. Series.

QL391.R8 R68 2001
592'.52--dc21

2001020395

ISBN 978-94-010-3820-1

---


Originally published by Kluwer Academic Publishers in 2001

# TABLE OF CONTENTS

## Part II. Reproduction and Life History



## Part III. Trophic Relations and Interspecific Interactions



## Part IV. Ecosystem Ecology

## Part VI. Genetics and Evolution



## Part VII. Aquaculture and Ecotoxicology

Hydrobiologia 446/447: xi–xiii, 2001.
L. Sanoamuang, H. Segers, R.J. Shiel & R.D. Gulati (eds), Rotifera IX.



# Preface

The IXth International Rotifer Symposium was held at Khon Kaen, Thailand, during 16–23 January 2000, in the Hotel Sofitel Raja Orchid. It was the first rotifer symposium held in Asia. One hundred and thirteen participants from 25 countries attended this triennial meeting. The Symposium was organised by La-orsri Sanoamuang and hosted by the Department of Biology, Faculty of Science, Khon Kaen University (KKU). It marked the celebration of three important events: the 72$^{nd}$ birthday anniversary of His Majesty King Bhumibol Adulyadej of Thailand, the 36$^{th}$ Anniversary of Khon Kaen University, and the end of the second Millennium. The symposium venue provided an excellent opportunity to the community of rotifer researchers to acquaint themselves with numerous scientists and research students of the Khon Kaen University.

After the opening ceremony Bob Wallace paid homage to Tommy Edmondson, who sadly had passed away just a few days prior to the beginning of the symposium. Birger Pejler, to whom the previous symposium proceedings had been dedicated, also had left us in the months preceding the symposium.[1] During the preparation of the present proceedings, we heard of the untimely demise of Tom Frost. In the name of the entire rotifer community, we extend our profound sympathy to the bereaved families of these colleagues, who were very dear to us all.

The scientific program followed the tradition of previous symposia – 8 invited review lectures, 43 oral and 43 poster presentations. The major theme of the Symposium was "How far have we come"? This chronicled the importance of a sense of history, especially that relating to the development of our knowledge of rotifers since these symposia were initiated in the seventies in Lunz, Austria. The contributions were grouped into the following sessions: autecology and population ecology; phylogeny and evolution; physiology, biochemistry and population genetics; aquaculture and ecotoxicology; taxonomy and zoogeography; biomanipulation and rotifers in man-made habitats; trophic interactions and behaviour; and community ecology. During one of the two evening sessions, a workshop on "Quantitative sampling of littoral and periphytic rotifers" was arranged and on the other, a video presentation was given.

Social activities began with the Welcome Party that was held on Monday evening at the Art and Culture Museum of KKU, which was marked by a consummate Thai atmosphere. Participants were invited to experience and join in the typical Thai dancing (with varying degrees of success!). On Wednesday afternoon, the participants enjoyed a cruise around the Ubolrat reservoir, and had a 'walk and climb' exercise to the large Buddha statue atop a hill near the reservoir, and ended with the BBQ party in the evening. On Friday evening, the Farewell ceremony took place at the Golden Jubilee Convention Hall of KKU with classical and serene Thai music and ambience. Accompanying guests made a series of day trips to places of cultural and historical interest in northeast Thailand. Finally, a two-day post-conference excursion to Phimai Historic Park, Khao Yai National Park and the ancient Thai capital city "Ayutthaya" was arranged for interested scientists to get an impression of the geography, history and natural beauty of Thailand.

As with all but one of the previous symposia, Kluwer Academic Publishers and Dr Henri J. Dumont (Editor-in-chief, *Hydrobiologia*) have agreed to publish the proceedings of this symposium as a special volume of *Hydrobiologia* as well as in the series *Developments in Hydrobiology*. Each of the manuscripts published in this volume has undergone at least two rigorous peer reviews and a major or minor revision, before its acceptance and final editing by us. For the final round of editing we gathered at Khon Kaen in September 2000, La-orsri again playing a (wonderful) host. We are pleased to present a volume that contains numerous contributions spanning a wide range of disciplines relating to the science of rotifers.

We wish to thank the 'local organizing committee' for its invaluable help during the symposium and during our final editorial work. We are grateful to Bob Wallace and Liz Wurdak for sharing with us their experience from the VIII$^{th}$rotifer symposium. We also wish to acknowledge Khon Kaen University, INVE Asia Services Limited and the Tourist Authority of Thailand, Northeastern Office for financial support.

*The Editors*
LA-ORSRI SANOAMUANG, HENDRIK SEGERS
RUSSELL J. SHIEL, RAMESH D. GULATI

[1] For an obituary to Professor Birger Pejler, see *Limnologica* **30**: 233–234 (2000).

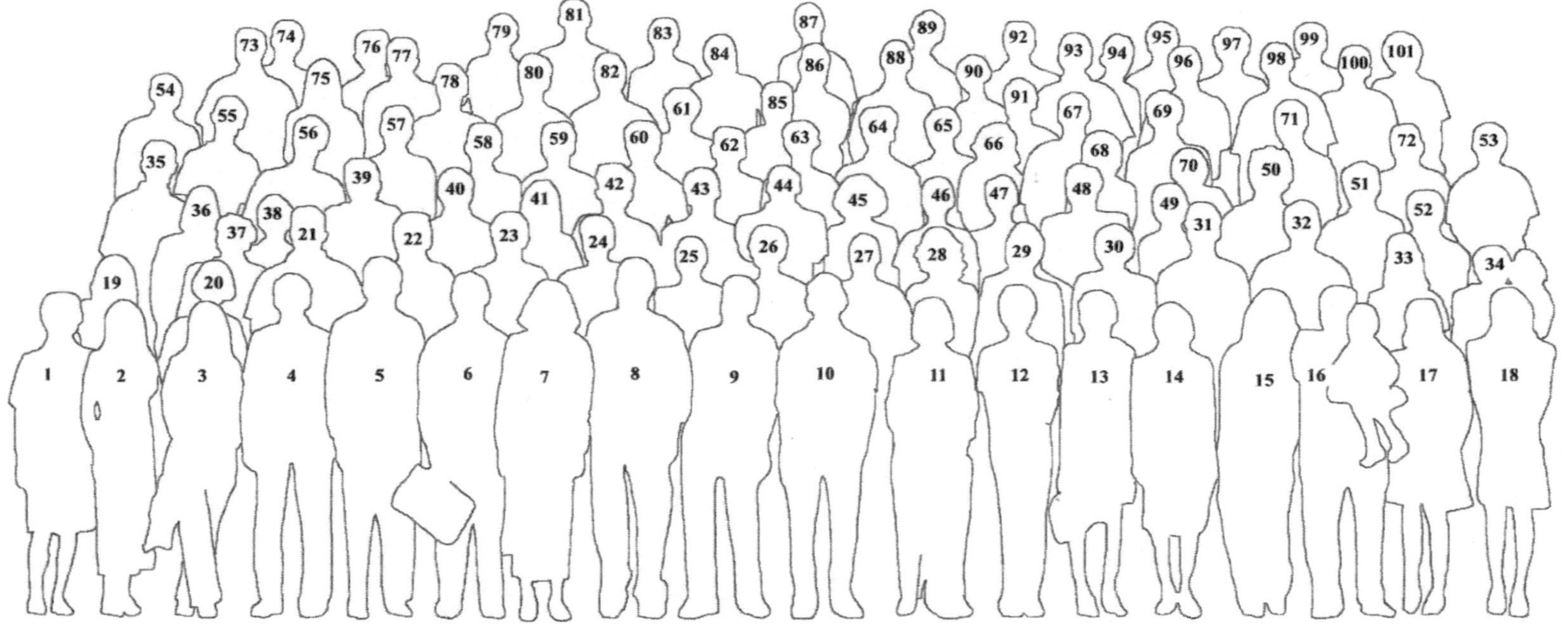

## Participants in the IXth International Rotifer Symposium

1. Tida Pechmanee
2. Raquel Ortells
3. Maria José Carmona
4. Cesar Alberto Velázquez-Rojas
5. B.K. Sharma
6. Mohamed M. Dorgham
7. La-orsri Sanoamuang
8. Russell J. Shiel
9. Ramesh D. Gulati
10. Ramakrishna T. Rao
11. Celia Joaquim-Justo
12. Melissa Anne Garcia
13. Jittra Teeramaethee
14. Nitaya Lauhachinda
15. Maria Guadalupe
16. Roberto Rico-Martínez
17. Sukontip Savatenalinton
18. Supenya Chittapun
19. Africa Gomez
20. Liliana Valle
21. Ignacio Alejandro Perez-Legaspi
22. Stanisław Radwan
23. Nobert Walz
24. Pornsilp Pholpunthin
25. Supawadee Chullasorn
26. Galina A. Galkovskaya
27. Mavit Assavaaree
28. Adriana Belem De Araujo
29. Thanathip Lamkom
30. Nuntaporn Charubhun
31. Claus-Peter Stelzer
32. Jun Zhu
33. Stephanie Hampton
34. S. Nandini
35. Brian J. Dingmann
36. Elizabeth J. Walsh
37. Esther Lubzens
38. Supiyanit Mypaa
39. Gustavo Emilio Santos-Medrano
40. Nukul Saengphan
41. Jin-qiu Wang
42. Kaseme Chetawan
43. Tatsuki Yoshinaga
44. Javier Armengol
45. Maria Sahuquillo
46. Manuela Caprioli
47. Nadia Santo
48. Jolanta Ejsmont-Karabin
49. Linda May
50. Tomonari Kotani
51. Li Zhou
52. S.S.S. Sarma
53. Hendrik Segers
54. Alexander K. Gorbunov
55. Laurent Viroux
56. Rainer Deneke
57. Peter Funch
58. Wei Bin Liu
59. Chun Shi
60. Miloslav Devetter
61. V. Ramasubramanian
62. Hachiro Hirata
63. Min-Min Jung
64. Heum Gi Park
65. Claudia Ricci
66. Natalia Kuczyńska-Kippen
67. Giulio Melone
68. Maria Rosa Miracle
69. Atsushi Hagiwara
70. Irena Bielańska-Grajner
71. Peter Starkweather
72. Parke Rublee
73. David Mark Welch
74. Udomsak Darumat
75. Jessica Mark Welch
76. Cally Gilbert
77. Willem H. De Smet
78. Ian C. Duggan
79. John J. Gilbert
80. Bennan Chen
81. Steven C. Fradkin
82. Manuel Yufera
83. Tania De Wolf
84. Eduardo Vicente
85. N. Munuswamy
86. Walter Kleinow
87. Thomas Schroeder
88. Wilko H. Ahlrichs
89. Guntram Weithoff
90. Sara Lapesa
91. Robert Lee Wallace
92. Martin Vinther Sorensen
93. Jorge Ciros-Perez
94. Manuel Serra
95. Terry Snell
96. Charles E. King
97. Yilong Xi
98. Taavi Virro
99. Jia-xin Yang
100. Jim Green
101. Rosalind M. Pontin

Hydrobiologia **446/447:** xv–xxi, 2001.
L. Sanoamuang, H. Segers, R.J. Shiel & R.D. Gulati (eds), Rotifera IX.



# In memoriam: W.T. Edmondson (1916–2000)

*Dr W.T. 'Tommy' Edmondson died in Seattle, Washington on Monday, January 10, 2000, just a few days before the start of the IXth International Rotifer Symposium. He is survived by Yvette Edmondson, his wife and coworker of over 50 years*

Following the receipt of his Ph.D. from Yale University, Edmondson's professional life included work at Woods Hole Oceanographic Institution (during WWII) and Harvard University (1946–1949), but the majority of his career was spent on the faculty of the Department of Zoology at the University of Washington in Seattle (1949–1986). Even after retirement from active teaching, Edmondson continued his dynamic research program on Lake Washington. While his contributions to the science of ecology in general and limnology in particular had a major impact on both fields, he retained an abiding interest in rotifers throughout his life while simultaneously viewing them with scientific rigor and imagination. His studies on rotifers are marked with an uncanny ability to tease out important aspects of their life histories in remarkably resourceful ways.

*Figure 1.* Edmondson as a young scientist. (18 years old, Osborn Laboratory, Yale University, New Haven; 29 May 1934; photographer, B.J. Kaston; photograph courtesy of Yvette Edmondson.)

## The research of W.T. Edmondson: the rotifer perspective

Edmondson was well known for his long-term, limnological studies of Lake Washington. These began in the mid 1950s and occupied most of his attention from the 1960s until his death. He also was widely respected for his pioneering and seminal contributions to the analysis of zooplankton population dynamics, which included his 1960 publication on the "Reproductive rate of rotifers in natural populations," his 1959 edition of Ward and Whipple's *Fresh-Water Biology*, and for his many papers on the ecology, distribution and taxonomy of rotifers, most of which were published between 1934 and 1968.

Edmondson, known as Tommy to his friends and colleagues, developed his affection for rotifers at a very early age (Fig. 1). He published papers on their taxonomy and distribution while still in high school in New Haven and as an undergraduate at Yale University. It was during his high school days that he was introduced to Frank Myers, who invited him to visit his home in Ventnor, New Jersey, to study the rotifers of the ponds and bogs of the nearby Pine Barrens. Because of his relationship with G. Evelyn Hutchinson, which continued through his undergraduate studies at Yale and later, Edmondson was able to work on the rotifer fauna collected from some very distant places, including North India and Hispaniola. During a year at the University of Wisconsin in Madison (1938–1939), he took Chancey Juday's limnology course and did research at the University's field station (Trout Lake Station in Vilas County, Wisconsin). While in Wisconsin, he met and later married Yvette Hardman who was working on a Ph.D. dissertation on the bacteria of bottom muds in Lake Mendota. He returned to Yale in 1939 to finish his Ph.D. with Hutchinson. Edmondson's dissertation on the ecology of sessile rotifers – published as two installments, one in 1944 and the other in 1945 – was extremely comprehensive and innovative. Ranging from descriptive observations to analysis of the dynamics of natural populations, this work represented a fundamental advance in the field of aquatic ecology.

His classic and highly original experiments on the growth, mortality and reproductive rates of the sessile rotifer *Floscularia conifera* are unrivaled to this day. He devised a method to determine the growth and survival rates of individuals in nature; then, by relating mortality rates to age, he was able to construct a complete life table for this natural rotifer population – a remarkable feat that has not been duplicated for any other group of freshwater microfauna. This study was extended by comparing age-specific mortality, and egg production, in both solitary and colonial individuals. He found that colonial animals grew at the same rate as solitary ones but survived and reproduced better.

Edmondson made a major breakthrough in analyzing the dynamics of natural zooplankton populations when he developed the 'egg ratio' method for estimating birth rate. His thinking about this technique actually began with his study of *Floscularia conifera* reproduction and was mentioned in his 1945 monograph on sessile rotifers. By knowing the ratio of eggs – individuals in a sample, and the development time of the eggs, finite and instantaneous birth rates could be calculated. Then from the instantaneous birth rate and population growth rate (determined from two sample dates), the instantaneous death rate could be calculated ($d=b-r$). This permitted investigators to relate birth rates to ecological conditions, such as temperature and food availability, and death rates to predator abundance. Edmondson's major work along these lines was with rotifers and described in his 1960 and 1965 papers. In the latter, he was able to show, for example, that the birth rate of *Polyarthra vulgaris* was strongly related to the abundance of *Cryptomonas*. These papers demonstrated, for the first time, how critical population parameters could be estimated from natural populations, and how ecological factors affecting them could be assessed.

## The research of W.T. Edmondson: the limnological perspective

When Edmondson joined the zoology faculty at the University of Washington in 1949, Lake Washington had a Secchi disk reading of about 4 m and the lake appeared to be in good health. However, in 1955 the cyanobacterium *Oscillatoria rubescens* (known to produce nuisance blooms in European lakes) was found in Lake Washington. This discovery stimulated Edmondson to initiate an intensive investigation of lake conditions in 1957 and 1958. The source of the eutrophication was no mystery; the Seattle area was experiencing a population increase of *Homo sapiens* and 10 waste-treatment facilities were discharging their nutrient-rich, primary-treated sewage directly into the lake. By the end of the decade thick, foul-smelling algal mats were accumulating along the shoreline, resulting in the closure of popular beaches and restricting the public's recreational use of the lake.

In 1958, the greater metropolitan Seattle area passed legislation authorizing diversion of the sewage inputs. Edmondson was the major scientific advisor to the commission, which had sought the authorization and funding from the public for the recovery plan. In that position he frequently provided information to the print media and appeared on television programs to supply facts and predictions about the lake, being careful to avoid partisan activity in the political campaign. Ultimately he was credited with a major role in passage of the legislation to fund the clean-up efforts. The Chairman of the Commission described him as their 'data bank' and his predictions that the deterioration could be reversed carried great weight with the public. In particular, Edmondson pointed to phosphates, rather than nitrates (a prediction that was later confirmed), as the most likely culprit for the eutrophication. He also predicted that if nothing were done to reverse this pattern, the water quality would continue to deteriorate at an increasing rate. Secchi disk readings reached a minimum depth of only 1 m in 1963 when the diversion of sewage from Lake Washington to Puget Sound began. However, by 1968, when the sewage diversion was completed, Secchi disk readings had increased to 4 m and *Oscillatoria* populations were decreasing, finally disappearing in 1976. That same year, Edmondson submitted his final report to the Environmental Protection Agency predicting that the lake would maintain a trophic equilibrium with its stable nutrient supply. However, as the density of the abundant crustacean, *Neomysis mercedis,* was reduced by predation from the longfin smelt (*Spirinchus thaleichthys*), there was a massive population increase of *Daphnia*. As a result, the Secchi disk transparency doubled. The quirks of this dynamic lake simply added to the pleasure Edmondson so obviously found in his research.

*Figure 2.* Edmondson's busy office in Kincaid Hall. (University of Washington, Seattle; 1982. photograph courtesy of Aida Infante.)

Unraveling the nexus of trophic interactions among the algae, zooplankton and fish was to occupy most of Edmondson's research for the remainder of his life. Lake Washington has become **the** model for reversal of cultural eutrophication and a tremendous body of basic scientific knowledge has resulted from the studies that were set up to investigate the dynamics of the recovery process (Fig. 2).

In 1973, Edmondson was elected to the U. S. National Academy of Sciences and received its Garner Cottrell Award for Environmental Quality. The years surrounding his 'retirement' in 1986 were both scientifically productive and filled with the sort of recognition that generally comes only late in the career to ecologists dealing with long-term environmental change. His numerous awards include the Eminent Ecologist Award (Ecological Society of America), and the Einar Naumann-August Thienemann Medal (International Society for Theoretical and Applied Limnology) (Schelske, 1990). In 1987, a Resolution of Respect was passed by the State of Washington House of Representatives, and in 1990, he received the G. Evelyn Hutchinson Medal of the American Society of Limnology and Oceanography. In his acceptance remarks for the latter award he stated:

> "Limnologists work on fascinating systems and find out things that interest them intensely. While that is personally gratifying and worth much effort, it is even better when other people find the results interesting too. Sometimes that reaction is expressed by the award of a prize or by election to something. It is a pleasure to have a signal that somebody else understands what you did and finds it good. But the prize or the election is secondary; they are symptoms of accomplishment, not accomplishment itself. The real thing is the excitement felt by the investigator, and the fact that the excitement has been transmitted to colleagues."

## Personal characteristics

While Edmondson influenced and inspired countless scientists indirectly through his research, he also affected many directly though his intellect and personality. This influence is worthy of celebration. He was one of those individuals who transform a place just by their presence and his wit always enlivened the conversation. Evidence of this humor is liberally sprinkled throughout much of Edmondson's writings and was manifested early in his career. He once remarked to one of us [RLW] that he sampled one particular lake in Massachusetts just so that he could fill an entire line of his Ph.D. dissertation with the lake's name (Lake Chaugogagogmanchaugagogchaubunagungamaug). (*nb*: Reference to this water body also appears in his 1944 monograph on sessile rotifers. This lake, whose spelling has many variants, is generally called Webster Lake.) Also, one never knew when Edmondson might offer some new and insightful analysis to his readers. His suggestion for an entirely new taxon within the Proboscidea (family Microelephantidae) was inspired by observations of unusual microfossils from sediment cores of Lake Washington (Edmondson, 1991: Fig. 9); and his subtle analysis of G. Evelyn Hutchinson Award recipients is a clear reflection of his attempt always to inform rather than muddle his readers (Schelske, 1990: 986–988). However, he was not all work. His interests included both astronomy and music, with a special fondness for music written for the pipe organ. The walls of his house overlooking Lake Washington are

*Figure 3.* Edmondson as teacher: a shipboard lecture to his limnology class about the use of the Clark-Bumpus zooplankton sampler. (Lake Washington, Seattle; 1963; photograph courtesy of John J. Gilbert.)

lined with hundreds of LPs and CDs of music, and one wall of his study has a large topographic lunar map. Edmondson also collected books on astronomy as well as facsimiles of early instruments for making astronomical measurements.

Edmondson's complete scientific legacy cannot be assessed in these few paragraphs. We are most grateful to have known him and worked with him, and we cannot help but realize how much his knowledge and enthusiasm have influenced our own efforts in teaching and research. Each time we step into a classroom or each time we direct a student in a research project, we carry some of the insights he imparted to us. And our experience is not unique; he had a major impact on many of those with whom he came in contact. Thus, on behalf of the international family of Rotiferologists and the organizers and editors of the IXth International Rotifer Symposium we offer this dedication to the memory of our friend and colleague, Tommy Edmondson (Figure 3). We do this in recognition of his remarkable contributions to the disciplines of freshwater ecology and rotifer biology.

ROBERT L. WALLACE[1]
JOHN J. GILBERT[2]
CHARLES E. KING[3]
[1] *Ripon College*
[2] *Dartmouth College*
[3] *Oregon State University*

## Acknowledgements

We are most grateful to Arni Litt and Sally Abella, both colleagues of W. T. Edmondson, for reading and improving this dedication. We are also very much in debt to Yvette Edmondson who applied her editorial skills to our efforts.

## References

For additional accounts of his extraordinary life, consult the chronicles presented in the volume edited by Hairston et al. (1988) or by Tommy himself in the volume of the Vth International Rotifer Symposium, held in Gargnano, Italy 1988 (Edmondson, 1989) or his book *The Uses of Ecology: Lake Washington and Beyond* (Edmondson, 1991).

Edmondson, W. T., 1989. Rotifer study as a way of life. Hydrobiologia 186/187: 1–9.

Edmondson, W. T., 1991. Sedimentary record of changes in the condition of Lake Washington. Limnol. Oceanogr. 36: 1031–1044.

Edmondson, W. T., 1991. The Uses of Ecology: Lake Washington and Beyond. Univ. Wash. Press, Seattle.

Hairston, N. G. Jr., J. T. Lehman & J. G. Stockner, 1988. W. T. Edmondson Celebratory Issue. Limnol. Oceanogr. 33(6, part 1): 1231–1429.

Schelske, C., 1990. Announcement: G. Evelyn Hutchinson Medal Award. Limnol. Oceanogr. 35: 985–988.

## Bibliography of W.T. 'Tommy' Edmondson

Edmondson, W. T., 1934. Investigations of some Hispaniolan lakes. 1. The Rotatoria. Arch. Hydrobiol. 26: 465–471.

Edmondson, W. T. & G. E. Hutchinson, 1934. Yale North India Expedition. Article IX. The Rotatoria. Mem. Conn. Acad. Arts Sci. 10: 153–186.

Edmondson, W. T., 1935. Some Rotatoria from Arizona. Trans. Am. Microsc. Soc. 54: 301–306.

Edmondson, W. T., 1936. Fixation of sessile Rotatoria. Science 84: 444.

Edmondson, W. T., 1936. New Rotatoria from New England and New Brunswick. Trans. Am. Microsc. Soc. 55: 214–222.

Edmondson, W. T. & J. L. Fuller, 1937. Food conditions in some New Hampshire lakes. New Hampshire Fish and Game Department Survey Report No. 2. Biological Survey of the Androscoggin, Saco and Coastal Watersheds: 95–99.

Edmondson, W. T., 1938. Notes on the plankton of some lakes in the Merrimack watershed. New Hampshire Fish and Game Department Survey Report No. 3. Biological Survey of the Merrimack Watershed: 107–210.

Edmondson, W. T., 1938. Three new species of Rotatoria. Trans. am. Microsc. Soc. 57: 153–157.

Edmondson, W. T., 1939. New species of Rotatoria with notes on heterogonic growth. Trans. am. Microsc. Soc. 58: 459–472.

Edmondson, W. T., 1940. The sessile Rotatoria of Wisconsin. Trans. am. Microsc. Soc. 59: 433–459.

Edmondson, W. T., 1941. Substrate limitations in sessile Rotatoria. [Abstract]. Anat. Rec. Suppl. 81: 109.

Edmondson, W. T., 1944. Ecological studies of sessile Rotatoria. Part I. Factors affecting distribution. Ecol. Monogr. 14: 31–66.

Edmondson, W. T., 1945. Ecological studies of sessile Rotatoria. Part II. Dynamics of populations and social structures. Ecol. Monogr. 15: 141–172.

Edmondson, W. T., 1946. Factors in the dynamics of rotifer populations. Ecol. Monogr. 16: 357–372.

Edmondson, W. T., 1946. Review: Coasts, waves and weather for navigators (John Q. Stewart). Rev. Sci. Instr. 17: 148–149.

Clarke, G. L., W. T. Edmondson & W. E. Ricker, 1946. Mathematical formulation of biological productivity. Ecol. Monogr. 16: 336–337.

Edmondson, W. T. & Y. H. Edmondson, 1947. Measurements of production in fertilized salt water. Sears J. mar. Res. 6: 228–246.

Bigelow, H. B. & W. T. Edmondson, 1947. Wind waves at sea, breakers and surf. U.S. Hydrographic Office Pub. No. 602: xi + 177.

Edmondson, W. T., 1948. Rotatoria from Penikese Island, Mass., with description of *Ptygura agassizi*, n. sp. Bio. Bull. 94: 169–173.

Edmondson, W. T., 1948. Two new species Rotatoria from sand beaches. Trans. am. Microsc. Soc. 67: 149–152.

Edmondson, W. T., 1948. Ecological applications of Lansing's physiological work on longevity in Rotatoria. Science 108: 123–126.

Edmondson, W. T., 1949. A formula key to the Rotatorian genus *Ptygura*. Trans. am. Microsc. Soc. 68: 127–135.

Edmondson, W. T., 1949. Wave action and effects on refraction. Shore and Beach 17: 35–36. (Note: Corrected revision of a stenographic report based on notes taken in a darkened room, published in an earlier issue. 16: 22–23.)

Edmondson, W. T., 1950. Centrifugation as an aid in examining and fixing rotifers. Science 112: 49.

Comita, G. W. & W. T. Edmondson, 1953. Some aspects of the limnology of an Arctic Lake. Stanford Univ. Pub. Univ. Ser. Biol. Sci. 11: 7–13.

Edmondson, W. T., 1954. Review: intertidal invertebrates of the central California coast. Science 120: 489–490.

Edmondson, W. T., 1955. Seasonal life history of *Daphnia* in an Arctic Lake. Ecology. 36: 439–455.

Nelson, P. R. & W. T. Edmondson, 1955. Limnological effects of fertilizing Bare Lake, Alaska. U.S. Fish & Wildlife Service. Fish. Bull. 102: 413–436.

Edmondson, W. T., 1955. Factors affecting productivity in fertilized salt water. Deep-Sea Res. 3: 451–464.

Edmondson, W. T., 1956. The relationship of photosynthesis by phytoplankton to light in lakes. Ecology 37: 161–174.

Edmondson, W. T., 1956. Measurement of conductivity of lake water *in situ*. Ecology. 37: 201–204.

Edmondson, W.T., G. C. Anderson & Donald R. Peterson, 1956. Artificial eutrophication of Lake Washington. Limnol. Oceanogr. 1: 47–53.

Edmondson, W. T., 1956. Biological aspects of the problem. In Sylvester O. R., W. T. Edmondson & R. H. Bogan (eds), A New Critical Phase of the Lake Washington Problem. Trend Eng. 8: 11–13.

Edmondson, W. T., 1957. Trophic relations of the zooplankton. Trans. am. Microsc. Soc. 76: 225–245.

Edmondson, W. T. (ed.), 1959. Ward and Whipple's Fresh-Water Biology, 2nd edn. Wiley. [Note: Tommy authored the Preface and Chapters 1 (Introduction), 18 (Rotifera), and 46 (Methods & Equipment).]

Edmondson, W. T., 1960. Reproductive rate of rotifers in natural populations. Mem. Ist. ital. Idrobiol. 12: 21–77.

Edmondson, W. T., 1960. Figures of rotifers (Plates XVII–XX). In American Public Health Association. Standard Methods for the Examination of Water and Waste Water. 11th edn. American Public Health Association.

Edmondson, W. T., 1960. Rotifera. Encyclopedia of the Biological Sciences. Reinhold.

Edmondson, W. T., 1960. Rotifera. Encyclopedia Britannica.

Edmondson, W. T., 1961. Changes in Lake Washington following an increase in the nutrient income. Verh. int. Ver. Limnol. 14: 167–175. (Note: reprinted In Ford R. F. & W. T. Hazen (eds), (1972) Readings in Aquatic Ecology. Saunders: 364–372.)

Edmondson, W. T., 1961. Secondary production and decomposition. Verh. int. Ver. Limnol. 14: 316–339.

Edmondson, W. T., 1962. Food supply and reproduction of zooplankton in relation to phytoplankton population. Rapp. Proc.-Verb. Cons. Internat. Explor. de la Mer. 153: 137–141.

Edmondson, W. T., G. W. Comita & G. C. Anderson, 1962. Reproductive rate of copepods in nature and its relation to phytoplankton population. Ecology 43: 625–634.

Edmondson, W. T., 1963. Pacific Coast and Great Basin. In Frey, D. G. (ed.), Limnology in North America. University of Wisconsin Press, Madison: 371–392.

Edmondson, W. T., 1964. The rate of egg production by rotifers and copepods in natural populations as controlled by food and temperature. Verh. int. Ver. Limnol. 15: 673–675.

Edmondson, W. T., 1965. Reproductive rate of planktonic rotifers as related to food and temperature in nature. Ecol. Monogr. 35: 61–111.

Edmondson, W. T. & G. C. Anderson, 1965. Some features of saline lakes in Central Washington. Limnol. Oceanogr. 10(suppl.): R87–R96.

Edmondson, W. T., 1966. Changes in the oxygen deficit of Lake Washington. Verh. int. Ver. Limnol. 16: 153–158.

Edmondson, W. T. (ed.), 1966. Ecology of Invertebrates. Marine Biology, Vol. 3. N.Y. Acad. Sci. Interdisciplinary Communications Program.

Oglesby, R. T. & W. T. Edmondson, 1966. Control of eutrophication: rehabilitation. J. Water Pollut. Control Fed. 38: 1452–1460.

Edmondson, W. T., 1967. Why study blue-green algae? In Environmental requirements of blue-green algae. Symp. Pub. Pacific Northwest Water Lab., Corvallis, Oregon: 1–6.

Edmondson, W. T., 1967. The present condition of Lake Washington. In Municipality of Metropolitan Seattle Quarterly, Summer: 4 pp.

Edmondson, W. T., 1968. Water quality management and lake eutrophication: The Lake Washington case. In Campbell, T. H. & R. O. Sylvester (eds), Water Resources Management and Public Policy. Univ. Wash. Press: 139–178. (Note: This paper has been translated into Hungarian by Dr Olga Sevestyen: Visminösegi ugyvetes es a tavak eutrofikacioja: A Lake Washington ügy.)

Edmondson, W. T., 1968. A graphical model for evaluating the use of the egg ratio for measuring birth and death rates. Oecologia 1: 1–37.

Edmondson, W. T., 1968. The history of the lake. In Metro – the first ten years. Municipality of Metropolitan Seattle: 22–23.

Edmondson, W. T., 1968. Review: Limnofauna Europaea. Limnol. Oceanogr. 13: 208.

Edmondson, W. T., 1969. Eutrophication in North America. In Eutrophication: Causes, Consequences, Correctives. Nat. Acad. Sci. Publ. No. 1700: 124–149.

Edmondson, W. T., 1969. Cultural eutrophication with special reference to Lake Washington. Mitt. int. Ver. Limnol. 17: 19–32.

Edmondson, W. T., 1969. The present condition of the saline lakes in the Lower Grand Coulee, Washington. Verh. int. Ver Limnol. 17: 447–448.

Edmondson, W. T., 1969. Ecology and weather modification. In Fleagle, R. G. (ed.), Weather Modification; Science and Public Policy. Univ. Wash. Press: 87–93.

Edmondson, W. T., 1969. Review: Introduction to paleolimnology (C. C. Reeves, Jr.). Limnol. Oceanogr. 14: 650.

Griffiths, M, P. S. Perrott & W. T. Edmondson, 1969. Oscillaxanthin in the sediment of Lake Washington. Limnol. Oceanogr. 14: 317–326.

Edmondson, W. T., 1970. Phosphorus, nitrogen and algae in Lake Washington after diversion of sewage. Science 196: 690–691. (Note: reprinted In Ford, R. F. & W. E. Hazen (eds), (1972) Readings in Aquatic Ecology. Saunders: 372–374.)

Edmondson, W. T., 1970. Lake Washington – history of success. University of Washington Daily, 20 April, 1970: B9. (Note: a reprinting of the Edmondson (1968) Metro article).

Edmondson, W. T., 1970. One Professor's Story. University of Washington Daily. 20 April, 1970: D6.

Edmondson, W. T., 1970. Review: Water management research (R. A. Vollenweider). Limnol. Oceanogr. 15: 169–170.

Edmondson, W. T. & D. E. Allison, 1970. Recording densitometry of X-radiographs for the study of cryptic laminations in the sediment of Lake Washington. Limnol. Oceanogr. 15: 138–144.

Edmondson, W. T., 1971. Freshwater pollution. In Murdoch W. W. (ed.), Environment: Resources, Pollution and Society. Sinauer Assoc.: 213–229. (Note: 2nd edn., 1975: 251–271).

Edmondson, W. T., 1971. Review: The water encyclopedia (D. K. Todd). Limnol. Oceanogr. 16: 597.

Edmondson, W. T. & G. G. Winberg (eds), 1971. A Manual on Methods for the Asseessment of Secondary Productivity in Fresh Waters. IBP handbook No. 17. Blackwell, Oxford.

Shapiro, J., W. T. Edmondson & D. E. Allison, 1971. Changes in the chemical composition of sediments of Lake Washington, 1958–1970. Limnol. Oceanogr. 16: 437–452.

Edmondson, W. T., 1972. Nutrients and phytoplankton in Lake Washington. In Likens, G. (ed.), Nutrients and Eutrophication. Am. Soc. Limnol. Oceanogr., Spec. Symp. No. 1: 172–193.

Edmondson, W. T., 1972. The present condition of Lake Washington. Verh. int. Ver. Limnol. 18: 284–291.

Edmondson, W. T., 1972. Instantaneous birth rates of zooplankton. Limnol. Oceanogr. 17: 792–795.

Edmondson, W. T., 1972. Review: New York Times encyclopedic dictionary of the environment (P. Sarnoff). Limnol. Oceanogr. 17: 502.

Beeton, A. M. & W. T. Edmondson, 1972. The eutrophication problem. J. Fish. Res. Bd Canada 19: 673–682.

Edmondson, W. T., 1973. Lake Washington. In Goldman C. R., James McEvoy III & Peter J. Richerson (eds), Environmental Quality and Water Development. Freeman: 281–298.

Edmondson, W. T., 1974. Secondary production. Mitt. int. Ver. Limnol. 20: 229–272.

Edmondson, W. T., 1974. The sedimentary record of the eutrophication of Lake Washington. Proc. natl. Acad. Sci. U.S.A. 71: 5093–5095.

Edmondson, W. T., 1974. Review: the environmental phosphorus handbook (E.J. Griffith, A. Beeton, J. M. Spencer & D. T. Mitchell, eds). Limnol. Oceanogr. 19: 369–375.

Edmondson, W. T., 1975. Microstratification of Lake Washington sediments. Verh. int. Ver. Limnol. 19: 770–775.

Griffiths, M. & W. T. Edmondson, 1975. Burial of oscillaxanthin in the sediment of Lake Washington. Limnol. Oceanogr. 20: 945–952.

Edmondson, W. T., 1977. Recovery of Lake Washington from eutrophication. In Cairns J. Jr., K. L. Dickson & E. E. Herricks (eds), Recovery and Restoration of Damaged Ecosystems. Univ. Press Virginia: 102–109.

Edmondson, W. T., 1977. Lake Washington. In North American Project – A Study of United States Water Bodies. A Report of the Organization for Economic Co-operation and Development. (EPA-600/3-77-086) Published by Environmental Research Laboratory, Environmental Protection Agency, Corvallis.

Edmondson, W. T., 1977. Trophic equilibrium of Lake Washington. Final Report on EPA Project R 8020 82-03-1. (EPA-600/3-77-087). Environmental Research Laboratory, Environmental Protection Agency, Corvallis, Oregon.

Edmondson, W. T., 1977. Population dynamics and secondary production. Arch. Hydrobiol. Beih. 8: 56–64.

Edmondson, W. T., 1977. Human influence on lakes as limnological experiments. [Abstract]. In Soltero, R. A. et al. (eds), Proc. Symp. Terrestrial and aquatic studies in the Northwest. Eastern Washington State College: 387 pp.

Edmondson, W. T., 1978. Formation of recent laminations in sediment of Lake Washington. [Abstract]. Verh. int. Ver. Limnol. 20: 361.

Edmondson, W. T., 1979. Lake Washington and the predictability of limnological events. Ergebn. Limnol. 13: 234–241.

Edmondson, W. T., 1979 Problems of zooplankton population dynamics. In. De Bernardi, R. (ed.), Proc. Symp. Biological and Mathematical Aspects in Population Dynamics. Me. ist. Ital. Idrobiol. (Suppl.) 37: 1–11.

Edmondson, W. T., 1980. Secchi disc and chlorophyll. Limnol. Oceanogr. 25: 378–379.

Edmondson, W. T., 1981. Review: Rotatoria. Die radertiere Mitteleuropas (W. Koste), Limnol. Oceanogr. 26: 400.

Edmondson, W. T., & J. T. Lehman, 1981. The effect of changes in the nutrient income on the condition of Lake Washington. Limnol. Oceanogr. 26: 1–29.

Edmondson, W. T., 1982. Review: Das Zooplankton der Binnengewässer (F. Kiefer & G. Fryer), Limnol. Oceanogr. 27: 196–197.

Edmondson, W. T., 1982. Review: Neusiedlersee. The limnology of a shallow lake in central Europe (H. Loffler [ed.]). Limnol. Oceanogr. 27: 198.

Edmondson, W. T., 1982. Review: Shallow lakes (M. Dokulil, H. Metx & D. Jewson [eds]). Limnol. Oceanogr. 27: 396.

Edmondson, W. T., 1982. Review: Rotatoria. Proceedings of the 2nd International Rotifer Symposium. (H. J. Dumont & J. Green). Limnol. Oceanogr. 27: 397.

Edmondson, W. T. & A. H. Litt, 1982. *Daphnia* in Lake Washington. Limnol. Oceanogr. 27: 272–293.

Lehman, J. T. & W. T. Edmondson, 1983. The seasonality of phosphorus deposition in Lake Washington. Limnol. Oceanogr. 18: 796–800.

Edmondson, W. T., 1984. Volcanic ash in lakes. Northwest Environ. J. 1: 139–150.

Edmondson, W. T., 1984. Review: Écologie du plancton des eaux continentales (R. Pourriot, J. Capblancq, P. Champ & J. A. Meyer). Limnol. Oceanogr. 29: 1349.

Edmondson, W. T., 1984. Review: Limnologia (R. Margalef). Limnol. Oceanogr. 84: 1349–1350.

Edmondson, W. T. & A. H. Litt, 1984. Mt. St. Helens ash in lakes in the Lower Grand Coulee, Washington State. Verh. int. Ver. Limnol. 22: 510–512.

Edmondson, W. T., 1985. Reciprocal relations between *Daphnia* and *Diaptomus* in Lake Washington. Ergeb. Limnol. 21: 475–481.

Edmondson, W. T., 1985. Recovery of Lake Washington from eutrophication. In Proc. Internat. Congress Lake pollution and recovery., Eur. Water Pollut. Control Assoc. Rome, April, 1985: 308–314. (Note: pp. 228–234 in preprints.)

Infante, A. & W. T. Edmondson, 1985. Edible phytoplankton and herbivorous zooplankton in Lake Washington. Ergebn. Limnol. 21: 161–171.

Wallace, R. L. & W. T. Edmondson, 1986. Mechanism and adaptive significance of substrate selection in the sessile rotifer *Collotheca gracilipes*. Ecology 67: 314–323.

Edmondson, W. T., 1987. *Daphnia* in experimental ecology: notes on historical perspectives. Mem. Ist. ital. Idrobiol. 45: 11–30.

Edmondson, W. T. & A. H. Litt, 1987. *Conochilus* in Lake Washington. Hydrobiolpgia. 147: 157–162.

Edmondson, W. T., 1987. Review: Lakes and Reservoirs (F. Taub [ed.]). Limnol. Oceanogr. 32: 776–777.

Edmondson, W. T., 1988. On the modest success of *Daphnia* in Lake Washington in 1965. In Round, F. E. (ed.), Algae and the Aquatic Environment. Contributions in honor of J.W.G. Lund, C.B.E., F.R.S. Biopress: 223–243.

Edmondson, W. T., 1988. Review of organ recital by Svend Prip in the Nikolaikriche in Plön, West Germany (in German). Ostholsteinisches Tageblatt.

Edmondson, W. T. & S. E. B. Abella, 1988. Unplanned biomanipulation in Lake Washington. Limnologica 19: 73–79.

Edmondson, W. T., 1988. Lessons from Washington lakes. In Popoff, G. I., C. R. Goldman, S. L. Loeb & L. B. Leopold (eds), Proc. Mountain Watershed Symp. Lake Tahoe. Tahoe Resource Conservation District: 457–463.

Edmondson, W. T., 1989. Rotifer study as a way of life. Hydrobiologia 186/187: 1–9.

Edmondson, W. T. & A. H. Litt, 1989. Morphological variation of *Kellicottia longispina*. Hydrobiologia 186/187: 109–117.

Edmondson, W. T., 1990. Perspectives in plankton studies. Mem. Ist. ital. Idrobiol. 47: 331–361.

Edmondson, W. T., 1990. Lake Washington entered a new state in 1988. Verh. int. Ver. Limnol. 24: 428–430.

Edmondson, W. T. & Y. H. Edmondson, 1990. Pallanza as a haven for visiting limnologists. Me,. Ist. ital. Idrobiol. 47: 45–55.

Edmondson, W. T., 1990. G. Evelyn Hutchinson Medal Award [Remarks by W.T. Edmondson]. Limnol. Oceanogr. 35: 986–988.

Edmondson, W. T., 1990. On prizes. Amer. Soc. Limnol. Oceanogr. Comm. to Members. Fall 1990: 9–10.

Edmondson, W. T., 1991. The uses of ecology: Lake Washington and beyond. Univ. Wash. Press. (Note: a Russian translation was published (1999) by MIR, Moscow. An abbreviated version of Chapter 10 on the Coulee lakes was published (1992) in the International Journal of Salt Lake Research 2: 9–20.)

Edmondson, W. T., 1991. Responsiveness of Lake Washington to human activity in the watershed. In Puget Sound Research '91. Puget Sound Water Quality Authority, Seattle, WA: 629–638.

Edmondson, W. T., 1991. Sedimentary record of changes in the condition of Lake Washington. Limnol. Oceanogr. 36: 1031–1044.

Edmondson, W. T., 1991. Resolution of respect. G. Evelyn Hutchinson. Bull. Ecol. Soc. am. 72: 212–216.

Edmondson, W. T., 1991. Homage to G. Evelyn Hutchinson (1903–1991). Mem. Ist. ital. Idrobiol. 49: 1–2.

Edmondson, W. T., 1992. Unsigned review: Homage to Ramon Margalef or, why there is so much pleasure in studying nature (J. Ros & N. Prat). Limnol. Oceanogr. 37: 1832–1833.

Edmondson, W. T., 1993. Eutrophication effects on the food chains of lakes: long-term studies. Mem. Ist. ital. Idrobiol. 52: 113–132.

Edmondson, W. T., 1993. Experiments and quasi-experiments in limnology. Bull. mar. Sci. 53: 65–83.

Edmondson, W. T., 1993. Eulogy for G. Evelyn Hutchinson (1903–1991). Verh. int. Ver. Limnol. 25: 49–55.

Edmondson, W. T., 1994. What is limnology? In Margalef, R. (ed.), Limnology Now: A Paradigm For Planetary Problems. Elsevier, Amsterdam: 547–553.

Edmondson, W. T., 1994. Sixty years of Lake Washington: a curriculum vitae. Lake Res. Manage. 10: 75–84.

Edmondson, W. T., 1994. Plankton. In Eblen R. A. & W. R. Eblen (eds), The Encyclopedia of the Environment. Houghton Mifflin, Boston: 537–538.

Edmondson, W. T., 1995. Eutrophication. In Nierenberg, W. A. (ed.), Encyclopedia of Environmental Biology Vol. 1. Academic Press, New York: 697–703.

Edmondson, W. T., 1996. Review: Freshwater Algae: their microscopic world explored (H. Canter-Lund & J. W. G. Lund). Limnol. Oceanogr. 41: 194–195.

Edmondson, W. T., 1996. Review: Lough Neagh: the ecology of a multipurpose water resource (R. B. Wood, & R. V. Smith [eds]). Limnol. Oceanogr. 41: 1373.

Edmondson, W. T., 1996. Review: Limnologie générale (R. Pourriot, M. Meybeck, P. Champ & J. Arcady-Meyer [eds]). Limnol. Oceanogr. 41: 1583.

Edmondson, W. T., 1996. Libraries and lakes. Lakeline (Published by NALMS). 16: 40–61.

Edmondson, W. T., 1997. Edward S. Deevey, Jr. Nat. Acad. Sci. U.S.A. Biographical Memoirs. 71: 1–15.

Edmondson, W. T., 1997. *Aphanizomenon* in Lake Washington. Arch. Hydrobiol. (Suppl.) 107 (Monographic Studies) 4: 409–446.

Hydrobiologia 446/447: xxiii–xxiv, 2001.
L. Sanoamuang, H. Segers, R.J. Shiel & R.D. Gulati (eds), Rotifera IX.



# Thomas M. Frost in memoriam (1950–2000)

*Figure 1.* Thomas M. Frost

On 25 August 2000, Tom Frost drowned in Lake Superior while saving his son Eliot. His untimely death is a tragedy for his family and a great loss to his many friends and colleagues. Tom is survived by his wife, Susan Knight, and sons Eliot, 9, and Peter, 6.

Tom was a limnologist with extremely broad interests in ecosystems, plankton communities and organisms, especially rotifers and sponges. After receiving his B.S. degree in Biology at Drexel University, Tom continued his studies at Dartmouth College. His Ph.D. dissertation on the biology and ecology of the freshwater sponge *Spongilla lacustris* included studies on population growth rates of both normal sponges with zoochlorellae and also experimentally induced aposymbiotic sponges, *in situ* feeding rates and selectivity; and grazing impact on the ecosystem. Tom's interest in sponges continued throughout his career and led to many papers and book chapters on these organisms. He was regarded as one of the world's experts on freshwater sponges.

After receiving his Ph.D. in 1978, Tom went to the University of Colorado, where he was a Research Associate with William Lewis and Director of the North American Field Group for the Lake Valencia Project in Venezuela. In 1981, Tom was appointed Associate Director for the Trout Lake Station of the Center for Limnology, University of Wisconsin-Madison. In 1990, he became Senior Lecturer in the Zoology Department at the Madison campus, where he taught Limnology. He remained at his post as Associate Director for the Trout Lake Station for the rest of his career, except for a 2 year leave (1997–1999) to become Program Director for Ecology at the National Science Foundation.

Tom's major research interests were zooplankton ecology and the effect of experimental acidification on zooplankton community structure. He collaborated with many students and colleagues on this research and on a variety of other projects involving snail migrations, temporary ponds, ultraviolet radiation, phytoplankton communities and population regulation, and sponge ecology, energetics and distribution. He also worked with his wife on carnivory by the vascular hydrophyte *Utricularia*. Tom advised six graduate students who completed their doctoral dissertations between 1988 and 1998.

Tom's research on rotifers began while he was at Dartmouth, when he collaborated with Peter Starkweather and John Gilbert on the ability of *Brachionus calyciflorus* to feed on bacteria. At Wisconsin, Tom worked on rotifers with his students, Michael Sierszen, María González and Daniel Schneider, and with his postdoctoral associate, Rita Adrian. Mike and Tom focused on the effect of lake acidification on Rotifer Feeding Selectivity. With María, Tom and others examined the effects of food limitation and acidification on rotifer population dynamics. As part of a study on temporary ponds, Daniel and Tom studied competition between *Daphnia* and *Keratella*. Rita and Tom investigated feeding in *Tropocyclops* and determined the ability of this small copepod to eat rotifers.

Tom published a total of 65 papers and 17 book chapters. During his time as Associate Director for the Trout Lake Station, Tom successfully obtained numerous research grants from NSF and EPA to conduct research as well as to improve the facilities at the station.

Tom was a very friendly and positive person. He constantly transmitted his optimism to his students, colleagues and friends. As an advisor, he cared deeply for his students. He provided constructive criticism of their proposals, manuscripts and presentations in a very gentle and positive manner. He very enthusiastically celebrated his students' accomplishments. He was well respected by everybody. His warm personality made all those he met feel welcome and appreciated. Even as a highly respected and established scientist, he always had time at professional meetings to lend an ear and give a kind word to young students just starting out in the field.

Tom was a great friend. He was always very grateful for life, and he could see the positive side of any situation. He frequently kept in touch with many of his friends in spite of distance and his busy schedule. It was always a joy to catch up with Tom at scientific meetings over a meal, a coffee or a hike. We all will miss his optimism and friendly smile!

MARÍA GONZÁLEZ
*Miami University*
*Department of Zoology*
*Oxford, OH 45056, U.S.A.*

J. J. GILBERT
*Dartmouth College*
*Department of Biology*
*Hanover, NH 03755, U.S.A.*

CRAIG E. WILLIAMSON
*Department of Earth and Env. Sciences*
*Lehigh University*
*Bethlehem, PA 18015, U.S.A.*

## Papers by T.M. Frost involving rotifers

Adrian, R. & T. M. Frost, 1992. Comparative feeding ecology of *Tropocyclops prasinus mexicanus* (Copepoda, Cyclopoeda). J. Plankton Res. 14: 1369–1382.

Adrian, R. & T. M. Frost, 1993. Omnivory in cyclopoid copepods: Comparisons of algae and invertebrates as food for three differently sized species. J. Plankton Res. 15: 643–658.

Frost, T. M. & P. K. Montz, 1988. Early zooplankton response to experimental acidification in Little Rock Lake, Wisconsin, U.S.A. Verh. int. Ver. Limnol. 23: 2279–2285.

Frost, T. M., S. R. Carpenter & T. K. Kratz, 1992. Choosing ecological indicators: Effects of taxonomic aggregation on sensitivity to stress and natural variability. In McKenzie, D. H., D. E. Hyatt & V. J. McDonald (eds), Ecological Indicators, 1. Elsevier Applied Science Publishers, Essex, England: 215–227.

Frost, T. M., P. K. Montz & T. K. Kratz, 1998. Zooplankton community responses during recovery from acidification: limited persistence by acid-favored species in Little Rock Lake, Wisconsin. Restor. Ecol. 6: 336–342.

Frost, T. M., P. K. Montz, T. K. Kratz, T. Badillo, P. L. Brezonik, M. J. González, R. G. Rada, C. J. Watras, K. E. Webster, J. G. Wiener, C. E. Williamson & D. P. Morris, 1999. Multiple stresses from a single agent: diverse responses to the experimental acidification of Little Rock Lake, Wisconsin. Limnol. Oceanogr. 44: 784–794.

Frost, T. M., P. K. Montz, M. J. González, B. L. Sanderson & S. E. Arnott, 1999. Rotifer responses to increased acidity: long-term patterns during the experimental manipulation of Little Rock Lake. In Wurdak, E., R. Wallace & H. Segers (eds), Rotifera VIII: A Comparative Approach. Hydrobiologia 387/388: 141–152.

González, M. J., T. M. Frost & P. K. Montz, 1990. Effects of experimental acidification on rotifer population dynamics in Little Rock Lake, Wisconsin, U.S.A. Verh. int. Ver. Limnol. 24: 449–456.

González, M. J. & T. M. Frost, 1992. Food limitation and the seasonal population dynamics of rotifers. Oecologia 89: 560–566.

González, M. J. & T. M. Frost, 1994. Comparisons of laboratory bioassays and a whole-lake experiment: rotifer responses to experimental acidification. Ecol. Appl. 4: 69–80.

Starkweather, P. L, J. J. Gilbert & T. M. Frost, 1979. Bacterial feeding by the rotifer *Brachionus calyciflorus*: clearance and ingestion rates, behavior and population dynamics. Oecologia 44: 26–30.

Sierszen M. E. & T. M. Frost, 1990. Effects of an experimental lake acidification on zooplankton feeding rates and selectivity. Can. J. Fish. aquat. Sci. 47: 772–779.

Williamson, C. E., R. S. Stemberger, D. P. Morris, T. M. Frost & S. G. Paulsen, 1996. Ultraviolet radiation in North American lakes: attenuation estimates from DOC measurements and implications for plankton communities. Limnol. Oceanogr. 41: 1024–1034.

*Hydrobiologia* **446/447:** 1–11, 2001.
*L. Sanoamuang, H. Segers, R.J. Shiel & R.D. Gulati (eds), Rotifera IX.*




# Dormancy patterns in rotifers

Claudia Ricci
*Dipartimento di Biologia, Università degli Studi di Milano, Italy*
*E-mail: rotiferi@mailserver.unimi.it*

*Key words:* quiescence, diapause, anhydrobiosis, resting eggs, habitat disturbance, coarse- and fine-grained environment

## Abstract

Dormancy is common among rotifers: monogononts produce resting eggs (diapause) commonly after switching to mictic phase, and bdelloids enter anhydrobiosis (quiescence) at any time during their life cycle. Monogononts are short-lived and inhabit coarse-grained environments; their dormancy is a long-lasting diapause, commonly initiated by indirect remote cues. Bdelloids live 3 times as long, live in fine-grained environments and enter short-lasting quiescence as a direct response to changing environment. The two dormancy forms of the rotifers can be related to the temporal variation of their environments and seem to represent diverse responses to disturbance occurring at different rates. The two strategies are alternative and mutually exclusive, as no single rotifer species seems capable of both diapause and quiescence. Dormancy has great ecological significance: it can carry the population through stressful conditions, promote species coexistence and serve as a biodiversity bank providing reliable colonization source.

## Introduction

Many animals live in ephemeral habitats, and when the conditions become unsuitable for their life, they may either move to a better habitat, or if unable to actively migrate, they may persist in their environment by entering into dormancy. Dormancy is a temporally suspended life, characterized by reduced metabolism and arrested development. Dormant forms often present enhanced resistance to environmental stress, such as drought or extreme temperature.

Onset and termination of dormancy are commonly under environmental control, although some autonomous control is apparent too. The temporally suspended life may be achieved through a number of diverse physiological states, and form and duration vary consistently among animals (e.g. Hand, 1991; Càceres, 1997a; Womersley et al., 1998). The term 'dormancy' includes both diapause and quiescence. The two phenomena are distinguished by being under endogenous and exogenous control, respectively.

**Diapause** is initiated in response to a variety of stimuli (population density, temperature, photoperiod) that are directly or indirectly predictive of environmental deterioration. Diapause is temporarily irreversible, and its duration may greatly exceed the duration of the harsh period (e.g. Hairston et al., 1995). Diapause is terminated by a specific trigger that does or does not correspond to the return of favourable conditions. This form of dormancy necessitates the presence of specific ontogenetic phases, like resting eggs in several invertebrates or larvae in nematodes and insects.

**Quiescence** refers to states such as hibernation, aestivation and all forms of cryptobiosis (i.e. cryobiosis, anhydrobiosis, osmobiosis, according to Keilin, 1959). Quiescence is directly induced and maintained by the occurrence of adverse environmental conditions and is promptly broken once the condition is removed; it lasts as long as the unfavourable situation. Animals can enter quiescence at any age, and do not need to produce specific dormant stages.

Both dormancy forms occur in rotifers. Phylum Rotifera comprises three classes, Seisonidea, Bdelloidea and Monogononta. Seisonidea are marine, their biology is mostly unknown, and the presence of dormancy in this class has never been investigated. The other classes, Monogononta and Bdelloidea, consist mostly of freshwater animals that reproduce parthenogenetically and commonly live in unstable habitats prone to temporal catastrophic events. To resist uninhabitable conditions rotifers of these classes undergo dormancy.

Dormancy of the rotifers was extensively reviewed by Gilbert (1974); more recently it has been dealt with by scientists who focused on peculiar aspects of the phenomenon, particularly the relevance of monogonont resting eggs as products of bisexual reproduction (e.g. Snell, 1987; Aparici et al., 1998; Serra & King, 1999), or the effect of dormancy on bdelloid demography (Ricci et al., 1987). Pourriot & Snell (1983) reviewed production, morphology, hatching and evolutionary significance of resting eggs. Since then, another resistant form, diapausing amictic egg, has been described in one monogonont species (Gilbert, 1995; Gilbert & Schreiber, 1995, 1998), adding to the already known dormancy forms in rotifers. Here, I analyse and compare the forms of dormancy occurring in these two rotifer classes, focusing on their ecological conditions and life history characteristics.

## Dormant stages of rotifera

### *Monogononta*

#### *Resting egg*

There is no evidence that monogononts can survive harsh conditions as adults. Their dormant stage is represented by the egg produced after mating (resting egg). Cued by some external factor, monogonont rotifers can interrupt their parthenogenetic reproduction (amictic cycle) and initiate a mictic cycle that leads to the formation of sexually-produced resting eggs. Occurrence of resting eggs (r.e.) produced without mating, called pseudo-sexual eggs, has been demonstrated in one rotifer species, *Keratella hiemalis*, by Ruttner-Kolisko (1946); these eggs have the same appearance and the same resistance properties as the common r.e. Recently, other eggs capable of dormancy and produced without mating have been found (see below).

Although current literature ascribes the production of r.e. to all monogonont species, and although the annual cycle of any 'average' monogonont is reported to consist of amictic plus mictic phases (see Wallace, 1999), males and r.e. have been found in a minority of species only. Many species, in which r.e. are unknown, are commonly cultivated in the laboratory and their life cycles thoroughly studied. Whether the absence of r.e. in these and other species is a matter of paucity of data or of lack of bisexual reproduction, needs further investigation.

Reasonably, the initiation of bisexual reproduction is influenced by population density, that must be high enough to locate a mate, and the production of the r.e. is affected by quantity and quality of the diet that should be sufficiently rich to provide resources for preparing a costly egg. Many studies have addressed the problem of 'how much and when' bisexual reproduction occurs in the monogononts, analysing costs (arrest of population growth) and benefits (genetic variation, mainly) (Wesenberg-Lund, 1930; Gilbert, 1963, 1977, 1993; King, 1980; Snell, 1986, 1987; Serra & Carmona, 1993; Aparici et al., 1998; Serra et al., 1998; Serra & King, 1999). Nevertheless, the outcome of bisexual reproduction in monogononts is the r.e. that is their common dormant stage (King & Schonfeld, 2001; Ruttner-Kolisko, 1963); it is destined to meet unfavourable conditions.

The sexually-produced r.e. is a diploid fertilized egg with a thick, often sculptured shell made of three layers (Wurdak et al., 1978). The r.e. is bigger and contains higher energy stores, mainly lipids and glycogen, than the unfertilized amictic egg, and its production is more costly to the mother (Gilbert, 1993). Females producing r.e. may have only about 10% the fecundity of females producing amictic eggs (King, 1970; Ruttner-Kolisko, 1974; see Gilbert, 1993). The r.e. contains a diapausing embryo, differentiated into an internal mass and an ectodermal layer, made of about forty to several hundred nuclei (Wurdak et al, 1978; Gilbert, 1993; Pourriot & Snell, 1983; Clément & Wurdak, 1991).

The r.e. can withstand several stresses, like low temperatures and drying (Gilbert, 1974; T. Schroeder, unpublished data). It generally sinks to the bottom and lays on top of the sediment where it remains dormant until the next growing season, usually months later. Pourriot & Snell (1983) listed hatching rates of a number of monogonont resting eggs and, despite variation among species, the rates are often close to 100%. Apparently, the r.e. viability is not significantly affected by increasing duration of dormancy; from a clutch of r.e. in a sediment sample, estimated to be 13 years old, up to 77% were found to hatch (King & Serra, 1998). Kotani et al. (2001) observed that about 80% of the resting eggs of *Brachionus rotundiformis* hatched after a dormancy lasting over 60 years.

Commonly, the dormancy of the resting embryo extends longer than the persistence of the harsh condition, is temporally irreversible and is broken independently from the cessation of the harsh period. The resumption of development is commonly influenced by a variety of factors – temperature, oxygen and light being the most common (Pourriot & Snell, 1983; Wallace, 1999). Hatching may be synchronous for a given group of eggs, or scattered over a wide time interval (Pourriot & Snell, 1983). Resting eggs hatch into amictic females.

*Diapausing amictic egg*

Recently, another type of monogonont dormant egg has been described in one rotifer species, *Synchaeta pectinata,* by Gilbert (1995). He called it a 'diapausing amictic egg' and found that its production was induced by starvation or low food availability (Gilbert & Schreiber, 1995, 1998). The diapausing amictic egg (d.a.e.) has characteristics similar to the subitaneous amictic egg, and its production does not require additional energetic costs to the mother (Gilbert, 1995). The shell of this egg is single-layered, like that of the subitaneous amictic egg, but thicker, and contains the embryo arrested at first cleavage divisions. It resumes development after a fixed time interval, regardless of the external conditions. Similarly to the r.e., these eggs undergo an obligatory dormant period before resuming development and hatching, but this period is much shorter than that usually required by the bisexually-produced resting eggs. As a response to different conditions, *S. pectinata* may be able to produce both d.a.e. after episodes of starvation and r.e. following the usual heterogonic reproductive cycle of monogononts. The species seems to be the only rotifer species, so far, capable of two forms of dormancy.

*Bdelloidea*

*Anhydrobiotic form*

In response to the evaporation of water in the environment, bdelloids lose internal water and enter a particular form of dormancy, called anhydrobiosis. Almost all bdelloids are capable of anhydrobiosis – either as eggs or embryos, juveniles or adults – but capacity differs among species and strains (Ricci, 1998). In preparation for the anhydrobiotic condition, a bdelloid follows a sequence of adjustments of morphology and metabolism. The animal contracts into a compact shape, called 'tun', by withdrawing its cephalic and caudal extremities into the trunk (Ricci & Melone, 1984). This form may help the animal to better control the rate of water loss during the process of desiccation, and also to reduce transpiration when the dormant condition is attained. Tissues and cells become packed and, in this form, can preserve their integrity (Dickson & Mercer, 1967; Schramm & Becker, 1987).

The series of adjustments that allow the rotifer to enter anhydrobiosis properly (preparation phase) require a few hrs, and duration of the period during which the rotifer is undergoing desiccation affects subsequent recovery success (Caprioli & Ricci, 2001). The time necessary to complete the preparation phase differs among species, and possibly this is related to the different desiccation velocity of the natural habitats. The metabolism of the desiccating animal is known to slow down until it is undetectable, and an osmo-protectant molecule, trehalose, is synthesized and stored. It stabilizes cellular structures and preserves macromolecular integrity. The presence of trehalose during dormancy has been assessed in the other taxa able to enter anhydrobiosis – tardigrades and nematodes (e.g. Crowe, 1971; Wright et al., 1992), as well as arthropods and plants (Crowe et al., 1992). The physiology and biochemistry of anhydrobiotic rotifers have never been explored, but it seems likely that trehalose also is produced by bdelloids.

In contrast to monogonont eggs, bdelloids entering dormancy are not passive forms. Commonly, before entering anhydrobiosis, the animals burrow actively and move into the sediment where they can be better protected from the deleterious effects of drying. Dormant bdelloids are known to withstand very severe conditions of temperature, pressure and radiations, but most existing reports are anecdotal (Gilbert, 1974). Recently, the resistance of a bdelloid species to stressful conditions during anhydrobiosis has rigorously been tested, and high temperature, vacuum, low and high humidity proved hazardous for the dormant rotifer (Caprioli & Ricci, 2001).

Dormancy of the anhydrobiotic rotifer, either egg or animal, is promptly broken by the return of water to the environment, and the animal resumes activity in a few hours. After addition of water, the rotifer swells up, unfolds, starts moving and, after a variable time, starts feeding. Possibly, the process of recovery tracks the modifications that occurred at induction of anhydrobiosis in reverse, and trehalose is degraded by the rotifer, as it is by nematodes and tardigrades.

Rahm (1923) observed an unspecified number of specimens of *Macrotrachela quadricornifera* become active after wetting a herbarium moss that was kept

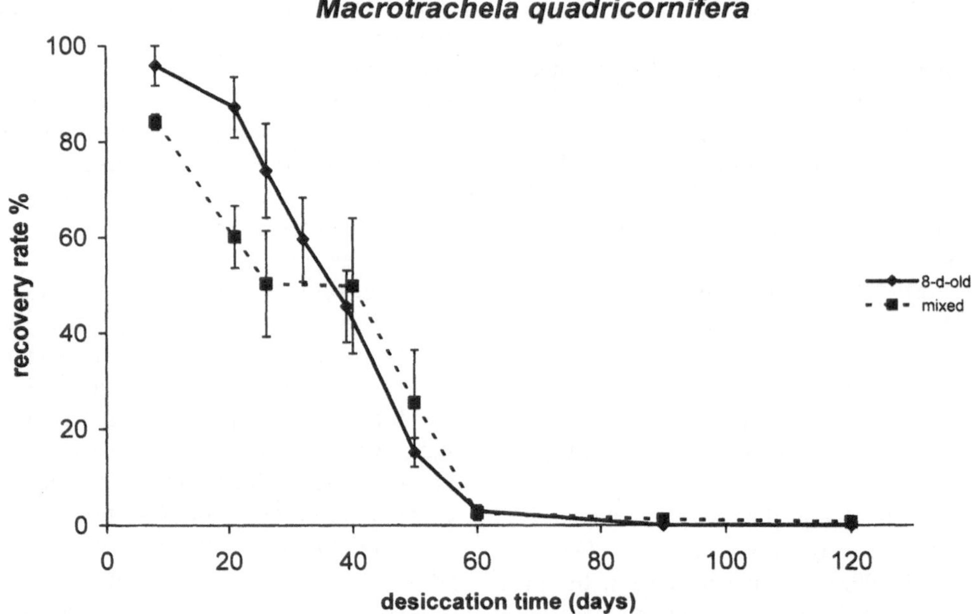


*Figure 1.* Rates of recovery of *Macrotrachela quadricornifera* (Rotifera, Bdelloidea) after different durations of anhydrobiosis. Solid line: all rotifers entered desiccation when 8-d-old. Broken line: the rotifers were not controlled for their age. Vertical bars: 1 standard deviation of the mean.

dry for 59 years. On the other hand, Caprioli & Ricci (2001) kept samples of *M. quadricornifera* anhydrobiotic for different time lengths under controlled conditions and found that the recovery percentages decreased drastically with increasing desiccation duration. Longer duration of anhydrobiosis seemed to be better tolerated by late embryos of this species, as apparent from the pattern of recovery rates of a mixed-age population; at desiccation times longer than 40 days, the population consisted mostly of newborns (Fig. 1).

The rate of bdelloid recovery is known to be affected by rotifer age at desiccation. Other things being equal, recovery is maximal for adults and late embryos, and minimal for juveniles and newly-laid eggs (Ricci et al., 1987; Örstan, 1995; Orsenigo et al., 1998). At present, we cannot offer any explanation for such difference that seems to be independent of resource availability (C. Ricci, unpublished). Whether short or long, dormancy represents a 'blind' interval in bdelloid life time, and the recovered rotifers ignore the duration of anhydrobiosis (according to the 'Sleeping Beauty' model, Ricci & Caprioli, 1998; Ricci et al., 1987). No difference was detected between age-specific fecundity and survival rates of rotifers maintained hydrated and those desiccated (Ricci et al., 1987). In contrast, considerable differences of fitness components were found between populations that had experienced dormancy and those that were kept under conditions of constant hydration (Ricci, 1987). Species with high rates of dormancy recovery, like *Abrochtha intermedia*, *Philodina vorax* and *P. rapida*, were found to decline in number if cultivated under constantly hydrated conditions for several generations (C. Ricci, pers. obs.). Dobers (1915) and Wesenberg-Lund (1930), remarked that anhydrobiosis could enhance life-history traits of the bdelloids that experience it. Thus, while the rotifer seem to be unaffected by anhydrobiosis if directly exposed, populations derived from recovered mothers could have some advantage in terms of life-history traits.

The three dormant forms occurring in the rotifers differ in modalities of initiation, duration and termination. The resting egg is the end-product of bisexual reproduction that necessitates two generations following the first signal, whereas the anhydrobiotic condition requires a few hrs to complete. An intermediate situation is represented by the diapausing amictic egg of *Synchaeta*, a parthenogenetic egg produced after starvation of the mother. The duration of dormancy of both monogonont eggs is independent of the duration of the environmental hardship, it is not terminated upon the return of favourable conditions, but necessitates some specific stimulus. These forms of dormancy fit the definition of 'diapause'. In contrast, bdelloid dormancy lasts as long as the drought of the habitat, does not require the formation of specific ontogenetic stages, but the animal recovery rate decreases at increasing desiccation times. The anhydrobiotic form corresponds to 'quiescence'.

## Environmental grain

Stability or variability of the environment are relative concepts and should be standardized to the organisms, to their biology and, more importantly, to their generation time. The same duration of time, say a year, is short for a long-lived animal, like man or an elephant, but is very long for a short-lived organism, like a rotifer. Thus, when speaking of environmental variability it is advisable to make use of the notion of 'grain' (*sensu* Levins, 1968). An environment is coarse-grained when the animal spends its whole life in a single alternative, that is, when an entire life time either includes, or does not include a harsh condition. In contrast, an environment is fine-grained when the time lapsing between two successive changes is shorter than the animal's generation time. Thus, if an animal in its life has a high probability of facing uninhabitable situations one or multiple times, it lives in a fine-grained environment (Levins, 1969). Therefore, the same environment can be fine-grained or coarse-grained according to an animal's life span, and the time scale for defining stability of environment should be related to the generation time of the organisms that live in it.

### *Monogononta*

Monogonont rotifers consistently occur in the plankton of lakes (e.g. Edmondson, 1959; Nogrady et al., 1993; Shiel et al., 1998), but many species dwell in the bottom (Schmid-Araya, 1998; Ricci & Balsamo, 2000) and in the periphyton (Duggan, 2001) of lentic and lotic habitats. Our knowledge of rotifer biology and ecology is chiefly based on data from planktonic rotifers in temperate zones, and relatively little is known about the traits of rotifers in other regions or habitats. I shall refer to planktonic temperate rotifers, aware that it will be wrong to assume that all species have similar characteristics.

Most monogonont rotifers are short-lived and complete their life cycle in 5–10 days, on average, at 19–25 °C (Table 1). The conditions in freshwater environments in temperate areas are likely to change gradually between cold unsuitable winters and hot dry summers. Thus, the short life-span of these rotifers either includes or does not include a winter or a dry season, and their environment can be defined as coarse-grained. The harsh condition of this sort of habitat occurs at more or less fixed time intervals, spaced by several rotifer generations. Some authors (e.g. Pourriot & Snell, 1983; Snell & Hoff, 1985; Gilbert, 1993) believe that r.e. of rotifers are not necessarily produced in anticipation of harsh conditions, but that mictic-female-production and r.e. formation, occurs under conditions that are favourable for the production of the energy-rich eggs. This interpretation does not account for the production of r.e. as dormant stages, but only as resource-rich eggs, that, as a side effect, will remain dormant for a long time. The widely-held opinion is that r.e. are produced in order to meet uninhabitable conditions (e.g. Wesenberg-Lund, 1930; see also Ruttner-Kolisko, 1974), but this requires that planktonic rotifers 'forecast' the future environment: after some stimulus, rotifers initiate the mictic cycle and after two generations release the resting eggs. Coupled to environmental conditions or not, the cues for sexual reproduction differ greatly between species and between strains (see e.g. Gilbert, 1977; Snell & Hoff, 1985; Snell, 1987; Serra & Carmona, 1993). The stimuli responsible for the dormancy of monogononts are remote and indirect, because do not initiate the production of dormant stages directly, but activate a cascade of events, which, once initiated, cannot be broken. Moreover, the massive production of resting eggs provides massive dormant stages that escape habitat selection, but that depress population growth rate. Thus, it could be disadvantageous for a rotifer to undergo diapause when the environmental conditions are still suitable for growth and reproduction, or, worst, not to have produced dormant stages at the right time to elude uninhabitable conditions.

Monogononts reduce the risk of variation of the catastrophe date by maintaining a variable fraction of amictic females while initiating the mictic phase and following a bet-hedging strategy, similarly to other animals (e.g. Hairston et al., 1985). Commonly, the rate of production of mictic females is a stochastic event, in that it takes a range of possible values in different populations and in different species, and its probability may be related to the environmental quality. It seems likely that the more regular the occurrence of the uninhabitable conditions, the larger the fraction of mictic females present at a given moment, and the more numerous the resting eggs in the sediment at a given

*Table 1.* Mean life span of rotifer species under controlled conditions

| Species | t°C | Approx. life span | Reference |
|---|---|---|---|
| **Monogononta** | | | |
| *A. priodonta* | 20° | 5 d | Kirk (1997) |
| *A. silvestri* | 20° | 8 d | Kirk (1997) |
| *Asplanchna brightwelli* | 25° | 4 d | Snell & King (1977) |
| *Brachionus angularis* | 20° | 5 d | Walz (1987) |
| *B. calyciflorus* | 20° | 11 d | Halbach (1973) |
| *B. caudatus* | 20° | 6 d | Kirk (1997) |
| *B. plicatilis* | 20° | 12 d | Schmid-Araya (1991) |
| *Encentrum linnhei* | 20° | 12 d | Schmid-Araya (1991) |
| *Euchlanis dilatata* | 22° | 7 d | King (1967) |
| *E. triquetra* | 20° | 7 d | Lansing (1947) |
| *Keratella cochlearis* | 20° | 9 d | Walz (1987) |
| *Lecana tenuiseta* | 20° | 26 d | Hummon & Bevelhymer (1980) |
| *Platyias quadricornis* | 20° | 7 d | Kirk (1997) |
| *Proales sordida* | 19° | 10 d | Jennings & Lynch (1928) |
| *Synchaeta pectinata* | 20° | 5 d | Kirk (1997) |
| **Bdelloidea** | | | |
| *Adineta grandis* | 4° | 40 d | Dartnall (1992) |
| *Adineta vaga* | 24° | 17 d | Ricci (1983) |
| *Embata laticeps* | 24° | 27 d | Ricci (1983) |
| *Habrotrocha constricta* | 24° | 38 d | Ricci (1983) |
| *H. elusa vegeta* | 24° | 32 d | Ricci (1983) |
| *H. sylvestris* | 24° | 40 d | Ricci (1983) |
| *H. tranquilla* | 22° | 29 d | Ricci (unpublished) |
| *Macrotrachela inermis* | 24° | 31 d | Ricci (1983) |
| *M. insolita* | 24° | 76 d | Ricci (1983) |
| *M. quadricornifera* | 22° | 25 d | Ricci & Fascio (1995) |
| *Otostephanos monteti* | 23° | 44 d | Ricci (unpublished) |
| *O. torquatus* | 24° | 45 d | Ricci (1983) |
| *Philodina acuticornis* | 24° | 23 d | Meadow & Barrows (1971) |
| *P. citrina* | 20° | 22 d | Lansing (1947) |
| *P. gregaria* | 4° | 37 d | Dartnall (1992) |
| *P. rapida* | 22° | 32 d | Ricci (unpublished) |
| *P. roseola* | 24° | 25 d | Ricci (1983) |
| *P. vorax* | 22° | 22 d | Ricci & Fascio (1995) |

time (Carmona et al., 1995; Serra & King, 1999). To have dormant stages available at the right time, monogonont populations can have three strategies. i) they can produce resting eggs at a low rate but continuously, while most of the population reproduces parthenogenetically (Fig. 2, 'polyphasic'); and ii) they can produce resting eggs in a single event while a minor part of the population keeps parthenogenetic reproduction, if the environment is predictable and the uninhabitable condition occurs in a cyclic, regular sequence (Fig. 2, 'monophasic', see also Carmona et al., 1995). (iii) They can have a massive production of resting eggs and then disappear from the habitat, if the environment changes quickly and suddenly (Fig. 2, 'bang-bang', see also Serra & King, 1999).

Monogonont rotifers are not known to elaborate fast dormancy responses, like quiescence, even though several species live in ephemeral habitats prone to sudden changes. The only dormant stage resulting from a direct cue is the 'diapausing amictic egg' of *S. pec-*

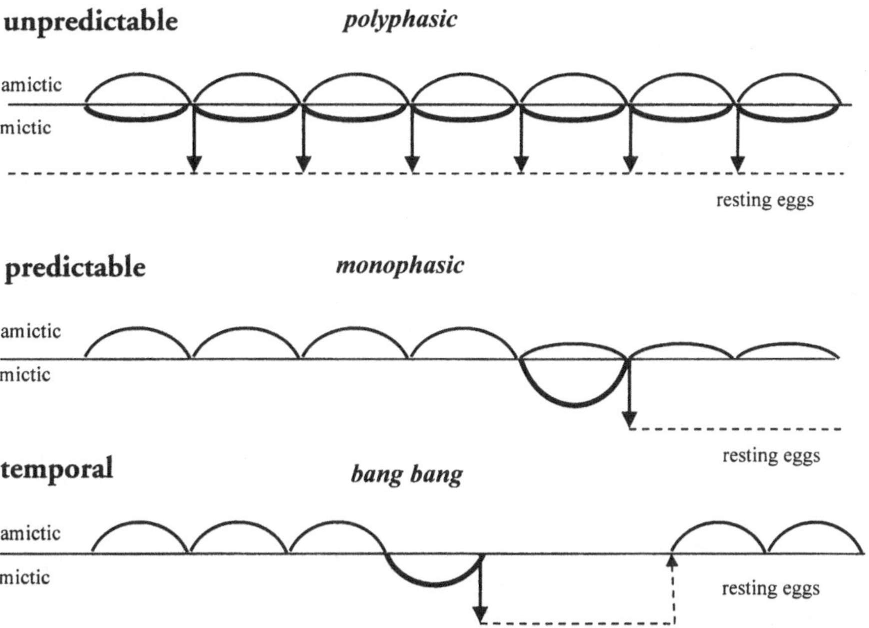


*Figure 2.* Theoretical patterns of resting egg production (arrows) of monogonont rotifers in response to three possible environmental conditions. Upper panel: unpredictable environment: a small part of the population switches continuously to mictic cycle (bold curve): the cycle is polyphasic. Middle panel: predictable environment: the rotifers switch once to mictic cycle: the cycle is monophasic. Lower panel: temporal environment: all rotifers switch to mictic cycle: the resting eggs are produced massively ('bang-bang').

*tinata*, induced by starvation (Gilbert, 1995). The species occurs in the plankton of lakes, an environment not exposed to sudden unpredictable changes. I am not aware of other reports of a fast dormancy response in the monogononts, but I would predict that such dormancy may exist, because many monogonont species inhabit puddles or shallow waters that are ephemeral habitats (Ruttner-Kolisko, 1974; Nogrady et al., 1993).

*Bdelloidea*

Bdelloid rotifers inhabit the sediments of water bodies, the shores of freshwater environments and the thin water film around moss leaves or soil particles (Donner, 1965; Ricci, 1987). Their habitat conditions shift from suitable to uninhabitable quickly, and the change in environmental parameters, like temperature, ion concentration, amount of $O_2$, can be very fast, because the habitats are prone to sudden desiccation. Bdelloid rotifers live 30–40 days under controlled conditions (Table 1). The disturbance of their habitat can be frequent, and the life span of these rotifers is long enough to expose each bdelloid to one or more episodes of desiccation in its life time, to which it responds by entering anhydrobiosis quickly. Following the definition given earlier, the habitat of the bdelloids is a fine-grained environment.

Time-consuming adaptations would not help the animals survive the fast changes of their environment. As their habitat dries up, all bdelloids must initiate morphological and metabolic adjustments, and, within hours, enter anhydrobiosis. With the same rapidity, the environment will get water and become suitable again. Bdelloids re-hydrate, emerge from their dormancy in few hrs and resume reproduction in about 24 h. The mechanism looks like an 'on-off switch' of metabolism, with complicated adjustments of physiology and morphology (Ricci, 1996). The on-off strategy may be of advantage when the favourable environment lasts for a short period of time and the recovered animals must be able to reproduce promptly. Adult animals act as fast colonizers that are able to establish an active community when conditions are favourable for growth and reproduction. The induction of, and emergence from, anhydrobiosis are deterministic events, all bdelloids enter dormancy at the same time and cannot delay the start or end of quiescence.

The difference between the duration of the life cycles of the two rotifer groups is remarkable, as well as the differences between their environments,

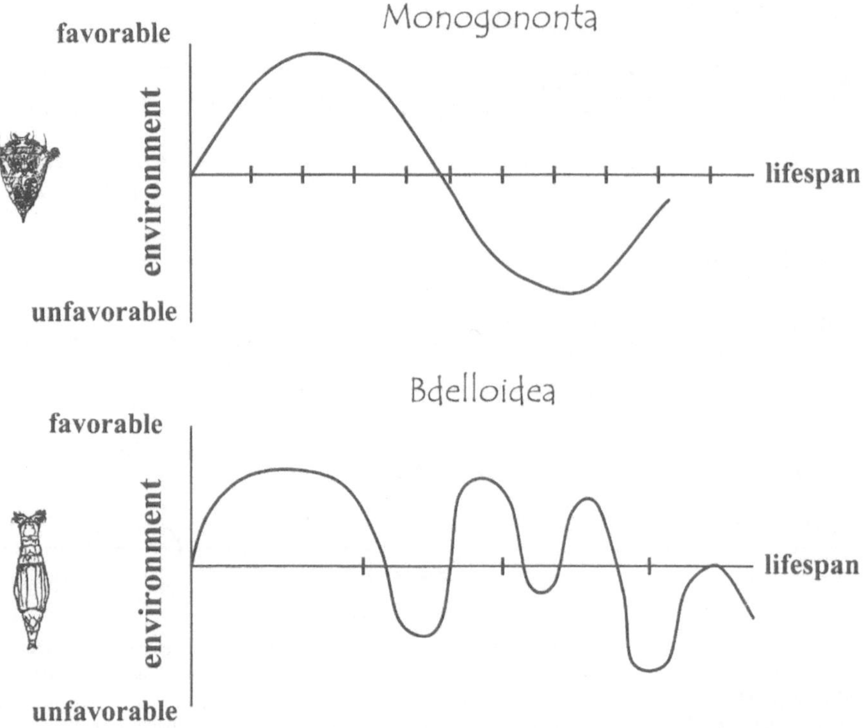


*Figure 3.* Environmental changes: below the horizontal line the conditions are unfavourable to the rotifers. The environment fluctuates regularly and the generation time (tick marks on x-axis) of the animal either includes or does not include an uninhabitable condition. The environment changes suddenly and quickly, and the whole lifespan of the rotifer is likely to include environmental changes.

especially in terms of environmental grain (Fig. 3). Monogononts are short-lived, live in a coarse-grained environment and their dormancy lasts long enough to carry populations to the next growing season that may occur several months later. Resting eggs can remain dormant for days, months, or years, and yet remain viable. Bdelloids live three times as long, inhabit fine-grained environments, and become quiescent directly in response to environment disturbance. Anhydrobiotic bdelloids have to survive until the next suitable environment – commonly after the next rain. While the diapause of monogononts permits survival through seasonal changes, anhydrobiosis has to carry bdelloids to the next suitable condition, often a few days later. This difference seems consistent with the remarkable difference between the duration of the rotifer dormancy.

No rotifer species is capable of both diapause (resting eggs) and quiescence (cryptobiosis). Resting eggs and anhydrobiotic animals seem two alternative dormancy forms, endowed with different characteristics and adapted to habitats where disturbance occurs at different rates (Fig. 3) (Ricci, 1992).

## Consequences of dormancy

Dormancy is of survival value when its duration spans unfavourable periods in the natural environment. Thus, its major significance is that it carries the population to the next suitable condition. The dormant forms represent a biodiversity bank as they survive through environmental disasters, preserving species diversity and provide a reliable colonization source when conditions improve. The biodiversity bank assures genetic continuity through periods of environmental adversity. I propose to use 'biodiversity' bank to extend the concept of 'seed' or 'egg' bank (Lampert, 1995) to all dormant forms, including quiescent stages like the anhydrobiotic animals.

No variability can be expected among the genotypes of the bdelloids that recover after anhydrobiosis, which faithfully reflect the genotypes present before the disaster, nor among the genotypes of *Synchaeta* before and after the d.a.e., because of the amictic production of the egg. In contrast, resting eggs of monogonont rotifers introduce genetic variability into species that reproduce chiefly through ameiotic parthenogenesis. The resting egg, resulting from a sexual reproduction, possesses the novelties of the genetic recombination, and its contribution to genetic variability through recombination seems relevant (e.g. Gilbert, 1963, 1977; King, 1980; Snell, 1987; Aparici et al., 1996, 1998; King & Serra, 1998; Serra & King, 1999; Gómez & Carvalho, 2000). The role of the biodiversity bank is more accurately performed by the anhydrobiotic state or by the diapausing egg, because these preserve the genotypes without any modification.

The dormant stages act as propagules that when conditions are suitable, can originate a new population, either in the same place already occupied, or in other habitats, if dispersed. Actually, some bdelloid and very few monogonont rotifers were found in the dust collected with windsocks and rain samplers, demonstrating that passive dispersal is possible (Jenkins & Underwood, 1998). Several possible mechanisms for bdelloid dispersal have been recognized, wind and running waters being the most likely (Örstan, 1998; Schmid-Araya, 1998).

The pattern of emergence after dormancy reflects the same trend as already seen for induction: an 'all-together' recovery for the bdelloids, and a great variability in timing of the emergence for the monogononts. Variability in the time of hatching for the monogononts is consistent with the emergence patterns of other taxa that produce resting eggs (Gilbert, 1974; Hairston, 1996), and can produce the succession of different genotypes derived from different egg clutches of a single species (King, 1972; Gómez et al., 1995). Such a hatching pattern influences both the genetic composition of a population and the species structure of a community (e.g. Càceres, 1998; King & Serra, 1998).

Any dormancy type would reduce the rate of elimination of a species, or of a strain, from a system when conditions for active life become unfavourable, and increases the rate of recolonization of a habitat when conditions improve. Dormancy can also promote species coexistence and regulate sib-competition; coexistence of competitors without competitive exclusion necessitates resource partitioning and spatial or temporal separation of the competitors. Several rotifer species apparently make use of the same resources, are not separate in time or space, and possibly compete, yet they co-occur (e.g. King, 1972; Snell, 1979; King & Serra, 1998). We can hypothesize that the coexistence of competitors could be made possible by different dormancy capacities, that might determine temporal fluctuations in recruitment, consistently with the 'storage effect' demonstrated in *Daphnia* (Càceres, 1997b).

## Conclusions and perspectives

Dormancy forms of the rotifers can be related to the temporal variation of their environments. Monogononts are short-lived, inhabit coarse-grained environments, and have a long-term diapause, commonly initiated by indirect, remote cues. Bdelloids live 3 times as long, live in fine-grained environments and can transform into short-term quiescent forms directly in response to the environment disturbance. The two dormancy forms seem to represent alternative responses to disturbance occurring at different rates, and no single rotifer species seems capable of both diapause and quiescence. *S. pectinata* can produce short-lasting 'diapausing amictic eggs' as a direct response to episodes of starvation and long-lasting resting eggs following the usual monogonont reproductive cycle. This species is capable of two different dormancy forms, both diapausing stages. The dormant state carries the population through stressful conditions, preserves it against dramatic selection operated by the uninhabitable environment, and contributes to a biodiversity bank providing a reliable colonization source. Dormancy is a very efficient time- and space-dispersal system that allows prompt recolonization (Templeton & Levin, 1979; Lampert, 1995). All these aspects have been ascertained in other animals capable of dormancy, are likely to fit rotifers as well, but need experimental tests.

Several areas need to be explored by rigorous experiments. Among the many questions that deserve an answer are:

1. How do monogononts cope with ephemeral habitats? Are they capable of fast dormancy similar to quiescence?
2. Do bdelloid rotifers use trehalose as a osmoprotective molecule, in analogy with anhydrobiotic nematodes and tardigrades?
3. Are the resting eggs of the monogononts making use of the same chemicals?
4. How does bdelloid age affect recovery after anhydrobiosis?
5. What causes mortality of bdelloids after anhydrobiosis, if they appear insensitive to the environmental conditions during dormancy?

6. Since the production of resting eggs is costly, is quiescence a costly condition as well?
7 Do bdelloid eggs synthesize protective chemicals?

## Acknowledgements

Many thanks are due to Maria José Carmona, Giulio Melone, Charles King and Terry Snell, who significantly improved this manuscript with their criticism and suggestions. John Gilbert and Elizabeth Wurdak commented on an earlier version of the manuscript. Manuela Caprioli gave technical and experimental support, Simona Orsenigo and Cesare Covino helped in preparing the manuscript. Financial support for the study and for attending the rotifer meeting came from ASI (Italian Space Agency) grant.

## References

Aparici, E., M. J. Carmona & M. Serra, 1996. Polymorphism in bisexual reproductive patterns of cyclical parthenogens. A simulation approach using a rotifer growth model. Ecol. Model. 88: 133–142.

Aparici, E., M. J. Carmona & M. Serra, 1998. Sex allocation in haplodiploid cyclical parthenogens with density-dependent proportion of males. Am. Nat. 152: 654–659.

Càceres, C. E., 1997a. Dormancy in invertebrates. Inver. Biol. 116: 371–383.

Càceres, C. E., 1997b. Temporal variation, dormancy and coexistence: a field test of the storage effect. Proc. natl. Acad. Sci. 94: 9171–9175.

Càceres, C. E., 1998. Interspecific variation in the abundance, production and emergence of *Daphnia* diapausing eggs. Ecology 79: 1699–1710.

Caprioli, M. & C. Ricci, 2001. Recipes for successful anhydrobiosis in bdelloid rotifers. Hydrobiologia 446/447 (Dev. Hydrobiol. 153): 13–17.

Carmona, M. J., A. Gómez & M. Serra, 1995. Mictic patterns of the rotifer *Brachionus plicatilis* Mûller in small ponds. Hydrobiologia 313/314: 365–371.

Clément, P. & E. Wurdak, 1991. Rotifera. In Harrison, F. W. & E. E. Ruppert (eds), Microscopic Anatomy of Invertebrates. Vol. 4: Aschelmintes. Wiley-Liss, New York: 219–297.

Crowe, J. H., 1971. Anhydrobiosis: an unsolved problem. Am. Nat. 105: 563–573.

Crowe, J. H., F. A. Hoeckstra & L. M. Crowe, 1992. Anhydrobiosis. Ann. Rev. Physiol. 54: 579–599.

Dartnall, H. J. G., 1992. The reproductive strategies of two Antartic rotifers. J. Zool. Lond. 227: 145–162.

Dickson, M. R. & E. H. Mercer, 1967. Fine structural changes accompanying desiccation in *Philodina roseola* (Rotifera). J. Microsc. 6: 331–348.

Dobers, E., 1915. Über die Biologie der Bdelloidea. Int. Rev. ges. Hydrobiol. Hydrogr. 7: 1–128.

Donner, J., 1965. Ordnung Bdelloidea. Akademie Verlag: 297 pp.

Duggan, I.C., 2001. The ecology of periphytic rotifers. Hydrobiologia 446/447 (Dev. Hydrobiol. 153): 139–148.

Edmondson, W. T., 1959. Rotifera. In Edmondson, W.T. (ed.), Freshwater Biology. Wiley & Sons.

Gilbert, J. J., 1963. Mictic female production in the rotifer *Brachionus calyciflorus*. J. exp. Zool. 153: 113–124.

Gilbert, J. J., 1974. Dormancy in rotifers. Trans. am. Microsc. Soc. 93: 490–513.

Gilbert, J. J., 1977. Mictic female production in monogonont rotifers. Arch. Hydrobiol. Beih. 8: 142–155.

Gilbert, J. J., 1993. Rotifera. In Adjiodi, K. G. & R. G. Adjiodi (eds), Reproductive Biology of Invertebrates. Vol VI, part A: 231–263.

Gilbert, J. J., 1995. Structure, development and induction of a new diapause stage in rotifers. Freshwat. Biol. 34: 263–270.

Gilbert, J. J. & D. K. Schreiber, 1995. Induction of diapausing amictic eggs in *Synchaeta pectinata*. Hydrobiologia 313/314: 345–350.

Gilbert, J. J. & D. K. Schreiber, 1998. Asexual diapause induced by food limitation in the rotifer *Synchaeta pectinata*. Ecology 79: 1371–1381.

Gómez, A. & G. R. Carvalho, 2000. Sex, parthenogenesis and genetic structure of rotifers: microsatellite analysis of contemporary and resting egg bank populations. Molec. Ecol. 9: 203–214.

Gómez, A., M. Temprano & M. Serra, 1995. Ecological genetics of a cyclical parthenogen in temporary habitats. J. Evol. Biol. 8: 601–622.

Hairston, N. G., 1996. Zooplankton egg banks as biotic reservoirs in changing environments. Limnol. Oceanogr. 41: 1087–1092.

Hairston, N. G., E. J. Olds & W. R. Munns, 1985. Bet-hedging and environmentally cued diapause strategies of diaptomid copepods. Verh. int. Ver. Limnol. 22: 3170–3177.

Hairston, N. G., R. A. Van Brunt, C. M. Kearns & D. R. Engstrom, 1995. Age and survivorship of diapausing eggs in a sediment egg bank. Ecology 76: 1706–1711.

Halbach, U., 1973. Life table data and population dynamics of the rotifer *Brachionus calyciflorus* Pallas as influenced by periodically oscillating temperatures. In Wieser, W. (ed.), Effects of Temperature on Ectothermic Organisms. Springer Verlag.

Hand, S. C., 1991. Metabolic dormancy in aquatic invertebrates. Comp. Environ. Physiol. 8: 2–50.

Hummon, W. D. & D. P. Bevelhymer, 1980. Life table demography of the rotifer *Lecane tenuiseta* under culture conditions, and various age distributions. Hydrobiologia 70: 25–28.

Jenkins, D. G. & M. O. Underwood, 1998. Zooplankton may not disperse readily in wind, rain or waterfowl. Hydrobiologia 387/388: 15–21.

Jennings, H. S. & R. S. Lynch, 1928. Age, mortality, fertility and individual diversities in the rotifer *Proales sordida* Gosse. I. Effect of age of the parent on characteristics of the offspring. J. exp. Zool. 50: 345–407.

Keilin, D., 1959. The problem of anabiosis or latent life: history and current concept. Proc. r. Soc. London 150: 149–191.

King, C. E., 1967. Food, age and the dynamics of a laboratory population of rotifers. Ecology 48: 111–128.

King, C. E., 1970. Comparative Survivorship and Fecundity of Mictic and Amictic Female Rotifers. Physiol. Zool. 43: 206–212.

King, C. E., 1972. Adaptation of rotifers to seasonal variation. Ecology 53: 408–418.

King, C. E., 1980. The genetic structure of zooplankton populations. In Kerfoot, W. C. (ed.), Evolution and Ecology of Zooplankton Communities Univ. Press of New England, Hanover: 315–328.

King, C. E. & J. Schonfeld, 2001. The approach to equilibrium of multilocus genotype diversity under clonal selection and cyclical parthenogenesis. Hydrobiologia 446/447 (Dev. Hydrobiol. 153): 323–331.

King, C. E. & M. Serra, 1998. Seasonal variation as a determinant of population structure in rotifers reproducing by cyclical parthenogenesis. Hydrobiologia 387/388: 361–372.

Kirk, K. L., 1997. Life-history responses to variable environments: starvation and reproduction in planktonic rotifers. Ecology 78: 434–441. on ectothermic organisms. Springer Verlag.

Kotani, T. M. Ozaki, K. Matsuoka, T. W. Snell & A. Hagiwara, 2001. Reproductive isolation among geographically and temporarily isolated marine *Brachionus* strains. Hydrobiologia 446/447 (Dev. Hydrobiol. 153): 283–290.

Lampert, W., 1995. Egg bank investment. Nature 377: 479.

Lansing, A. I., 1947. A transmissible, cumulative and reversible factor in aging. J. Geront. 2: 228–239.

Levins, R., 1968. Evolution in changing environments. Princeton Univ. Press.

Levins, R., 1969. Dormancy as an adaptive strategy. Sym. Soc. Exp. Biol. 23: 10.

Meadow, N. D. & C. H. Barrows, 1971. Studies on aging in a Bdelloid Rotifer. II The effects of various environmental conditions and maternal age on longevity and fecundity. J. Geront. 26: 302–309.

Nogrady, T., R. L. Wallace & T. W. Snell, 1993. Rotifera. Volume 1: Biology, ecology and systematics. SPB Academic Publishing bv, The Hague.

Orsenigo, S., C. Ricci & M. Caprioli, 1998. The paradox of bdelloid egg size. Hydrobiologia 387/388: 317–320.

Örstan, A., 1995. Desiccation survival of the eggs of the rotifer *Adineta vaga* (Davis, 1873). Hydrobiologia 313/314: 373–375.

Örstan, A., 1998. Microhabitats and dispersal routes of bdelloid rotifers. Sci. Nat. 1: 27–36.

Pourriot, R. & T. W. Snell, 1983. Resting weggs in rotifers. Hydrobiologia 104: 213–224.

Rahm, P. G., 1923. Biologische und physiologische Beiträge zur Kenntnis der Moosfauna. Z. Allg. Physiol. 20: 1–34.

Ricci C., 1983. Life histories of some species of Rotifera Bdelloidea. Hydrobiologia 104: 175–180.

Ricci, C., 1987. Ecology of bdelloids: how to be successful. Hydrobiologia 147: 117–127.

Ricci, C., 1992. Rotifers: parthenogenesis and heterogony. In Dallai, R. (ed.), Sex Origin and Evolution. Selected Symposia and Monographs UZI 6.: 329–341.

Ricci, C., 1996. Desiccation as a switch for some microinvertebrates. Proc. VI Eur. Symp. Life Sciences Research in Space. ESA SP-390: 273–275.

Ricci. C., 1998. Anhydrobiotic capabilities of bdelloid rotifers. Hydrobiologia 387/388: 321–326.

Ricci C. & M. Balsamo, 2000. The biology and ecology of lotic rotifers and gastrotrichs. Freshwat. Biol. 44: 15–28.

Ricci C. & M. Caprioli, 1998. Stress during dormancy: effect on recovery rates and life-history traits of anhydrobiotic animals. Aquat. Ecol. 32: 353–359.

Ricci, C. & U. Fascio, 1995. Life-history consequences of resource allocation of two bdelloid rotifer species. Hydrobiologia 299: 231–239.

Ricci, C. & G. Melone, 1984. *Macrotrachela quadricornifera* (Rotifera Bdelloidea): a SEM study on active and cryptobiotic animals. Zool. Scr. 13: 195–200.

Ricci, C., L. Vaghi & M. L. Manzini, 1987. Desiccation of rotifers (*Macrotrachela quadricornifera*): survival and reproduction. Ecology 68: 1488–1494.

Ruttner-Kolisko, A., 1946. Über das Auftreten unbefruchteter 'Dauereier' bei *Keratella quadrata*. Öst. Zool. Z. 1: 179–191.

Ruttner-Kolisko, A., 1963. The interrelationships of the Rotatoria. In Dougherty, E. C. (ed.), The Lower Metazoa. Univ. Calif. Press.

Ruttner-Kolisko, A., 1974. Planktonic rotifers: biology and taxonomy. Die Binnengewässer 26: 1–146.

Schmid-Araya, J. M., 1991. The effect of food concentration on the life histories of *Brachionus plicatilis* (O.F.M.) and *Encentrum linnhei* Scott. Arch. Hydrobiol. 121: 87–102.

Schmid-Araya, J. M., 1998. Rotifers in interstitial sediments. Hydrobiologia 387/388: 231–240.

Schramm, U. & W. Becker, 1987. Anhydrobiosis of the Bdelloid Rotifer *Habrotrocha rosa* (Aschelminthes). Z. mikrosk.-anat. Forsch. 101: 1–17.

Serra, M. & M. J. Carmona, 1993. Mixis strategies and resting egg production of rotifers living in temporally-varying habitats. Hydrobiologia 255/256: 117–126.

Serra, M., A. Gómez & M. J. Carmona, 1998. Ecological genetics of *Brachionus*. Hydrobiologia 387/388: 373–384.

Serra, M. & C. E. King, 1999. Optimal rates of bisexual reproduction in cyclical parthenogens with density-dependent growth. J. Evol. Biol. 12: 263–271.

Shiel, R. J., J. D. Green & D. L. Nielsen, 1998. Floodplain biodiversity: why are there so many species? Hydrobiologia 387/388: 39–46.

Snell, T. W., 1979. Intraspecific competition and population structure in rotifers. Ecology 60: 494–502.

Snell, T. W., 1986. Effects of temperature, salinity and food levels on sexual and asexual reproduction in *Brachionus plicatilis* (Rotifera). Mar. Biol. 92: 157–162.

Snell, T. W., 1987. Sex, population dynamics and resting egg production in rotifers. Hydrobiologia 144: 105–111.

Snell, T. W. & F. H. Hoff, 1985. The effect of environmental factors on resting egg production in the rotifer *Brachionus plicatilis*. J. World Mar. Soc. 16: 484–497.

Snell, T. W. & C. E. King, 1977.Lifespan and fecundity patterns of rotifers: the cost of reproduction. Evolution 31: 882–890.

Templeton, A. R. & D. A. Levin, 1979. Evolutionary consequences of seed pools. Am. Nat. 114: 232–249.

Wallace, R. L., 1999. Rotifera. In Encyclopedia of reproduction. Vol. 4. Academic Press.

Walz, N., 1987. Comparative population dynamics of the rotifers *Brachionus angularis* and *Keratella cochlearis*. Hydrobiologia 147: 209–213.

Wesenberg-Lund, C., 1930. Contributions to the biology of the Rotifera. Part II. The periodicity and sexual periods. K. Danske Videnske Selsk. Skrifter, Nat. Afd. 9: 1–230.

Womersley, C. Z., D. A. Wharton & L. M. Higa, 1998. Survival biology. In Perry, R. N. & D. J. Wright (eds), The Physiology and Biochemistry of Free-living and Plant-parasitic Nematodes: 271–302.

Wright, J. C., P. West & H. Ramløv, 1992. Cryptobiosis in Tardigrada. Biol. Rev. 67: 1–29.

Wurdak, E., J. J. Gilbert & R. Jagels, 1978. Fine structure of the resting eggs of the rotifers *Brachionus calyciflorus* and *Asplanchna sieboldi*. Trans. am. Microsc. Soc. 97: 49–72.

Hydrobiologia **446/447:** 13–17, 2001.
L. Sanoamuang, H. Segers, R.J. Shiel & R.D. Gulati (eds), Rotifera IX.



# Recipes for successful anhydrobiosis in bdelloid rotifers

Manuela Caprioli & Claudia Ricci
*Department of Biology, State University of Milan, via Celoria 26, 20133 Milan, Italy*
*E-mail: rotiferi@mailserver.unimi.it*

*Key words:* bdelloid rotifers, desiccation, dormancy, recovery, humidity

## Abstract

We tested the effect of several environmental variables on the ability of three bdelloid rotifers (*Macrotrachela quadricornifera, Philodina roseola* and *Adineta oculata)* to recover from the anhydrobiotic state. The variables we examined were (1) rate of water evaporation, (2) relative humidity during anhydrobiosis, (3) temperature during anhydrobiosis, (4) duration of anhydrobiosis, and (5) rehydration rate. Our results indicate that bdelloids can regulate to some degree net water balance during onset and termination of anhydrobiosis.

## Introduction

Of the numerous freshwater metazoa, only bdelloid rotifers, nematodes and tardigrades can live in environments prone to unpredictable periods of drought. Such ephemeral environments include soil and the surface films on mosses and lichens. Persistence of these organisms in such harsh habitats is due to their ability to enter a dormant state called anhydrobiosis (Keilin, 1959; Crowe, 1971; Cáceres, 1997). Anhydrobiosis is a reversible process that allows the animals to escape unfavourable conditions and to resume active life when water again becomes available (Keilin, 1959). Anhydrobiotic ability implies that the animals must be capable of undergoing complex morphological and physiological adjustments, such as synthesising protective chemicals to preserve their structures from injury (Crowe, 1971; Higa & Womersley, 1993). Because anhydrobiotic animals have to face loss of water, it is reasonable to predict that the availability of water has a key role during all steps of the anhydrobiotic process.

Studies on the effect of drying rate on nematode recovery indicated that two groups of species can be distinguished; the highest recovery was obtained by one group after slow desiccation (slow-dehydration strategist) and by the other after fast desiccation (fast-dehydration strategists) (Womersley, 1987). The capacity of tardigrades and bdelloid rotifers to withstand different desiccation rates was not investigated in detail; nevertheless, most authors considered that both taxa are slow-dehydration strategists (Jacobs, 1909; Wright et al., 1992).

The recovery capacity of the animals also may be affected by the conditions during the anhydrobiotic period. One important parameter is the duration of the dry period; longer dryness causes higher mortality rates in the rotifers (Jacobs, 1909; Ricci & Pagani, 1997; Ricci & Caprioli, in press). The increase of mortality rate with longer desiccation time apparently contrasts with the anecdotal resistance of anhydrobiotic bdelloids to external conditions. Anhydrobionts are said to tolerate conditions that cannot be successfully tolerated by the active forms (Gilbert, 1974). Anhydrobiotic bdelloids were found to resist extreme temperature and pressure (Rahm, 1923; Becquerel, 1950), and to better withstand low rather than high humidity. However, the experimental conditions were not described in sufficient detail to make meaningful generalizations (Jacobs, 1909). For example, these older studies often did not report the number of animals used or the experimental conditions. In some cases, the results have only anecdotal value (see Gilbert, 1974).

The rehydration process, too, may be critical for the recovery of anhydrobiotic organisms because, when placed in water, they imbibe rapidly and might passively lose solutes in the water medium (Crowe et al., 1992). Such leakage could damage the organism, eventually causing its death (Crowe et al., 1992).

Thus, a gradual increase of humidity can limit sudden imbibition and may provide the animals with a greater opportunity to adjust their internal structures more gradually. Increased viability after slow rehydration is reported for some nematode species (Womersley et al., 1998).

In this study, we tested the effect of environmental parameters on the recovery rates of bdelloid rotifers. Our larger goal is to establish a protocol for inducing anhydrobiosis with reproducible results. Here, we separate anhydrobiosis into three major phases (initiation of desiccation, dry state and rehydration), and focus on the importance of the availability of water to the animals during these phases.

## Materials and methods

Three bdelloid species, *Macrotrachela quadricornifera (M.q.)*, *Philodina roseola (P.r.)* and *Adineta oculata (A.o.)*, were dried, exposed to different conditions and their recovery rate was assessed. The species were collected from terrestrial mosses (*M.q.* and *A.o.*) and from aquatic habitats (*P.r.*), and are known to have different anhydrobiotic capabilities (Ricci, 1998). Rotifers were cultured at 22 °C in embryo dishes (5 ml capacity) for several generations, with deionized water as medium and a suspension of powdered fish food (Friskies® for goldfish). Anhydrobiosis was induced to 8-d-old rotifers, because this age gave best recovery after desiccation in previous experiments (Ricci et al., 1987; Ricci, 1998). Desiccation was always initiated by (1) adding small pieces of filter paper to the rotifer culture, (2) removing most water medium with a pipette and (3) allowing the rest to evaporate in a humido-thermostatic chamber (VC 0020-Votsch) at 22 °C and fixed evaporation rate. The filter paper was used to retard evaporation and to provide the animals a substratum during anhydrobiosis (Ricci et al., 1987). The amount of water left in the culture dish was 0.3–0.5 ml. The humido-thermostatic chamber controls both temperature and rate of relative humidity (RH) decrease. These variables may be adjusted to achieve different values at start and end of the dehydration regime. Unless stated otherwise, the samples were dried by dropping RH from 95 to 40% in 12 h. At the end of the anhydrobiotic period, the animals were rehydrated by adding water and food. The rotifers were considered alive if they became active within 24 h after rehydration. Each experiment was coupled to control samples.

For each experiment, several replicates were run, and the average (±S.E.) of the replicates was reported. Each replicate consisted of 20–50 rotifers. The different experimental conditions were compared by $\chi^2$ test.

### *Rate of water evaporation*

Effect of evaporation rate was tested on *M.q.*, *P.r.* and *A.o.* The rotifers were desiccated in the chamber dropping RH from 95 to 40% in 1, 12, 24 or 46 h. Because the faster treatment was not long enough to dry the rotifer samples completely, this and the other treatments were maintained at 40% RH for 48 h. For each treatment, 6 replicates and 2 controls were tested. The controls were desiccated at room conditions. The total duration of the anhydrobiosis period was 7 days.

### *Relative humidity during anhydrobiosis*

Different RH during a 7-d anhydrobiosis were tested on dry *M.q.* Very low humidity (~0% RH) was attained in a desiccator using $CaCl_2$, Si gel or under vacuum ($10^{-3}$ Tor). From three to nine replicate samples plus two controls were run for each treatment. One control was exposed to room humidity (40–55% RH), while the other one was maintained at 40% RH.

Two experiments tested the effect of high RH during long dryness. Dry samples were maintained at 75% RH for 20 days (5 samples) and for 40 days (10 samples). The controls were kept at 40% RH for either 20 or 40 days.

### *Temperature during anhydrobiosis*

More than 10 *M.q.* samples were used to test the effect of temperature on anhydrobiotic rotifers. In one experiment, three samples were kept at −80 °C for 5 days, and then rehydrated. In a second experiment, dry rotifers were kept at 60 °C for 1, 2, 3 and 4 days (two samples per treatment), and then were transferred at 22 °C. Controls were kept at 22 °C. All rotifers remained dry for 7 days; humidity was not controlled in these experiments.

### *Duration of anhydrobiosis*

*M.q.* samples were kept dry for 7, 21, 26, 32, 40, 50, 60, 90 and 120 days. Each treatment was run on 6–8 replicate samples. The humidity during this anhydrobiotic period ranged between 40 and 55% RH.

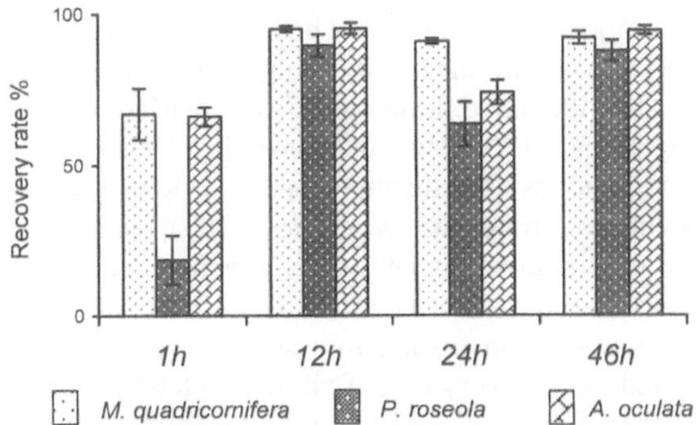


*Figure 1.* Percent recovery of three bdelloid rotifers desiccated at different dehydration rates. The abscissa notes the length of time over which the relative humidity was decreased from 95 to 40%. Rotifers in each treatment were kept dry for 7 days. Error bars are ±1SE.

*Rehydration rate*

Dry *M.q.* samples were maintained at ~0% RH (Si gel) for 5 days, then Si gel was replaced with deionized water in the Petri dish of 2 samples to increase RH 24 h before re-hydration. Other samples were re-hydrated directly without this pre-treatment. The control was kept at 40% RH and then re-hydrated directly.

## Results

*Rate of water evaporation*

The recovery rates of all species under tested conditions followed a similar trend (Fig. 1). The lowest survival was recorded when the relative humidity dropped from 95 to 40% in 1 h. Slower dehydration gave higher recovery; nevertheless, when the evaporation rate occurred over 24 h, two out of three species showed unexpectedly low viability. Among the tested species, *P.r.* had the lowest survival under all treatments.

*Relative humidity during anhydrobiosis*

This experiment was performed on *M.q.* Both controls gave similar recovery rates (~95%) (Fig. 2 upper panel). Exposure to very low RH levels (~0%), whether obtained with $CaCl_2$ or Si gel, significantly decreased the rotifer survival (~35%). The lowest recovery (~25%) was obtained by keeping the dry rotifers under vacuum but the three treatments did not differ ($\chi^2_{2df}$=2.8; $P$>0.1). Recovery rates after all treatments were significantly different from those of the controls ($\chi^2_{1df}$=200.33; $P$<0.0001).

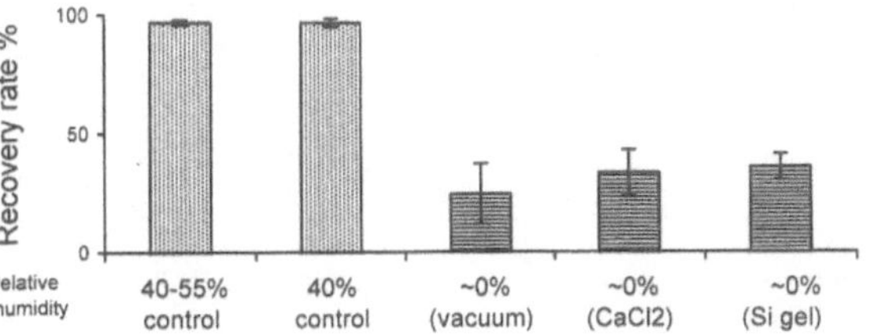


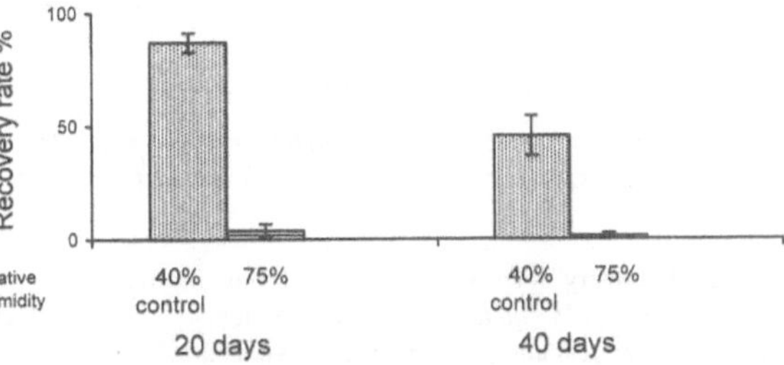


*Figure 2.* Upper panel: recovery rates of *M. quadricornifera* experiencing low relative humidity (~0%) during 7-d anhydrobiosis. Bars are ±1SE; $\chi^2$ compares all treatments to controls ($\chi^2_{1df}$=200.33; $P$<0.0001). Lower panel: recovery rates of *M. quadricornifera* experiencing high relative humidity (40%; 75%) during 20 and 40-d of anhydrobiosis. Bars are ±1SE.

*Table 1.* Recovery rates (%, average among replicates; ±S.E.) of anhydrobiotic *M. quadricornifera* after exposure to high (60 °C) and low (−80 °C) temperatures for different time lengths. $\chi^2$test compares − 80 °C treatments to control

| | # (Replicates) | Recovered | %±S.E. |
|---|---|---|---|
| | 60 °C | | |
| Control | 49 (2) | 49 | 100 - |
| 1 day | 47 (2) | 0 | 0 - |
| 2 days | 51 (2) | 0 | 0 - |
| 3 days | 49 (2) | 0 | 0 - |
| 4 days | 52 (2) | 0 | 0 - |
| | −80 °C | | |
| Control | 36 (1) | 34 | 94.4 - |
| 5 days | 90 (3) | 79 | 88.6 ±6.5 |
| $\chi^2_{1df}$=0.6; n.s. | | | |

Exposure to 75% RH during anhydrobiosis for 20 and 40 d caused a dramatic mortality rate of the rotifers (~100%) (Fig. 2 lower panel).

*Temperature during anhydrobiosis*

The dry rotifers experiencing 60 °C died regardless of the duration to the exposure (Table 1). Only a few eggs, laid during the evaporation process, were vi-

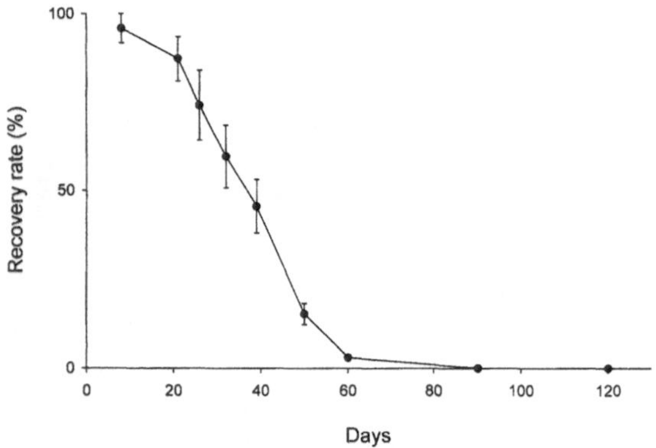


*Figure 3.* Recovery rates of *M. quadricornifera* kept dry for different time lengths. Bars are ±1SE of no fewer than 6 replicates.

*Table 2.* Recovery rates (%, average among replicates; ±S.E.) of anhydrobiotic *M. quadricornifera* experiencing slow and fast rehydration. $\chi^2$ test compares treatments vs. control

| | # (Replicates) | Recovered | % ±S.E. |
|---|---|---|---|
| Control | 34 (2) | 33 | 96.4 ±3.6 |
| Slow rehydration | 30 (2) | 19 | 63.1 ±1.6 |
| Fast rehydration | 38 (2) | 35 | 91.9 ±3.1 |
| $\chi^2{}_{2df}$=16.7; $P$<0.001 | | | |

able after exposure to 60 °C for 1 day. In contrast, the dry rotifers kept at −80 °C for 5 days recovered at percentages similar to the controls ($\chi^2{}_{1df}$=0.62; $P$>0.1).

*Duration of anhydrobiosis*

Increasing duration of anhydrobiosis decreased the recovery rates of 8-d-old *M.q.* dramatically. Viability was approximately halved when anhydrobiosis lasted 40 days, but no survivors were found when this period was extended to 90 d or longer (Fig. 3).

*Rehydration rate*

Addition of water in the Petri dish before rehydration resulted in increased mortality of the rotifers ($\chi^2{}_{2df}$=16.7; $P$<0.001) (Table 2). Each treatment was tested on two replicate samples.

## Discussion

Given the results obtained by other anhydrobionts, we expected that viability of these bdelloid species to anhydrobiotic conditions would be influenced by the rate of drying, conditions faced during the dry period, and probably, the rehydration rate (Crowe & Madin, 1975; Wright et al., 1992; Higa & Womersley, 1993).

Nematodes were distinguished into slow- and fast-dehydration strategists (Womesley, 1987), but no evidence of a similar response was apparent among the bdelloids tested. Consistently with the observations of Jacobs (1909), fast water loss during desiccation decreased the recovery rate of all three bdelloids tested. The three species came from different habitats and are likely to be exposed to different desiccation frequencies in their environment (Ricci, 1998). For example, *P.r.* lives in aquatic habitats and had the poorest recovery at all desiccation protocols. On the other hand, the relationship between the viability and the desiccation protocols followed a similar trend for the three species. The evaporation rate of 12 h gave the best results as a compromise between relatively fast process and high viability of the animals, and was utilised in the other experiments.

After attaining the anhydrobiotic state, the rotifers were assumed to be insensitive to environmental conditions (Rahm, 1923; Becquerel, 1950; Ricci & Caprioli, 1998), although RH was hypothesised to have some role in the survival of the rotifers (Jacobs, 1909). Under the hypothesis that low humidity could better preserve the integrity of the structures of dry animals, we exposed the rotifers to very low RH. However, anhydrobiotic *M.q.* could withstand low humidity very poorly. The high mortality we recorded could be ascribed to a passive loss of water that cannot be controlled by the anhydrobionts. In contrast, we should predict that high RH could better preserve the water content of the anhydrobionts. Thus, we exposed the dry rotifers to high RH (75%) for 20 and 40 days. But, the treatments caused the mortality of almost all rotifers. We cannot offer any plausible explanation for these results. It seems that both high and low RH are unfavourable conditions for viability of anhydrobiotic animals, while intermediate (40–50%) RH gave high recovery rates.

Temperature also affected rotifer recovery after anhydrobiosis, possibly influencing the amount of water present in the dry animal. Low temperature enhanced survival, but relatively higher temperatures were lethal. Rahm (1923) and Becquerel (1950) reported that dry rotifers could withstand both low (close to 0 °K) and high temperatures (110–151 °C) for a short time (about 30 min). In this study, the rotifers were exposed to the harsh conditions for several days, so it is possible that a long exposure to these conditions caused the decrease in survival.

The length of the anhydrobiotic period is known to affect viability of anhydrobiotic animals (Jacobs, 1909; Crowe et al., 1992; Ricci & Pagani, 1997), but specimens of several bdelloid species were reported to recover after very long periods of desiccation (Gilbert, 1974). Most of these records referred to occasional observations and not to rigorous experimental results. Under controlled conditions, *M.q.* was kept dry for different periods, its viability was strongly affected by the duration of anhydrobiosis, and no recovery was observed after 60 d dryness. It should be noted that different bdelloid species have different recovery capabilities (Ricci, 1998), and results obtained with one species cannot be extended to other species.

Slow rehydration gave lower viability of anhydrobiotic *M.q.,* in contrast with the results on the nematodes (Womersley et al., 1998). Possibly, the rotifers can better control the uptake of water during rehydration.

In conclusion, results obtained from these experiments represent a starting point for working out a reliable protocol for anhydrobiosis studies. Our results indicated that bdelloid rotifers can exert an efficient control over the rate of water loss and gain during the onset and termination of anhydrobiosis. How the dry rotifers can control their water content at any step of this process represents a critical feature which remains to be fully understood.

## Acknowledgements

Claudia Galimberti, Giulio Melone and Stefania Pellegrini assisted us with several technical aspects of the research. John J. Gilbert gave us access to some literature that was not available to us. Financial support came from ASI grant to C.R.

## References

Becquerel, P., 1950. La suspension de la vie au-dessous de 1/20 °K absolu par démagnétisation adiabatique de l'alun de fer dans le vide le plus élevé. Compt. Rend. Acad. Sci. 231: 261–263.

Cáceres, C. E., 1997. Dormancy in invertebrates. Invert. Biol. 116(4): 371–383.

Crowe, J. H., 1971. Anhydrobiosis: an unsolved problem. Am. Nat. 105: 563–573.

Crowe, J. H. & K. A. C. Madin, 1975. Anhydrobiosis in nematodes: evaporative water loss and survival. J. exp. Zool. 193: 323–334.

Crowe, J. H., F. A. Folkert & L. M. Crowe, 1992. Anhydrobiosis. Ann. Rev. Physiol. 54: 579–599.

Gilbert, J. J., 1974. Dormancy in rotifers. Trans. am. Microsc. Soc. 93: 490–513.

Higa, L. M. & C. Z. Womersley, 1993. New insights into the anhydrobiotic phenomenon: the effect of trehalose content and differential rates of evaporative water loss on the survival of *Aphelenchus avenae*. J. exp. Zool. 267: 120–129.

Jacobs, M. H., 1909. The effects of desiccation on the rotifer *Philodina roseola*. J. exp. Zool. 6: 207–263.

Keilin, D., 1959. The problem of anabiosis or latent life: history and current concept. Proc. R. Soc. London 150: 149–191.

Rahm, P. G., 1923. Biologische und physiologische Beiträge zur Kenntnis der Moosfauna. Z. Allg. Physiol. 20: 1–34.

Ricci, C., 1998. Anhydrobiotic capabilities of bdelloid rotifers. Hydrobiologia 387/388: 321–326.

Ricci, C. & M. Caprioli, 1998. Stress during dormancy: effect on recovery rates and life-history traits of anhydrobiotic animals. Aquat. Ecol. 32: 353–359.

Ricci, C. & M. Caprioli. Anhydrobiosis as a fast response to environmental stress. Proceedings SIL, in press.

Ricci, C. & M. Pagani, 1997. Desiccation of *Panagrolaimus rigidus* (Nematoda): survival, reproduction and the influence on the internal clock. Hydrobiologia 347: 1–13.

Ricci, C., L. Vaghi & M. L. Manzini, 1987. Desiccation of rotifers (*Macrotrachela quadricornifera*): survival and reproduction. Ecology 68: 1488–1494.

Womersley, C. Z., 1987: A reevaluation of strategies employed by nematode anhydrobiotes in relation to their natural environment. In Veech, J. & D. W. Dickson (eds), Vistas on Nematology. Soc. Nematol.: 165–173.

Womersley, C. Z., D. A. Wharton & L. M. Higa, 1998. Survival Biology. In Perry, R. N. & D. J. Wright (eds), The Physiology and Biochemistry of Free-living and Plant-parasitic Nematodes. CABI Publishing; Wallingford: 271–302.

Wright, J. C., P. West & H. Ramlov, 1992. Cryptobiosis in Tardigrada. Biol. Rev. 67: 1–29.

Hydrobiologia **446/447:** 19–28, 2001.
L. Sanoamuang, H. Segers, R.J. Shiel & R.D. Gulati (eds), Rotifera IX.




# Spine development in *Brachionus quadridentatus* from an Australian billabong: genetic variation and induction by *Asplanchna*

John J. Gilbert
*Department of Biological Sciences, Dartmouth College, Hanover, New Hampshire 03755, U.S.A.*
*E-mail: John.J.Gilbert@Dartmouth.EDU*

*Key words:* *Asplanchna*, *Brachionus*, genetic variation, resting eggs, spines, spine induction

**Abstract**

Fertilized resting eggs of Australian *Brachionus quadridentatus* hatched 2–3 days after hydration into females with or, more frequently, without posterior lateral spines. These females then produced clones with short-spined or long-spined phenotypes. *Asplanchna girodi* induced females from two short-spined clones and one long-spined clone to produce daughters with significantly longer posterior lateral spines. In all clones, there were significant differences in spine development among offspring of mothers within *Asplanchna* and control treatments. The range of phenotypes reported in one short-spined clone is observed in the billabong and includes much of the variation described for the species, with *mehleni* (long-spined) phenotypes occurring with *Asplanchna*. In *B. quadridentatus*, the ecological significance of long-spined, basic phenotypes, and of the spine-development response to *Asplanchna*, is unclear. In laboratory cultures, females of all clones were attached to the substratum or water surface, and were safe from *Asplanchna*; in nature, females are epiphytic and probably rarely susceptible to *Asplanchna*. Most (96%) resting eggs produced in cultures and kept under culture conditions hatched after a 7-day latent period. This raises questions regarding natural conditions which might prevent hatching and allow accumulation of resting eggs in a sediment egg bank. Hatching of resting eggs in nature may be enhanced in sediments which dry and then become flooded after rains.

## Introduction

The availability of fertilized resting eggs of *Brachionus quadridentatus* Hermann in dried sediment from a billabong in southeastern Australia provided an opportunity to investigate several questions regarding the phenotype of this species. Spine development of this species is very variable and is used to determine taxonomic status below the species level (Koste & Shiel, 1987).

The major goals of this study are to assess the extent of genetic variation in phenotype among clones hatching from resting eggs, and then to test the hypothesis that the phenotypes of these clones are influenced by the predatory rotifer *Asplanchna*. *Asplanchna* produces a kairomone that affects spine development in other species of *Brachionus* and in some species of other brachionid genera (Gilbert, 1999). *Asplanchna*-induced phenotypes may have spines not present in the basic phenotype, longer spines than those in the basic phenotype, or both. The effect of *Asplanchna* on *B. quadridentatus* has not been determined. The *B. quadridentatus* in this study comes from a population that co-occurs with *Asplanchna* (R. Shiel, unpublished) and so may have evolved the potential to respond to the predator.

In addition, this study compares phenotypes of *B. quadridentatus* hatching from resting eggs with those in subsequent parthenogenetic generations. Some strains of *B. calyciflorus* from North America and Australia have a basic phenotype with posterior lateral spines, but females hatching from the resting eggs of these strains do not have them (J.J. Gilbert, unpublished). These spines are expressed only in later generations.

Finally, this study includes some observations on the hatching of *B. quadridentatus* resting eggs present in the sediment and produced in culture. These observations are of interest regarding time to hatching following the flooding of dried sediments, minimum duration of dormancy and accumulation of resting eggs in sediments.

## Materials and methods

All clones of *Brachionus quadridentatus* were derived initially from resting eggs in dried sediment collected on 28 May 1997 from above the water level of Ryan's 2 billabong – a River Murray billabong near Wodonga, Victoria (see Hillman & Shiel, 1991). The sediment was transported to Hanover, New Hampshire and kept in a closed, clear plastic jar exposed to light at room temperature. *Asplanchna girodi* de Guerne was collected from Star Lake (Norwich, Vermont).

*B. quadridentatus* was cultured in lake water containing *Cryptomonas erosa* or a mixture of *C. erosa* and *C.* sp. in plastic Petri dishes (35 × 10 mm or 60 × 15 mm) or concavities of glass or plastic well plates kept at 20 °C in a photoperiod (L:D 16:8). *A. girodi* was cultured the same way but using only *C. erosa* as food. Rotifers were transferred to fresh culture medium and food every 2–4 days. The algae were cultured in modified MBL medium (Stemberger, 1981) at 20 °C in a photoperiod (L:D 16:8). Approximately 1-week-old cultures were used to feed rotifers and to inoculate new cultures.

The lake water used for rotifer cultures was collected in late fall from the surface of Storrs Pond (Hanover), filtered (0.45 $\mu$m), and stored at 4 °C in the dark. Before use, the water was adjusted to pH 7.5–8.0 with 1 N NaOH. The water was assumed to be free of *Asplanchna* kairomone because any kairomone that might have been present probably would have degraded during storage (see Gilbert, 1999).

To hatch resting eggs of *B. quadridentatus* from the dried sediment, a small amount of the material (ca. 5 ml) was placed in a glass bowl and mixed with about 200 ml lake water containing *Cryptomonas*. To obtain individuals just hatching from eggs in the sediment, once or twice daily for about a week all the water in a bowl was removed in aliquots, examined with a stereomicroscope, and returned to the bowl.

The effect of storage conditions on the hatching of resting eggs produced in culture was assessed using a population of *B. quadridentatus* derived from several females hatching from the billabong sediment and maintained at a high population density. When in culture medium with *Cryptomonas* at 20 °C, resting eggs were transferred to fresh medium and food every 2 – 4 days.

Body (lorica) and spine measurements of *B. quadridentatus* were made at 500 magnifications to the nearest 1.9 $\mu$m on specimens preserved in acid Lugol's solution. Spine terminology was that of Koste & Shiel (1987, Fig. 12). Body length was defined as the distance between the notch at the base of the dorsal anterior median spines and the dorsal median posterior margin at the midline. Lengths of one of each of the three pairs of anterior spines were measured from the base of the spine. Anterior median spines were measured from the notch at the base to the tip along the angle of the spine; their lengths were underestimated when they were long and curved over ventrally. Lengths of both posterior lateral spines were measured from the posterior margin of the lorica on the dorsal surface, and then averaged. Relationships between body and spine lengths, or between lengths of different spines, among individuals in a given *B. quadridentatus* population were tested using correlation analysis.

The hypothesis that *Asplanchna* promotes spine development in *B. quadridentatus* was tested using three clones – two short-spined clones in which the basic (non-induced) phenotype had short or no posterior lateral spines, and one long-spined clone in which the basic phenotype had pronounced posterior lateral spines. Depending on the experiment, from 7 to 18 young *B. quadridentatus* were placed individually in 2–3 ml medium with *C. erosa* or a mixture of *C. erosa* and *C.* sp. at a measured or estimated density of 3 × $10^4$ cells $ml^{-1}$. In each experiment, about half of the vessels were designated for the experimental treatment and received one *Asplanchna*; the other vessels served as control treatment. The *Asplanchna* did not appear to eat *B. quadridentatus*, which attached by its foot to the vessel or the water surface. On each of the next 2 or 3 days, all offspring of each of the parental females were removed, preserved in acid Lugol's solution, and measured for body and spine lengths. Also, on each day, the number of *Asplanchna* in the experimental vessels was adjusted to 1 or 2 individuals, and any mictic-female *Asplanchna* were removed or replaced with amictic ones. Cryptomonad food generally remained high throughout the experiment, and additional cells were added after the first day only in the experiment involving the long-spined clone.

Spine development of different-sized *B. quadridentatus* offspring born in treatments with and without *Asplanchna* was compared by standardizing spine lengths as a fraction of body lengths (S/B ratios). Use of absolute spine lengths for different-sized rotifers would be inappropriate, since spine length probably increases with body length as juveniles grow to adults. However, it is not known if spine growth in *B. quadridentatus* is isometric, with S/B ratios remaining the same as body length increases. In *B. calyciflorus*, growth of the posterior lateral spines is negatively allometric (Gilbert, 1967; Halbach, 1970). S/B ratios were compared across treatments and among mothers using nested ANOVAs.

## Results

### *Resting eggs: hatching and occurrence of posterior lateral spines*

After dried sediment was added to lake water, *B. quadridentatus* females hatching from resting eggs appeared and were collected 2–3 days later. Of these, 21 had no posterior lateral spines, and 3 had short posterior lateral spines.

Ten females, one of which had posterior lateral spines, were used to initiate clones that were cultured for up to 4 generations. Six clones starting from females with no posterior lateral spines produced some females with short posterior lateral spines by the F1 and F2 generations and thereafter. Three clones starting from females with no posterior lateral spines produced only females without these spines through the F1, F3 or F4 generations. One of these clones was cultured in 15-ml volumes for 5 weeks and produced some individuals with short posterior lateral spines and some without these spines; this short-spined clone was designated SS 3 (see below). The one clone starting with a female with posterior lateral spines produced females that all had long posterior lateral spines. This clone was cultured in 15-ml volumes for 5 weeks and continued to produce individuals with long posterior lateral spines; this long-spined clone was designated LS 1 (see below).

On 14 June 1999, a large number of resting eggs was collected from a culture originating from about 5 females that had hatched from dried sediment. These eggs hatched at a very high rate under the three conditions tested. Of the 48 kept in culture medium at 20 °C (L:D 16:8), 96% hatched after 7–23 days. Of the 44 kept in culture medium at 5 °C in the dark for 7 days after collection, 89% hatched when returned to 20 °C (L:D 16:8) after 4–20 days. Of the 14 resting eggs allowed to gradually dry and kept dry for 4 days at 20 °C (L:D 16:8), 76% hatched following immersion in culture medium after 4–9 days. No more of these eggs hatched during the next 12 days.

*Table 1.* Body and spine lengths of basic (non-induced) phenotypes of ovigerous females from a short-spined (SS 3) and a long-spined (LS 1) clone of *Brachionus quadridentatus*. Samples taken at two separate times – after 3 weeks (time 1) and 5 weeks (time 2) of culture. Spines are posterior lateral (PL), anterior median (AM), and anterior lateral (AL). Values are means (1 SD)

| Clone | Sample time | N | Length (μm) | | | |
|---|---|---|---|---|---|---|
| | | | Body | PL spine | AM spine | AL spine |
| SS 3 | 1 | 31 | 149 (6) | 2 (5) | 27 (3) | 17 (2) |
| | 2 | 28 | 152 (5) | 4 (6) | 27 (3) | 17 (2) |
| LS 1 | 1 | 20 | 116 (4) | 42 (5) | 36 (3) | 20 (2) |
| | 2 | 23 | 116 (4) | 34 (8) | 33 (4) | 18 (3) |

All 98 of the females hatching from these eggs had short posterior lateral spines. Clones starting from 13 of these females were cultured in 1-ml volumes through 5 generations. Females from all clones and generations had short or no posterior lateral spines.

### *Clones with major differences in body size and spine development*

Females from the SS 3 and LS 1 clones were haphazardly collected and preserved after about 3 weeks (time 1) and 5 weeks (time 2) of culture. The relationships between body length and posterior lateral spine length for all individuals measured are shown in Figure 1. SS 3 females had considerably greater maximal body lengths than LS 1 females. Only some SS 3 females had posterior lateral spines, and these spines were short. All LS 1 females had posterior lateral spines, and these generally were long. In SS 3 females, there was no correlation between body length and spine length at either time 1 (product-moment correlation coefficient, $r$=−0.235, $p$=0.10) or time 2 ($r$=−0.192, $p$=0.23). Females either did or did not have posterior lateral spines, whether they were small or large, and there was no tendency for larger females to have longer spines. In LS 1 females, there was a significant tendency for larger females to have longer posterior lateral spines at time 1 ($r$=0.545, $p$<0.0001) but not at time 2 ($r$=0.207, $p$=0.25), when very few small animals were measured.

Spine and body lengths for all ovigerous, and hence fully grown, females in the samples of the SS 3 and LS 1 clones are shown in Table 1. LS 1 females were much smaller and had longer spines, both

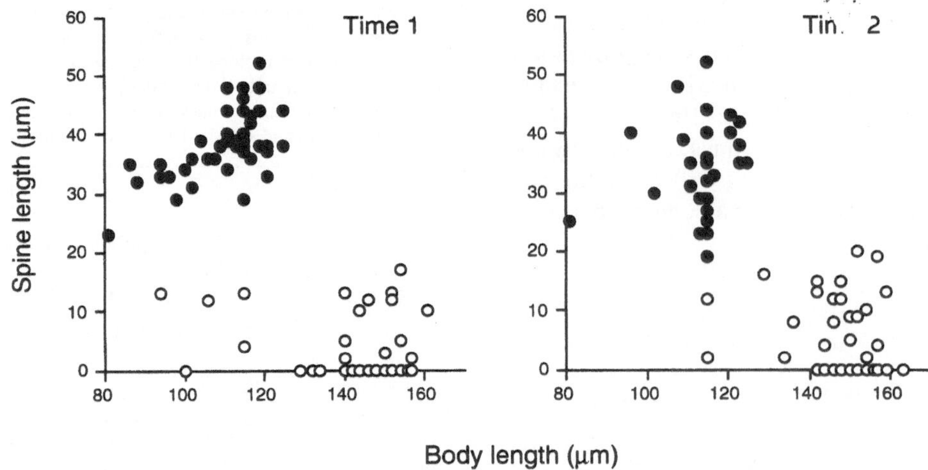


*Figure 1.* Spine and body (lorica) lengths of basic phenotypes of females of various ages from a short-spined clone (SS 3), open circles, and a long-spined clone (LS 1), closed circles, of *Brachionus quadridentatus*. Measurements taken on two separate dates – about 3 weeks (Time 1) and 5 weeks (Time 2) after initiation of culture.

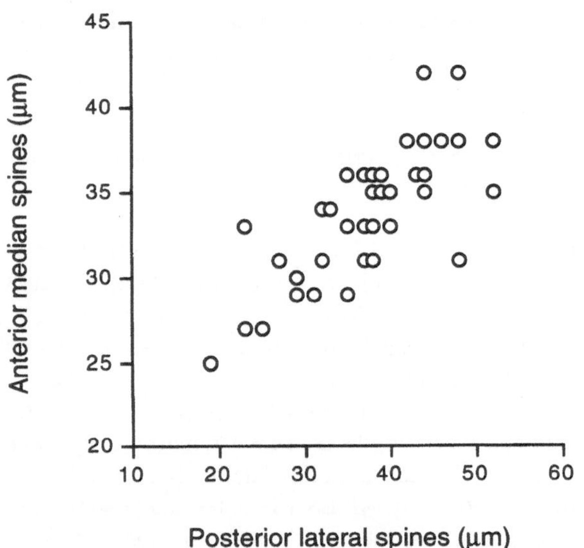


*Figure 2.* Correlation between lengths of posterior lateral and anterior median spines of basic phenotypes of ovigerous females from a long-spined clone (LS 1) of *Brachionus quadridentatus*. Correlation coefficient is 0.763 ($p$<0.0001).

in absolute length and especially in proportion to body length. Posterior lateral spines of LS 1 females ranged from 19 to 52 $\mu$m, while those of SS 3 females ranged from 0 to 19 $\mu$m. Anterior median spines of LS 1 females also were consistently longer than those of SS 3 females, and the longest ones curved ventrally. Anterior lateral spines of LS 1 females were slightly longer than those of SS 3 females. The shorter, anterior submedian spines were not measured.

*Table 2.* Statistical analysis (nested ANOVA) of effects of *Asplanchna* and mother on spine development of offspring in *Brachionus quadridentatus*

| Clone | Spine | Effect | df | *F* ratio | *p* |
|---|---|---|---|---|---|
| SS 1 | posterior lateral | *Asplanchna* | 1 | 606.4 | <0.0001 |
| | | mother | 5 | 10.6 | <0.0001 |
| SS 2 | posterior lateral | *Asplanchna* | 1 | 341.3 | <0.0001 |
| | | mother | 16 | 3.4 | 0.0001 |
| | anterior median | *Asplanchna* | 1 | 16.8 | 0.0001 |
| | | mother | 16 | 0.8 | 0.69 |
| | anterior submedian | *Asplanchna* | 1 | 23.6 | <0.0001 |
| | | mother | 16 | 1.8 | 0.048 |
| | anterior lateral | *Asplanchna* | 1 | 26.4 | <0.0001 |
| | | mother | 16 | 1.9 | 0.038 |
| LS 2 | posterior lateral | *Asplanchna* | 1 | 19.7 | 0.0006 |
| | | mother | 10 | 3.3 | 0.0015 |

In adult LS 1 females, there was a highly significant, positive correlation between the lengths of the posterior lateral and anterior median spines (Fig. 2, $r$=0.763, $p$<0.0001). Thus, while there was considerable variation in spine development among females of this clone, the lengths of these two most pronounced pairs of spines were closely coupled.

The SS 3 and LS 1 clones produced resting eggs that hatched into females reflecting the phenotype of their parental clones. Ten SS 3-clone resting eggs all hatched into females with short posterior lateral spines, while 7 LS 1-clone resting eggs all hatched into females with much longer posterior lateral spines.

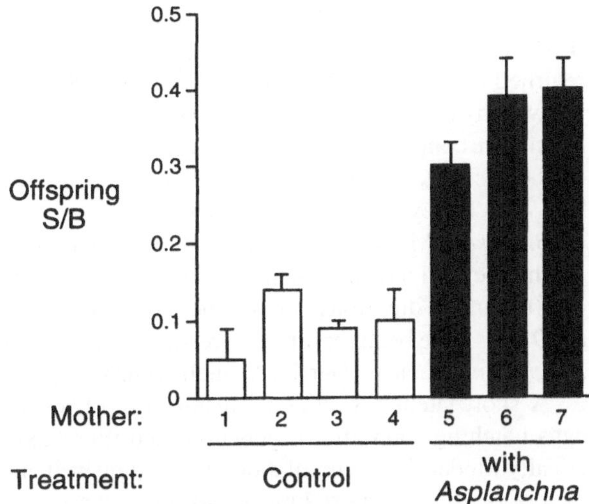


*Figure 3.* Mean spine to body (lorica) length ratios (S/B) of posterior lateral spines of offspring of mothers from a short-spined clone (SS 1) of *Brachionus quadridentatus* cultured with (closed bars) and without (open bars) *Asplanchna girodi*. Error bars are 1 SD. See Table 2 for statistics.

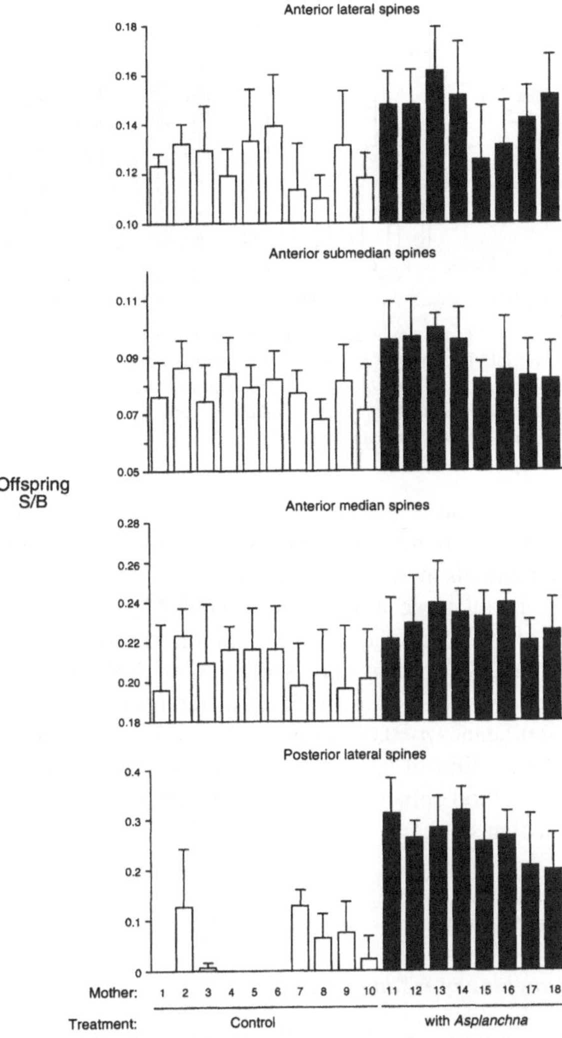


*Figure 4.* Mean spine to body (lorica) length ratios (S/B) of posterior lateral and three pairs of anterior spines of the offspring of mothers from a short-spined clone (SS 2) of *Brachionus quadridentatus* cultured with (closed bars) and without (open bars) *Asplanchna girodi*. Error bars are 1 SD. See Table 2 for statistics.

### *Effect of* Asplanchna *on spine development*

Two short-spined clones, clones with females having a basic phenotype with no or short posterior lateral spines, were derived from resting eggs in the sediment – one in January 1998 (clone SS 1) and one in February 1998 (clone SS 2). In both of these clones, the presence of *Asplanchna* induced females to produce offspring with longer spines (Figs. 3 and 4, Table 2). Mean posterior lateral spine to body length ratios of offspring from females in the *Asplanchna* treatments were much greater (ANOVA, $p<0.0001$) than those from females in the control treatments – on average, 3.8 times greater in the SS 1 clone and 6.2 times greater in the SS 2 clone, in which 4 of the 10 mothers produced only offspring with no posterior lateral spines. In clone SS 2, the three pairs of anterior spines also were measured. Spine to body length ratios of these spines were significantly greater (ANOVA, $p=0.0001$) in offspring of females cultured with *Asplanchna*.

A long-spined clone, with females having a basic phenotype with long posterior lateral spines, was derived from a female hatching from a resting egg produced by the LS 1 clone (see above). The presence of *Asplanchna* induced females of this LS 2 clone to produce offspring with even longer posterior lateral spines (Fig. 5, Table 2). On average, mean spine to body length ratios of offspring in the *Asplanchna* treatment were 0.39, while those of offspring in the control treatment were 0.27. This 1.4-fold difference was highly significant (ANOVA, $p=0.0006$).

The effect of *Asplanchna* on the long-spined clone was primarily on young individuals. Without *Asplanchna*, there was a highly significant correlation between spine length and body length, with larger individuals having longer spines (product-moment correlation coefficient, $r=0.654$, $p<0.0001$). With *Asplanchna*, there was a trend in this direction, but the relationship was not significant ($r=0.305$, $p=0.102$). Spines of small individuals were almost as long as

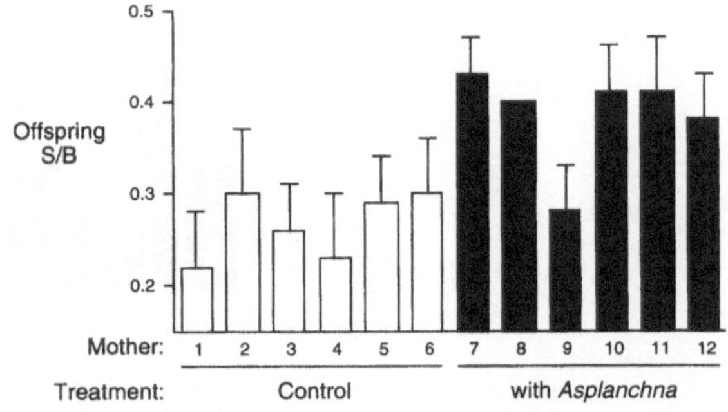


*Figure 5.* Mean spine to body (lorica) length ratios (S/B) of posterior lateral spines of the offspring of mothers from a long-spined clone (LS 2) of *Brachionus quadridentatus* cultured with (closed bars) and without (open bars) *Asplanchna girodi*. Error bars are 1 SD. See Table 2 for statistics.

those of large individuals, and spines of large individuals were not appreciably longer than those of large individuals in the control treatment.

In all three clones, the nested ANOVAs (Table 2) showed that there were pronounced differences in spine development among offspring of different mothers in a given treatment. These differences were highly significant ($p<0.001$) for the posterior lateral spines, just significant ($p<0.05$) for the anterior submedian and lateral spines, and not significant for the anterior median spines.

## Discussion

### *Hatching of resting eggs*

Resting eggs of *B. quadridentatus* in the dried sediment hatched as soon as 2–3 days after mixing the sediment with culture medium. Therefore, natural flooding of dried sediment in billabongs which have partially or completely dried should quickly lead to the hatching of this and probably other rotifer species. Introductions of new rotifers into pre-existing communities from the hatching of resting eggs in recently flooded sediment are immigration events that may facilitate the maintenance of a diversity of genotypes within a species and of species within a community. While resting eggs certainly would be present in water-covered sediments from the relatively deep basins of incompletely dried billabongs, those that had dried and were recently flooded in the more shallow areas may be more likely to hatch.

Resting eggs of *B. quadridentatus* produced by laboratory cultures hatched at a very high rate (76–96%) whether they were kept in culture medium at 20 °C (L:D 16:8), or whether they were dried at 20 °C (L:D 16:8) or stored at 5 °C in the dark before being returned to culture medium at 20 °C (L:D 16:8). It is noteworthy that 96% of the resting eggs maintained in the culture conditions under which they were produced (20 °C, L:D 16:8) hatched after 7–23 days. Similarly, Lite & Whitney (1925) noted that most resting eggs of *B. bakeri* Müller (=*B. quadridentatus* Hermann) hatched several days after being produced. Latent periods (minimum durations of dormancy) ranging from 2 to 29 days have been reported for resting eggs of other *Brachionus* species (Pourriot & Snell, 1983).

A short latent period could result in most resting eggs hatching soon after production, and thus theoretically hinder the accumulation of an extensive bank of resting eggs in the sediment. However, the occurrence of many resting eggs of *B. quadridentatus* in the dried sediment from Ryan's 2 billabong indicates that this species does have a large bank of resting eggs in this system, just as other rotifers do in other systems (Gilbert, 1974; Pourriot & Snell, 1983; Snell et al., 1983). This apparent paradox may be explained if resting eggs quickly settle to the bottom where conditions are unsuitable for hatching (Gilbert, 1993). In Ryan's 2 billabong, temperature and light conditions at the bottom probably should permit hatching (Blanchot & Pourriot, 1982; Pourriot et al., 1982, 1983; Pourriot & Snell, 1983), but concentrations of dissolved oxygen may be too low for hatching. For example, resting eggs of *B. bakeri* hatched when transferred from stale to aerated medium (Lite & Whitney, 1925).

### *Genetic variation in spine development and body size*

Genotypes with very different basic phenotypes clearly exist in Ryan's 2 billabong. In clones from 3 resting eggs, females had no or only short posterior lateral spines. In clones from 2 resting eggs, females always had long posterior lateral spines. Measurements of females from one long-spined clone (LS 1) and one short-spined clone (SS 3) from samples taken at two times separated by about 2 weeks showed that the long-spined clone, on two separate occasions, had much longer posterior lateral spines, longer anterior median and lateral spines, and much smaller bodies (Fig. 1, Table 1). Thus, the spines of the long-spined clone were especially long relative to body length. Mean spine to body length ratios for adult females from the long-spined and short-spined clones were, respectively, 0.33 and 0.02 for posterior lateral spines and 0.30 and 0.18 for anterior median spines (see Table 1).

In adult females from the long-spined clone (LS 1), there was considerable individual variation in the lengths of the different spines (Table 1). However, the length of the different spines did not vary independently. Females with relatively long posterior lateral spines also had relatively long anterior spines. This was particularly true for the anterior median spines, where the correlation was highly significant ($p<0.0001$, Fig. 2).

It is not known if there typically are two types of *B. quadridentatus* clones in Ryan's 2 billabong – one like LS 1 with small (ca. 116 $\mu$m) bodies and long spines, and one like SS 3 with large bodies (ca. 151 $\mu$m) and short spines. If this were the case, it might indicate either a trade-off between body size and spine development, or a selection for long spines in genotypes with small bodies – perhaps to reduce mortality from invertebrate predators that prefer small prey. The presence of two categories of phenotypes could reflect a genetic polymorphism within a single species, or the existence of two separate, sibling species. In *B. plicatilis*, small and large phenotypes are now known to be distinct species – *B. rotundiformis* and *B. plicatilis*, respectively (see Segers, 1995).

It is noteworthy that *B. quadridentatus* hatching from resting eggs may have no or pronounced posterior lateral spines. Whether or not resting eggs hatch into females with these spines probably depends on whether the mictic females and males producing them belonged to clones with genotypes coding for long spines. This is supported by the fact that resting eggs produced by a long-spined clone (LS 1) hatched into females with longer spines than those hatching from resting eggs produced by a short-spined clone (SS 3) (J.J. Gilbert, unpublished).

The finding that resting eggs may hatch into females with posterior lateral spines in *B. quadridentatus* appears to differ from the situation in *B. calyciflorus*. Females hatching from *B. calyciflorus* resting eggs have never been observed to have posterior lateral spines, even though they may be the progenitors of clones with a basic phenotype having such spines. The pattern where *B. calyciflorus* clones develop posterior lateral spines only after one to several generations from the parental resting-egg generation has been noticed many times with resting eggs from a Gainesville, Florida population (Aquaculture Supply, Dade City, Florida) and on two occasions with resting eggs hatching from dried sediment from Ryan's 2 billabong (J.J. Gilbert, unpublished).

The genetic variation for spine length observed among *B. quadridentatus* clones in Ryan's 2 billabong may be maintained by fluctuating or disruptive selection. Long-spined clones may be more fit than short-spined clones only at some times or in some microhabitats. Of the 24 females hatching from resting eggs in two batches of dried sediment from this billabong, 3 or 12% had posterior lateral spines, while 21 or 88% did not. Some of these females without posterior lateral spines were cultured and found to produce clones with no or only short posterior lateral spines. Thus, it seems likely that all 88% of the females that hatched from the sediment and had no posterior lateral spines would be progenitors of short-spined clones. The preponderance of short-spined clones in the sediment suggests selection for these clones.

### *Induction of spine development by* Asplanchna

*Asplanchna girodi* induced females from two short-spined clones (SS 1 and 2) and one long-spined clone (LS 2) to produce offspring with significantly longer posterior lateral spines (Figs. 3–5, Table 2). The *Asplanchna* effect was most dramatic in the short-spined clones where the basic phenotype had no or only short spines. In the long-spined clone, spine to body length ratios were higher in the *Asplanchna* treatment, but the effect appeared to be primarily on juveniles. While the spines of juveniles born in the *Asplanchna* treatment were relatively long, they appeared to grow little as body length increased.

*Asplanchna* also induced females from a short-spined clone (SS 2) to produce offspring with significantly longer anterior spines (Fig. 4, Table 2). This is the third type of comparison in the study showing that *B. quadridentatus* with relatively long posterior lateral spines has relatively long anterior spines as well: (1) basic phenotypes of adults from short- and long-spined clones, (2) basic phenotypes of long-spined adults with posterior lateral spines of different lengths, and (3) basic and *Asplanchna*-induced phenotypes from a short-spined clone. Clearly, both genetic and environmental factors that affect spine development in this species affect the growth of several sets of spines simultaneously. Most notably, individuals with much longer posterior lateral spines also have much longer anterior median spines. The ability of *Asplanchna* to induce longer anterior as well as posterior

*Table 3.* Reports of predator-induced morphological defenses in rotifers

| Rotifer | | Predator(s) | Reference(s) |
|---|---|---|---|
| Genus | Species | | |
| *Brachionus* | *bidentatus* | *Asplanchna* | 9 |
| | *calyciflorus* | | 1, 2, 6, 11 |
| | *quadridentatus* | | 3 |
| | *sericus* | | 8 |
| *Plationus* | *patulus* | *Asplanchna* | 10 |
| *Keratella* | *cochlearis* | *Asplanchna* | 4, 13 |
| | | copepods | |
| | | cladocerans | |
| | *quadrata* | *Diacyclops thomasi* | 12 |
| | *slacki* | *Asplanchna* | 5 |
| | *testudo* | *Asplanchna* | 14 |
| | | copepods | |
| | | cladocerans | |
| | *tropica* | copepods | 7, 15 |
| | | cladocerans | |
| *Filinia* | *mystacina* | *Asplanchna* | 8 |

**References:** (1) Beauchamp, 1952; (2) Gilbert, 1966, 1967; (3) Gilbert, this study; (4) Gilbert & MacIsaac, 1989; (5) Gilbert & Stemberger, 1984; (6) Halbach, 1970; (7) Marinone & Zagarese, 1991; (8) Pourriot, 1964; (9) Pourriot, 1974; (10) Sarma, 1987; (11) Stemberger, 1990; (12) Stemberger, unpublished; (13) Stemberger & Gilbert, 1984; (14) Stemberger & Gilbert 1987; (15) Zagarese & Marinone, 1992.

lateral spines has been demonstrated previously in *B. calyciflorus* (Gilbert, 1967), *B. bidentatus* (Pourriot, 1974), *Keratella slacki* (Gilbert & Stemberger, 1984) and *K. testudo* (Stemberger & Gilbert, 1987).

In all three experiments testing the effect of *Asplanchna* on spine development, there was significant variation in spine development among offspring of different mothers within treatments (Table 2). These differences were greatest for the posterior lateral spine. The basis for such differences among mothers from the same clone and under the same conditions is not known. Comparable non-genetic variation has been noted regarding the tendency of females to be mictic. In *Asplanchna brightwelli*, amictic females differed significantly from one another in the proportion of mictic daughters they produced (Gilbert, 1968). Variation in this trait in *Notommata copeus* is known to be associated with the age of the mothers of these amictic females (Clément & Pourriot, 1979).

*Asplanchna*, and some crustacean predators and competitors, can induce spine development in a variety of rotifers (Gilbert, 1999). All reports of such induction known to date are listed in Table 3. The induction is mediated by kairomones that are released by the predator or competitor and that somehow cause females to produce eggs that develop into offspring with longer and sometimes new spines (Gilbert, 1999). This study on *B. quadridentatus* documents the fourth *Brachionus* species, and the ninth rotifer species, so affected by *Asplanchna*. Two species of *Keratella* that develop longer spines with crustaceans (*K. quadrata* and *K. tropica*) have not been tested with *Asplanchna*.

The variation in spine development observed in one clone of *B. quadridentatus* in the present study includes much of the variation known for the species (Koste & Shiel, 1987, Fig. 16). Thus, phenotypes with no posterior lateral spines, one or two short posterior lateral spines, or long, *Asplanchna*-induced posterior lateral and anterior median spines – as in the subspecies *mehleni* – were all expressed by females from a short-spined clone. Phenotypes spanning this range of variation have been collected from Ryan's 2 billabong over a 2-year period (R. Shiel, pers. comm.). Of particular interest is the observation that the very long-spined, *melheni* phenotypes typically were associated with the presence of *Asplanchna*, either *A. asymmetrica*, *A. brightwelli* or *A. sieboldi* (R. Shiel, pers. comm.). A strong association between long-spined *B. calyciflorus* and *Asplanchna* has been reported by several investigators (Gilbert, 1967; Gilbert & Waage, 1967; Halbach, 1970).

It is tempting to believe that long-spined phenotypes of *B. quadridentatus*, as in *B. calyciflorus*, are adaptations to reduce predation by *Asplanchna*. Certainly the pronounced developmental response of short-spined clones of *B. quadridentatus* to *Asplanchna* observed in the present study is consistent with this idea. However, there is no evidence that *Asplanchna* actually eats *B. quadridentatus* or is more likely to ingest short-spined phenotypes than long-spined ones. In the present study, females from laboratory cultures of all *B. quadridentatus* clones were almost always attached to the bottom of the vessel or to the water surface, and the presence of *Asplanchna girodi* did not appear to cause any mortality (J.J. Gilbert, unpublished). The firm attachment of *B. quadridentatus* to a substratum may prevent *Asplanchna* from capturing this rotifer.

In Ryan's 2 billabong, and certainly other ecosystems as well, *B. quadridentatus* probably is epiphytic on littoral and floating vegetation and only planktonic when dislodged from its substratum (Shiel, pers. comm.). For example, it is commonly found on the roots of *Azolla* or within the rolled up scales of *Ricciocarpus* (J. De Manuel & R. Shiel, unpublished). Accordingly, it seems unlikely that *B. quadridentatus* would be susceptible to *Asplanchna* predation. It is interesting to note that the stomachs of *A. brightwelli* collected from a Javanese sewage pond in which *B. quadridentatus* occurred contained no individuals of this species (Green & Lan, 1974), and that stomach content analyses of *A. intermedia* from Delhi ponds showed that this rotifer strongly avoided the periphytic rotifers *Brachionus budapestinensis* and *Lecane bulla* (Iyer & Rao, 1996). Thus, in *B. quadridentatus*, the ecological significance of long-spined, basic phenotypes, and of the spine-development response to *Asplanchna*, is unclear. Notably, *B. quadridentatus* was the only surviving rotifer found in nearby Ryan's 3 billabong (Shiel et al. 1998) in the presence of the carnivorous centropagid calanoid *Boeckella major*, although numerous other rotifer species had been preyed upon by the copepod (Green & Shiel 2000). The long-spined phenotype may provide some deterrence to handling by copepod predators.

Of the nine rotifer species known to have developmental responses to *Asplanchna* (Table 3), seven are planktonic and thus clearly susceptible to predation by *Asplanchna*. In addition to *B. quadridentatus*, which is epiphytic, the only other non-planktonic species is *Plationus patulus*, which is primarily associated with sediments, detritus and water plants (Koste, 1978; Koste & Shiel, 1987) and which, in laboratory cultures, attaches to the substratum by a mucus strand (Hampton & Gilbert, 2001). Of these two species, *B. quadridentatus* has the most pronounced response. In *P. patulus*, *A. brightwelli* induces an elongation of the right posterior and posterior lateral spines (Sarma, 1987).

## Acknowledgements

The author thanks María Diéguez, Robert L. Wallace and an anonymous referee for reviewing the manuscript, is most grateful to Stephanie Hampton for improving the manuscript and conducting the nested ANOVAs, and is indebted to Russell J. Shiel for enabling his visit to the Murray-Darling Freshwater Research Centre (Albury, New South Wales, Australia) and for providing unpublished observations on *Brachionus quadridentatus* and *Asplanchna* in Ryan's 2 billabong.

## References

Blanchot, J. & R. Pourriot, 1982. Effets de l'intensité d'eclairement et de la longueur d'onde sur l'éclosion des œufs de durée de *Brachionus rubens* (Rotifère). C.R. Acad. Sci. 295: 123–125.

Clément, P. & R. Pourriot, 1979. Influence de l'âge des grand-parents sur l'apparition des mâles chez le Rotifère *Notommata copeus* Ehr. Internat. J. Invert. Reproduction 1: 89–98.

De Beauchamp, P., 1952. Un facteur de la variabilité chez les rotifères du genre *Brachionus*. C. R. Acad. Sci. 234: 573–575.

Gilbert, J. J., 1966. Rotifer ecology and embryological induction. Science 151: 1234–1237.

Gilbert, J. J., 1967. *Asplanchna* and posterolateral spine induction in *Brachionus calyciflorus*. Arch. Hydrobiol. 64: 1–62.

Gilbert, J. J., 1968. Dietary control of sexuality in the rotifer *Asplanchna brightwelli* Gosse. Physiol. Zool. 41: 14–43.

Gilbert, J. J., 1974. Dormancy in rotifers. Trans. am. Microsc. Soc. 93: 490–513.

Gilbert, J. J., 1993. Rotifera. In K. G. & R. G. Adiyodi (eds), Reproductive Biology of Invertebrates. Vol. VI, Part A. Asexual Propagation and Reproductive Strategies. Oxford & IBH Publishing Co., New Delhi: 231–263.

Gilbert, J. J., 1999. Kairomone-induced morphological defenses in rotifers. In Tollrian, R. & C. D. Harvell (eds), The Ecology and Evolution of Inducible Defenses. Princeton University Press, Princeton, N.J.: 127–141.

Gilbert, J. J. & H. J. MacIsaac, 1989. The susceptibility of *Keratella cochlearis* to interference from small cladocerans. Freshwat. Biol. 22: 333–339.

Gilbert, J. J. & R. S. Stemberger, 1984. *Asplanchna*-induced polymorphism in the rotifer *Keratella slacki*. Limnol. Oceanogr. 29: 1309–1316.

Gilbert, J. J. & J. K. Waage, 1967. *Asplanchna*, *Asplanchna*-substance and posterolateral spine length variation of the rotifer *Brachionus calyciflorus* in a natural environment. Ecology 48: 1027–1031.

Green, J. & O. B. Lan, 1974. *Asplanchna* and the spines of *Brachionus calyciflorus* in two Javanese sewage ponds. Freshwat. Biol. 4: 223–226.

Green, J. D. & R. J. Shiel, 2000. Predation by the centropagid calanoid, *Boeckella major*, structuring microinvertebrate communities in the absence of fish. Verh. Int. Ver. Limnol, 27, in press.

Halbach, U., 1970. Die Ursachen der Temporalvariation von *Brachionus calyciflorus* Pallas (Rotatoria). Oecologia 4: 262–318.

Hampton, S. E. & J. J. Gilbert, 2001. Observations of insect predation on rotifers. Hydrobiologia, in press.

Hillman, T. J. & R. J. Shiel, 1991. Macro- and microinvertebrates in Australian billabongs. Verh. int. Ver. Limnol. 24: 1581–1587.

Iyer, N. & T. R. Rao, 1996. Responses of the predatory rotifer *Asplanchna intermedia* to prey species differing in vulnerability: laboratory and field studies. Freshwat. Biol. 36: 521–533.

Koste, W., 1978. Rotatoria. Die Rädertiere Mitteleuropas (Überordnung Monogononta), Bestimmugswerk begründet von Max Voigt. 2 Vols. Gebrüder Borntraeger Stuttgart.

Koste, W. & R. J. Shiel, 1987. Rotifera from Australian inland waters. I. Epiphanidae and Brachionidae (Rotifera: Monogononta). Invert. Taxon. 1: 949–1021.

Lite, J. C. & D. D. Whitney, 1925. The role of aeration in the hatching of fertilized eggs of rotifers. J. exp. Zool. 43: 1–9.

Marinone, M. C. & H. E. Zagarese, 1991. A field and laboratory study on factors affecting polymorphism in the rotifer *Keratella tropica*. Oecologia 86: 372–377.

Pourriot, R., 1964. Étude experimentale de variations morphologiques chez certaines espéces de rotifères. Bull. Soc. Zool. Fr. 89: 555–561.

Pourriot, R., 1974. Relations prédateur-proie chez les Rotifères: influence du prédateur (*Asplanchna brightwelli*) sur la morphologie de la proie (*Brachionus bidentata*). Ann. Hydrobiol. 5: 43–55.

Pourriot, R., D. Benest & C. Rougier, 1982. Processus d'éclosion des œufs de durée de *Brachionus calyciflorus* Pallas (Rotifère). Comparaison de deux clones. Vie Milieu 32: 83–87.

Pourriot, R., D. Benest & C. Rougier, 1983. Effet de la température sur l'éclosion d'œufs de durée provenant de populations naturelles de Brachionidae (Rotifères). Bull. Soc. Zool. Fr. 108: 59–66.

Pourriot, R. & T. W. Snell, 1983. Resting eggs in rotifers. Hydrobiologia 104: 213–224.

Sarma, S. S. S., 1987. Experimental studies on the ecology of *Brachionus patulus* (Müller) (Rotifera) in relation to food, temperature and predation. PhD thesis, University of Delhi, India.

Segers, H., 1995. Nomenclatural consequencesof some recent studies on *Brachionus plicatilis* (Rotifera, Brachionidae). Hydrobiologia 313/314: 121–122.

Shiel, R. J., J. D. Green & D. L. Nielsen, 1998. Floodplain biodiversity: why are there so many species? Hydrobiologia 387/388: 39–46.

Snell, T. W., B. E. Burke & S. D. Messur, 1983. Size and distribution of resting eggs in a natural population of the rotifer *Brachionus plicatilis*. Gulf Res. Rep. 7: 285–287.

Stemberger, R. S., 1981. A general approach to the culture of planktonic rotifers. Can. J. Fish. aquat. Sci. 38: 721–724.

Stemberger, R. S., 1990. Food limitation, spination and reproduction in *Brachionus calyciflorus*. Limnol. Oceanogr. 35: 33–44.

Stemberger, R. S. & J. J. Gilbert, 1984. Spine development in the rotifer *Keratella cochlearis*: induction by cyclopoid copepods and *Asplanchna*. Freshwat. Biol. 14: 639–647.

Stemberger, R. S. & J. J. Gilbert, 1987. Multiple species induction of morphological defenses in the rotifer *Keratella testudo*. Ecology 68: 370–378.

Zagarese, H. E. & M. C. Marinone, 1992. Induction and inhibition of spine development in the rotifer *Keratella tropica*: evidence from field observations and laboratory experiments. Freshwat. Biol. 28: 289–300.

Hydrobiologia **446/447:** 29–34, 2001.
L. Sanoamuang, H. Segers, R.J. Shiel & R.D. Gulati (eds), Rotifera IX.




# The relationship between *Trichocerca pusilla* (Jennings), *Aulacoseira* spp. and water temperature in Loch Leven, Scotland, U.K.

Linda May, A.E. Bailey-Watts & A. Kirika
*Centre for Ecology and Hydrology, Edinburgh Research Station, Bush Estate, Penicuik, Midlothian EH26 0QB, Scotland, U.K.*
*E-mail: lmay@ceh.ac.uk*

*Key words: Trichocerca pusilla*, rotifer, *Aulacoseira*, diatom, grazing, silica, water temperature, Loch Leven

## Abstract

Loch Leven is a shallow, eutrophic lake in the Scottish lowlands that is famous for its brown trout (*Salmo trutta* L.) fishery. Studies of planktonic rotifer populations began here in January 1977. Since then, samples have been collected and analysed at more or less weekly intervals. Additional information on the composition and abundance of phytoplankton and crustacean zooplankton species, and on a variety of physical and chemical determinants, has been recorded on each sampling occasion.

Long-term datasets, such as that described above, are invaluable for identifying interactions between components of the plankton that only appear for short periods each year, as these interactions would probably be overlooked in data spanning a shorter period of time. This study uses the long-term data from Loch Leven to examine the food and temperature requirements of the summer rotifer species *Trichocerca pusilla* (Lauterborn). The results suggest that *T. pusilla* prefers water temperatures above 12 °C and that it feeds, primarily, on the filamentous diatom *Aulacoseira* spp. During the summer months, its abundance was closely related to the availability of this diatom. When filaments of *Aulacoseira* spp. were abundant, rotifer densities reached 1000–3000 ind. $l^{-1}$ and when they were scarce (e.g. 1980, 1997 and 1998) *T. pusilla* densities also remained low (i.e. less than 100 ind. $l^{-1}$). The reason for the success or failure of *Aulacoseira* during the summer months each year is unclear but, in general, its abundance was related to the availability of dissolved silica in the water.

## Introduction

Rotifers are an important and numerically abundant component of the plankton of most freshwater lakes, but relatively little is known about the ecology of individual species in the natural environment. This is particularly true of seasonal species that appear in the plankton for very short periods each year. This combined with the relatively long sampling intervals used in many studies (e.g. 1 month or more) means that very little data on such species are collected in studies lasting only a year or two.

The plankton of Loch Leven has been routinely monitored at weekly or fortnightly intervals since 1968, and rotifer studies began here in January 1977. This has provided a long series of data for zooplankton, algae and water chemistry that can be examined for trends, especially among seasonal species. This paper summarises long-term data and other observations from Loch Leven that contribute to our understanding of the ecology of *Trichocerca pusilla* (Lauternborn). As the long-term data are incomplete between January 1983 and December 1994 (due to a lack of funding for the analysis of samples collected), this study reports data for 1977–1982 and 1994–1998, only.

## Site description

Loch Leven (56° 12′ N, 3° 22′ W) is a shallow lake in lowland Scotland that lies in a morainic basin over Old Red Sandstone. It has a surface area of 13.3 km$^2$, and mean and maximum depths of 3.9 m and 25.5 m, respectively. The loch is eutrophic due to a relatively high phosphorus load (≈9 tonnes yr$^{-1}$) from the catchment. This arises from nutrient laden inflows draining a mainly agricultural landscape, and phosphorus-rich discharges from local sewage works and an industrial source. This, combined with the shallowness of the site and an average retention time of about 5.2 months (Smith, 1974) makes the loch very prone to algal blooms. As phytoplankton crops here are generally phosphorus (P) limited, efforts have been made to reduce the P load from the catchment first by targeting point sources and, more latterly, by targeting diffuse sources (LLCMP, 1999). The overall effect of these measures has been to reduce the P-load from about 20 tonnes per year (1985) to about 8.5 tonnes per year (1995). In addition, annual stocking with rainbow trout to improve the sport fishery began in 1993.

## Methods

Plankton and water chemistry samples have been collected from Loch Leven at more or less weekly intervals since January 1977, with little change in sampling procedures. If weather conditions were good, samples were usually collected by boat at the Reed Bower site (Fig. 1) with a weighted polythene tube (Lund, 1949). This provided an integrated water sample over the entire water column, from the surface to 0.25 m above the sediment. This was usually to a depth of 3–3.5 m. However, during adverse weather conditions (strong winds, ice cover, etc.), samples were collected from the shore by dipping a 10 l bucket into deep water close to outflow (Site SL, Fig. 1). Long-term records have shown that, because Loch Leven is well-mixed, the abundances of small planktonic organisms (such as algae and rotifers) and the water chemistry at these two sites are usually very similar at any given time (May, 1980). Surface water temperature was recorded on each sampling occasion.

The raw water samples were mixed well and subsampled for the analyses of water chemistry (250 ml), algal abundance (250 ml) and rotifer numbers (1 l or 500 ml). Dissolved silica analyses were carried out in accordance with the methods given by Golterman et al. (1978). Algal samples were preserved in Lugol's iodine, concentrated by sedimentation in a 250 ml glass measuring cylinder and counted under the microscope in a calibrated glass nannoplankton chamber (Lund, 1959, 1962; Youngman, 1971). Counts of *Aulacoseira* spp. filaments were converted to an estimate of biomass using a conversion factor of 7000 $\mu m^3$ for *per* filament. This average value was based on the measured linear dimensions of these algae in Loch Leven, approximating each filament to a simple cylindrical shape. Rotifer samples were narcotised by adding sufficient procaine hydrochloride ($NH_2.C_6H_4.COO.CH_2.CH_2.N(C_2H_5)_2.HCl$) to give a final concentration of 0.2 g $l^{-1}$ (May, 1985). The samples were then preserved in 4% formalin, concentrated using a sedimentation technique, sub-sampled and counted under $\times 100$ magnification in the counting chamber of a Wild® inverted microscope. Several sub-samples, each 3 ml in volume, were examined until at least 200 rotifers had been recorded from each sample. Species identifications were carried out on live specimens in accordance with Koste (1978).

Data on the hatching of *T. pusilla* from lake sediments collected in winter when incubated at 5 °C, 15 °C and 20 °C are also presented. The methods for, and further details of, this experiment are described in detail by May (1987).

## Results

Surface water temperature in Loch Leven ranged from 0.15 °C (on 29 December 1981) to 21.4 °C (on 11 July 1977) over the period of study (Fig. 2, upper panel). Although temperatures regularly fell below 1 °C for short periods each winter, summer maxima above 20 °C were less common. These only occurred in 1977, 1982, 1995 and 1997. The coolest summer was 1978 when the highest water temperature recorded was only 16.7 °C. In general, however, seasonal increases and decreases in water temperature followed a smooth and regular pattern each year, with the warmest period (>10 °C) occurring from the beginning of May to the end of October.

*T. pusilla* was recorded in Loch Leven during the summer months only (Fig. 2, lower panel), and its ability to achieve high population densities seemed to be related to water temperature (Fig. 3). Abundance rarely exceeded 100 ind. $l^{-1}$ when water temperatures were below 12 °C, except occasionally in autumn when the summer population was declining gradually as water temperatures began to fall, e.g. October 1981 and October 1994. However, at higher water temperatures, the species could become very abundant. Figure 3 indicates the range of values recorded in Loch Leven over the study period. The data suggest an upper boundary on abundance that is related to water temperature and below which lower values probably reflect other constraints on population growth – such as food limitation.

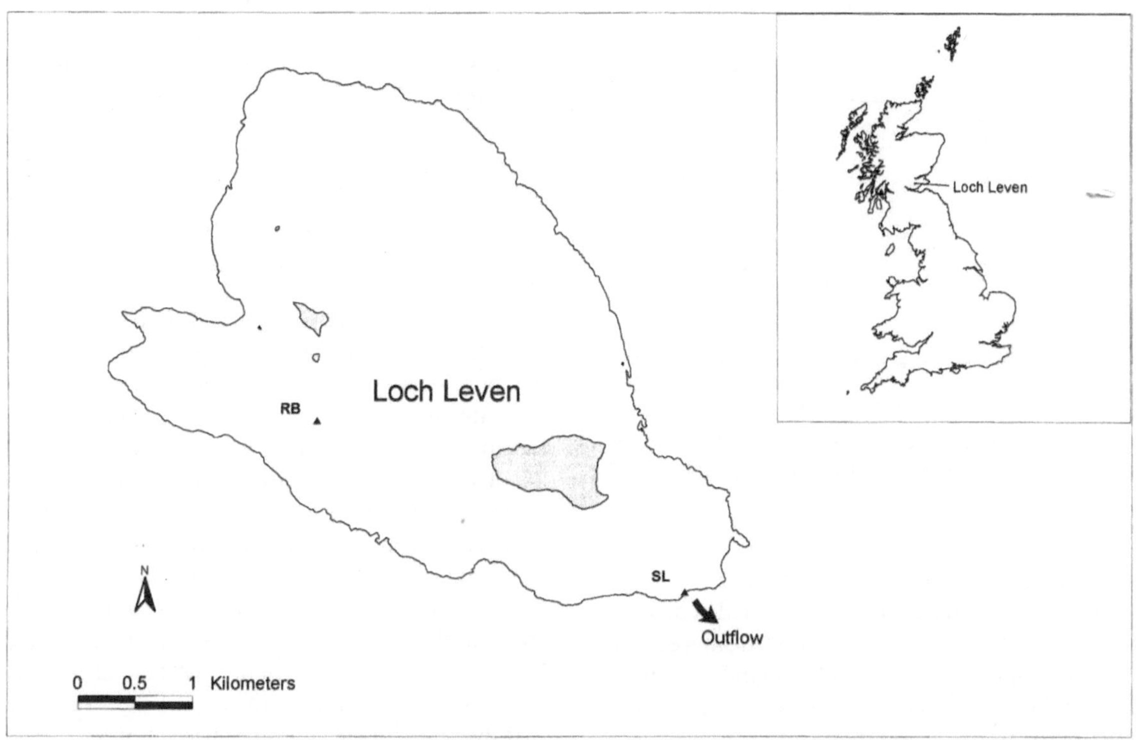


*Figure 1.* Map of Loch Leven showing the Reed Bower (RB) and Sluices (SL) sampling sites. Inset shows the location of the loch in Great Britain.

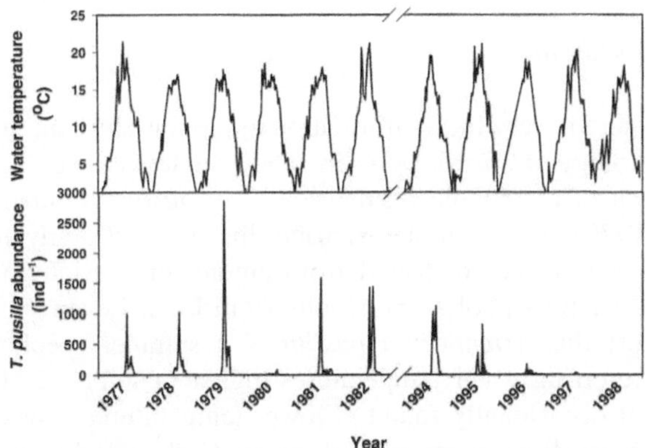


*Figure 2.* Surface water temperature (upper panel) and abundance of *Trichocerca pusilla* (lower panel) in Loch Leven, 1977–1982 and 1994–1998.

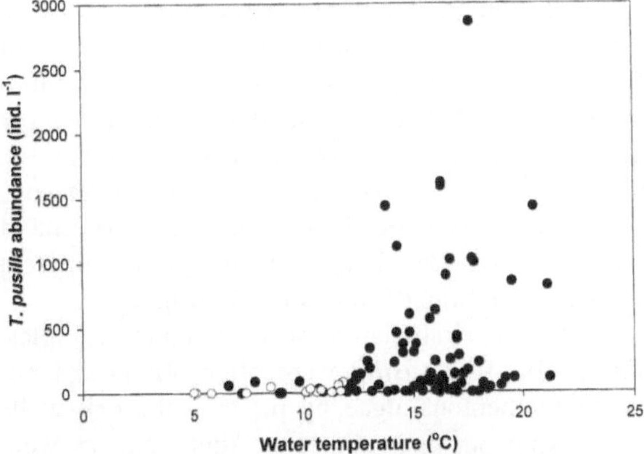


*Figure 3.* Abundance of *Trichocerca pusilla* in relation to water temperature in Loch Leven, 1977–1982 and 1994–1998. Grey circles indicate samples collected during low temperatures in autumn (see text for details).

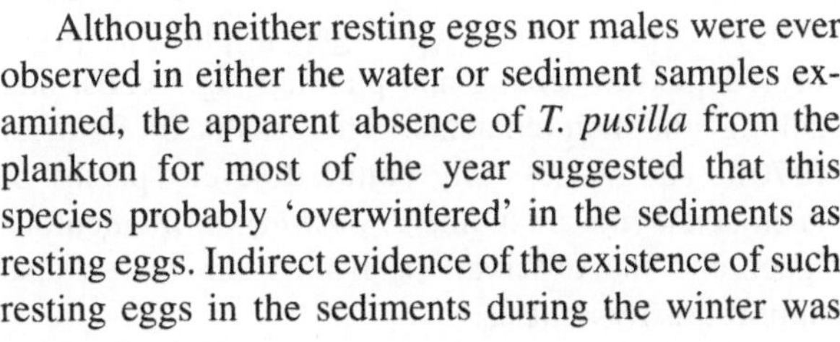

Although neither resting eggs nor males were ever observed in either the water or sediment samples examined, the apparent absence of *T. pusilla* from the plankton for most of the year suggested that this species probably 'overwintered' in the sediments as resting eggs. Indirect evidence of the existence of such resting eggs in the sediments during the winter was also provided by a sediment incubation experiment carried out in February 1984 and described in full by May (1987). In this experiment, *T. pusilla* individuals emerged from these 'winter' sediments when they were incubated at 'summer' temperatures (Fig.

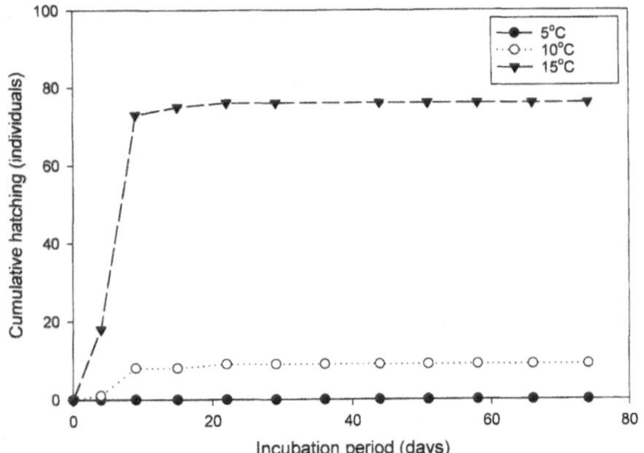


*Figure 4.* Cumulative emergence of *Trichocerca pusilla* from loch sediments incubated at different temperatures in the laboratory.

4). Hatching success was markedly greater at 15 °C than at 10 °C, and no hatching was recorded at 5 °C.

The abundance of *T. pusilla* varied from year to year. The study-period maximum (2860 ind. $l^{-1}$) was recorded on 30 July 1979 (Fig. 2). However, most annual maxima were between 1000 ind. $l^{-1}$ and 1500 ind. $l^{-1}$. Only in 1980 and from 1996 to 1998 were these values especially low (max. <100 ind. $l^{-1}$). The low population maximum recorded in 1980 appeared to be an isolated event, rather than part of a long term trend, as it was followed by population maxima of 1590 ind. $l^{-1}$ in 1981, and two similar peaks in abundance of about 1440 ind. $l^{-1}$ in 1982. The most likely explanation for this was food limitation (see below). However, from 1994 onwards, a more progressive, long-term decline in the annual population maxima of this species was recorded, culminating in its apparent disappearance from the loch in 1998 (Fig. 2).

During the examination of live plankton samples, individuals of *T. pusilla* were often observed feeding on filamentous algae by piercing the cell walls and sucking out their contents. Adult females were also seen depositing their eggs on these filaments. These observations suggested that *T. pusilla* both fed on, and attached its eggs to, filamentous algae in the loch. As the dominant filamentous algae in the loch when *T. pusilla* was abundant were *Aulacoseira* spp., and *T. pusilla* eggs had been seen attached to these filaments during the examination of the preserved samples, the monitoring data were examined for evidence of interactions between these species.

Comparison of the population dynamics of *Trichocerca* and *Aulacoseira* spp. suggested that their abundances were closely linked (Fig. 5). *T. pusilla* tended to achieve much higher population densities in years when the biomass of *Aulacoseira* in July/August was high (e.g. 1977, 1978, 1979, 1994) than in years when there was no significant summer population of *Aulacoseira* (e.g. 1980, 1997 and 1998). However, this relationship did not hold for 1996, when *Aulacoseira* was very abundant in July/August but rotifer numbers remained low, or in 1982 when *T. pusilla* abundance was higher than might have been expected given the relatively low abundance of *Aulacoseira* during this period. The reasons for these apparent deviations from the more usual patterns of abundance are unclear.

It has been shown that diatom abundance in Loch Leven tends to be closely related to the availability of dissolved silica (Bailey-Watts & Lund, 1973; Bailey-Watts, 1976a,b). This relationship is illustrated for *Aulacoseira* abundance in Figure 6, which clearly shows that it declined rapidly once the supply of dissolved silica had become limiting and open water concentrations were on the decline (Fig. 6). Figure 6 also shows that, in general, the annual maxima of *Aulacoseira* tended to increase over the study period. However, it should be noted in relation to this study that these were spring maxima that occurred outside the rotifer's preferred temperature range.

## Discussion

There is very little information in the literature on the ecology of *Trichocerca* spp., and even less concerning that of planktonic forms such as *T. pusilla*. Pourriot (1970), for example, restricts his detailed study to seven species collected from among the periphyton of shallow pools. The results from Loch Leven suggest that *Trichocerca pusilla* is a summer species, preferring water temperatures higher than 11–12 °C but occasionally found at lower temperatures. These observations agree with those of Carlin (1943) and Bērzinš & Pejler (1989).

Jennings (1903) suggested that the main food of *Trichocerca* spp. was probably small, suspended particles in the water or flocculant material on the surface of aquatic plants. However, De Beauchamp (1909) recognised that their virgate trophi were probably adapted to seizing and piercing algal cells, and sucking out their contents. The results of Pourriot (1970), who successfully cultured seven species of *Trichocerca* (not including *T. pusilla*) on filamentous algae, and the observations from Loch Leven also support this hypothesis.

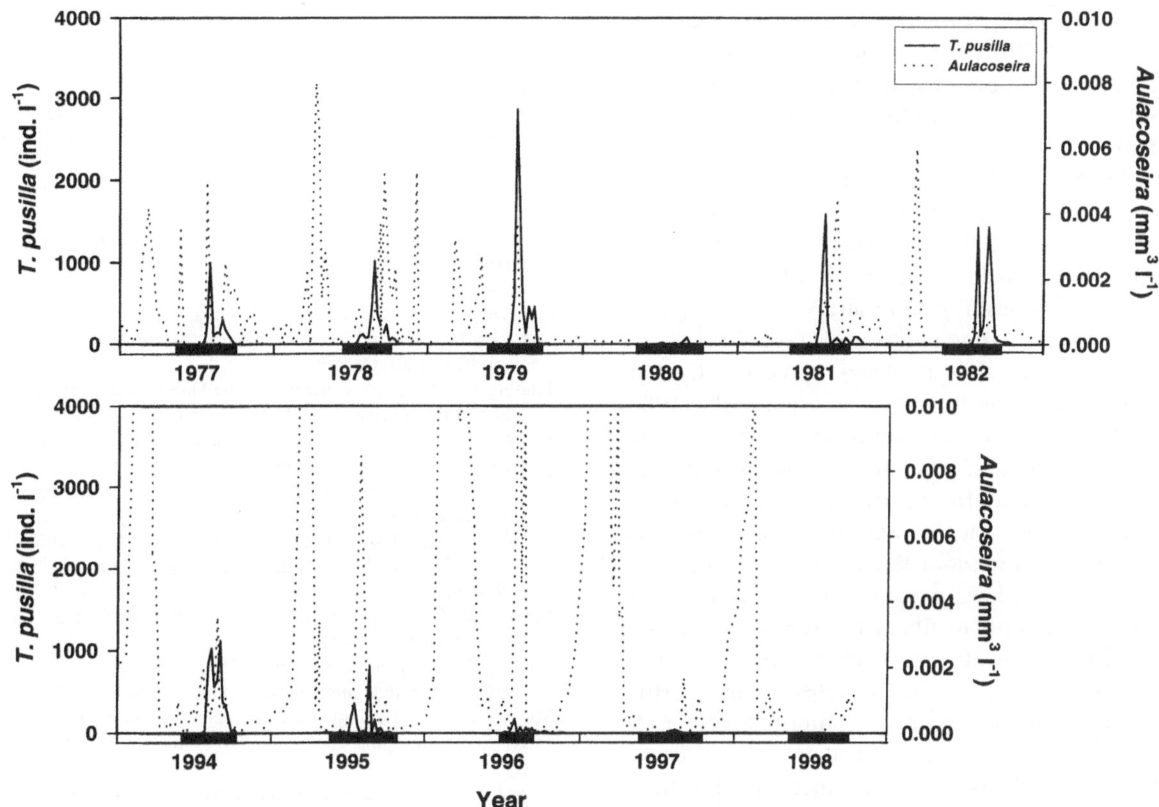


*Figure 5.* The relationship between *Aulacoseira* spp. biomass and *Trichocerca pusilla* abundance in summer, 1977–1982 and 1994–1998. Periods when the temperature was ≥ 12 °C are indicated by the thick bars along the horizontal axes.

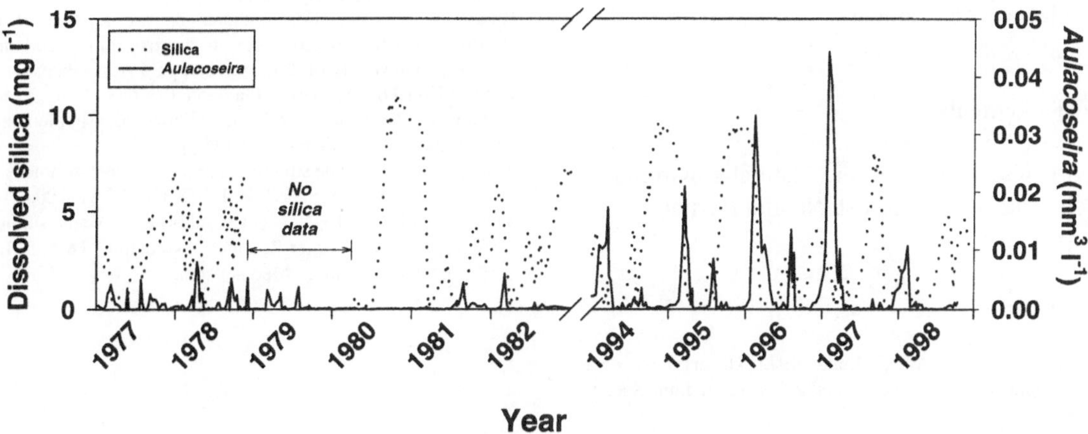


*Figure 6.* The relationship between dissolved silica concentrations and *Aulacoseira* spp. biomass, 1977–1982 and 1994–1998.

The literature also suggests that, in general, planktonic *Trichocerca* species attach their eggs to other rotifers, especially *Brachionus angularis* (Ruttner-Kolisko, 1974; Koste, 1976). This does not seem to be the case in Loch Leven where *B. angularis* was rarely recorded. Here, eggs were frequently seen attached to filaments of *Aulacoseira* spp., and other filamentous algae, but not to other rotifers. There is little information on the sexual periods or resting eggs of *T. pusilla* in the literature (Ruttner-Kolisko, 1974). However, evidence from Loch Leven suggests that, like many other rotifer species, *T. pusilla* survives as resting eggs in the sediments during unfavourable conditions.

*T. pusilla* is also predated upon by *Cyclops abyssorum* (Sars) in Loch Leven. Rutkowski (1980) found the remains of this rotifer among the gut contents of copepodite stages 3 to adult during her study of plankton samples from Loch Leven in 1978. However, recognisable remains were found in <2% of the guts examined, suggesting that losses due to predation from *Cyclops* are low. In addition,*C. abyssorum* feeds heavily on *Aulacoseira* filaments (Rutkowski, 1980). Some of the filaments grazed by *Cyclops* will probably have *Trichocerca* eggs attached, thus exerting some indirect predation on the rotifer population. As both *Trichocerca* and *Cyclops* feed on *Aulacoseira* filaments, there will also be some effects of competition for food between these species.

The reason for the general decline in *T. pusilla* abundance from 1994 onwards seemed to be, at least partly, related to changes in the timing and abundance of *Aulacoseira* populations. The reason for these changes in *Aulacoseira* abundance is unclear, but it may be related to recent changes in the management of the loch and its catchment.

## Acknowledgements

This work was funded by the Natural Environment Research Council and Scottish Natural Heritage.

## References

Bailey-Watts, A. E. & J. W. G. Lund, 1973. Observations on a diatom bloom in Loch Leven, Scotland. Biol. J. linn. Soc. 5: 235–253.

Bailey-Watts, A. E., 1976a. Planktonic diatoms and silica in Loch Leven, Scotland: a one month silica budget. Freshwat. Biol. 6: 203–213.

Bailey-Watts, A. E., 1976b. Planktonic diatoms and some diatom-silica relations in a shallow eutrophic Scottish loch. Freshwat. Biol. 6: 69–80.

De Beauchamp, P., 1909. Recherches sur les Rotifères: les formations tégumentaires at l'appareil digestif. Arch. Zool. exp. gJn. 10: 1–410.

Bērziņš, B. & B. Pejler, 1989. Rotifer occurrence in relation to temperature. Hydrobiologia 175: 223–231.

Carlin, B., 1943. Die Planktonrotatorien des Motalaström: zur taxonomie und Ökologie der Planktonrotatorien. Medd. Lunds Univ. limnol. Inst. 5: 1–255.

Golterman, H. L., R. S. Clymo & M. A. M. Olmstad, 1978. Methods for the Physical and Chemical Analysis of Freshwaters. 2nd edn. Publ. Blackwell, Oxford.

Jennings, H. S., 1903. Rotatoria of the United States. II. A monograph of the *Rattulidae*. Bull. U.S. Fish Comm.: 273–352.

Koste, W., 1976. Rotatoria: Die Rädertiere Mitteleuropas. Publ. Gebrhder Borntraeger, Berlin: 673 pp.

LLCMP (Loch Leven Catchment Management Project), 1999. The Loch Leven Catchment Management Plan: 93 pp.

Lund, J. W. G., 1949. Studies on *Asterionella* I. The origin and nature of the cells producing seasonal maxima. J. Ecol. 37: 389–419.

Lund, J. W. G., 1959. A simple counting chamber for nannoplankton. Limnol. Oceanogr. 4: 5–75.

Lund, J. W. G., 1962. Concerning a counting chamber for nannoplankton described previously. Limnol. Oceanogr. 7: 261–262.

May, L., 1980. Ecology of planktonic rotifers at Loch Leven, Kinross-shire. Ph.D Thesis, Paisley College of Technology: 180 pp.

May, L., 1983. Rotifer occurrence in relation to water temperature in Loch Leven, Scotland. Hydrobiologia 104: 311–315.

May, L., 1985. The use of procaine hydrochloride in the preparation of rotifer samples for counting. Verh. int. Ver. Limnol. 22: 2987–2990.

May, L., 1987. Effect of incubation temperature on the hatching of resting eggs collected from sediments. Hydrobiologia 147: 335–338.

Pourriot, R., 1970. Quelques *Trichocerca* (Rotifères) et leurs régime alimentaires. Ann. Hydrobiol. 1: 155–171.

Rutkowski, E. W., 1980. Studies on the feeding of *Cyclops abyssorum* Sars in Loch Leven, Kinross-shire, Scotland. MSc. Thesis, University of Stirling: 126 pp., 4 Appendices.

Ruttner-Kolisko, A., 1974. Plankton Rotifers: biology and Taxonomy. Die Binnengewässer, volume 26, supplement. Publ. Gebrüder Ranz, Dietenkeim: 146 pp.

Smith, I. R., 1974. The structure and physical environment of Loch Leven, Scotland. Proc. r. Soc. Edinb. (B) 74: 81–100.

Youngman, R. E., 1971. Algal monitoring of water supply reservoirs and rivers. Water Research Association TM63. The Water Research Association, Medmenham.

Hydrobiologia **446/447**: 35–38, 2001.
L. Sanoamuang, H. Segers, R.J. Shiel & R.D. Gulati (eds), Rotifera IX.



# Swimming speed and Reynolds numbers of eleven freshwater rotifer species

Gustavo Emilio Santos-Medrano, Roberto Rico-Martinez & César Alberto Velázquez-Rojas
*Universidad Autónoma de Aguascalientes, Centro de Ciencias Básicas, Departamento de Química, Avenida Universidad 940, C.P. 20100, Aguascalientes, Ags., México*
*E-mail: rrico@aqua.uaa.mx*

*Key words:* swimming speed, Rotifera, fluid dynamics, viscosity

## Abstract

We obtained data on the swimming speed of 11 freshwater rotifer species. These data were analyzed in terms of Reynolds numbers given the fact that rotifers are in the evolutionary boundary between the use of cilia and swimming appendages. Swimming speed ranged from 0.174 to 0.542 mm $^{-1}$. *Philodina acuticornis odiosa* was the fastest rotifer in terms of absolute speed. However, *Gastropus hyptopus*, the smallest of the rotifers analyzed, was the fastest (2.849) in terms of body-lengths $s^{-1}$. Reynolds numbers (Re) among the 11 species analyzed varied from 0.023 to 0.301. *Lecane furcata* had the smallest Re, while *Epiphanes senta* had the highest Re. The two bdelloid rotifers analyzed in this work swam five times faster than they crept. The importance of Reynolds numbers and drag coefficients are discussed in view of the present results and data found in the literature.

## Introduction

Most zooplankton including Rotifera are small, hence live at low Reynolds numbers and, therefore, inhabit a viscous environment (Purcell, 1977). Different alternatives, such at using flagella, cilia or muscles, exist as adaptations to propel through a viscous medium (e.g. Brennen & Winet, 1977), but there is not a single species that uses more than one of these in the adult stage (Sleigh & Blake, 1977). Among the few phyla that use cilia for swimming are protozoans, mollusc larvae, flatworms, ctenophores and rotifers. Epp & Lewis (1984) demonstrated that ciliary locomotion is energetically costly, and concluded that this may be the reason why ciliary locomotion is not commonly found in the animal kingdom (Epp & Lewis, 1979, 1984). Animals using cilia mostly range in size from 20 $\mu$m up to 2 mm (Sleigh & Blake, 1977), with exceptions (e.g. *Asplanchna sieboldi*: see Koste, 1978), but the efficiency of swimming by cilia is optimal in by animals smaller than 350 $\mu$m long (Sleigh & Blake, 1977). This restriction of size for the effective use cilia for swimming results from the fact that at low Re, viscous mechanism rather than inertial forces are important (Sleigh, 1989).

Drag is defined as the "rate of removal of momentum from a moving fluid by an immersed body." However, there is no universal and practical way to predict drag of even simple objects. Therefore, Reynolds number has become the most important fluid mechanics index used in biological studies to explain the interaction between solid bodies moving in liquids (Vogel, 1981). In order to quantify the energy utilization during movement of an object of known shape through a liquid, it is necessary to know the relation between Reynolds numbers (Re) and drag coefficients ($C_D$) (Vlymen, 1970).

Data on Re and $C_D$ in rotifers is scarce, and restricted to three genera: *Asplanchna, Brachionus* and *Polyarthra* (Epp & Lewis, 1984; Gilbert, 1985). Therefore, the goal of this contribution is to study the relationship between swimming speeds, Re and $C_D$ in 11 species of rotifers previously uncharacterized for these features.

## Materials and methods

Eleven rotifer species, all originating from wild populations collected in different localities in Mexico, were used in this work. All were raised in EPA medium

*Table 1.* Morphometric characterization of the 11 species used in this work, means ± SD of measurements in micrometers of 20 specimens. Abbreviations: ML = Maximum Length; MW = Maximum Width; FL = Foot Length. In the case of bdelloids, maximum length includes foot length. ML was used to calculate Re and $C_D$ in all cases

| Species: | ML | MW | FL |
|---|---|---|---|
| *Cephalodella gibba* | 133.12±17.53 | 52.12±13.90 | 25.25±2.72 |
| *Epiphanes senta* | 671.50±71.15 | 314.00±34.84 | 44.50±4.97 |
| *Gastropus hyptopus* | 109.50±18.95 | 68.12±8.79 | 13.25±1.60 |
| *Lecane furcata* | 121.87±7.97 | 76.00±4.76 | 32.12±2.40 |
| *Lecane hamata* | 141.52±6.15 | 68.12±5.74 | 33.75±4.50 |
| ∗*Lecane luna* | 201.40±9.10 | 134.50±9.10 | 67.70±5.50 |
| *Lecane pyriformis* | 133.50±8.92 | 75.62±4.99 | 37.25±3.78 |
| *Lepadella patella* | 127.35±12.97 | 68.00±3.67 | 36.37±7.00 |
| *Philodina acuticornis* | 291.66±34.58 | — | — |
| *Platyias quadricornis* | 289.00±21.42 | 226.50±18.51 | — |
| *Rotaria megaceros* | 394.43±49.37 | — | — |

*After Pérez-Legaspi & Rico-Martínez, 1998

*Table 2.* Swimming speed, Reynolds Numbers and Drag Coefficients of the eleven species used in this work

| Species | Swimming speed (mm s$^{-1}$) | Swimming speed (body length s$^{-1}$) | Reynolds Number | Drag Coefficient (Newtons) |
|---|---|---|---|---|
| *Cephalodella gibba* | 0.174±0.064 | 1.307 | 0.024 | $2.11\times10^{-8}$ |
| *Epiphanes senta* | 0.421±0.142 | 0.626 | 0.301 | $3.78\times10^{-6}$ |
| *Gastropus hyptopus* | 0.312±0.071 | 2.849 | 0.036 | $7.35\times10^{-8}$ |
| *Lecane furcata* | 0.183±0.039 | 1.501 | 0.023 | $3.14\times10^{-8}$ |
| *Lecane hamata* | 0.315±0.080 | 2.225 | 0.047 | $9.69\times10^{-8}$ |
| *Lecane luna* | 0.417±0.101 | 2.070 | 0.087 | $4.77\times10^{-7}$ |
| *Lecane pyriformis* | 0.198±0.080 | 1.483 | 0.028 | $4.01\times10^{-8}$ |
| *Lepadella patella* | 0.262±0.071 | 2.057 | 0.035 | $6.02\times10^{-8}$ |
| *Philodina acuticornis* | 0.542±0.140 | 1.858 | 0.168 | $2.88\times10^{-7}$ |
| *Platyias quadricornis* | 0.529±0.151 | 1.830 | 0.163 | $1.85\times10^{-6}$ |
| *Rotaria megaceros* | 0.490±0.178 | 1.242 | 0.206 | $3.06\times10^{-8}$ |

(U.S. EPA, 1985) and fed *Nannochloropsis oculata* (UTEX Collection LB 2164) grown in Bold's Basal Medium (Nichols, 1973) at ambient temperature in the laboratory (~28.8 °C) for at least 1 month prior to the experiments.

The density of deionized (DI) water and EPA medium was recorded using a density meter, and the dynamic viscosity with a Brookfield Viscometer Model DV-II+. Readings were taken at room temperature (~28.8 °C) in a closed room, similar to that where the experiments were conducted.

Swimming speed was estimated by videotaping ten different females at 50× magnification for several minutes in a small 100 $\mu$l chamber. We used a Hitachi KP50 Colour Digital Camera attached to an Olympus dissection microscope and a Toshiba HQ VCR in VHS format. Then, the video was replayed with a clear acetate sheet taped to the monitor and the swimming path of a rotifer was traced for 10 s. A cartometer was then used to measure the length of the path. Ten replicates were used to calculate swimming speed. This methodology for calculating swimming speed yields results comparable to when using high

*Table 3.* Creeping speed of the two bdelloid species

| Species | mm $s^{-1}$ | Body length $s^{-1}$ |
|---|---|---|
| *Philodina acuticornis* | 0.090±0.024 | 0.308 |
| *Rotaria megaceros* | 0.100±0.027 | 0.254 |

resolution cinematography (Rico-Martínez & Snell, 1997) or image analysis (Rico-Martínez, unpublished data). The mean of the body lengths of 20 individuals of each species was used to determine swimming speed in body-lengths $s^{-1}$. Morphometric data were obtained using the guidelines proposed by Stemberger (1979). Experiments were conducted at room temperature (approximately 28.8 °C). We used EPA medium values in all calculations of Re and $C_D$. Reynolds numbers and drag coefficients were calculated according to Vlymen (1970).

## Results

At 28.8 °C, the density of DI water was 1.0023 g $cm^{-3}$ and that of EPA medium was 1.0135 g $cm^{-3}$. The dynamic viscosity of DI water was 1.18 centipoisse and that of EPA medium was 0.95 centipoisse.

Morphometric data on the eleven rotifers species are as in Table 1. These, together with swimming speeds and density and dynamic viscosity data, were used to calculate Re and $C_D$ (Table 2). Swimming speeds in absolute terms ranged from 0.174 mm $s^{-1}$ in *C. gibba* to 0.542 mm $s^{-1}$ in *P. acuticornis odiosa* (Table 1). Re varied from 0.023 in *L. furcata* to 0.301 in *E. senta* (Table 2). However, the lowest drag coefficient belonged to *C. gibba*, which was also the slowest species. *Epiphanes senta*, the largest species used, had both the highest *Re* and drag coefficient (Table 2). Finally, we also recorded the creeping speed in the two bdelloids (Table 3).

## Discussion

Most of the rotifer species here investigated are smaller than the 350 $\mu$m size treshold suggested by Sleigh & Blake (1977), and only *E. senta* and *R. megaceros* exceed this value. Interestingly, both the large species have the highest Re and the slowest swimming speeds as expressed in body lengths $s^{-1}$ of all species reported, whereas *E. senta* also has the highest $C_D$ (Table 2). In contrast, *Cephalodella* sp., one of the smallest species investigated, also has the smallest $C_D$, yet swims faster, in terms of body lengths $s^{-1}$, than *E. senta* and *R. megaceros*. Therefore, our results suggest that the threshold value proposed by Sleigh & Blake (1977) is valid for rotifers, too.

Although we did not obtain direct measurements of energy expenditure for any species investigated, drag coefficients are directly related to energy expenditure and, therefore, to swimming efficiency (Vlymen, 1970). Therefore, we conclude that *E. senta* and *R. megaceros* spend more energy to move than all the other species investigated, at least with regard to theoretical power required for movement. However, some authors warn about the relationship between theoretical and actual power for movement (Epp & Lewis, 1984).

When creeping *versus* swimming speed are compared in the two bdelloid species investigated, it can be seen that bdelloids swim 5 times faster than they creep. This may indicate that swimming is a better strategy for fast swimming in bdelloids, whereas creeping may confer special advantages to bdelloids travelling at low speeds.

## Acknowledgements

We thank Hendrik Segers for identifying *Lecane furcata*, and Aydin Örstan for identifying *Philodina acuticornis odiosa*.

## References

Brennen, C. & U. Winet, 1977. Fluid mechanics of propulsion by cilia and flagella. Ann. Rev. Fluid Mech. 9: 339–398.

Epp, R. W. & W. M. Lewis, Jr., 1979. Sexual dimorphism in *Brachionus plicatilis* (rotifera): its evolutionary and adaptive significance. Evolution 33: 919–928

Epp, R. W. & W. M. Lewis, Jr., 1984. Cost and speed of locomotion for rotifers. Oecologia 61: 289–292.

Gilbert, J. J., 1985. Escape response of the rotifer *Polyarthra*: a high-speed cinematographic analysis. Oecologia 66: 322–331.

Koste, W., 1978. Rotatoria. Die Rädertiere Mitteleuropas. 2 vols., Gebrüder Borntraeger, Berlin: 673 pp.

Nichols, H. W., 1973. Growth media – freshwater. In Stein, J. R. (ed.), Handbook of Phycological Methods. Cambridge University Press, Cambridge, MA. U.S.A.: 7–24.

Nogrady, T., R. L., Wallace & T. W. Snell, 1993. Guides to the Identification of the Microinvertebrates of the Continental Waters of the World: Volume 4: Rotifera. SPB Academic Publishing, Amsterdam: 142 pp.

Pérez-Legaspi, I. A. & R. Rico-Martínez, 1998. Effect of temperature and food concentration in two species of littoral rotifers. In Wurdak, E., R. Wallace & H. Segers (eds), Rotifera VIII: A Comparative Approach. Developments in Hydrobiology 134. Kluwer Academic Publishers, Dordrecht: 341–348. Reprinted from Hydrobiologia 387/388.

Purcell, E. M., 1977. Life at low Reynolds numbers. Am. J. Phys. 45: 3–11.

Rico-Martínez, R. & T. W. Snell, 1997. Mating behavior in eight rotifer species: using cross-mating tests to study species boundaries. Hydrobiologia 356: 165–173.

Sleigh, M. A., 1989. Adaptations of ciliary systems for the propulsion of water and mucus. Comp. Biochem. Physiol. 94A: 359–364.

Sleigh, M. A. & J. R. Blake, 1977. Methods of ciliary propulsion and their size limitations. In Pedley, J. R. (ed.), Scale Effects in Animal Locomotion. Academic Press, New York: 243–256.

Stemberger, R. S., 1979. A Guide to Rotifers of the Laurentian Great Lakes. U.S. EPA publication: EPA/600/4-79/021, Washington, DC: 185 pp.

United States Environmental Protection Agency, 1985. Methods for measuring the acute toxicity of effluents to freshwater and marine organisms. In Peltier, W. H. & C. I. Weber (eds), EPA-600/4-85-013. U.S. Environmental Protection Agency, Washington, D.C.: 600 pp.

Vlymen, W. J., 1970. Energy expenditure of swimming in copepods. Limnol. Oceanogr. 15: 348–356.

Vogel, S., 1981. Life in Moving Fluids. Princeton University Press, Princeton, New Jersey, U.S.A.: 352 pp.

Hydrobiologia **446/447:** 39–44, 2001.
L. Sanoamuang, H. Segers, R.J. Shiel & R.D. Gulati (eds), Rotifera IX.




# Density-dependent regulation of natural and laboratory rotifer populations

Terry W. Snell[1], Brian J. Dingmann[1] & Manuel Serra[2]
[1]*School of Biology, Georgia Institute of Technology, Atlanta, GA 30332-0230, U.S.A.*
[2]*Departament de Ecologia i Microbiologia, Universitat de Valencia, E-46100 Burjassot, Valencia, Spain*
*E-mail: terry.snell@biology.gatech.edu*

*Key words:* population regulation, density-dependent, growth rates, models, food limitation, autotoxicity

**Abstract**

Density-dependent regulation of abundance is fundamentally important in the dynamics of most animal populations. Density effects, however, have rarely been quantified in natural populations, so population models typically have a large uncertainty in their predictions. We used models generated from time series analysis to explore the form and strength of density-dependence in several natural rotifer populations. Population growth rate ($r$) decreased linearly or non-linearly with increased population density, depending on the rotifer species. Density effects in natural populations reduced $r$ to 0 at densities of 1–10 $l^{-1}$ for 8 of the 9 rotifer species investigated. The sensitivities of these species to density effects appeared normally distributed, with a mean $r$=0 density of 2.3 $l^{-1}$ and a standard deviation of 1.9. *Brachionus rotundiformis* was the outlier with 10–100× higher density tolerance. Density effects in laboratory rotifer populations reduced $r$ to 0 at population densities of 10–100 $ml^{-1}$, which is $10^4$ higher than densities in natural populations. Density effects in laboratory populations are due to food limitation, autotoxicity or to their combined effects. Experiments with *B. rotundiformis* demonstrated the absence of autotoxicity at densities as high as 865 $ml^{-1}$, a much higher density than observed in natural populations. It is, therefore, likely that food limitation rather than autotoxicity plays a major role in regulating natural rotifer populations.

## Introduction

Density-dependence is the basis of population regulation, yet there are few data on how increased density regulates abundance, especially in natural populations. This is probably because it is difficult in most animals to measure how survival and fecundity change with density. The fundamental importance of quantifying population density effects in natural populations has been well recognized (Sibly, 1999). However, the long time series of natural population abundance required to detect density-dependence are scarce. As a result, there has been a great deal of uncertainty in animal population models, constraining their reliability and ultimately their usefulness in predicting the fate of populations. This is particularly problematic since predictive population models are becoming more essential for making good resource management decisions (Groom & Pascual, 1997).

There are several key unanswered questions about density-dependence: what is the strength of density-dependence? over what densities does it operate? how long before the effects of previous densities are removed (time lag)? what are the mechanisms of density-dependence – the relative contributions of food limitation, autotoxicity, predation, parasitism and interference competition? what kind of dynamics it generates – monotonic damping, oscillatory damping, limit cycles or chaos? Recent advances in the analysis of short, noisy time series (Turchin & Taylor, 1992; Ellner & Turchin, 1995) have made it possible to investigate density-dependent population processes in natural rotifer populations. Snell & Serra (1998) have used this approach to develop models of the dynamics of several rotifer species and to explore how perturbations affect their probabilities of extinction (Snell & Serra, 2000).

Building on these observations, in this paper we use models generated from time series analysis to explore the form and strength of density-dependence in several natural rotifer populations. We describe the densities where regulation occurs, comparing among species and contrasting field and laboratory populations. The mechanisms of density-dependence are explored in experiments on food limitation and autotoxicity.

## Methods

Density-dependence was investigated in several natural rotifer populations using time series analysis of published data. For *Asplanchna girodi*, population dynamics data for Golf Course Pond, Florida, was obtained from King & Snell (1980). Data for *Brachionus rotundiformis* were obtained for Westshore Pond, Florida, from Snell & Serra (2000). Data for *B. angularis, B. dichotomus, B. lyratus, B. budapestinensis, Filinia pejleri, Keratella tropica* and *Monostylla bulla* were obtained for a billabong on the River Murray floodplain in northeastern Victoria, Australia, from Tan & Shiel (1993).

We used a commercial computer program called Ramas/time from Applied Biomathematics (Setauket, New York) to extract deterministic dynamics from short rotifer time series data (Snell & Serra, 1998). The general model of population dynamics is:

$$N_t = F(N_{t-1}, N_{t-2}\ldots, N_{t-d}, \epsilon_t),$$

where $N_t$ is the population density at time $t$, $F$ is a function approximated using a general method of response surface fitting similar to regression analysis (Box & Draper, 1987). Several types of models were evaluated to approximate the shape of $F$, but first or second order polynomials typically yielded the best fit. Ramas/time calculated the autocorrelation function, which indicates the number of time lags ($d$) having significant effect. $\epsilon_t$ is the parameter accounting for environmental stochasticity, the exogenous, random and density-independent component of population dynamics. It is assumed here to be a random variable, normally distributed with mean 0 and variance $\epsilon_t$. This $\epsilon_t$ was estimated empirically from the observed population dynamics. Examples of exogenous factors contributing to population variance include storms, temperature extremes, cyanobacterial blooms and pollution. Ramas/time approximates the per capita rate of change ($r$) rather than population density itself. This is because a function relating $r$ to lagged densities is usually simpler and easier to approximate than a function for $N_t$ (Snell & Serra, 1998).

A similar approach was used for modeling the dynamics of laboratory rotifer populations. Time series data for laboratory cultures of *Asplanchna brightwelli* was obtained from Snell (unpublished), *B. rotundiformis* from Fu et al. (1997), *Brachionus angularis* from Walz (1983) and *Synchaeta pectinata* from Kirk (1998).

The Ramas/time calculated functions relating $r$ to lagged densities of natural and laboratory rotifer populations are presented in Table 1. These functions were used to calculate $r$'s over all intervals from $N_{t-1}$ to $N_t$ for each species, where $t = 1$ or 2 days. Then $r_{t-1}$ was plotted against log $N_t$ to test for density-dependence as suggested by Pollard et al. (1987).

Experiments to detect autotoxicity were conducted in batch culture using 5-l plastic bags. The bag was inoculated with *B. rotundiformis* at a density of 2.5 ind.ml$^{-1}$, with $10^6$ cells ml$^{-1}$ of *Tetraselmis suecica,* at 25 °C in constant light of 2000 lux. Synthetic seawater at 15 ppt was prepared with de-ionized water and Instant Ocean$^{TM}$ sea salts. Rotifer density was determined daily and 100 ml samples of the culture medium were removed, 0.45 $\mu$m filtered, and tested employing the two-day reproductive autotoxicity test on days 1, 3, 5, 7 and 8 (Snell & Moffat, 1992). For this, 6 neonate rotifers, less than 4 h old, were placed into 12 ml of filtered culture medium in each of 7 replicate test tubes containing $10^6$ *T. suecica* cells ml$^{-1}$. This design allowed us to avoid food limitation and test the culture medium for toxicity. Test tubes were incubated at 25 °C in darkness for 48 h. The contents of each tube were then emptied into a petri-dish and the number of rotifers was counted. All reproduction was parthenogenetic and the instantaneous population growth rate ($r$) was estimated as: $r = (\ln N_T - \ln N_0)/T$, where ln $N_T$ and ln $N_0$ are the natural logs of the final and initial population sizes, and $T$ is 2 days.

## Results

A test of density-dependence is to plot $r_{t-1}$ against log-transformed $N_t$. Pairs of $N_t$ and $N_{t-1}$ were taken from the time series and $r_{t-1}$ without the random component was calculated from the species-specific models. We explored this relationship for several natural populations of rotifers (Fig. 1a,b). Population growth rate declined with increasing density, nonlinearly for *Asplanchna girodi* and *B. rotundiformis* and linearly for *B. angularis, Filina pejleri, Lecane bulla* and *Keratella tropica*. A critical value is where density effects reduce $r$ to 0. The $r = 0$ density falls between 1 and 10 rotifers l$^{-1}$ for natural populations of all these

*Table 1.* Ramas/time calculated functions relating $r$ to lagged densities of natural and laboratory rotifer populations. The time unit is 1–2 days

| Species | Population growth equation |
|---|---|
| **Natural Pops** | |
| *Asplanchna girodi* | $r_t = -0.030 + 0.061N_{t-1}^{-1} + 0.0088N_{t-2}^{-1} + \epsilon_t$ |
| *Brachionus rotundiformis* | $r_t = -1.51 + 0.63N_{t-1}^{-0.5} + 0.023N_{t-2}^{-1} + \epsilon_t$ |
| *Brachionus angularis* | $r_t = -2.69 + 3.54N_{t-1}^{-1} - 0.80N_{t-1}^{-2} + \epsilon_t$ |
| *Filinia pejleri* | $r_t = -0.94 + 0.36N_{t-1}^{-1} + 0.31N_{t-2}^{-0.5} + \epsilon_t$ |
| *Keratella tropica* | $r_t = -1.42 + 0.085N_{t-1}^{-1} + 1.15N_{t-2}^{-0.5} + \epsilon_t$ |
| *Lecane bulla* | $r_t = 1.15 - 2.00N_{t-1} + 0.062N_{t-2}^{-1} + 0.98N_{t-3} + \epsilon_t$ |
| **Laboratory pops** | |
| *Asplanchna brightwelli* | $r_t = 0.036 - 0.68N_{t-1} + 0.66N_{t-2}^{-1} - 0.15\log N_{t-3} + \epsilon_t$ |
| *Brachionus rotundiformis* | $r_t = 3.26 - 3.03N_{t-3}^{0.5} + \epsilon_t$ |
| *Brachionus angularis* | $r_t = -0.16 - 0.76\log N_{t-3} + \epsilon_t$ |
| *Synchaeta pectinata* | $r_t = 0.037 - 0.64\log N_{t-3} + \epsilon_t$ |

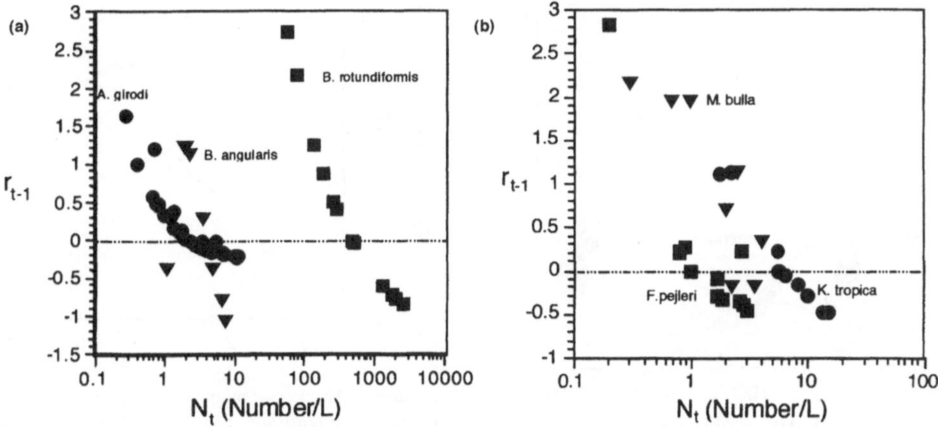


*Figure 1.* a,b) Density-dependence in natural populations of rotifers. $r_{t-1}$ is the instantaneous population growth rate in one time step ($t$), from density $N_{t-1}$ to $N_t$. Equations for $r$ for each species are given in the Methods section.

rotifers except for *B. rotundiformis*, which are clearly outliers, with the $r = 0$ density at 482 rotifers $l^{-1}$. A probability plot of the $r = 0$ densities illustrates how distinct *B. rotundiformis* is from all other rotifer species (Fig. 2). The percentile rank of the $r$=0 density of nine species is plotted against the log of the $r$=0 density. The linearity of these data, suggests that the $r$=0 density for all rotifer species except *B. rotundiformis* is normally distributed with a mean of 2.3 rotifers $l^{-1}$ and standard deviation of 1.9. In contrast to the other rotifer species, *B. rotundiformis* is about 10–100 times more tolerant of density effects.

Laboratory populations of *B. angularis, A. brightwelli, S. pectinata*, and *B. rotundiformis* are density dependent (Fig. 3), i.e. $r$ declines linearly with increasing density. Density effects reduce $r$ to 0 at 5–100 rotifers $ml^{-1}$ in the first three species. Once again, *B. rotundiformis* is 10–100 times more tolerant of density effects, with an $r = 0$ at density of about 5000 animals $ml^{-1}$. A probability plot of $r$=0 densities for laboratory rotifer populations is shown in Fig. 4. The mean density for all species except *B. rotundiformis* was 23.3±25.1 rotifers $ml^{-1}$.

In comparing density-dependence in natural and laboratory rotifer populations, one obvious difference is that $r$=0 densities in lab populations are $10^3$–$10^4$ times higher than in natural populations. For *B. rotundiformis*, for example, $r$=0 in natural populations at 482 $l^{-1}$ as compared to 5000 ind. $ml^{-1}$ in laboratory populations. The two factors most likely to

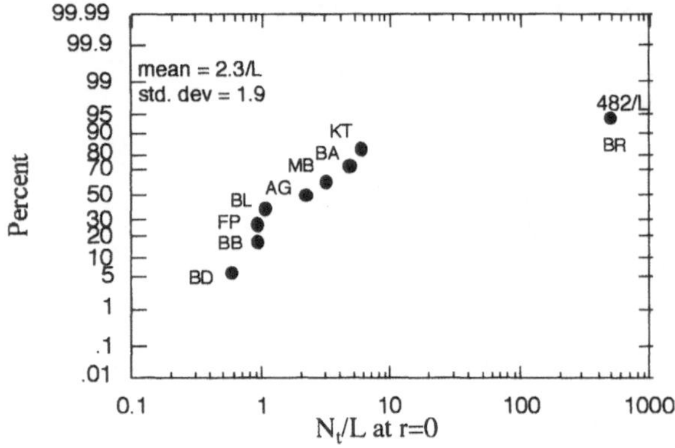


*Figure 2.* Probability plot of $r$=0 densities for natural populations of nine rotifer species. Percent on the Y axis is the percentage rank of the $r$ = 0 density. Species are abbreviated by the first letters of their genus and species names.

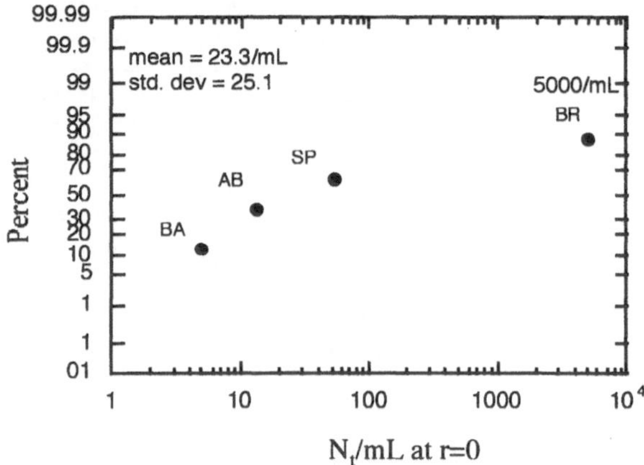


*Figure 4.* Probability plot of $r$=0 densities for laboratory populations of four rotifer species. Percent on the Y axis is the percentage rank of the $r$=0 density. Species are abbreviated by the first letters of their genus and species names.

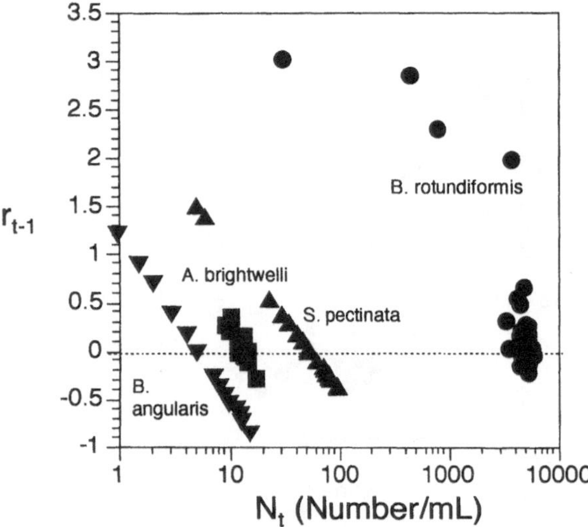


*Figure 3.* Density-dependence in laboratory populations of rotifers. $r_{t-1}$ is the instantaneous population growth rate in one time step ($t$), from density $N_{t-1}$ to $N_t$. Equations for $r$ for each species are given in the 'Methods' section.

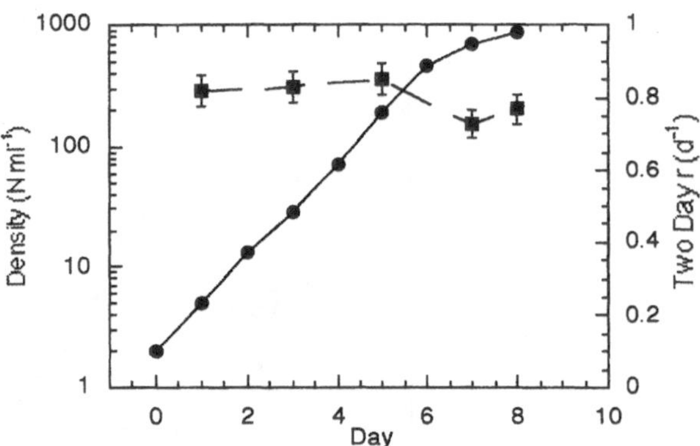


*Figure 5.* Absence of autotoxicity in laboratory populations of *Brachionus rotundiformis*. Day on the X axis is the age of a 5 l batch culture. Density ($N$ ml$^{-1}$) of this population is given on the left Y axis (circles). Population growth rate ($r$) per day from a two day reproductive test using filtered batch culture medium is given on the right Y axis (squares). Vertical lines on squares indicate one standard deviation.

cause density-dependent reductions in the growth of laboratory rotifer populations are food limitation and accumulation of an autotoxin resulting from metabolic waste products. The density-dependent effects observed in natural populations are not likely due to autotoxicity since lab populations grow at similar $r$ values at $10^3$–$10^4$ times higher densities. For example, natural populations of *B. rotundiformis* grew at $r$=2 at a density of about 100 ind. l$^{-1}$ (Fig. 1a) as compared with lab populations which grew at $r$=2 at a density of about 1000 ml$^{-1}$ (Fig. 3). Consequently, natural rotifer populations probably are not realizing their potential growth rate primarily due to food limitation, not autoxicity which does not occur until $10^3$–$10^4$ higher densities.

We demonstrated the ineffectiveness of autotoxicity to cause density effects in laboratory populations of *B. rotundiformis* (Fig. 5). Filtrates isolated from a population at various densities were tested for their ability to inhibit rotifer population growth. No evidence of medium conditioning to produce autotoxicity was detected in populations with densities as high as 865 animals ml$^{-1}$. This is despite signs that population growth began to slow at a density of about 500/ml, probably due to food limitation.

## Discussion

The population density where $r$=0 can be considered the carrying capacity, which is defined as "the dynamic equilibrium around which a population fluctuates ... when population density is above it, the number of individuals will tend to decrease, and when the population density is below it the number of individuals will tend to increase ..." (Calow, 1998). This concept of carrying capacity is independent of the logistic equation and can be defined even for populations with chaotic dynamics. Notably, carrying capacity and steady-state are different concepts. If density-dependent regulation is to be analyzed using models, it is important to chose from a large array of model types. Classical population dynamics models like the logistic or time-lagged logistic models cannot describe the complex dynamics observed in many laboratory and natural populations. Therefore, key features of regulation could be missed using these classical models. The contribution of this work is that we have identified a small set of time series models capable of describing the complex dynamics of natural rotifer populations from a large array of possible models.

Short time series of abundance for nine natural rotifer populations were described by Snell & Serra (1998). All time series were stationary, exponentially damped, fluctuating around a constant mean population size with constant variance. Natural rotifer populations appear to be tracking very recent perturbations and their dynamics cannot be predicted from perturbations from the more distant past. Effects of a perturbation are rapidly removed within 1–2 days as evidenced by the strongly negative Lyapanov exponents, indicating that the populations are stable and rapidly return to an equilibrium point after disturbance (Snell & Serra, 1998). Snell et al. (1999) further explored time series models of rotifer populations, examining effects on population viability of reduced resting egg production and heterozygosity loss. Snell & Serra (2000) extended these models to examine the probability of extinction associated with reduced resting egg production.

In all of these models, density-dependence is effective in population regulation, yet little is known about the population densities when it begins to operate and how population growth rate is reduced with increasing density. Density effects reduce $r$ to 0 at densities of 1–10 $l^{-1}$ in natural populations of most rotifer species. *B. rotundiformis* is especially tolerant of density effects, responding at 10–100 times higher densities.

The observation that density induced reductions in $r$ occurs at $10^4$-fold higher densities in laboratory populations requires further explanation. Natural populations experience a variety of factors that could cause density-dependent effects. These include food limitation, autotoxicity, mixis, predation, parasitism and interference competition. Density-dependence in lab populations is due only to the first three factors. The key to obtaining high population densities in laboratory is probably the provision of abundant, high quality food. Food availability may account for as much as $10^3$ of the $10^4$-fold difference in density effects between lab and natural populations. Autotoxicity may also play a regulatory role in laboratory populations of some rotifer species, but its effects are not felt until densities of 10–100 $ml^{-1}$. Kirk (1998) demonstrated chemically mediated, autotoxicity regulation of laboratory *Synchaeta pectinata* populations at densities of 100 $ml^{-1}$. Similar inhibitory effects of autotoxins have been reported for laboratory populations of *Daphnia* at densities as low as 40–150 $l^{-1}$ (Goser & Ratte, 1994; Burns, 1995). In contrast, we found no autotoxicty in *B. rotundiformis* at densities up to 865 $ml^{-1}$. The requirement of very high rotifer densities before autotoxicity can be detected raises questions about whether autotoxicity is simply an artifact of the laboratory. The possibility that autotoxicity is lower in the synthetic seawater used in laboratory experiments than the complex organic mixture present in natural seawater needs investigation.

The extraordinary tolerance of *B. rotundiformis* to high population densities distinguishes it from all other rotifer species tested. It would be interesting to explore the ecological factors in its life history that impart this tolerance. Moreover, it would be interesting to investigate the relative density tolerance of closely related *B. plicatilis*, which like *B. rotundiformis* is grown in mass cultures as larval fish feed in aquaculture. Examination of variation in density tolerance among *B. rotundiformis* and *B. plicatilis* strains may prove useful in aquaculture.

## Acknowledgements

We thank David Dusenbery, Nobert Walz and Claudia Ricci for comments that improved this paper.

## References

Box, G. E. P. & N. R. Draper, 1987. Empirical Model Building and Response Surfaces. John Wiley & Sons, New York, NY.

Burns, C. W., 1995. Effects of crowding and different food levels on growth and reproductive investment of *Daphnia*. Oecologia101: 234–244.

Calow, P. (ed.), 1998. The Encyclopedia of Ecology and Environmental Management. Blackwell Science, Oxford.

Ellner, S. & P. Turchin, 1995. Chaos in a noisy world: New methods and evidence from time-series analysis. Am. Nat. 145: 343–375.

Fu, Y., A. Hada, T. Yamashita, Y. Yoshida & A. Hino, 1997. Development of a continuous culture system for stable mass production of the marine rotifer *Brachionus*. Hydrobiologia 358: 145–151.

Goser, B. & H. T. Ratte, 1994. Experimental evidence of negative interference in *Daphnia magna*. Oecologia 98: 354–361.

Groom, M. J. & M. A. Pascual, 1997. The analysis of population persistence: an outlook on the practice of viability analysis. In Fielder, P. L. & P. M. Karevia (eds), Conservation Biology for the Coming Decade. Chapman-Hall, New York, NY: 4–27.

King, C. E. & T. W. Snell, 1980. Density-dependent sexual reproduction in natural populations of the rotifer *Asplanchna girodi*. Hydrobiologia 73: 149–152.

Kirk, K. L., 1998. Enrichment can stabilize population dynamics: autotoxins and density dependence. Ecology 79: 2456–2462.

Pollard, E., K. H. Lakhani & P. Rothery, 1987. The detection of density-dependence from a series of annual censuses. Ecology 68: 2046–2055.

Sibly, R. M., 1999. Efficient experimental designs for studying stress and population density in animal populations. Ecol. Appl. 9: 496–503.

Snell, T. W. & B. D. Moffat, 1992. A two day life cycle test with the rotifer *Brachionus calyciflorus*. Env. Tox. Chem. 11: 1249–1257.

Snell, T. W. & M. Serra, 1998. Dynamics of natural rotifer populations. Hydrobiologia 368: 29–35.

Snell, T. W. & M. Serra, 2000. Using probability of extinction to evaluate the ecological significance of toxicant effects. Environ. Toxicol. Chem., 19: 2357–2363.

Snell, T. W., M. Serra & M. J. Carmona, 1999. Toxicity and sexual reproduction in rotifers: reduced resting egg production and heterozygosity loss, In Forbes, V. E. (ed.), Genetics and Ecotoxicology. Taylor and Francis, Philadelphia, PA: 169–185.

Tan, L.-W. & R. J. Shiel, 1993. Responses of billabong rotifer communities to inundation. Hydrobiologia 255/256: 361–369.

Turchin, P. & A. D. Taylor, 1992. Complex dynamics in ecological time series. Ecology 73: 289–305.

Walz, N., 1983. Continuous culture of the pelagic rotifer *Keratella cochlearis* and *Brachionus angularis*. Arch. Hydrobiol. 98: 70–92.

*Hydrobiologia* **446/447:** 45–50, 2001.
*L. Sanoamuang, H. Segers, R.J. Shiel & R.D. Gulati (eds), Rotifera IX.*




# Variability for mixis initiation in *Brachionus plicatilis*

Eduardo Aparici, Maria José Carmona & Manuel Serra*
*Institut "Cavanilles" de Biodiversitat i Biologia Evolutiva, Universitat de València, Apartat 22085, E46071-València, Spain*
*E-mail: Eduardo.Aparici@uv.es Maria.J.Carmona@uv.es Manuel.Serra@uv.es*
*(*Author for correspondence)*

*Key words:* rotifers, sexual reproduction, mixis initiation, intraspecific variability

## Abstract

Deductions from both evolutionary models and inductive argumentation from empirical data support the notion of intraspecific variability for the initiation of sexual reproduction (mixis) within rotifer populations. In this study, we focus on the time and density at which mixis is initiated in a growing population. Cyclical parthenogenetic clones of *Brachionus plicatilis* established by hatching of resting eggs, isolated from a natural habitat, have been tested at the start of their sexual phase. Clones exhibited great variation for this trait, their time of switching to sexual reproduction being correlated with population density. Most of the variation for mixis initiation has either low or no heritability and is caused by individual environmental factors.

## Introduction

Mictic pattern is a critical component of the life cycle of the cyclically parthenogenetic rotifers, since the resting eggs which are a product of the sexual reproduction, are only a diapausing stage in most species – the rotifer populations being seasonal, sometimes ephemeral and inhabiting isolated habitats. The number of resting eggs, produced by a particular genotype in a growth season, which will hatch in future growth seasons, strongly determine the success of this genotype, and is a measure of its fitness.

As an important component of mictic patterns, mixis initiation should be under strong selection. A late initiation of sexual reproduction might result in low efficiency, as environmental conditions could already be so unfavourable that mictic phase would not be completed, or the period for resting egg production would be shortened. An early initiation of sexual reproduction implies a decrease of resources allocated in current parthenogenetic growth, which would result in lower resting egg production in the complete current growth season, because the genotypes with a too early mixis would not achieve the maximum possible density.

The above arguments provide a rationale for the expectation of a strong stabilizing selection for mixis initiation and a lack of variation within population for this trait. However, other arguments have been proposed to expect within-population diversity for the timing of mixis initiation. Firstly, Carmona et al. (1995) observed extended mictic patterns in field populations and proposed that this pattern would be adaptive in unpredictable habitats, as a bet-hedging strategy to face the uncertain length of the growth season. An extended mictic pattern can be achieved by intermediate mictic ratios (the amictic females producing both sexual and asexual daughters) and/or by variation among individuals for the timing of mixis. Secondly, Aparici et al. (1996) studied theoretical populations growing in predictable habitats, without competition, in which mixis is expected to occur at the very end of the growth season (Serra & Carmona, 1993). Aparici et al. (1996) found that a low level of variation in mixis initiation could be expected because it improves resting egg production through the overlap of male and immature mictic female occurrence. Their observation is quite puzzling that mating competition could affect evolution of mixis initiation, promoting variation in this trait and a decrease in resting egg production at the population level. In the models constructed by Aparici et al. (1996), mixis initiation was assumed to be under genetic control with no phenotypic plasticity, and so mixis initiation variation is necessarily associated with genetic variation.

Since the pioneering works of Maupas, Nusbaum, Whitney, Shull and others (reviewed by Birky & Gilbert, 1971), a number of studies have addressed what cues allow rotifers to detect the status of the environment in order to initiate mixis optimally. Besides, and there is a long list of factors proposed that trigger as mixis (see Gilbert, 1977; Pourriot & Clément, 1981; Pourriot & Snell, 1983). A factor often directly related to mixis in the genus *Brachionus* is population density, both in field studies (Bogoslavsky, 1963; King & Snell, 1980) and in laboratory and field experiments (Snell & Boyer, 1988; Hagiwara et al., 1989; Carmona et al., 1993). Nevertheless, some studies have also shown the existence of intraspecific genetic variation in the mictic patterns of different species (e.g. Pourriot, 1969; Gilbert 1975, 1976; Clément & Pourriot, 1976; Snell & Hoff, 1985; Carmona et al., 1994), which is anticipated due to the diverse selective regimes experienced by the different populations of one species. For instance, Carmona et al. (1994) studied 13 clones of *B. plicatilis* from different geographical origins, and found high variability in the mictic response to population density, which was interpreted as due to genetic differences. However, a different but relevant question, in accord with the evolutionary arguments is if there is variation in mixis initiation within a single population, and if this variation is genetic, or not. Here, we address this question in a population of *Brachionus plicatilis sensu stricto* (old L-morphotype) inhabiting a small pond. According to Carmona et al. (1995), this species has extended mictic patterns.

First, we estimated the within-population variability for mixis initiation in a sample of clones, based on response to changes in population density. These clones were generated by hatching resting eggs collected in a sediment sample. Second, we studied mixis initiation in the $F_1$-clones sexually produced by the clones that showed early and late mixis initiation in the first part of the study. The correlations between the parental clones and their sexual offspring were examined.

## Materials and methods

For estimating the within population variability for mixis initiation, we sampled the upper layer (1–2 cm) of the sediment in the *Poza Sur* pond, Prat de Cabanes-Torreblanca Marsh (Castelló de la Plana, East Spain) in summer 1997, when the pond was nearly dry. For a description of the limnological parameters of the pond, see Carmona et al. (1995). We obtained rotifer hatchlings from the resting eggs according to Snell et al. (1983) and Marcus (1990). About 10 $cm^3$ of the dried (25 °C) and sieved (250 $\mu$m mesh net) sediment sample were mixed with 1.75 M sucrose solution and made up to 40 ml. It was centrifuged at 600 rpm for 5 min. Suspended particles in the supernatant (including resting eggs) were collected and rinsed with distilled water and placed in petri dishes with diluted artificial seawater (2 g $l^{-1}$; Instant Ocean®, Aquarium Systems). Dishes were checked every 8 h for hatchling *B. plicatilis*. From the hatchlings, 100 newborns were individually isolated in 1 ml wells containing 0.5 ml of culture medium consisting of diluted artificial sea water (12 g $l^{-1}$) enriched with f/2 solution (Guillard & Ryther, 1962), and with the unicellular algae *Tetraselmis suecica* (density 500 000–700 000 cells/ml) as food. Cultures of rotifers and algae were kept at constant temperature (21 °C) and illumination (PAR: approx. 35 $\mu$E $m^{-2}$ $s^{-1}$). Culture medium was not renewed during the experiment. As a result of parthenogenetic growth in the wells from the 100 resting egg hatchlings, clones (hereafter called P-clones) were obtained. The wells were monitored every 8 h and when males were first observed, the time and population density were recorded. The individuals in the well were then transferred into a 20-ml flask with culture medium and excess of food. The resulting clones were maintained in the laboratory.

For studying the mixis initiation in the $F_1$-clones sexually produced, the resting eggs were obtained in the laboratory by inbreeding P-clones. Resting eggs were collected from the bottom of the culture flasks containing the P clones using a micropipette and a stereomicroscope. The isolated eggs were dried and later hatched as described above. From these hatchlings, filial clones ($F_1$-clones) were established. Both density and time at mixis initiation were determined as for the P-clones. Only the $F_1$-clones that showed mixis initiation were used for studying heritability. We thus tested if the largest differences observed in the P-clones are maintained in their descendents. It should be noted that more than one $F_1$-clone could be obtained for each P-clone. Therefore, the differences in the mictic response of the $F_1$-clones can be analyzed as related to two nesting factors: (1) P-clone from which

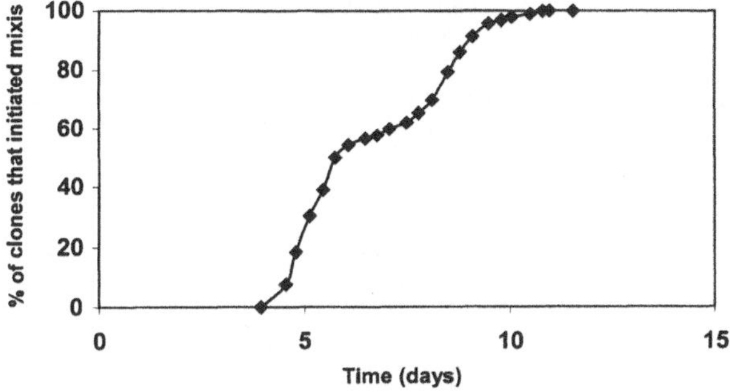


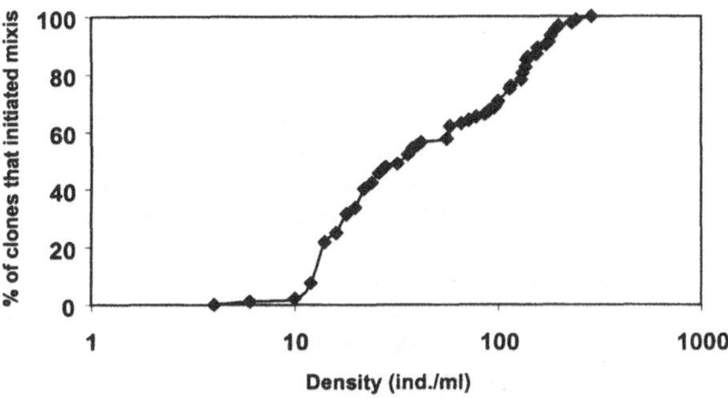


*Figure 1.* Cumulative distributions of the time (days) of mixis initiation (top) and mixis initiation population density (bottom) corresponding to the clones obtained from the resting eggs collected in the field. The percentage is expressed relative to the total number of clones in which mixis was observed ($n$ = 92).

the $F_1$-clone was obtained and (2) timing of their P-clone mictic response (early vs. late mixis initiation). A two-level nested ANOVA was performed, using SPSS 8.0 for Windows. We also estimated resting-egg-viability produced by P-clones with early and late mixis initiation, the differences being statistically analyzed by $X^2$ test, using the Yates correction.

## Results and discussion

From the 100 resting egg hatchlings obtained from eggs collected in the field, a total of 95 P-clones were founded. The other five hatchlings obtained from resting eggs failed to reproduce. Three of these 95 P-clones did not initiate mixis after 12 days in culture. The experiment was terminated at this stage due to the medium deterioration in the culture wells. Two of these three clones did not initiate mixis during the experiment, and the third one was lost. The remaining 92 P-clones exhibited a wide range for mixis initiation time (Fig. 1). The earliest clones produced their males as early as 4 days after start of the experiment at population densities <5–10 ind per well, i.e. 10–20 ind ml$^{-1}$. About 90% of clones had produced males at population densities of about 100 ind per well (200 ind ml$^{-1}$) only after 9 days after the start of the experiment.

The first phases of clonal culture showed that the age at maturity (first egg laid) is ca. 1 day, and eggs hatched about 1 day thereafter. If we assume that the type of female (mictic or amictic) is determined in early developmental stages, before the extrusion of the egg (Buchner, 1941; Ruttner-Kolisko, 1964), means that cues for mixis initiation acted about 3 days before the first males were produced. In fact, we observed that the first daughter of the female hatched from a resting egg could be a mictic female. However, a mictic female hatched from a resting egg was never observed, which agrees with the findings of other authors (Gilbert, 1974; Wallace & Snell, 1991). Therefore, we conjecture that cues for mixis initiation worked on the first day for earliest clones and on the 6th day (or later) for the latest ones. The corresponding densities are 2 and 40 ind ml$^{-1}$, assuming that the clones grew exponentially ($r \approx 0.53$ d$^{-1}$, estimated from our experimental cultures). Notably, the latter density, although high, is not unlikely in wild populations. Iltis & Riou-Duwat (1971) reported 65 females ml$^{-1}$ for *B. plicatilis* in a Chadian lake. By contrast, the lowest density associated with mixis induction can be considerably lower than 2 ind ml$^{-1}$ in the field. Carmona et al. (1995) estimated a mixis density threshold of 0.007 ind ml$^{-1}$ for *B. plicatilis*. Unfortunately, our estimation of mixis induction densities for individual clones implies practical constraints (e.g. the volume of the wells we used) that probably led to an overestimate of the lowest density at mixis induction and thus to an underestimate of within-population variation in this trait.

Despite the methodological shortcomings, we found a wide and continuous range for mixis initiation, while some clones responded quickly to the density, the others were less sensitive, or lagged in their response to population density, or both. The variation is possibly due to either genetic differences among clones or non-genetic factors as maternal effects, or to micro-environmental differences in the culture conditions. It should be noted that, because we used resting eggs collected from the field, and the measurement of mictic response is necessary shortly after resting egg hatching, within-clone variation could not be measured.

For addressing this question, the $F_1$-clones obtained by sexual reproduction of the P-clones were

*Table 1.* ANOVA on the log-transformed density for mixis initiation of the $F_1$-clones. Measurements taken of $F_1$-clones are grouped by their parental clone (Grouping factor: P-clones) and then are grouped by the type of parental P-clone, either early or late mixis initiation (Grouping factor: Early vs. Late P-clones)

| Grouping factor | | Sum of squares | df | Mean square | F | *P* |
|---|---|---|---|---|---|---|
| P-clones | Hypothesis | 26.20 | 11 | 2.38 | 3.28 | 0.003 |
| | Error | 29.06 | 40 | 0.73 | | |
| Early *vs.* Late P-clones | Hypothesis | 0.16 | 1 | 0.16 | 0.09 | 0.766 |
| | Error | 26.71 | 14.89 | 1.79 | | |

analyzed. The experimental design was focused on demonstrating heritable differences between early and late P-clones. Thus, we collected only the resting eggs of 14 P-clones with earliest mixis initiation (males observed before 5 days) and of 12 P-clones with latest mixis initiation (males observed after 9 days). Hatching was observed for the resting eggs of only the 9 early P-clones and for 4 late P-clones. The hatching differences were statistically not significant ($X^2$ = 1.406; df = 1; $P > 0.05$). Inbreeding depression, the requirement of a longer dormancy, or the lack of proper hatching cues, could explain hatching failure in 36% of early P-clones and 67% of late P-clones. Forty one $F_1$-clones (from 51 resting eggs hatched) were found from early P-clones. However, no males were observed in three of these clones during the experiment that had to be terminated after 15 days because of the deteriorating medium in the individual culture wells). Seventeen $F_1$-clones (from 19 resting eggs hatched) were founded from late P-clones, but mixis was not observed in two of them.

Patterns of mixis initiation of $F_1$-clones from early P-clones and from late P-clones were similar, despite delayed mixis in the latter, as expected if the trait is heritable, differences were not significant (Fig. 2; Table 1: $P$ = 0.766). Moreover, both groups of $F_1$-clones showed a wide range of time and density for mixis initiation, very similar to that found for the clones established from resting eggs based on field collections (P-clones; compare Figs. 1 and 2). Within-population variation in mixis initiation, which seems to be important in the population studied, has low heritability. In other words, there is little or no additive genetic variation because no correlated response to the selection for early and late mixis initiation clones could be detected. Surprisingly, our statistical analysis shows that variation within the $F_1$-clones obtained from the same P-clone is lower than variation among the $F_1$-clones obtained from different P-clones (Table 1; $P$ = 0.003; see also Table 2). The resemblance between sister $F_1$-clones suggests the existence of genetic variation, which according to the other results would be non-additive. In fact, regard-

*Table 2.* Mean values and standard deviation for population density (ind./ml) and time (days) at which mixis is initiated for early and late $F_1$-clones. Components of variance estimated from ANOVA (Table 1) are also shown (*n*: number of studied clones)

| Clones | *n* | Density (s.d.) | Time (s.d.) |
|---|---|---|---|
| Early | | | |
| C40 | 1 | 14.0 (–) | 5.0 (–) |
| C57 | 3 | 15.3 (9.45) | 4.5 (0.5) |
| C86 | 3 | 24.0 (10.4) | 4.8 (0.9) |
| C35 | 3 | 206 (184) | 9.9 (3.8) |
| C95 | 4 | 40.5 (36.2) | 5.9 (0.4) |
| C82 | 5 | 41.2 (34.2) | 6.8 (3.2) |
| C39 | 7 | 36.3 (17.8) | 5.8 (1.3) |
| C74 | 6 | 45.3 (34.3) | 6.6 (2.0) |
| C72 | 6 | 154 (97.2) | 9.0 (2.2) |
| Late | | | |
| C9 | 1 | 28.0 (–) | 4.5 (–) |
| C76 | 3 | 18.7 (16.8) | 6.3 (2.8) |
| C97 | 5 | 102 (98.9) | 7.0 (2.0) |
| C46 | 6 | 141 (133) | 8.5 (3.2) |
| $\sigma^2_{\text{amongclones}} = 0.421$ | | | |
| $\sigma^2_{\text{withinclones}} = 0.726$ | | | |

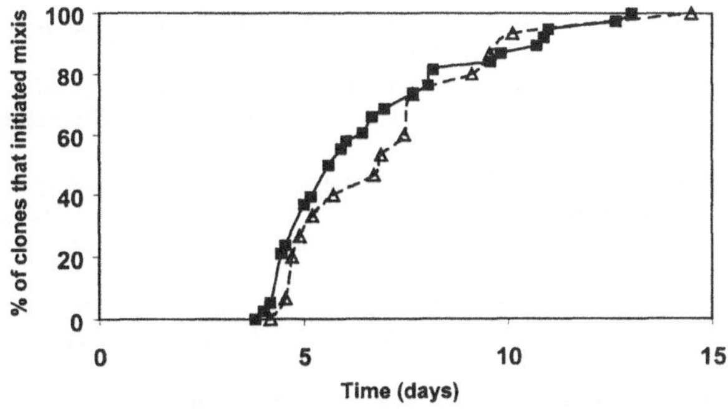


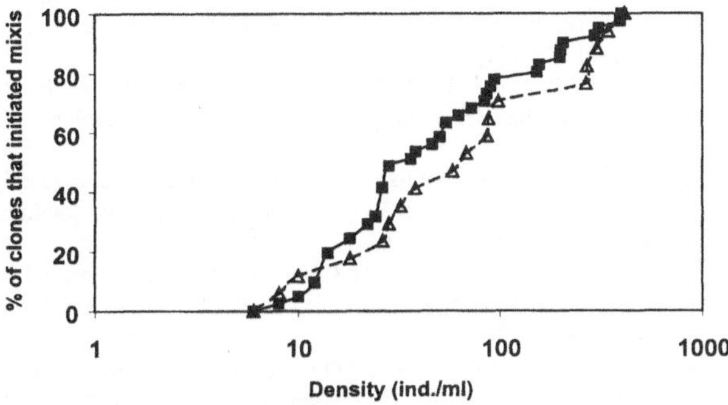


*Figure 2.* Cumulative distributions of the time (days) of mixis initiation (top) and mixis initiation population density (bottom) corresponding to the clones obtained from the resting eggs produced in the laboratory. The solid line and squares show the results for the group of $F_1$-clones originated from early P-clones ($n$ = 38). The dashed line and open triangles show the results for the group of $F_1$-clones originated from late P-clones ($n$ = 15). The percentage is expressed relative to the total number of clones in which mixis was observed.

less of whether they had early or late mixis initiation, some P-clones produced only early $F_1$-clones, while others produced mixtures of early and late $F_1$-clones (Table 2). Unfortunately, the experimental design that was focused on demonstrating heritable differences between early and late P-clones, does not allow estimation of the phenotypic variance components by standard methods of quantitative genetics.

Our results suggest that most of the phenotypic variation for mixis initiation within the studied population is environmental. This is due to micro-environmental differences or to the maternal environment (Rougier & Pourriot, 1977; Carmona et al., 1994) experienced during the development of the individual. In addition, the environmental differences explain the sensitivity differences for the mixis initiation cues. This result contrasts with the well-supported view of assigning the observed variation in mictic patterns among populations within a species to genetic differences (for *B. plicatilis*: Snell & Hoff, 1985; Carmona et al., 1994; for other species: Gilbert, 1975, 1976; Clément Pourriot, 1976). A high level of variation in mixis initiation is likely in the studied population in natural conditions, because micro-environmental effects (e.g. spatial heterogeneity caused by depth, littoral effects, etc.) should be higher than in controlled laboratory experiments. *B. plicatilis* populations inhabiting the Prat de Cabanes-Torreblanca Marsh experience an unpredictable environment in fall, winter and spring, when floods may suddenly and dramatically change ecological conditions. A wide range in the sensitivity to mixis induction cues based on phenotypic plasticity, rather than in genetic differences, would promote besides intermediate mictic ratios, an extended period of mixis. This is because in some individuals mixis is induced earlier than others during the growth season. This bet-hedging strategy ensures that some resting eggs would be produced if the growth season is short, but because parthenogenetic growth continues, advantages would be obtained from a long growth season (Carmona et al., 1995; Serra & King, 1999). Because the variation could depend on the selective regimen experienced by this population, extrapolation of our results to other populations should be done with caution. Lastly, for the purpose of generalisation work is needed on other populations.

## Acknowledgements

This study was supported by Grant GV-2543/94 from the Conselleria de Cultura, Educacio i Ciencia (Generalitat Valenciana), and by a Fellowship granted to E. Aparici by the Conselleria de Cultura, Educacio i Ciencia (Generalitat Valenciana, Pla Valencia de Ciencia i Tecnologia). We also wish to thank to Marta Briasco and Maria Teresa Ibanez for their collaboration in the field and laboratory work, and Africa Gomez for her helpful review of the manuscript. The authors are grateful to D. Clifton-Sewell who improved the language.

## References

Aparici, E., M. J. Carmona & M. Serra, 1996. Polymorphism in bisexual reproductive patterns of cyclical parthenogens. A simulation approach using a rotifer growth model. Ecol. Mod. 88: 133–142.

Birky, C. W. & J. J. Gilbert, 1971. Parthenogenesis in rotifers: the control of sexuality and asexuality. Am. Zool. 11: 245–266.

Buchner, H., 1941. Entwicklungsphysiologische Untersuchungen über den Determinationspunkt. Arch. Entwicklungsmech. Organ. 141: 145–158.

Carmona, M. J., M. Serra & M. R. Miracle, 1993. Relationships between mixis in *Brachionus plicatilis* and preconditioning of culture medium by crowding. Hydrobiologia 255/256: 145–152.

Carmona, M. J., M. Serra & M. R. Miracle, 1994. Effect of population density and genotype on life-history traits in the rotifer *Brachionus plicatilis* O.F. Mûller. J. exp. mar. Biol. Ecol. 182: 223–235.

Carmona, M. J., A. Gómez & M. Serra, 1995. Mictic patterrns of the rotifer *Brachionus plicatilis* Müller in small ponds. Hydrobiologia 313/314: 365–371.

Clément P. & R. Pourriot, 1976. Influences du groupement et de la densité de population sur le cycle de reproduction des rotifères. II. Comparaison de deux souches de *Notommata copeus* Ehrb. Arch. Zool. exp. gén.117: 5–13.

Gilbert, J. J., 1974. Dormancy in rotifers. Trans. am. Micros. Soc. 93: 490–513.

Gilbert, J. J., 1975. Polymorphism and sexuality in the rotifer *Asplanchna*, with special reference to the effects of prey-type and clonal variation. Arch. Hydrobiol. 75: 442–483.

Gilbert, J. J., 1976. Polymorphism in the rotifer *Asplanchna sieboldi*. Variability in the body-wall-outgrowth response to dietary tocopherol. Physiol. Zool. 48: 404–419.

Gilbert, J. J., 1977. Mictic-female production in monogonont rotifers. Arch. Hydrobiol. Beih. Ergebn. Limnol. 8: 142–155.

Guillard, R. R. L. & J. H. Ryther, 1962. Studies of marine planktonic diatoms. I. *Cyclotella nana* Hustedt, and *Detonula confervacea* (Cleve). Can. J. Microbiol. 8: 229–239.

Hagiwara, A., C. S. Lee, G. Miyamoto & A. Hino, 1989. Resting egg formation and hatching of the S-type rotifer *Brachionus plicatilis* at varying salinities. Mar. Biol. 103: 327–332.

Iltis, A. & S. Riou-Duwat, 1971. Variations saissonniéres du peuplement en Rotiféres des eaux natron?es du Kanem (Tchad). Cahiers O.R.S.T.O.M. Seriés Hydrobiologie V, 2: 101–112.

King, C. E. & T. W. Snell, 1980. Density-dependent sexual reproduction in natural populations of the rotifer *Asplanchna girodi*. Hydrobiologia 73: 149–152.

Lubzens, E., G. Minkoff & S. Marom, 1985. Salinity dependence of sexual and asexual reproduction in the rotifer *Brachionus plicatilis*. Mar. Biol. 85: 123–126.

Marcus, N. H., 1990. Calanoid copepod, cladoceran, and rotifer eggs in sea-bottom sediments of northern Californian coastal waters: identification, ocurrence and hatching. Mar. Biol. 105: 413–418.

Pourriot, R. & P. Clément, 1981. Action de facteurs externes sur la reproduction et le cycle reproducteur des Rotifers. Acta Oecol. Gen. 2: 135–151.

Pourriot, R. & T. W. Snell, 1983. Resting eggs in rotifers. Hydrobiologia 104: 213–224.

Rougier, C. L. & R. Pourriot, 1977. Aging and control of the reproduction in *Brachionus calyciflorus* (Pallas) (Rotatoria). Exp. Gerontol. 12: 137–151.

Ruttner-Kolisko, A., 1964. Über die labile Periode im Fortpflanzungszyklus der Rádertiere. Inv. Rev. ges. Hydrobiol. 49: 473–482.

Serra, M. & M. J. Carmona, 1993. Mixis strategies and resting egg production of rotifers living in temporally-varying habitats. Hydrobiologia 255/256: 117–126.

Serra, M. & C. E. King, 1999. Optimal rates of bisexual reproduction in cyclical parthenogens with density-dependent growth. J. Evol. Biol. 12: 263–271.

Snell, T. W. & E. M. Boyer, 1988. Thresholds for mictic female production in the rotifer *Brachionus plicatilis* (Müller). J. exp. mar. Biol. Ecol. 124: 73–85.

Snell, T. W. & F. H. Hoff, 1985. The effect of environmental factors on resting egg production in the rotifer *Brachionus plicatilis*. J. World Marincult. Soc. 16: 484–497.

Snell, T. W., B. E. Burke & S. D. Messur, 1983. Size and distribution of resting eggs in a natural population of the rotifer *Brachionus plicatilis*. Gulf Res. Rep. 7: 285–287.

Wallace, R. L. & T. W. Snell, 1991. Rotifera. In Ecology and classification of North American freshwater invertebrates. Academic Press: 187–248.

Hydrobiologia 446/447: 51–55, 2001.
L. Sanoamuang, H. Segers, R.J. Shiel & R.D. Gulati (eds), Rotifera IX.




# Parasites in rotifers from the Volga delta

Alexander K. Gorbunov & Anna A. Kosova
*Astrakhanskiy Biosphere Reserve, Astrakhan 414021, Russia*
*E-mail: abnr@astranet.ru*

*Key words:* rotifers, internal parasites, bacteria, fungi, protozoa

## Abstract

Internal parasites were recorded in more than 20 species of rotifers from water bodies of the Volga delta. The most abundant species, *Asplanchna priodonta* Gosse and *Brachionus calyciflorus* Pallas, were infested every year. The infestation was usually recorded in May and June during the spring-summer maximum in abundance. *B. calyciflorus* was usually infested by *Microsporidium asperospora* (Fritsch, 1895), while *A. priodonta* was more often infested by *Pythium* sp. and bacteria. At times, the incidence of infestation was as high as 20–40%. The main developmental stages of some parasites were examined and photographed on living material.

## Introduction

At the end of the 19th century, the infestation of rotifer by internal parasites attracted the attention of several investigators including Bertram (1892) and Fritsch (1895). Later, Budde (1927) and Voigt (1956, 1957) presented overviews of the parasites of rotifers. More recently, it was reported that parasites could greatly affect the population dynamics of some rotifer species (Miracle, 1977; Ruttner-Kolisko, 1977).

Our aim was to investigate the problem of rotifer infestation by parasites in the water bodies of the Volga delta. This investigation was carried out between 1981 and 1999, in parallel with a study of the species composition and quantitative dynamics of zooplankton. The investigations showed that many species of rotifers inhabiting these water bodies are subjected to infestation by internal parasites (Gorbunov & Kosova, 1990, 1993).

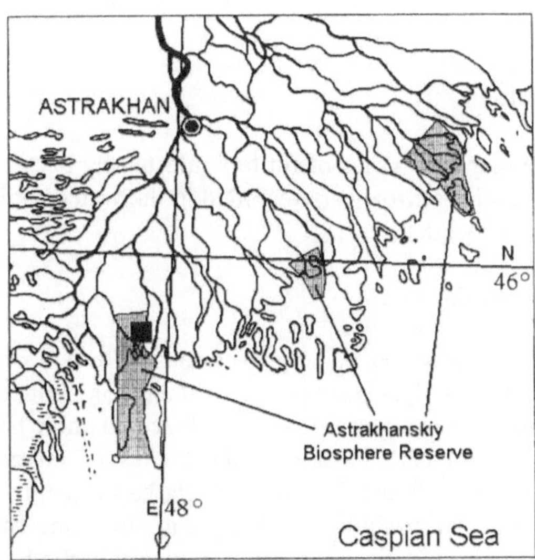


*Figure 1.* Map of the Volga delta. Black square shows location of the main sampling point.

## Description of sites

The investigation was carried out in the water bodies of the Volga delta. These included the main sampling point in the Bystraya channel in the lower zone of the western part of the delta and other sample points located in a number of other channels of the delta, in temporary high-water bodies, and in the fore-delta mainly within the bounds of the Astrakhanskiy Biosphere Reserve (see Fig. 1).

The channels of the delta are 2–5 m deep and 10–70 m wide. The temporary high-water bodies are formed on islands in the delta during spring-summer floods, and remain for 1.5–2 months. The fore-delta is a shallow fresh water area inclined towards the Caspian Sea with depths of 0.5–2 m. In winter, all of these water bodies are covered by ice. In summer, the temperature of the water in the channels rises to 25–26 °C, and in the temporary high-water bodies and fore-delta up to 25–30 °C.

## Materials and methods

Samples of zooplankton were obtained from the water bodies in the delta by filtering 100 l of water through a plankton net (mesh – 70 $\mu$m). The samples were fixed with 4% formalin. Sampling was performed once every month, with increasing frequency to several times a month in the spring and summer periods. The study of rotifer infestation was carried out simultaneously with the processing of zooplankton samples for quantitative analysis. In addition, qualitative samples were also taken. From time to time, living material was examined under stereomicroscopes at the magnification from 32 to 56× and under light microscopes at the magnification from 56 to 1350×. Infected rotifers were photographed under a light microscope with flash.

## Results

More than 20 species of rotifers infested by bacteria, fungi and protozoa were recorded in the water bodies of the Volga delta.

### *Bacterial infections*

Bacterial infection was observed mainly in *A. priodonta*. Under light microscope examination, infected rotifers had recognisable fawn-colour and short bacteria were clearly visible in the bodies of infected rotifers (Figs 2 and 3). The bacteria had a length of 2–6 $\mu$m and a width of 0.3–0.5 $\mu$m. In living rotifers, the bacterial cells moved freely within the body cavity together with the interstitial fluid when the rotifer contracted. Bacterial infection in *A. priodonta* occurred from the end of May to the beginning of June. In rare instances, this infection was also recorded in *A. henrietta* (Langhans), *Brachionus calyciflorus* Pallas, *B. plicatilis* (Müller), *Polyarthra longiremis* (Carlin), *Synchaeta pectinata* Ehrenberg and *S. stylata* Wierzejski.

Similar bacteria were described in *B. calyciflorus* by Budde (1927), who named them *Entobacter inflans*. This author found infected rotifers in water bodies in Germany in June.

### *Fungal infections*

Several fungi were found infesting rotifers. *Pythium* sp. *(Oomycetes)* occurred in *A. priodonta*, which was the rotifer most frequently parasitised by fungi. Under the light microscope, infected rotifers could be clearly seen to have hyphae inside their bodies. During the early stages of infection, affected rotifers swam normally. They did not differ from healthy rotifers, except for the presence of hyphae in the body (Fig. 4). However, during the next stages, affected rotifers had low swimming activity, followed by death. Shortly after the death of the rotifer, hyphae began to grow through the rotifer cuticle (Fig. 5). On the top of the hyphae, vesicles started to appear with zoospores developing in the vesicles after about 45 min (Figs 6–8). The thin outer coat of the vesicles then disappeared and biflaggelated zoospores escaped into the surrounding water (Fig. 9). These zoospores had a length of 10–12 $\mu$m and a width of 7–10 $\mu$m (Fig. 10). At first, the zoospores swam very actively. Later, they settled down on the glass and changed into motionless round cysts about 9–10 $\mu$m in diameter (Fig. 11). Then, the cysts produced new hyphae (Fig. 12). In some cases, oospores were seen in rotifers as well (Fig. 13). These had diameter about 22–33 $\mu$m and a thick coat of about 2–5 $\mu$m. This fungus was identified as *Pythium* sp. Fungus epidemics in *A. priodonta* were recorded in the water bodies of the Volga delta every year from May to June, with maximum infection percentages of 5–20%.

Two other fungus species belonging to the *Chytridiomycetes* were identified. *Catenaria anguillulae* Sorokin, 1876, was recorded in *B. calyciflorus*, *B. diversicornis* (Daday), *Eosphora thoides* Wulfert, *Filinia longiseta* (Ehrb.), *Lecane luna* (Müller) and *Sinantherina semibullata* (Thorpe). *Olpidium gregarium* Nowakowski, 1876, was found in the eggs of *F. longiseta* (Fig. 14). Some other types of internal parasitic fungi were also observed in rotifers, but these have not yet been identified.

### *Protozoan parasites*

The most frequently found protozoan parasites of rotifers were *Microsporidium asperospora* (Fritsch, 1895). These parasites were first described by Bertram (1892) from water bodies in Germany, but were not assigned a name. Later, Fritsch (1895) found similar parasites of rotifers in Czechia and named them as *Glugea asperospora*. Sprague (1977) considered these parasites to be part of the *Microsporidium* group. As

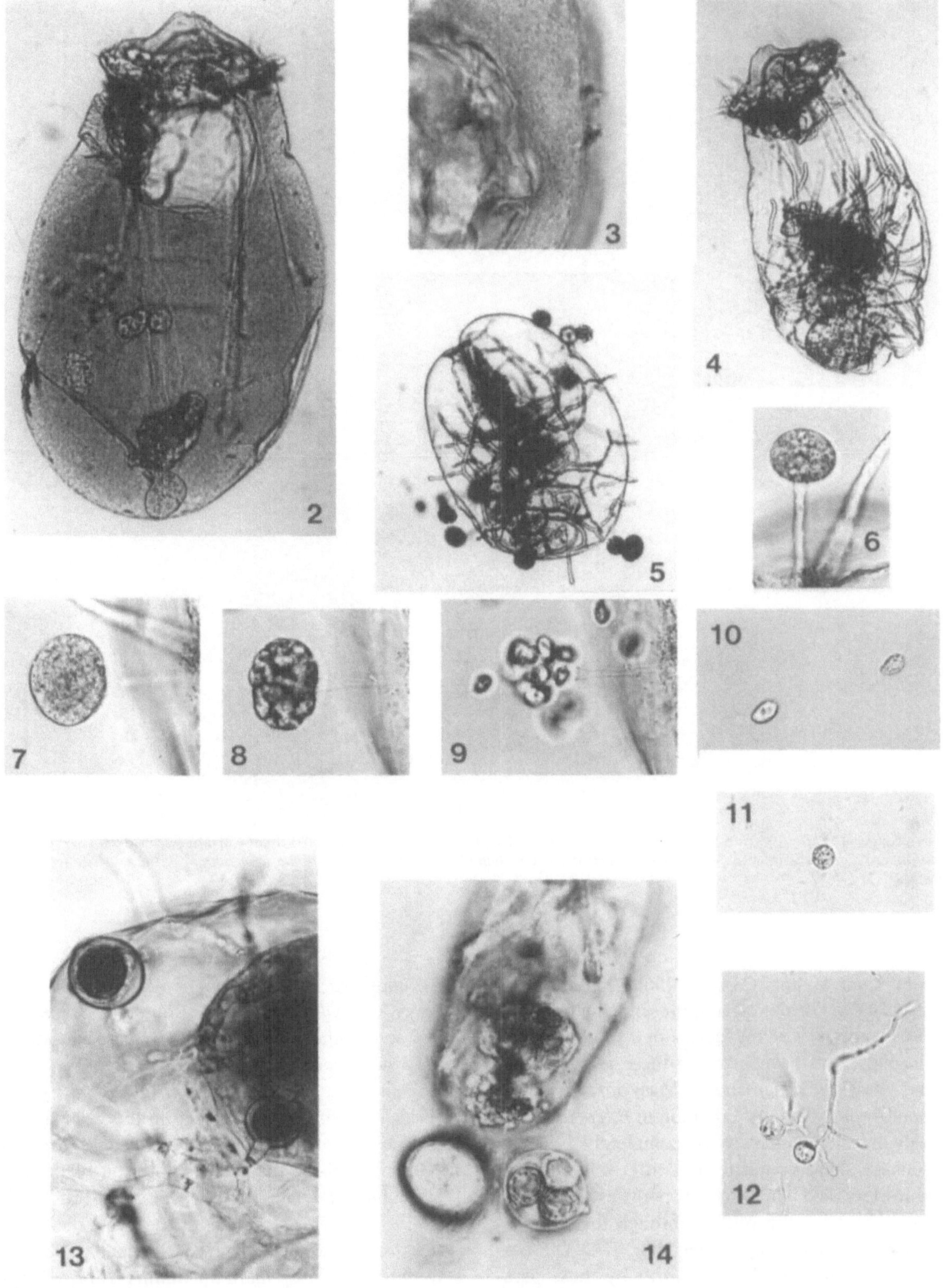

*Figures 2–14.* (2) *Asplanchna priodonta* infected with *Entobacter inflans.* (3) Cells of *Entobacter inflans* inside the rotifer body. (4) *Asplanchna priodonta* infected with *Pythium* sp. (5) Dead *Asplanchna priodonta* infected with *Pythium* sp. Appearance of vesicles. (6) Formation of vesicle of *Pythium* sp. (7) Formed vesicle of *Pythium* sp. (8) Formation of zoospores inside vesicle of *Pythium* sp. (9) Appearance of zoospores from vesicle of *Pythium* sp. (10) Zoospore of *Pythium* sp. (11) Cyst of *Pythium* sp. (12) Development of new hyphae of *Pythium* sp. from cysts. (13) Oospores of *Pythium* sp. in *Asplanchna priodonta.* (14) Zoosporangia of *Olpidium gregarium* in the egg of *Filinia longiseta.*

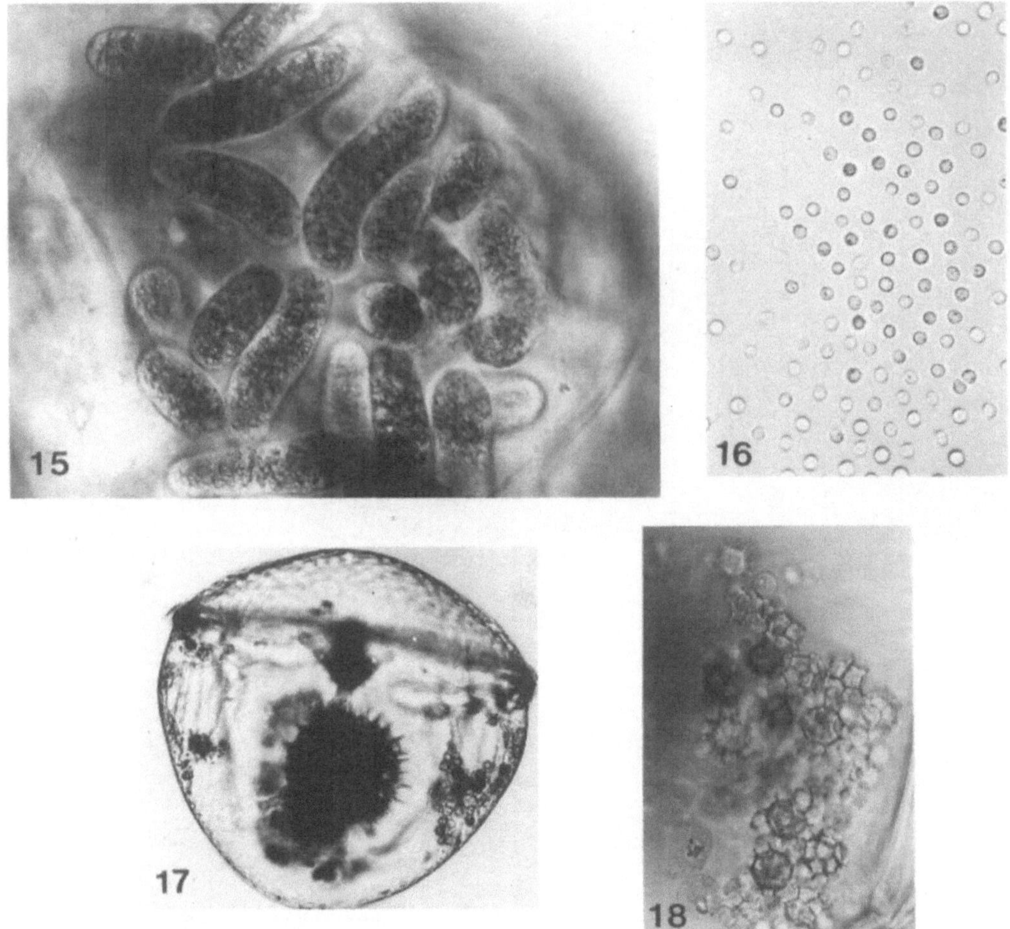

*Figures 15–18.* (15) *Microsporidium asperospora* in the body of *Brachionus calyciflorus.* (16) Spores of *Microsporidium asperospora.* (17) *Trochosphaera solstitialis* infected with cysts of *Bertramia beauchampi.* (18) Cysts of *Bertramia beauchampi* in the body of *Trochosphaera solstitialis.*

noted by some authors (Voigt, 1956, 1957; Ruttner-Kolisko, 1977), it is not clear at present whether these parasites comprise several species or a single species. In this paper, we have grouped these parasites under the name *M. asperospora.* Mass infestation by *M. asperospora* was most common in *B. calyciflorus.* Severely infested rotifers were completely filled by this sausage-shaped parasite (Fig. 15). When the infested rotifers died, small round spores escaped from the body cavity of the rotifers through the head region into the surrounding water (Fig. 16). The percentage infestation of *B. calyciflorus* by these parasites sometimes amounted to 30–40%. Single cases infestation of *Brachionus urceus* (L.). *B. diversicornis, B. quadridentatus* Hermann, *P. longiremis, P. dolichoptera* Idelson, *Epiphanes brachionus* (Ehrenberg), *Conochilus unicornis* Rousselet, *S. semibullata* and some other rotifer species by parasites similar to *M. asperospora* were recorded.

Another parasite, *Microsporidium polygona* (Fritsch, 1895), was recorded in the water bodies of the Volga delta in *A. priodonta* and *A. brightwelli* (Gosse). This is the first observation of these parasites since they were first described by Fritsch (1895) in Czechia. Cysts similar to the cysts of *Bertramia beuchampi,* as described by Stempell (1921), were found in *Trochosphaera solstitialis* Thorpe (Figs. 17 and 18) (Gorbunov & Kosova, 1993). Some other protozoan parasites were observed in rotifers from the water bodies of the Volga delta, but have not been identified.

## Discussion

The investigations have shown that many species of rotifers inhabiting the water bodies of the Volga delta are subjected to infestation by internal parasites. For most rotifer species, only single cases of infestation were recorded. However, mass infestation took place every year in some abundant species, usually at the population maximum. Sometimes the level of infestation reached 20–40%. For these species, infestation appears to be a factor regulating their population dynamics. This observation agrees with the data of Miracle (1977) and Ruttner-Kolisko (1977) on the study of infestation of *Filinia terminalis*, *S. pectinata* and *Conochilus unicornis*, and with the general theory of epidemiology.

## References

Bertram, C., 1892. Beiträge zur Kenntniss der Sarcosporidien nebst einem Anhange über parasitische Schläuche in der Leibeshöhle von Rotatorien. Zool. Jb. Abt. Anat. Ontog. 5: 581–604.

Budde, E., 1927. Über die in Rädertieren lebenden Parasiten. Arch. Hydrobiol. 18: 442–459.

Gorbunov, A. K. & A. A. Kosova, 1990. Infection of rotifers of bodies of water of delta of the Volga River. Proceedings of the All-Union Rotifer Symposium: 50–51 (in Russian).

Gorbunov, A. K. & A. A. Kosova, 1993. *Trochosphaera solstitialis* (Thorpe, 1893) in the Volga River delta, and the first recorded instance of its infestation. Zoosystematica Rossica 2 (2): 223–224.

Fritsch, A., 1895. Über Parasiten bei Crustaceen und Rädertieren der süssen Gewässer. Bull.int II Acad. Sci. Prag.: 79–85.

Miracle, M. R., 1977. Epidemiology in rotifers. Arch. Hydrobiol. Beih. Ergebn. Limnol. H. 8: 138–141.

Ruttner-Kolisko, A., 1977. The effect of the microsporid *Plistophora asperospora* on *Conochilus unicornis* in Lunzer Untersee (LUS). Arch. Hydrobiol. Beih. Ergebn. Limnol. H. 8: 135–137.

Sprague, V., 1977. Systematics of the Microsporidia. Comparative Pathobiology. New York 2: 1–510.

Stempell, W., 1921. Haplosporidienstudien. II. Über *Bertramia beauchampi* n. sp. aus *Conochilus volvox* Ehrbg. Arch. Protistenk. 43: 355–360.

Voigt, M., 1956/57. Rotatoria. Die Rädertiere Mitteleuropas. Berlin-Nikolassee. I Textband: 1–508; II Tafelband: 115.

Hydrobiologia 446/447: 57–61, 2001.
L. Sanoamuang, H. Segers, R.J. Shiel & R.D. Gulati (eds), Rotifera IX.




# Use of microparticulate markers in examination of rotifer physiology: results and prospects

N. Lindemann, L. Discher & W. Kleinow*
*Zoologisches Institut der Universität zu Köln, Weyertal 119, D-50923 Köln, Germany*
*E-mail: Walter.Kleinow@Uni-Koeln.De*
*(*Author for correspondence)*

*Key words:* pH, intestine, stomach, feeding, filtration rate, microcapsules

## Abstract

A short overview is given of some possible ways of using microparticles for the study of rotifer physiology. Besides synthetic microcapsules, some 'natural' microparticles have been found to be appropriate for this purpose. Using yeast cells, erythrocytes and erythrocyte ghosts, we were able to examine feeding rates, time course and control of nutrient passage, to determine the pH milieu in the digestive tract and to demonstrate absorptive processes. Preliminary results exist of studies on the mechanical efficiency of the mastax and on the secretion of $H^+$-ions by the intestine. Different types of synthetic microcapsules can be produced for an array of applications. Some of these are suitable for use with rotifers. Problems exist in producing them in a suitable uniform size and in loading them efficiently.

## Introduction

Some physiological processes within living rotifers may be observed through the light microscope. The uptake of particulate food for example, and its passage through the digestive tract, can easily be examined by providing the rotifers with several kinds of microparticles (for reviews, see Starkweather, 1980; Salt, 1987, see also Lindemann & Kleinow, 2000). Microparticulate markers are also suitable for experimental approaches. Several types of synthetic microparticles have already been used to study rotifer physiology besides natural food particles. Thus, latex beads have been used to characterize size preferences of rotifers in uptake of nutrient particles (see Ronneberger, 1998 and literature cited therein). Rotifers have also been cultured for a restricted time on nutrients enclosed in microcapsules (Teshima et al., 1981). This paper will present additional ways of using microparticulate markers for the study of rotifer physiology. In a stage, intermediate between use of natural food organisms or purely synthetic microparticles, these examinations were performed with artificial microparticles of biological origin. Some of these techniques are quite simple and direct and, therefore, also suitable for teaching purposes.

## Materials and methods

*Brachionus plicatilis* (O.F.Müller, 1786) was reared in xenic culture as described previously (Kühle & Kleinow, 1985, 1989; Wethmar & Kleinow, 1993). Staining of yeast cells by bromothymol blue (change from yellow at pH 6.0 to blue-green at pH 7.6) was performed as follows: 250 mg baker's yeast were suspended in 5 ml of 0.05 m phosphate buffer, pH 7.0. After adding 25 mg bromothymol blue, the suspension was heated at 100 °C for 60 min. The stained yeast cells were recovered by centrifugation (10 min, 800 × ***g***), washed three times in 10 ml and finally suspended in 5 ml phosphate buffer (ca. $10^7$ cells/ml).

Erythrocyte suspensions were obtained by diluting blood with seawater. The density of erythrocytes in a suspension was controlled by reading its absorbancy at 630 nm and 420 nm on a Beckman DU 600 spectrophotometer and calibrated by counting them in a haemocytometer. Erythrocytes with fixed membranes were obtained by the method of Bing et al. (1967).

The feeding rate of rotifers was determined by mixing specified concentrations of rotifers and erythrocytes, contained in seawater. Samples were taken at different times from a compartment in the experimental solution which was accessible to erythrocytes but not to rotifers. This compartment consisted of three 'Tissue Baskets' (Plano GmbH, Wetzlar), covered with gauze with a mesh of 75 $\mu$m, which were mounted together (Lindemann & Kleinow, 2000).

The percentage of rotifers with digestive tracts filled with haemoglobin was determined as follows: at the indicated times after mixing, aliquots of 0.5 ml were taken out and pipetted into 0.5 ml solution containing 20 mg ml$^{-1}$ sodium-dithionite in order to immobilize the animals in an extended state. The whole resulting mixture was examined through the microscope and counted in 0.1 ml samples on an excavated slide over the next 20 min (from about this time, after treatment by dithionite, the red colour starts to diminish in the digestive tract and to spread through the bodies of the animals).

## Results and discussion

We have found some kinds of 'natural' microparticles to be useful as markers for examining feeding and digestive processes in *B. plicatilis*. A first example is the use of yeast cells for determining the pH values in different parts of the digestive tract (Fok et al., 1982). Figure 1 shows the result of an experiment in which yeast cells stained by bromothymol blue were fed to *B. plicatilis*. Through the microscope, yeast cells in the stomach can be seen to display a blue-green (alkaline) colour. This colour is shown also by yeast cells which have just been propelled into the intestine (Fig. 1a). Three minutes later (Fig. 1b), however, the yeast cells in the intestine become more bright, due to a colour change from blue-green to yellow indicating a respective shift of pH. From this experiment, two things may be concluded: (1) stomach and intestine are separable compartments with different conditions and functions and (2) since the pH of the intestine is more acidic than that of the stomach $H^+$-ions are secreted by intestine wall cells. These results (Kleinow & Kühle, 1984) have recently been confirmed by using fluorescent pH markers and confocal laser microscopy by which the secretion of $H^+$-ions can also be examined more closely (work in progress).

Another useful type of natural microcapsules are mammalian erythrocytes (Kleinow et al., 1997). They have an appropriate size (3–9 $\mu$m), are intensely coloured and can be modified in several ways. In contrast to yeast cells, untreated mammalian erythrocytes are completely broken down by the mastax of *B. plicatilis*. Since erythrocytes are also concentrated during feeding, the uptake of erythrocytes turns the digestive tracts of rotifers bright red. By comparing the number of rotifers with red digestive tracts with the total number of rotifers in a sample, it is possible to calculate the percentage of animals which at different times have their digestive tract completely filled. With this information, it is possible to estimate how fast nutrient particles are taken up by the rotifers and how long they stay in the digestive tract under different conditions. The data can also be used to check the percentage of rotifers in a sample that is actively feeding.

Moreover, changes in the concentration of erythrocytes in the medium can be monitored photometrically by measuring the absorbancy caused by light scattering of the particles (at 630 nm) and the absorbancy of hemoglobin at 416–420 nm (Soretband). By centrifuging the medium, a distinction can be made between hemoglobin still contained in erythrocytes and free hemoglobin. It is, therefore, possible to determine aspects of filtration, feeding and passage of nutrient particles by rotifers in the same experiment. Parallel determinations of this type are difficult to obtain using food organisms.

In Figure 2, the results of a representative experiment are shown in which the uptake of ovine erythrocytes was monitored for two different densities of rotifers. Figure 2a shows the percentage of rotifers having a red stained digestive tract. Erythrocytes are readily taken up, thus after 5 min nearly all rotifers have their digestive tracts filled. This condition remains constant for about 60–90 min. Then the percentage of stained rotifers diminishes. The number of stained rotifers is reduced, since as the supply of erythrocytes is depleted, fewer and fewer rotifers are able to fill their digestive tracts completely. Thus, a significant decrease of stained rotifers below the maximal percentage (found at about time 5 min), indicates conditions when the supply of erythrocytes becomes rate limiting for the ingestion rate of rotifers. This exhausting of the red blood cell supply occurs more quickly when higher numbers of rotifers are ingesting the cells.

From Figure 2b it can be seen that the particle density (absorbancy at 630 nm) decreases steadily by a

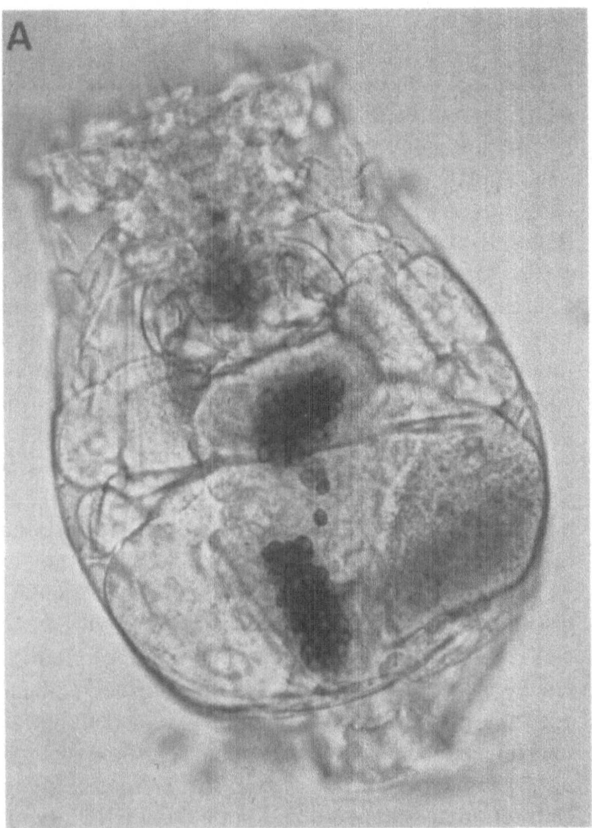


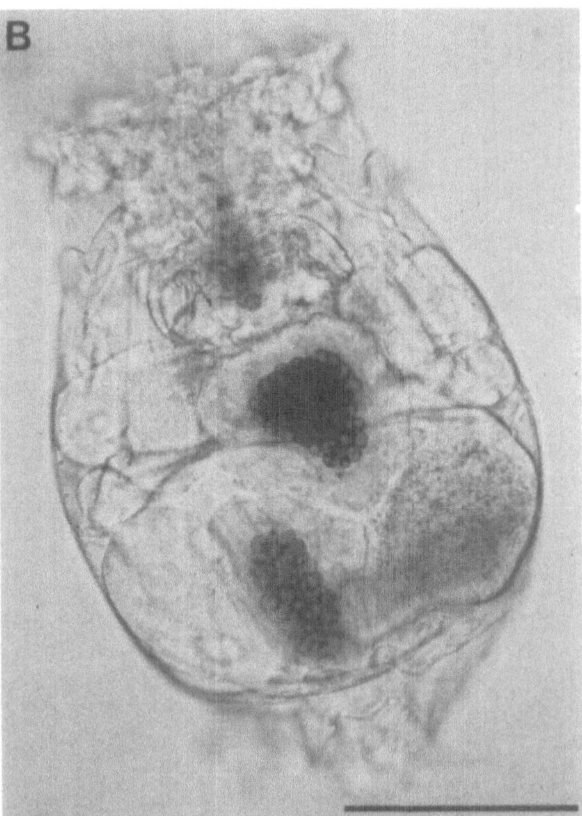


*Figure 1.* Light microscopic pictures of the same specimen of *B. plicatilis* after providing it with yeast cells which had been stained by bromothymol blue. The yeast cells within the stomach appear to be dark in both pictures, corresponding to the blue-green alkaline colour of bromothymol blue. Some of the yeast cells in Figure 1 a have just been propelled from the stomach into the intestine and are still of dark appearance. Figure 1b. was taken 3 min later. The yeast particles in the intestine are brighter, indicating a change to yellow, respectively, to a pH of 6 or lower. Calibration bar = 0.1 mm.

function which can be described quite well by first order kinetics. For the experiment with 296 rotifers/ml, it may furthermore be noted that the percentage of rotifers with stained digestive tracts (Fig. 2a) is starting to decrease steeply at about the time when the absorbancy at 630 nm in the medium approaches zero.

The absorbancy curve for hemoglobin at 420 nm complements this information in other respects. During the first 10 min, this absorbancy can be seen to diminish since erythrocytes are taken up by the rotifers and, therefore, disappear from the aliquots which are taken for measurements. This decrease of absorbancy at 420 nm continues until all actively feeding rotifers present in the experiment have their digestive tracts filled. Then it increases again since now hemoglobin is given up by the rotifers and, therefore, reappears in the medium. Therefore, from the course of this curve, one may estimate the minimum gut passage time of nutrient particles in the rotifers digestive tract. In this experiment, the gut passage time was about 10 min (in other experiments up to 20 min). The extent of the decrease is clearly related to the number of rotifers present in the sample. It seems reasonable to suggest that under these conditions, the rotifer digestive tracts are completely filled by ovine erythrocytes (volume 0.031 pl). Therefore, from the decrease of the number of erythrocytes per number of rotifers, it is possible to estimate the mean volume of their digestive tracts. In the experiment shown, the respective volumes were calculated to be 102, respectively 145 pl. These values correspond quite well with estimates obtained by evaluation of electron microscopic pictures (Kleinow et al., 1991).

The supply of particles also influences the passage time of nutrients through the digestive tracts. If remaining food particles are completely removed from the medium, after the rotifers have filled their digestive tracts, the residence time of particles becomes signific-

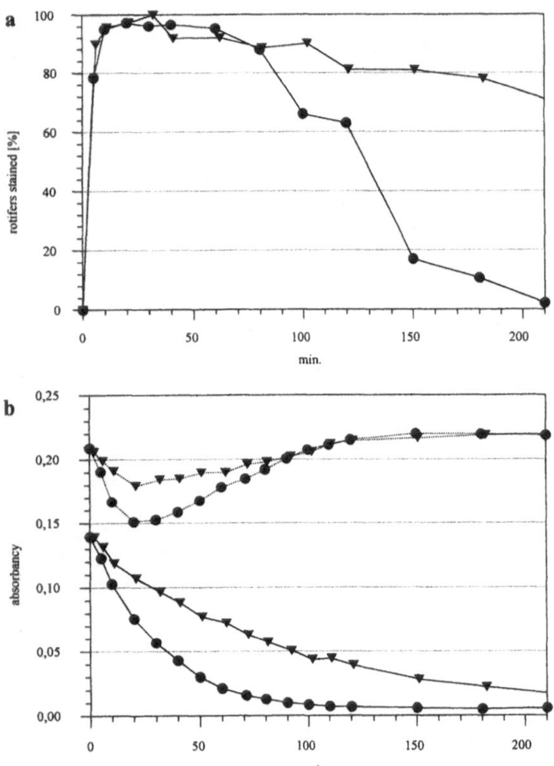


*Figure 2.* Percentages of *B. plicatilis* with digestive tracts filled with erythrocytes (a) and absorbancy at 420 nm ( ·········· ) and at 630 nm ( ———— ) (b) at different times after adding erythrocytes to the medium. Explanations: rotifer densities ● 296 ind. $ml^{-1}$.; ▼ 96 ind. $ml^{-1}$.

antly longer. If a new source of particles is added to the medium, they will be taken up very quickly, replacing the original particles in the digestive tract (Lindemann & Kleinow, 2000).

This provides a technique for quick exchange of the contents of the digestive tract of rotifers. Replacement of food organisms or marker particles in the digestive tract of rotifers with other sorts of particles may be useful in several types of experiments. Such replacement might be needed before analytical determinations to control or remove contaminating material contained in the digestive tract. Mammalian erythrocytes are useful in this respect since they do not contain glycogen, RNA or DNA. By replacing nutrient particles with erythrocytes, it is possible to evaluate the effects of dietary glycogen in rotifers (work in progress). Similarly, it may be important to replace nutrient particles from the digestive tract of rotifers by mammalian erythrocyte before determination or isolation of RNA or DNA. This is particularly true when an investigator wishes to PCR amplify rotifer DNA while avoiding the DNA of food species in the digestive tract.

Replacement of particles was also used to demonstrate absorptive processes: Albumin was labelled by fluorescent stain and enclosed in erythrocyte ghosts. After uptake of these particles, the digestive tract showed a bright fluorescence. This material was then replaced by erythrocyte ghosts containing nonfluorescent albumin. After this treatment some residual fluorescence was found in the rotifer caused by labelled protein which was no longer contained in the lumen of the digestive tract but had already been absorbed by the cells of the stomach (Lindemann & Kleinow, 2000).

A further possible application of erythrocytes is to measure the mechanical efficiency of the mastax by determining the amount of free hemoglobin given up by the rotifers. Compared to untreated erythrocytes, we found a strong decrease in the free hemoglobin given up, if rotifers were fed with erythrocytes which had been fixed by different concentration of glutardialdehyde. Treatment of erythrocytes by 0.5% glutardialdehyde inhibited the release of free hemoglobin thus it has to be assumed that treated erythrocytes cannot be broken down by the *B. plicatilis* mastax. After calibration and standardization of this method (work in progress), it will be possible to make quantitative assessments of the mechanical effectiveness of the mastax. This would clarify the relative contributions of mechanical breakdown and enzymatical degradation of nutrient particles in *B. plicatilis*.

We have found no satisfactory method for preparing synthetic microcapsules for examination of rotifers. Problems exist in producing microcapsules in a suitable (2–10 $\mu$m), uniform size, in large amounts and in loading them efficiently. Moreover, a suitable type of material has yet to be identified for constructing microcapsules. When rotifers are provided with alginate capsules, their cilia become stuck together.

It seems, however, worthwhile examining these possibilities further since several problems of rotifer physiology might be solved by using synthetic microcapsules. These problems include localization of digestive enzymes and determination of nutritional demands of rotifers under more or less axenic conditions. Such studies would be a useful complement to studies in which the nutritional demands of rotifers are examined by feeding them with different algae (Rothhaupt, 1990).

## Acknowledgements

We wish to thank Mr Helmut Wratil for excellent technical assistance and Mrs Frances Wharton for correcting the English in the manuscript.

## References

Bing, D. H., J. G. M. Weyand & A. B. Stavitsky, 1967. Hemagglutination with aldehyde-fixed erythrocytes for assay of antigens and antibodies. Proc. Soc. Exp. Biol. Med. 124: 1166–1170.

Fok, A. K., Y. Lee & R. D. Allen, 1982. The correlation of digestive vacuole pH and size with the digestive cycle in *Paramecium caudatum*. J.Protozool. 29(3): 409–414.

Kleinow, W. & K. Kühle, 1984. Zum pH-Optimum hydrolytischer Enzyme und zum pH-Milieu im Verdauungstrakt bei *Brachionus plicatilis* (Rotatoria). Verh. Dtsch. Zool. Ges. 77: 299.

Kleinow, W., N. Lindemann & S. Köhler, 1997. Investigation of food uptake of *Brachionus plicatilis* (Rotatoria) by means of suspended erythrocytes (in German). Verh. Dtsch. Zool. Ges. 90.1: 112.

Kleinow, W., H. Wratil, K. Kühle & B. Esch, 1991. Electron microscope studies of the digestive tract of *Brachionus plicatilis*. Zoomorphology 111: 67–80.

Kühle, K. & W. Kleinow, 1985. Measurements of hydrolytic enzymes in homogenates from *Brachionus plicatilis* (Rotifera). Comp. Biochem. Physiol. [B] 81 B: 437–442.

Kühle, K. & W. Kleinow, 1989. Localization of hydrolytic enzyme activities within cellular fractions from *Brachionus plicatilis* (Rotatoria). Comp. Biochem. Physiol. [B] 93 B: 565–574.

Lindemann, N. & W. Kleinow, 2000. A study of rotifer feeding and digestive processes using erythrocytes as microparticulate markers. Hydrobiologia 435: 27–41.

Ronneberger, D., 1998. Uptake of latex beads as size-model for food of planctonic rotifers. Hydrobiologia, 387/388: 445–449.

Rothhaupt, K. O., 1990. Population growth rates of two closely related rotifer species: effects of food quantity, particle size and nutritional quality. Freshwat. Biol. 23: 561–570.

Salt, G. W., 1987. The components of feeding behavior in rotifers. Hydrobiologia 147: 271–281.

Starkweather, P. L.,1980. Aspects of the feeding behaviour and trophic ecology of suspension-feeding rotifers. Hydrobiologia 73: 63–72.

Teshima, S., A. Kanazawa & M. Sakamoto, 1981. Attempt to culture the rotifers with microencapsulated diets. Bull. Japan. Soc. Sci. Fish. 47: 1575–1578.

Wethmar, C. & W. Kleinow, 1993. Characterization of Proteolytic Activities Stimulated by SDS or Urea and 2-Dimensional Gel Electrophoresis of Proteins from *Brachionus plicatilis* (Rotifera). Comp. Biochem. Physiol. [B] 106: 349–358.

Hydrobiologia **446/447**: 63–69, 2001.
L. Sanoamuang, H. Segers, R.J. Shiel & R.D. Gulati (eds), Rotifera IX.




# Population growth of *Lepadella patella* (O. F. Müller, 1786) at different algal (*Chlorella vulgaris*) densities and in association with *Philodina roseola* Ehrenberg, 1832

S. Nandini & S. S. S. Sarma
*Division of Interdisciplinary Research, National Autonomous University of Mexico (UNAM – Campus Iztacala) AP 314, CP 54090, Los Reyes, Tlalnepantla, State of Mexico, Mexico*
*E-mail: nandini@servidor.unam.mx*

*Key words:* population growth, rotifers, littoral species, food density, Bdelloidea

## Abstract

Population growth of *Lepadella patella* was studied using *Chlorella* as the sole food at five concentrations ranging from $0.25 \times 10^6$ to $4.0 \times 10^6$ cells $ml^{-1}$ at 25 °C for 22 days. The population densities increased with increasing algal concentration up to $1.0 \times 10^6$ cells $ml^{-1}$. The population growth of *L. patella* was lower at algal concentration of $2.0 \times 10^6$ cells $ml^{-1}$ and above. In a separate experiment, we tested the influence of the bdelloid rotifer *Philodina roseola* on the population growth of *L. patella* at different ratios of initial inoculation densities using $1.0 \times 10^6$ cells $ml^{-1}$ of *Chlorella* at 28 °C. Despite lower initial inoculation densities compared with those in the controls, both *L. patella* and *P. roseola* showed higher peak abundances when grown together. The maximum peak abundance values recorded for *L. patella* and *P. roseola* were 830 and 230 ind. $ml^{-1}$, respectively, at an inoculation ratio of 1:1.

## Introduction

Population growth of planktonic rotifers is strongly controlled by several factors including food availability, temperature, interactions with co-occurring species and initial inoculation density (Edmondson, 1965; DeMott, 1989; Sarma et al., 1996). A vast majority of the population growth studies on brachionids and some littoral species have shown a positive relation between the growth rates of rotifers and the availability of food (Stemberger & Gilbert, 1985a; Ooms-Wilms, 1998). Population growth studies on littoral and benthic species are less common than those on planktonic rotifers. Studies on bdelloids and other non-planktonic species have so far shown food-density related abundance until a certain food level, usually lower than those for planktonic rotifers (Ricci, 1984; Pérez-Legaspi & Rico-Martinez, 1998).

The relation between rotifer body size and the optimal concentration of food required to produce a maximal abundance have yielded inconclusive results. Based on some studies, larger species require more food to reproduce than smaller ones and species may be inhibited by high algal concentrations (Stemberger & Gilbert, 1985b). Conversely, some small sized rotifer species do well under very high food concentrations (Dumont et al., 1995). Most of these studies have been conducted using planktonic rotifer species. Littoral and benthic rotifers are equally important in freshwater ecosystems as an intermediate step for the transfer of energy from detrius and benthic algae to invertebrate and vertebrate predators (Lampert & Sommer, 1997). Furthermore, compared to planktonic rotifers, littoral and benthic species are much more diverse (Koste, 1978; Nogrady et al., 1993) and thus are more intricately linked in the trophic web.

Since the planktonic system is relatively homogenous, there is an intense competition as a result of which only a few species can thrive. Competition is a much slower process than predation. In natural systems, competitive exclusion in the planktonic community can be detected in ancient lakes. This has been elegantly documented with reference to Lake Baikal and Lake Tanganyika (Dumont, 1994). However, in the benthic community, the situation is different. Due to the high complexity of benthic systems leading to a variety of niches, inter- and intra-specific competition among benthic organisms may be less intense than among planktonic ones. At times, co-occurrence of two benthic species may actually facilitate their growth (Nandini et al., 1998). This has, however, not been experimentally well established under laboratory conditions.

In this study, we present some quantitative information on the population growth of the benthic species *Lepadella patella* under different food concentrations and in association with *Philodina roseola* at varying inoculation densities.

## Materials and methods

We isolated the test zooplankton, *Lepadella patella* and *Philodina roseola,* from local water bodies in Mexico City. *L. patella* is a littoral monogonont rotifer (Koste, 1978), while *P. roseola* is a bdelloid, predominantly found attached to littoral zone vegetation (Donner, 1965). Both species were cloned separately and maintained on the unicellular green alga *Chlorella vulgaris* at concentrations of 1–2 $\times$ $10^6$ cells $ml^{-1}$.

We tested the effect of five algal food densities (0.25 $\times$ $10^6$, 0.5 $\times$ $10^6$, 1.0 $\times$ $10^6$, 2.0 $\times$ $10^6$ and 4.0 $\times$ $10^6$ cells $ml^{-1}$) on the population growth of *L. patella*. For all experiments, we used EPA medium (prepared by dissolving 96 mg $NaHCO_3$, 60 mg $CaSO_4$, 60 mg $MgSO_4$ and 4 mg KCl in 1 l of distilled water) (Anon., 1985) and *Chlorella vulgaris* was mass cultured using Bold Basal medium (Borowitzka & Borowitzka, 1988). The alga was harvested during log phase, centrifuged at 3000 rpm for 5 min, and resuspended in EPA medium; the density was determined using a haemocytometer. Into 25 ml capacity beakers with 20 ml EPA medium, rotifers were introduced at a density of 1 ind. $ml^{-1}$. Four replicates were set up for each treatment. All beakers were kept in diffused and continuous fluorescent illumination (200 lux) at 25$\pm$2 °C. Observations were taken every day on the population density of the test organisms. After every 24 h, rotifers were filtered using a small net of 25 $\mu$m mesh and food suspension medium was renewed 100%. When densities of rotifers were greater than 10 $ml^{-1}$, counts from 3 to 4 aliquot samples of 1 ml each were used to estimate population density. Population growth was followed for 22 days, by which time a declining trend in the growth curve was observed for all the food concentrations tested.

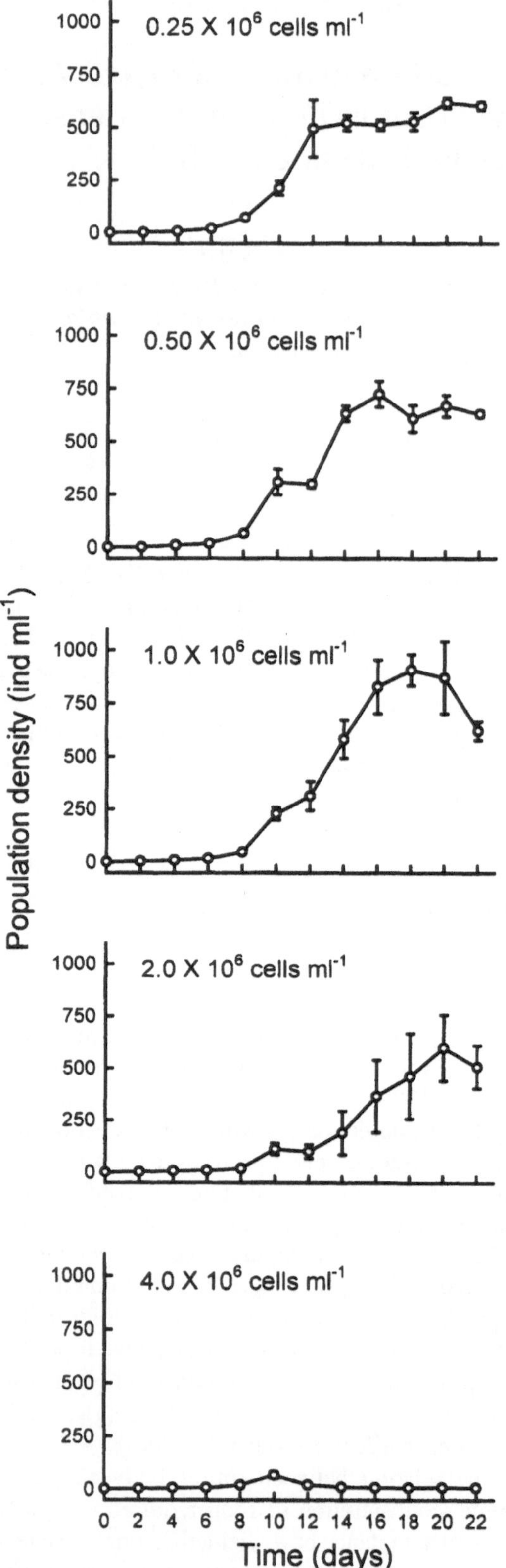


*Figure 1.* Population growth of *Lepadella patella* at various concentrations of *Chlorella vulgaris*. Values are mean ($\pm$ 1 standard error).

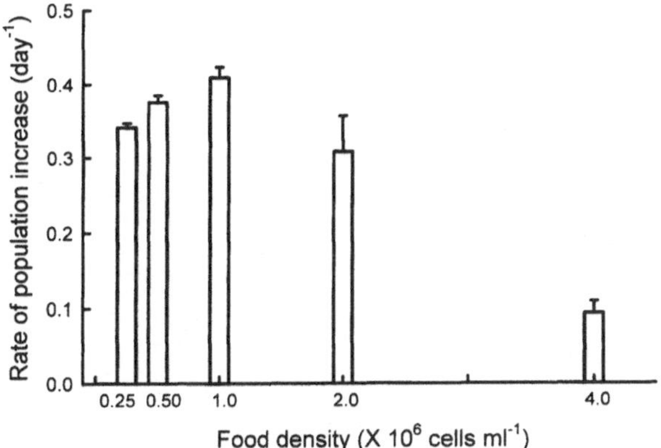


*Figure 2.* Rate of population increase of *L. patella* in relation to food density. Values are mean (± 1 SE).

Based on the results of the above experiment, we selected a food concentration of 1.0 × 10$^6$ cells ml$^{-1}$ of *Chlorella vulgaris* to study interactions between *P. roseola* and *L. patella.* The experiment was conducted at 28 °C, which was suitable for both the species. Into 25 ml beakers we introduced 20 ml of EPA medium with the appropriate algal concentration. We introduced the rotifers at varying intial densities: *Lepadella* control (20 *Lepadella* individuals), *Lepadella:Philodina* (15:5), *Lepadella:Philodina* (10:10), *Lepadella:Philodina* (5:15) and *Philodina* control (20 *Philodina* individuals). *Philodina* had a tendency to stick to the walls of the beaker; therefore, a delicate brush was used to detach them before counting and transferring them to fresh medium. Population growth of rotifers in all treatments was followed for 22 days, by which time a declining trend in the growth curve was observed.

Population growth rate ($r$) was calculated from the exponential phase using the formula:

$$r = (\ln N_t - \ln N_{\text{o}})/t,$$

where $N_{\text{o}}$ = initial population density and $N_t$= population density after time $t$. The $r$ was calculated for each replicate separately, for different time intervals, and mean values were used for analysis. The maximal population densities and time required to reach them were derived from each replicate separately following Sarma et al. (1998).

The various population characteristics were compared using Analysis of Variance and multiple comparison tests (Students–Newmans–Keuls test) (Sokal & Rohlf, 1981).

## Results

### *Population growth of* L. patella *in relation to food density*

Population growth curves of *L. patella* offered five different concentrations of *Chlorella vulgaris* which are presented in Figure 1. In general, the population abundance of *L. patella* increased with increasing food concentration from 0.25 × 10$^6$ to 1.0 × 10$^6$ cells ml$^{-1}$ after which a further increase in food concentration resulted in a reduced population growth. The growth curves had a long lag phase (6–8 days) followed by an exponential phase for the next 8–12 days and thereafter remained stationary or began to decline. The maximal population density reached by *L. patella* was 1000±96 (mean±se) ind. ml$^{-1}$ under a food concentration of 1.0 × 10$^6$ cells ml$^{-1}$. There was a significant difference in both, the maximum population density and the day at which this was reached ($P$<0.001, Table 1, ANOVA) with increasing food concentration. Maximum population density was reached earliest at the highest food concentration (4.0 × 10$^6$ cells ml$^{-1}$). The rate of population increase per day (mean±standard error) varied from 0.09±0.02 at 4.0 × 10$^6$ cells ml$^{-1}$ to 0.41±0.01 at 1.0 × 10$^6$ cells ml$^{-1}$ (Fig. 2). There was a significant difference in the population growth rates at the various food concentrations tested (Table 1).

### *Growth of* L. patella *and* P. roseola *under conditions of co-existence*

Population growth curves of *L. patella* and *P. roseola* grown together at a food concentration of 1.0 × 10$^6$ cells ml$^{-1}$ using different inoculation densities are presented in Figure. 3. Regardless of the inoculation ratio, both *L. patella* and *P. roseola* showed a significantly higher population abundance, in the presence of each other, when compared to controls ($P$< 0.05, SNK-test). The maximum population abundance of *L. patella* (920±230 ind. ml$^{-1}$) was obtained when grown with *P. roseola* in an equal inoculation ratio. There was no negative influence of the presence of *L. patella* on *P. roseola* or vice versa; instead both species benefitted from their association as reflected in their growth rates. There was no significant difference in the peak population density reached by *Lepadella* in control or in association with *Philodina*; on the other hand, those reached by *Philodina* were significantly higher in association with *Lepadella* than in controls ($P$ < 0.01; Table 1). The rate of population increase of

*Table 1.* Analysis of variance (ANOVA) performed on population growth variables of *Lepadella patella*. DF = degrees of freedom, SS = sum of squares; MS = mean square, *F* = *F*-ratio

| Source | DF | SS | MS | *F* |
|---|---|---|---|---|
| **Population growth of** ***L. patella*** | | | | |
| *Peak population density* | | | | |
| Among food levels | 4 | 1888093.0 | 47202 | 3.013.98*** |
| Error | 15 | 506637.0 | 337775.0 | – |
| *Day at peak population density* | | | | |
| Among food levels | 4 | 246.00 | 61.5 | 12.47*** |
| Error | 15 | 74.00 | 4.93 | – |
| *Rate of population increase* | | | | |
| Among food levels | 4 | 0.252 | 0.063 | 23.62*** |
| Error | 15 | 0.04 | 0.002 | – |
| **Population growth of** ***L. patella*** **and** ***P. roseola*** **in single- and mixed-species cultures** | | | | |
| *Peak population density (L. patella)* | | | | |
| Among inoculation densities | 1 | 141376.0 | 141376.0 | 1.76ns |
| Error | 14 | 1124760 | 80340 | – |
| *Peak population density (P. roseola)* | | | | |
| Among inoculation densities | 1 | 0.014 | 0.014 | 12.25** |
| Error | 14 | 0.016 | 0.0011 | – |
| *Rate of population increase of L. patella* | | | | |
| Among inoculation densities | 3 | 0.01 | 0.003 | 5.71* |
| Error | 12 | 0.007 | 0.0006 | – |
| *Rate of population increase of P. roseola* | | | | |
| Among inoculation densities | 3 | 0.014 | 0.0047 | 3.50* |
| Error | 12 | 0.016 | 0.0013 | – |

* = $p< 0.05$; ** = $p< 0.01$; *** = $p<0.001$; ns = non-significant ($p>0.05$).

both test rotifers was significantly higher in the mixed-species treatments than in their respective controls ($P < 0.05$; ANOVA; Table 1). However, on comparing the three inoculation ratios tested, we found that they had no significant effect on the population growth rate of either species ($P > 0.05$, SNK-test). There was no significant difference in the day at which the maximum population density was reached by both species. Similar trends were observed with the population growth rate values of both rotifers which were significantly higher in mixed species culture than in single-species cultures ($p <0.05$, Table 1). The rate of population increase per day varied between 0.35±0.01 (*L. patella* = 100%) and 0.42±0.02 (*L. patella* = 50%), while for *P. roseola* the range was 0.24±0.03 (*P. roseola* = 100%) and 0.32±0.01 (*P. roseola* = 25%) (Fig. 4).

## Discussion

In this study, both *Lepadella* and *Philodina* showed typical growth curves similar to those in other studies on non-planktonic rotifers (Ricci, 1984; Nandini et al., 1998). These growth curves differ from those ob-

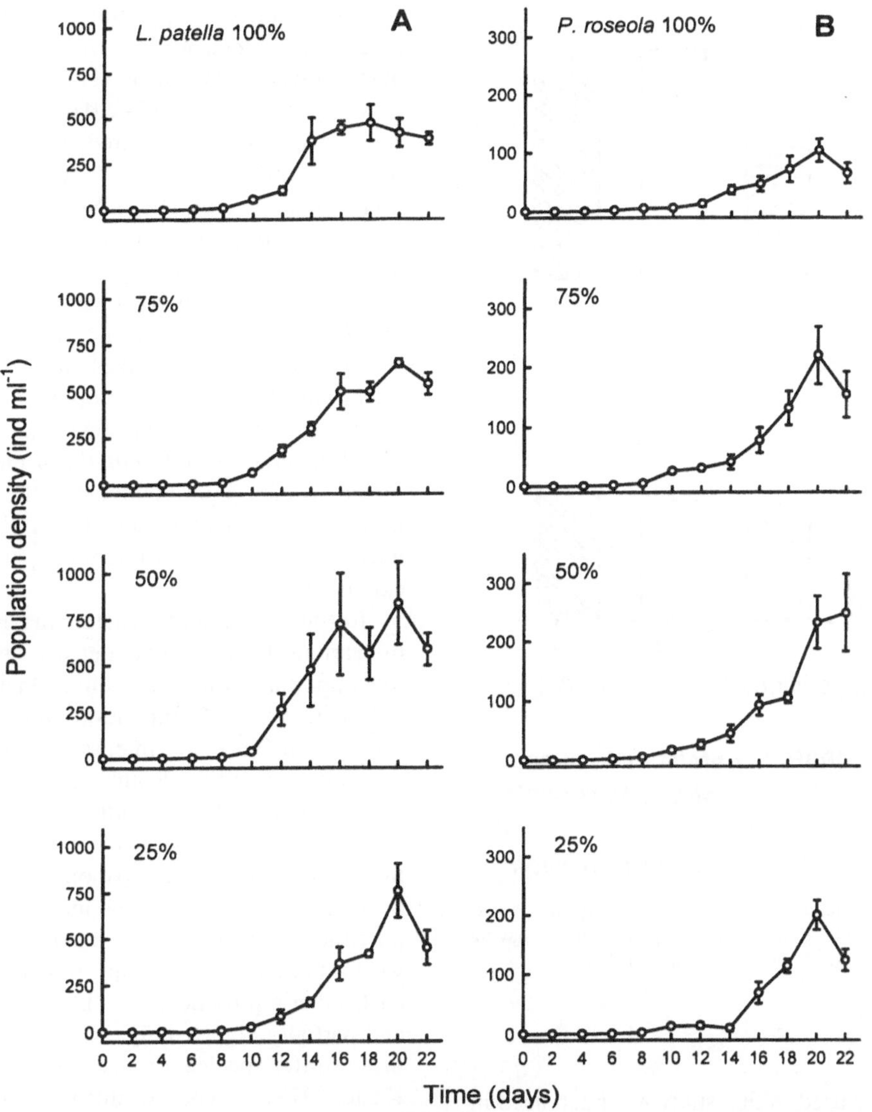


*Figure 3.* Population growth of *L. patella* and *P. roseola* alone and when grown together at a *Chlorella* concentration of $1 \times 10^6$ cells ml$^{-1}$. Shown in column A are 100% *L. patella*; 75% *L. patella* + 25% *P. roseola*; 50% *L. patella* + 50% *P. roseola* and 25% *L. patella* + 75% *P. roseola*. Values are mean (± 1 SE). Shown in column B are 100% *P. roseola*; 75% *P. roseola* + 25% *L. patella*; 50% *P. roseola* + 50% *L. patella* and 25% *P. roseola* + 75% *L. patella*. Values are mean (± 1 SE).

served in planktonic species in the prolonged lag phase of almost 8 days. Planktonic rotifers, on the other hand, begin to show increases in population density within 3 days at comparable temperatures (Walz, 1995). Among the littoral rotifers, data on population growth have concentrated on *Euchlanis* (Gulati et al., 1987) and *Lecane* (Hummon & Bevelhymer, 1980). *Lepadella patella*, a common rotifer species often found in high abundance in many freshwater bodies, has not previously been cultured for a prolonged period of time. In this study, this species was cultured exclusively on *Chlorella* for several months before the experiments were commenced. The fact that *L. patella* could grow well on *Chlorella* indicates the possible utilization of planktonic algae by benthic rotifers. In our experimental containers, we noticed that *L. patella* individuals were often present in great numbers on the surface of the water column. *Philodina roseola* has been cultured earlier on green algae (Lebadeva & Gerasimova, 1985, 1987). The range of

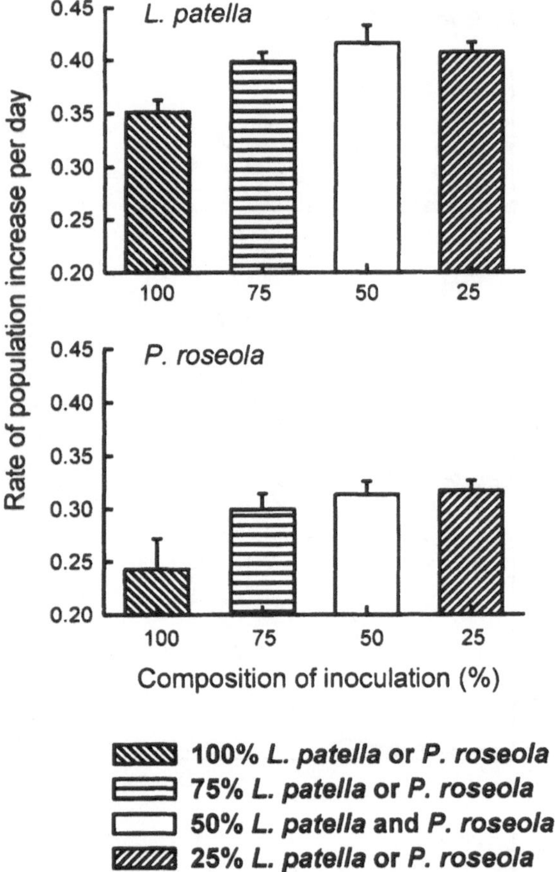


*Figure 4.* Rate of population increase of *L. patella* and *P. roseola* in single species cultures and in mixed-species cultures having different inoculation ratios. Values are mean (± 1 SE).

algal food density used in this study has been used in the past for other rotifers including planktonic, littoral and benthic species (Sarma et al., 1996; Nandini et al., 1998).

The patterns of population growth of *L. patella* indicate a preference for lower algal levels. In general, smaller species can survive and reproduce at lower algal levels compared to larger species (Stemberger & Gilbert, 1985a). This has been documented also for *Anuraeopsis fissa* (70 μm body size) vs *Brachionus calyciflorus* (255 μm) (Sarma et al., 1996) and *Brachionus patulus* (97 μm) vs *B. calyciflorus* (175 μm) (Sarma et al., 1999). The maximal abundance reached by a species is also influenced by the body size. Thus, for any given food concentration, the numerical abundance of *L. patella* was significantly higher than that of *P. roseola*.

The range of population growth rates recorded here for *L. patella* was within the range recorded for many other rotifers. In general, most rotifer species have a growth rate of 0.2–2.0 depending on food type and density. When compared to monogonont rotifers, bdelloid rotifers have much lower rates of population increase (Ricci, 1984). Similarly, we found much higher rates of population increase in *Lepadella* compared to *Philodina* (Fig. 4), and the maximal abundance reached by *L. patella* was much higher than that of *Philodina* (Fig. 3). Since the two species differ strongly in their body sizes, the capacity of *L. patella* to reach higher mean population abundance may be attributed to its smaller body size. The differences in the abundance of *L. patella* in the controls of the mixed-species experiment and in the food density experiment (especially under $1.0 \times 10^6$ cells $ml^{-1}$) may have been due to the different temperatures of the two experiments.

Rotifer community structure in nature is controlled by various factors, food density and predation being important among the biotic factors (Lampert & Sommer, 1997). Competition among planktonic rotifers is known to be intense and is also influenced by food availability (Rothhaupt, 1990). A higher diversity of littoral and benthic rotifers (Koste, 1978) is probably due to less intense competitive interactions (Dumont, 1994). Among littoral and benthic rotifers, competition is less intense and, therefore, a possible co-existence can be expected. In the present work, both *L. patella* and *P. roseola* benefited by each other's presence, possibly by utilizing excreted and partially digested alga; it is established that both these rotifer species are detritivorous (Donner, 1965; Koste, 1978). The best results in terms of growth were obtained when both species were initially in equal proportions (Fig. 4). Nandini et al. (1998) have shown that *Philodina roseola* has an antagonistic influence on the population growth of planktonic monogonont rotifers. It is thus possible that the influence of *Philodina* on other rotifers is species-specific.

The present study indicates that *Lepadella patella* is well adapted to low food levels and that *P. roseola* had a positive effect on the population growth of *L. patella*. It thus appears that the ability of littoral and benthic species to co-exist is related to their low food requirements as well as species-specific interactions. Further research on interactions among other littoral species may help clarify the reason for the high species diversity observed in littoral and benthic regions of lakes and ponds.

## Acknowledgements

We thank The National System of Investigators (Mexico) (Ref. No. 20520 & 18723) for support.

## References


Anonymous, 1985. Methods of measuring the acute toxicity of effluents to freshwater and marine organisms. U.S. Environment Protection Agency EPA/600/4-85/013.

Borowitzka, M. A. & L. J. Borowitzka, 1988. Micro-algal Biotechnology. Cambridge University Press, London.

DeMott, W. R., 1989. The role of competition in zooplankton succession. In Sommer, U. (ed.), Plankton Ecology: Succession in Plankton Communities. Springer, New York: 195–252.

Donner, J., 1965. Ordnung Bdelloidea. Bestimmungsbücher zur Bodenfauna Europas. Akademie Verlag, Berlin. 6: 1–267.

Dumont, H. J., 1994. Ancient lakes have simplified pelagic food webs. Arch. Hydrobiol. Beih. 44: 223–234.

Dumont, H. J., S. S. S. Sarma & A. J. Ali, 1995. Laboratory studies on the population dynamics of *Anuraeopsis fissa* (Rotifera) in relation to food density. Freshwat. Biol. 33: 39–46.

Edmondson, W. T., 1965. Reproductive rate of planktonic rotifers as related to food and temperature in nature. Ecol. Monogr. 35: 61–111.

Gulati, R. D., J. Rooth & J. Ejsmont-Karabin, 1987. A laboratory study of feeding and assimilation in *Euchlanis dilatata lucksiana*. Hydrobiologia 147: 289–296.

Hummon, W. D. & D. P. Bevelhymer, 1980. Life table demography of the rotifer *Lecane tenuiseta* under culture conditions and various age distributions. Hydrobiologia 70: 25–28.

Koste, W., 1978. Rotatoria. Die Rädertiere Mitteleuropas. Ein Bestimmungswerk begründet von Max Voigt. Bornträger, Stuttgart. Vol. 1: Textband 673 pp., Vol. 2: Tafelband 234 pp.

Lampert, W. & U. Sommer, 1997. Limnoecology: The Ecology of Lakes and Streams. (Translated by J.F. Haney). Oxford Univeristy Press, New York: 382 pp.

Lebedeva, L. I. & T. N. Gerasimova, 1985. Peculiarities of *Philodina roseola* (Ehrbg.) (Rotatoria, Bdelloida) – growth and reproduction under various temperature conditions. Int. Rev. ges. Hydrobiol. 70: 509–525.

Lebedeva, L. I. & T. N. Gerasimova, 1987. Survival and reproduction potential of *Philodina roseola* (Ehrenberg) (Rotatoria, Bdelloida) under various temperature conditions. Int. Rev. ges. Hydrobiol. 72: 695–707.

Nandini, S., S. S. S. Sarma & T. R. Rao, 1998. Effect of co-existence on the population growth of rotifers and cladocerans. Russ. J. Aquat. Ecol. 8: 1–10.

Nogrady, T., R. L. Wallace & T. W. Snell, 1993. Rotifera: Vol. 1: Biology, ecology and systematics. SBP Academic Publishers, The Hague: 142 pp.

Ooms-Wilms, A., 1998. On the food uptake and population dynamics of rotifers in a shallow eutrophic lake. Ph.D thesis, Universiteit van Amsterdam, The Netherlands: 153 pp.

Pérez-Legaspi, I. A. & R. Rico-Martínez, 1998. Effect of temperature and food concentration on two species of littoral rotifers. Hydrobiologia 387/388: 341–348.

Ricci, C., 1984. Culturing of some bdelloid rotifers. Hydrobiologia 112: 42–51.

Rothhaupt, K. O., 1990. Resource competition of herbivorous zooplankton: a review of approaches and perspectives. Arch. Hydrobiol. 118: 1–29.

Sarma, S. S. S., R. A. A. Stevenson & S. Nandini, 1998. Influence of food (*Chlorella vulgaris*) concentration and temperature on the population dynamics of *Brachionus calyciflorus* Pallas (Rotifera). Ciencia Ergo Sum 5: 77–81.

Sarma, S. S. S., M. A. Fernández-Araiza & S. Nandini, 1999. Competition between *Brachionus calyciflorus* Pallas and *Brachionus patulus* (Müller) (Rotifera) in relation to algal food concentration and initial population density. Aquat. Ecol. 33: 339–345.

Sarma, S. S. S., N. Iyer & H. J. Dumont, 1996. Competitive interactions between herbivorous rotifers: importance of food concentration and initial population density. Hydrobiologia 331: 1–7.

Sokal, R. R. & F. J. Rohlf, 1981. Biometry (2nd edn). W.H. Freeman and Company, San Francisco: 859 pp

Stemberger, R. S. & J. J. Gilbert, 1985a. Body size, food concentration and population growth in planktonic rotifers. Ecology 66: 1151–1159.

Stemberger, R. S. & J. J. Gilbert, 1985b. Assessment of threshold food levels and population growth in planktonic rotifers. Arch. Hydrobiol. Beih. 21: 269–275.

Walz, N., 1995. Rotifer populations in plankton communities: Energetics and life history strategies. Experientia 51: 437–453.

Hydrobiologia **446/447:** 71–74, 2001.
L. Sanoamuang, H. Segers, R.J. Shiel & R.D. Gulati (eds), Rotifera IX.




# Egg size and offspring fitness in a bdelloid rotifer

Nadia Santo, Manuela Caprioli, Simona Orsenigo & Claudia Ricci
*Department of Biology, State University of Milan, via Celoria 26, 20133 Milano, Italy*
*E-mail: rotiferi@mailserver.unimi.it*

*Key words:* bdelloid rotifers, egg volume, developmental time, age at maturity, anhydrobiosis

## Abstract

To test if the quality of offspring is affected by egg size and whether it increases with the amount of resources allocated to an individual egg, we compared the offspring hatched from small and large eggs of *Macrotrachela quadricornifera* (Rotifera, Bdelloidea). Differently sized eggs were obtained by feeding mothers with different food concentrations. Large eggs were expected to provide better offspring in terms of (1) development time of the embryos, (2) age at first reproduction of the juveniles and (3) recovery after desiccation of the newborns. The comparison between offspring hatched from large vs. small eggs of the same bdelloid species revealed that animals hatched from large eggs had shorter embryonic development and earlier age at first reproduction than animals hatched from small eggs. In contrast, the capacity to survive stress, like desiccation, was not affected by egg size. Thus, offspring hatched from larger eggs had shorter generation times, but were not favoured under harsh circumstances.

## Introduction

Offspring quality is expected to be positively correlated with egg size (Tessier & Consolatti, 1991), and the fitness of an individual offspring is expected to increase with the energy invested into it (Boersma, 1997). For instance, larger eggs of cladocerans and rotifers result in higher quality progeny, in terms of increased resistance to starvation (Tessier & Consolatti, 1989; Guisande & Gliwicz, 1992; Kirk, 1997a).

Genotype, age and size of the mother can influence egg size. Differences in egg size resulting from genetic differences are reported for several species and clones (Tessier & Consolatti, 1991; Glazier, 1992; Guisande & Gliwicz, 1992). In addition, egg size is positively correlated to maternal age and size (Glazier, 1992; Kirk, 1997a), and is affected by food quality and quantity, either positively or negatively (Boersma & Vijverberg, 1995; Kirk, 1997a).

Factors affecting egg size of bdelloid rotifers were studied in *Macrotrachela quadricornifera*. In this species egg size was influenced by genotype, size and age of the mother (Ricci, 1995), and increased with food concentration (Orsenigo & Ricci, 1997). Orsenigo et al. (1998) compared egg sizes in seven bdelloid species. Although genetic factors could have biassed the results as the large and small eggs used in the study belonged to different species, it was found that increased egg size was correlated to decreased fecundity, longevity, and to delayed maturity of the mother, and to longer developmental time of the egg. Thus, it appears that the production of large eggs is disadvantageous to the mothers. However, the production of large eggs, even if costly to the mothers in terms of decreased fertility (e.g. Glazier, 1992), could provide advantages to the offspring hatching from these eggs, e.g. endowing them with higher fitness. The possibility to manipulate the size of *M. quadricornifera* eggs by feeding the mothers different diets gave the opportunity to compare genetically identical eggs and to analyze the consequences of hatching from large vs. small eggs. Assuming that egg size affects offspring quality, we studied the effect of egg size on (1) development time of embryos, (2) age at first reproduction of juveniles, and (3) recovery after desiccation of newborns in *M. quadricornifera*.

Two traits, developmental time and age at maturity, are important fitness components and strongly affect population growth rate. Viability after desiccation was tested because the adaptations required to survive drying, a frequent stress factor in bdelloid habitats, are expected to be costly (Jacobs, 1909; Ricci et al., 1987). Our experiment was carried out on non-fed neonates, hypothesizing that survival could depend upon the amount of resources present in the egg, consequently on egg size.

## Materials and methods

The strain of *Macrotrachela quadricornifera* Milne, 1886 used in the present study has been reared under laboratory conditions (22–24 °C) for several years. Experiments were conducted at 24 °C, and rotifers were cultured according to the life table experimental protocol, in deionized water as medium. A suspension of powdered trout pellets was used as food, provided at two concentrations (0.06 and 6 mg ml$^{-1}$). Similarly to a previous study (Orsenigo & Ricci, 1997), rotifers fed the two diets produced eggs of different sizes. Volumes of ca. 30 eggs for each feeding regime were measured using an image analysis program (details in Ricci & Fascio, 1995; Orsenigo & Ricci, 1997). Viability of the eggs was randomly checked during all experiments. Life-history traits of the rotifers hatched from the differently sized eggs were compared by means of one-way ANOVA, and recovery rates after desiccation with a $\chi^2$-test.

## Experimental design

From a stock population, about 200 synchronous laid eggs were isolated and used to establish the pre-parental generation. This was reared on high food concentration (6 mg ml$^{-1}$). Eggs laid by 9 to 12 days old animals were collected and randomly divided into two groups representing the parental cohorts. One cohort, with approx. 300 rotifers, was fed a low diet (L, 0.06 mg ml$^{-1}$). A second cohort of 160 individuals was fed a high diet (H, 6 mg ml$^{-1}$). As egg size is known to depend on size of the mother (Ricci, 1995), we collected eggs laid by rotifers that already completed their somatic growth. This occurs at different ages for the two cohorts (Orsenigo & Ricci, 1997). Thus, eggs were collected from 9 to 10 days old rotifers in the H cohort, and from 11 to 12 days old rotifers in the L cohort, and assigned to the different experiments.

### *Development time*

The duration of embryonic development was assessed on about 200 eggs from either L or H cohorts. The eggs were laid synchronously and were inspected for hatching at intervals of 12 h.

### *Age at first reproduction*

About 35 eggs from L cohort and 80 eggs from H cohort were used for this experiment. All were isolated, allowed to hatch and fed with intermediate food concentration (0.6 mg/ml) until the onset of reproduction. The age at first reproduction was registered by checking the animals at 24 h intervals. The sample size of the two groups differed because of the different age-specific fecundity of the rotifers fed high vs. low diet, and therefore of the availability of freshly laid eggs.

### *Recovery after desiccation*

About 150 neonates born from either L and H cohorts were dried within a few hours after birth under controlled conditions, following the protocol of Caprioli & Ricci (2001). The rotifers were kept dry for a week and then were rehydrated. Recovery rates of the two groups were recorded by counting the living rotifers 24 h after rehydration.

## Results and discussion

*M. quadricornifera* produces eggs of different volumes if fed different diets (Orsenigo & Ricci, 1997), similar to other rotifer species (e.g. Kirk, 1997a). The increased egg size has been interpreted as a strategy to increase progeny fitness by equipping the newborns with greater amount of resources (Tessier & Consolatti, 1989; Guisande & Gliwicz, 1992; Kirk, 1997a). In this study, rotifers fed high diet (6 mg ml$^{-1}$) laid eggs that were about 30% larger than those produced by rotifers at low diet (0.06 mg ml$^{-1}$), and this difference was statistically significant (ANOVA, $F$= 39.5; 0.05>$P$>0.01; Table 1). Both egg groups were 100% viable, and the offspring survived till reproduction.

The duration of embryonic development differed between the two egg groups, and, unexpectedly, the embryos in the small eggs required a longer time to complete their development than the embryos in the large eggs. This contrasts with other studies, which report that increased egg size is associated with longer

*Table 1.* Offspring characteristics of *Macrotrachela quadricornifera* and statistical analysis. Sample sizes, mean values (± standard error) of each parameter and *P* values are reported

| | Small | | Large | | Small vs. large |
|---|---|---|---|---|---|
| | Sample size | Mean (± S.E.) | Sample size | Mean (± S.E.) | Test, Probability |
| Egg volume ($\mu m^3 \times 103$) | 30 | 177.41 (±3.52) | 30 | 241.77 (±3.72) | ANOVA, $P$=0.025 |
| Developmental time (days) | 203 | 5.16 (±0.06) | 205 | 4.73 (±0.05) | ANOVA, $P$<0.0001 |
| Age at first reproduction | 35 | 5.54 (±0.11) | 82 | 4.67 (±0.05) | ANOVA, $P$<0.0001 |
| Recovery rate after desiccation (%) | 140 | 93.33 (±2.21) | 146 | 95.36 (±1.31) | $\chi^2_{1df}$=0.737, $P$=0.391 |

embryogenesis time, e.g. in *Daphnia* spp. (Guisande & Gliwicz, 1992) or *Brachionus calyciflorus* and *B. caudatus* (Walz, 1997). The difference in development time between large and small eggs is not large (4.7 vs. 5.2 days), but statistically significant (ANOVA, $F$=27.5; $P$<0.0001; Table 1).

A similar trend was found when comparing age at first reproduction of juveniles from large and small eggs. Juveniles from large eggs reached maturity about 1 day earlier than those from small eggs (ANOVA, $F$=14.8; $P$<0.0001; Table 1). The juveniles from both egg groups were genetically identical and were reared at the same temperature and with the same food concentration, and the major difference between them was the volume of the eggs from which they hatched, that corresponded to the amount of resources available to each embryo during its development.

Increased egg volume decreased the duration of embryonic development and the time necessary to reach sexual maturity. Both traits are relevant in determining fitness as they result in a shorter generation time, and consequently in a greater contribution to population growth.

Offspring born from larger eggs were expected to be more resistant to conditions of stress (e.g. Kirk, 1997a,b). We subjected newborns from large and small eggs to desiccation in an attempt to obtain an indirect measure of the amount of resources present in the eggs. Bdelloid rotifers are able to withstand drought by entering anhydrobiosis, a particular form of dormancy. Anhydrobiosis induction requires a series of adjustments which are likely to be metabolically costly. Thus, the recovery capacity of the rotifers is assumed to be influenced by the amount of stored resources (Jacobs, 1909). However, the recovery rate from 7-d anhydrobiosis of neonates from large eggs was similar to that from small eggs (Table 1), indicating that the recovery rate of the neonates from anhydrobiosis was independent of egg size ($\chi^2_{1df}$=0.7; $P$>0.1). As the difference in egg size is related to a different level of resource availability (Guisande & Gliwicz, 1992; Boersma, 1997), then our results would suggest that the ability to survive desiccation be not affected by this factor.

The attempt to assess the costs and benefits of producing large vs. small eggs in different species of bdelloid rotifers revealed that producing large eggs implies costs rather than benefits to the mother animals (Orsenigo et al., 1998). Our comparison of offspring hatched from large vs. small eggs of the same species showed that animals hatched from large eggs were at advantage over animals hatched from small eggs, in terms of two traits relevant to fitness (embryonic developmental time and age at first reproduction). In contrast, the capacity to survive desiccation did not differ between the two groups of rotifers.

Life-history theory predicts that large offspring are favoured under harmful circumstances (Glazier, 1998). Assuming that larger eggs produce larger newborns, the larger offspring of *M. quadricornifera* have shorter generation times, and thus are favoured during population growth, but are not favoured under harsh circumstances.

## Acknowledgements

Financial support came from ASI grant to C.R.

## References

Boersma, M., 1997. Offspring size in *Daphnia*: does it pay to be overweight? Hydrobiologia 360: 79–88.

Boersma, M. & J. Vijverberg, 1995. Synergistic effects of different food species on life-history traits of *Daphnia galeata*. Hydrobiologia 307: 109–115.

Caprioli, M. & C. Ricci, 2001. Recipes for successful anhydrobiosis in bdelloid rotifers. Hydrobiologia 446/447 (Dev. Hydrobiol. 153): 13–17.

Glazier, D. S., 1992. Effects of food, genotype and maternal size and age on offspring investment in *Daphnia magna*. Ecology 73: 910–926.

Glazier, D. S., 1998. Does body storage act as a food-availability cue for adaptive adjustment of egg size and number in *Daphnia magna*? Freshwat. Biol. 40: 87–92.

Guisande, C. & Z. M. Gliwicz, 1992. Egg size and clutch size in two *Daphnia* species grown at different food level. J. Plankton Res. 14: 997–1007.

Jacobs, M. H., 1909. The effects of desiccation on the rotifer *Philodina roseola*. J. exp. Zool. 6: 207–263.

Kirk, K. L., 1997a. Egg size, offspring quality and food level in planktonic rotifers. Freshwat. Biol. 37: 515–521.

Kirk, K. L., 1997b. Life-history responses to variable environments: starvation and reproduction in planktonic rotifers. Ecology 78: 434–441.

Orsenigo, S. & C. Ricci, 1997. To grow or to reproduce: what bdelloid rotifers do choose. S. It. E. Atti 18: 139–142.

Orsenigo, S., C. Ricci & M. Caprioli, 1998. The paradox of bdelloid egg size. Hydrobiologia 387/388: 317–320.

Ricci C., 1995. Growth pattern of four strains of a bdelloid rotifer species: egg size and numbers. Hydrobiologia 313/314: 157–163.

Ricci C. & U. Fascio, 1995. Life-history consequences of resource allocation of two bdelloid rotifer species. Hydrobiologia 299: 231–239.

Ricci, C., L. Vaghi & M. L. Manzini, 1987. Desiccation of rotifers (*Macrotrachela quadricornifera)*: survival and reproduction. Ecology 68: 1488–1494.

Tessier, A. J. & N. L. Consolatti, 1989. Variation in offspring size in *Daphnia* and consequences for individual fitness. Oikos 56: 269–276.

Tessier, A. J. & N. L. Consolatti, 1991. Resource quantity and offspring quality in *Daphnia*. Ecology 72: 468–478.

Walz, N., 1997. Rotifer life history strategies and evolution in freshwater plankton communities. In Streit, B., T. Städler & C. M. Lively (eds), Evolutionary Ecology of Freshwater Animals. Birkhäuser Verlag, Basel, Switzerland: 119–149.

*Hydrobiologia* **446/447:** 75–83, 2001.
*L. Sanoamuang, H. Segers, R.J. Shiel & R.D. Gulati (eds), Rotifera IX.*




# Life table demography and population growth of *Brachionus variabilis* Hempel, 1896 in relation to *Chlorella vulgaris* densities

S.S.S. Sarma & S. Nandini
*Division of Research and Postgraduate Studies, National Autonomous University of Mexico, Campus Iztacala, AP 314, CP 54090, Los Reyes, Tlalnepantla, State of Mexico, Mexico*
*E-mail: sarma@servidor.unam.mx*

*Key words: Brachionus variabilis*, population growth, demography, *Chlorella*, Rotifera

**Abstract**

We studied the life history variables and population growth characteristics of *Brachionus variabilis*, which was recorded for the first time from Mexico. The animals were fed *Chlorella*, using five concentrations (0.25, 0.5, 1, 2 and 4 × $10^6$ cells $ml^{-1}$) at 25 °C. Food density was observed to have significant effect on life expectancy, average lifespan, gross reproductive rate, net reproductive rate, generation time and population growth rate. The average lifespan ranged from 3 to 6 days depending on the food density. The net reproductive rate ranged from 2 to 7 neonates $female^{-1}$ $d^{-1}$. The rate of population increase per day varied from 0.14 to 0.35. The highest net reproductive rate and average lifespan and life expectancy were recorded at *Chlorella* concentrations of 1 × $10^6$ and 2 × $10^6$ cells $ml^{-1}$.

## Introduction

The genus *Brachionus* has a tropical and subtropical distribution with more than 40 species described so far (Koste & Shiel, 1987). Much of the work concerning the population growth and life table demographic aspects is concentrated on a few species of this genus, for example *B. angularis* (Wang & Li, 1997), *B. calyciflorus* (Sarma et al., 1996), *B. dimidiatus* (Pourriot & Rougier, 1975), *B. patulus* (Sarma & Rao, 1991), *B. plicatilis* (Okauchi & Fukusho, 1984), *B. quadridentatus* (Heerkloss & Hlawa, 1995), *B. rubens* (Rothhaupt, 1990) and *B. urceolaris* (Wang & Li, 1997). All these species have a capacity to utilize relatively high algal food concentrations. Thus, a direct relation between food density and the abundance of brachionids is generally known. The threshold food hypothesis states that smaller rotifer species are well adapted to low food levels while larger species cannot survive and reproduce under these conditions (Stemberger & Gilbert, 1985a, b). Cladocerans, on the other hand, appear to have an inverse relation between the body size and the threshold food requirement (Gliwicz, 1990). The conclusions on the rotifer body size and threshold food requirements are based on diverse genera (*Keratella*, *Brachionus*, etc.). It, thus, remains to be tested whether all the species of *Brachionus* would behave similarly under changing food levels. Little information is available on some of the common species of brachionids often occurring in large numbers such as *Brachionus budapestinensis* and *Brachionus variabilis*.

The effect of varying food concentrations on brachionids may be quantified using population growth studies or lifetable demography aspects. Population growth studies provide information on the effect of food level on individuals of various generations simultaneously occurring in a growing culture. Although through this approach an overall effect of food density is known, the method cannot resolve the adverse effects into: (a) those related to survivorship and (b) those related to reproduction (Krebs, 1985). On the other hand, life table demographic studies of a cohort of a brachionid population can provide information on age-specific mortality and fecundity. However, such studies do not provide information on the possible influence of offspring on the growth of the population, nor on the maximal densities reached by a rotifer species when continuously grown under defined conditions of food quality, quantity and temperature. Thus, life table demographic studies and population growth studies are complementary to give a complete picture of the effects of different food levels on a rotifer species. Most studies have so far considered these two aspects but one at a time (see e.g. Dumont & Sarma, 1995). Rarely have these two aspects been evaluated in a single study (King, 1982).

In the present work, we provide information on the population growth and life table demography of *B. variabilis* offered five concentrations of *Chlorella vulgaris*.

## Material and methods

The test rotifer species *Brachionus variabilis*, not previously reported from Mexico (Sarma, 1999), was isolated from a local pond in the State of Aguascalientes (Mexico) and was cultured in 1 l glass beakers using the green alga, *Chlorella vulgaris,* as the exclusive food. Rotifers from culture vessels were transferred every alternate day to fresh medium, containing *Chlorella* at a density of $0.5 \times 10^6$ cells ml$^{-1}$. For maintenance of cultures and for the experiments, we used EPA medium (Anon., 1985), which is used extensively for rotifer cultures. This medium was prepared by dissolving 96 mg $NaHCO_3$, 60 mg $CaSO_4$, 60 mg $MgSO_4$ and 4 mg KCl in one litre of distilled water. *C. vulgaris* was mass cultured using Bold Basal medium (Borowitzka & Borowitzka, 1988). For population growth and demography experiments, we used the alga harvested during the logarithmic phase of growth, centrifuged at 3000 rpm for 5 min and resuspended in EPA medium. This process eliminated the nutrient-rich algal medium, which was not suitable for survival and reproduction of *B. variabilis*. The density of *C. vulgaris* was estimated using a haemocytometer. For both life table demography studies and population growth experiments, we selected five levels of *Chlorella*: $0.25 \times 10^6$, $0.5 \times 10^6$, $1.0 \times 10^6$, $2.0 \times 10^6$ and $4.0 \times 10^6$ cells ml$^{-1}$. Both the life-table demography and population growth studies were conducted at a of pH 7.5, an incubation temperature of 25 °C and under a photoperiod of 12:12 light:dark.

### *Life table demographic studies*

For the life table demographic studies, we obtained a large number of neonates (age ca. 6 h) hatched from parthenogenetic eggs from the large culture vessels. The experimental design consisted of a total of 15 (= 5 food concentrations × 3 replicates) transparent jars of 50 ml capacity, containing 20 ml EPA medium with specified algal type and density. Into each of these vessels we introduced 20 neonates of *B. variabilis* using a Pasteur pipette under a stereomicroscope at 30× magnification. Following inoculation, every day we counted the number of original test individuals alive from each cohort and the number of neonates produced. The surviving adults were then transferred to fresh jars containing the appropriate food level. The neonates were enumerated and eliminated every day. The life table demography experiments were terminated when each individual of every cohort died. Based on the data collected, we derived the following variables: age-specific survivorship, age-specific reproduction, average lifespan, age-specific life expectancy, gross reproductive rate, net reproductive rate, generation time, stable age distribution and rate of population growth. The following formulae were used (Krebs 1985; Pianka 1988). Jackknife and Bootstrap methods were not required because the test cohorts contained a population, not an individual (see also Meyers et al., 1986).

$$\text{Gross reproductive rate} = \sum_{o}^{\infty} m_x$$

$$\text{Net reproductive rate } R_o = \sum_{o}^{\infty} l_x m_x$$

$$\text{Generation time (T)} = \frac{\sum l_x m_x . x}{R_0}$$

$$\text{Rate of population increase} = \sum_{x=0}^{n} e^{-rx} l_x m_x = 1$$

Also from the equation: $\ln (R_0)/T$

$$\text{Stable age distribution } (C_x) = \frac{\lambda^{-x} l_x}{\Sigma \lambda^{-i} l_i} \quad i = 0$$

### *Population growth studies*

For population growth experiments, we used the same

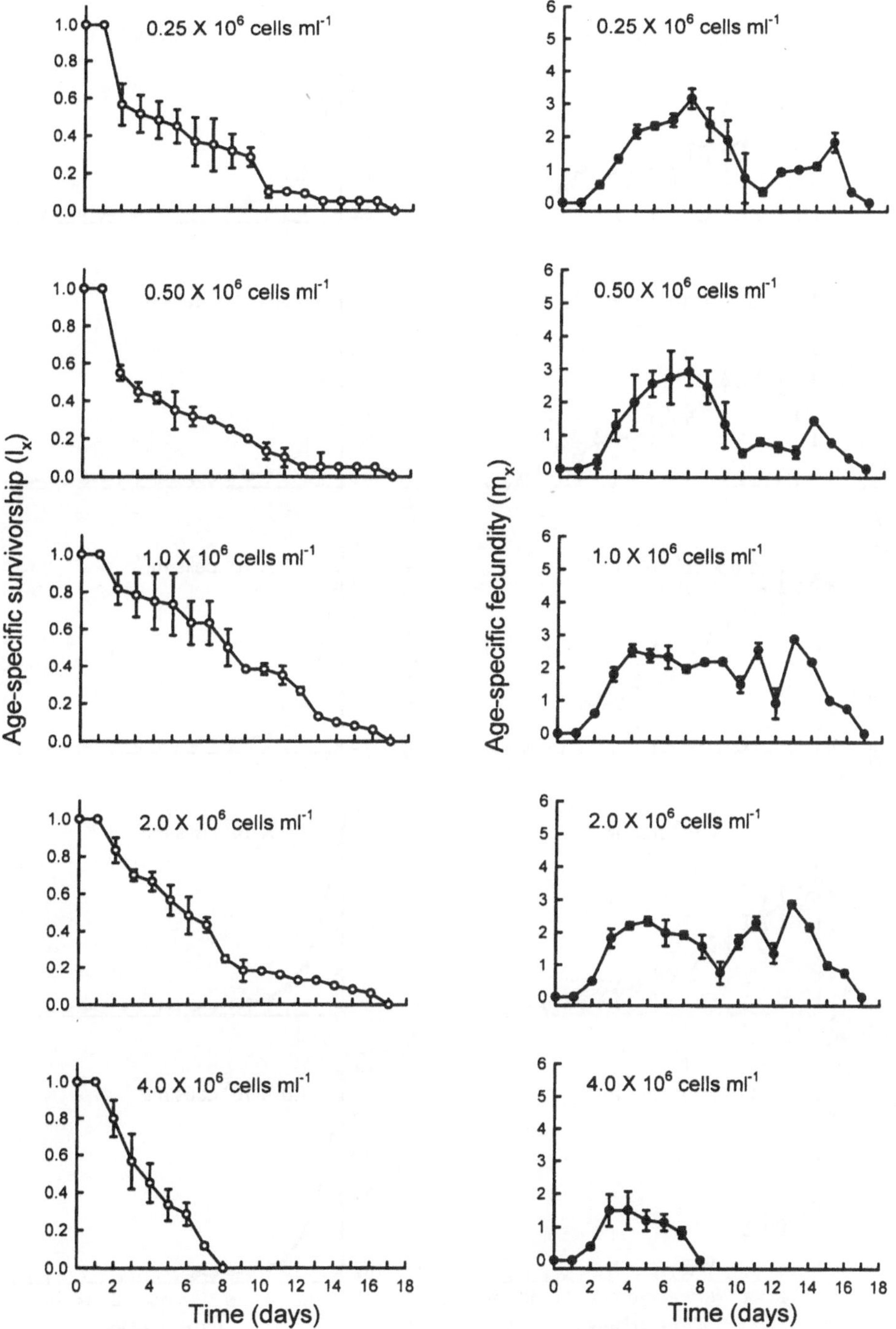


*Figure 1.* Age-specific survivorship (open circles) and fecundity (closed circles) curves of *Brachionus variabilis* offered *Chlorella vulgaris* at five different food concentrations. Shown are mean±standard error values based on 3 replicates.

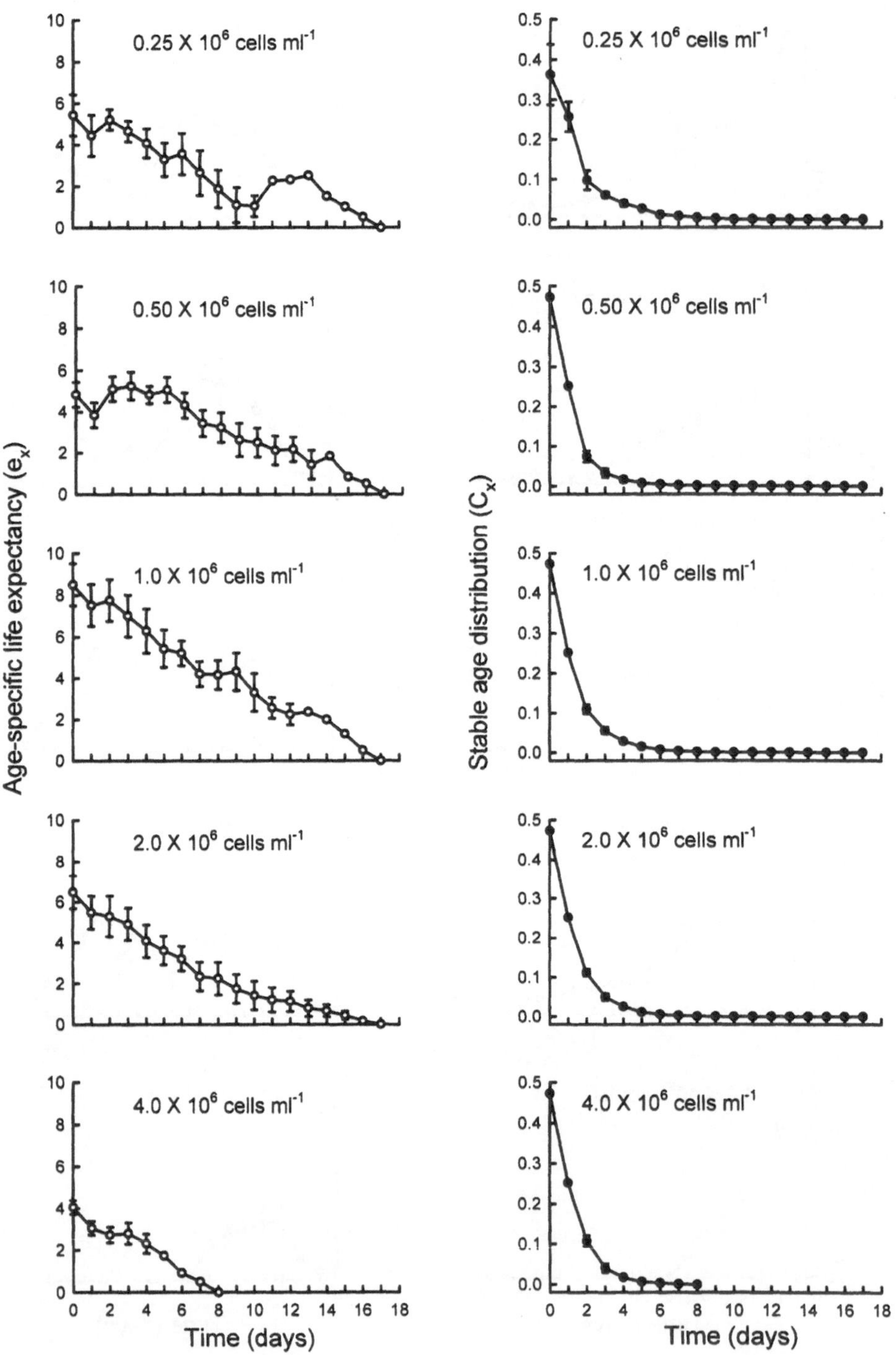


*Figure 2.* Age-specific life expectancy (open circles) and stable age distribution (closed circles) curves of *B. variabilis* offered *Chlorella* at five different food concentrations. Shown are mean±standard error values based on 3 replicates.

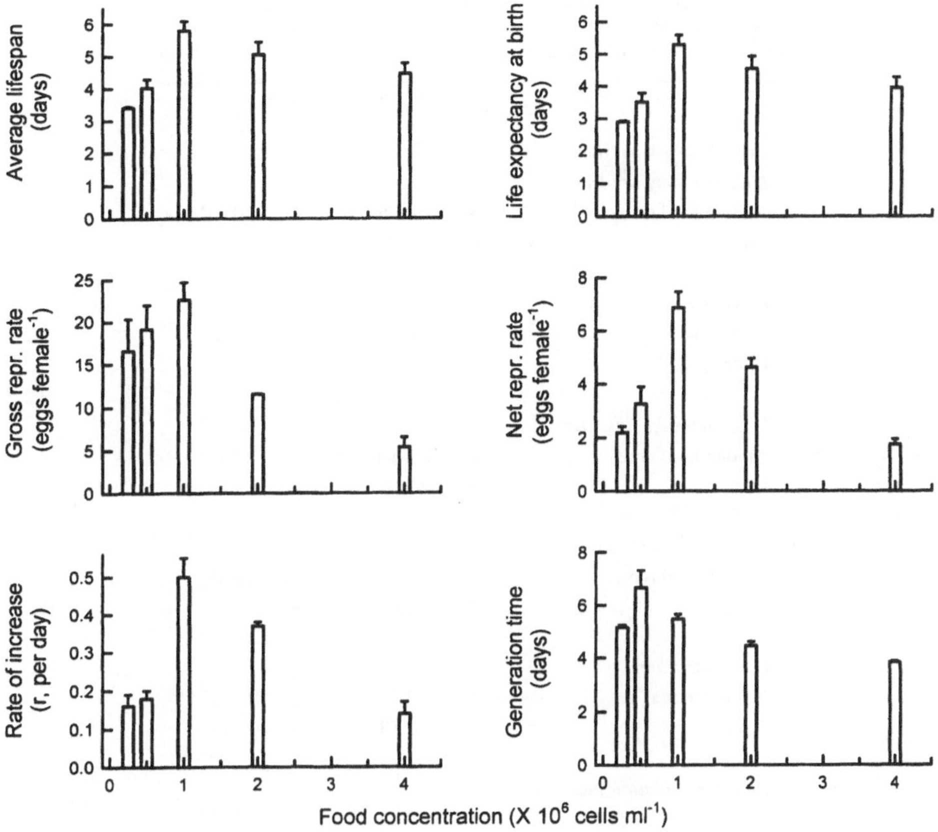


*Figure 3.* Data (mean±standard error values based on 3 replicates) on the life history variables of *B. variabilis* offered *Chlorella* at five different food concentrations. Shown are average lifespan (days), life expectancy at birth (days), gross reproductive rate (offspring female$^{-1}$), net reproductive rate (offspring female$^{-1}$), rate of population increase per day and generation time (days).

*Chlorella* concentrations as in the demography studies. Thus, the experimental design consisted a total of 15 test jars (= 5 food concentrations × 3 replicates) of 50 ml capacity containing 20 ml medium with a specified *Chlorella* density into which we introduced 20 individuals of *B. variabilis* of mixed age (non-ovigerous females). Following inoculation, every day we counted the total number of living individuals per container. Aliquot sampling was not used to quantify the number of rotifers per container. The living rotifers (including the loose eggs if present) were transferred to fresh EPA medium containing appropriate food density. The population growth experiments were terminated after 12 days when growth of most animals began to decline. We obtained the rate of population growth using the following equation:

$$r = (\ln N_t - \ln N_0)/t,$$

where $N_0$ and $N_t$ are the initial and final population densities; $t$ is time in days.

We used varying data points along the growth curve to calculate the mean per replicate; in general, we took 4–6 data points during the exponential stage of the population.

## Results

### *Life table demographic studies*

Among all the life history variables of *B. variabilis* fed different concentrations of *C. vulgaris*, life expectancy at birth ($e_0$), average lifespan, gross reproductive rate, net reproductive rate ($R_o$), generation time (T), and the rate of population increase per day ($r$) were significantly affected by the algal density ($p<0.05$, ANOVA, Table 1). Data on the selected life history variables (mean±standard error) of *B. variabilis* under different algal concentrations are presented in Figures 1–3. The age-specific survivorship curves ($l_x$) showed heavy mortalities in the first few days under low (0.25

*Table 1.* Analysis of variance (ANOVA) performed on various life history and population growth variables of *Brachionus variabilis*. DF = degrees of freedom, SS = sum of squares; MS = mean square, $F$ = $F$-ratio

| Source | DF | SS | MS | $F$ |
|---|---|---|---|---|
| **Life history variables** | | | | |
| *Life expectany at birth* | | | | |
| Among food types | 4 | 10.26 | 2.56 | 10.67** |
| Error | 10 | 2.40 | 0.24 | – |
| *Average lifespan* | | | | |
| Among food types | 4 | 10.27 | 2.57 | 10.66** |
| Error | 10 | 2.41 | 0.24 | – |
| *Gross reproductive rate* | | | | |
| Among food types | 4 | 547.10 | 136.77 | 8.19** |
| Error | 10 | 167.00 | 16.70 | – |
| *Net reproductive rate* | | | | |
| Among food types | 4 | 52.55 | 13.14 | 22.38*** |
| Error | 10 | 5.87 | 0.59 | – |
| *Generation time* | | | | |
| Among food types | 4 | 13.96 | 3.49 | 12.33*** |
| Error | 10 | 2.83 | 0.28 | – |
| *Rate of population increase* | | | | |
| Among food types | 4 | 0.14 | 0.035 | 35.00*** |
| Error | 10 | 0.01 | 0.001 | – |
| **Population growth studies** | | | | |
| *Peak population density* | | | | |
| Among food types | 4 | 73.98 | 18.50 | 3.97* |
| Error | 10 | 46.62 | 4.66 | – |
| *Rate of population increase* | | | | |
| Among food types | 4 | 134.27 | 33.57 | 17.98*** |
| Error | 10 | 18.67 | 1.87 | – |

* = $p< 0.05$; ** = $p< 0.01$; *** = $p<0.001$.

× $10^6$ and 0.5 × $10^6$ cells ml$^{-1}$) and high (4.0 × $10^6$ cells ml$^{-1}$) food concentrations. At an intermediate food density (1.0 × $10^6$ cells ml$^{-1}$), a higher survivorship was obtained. Age-specific fecundity ($m_x$), which were not survival weighted, had a more or less normal distribution pattern but much lower production was observed at the highest food density (Fig. 1). Age-specific life expectancy ($e_x$) was higher at 1.0 × $10^6$ cells ml$^{-1}$ food density, while, the stable age distribution curves were similar at intermediate algal concentration (Fig. 2). Data on the life expectancy at birth ranged from 2.9±0.03 to 5.3±0.29 days depending on the food density. The highest average lifespan (6 d) of *B. variabilis* was obtained when fed *Chlorella* at a food density 1.0 × $10^6$ cells ml$^{-1}$, while the lowest value (3 d) was recorded at the lowest food concentration. Gross-reproductive rate varied from 16.6±3.8 to 22.6±2.1 offspring female$^{-1}$, the highest being at a *Chlorella* concentration of 1.0 × $10^6$ cells ml$^{-1}$. The survival weighted net reproductive rates on the other hand varied from 2.2±0.2 to 6.9±0.6 offspring female$^{-1}$. The shortest generation time (4 d) was observed under the highest algal food level. The highest rate of population increase (0.50±0.05) was observed at $1.0 \times 10^6$ cells $ml^{-1}$ food level (Fig. 3).

*Population growth studies*

Data on the population growth of *B. variabilis* in relation to algal concentrations are shown in Figure 4. Food concentration had a significant effect on the peak population abundance of *B. variabilis* ($p<0.05$, ANOVA, Table 1). Similarly, day at which peak population abundance occurred was also significantly affected by the algal food concentration ($p<0.01$, ANOVA, Table 1). Throughout the population growth study, peak population abundance values of the rotifer varied from 2.0±0.3 to 9.0±2.0 ind $ml^{-1}$ depending on the food concentration. The highest rate of population increase ($r$) was 0.13±0.01 under $0.5 \times 10^6$ cells $ml^{-1}$, the rates being related inversely to the food abundance (Fig. 5).

## Discussion

Among the brachionid rotifers, population densities exceeding 1000 ind $ml^{-1}$ have been reported for *B. calyciflorus* and *B. plicatilis*, but rarely other species such as *B. patulus* (reviewed in Sarma, 1991). Such high densities are normally obtained employing algal food concentrations in the range from $0.5 \times 10^6$ to $25 \times 10^6$ cells $ml^{-1}$ of phytoplankton (depending on the algal type, cell diameter etc). For *Chlorella*, most investigators have normally used algal concentrations ranging from $0.1 \times 10^6$ to $10 \times 10^6$ cells $ml^{-1}$ for *Brachionus* cultures (see Sarma et al., 2001). However, it is not known if algal concentrations in this range are suitable for all the *Brachionus* species. From the present study, it is evident that an algal concentration $>1 \times 10^6$ cells $ml^{-1}$ is unsuitable for successful growth of *B. variabilis.* From the data on the life table demography, the optimal range of algal food concentration lies close to $1 \times 10^6$ cells $ml^{-1}$, and for population growth to $0.5 \times 10^6$ cells $ml^{-1}$. From the literature, it appears that larger brachionid species require higher food abundance to maximize the rate of population increase. In the present study, although *B. variabilis* (length without spines, 216±26 $\mu$m, $n$=50) was larger than *B. calyciflorus* (170±10 $\mu$m), it could not utilize higher algal availability. Simultaneously, we observed the production of male eggs in the culture vessels resulting in a decline of the population, even

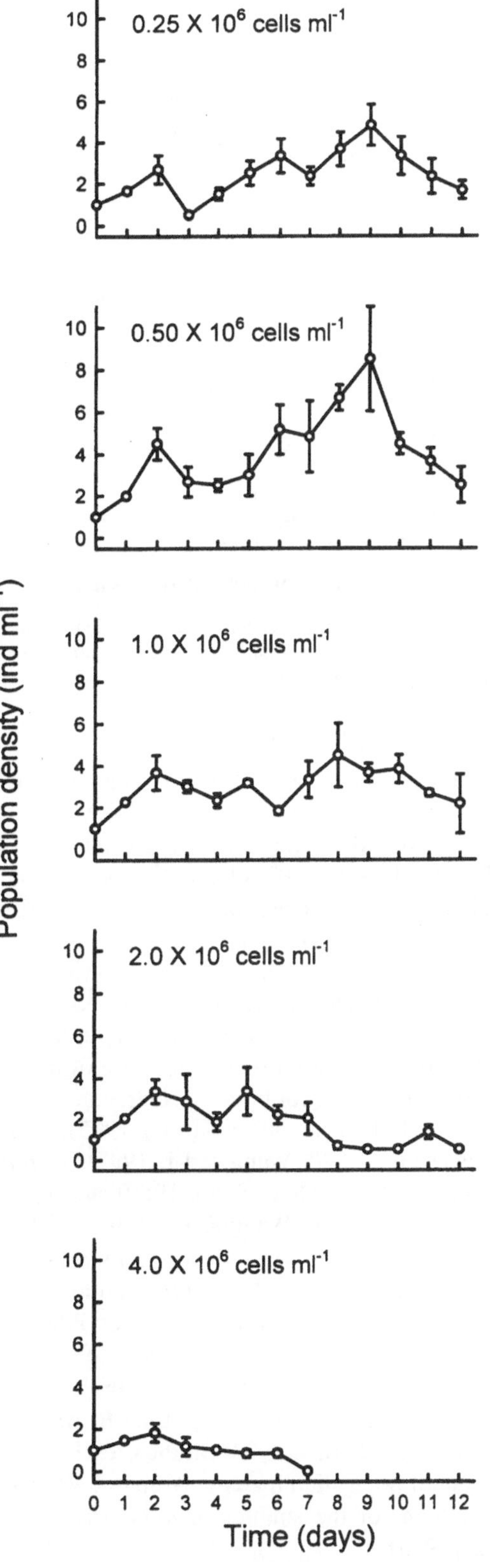


*Figure 4.* Population growth curves of *B. variabilis* offered *Chlorella* under five food concentrations. Shown are mean ± standard error values based on 3 replicate recordings.

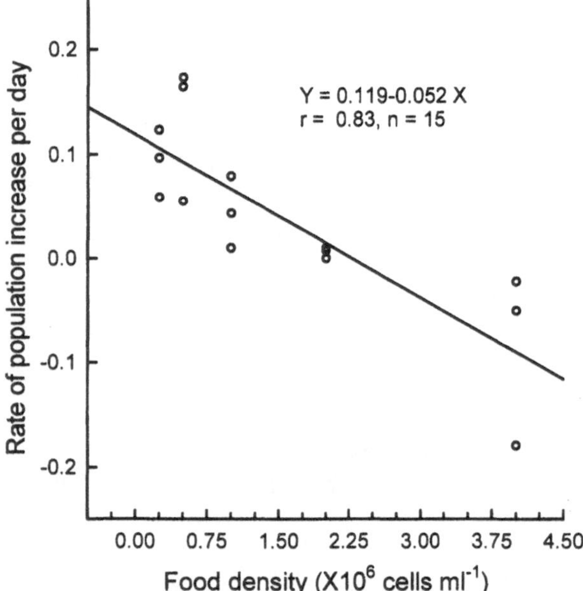


*Figure 5.* Relation between *Chlorella* concentration and the rate of population growth of *B. variabilis*. At each food concentration, all the three replicates were plotted.

though we used in our experiments a clonal population that was maintained at least 6 months. Moreover, the fact that we did not encounter males and *B. variabilis* did well only at $1 \times 10^6$ cells $ml^{-1}$ food concentration implied that this species was well adapated to living under low food levels.

The relation between length and *r* in rotifers species has not been definitively established. Under optimal food concentrations, it is often thought that large rotifers have higher *r* values than those of the smaller species. This is true for larger species, e.g. *Asplanchna* spp. ($r > 1.50$, Dumont & Sarma, 1995), *B. calyciflorus* (*r* up to 2.20, Wang & Li, 1997), *B. plicatilis* (*r* up to 2.00, Miracle & Serra, 1989), and *Epiphanes brachionus* ($r > 0.9$; Wang & Li, 1997), and smaller rotifer species such as *Anureaopsis fissa* (<0.8; Dumont et al., 1995) and *B. patulus* (*r* up to 0.6; Sarma & Rao, 1990). However, this condition is not always observed. For example, Iyer & Rao (1996) have documented lower *r* values (0.11–0.68) for *A. intermedia,* which is a much larger species than *B. rubens* (*r* up to 0.80) (Iyer & Rao, 1993). In the present work, too, the rate of population increase (*r* up to 0.2) was much lower than for the smaller *B. angularis* ($r > 0.58$; Walz, 1995).

The range of life history variables observed for *B. variabilis* in this study were in general agreement with the range expected for other *Brachionus* spp (Sarma & Rao, 1991; Wang & Li, 1997). When survivorship curves show rapidly declining trends, the differences in the gross and net reproductive rates become higher (Krebs, 1985; see also Fig. 4). As in the case of population growth experiments, life table demography study of *B. variabilis* showed consistently higher values for the different life history variables at relatively lower food levels, indicating that this species is adapted to lower *Chlorella* concentrations.

## Acknowledgements

The authors thank National System of Investigators (Mexico) (Ref. No. 18723 & 20520) for support.

## References

Anonymous, 1985. Methods of measuring the acute toxicity of effluents to freshwater and marine organisms. U.S. Environment Protection Agency EPA/600/4-85/013.

Borowitzka, M. A. & L. J. Borowitzka, 1988. Micro-algal Biotechnology. Cambridge University Press, London.

Dumont, H. J. & S. S. S. Sarma, 1995. Demography and population growth of *Asplanchna girodi* (Rotifera) as a function of prey (*Anuraeopsis fissa*) density. Hydrobiologia 306: 97–107.

Dumont, H. J., S. S. S. Sarma & A. J. Ali, 1995. Laboratory studies on the population dynamics of *Anuraeopsis fissa* (Rotifera) in relation to food density. Freshwat. Biol. 33: 39–46.

Gliwicz, Z. M., 1990. Food thresholds and body size in cladocerans. Nature 343: 638–640.

Heerkloss, R. & S. Hlawa, 1995. Feeding biology of two brachionid rotifers: *Brachionus quadridentatus* and *Brachionus plicatilis*. Hydrobiologia 313/314: 219–221.

Iyer, N. & T. R. Rao, 1993. Effect of the epizoic rotifer *Brachionus rubens* on the population growth of three cladoceran species. Hydrobiologia 255/256: 325–332.

Iyer, N. & T. R. Rao, 1996. Responses of the predatory rotifer *Asplanchna intermedia* to prey species differing in vulnerability: Laboratory and field studies. Freshwat. Biol. 36: 521–533.

King, C. E., 1982. The evolution of lifespan. In Dingle, H. & J. P. Hegmann (eds), Proceedings in Life Sciences: Evolution and Genetics of Life Histories. Springer-Verlag, New York: 121–138.

Koste, W. & R. J. Shiel, 1987. Rotifera from Australian inland waters. 2. Epiphanidae and Brachionidae (Rotifera: Monogononta). Invertebr. Taxon. 7: 949–1021.

Krebs, C. J., 1985. Ecology. The Experimental Analysis of Distribution and Abundance. 3rd edn. Harper & Row, New York. pp

Meyers, J. S., C. G. Ingresol, L. L. McDonald & M. S. Boyce, 1986. Estimating uncertainity in population growth rates: Jackknife vs bootstrap techniques. Ecology 67: 1156–1166.

Miracle M. R. & M. Serra, 1989. Salinity and tempertature influence in rotifer life history characteristics. Hydrobiologia 186/187: 81–102.

Okauchi, M. & K. Fukusho, 1984. Food value of minute alga, *Tetraselmis tetrathele*, for the rotifer *Brachionus plicatilis* culture:

1. Population growth with batch culture. Bull. Nat. Res. Inst. Aquacult. 5: 13–18.

Pianka, E. R., 1988. Evolutionary Ecology. Harper & Row, New York, 3rd edn.

Pourriot, R. & C. Rougier, 1975. The dynamics of a laboratory population of *Brachionus dimidiatus* (Bryce) (Rotatoria) in relation to food and temperature. Ann. Limnol. 11: 125–143.

Rothhaupt, K. O., 1990. Population growth rates of two closely related rotifer species effects of food quantity particle size and nutritional quality. Freshwat. Biol. 23: 561–570.

Sarma, S. S. S., 1991. Rotifers and aquaculture (Review). Envir. Ecol. 9: 414–428.

Sarma, S. S. S., 1999. Checklist of rotifers (Rotifera) from Mexico. Envir. Ecol. 17: 978–983.

Sarma, S. S. S. & T. R. Rao, 1990. Population dynamics of *Brachionus patulus* Müller (Rotifera) in relation to food and temperature. Proc. Indian Acad. Sci. (Anim. Sci.) 99: 335–343.

Sarma, S. S. S. & T. R. Rao, 1991. The combined effects of food and temperature on the life history parameters of *Brachionus patulus* Müller (Rotifera). Int. Rev. ges. Hydrobiol. 76: 225–239.

Sarma, S. S. S., N. Iyer & H. J. Dumont, 1996. Competitive interactions between herbivorous rotifers: importance of food concentration and initial population density. Hydrobiologia 331: 1–7.

Sarma, S. S. S., P. S. Larios-Jurado & S. Nandini, 2001. Effect of three food types on the population growth of the rotifers *Brachionus calyciflorus* and *Brachionus patulus* (Rotifera: Brachionidae). Rev. Biol. Trop. 49: (In press)

Stemberger, R. S. & J. J. Gilbert, 1985a. Body size, food concentration and population growth in planktonic rotifers. Ecology 66: 1151–1159.

Stemberger, R. S. & J. J. Gilbert, 1985b. Assessment of threshold food levels and population growth in planktonic rotifers. Arch. Hydrobiol. Beih. 21: 269–275.

Walz, N., 1995. Rotifer populations in plankton communities: Energetics and life history strategies. Experientia 51: 437–453.

Wang, J. & D. Li, 1997. Comparative studies on principal parameters of population growth of five freshwater rotifers. Acta Hydrobiol. Sinica 21: 131–136.

Hydrobiologia 446/447: 85–93, 2001.
L. Sanoamuang, H. Segers, R.J. Shiel & R.D. Gulati (eds), Rotifera IX.




# Life cycle patterns of rotifers in Lake Peipsi

Taavi Virro
*Department of Zoology and Hydrobiology, University of Tartu, 46 Vanemuise Street, 51014 Tartu, Estonia*
*E-mail: tvirro@ut.ee*

*Key words:* Rotifera, bisexual reproduction, life cycle patterns, egg ratio, mictic ratio

## Abstract

Life cycle strategies of rotifers in Lake Peipsi (Estonia) were examined. Bisexual (mictic) reproduction was detected in 26 species. Life cycle patterns were determined for 17 species. All three basic life cycle patterns were represented, with some evidence of intraspecific variability. The midcycle pattern, with mixis occurring near the population maximum, prevailed. Extended mixis was observed in littoral populations of *Polyarthra luminosa* and *P. remata*. Most species probably are monocyclic. *Polyarthra dolichoptera* and *Synchaeta oblonga* may be dicyclic. Mictic periods were not detected in *Conochilus hippocrepis* and *C. unicornis*, which probably are acyclic in L. Peipsi. Mictic ratios and egg ratios were calculated for some of the dominant species. The highest amictic egg ratios were observed slightly before or at the population maxima. High mictic ratios (1.0) indicated very intense bisexual reproduction in populations of *A. fissa*, *P. dolichoptera* and *S. verrucosa*. Definite periods of bisexual reproduction could not be distinguished in the rotifer community. During May–October, the spread of mictic reproduction merely reflects the general seasonal distribution of rotifers. Their life cycle strategies represent specific adaptations to unpredictability of their habitat. The results of the study confirm that mixis is an anticipatory event, and not a response to environment deterioration, or an ending of a rotifer population cycle.

## Introduction

Specific features of the rotifer life cycle play an important role in the determination of their population dynamics. As a rule, the life cycle of monogonont rotifers is characterised by cyclical parthenogenesis (heterogony), where parthenogenetic (amictic) and bisexual (mictic) modes of reproduction alternate during population development (e.g. Nogrady et al., 1993).

The factors inducing transition to mictic female production are not yet thoroughly understood. It is evident that specific controlling stimuli are involved (Birky & Gilbert, 1971). A number of laboratory studies on some species of *Asplanchna*, *Notommata* and *Brachionus* have revealed several external and internal factors initiating mictic female production (Birky & Gilbert, 1971; Snell, 1998 and references therein). Dietary tocopherol (vitamin E), photoperiod and population density have been stressed as external inducing stimuli. The rate of mictic female production has been found to be modulated by the age of parental female, food condition, water temperature, salinity, population density and photoperiod (references above, also Lubzens et al., 1993; Gómez et al., 1997). Genetic factors play an important role in determining the nature of the switch from amictic to mictic reproduction, sensitivity to mictic stimuli, and the extent of mictic response (Buchner, 1977; Lubzens et al., 1985; Walsh, 1993; Carmona et al., 1995; Snell, 1998 and references therein). Previous studies have shown that both the inducing stimuli and their responses are species, population and even clone specific.

Transition to bisexual reproduction may occur in different phases of the population cycle. Timing is important for its consequences on the genetic structure of the population and on its dynamics (Birky & Gilbert, 1971; King, 1980).

Although many field investigations deal with the dynamics of rotifer populations, studies addressing the timing of bisexual reproduction in natural populations of rotifers are not numerous (e.g. Carmona et al., 1995; Miracle & Armengol-Díaz, 1995; Gómez et al., 1997). To date, data on life cycle patterns are known for only 20 or 30 species.

The aim of the present work was to determine the life cycles of the dominant planktonic rotifers of Lake Peipsi with emphasis on the alternation of parthenogenetic and bisexual phases of reproduction in their life cycles. In addition, an attempt was made to quantify intensity of different modes of reproduction.

## The lake

Lake Peipsi (58° 22′ N–59 °00′ N, 26° 57′ E–27 °59′ E, surface area 2611 km$^2$, average depth 8.3 m, maximum depth 12.9 m) is the northern part of the compound Lake Peipsi–Pihkva on the eastern border of Estonia. It is eutrophic, with mesotrophic features in the northern part. A general description of L. Peipsi–Pihkva is given in Nõges et al. (1996). Physical and chemical characteristics of the lake are detailed in Jaani (1996) and Möls et al. (1996). Data on water temperature and transparency for the study period are presented in Virro (1996). Further information on phytoplankton can be found in Laugaste et al. (1996); on zooplankton in Haberman (1996) and Mäemets et al. (1996) and on rotifers in Virro & Haberman (1993) and Virro (1996).

## Materials and methods

Zooplankton samples were collected during 1986–1988 in the north-western part of Lake Peipsi. The sedimentation method was used to concentrate quantitative samples (Virro, 1996). Periods of bisexual reproduction were determined by the occurrence of haploid eggs, males or resting eggs.

In conventional sample analysis individuals and eggs (attached and detached) were counted differentiating: (1) females with amictic eggs (amictic females); (2) females with haploid eggs (mictic females); (3) females with resting eggs (mictic females); (4) non-ovigerous females; (5) males; (6) amictic eggs; (7) haploid eggs; and (8) resting eggs. From these counts, egg ratios and mictic ratios were calculated. Egg ratios for the three different types of eggs were calculated as the number of eggs per female (by including detached eggs and the total number of females). Mictic ratios were calculated by dividing the number of mictic females by the total number of females whose reproductive category could be determined (i.e. excluding non-ovigerous females). The number of mictic females was estimated as the sum of females with haploid eggs and those carrying resting eggs. Non-ovigerous females may be of any reproductive type. In case of fixed field samples, their types are not ascertainable. This may induce some error in these calculations. In the present material, the proportion of ovigerous females of the examined species constituted 0–75% of the total of females. In addition, at low population densities the calculations are less representative due to small sample sizes. However, rough estimation of these ratios on field material is possible for more abundant species.

## Results and discussion

### *Occurrence of bisexual reproduction*

Occurrence of bisexual (mictic) reproduction was detected in 26 species (Table 1). Only resting eggs attached to females were considered, because detached resting eggs, of *Filinia terminalis* and *Synchaeta verrucosa*, for example, often encountered outside their developmental cycles, could have originated from previous cycles. In the case of *Polyarthra remata*, amphoteric females carrying amictic and resting eggs were found. To date, amphoteric females have been described for only a few species of *Asplanchna, Conochiloides* and *Sinantherina* (Bogoslovsky, 1960 King & Snell, 1977; Nogrady et al., 1993 and references therein). The amphoteric type of females can be regarded as a peculiar 'bet-hedging' adaptation, which may be of use in unpredictable habitats.

Analysis of the life cycles of rotifers (Table 1; Figs 1–3), does not permit discrimination of definite periods of bisexual reproduction as was established in earlier papers (Virro & Haberman, 1993; Virro, 1995). Concentrating mostly on the vegetation period (in L. Peipsi from May to October), spread of mictic reproduction merely reflects the general seasonal distribution of rotifers. Bisexual reproduction can be initiated at any time of the year and can appear concurrently with parthenogenesis (Bogoslovsky, 1958; Nipkow, 1961; Nogrady et al., 1993).

### *Life cycle patterns*

Life cycle patterns were determined for 17 species (Table 2). Although mictic reproduction was recorded for nine more species, their population maxima could not be detected because of their sporadic occurrence and low numbers; their life cycle patterns were impossible to establish. All three basic life cycle

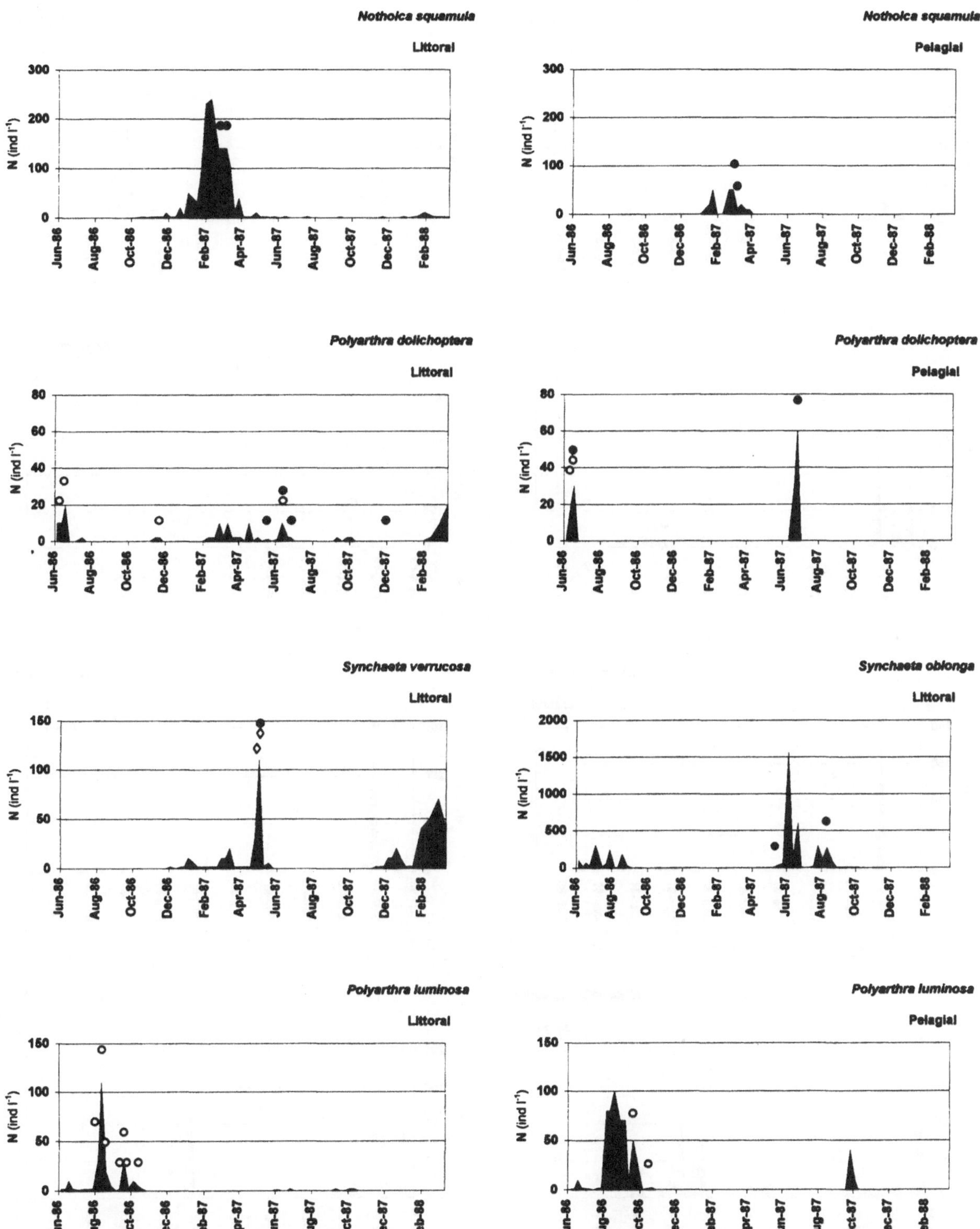


*Figure 1.* The seasonal dynamics of *Notholca squamula*, *Polyarthra dolichoptera*, *P. luminosa*, *Synchaeta verrucosa* and *S. oblonga* in Lake Peipsi in 1986–1988. N = population density (ind $l^{-1}$); ○ = haploid eggs; ● = resting eggs; ◇ = males.

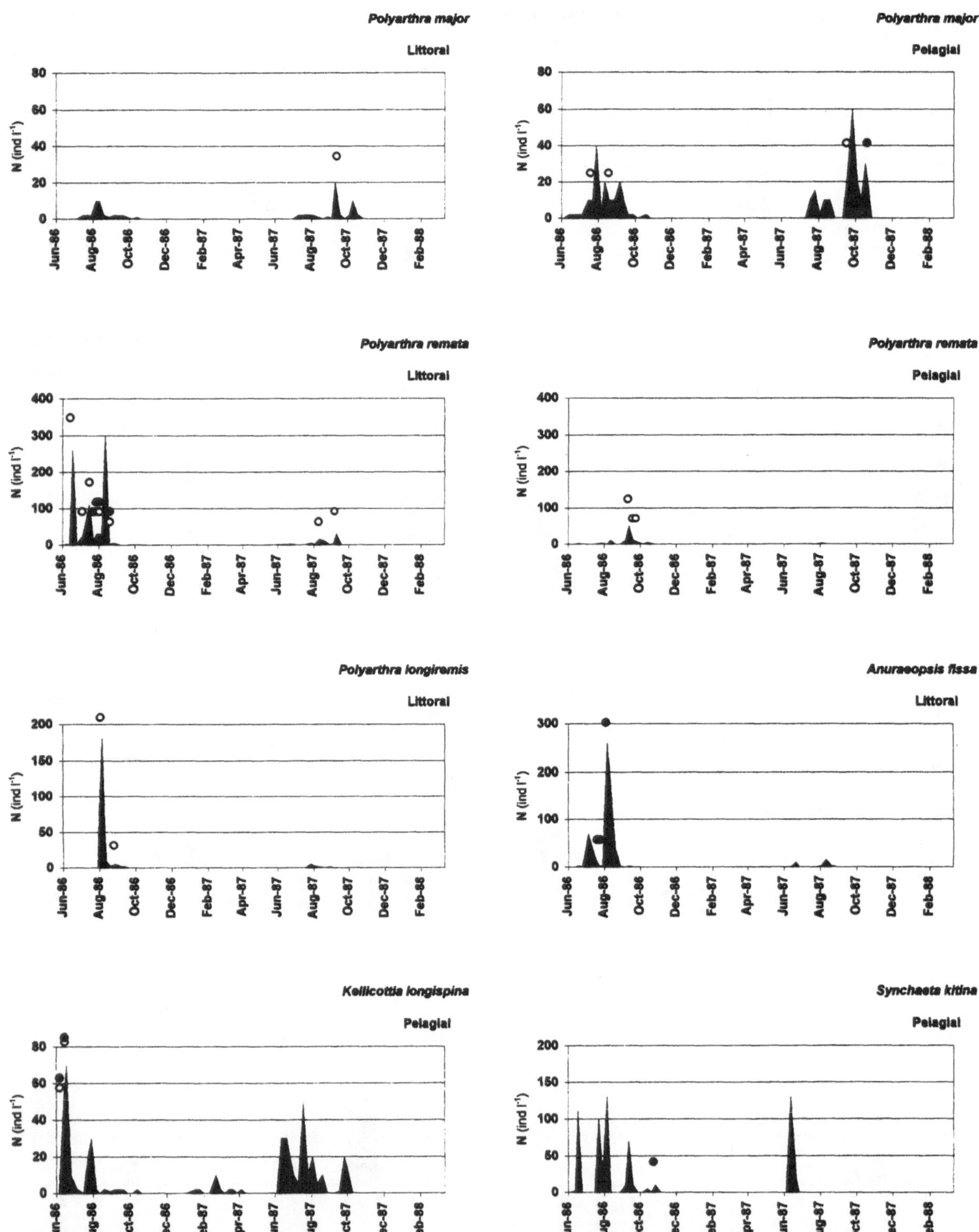


*Figure 2.* The seasonal dynamics of *Polyarthra major*, *P. remata*, *P. longiremis*, *Anuraeopsis fissa*, *Kellicottia longispina* and *Synchaeta kitina* in Lake Peipsi in 1986–1988. N = population density (ind $l^{-1}$); ○ = haploid eggs; ● = resting eggs.

*Table 1.* Occurrence of bisexual reproduction in Lake Peipsi 1986–1988. ○ = haploid eggs; ♂ = males; ☼ = resting eggs

| | J | F | M | A | M | J | J | A | S | O | N | D |
|---|---|---|---|---|---|---|---|---|---|---|---|---|
| *Anuraeopsis fissa* | | | | | | | ☼ | ☼ | | | | |
| *Asplanchna priodonta* | | | | | | ♂ | | | | | | |
| *Collotheca mutabilis* | | | | | | | | | ☼ | | | |
| *C. ornata* | | | | | | | | | | ☼ | | |
| *Euchlanis deflexa* | | | | | | | | | ♂ | ♂ | | |
| *E. dilatata* | | | | | | | | | ♂ | ♂ | | |
| *E. lyra* | | | | | | | | | | | ♂ | |
| *Filinia longiseta* | | | | | | | | ○☼ | ○ | ○ | | |
| *F. terminalis* | | | | | | ☼ | | | | | | |
| *Kellicottia longispina* | | | | | | ○☼ | | | | | | |
| *Keratella cochlearis* | | | | | | ☼ | | | | | | |
| *K. irregularis* | | | | | | ☼ | | ☼ | | | | |
| *K. quadrata* | | | | | | ○ | | | | | | |
| *Lecane lunaris* | | | | | | | | | ☼ | | | |
| *Notholca squamula* | | ☼ | ☼ | | | | | | | | | |
| *Polyarthra dolichoptera* | | | | | ☼ | ○☼ | | | | | ○☼ | |
| *P. longiremis* | | | | | | | | ○ | | | | |
| *P. luminosa* | | | | | | | | ○ | ○ | ○ | | |
| *P. major* | | | | | | | ○ | ○ | ○ | ☼ | | |
| *P. remata* | | | | | | ○ | ○☼ | ○☼ | ○ | | | |
| *Synchaeta kitina* | | | | | | | | | | ☼ | | |
| *S. lakowitziana* | | | | | | | | | | ☼ | | |
| *S. oblonga* | | | | | ☼ | | | ☼ | | | | |
| *S. pectinata* | | | | | | ☼ | ☼ | | | | | |
| *S. stylata* | | | | | | | ☼ | | | | | |
| *S. verrucosa* | | | | ♂ | ♂☼ | | | | | | | |

patterns are represented in L. Peipsi (Table 2; Figs 1–3). According to King (1980), one can distinguish: (1) early cycle species, initiating bisexual reproduction in the early phase of the population cycle; (2) midcycle species whose bisexual reproduction occurs near the population maximum; and (3) late cycle species whose bisexual reproduction occurs toward the end of the population cycle. Concerning duration of mictic period, Serra et al. (1998) have suggested two alternative strategies – extended and punctuated mixis patterns. Among the dominant rotifers in L. Peipsi, the punctuated mixis pattern seems to prevail, while the littoral populations of *Polyarthra remata* and *P. luminosa* (Figs 1–2) can be considered to have extended mixis patterns. In the two dominants, *Conochilus hippocrepis* and *C. unicornis*, mictic periods have not been detected. As a rule, seasonally occurring monogononts must have bisexual reproduction to survive in a changing environment. When the mictic period was very short, it might have remained unnoticed. In some years, owing to unfavourable conditions, mictic reproduction may not take place (Pozuelo & Lubian, 1993). Several, often perennial, species (e.g. *C. unicornis, K. longispina, K. cochlearis*) can be acyclic, with no periodicity in bisexual reproduction. In larger lakes, they can reproduce throughout the year or even in many years only parthenogenetically, omitting mixis altogether (Carlin, 1943; Pejler, 1957; Bogoslovsky, 1963, 1967; Kutikova, 1970). Perennial occurrence of *C. unicornis* has been reported earlier from L. Peipsi by Haberman (1976). Most likely, the majority of the species discussed here are monocyclic, having one mictic period per year. *Polyarthra dolichoptera* and *Synchaeta oblonga* may be dicyclic with two bisexual periods (Fig. 1). However, resting eggs, causing the second cycle, may originate from previous years, but not from the first cycle of the current year, thus,

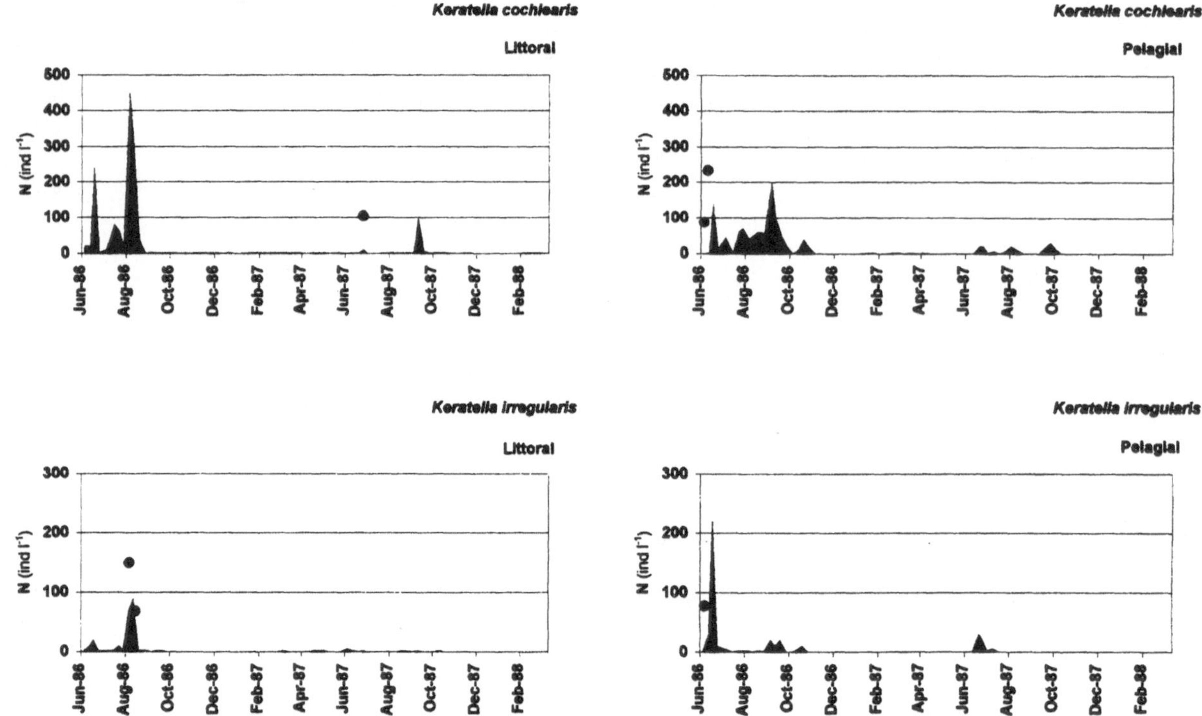


*Figure 3.* The seasonal dynamics of *Keratella cochlearis* and *K. irregularis* in Lake Peipsi in 1986–1988. N = population density (ind $l^{-1}$); ● = resting eggs.

the above-mentioned species can be pseudodicyclic (Bogoslovsky, 1969; Kutikova, 1970).

Both the present and literature data (Table 2) demonstrate that life cycle patterns are not fixed at the species level. All data reveal extensive intraspecific variability. The same species can have different patterns in different habitats, and further, spatially or temporally differentiated sub-populations behave differently within one and the same habitat. Available data from both natural (e.g. Carlin, 1943; Gilbert, 1977; Carmona et al., 1995) and experimental rotifer populations (e.g. Pozuelo & Lubian, 1993) confirm that the high intraspecific and intrapopulational variability is typical of life cycle patterns. Differences in life cycle patterns can favour the ecological and genetic divergence, and eventual speciation of rotifers in varying environments (e.g. Gómez et al., 1997; King & Serra, 1998).

Although bisexual reproduction can occur in any phase of the population cycle, it has most often been observed during population maxima (midcycle pattern) (Table 2), i.e. in the most favourable conditions. For two reasons, it would be adaptively advantageous for a rotifer population to have mixis at times of high population density (Birky & Gilbert, 1971; Gilbert, 1977; Serra & Carmona, 1993). First, mixis must lead to effective production of energy-rich resting eggs. This requires resources – abundant food supply. Second, resting egg production requires fertilisation, which may be more effective at higher population densities, when male-female encounter probability is higher. It can be hypothesized that early cycle rotifers have an advantage in highly unpredictable environments. On the contrary, late cycle species may have an advantage in stable environments. Midcycle rotifers, following the optimal strategy, are best suited for 'normal', moderately varying habitats. The effect of different life cycle patterns on the genetic structure of rotifer populations has been analysed by King (1980) and King & Serra (1998). Mixis in the early phase of the population cycle must preserve more genetic variability because it occurs before selection has eliminated most of the diverse genotypes hatched from resting eggs. Genotypic heterogeneity provides early cycle species with an evolutional advantage in unstable environments or when colonising new habitats. Bisexual reproduction in the middle or late cycle should fix homogeneity of the population,

*Table 2.* Life cycle patterns of rotifers in Lake Peipsi 1986–1988 and in other water-bodies (literature data). EC = early cycle; MC = midcycle; LC = late cycle; AC = acyclic; L = littoral; P = pelagial

| Species | Lake Peipsi | Other water-bodies (reference) |
|---|---|---|
| *Anuraeopsis fissa* | MC (L)[a] | MC (4; 7), LC (7) |
| *Asplanchna priodonta* | MC (P)[d] | EC (3; 5), MC (2) |
| *Conochilus hippocrepis* | AC (L, P) | |
| *C. unicornis* | AC (L, P) | MC (2; 6), LC (2), AC (2) |
| *Filinia longiseta* | MC (L)[e] | MC (3), LC (3; 5) |
| *Kellicottia longispina* | MC (P)[b] | AC (2; 6) |
| *Keratella cochlearis* | EC (L, P) | EC (5), MC (3; 4) |
| *K. irregularis* | MC (L, P) | |
| *Notholca squamula* | LC (L, P) | AC (2) |
| *Polyarthra dolichoptera* | MC (P), LC (L) | EC (2; 5), MC (1; 2; 4) |
| *P. longiremis* | MC (L)[a] | |
| *P. luminosa* | MC (L), LC (P) | MC (4), LC (5) |
| *P. major* | MC (L, P) | EC (2), MC (2), LC (2), AC (2) |
| *P. remata* | EC (L), MC (P) | MC (2), LC (2) |
| *Synchaeta kitina* | LC (P)[b] | AC (2) |
| *S. oblonga* | EC (L)[c] | AC (2) |
| *S. pectinata* | EC (P)[d] | AC (2) |
| *S. stylata* | EC (L)[c] | AC (2) |
| *S. verrucosa* | MC (L)[c] | MC (4) |

[a]Occurred only in littoral.
[b]Bisexual reproduction did not occur in the littoral.
[c]Bisexual reproduction did not occur in the pelagial.
[d]In the littoral, population density was too low to determine the life cycle pattern.
[e]In the pelagial, population density was too low to determine the life cycle pattern.
References (water-body):
1 = Amrén, 1964 (coastal ponds, Spitzbergen); 2 = Carlin, 1943 (Motalaström, Sweden); 3 = Dieffenbach, 1912 (ponds, Germany); 4 = Haberman, 1995 (L. Võrtsjärv, Estonia); 5 = Kutikova & Haberman, 1986 (L. Võrtsjärv, Estonia); 6 = Larsson, 1978 (L. Øvre Heimdalsvatn, Norway); 7 = Miracle & Armengol-Díaz, 1995 (L. Arcas-2, Spain).

a suitable adaptation to occupying narrower niches. Populations with extended mixis pattern have an advantage in randomly varying environments, whereas punctuated mixis pattern is favourable in predictable habitats (Serra et al., 1998).

Timing of bisexual reproduction has also general adaptive significance, related to environmental stability. In any case (early, mid-, late cycle), formation of resting eggs must be completed before a critical change of the environment (Pourriot & Snell, 1983).

### *Egg ratios and mictic ratios*

To quantify intensity of mictic and amictic reproduction, mictic ratios and egg ratios were calculated for *Anuraeopsis fissa*, *Kellicottia longispina*, *Notholca squamula*, *Polyarthra dolichoptera*, *P. luminosa* and *Synchaeta verrucosa*. In general, the highest egg ratios for amictic eggs were observed slightly before or at the population maxima, i.e. at the time of rapid growth. *A. fissa* and *P. luminosa* had the highest production of amictic eggs (egg ratios 0.1–2.0 and 0.2–2.2, respectively). Amictic egg ratios of *A. fissa* from L. Peipsi are rather high as compared with *A. fissa* from Lake Arcas–2 (Miracle & Armengol-Díaz, 1995). High mictic ratios (1.0) indicate very intense bisexual reproduction in the populations of *A. fissa*, *P. dolichoptera* and *S. verrucosa*. In the midcycle species (*A. fissa*, *K. longispina*, *S. verrucosa* and *P. luminosa* in littoral), maximum mictic ratios coincided with the population maxima. *N. squamula* and *P. dolichoptera* (littoral population), determined in L. Peipsi as late cycle species, had their highest mictic levels towards the end of population development. Additional measures of mictic activity, egg ratios for haploid eggs (*P.*

*dolichoptera* 0.1–0.2; *P. luminosa* 0.1–2.1) and resting eggs, and male numbers (*S. verrucosa*, maximum density 70 ind $l^{-1}$), show the same pattern. *P. luminosa*, which represents extended mixis pattern, has almost stable level of bisexual reproduction (at mictic ratio 0.5) from the moment of its induction till the end of population cycle. This should be characteristic of extended mixis strategy (Serra et al., 1998).

## Conclusions

Definite periods of bisexual reproduction can not be distinguished in Lake Peipsi. When concentrating mostly on the vegetation period, spread of mictic reproduction merely reflects the seasonal distribution of rotifers.

The results of the study demonstrate intraspecific variability of life cycle patterns. Expressing adaptive trade-off between the advantages of parthenogenetic and bisexual reproduction, the patterns reflect conditions of the habitat. It is apparent that mixis is an anticipatory event, and not a response to environment deterioration, or an ending of a rotifer population cycle.

## Acknowledgements

This research was supported by Estonian Science Foundation grant No. 728. I am very grateful to Mr Harri Virro for financial support to attend the 9th International Rotifer Symposium at Khon Kaen (Thailand).

## References

Amrén, H., 1964. Ecological and taxonomical studies on zooplankton from Spitsbergen. Zool. Bidr. Upps. 36: 209–276.

Birky, C. W. & J. J. Gilbert, 1971. Parthenogenesis in rotifers: The control of sexual and asexual reproduction. Am. Zool. 11: 245–266.

Bogoslovsky, A. S., 1958. Novye dannye po razmnozheniyu geterogonnykh kolovratok. Nablyudeniya za razmnozheniem *Sinantherina socialis* (Lin.). Zool. Zh. 37: 1616–1623 (New data on the reproduction of heterogonous rotifers. Observations on the reproduction of *Sinantherina socialis* (Lin.). In Russian).

Bogoslovsky, A. S., 1960. Nablyudeniya za razmnozheniem *Conochiloides coenobasis* Skorikov i ustanovlenie novoj dlya geterogonnykh kolovratok fiziologicheskoj kategorii samok. Zool. Zh. 39: 670–677 (Observations on the reproduction of *Conochiloides coenobasis* Skorikov and the statement of a physiological category of females new to heterogonous Rotifera. In Russian).

Bogoslovsky, A. S., 1963. Materialy k izucheniyu pokoyashchikhsya yaits kolovratok (soobshchenie 1). Byull. Mosk. Obshch. Isp. Prirody. Ot. Biol. 68 (6): 50–67 (Materials to the study of the resting eggs of rotifers (Communication 1). In Russian).

Bogoslovsky, A. S., 1967. Materialy k izucheniyu pokoyashchikhsya yaits kolovratok. Soobshchenie 2. Byull. Mosk. Obshch. Isp. Prirody. Ot. Biol. 72 (6): 46–67 (Materials to the study of the resting eggs of rotifers. Communication 2. In Russian).

Bogoslovsky, A. S., 1969. Materialy k izucheniyu pokoyashchikhsya yaits kolovratok. Soobshchenie 3. Byull. Mosk. Obshch. Isp. Prirody. Ot. Biol. 74 (3): 60–79 (Materials to the study of the resting eggs of rotifers. Contribution 3. In Russian).

Buchner, H., 1977. Physiological basis of the reproduction of heterogonous Rotatoria. Arch. Hydrobiol. Beih. Ergebn. Limnol. 8: 167–168.

Carlin, B., 1943. Die Planktonrotatorien des Motalaström. Zur Taxonomie und Ökologie der Planktonrotatorien. Medd. Lunds Univ. limnol. Inst. 5: 1–256.

Carmona, M. J., A. Gómez & M. Serra, 1995. Mictic patterns of the rotifer *Brachionus plicatilis* Müller in small ponds. Hydrobiologia 313/314 (Dev. Hydrobiol. 109): 365–371.

Dieffenbach, H., 1912. Biologische Studien an pelagischen Rädertieren. Int. Rev. ges. Hydrobiol. Hydrogr. Biol. Suppl. 3: 9–47.

Gilbert, J. J., 1977. Mictic female production in monogonont rotifers. Arch. Hydrpbiol. Beih. Ergebn. Limnol. 8: 142–155.

Gómez, A., M. J. Carmona & M. Serra, 1997. Ecological factors affecting gene flow in the *Brachionus plicatilis* complex (Rotifera). Oecologia 111: 350–356.

Haberman, J., 1976. An ecological characterization of the rotifers dominating in the pelagic region of Lakes Peipsi–Pihkva and Võrtsjärv. In Haberman, H., J. Haberman & K. Elberg (eds), Productivity of Estonian fresh waters, Estonian Contributions to the International Biological Programme, 10. Acad. Sci. Estonian SSR, Tartu: 35–59.

Haberman, J., 1995. Dominant rotifers of Võrtsjärv (Estonia). Hydrobiologia 313/314 (Dev. Hydrobiol. 109): 313–317.

Haberman, J., 1996. Contemporary state of the zooplankton in Lake Peipsi. Hydrobiologia 338: 113–123.

Jaani, A., 1996. Hydrology and water balance of Lake Peipsi. Hydrobiologia 338: 11–23.

King, C. E., 1980. The genetic structure of zooplankton populations. In Kerfoot, W. C. (ed.), Evolution and Ecology of Zooplankton Communities. The University Press of New England, Hanover (N. H.): 315–328.

King, C. E. & M. Serra, 1998. Seasonal variation as a determinant of population structure in rotifers reproducing by cyclical parthenogenesis. Hydrobiologia 387/388 (Dev. Hydrobiol. 134): 361–372.

King, C. E. & T. W. Snell, 1977. Genetic basis of amphoteric reproduction in rotifers. Heredity 39: 361–364.

Kutikova, L. A., 1970. Kolovratki fauny SSSR (Rotatoria). Podklass Eurotatoria (otryady Ploimida, Monimotrochida, Paedotrochida). Nauka, Leningrad, 744 pp (rotifers of the fauna of the USSR. In Russian).

Kutikova, L. & J. Haberman, 1986. Rotifers (Rotatoria) of Lake Võrtsjärv. 1. Taxonomical and ecological survey. Proc. Acad. Sci.. Estonia SSR, Biol. 35: 113–121.

Larsson, P., 1978. The life cycle dynamics and production of zooplankton in Øvre Heimdalsvatn. Holarct. Ecol. 1: 162–218.

Laugaste, R., V. V. Jastremskij & I. Ott, 1996. Phytoplankton of Lake Peipsi–Pihkva: species composition, biomass and seasonal dynamics. Hydrobiologia 338: 49–62.

Lubzens, E., G. Minkoff & S. Marom, 1985. Salinity dependence of sexual and asexual reproduction in the rotifer *Brachionus plicatilis*. Mar. Biol. 85: 123–126.

Lubzens, E., Y. Wax, G. Minkoff & F. Adler, 1993. A model evaluating the contribution of environmental factors to the production of resting eggs in the rotifer *Brachionus plicatilis*. Hydrobiologia 255/256 (Dev. Hydrobiol. 83): 127–138.

Mäemets, A., M. Timm & T. Nõges, 1996. Zooplankton of Lake Peipsi–Pihkva in 1909–1987. Hydrobiologia 338: 105–112.

Miracle, M. R. & X. Armengol-Díaz, 1995. Population dynamics of oxiclinal species in lake Arcas-2 (Spain). Hydrobiologia 313/314 (Dev. Hydrobiol. 109): 291–301.

Möls, T., H. Starast, A. Milius & A. Lindpere, 1996. The hydrochemical state of Lake Peipsi–Pihkva. Hydrobiologia 338: 37–47.

Nipkow, F., 1961. Die Rädertiere im Plankton des Zürichsees und ihre Entwicklungsphasen. Schweiz. Z. Hydrol. 23: 398–461.

Nogrady, T., R. L. Wallace & T. W. Snell, 1993. Rotifera. Vol. 1: Biology, ecology and systematics. Guides to the identification of the microinvertebrates of the continental waters of the world 4. SPB Academic Publishing bv, The Hague, 142 pp.

Nõges, T., J. Haberman, A. Jaani, R. Laugaste, S. Lokk, A. Mäemets, P. Nõges, E. Pihu, H. Starast, T. Timm & T. Virro, 1996. General description of Lake Peipsi–Pihkva. Hydrobiologia 338: 1–9.

Pejler, B., 1957. Taxonomical and ecological studies on planktonic Rotatoria from northern Swedish Lapland. K. svenska Vetensk. Akad. Hand. 6 (5): 1–68.

Pourriot, R. & T. W. Snell, 1983. Resting eggs in rotifers. Hydrobiologia 104 (Dev. Hydrobiol. 14): 213–224.

Pozuelo, M. & L. M. Lubian, 1993. Asexual and sexual reproduction in the rotifer *Brachionus plicatilis* cultured at different salinities. Hydrobiologia 255/256 (Dev. Hydrobiol. 83): 139–143.

Serra, M. & M. J. Carmona, 1993. Mixis strategies and resting egg production of rotifers living in temporally-varying habitats. Hydrobiologia 255/256 (Dev. Hydrobiol. 83): 117–126.

Serra, M., A. Gómez & M. J. Carmona, 1998. Ecological genetics of *Brachionus* sympatric sibling species. Hydrobiologia 387/388 (Dev. Hydrobiol. 134): 373–384.

Snell, T. W., 1998. Chemical ecology of rotifers. Hydrobiologia 387/388 (Dev. Hydrobiol. 134): 267–276.

Virro, T., 1995. The genus *Polyarthra* in Lake Peipsi. Hydrobiologia 313/314 (Dev. Hydrobiol. 109): 351–357.

Virro, T., 1996. Taxonomic composition of rotifers in Lake Peipsi. Hydrobiologia 338: 125–132.

Virro, T. & J. Haberman, 1993. The rotifers of Lake Peipus. Hydrobiologia 255/256 (Dev. Hydrobiol. 83): 389–396.

Walsh, E. J., 1993. Rotifer genetics: integration of classic and modern techniques. Hydrobiologia 255/256 (Dev. Hydrobiol. 83): 193–204.

Hydrobiologia **446/447:** 95–98, 2001.
L. Sanoamuang, H. Segers, R.J. Shiel & R.D. Gulati (eds), Rotifera IX.




# Life history characteristics of three types of females in *Brachionus calyciflorus* Pallas (Rotifera) fed different algae

Yi-Long Xi[1,2,3], Xiang-Fei Huang[2] & Hong-Jun Jin[1]
[1]*Institute of Environmental Science, Nanjing University, Nanjing, Jiangsu 210093, P. R. China*
[2]*Institute of Hydrobiology, Chinese Academy of Sciences, Wuhan, Hubei 430072, P. R. China*
[3]*Present address: Department of Biology, Anhui Normal University, Wuhu, Anhui 241000, P. R. China*

*Key words: Brachionus calyciflorus*, rotifera, life history, food

## Abstract

This study describes the life history characteristics of amictic, unfertilized mictic and fertilized mictic females of the rotifer *Brachionus calyciflorus* cultured individually on two different algae at 0.1 mg ml$^{-1}$ food concentration and 27 °C. The duration of the juvenile period of amictic females was significantly shorter on *Chlorella pyrenoidosa* Chick than on *Scenedesmus obliquus* Kütz or both algae together. The duration of the juvenile period of unfertilized mictic females was significantly longer, and the number of eggs produced by amictic females was significantly larger on *Chlorella pyrenoidosa* than on *S. obliquus*. When fed the same type of alga, the duration of the juvenile period of the fertilized mictic females was the longest among the three types of females, and the durations of the reproductive period of the amictic females and the post-reproductive period of the fertilized mictic females were longer than, or equal to those of the other two types of females, respectively. The number of eggs produced by an unfertilized mictic female was the largest among the three types of females, and that of amictic females was larger than or equal to that of fertilized mictic females, depending on the type of diet.

## Introduction

Many studies of rotifers have examined life history characteristics of amictic females. A few researchers studied aspects of the life history of amictic and unfertilized mictic females (Pourriot, 1973; Snell, 1986; Pourriot & Rougier, 1991; Galindo & Guisande, 1993; Dahril, 1997). Comparative studies on the influence of various environmental factors on the reproduction and survival of amictic, unfertilized mictic and fertilized mictic females are less numerous (King, 1970; Pilarska, 1972; Mitchel & Joubert, 1986). The present paper was designed to investigate the effect of algal diet on the life history characteristics of the three types of females in *B. calyciflorus*.

## Materials and methods

A clonal culture of *Brachionus calyciflorus* was obtained from a resting egg collected from sediment of Lake Donghu. Stock rotifer cultures were kept at 27±1 °C on a 16:8h light:dark photoperiod at 130 l× provided by natural light in a growth chamber. Rotifers were fed daily on *Chlorella pyrenoidosa*. Before an experiment commenced, the rotifers were cultured on 0.1 mg ml$^{-1}$ of one of the following algae: *C. pyrenoidosa*, *S. obliquus*, or a 1:1 mixture (wet weight) of the above two species of algae for at least 2 weeks.

Algae were grown in semi-continuous culture using HB-4 medium (Li et al ., 1959) replenished daily at 40%. Algae in exponential growth phase were concentrated by centrifugation and then resuspended in rotifer medium (Gilbert, 1963). Algal concentrations were measured with a hemocytometer and diluted to 0.1 mg ml$^{-1}$ (equivalent to 2.0×10$^6$ cells ml$^{-1}$ of *S. obliquus* or 5.0×10$^6$ cells ml$^{-1}$ of *C. pyrenoidosa*. The mean volume per cell of these algae was 500 $\mu$m$^3$ and 200 $\mu$m$^3$, respectively; Zhang, 1991).

Approximately 100 animals with amictic eggs were randomly selected from stock cultures and individually placed into 3.5 ml tissue culture plate wells (made of Plexiglas) containing 1.5 ml rotifer medium with each type of alga at a concentration of 0.1 mg ml$^{-1}$. Half of these animals were individually cultured and the others were cultured with 2~3 2-h-old males. These cultures were observed every 1~2 h under a Wild dissecting microscope at 25× magnification over a 10 h period. The time of birth of the offspring was recorded, and each neonate transferred to a new well. On the following day, once the neonates reached maturity, 12 amictic females (AF), unfertilized mictic females (UMF) and fertilized mictic females (FMF) were randomly selected. Thereafter, observations were made every 2~6 h for the occurrence of eggs and neonates, which were counted and removed. Rotifers were transferred to a new ration of food daily. Between observations, the culture plates were immersed in a water bath at 27±1 °C. Light intensity was approximately 300 lx (16L:8D). These observations continued until the parental rotifers died, the time of death of each animal was recorded.

Differences between means for each life history parameter were analyzed with one-way ANOVA. Significant differences ($P$<0.05) were further analyzed with the Least Significant Rank multiple comparisons test (Statgraphics, Statistical Graphics Corp.).

## Results

### *Effect of food type on life history characteristics*

Table 1 shows the life history parameters of amictic, unfertilized mictic and fertilized mictic females of *B. calyciflorus* fed different algae. There were significant effects of food type on the duration of juvenile period of AF and UMF, and the number of eggs produced by an AF ($P$<0.05). There were no significant effects of algal diet on the duration of other principal developmental stages in these three types of females, and the number of eggs produced by an UMF as well as a FMF ($P$>0.05). Further analysis (LSR) revealed that durations of the juvenile period of AF fed *S. obliquus* and mixed algae were not significantly different, but were longer than that of AF fed *C. pyrenoidosa*. The juvenile period of UMF and the number of eggs produced by AF fed *C. pyrenoidosa* were longer and larger than those fed *S. obliquus*; there were no significant differences in these parameters between mixed algae and either of two species of algae.

### *Comparison of life history characteristics of three types of females fed the same alga*

Whether the rotifers were fed *C. pyrenoidosa*, *S. obliquus* or mixed algae, lifespan did not differ significantly among the three types of females. The duration of reproductive period for these females did not differ significantly when they were fed mixed algae. Other life history traits showed significant or highly significant differences among the three types of females fed one of the algae; such parameters included the duration of juvenile period, reproductive period, post-reproductive period and the total number of eggs produced by a female (Table 1).

FMF had the longest juvenile period among the three types of females, regardless of the food type. The duration of juvenile period for AF was longer than or equal to that of UMF, when they were fed *S. obliquus* or mixed algae, and *C. pyrenoidosa*, respectively.

When rotifers were fed *C. pyrenoidosa*, the reproductive periods of the two types of mictic females were similar, but shorter than that of AF. Fed *S. obliquus*, the reproductive period of UMF was significantly longer than that of FMF, and both were similar to that of AF. When rotifers were fed *C. pyrenoidosa*, the post-reproductive period of FMF was the longest, and those of the other two types of females were similar. When they were fed *S. obliquus*, the post-reproductive period of FMF was much and significantly longer than that of AF, but both of them was similar to that of the UMF. Fed mixed algae, the post-reproductive period of FMF was significantly longer than that of the UMF, and both of them were similar to that of AF.

When they were fed *C. pyrenoidosa* and mixed algae, the number of eggs produced by UMF was the largest, and by FMF the smallest. This also was true for animals fed *S. obliquus*, except that there was no significant difference between AF and FMF.

## Discussion

### *Algal diets and life history characteristics of different types of females*

Some researchers have investigated the effect of different algal diets on the life history of amictic rotifers. Ruttner-Kolisko (1984) recorded a prolongation of juvenile period, as well as a shortening of reproductive period for *B. plicatilis* on *Dunaliella* sp. than on *Chlorella* sp. at 15 °C. Korstad et al. (1989) found similar effects when they used *Isochrysis galbana* Tahiti

*Table 1.* Duration (h) of principal developmental stages and number (ind.) of eggs of three types of females of *B. calyciflorus* fed different algae. Numbers in parentheses are ±SD. When one-way ANOVA for female type revealed significant ($P<0.05$), a Least Significant Rank multiple comparison test was done. Numbers under multiple comparison indicated sample means that are similar (same number) or different (different number)

| Type of | Dj | | | Dr | | | Dp | | |
|---|---|---|---|---|---|---|---|---|---|
| females | Chl. | Sce. | Mix. | Chl. | Sce. | Mix. | Chl. | Sce. | Mix. |
| A.F. | 20.42 | 24.21 | 23.96 | 30.00 | 25.88 | 29.67 | 21.75 | 19.58 | 23.38 |
| | (2.28) | (3.86) | (3.63) | (9.79) | (7.21) | (12.25) | (8.08) | (6.30) | (11.79) |
| U.M.F. | 23.88 | 18.29 | 19.54 | 23.54 | 33.54 | 29.08 | 19.13 | 22.71 | 20.29 |
| | (6.87) | (3.33) | (4.43) | (10.78) | (12.74) | (10.48) | (8.90) | (11.68) | (10.05) |
| F.M.F. | 29.63 | 31.08 | 29.88 | 13.92 | 11.63 | 20.63 | 41.94 | 32.92 | 32.88 |
| | (1.54) | (2.39) | (2.36) | (9.25) | (11.28) | (15.47) | (13.64) | (9.90) | (8.13) |
| ANOVA: | $P<0.01$ | <0.01 | <0.01 | <0.01 | <0.01 | =0.17 | <0.01 | <0.01 | <0.05 |
| LSR multiple comparison | | | | | | | | | |
| A.F. | 1 | 2 | 2 | 2 | 3,1 | – | 1 | 1 | 2,3 |
| U.M.F. | 1 | 1 | 1 | 1 | 3 | – | 1 | 1,3 | 2 |
| F.M.F. | 2 | 3 | 3 | 1 | 1 | – | 2 | 3 | 3 |

| Type of | L | | | Ne | | |
|---|---|---|---|---|---|---|
| females | Chl. | Sce. | Mix. | Chl. | Sce. | Mix. |
| A.F. | 72.17 | 69.67 | 77.08 | 5.83 | 3.58 | 4.75 |
| | (10.39) | (9.24) | (19.73) | (1.40) | (1.00) | (1.86) |
| U.M.F. | 66.54 | 74.54 | 68.92 | 10.92 | 10.42 | 12.25 |
| | (15.76) | (10.88) | (12.99) | (2.71) | (4.52) | (3.25) |
| F.M.F. | 85.46 | 75.63 | 83.38 | 1.75 | 1.58 | 2.00 |
| | (14.48) | (9.57) | (15.74) | (0.45) | (0.51) | (0.60) |
| ANOVA: $P=$ | 0.059 | 0.31 | 0.11 | <0.01 | <0.01 | <0.01 |
| LSR multiple comparison | | | | | | |
| A.F. | – | – | – | 2 | 1 | 2 |
| U.M.F. | – | – | – | 3 | 2 | 3 |
| F.M.F. | – | - | – | 1 | 1 | 1 |

Dj–duration of juvenile period; Dr–duration of reproductive period; Dp–duration of post- reproductive period; L–mean life-span; Ne–total number of eggs produced by a single female per life cycle; Chl–*Chlorella pyrenoidosa*; Sce–*Scenedesmus obliquus*; Mix–mixed algae; A.F.–amictic female; U.M.F.–unfertilized mictic female; F.M.F.–fertilized mictic females.

and *Tetraselmis* sp. instead of *Nannochloris atomus* as the rotifer's food at 20 °C and at 90 mg C ml$^{-1}$. Xi & Huang (1999) obtained similar results when *B. urceolaris* was fed *Scenedesmus accuminatus* Chod other than *Chlorella ellipsoidea* Gern. at 26 °C and at 0.3 mg ml$^{-1}$. In the present study, we found no significant effects of algal food type on the durations of the reproductive period of the three types of females, but a significant effect on the durations of the juvenile period of AF and UMF. The duration of the juvenile period of AF fed *C. pyrenoidosa* was shorter than that of those fed *S. obliquus* or a mixture of both species, and that of UMF fed *C. pyrenoidosa* was longer than that of those fed *S. obliquus*. Thus, these two types of females appear to respond differently to these algae.

The effect of algal diet on rotifer lifetime fecundity was found to be significant only in AF. The total number of eggs produced by AF fed *C. pyrenoidosa* was larger than when fed *S. obliquus*, which was similar to the results obtained in *B. urceolaris* by Xi & Huang (1999).

### *Comparison of life history characteristics of different types of females fed the same alga*

The durations of the juvenile periods of AF and UMF females of *Asplanchna brightwelli* were similar (Pourriot, 1973), as were those of AF, UMF and FMF of *B. rubens* (Pilarska, 1972). However, in the present study, we found that the duration of the juvenile period of FMF was the longest among the three types of females, and that of AF was longer than or equal to that of UMF, depending on the type of diet.

In Pourriot's (1973) study, the reproductive period of UMF in *B. calyciflorus* was longer than that of AF, but in our clone of this species that of the former was longer than or equal to that of the latter, depending on the food type. In Pourriot's (1973) study, and in ours, the post-reproductive periods of AF and UMF of *B. calyciflorus* were similar. Galindo & Guisande (1993) found that the lifespan of AF was longer than that of UMF and that the post-reproductive period was shorter. However, in the present study, our results showed that there were no significant differences either in lifespan or in post-reproductive period between these two types of females on the food types we used.

Pourriot (1973) showed that AF and UMF females of *B. calyciflorus* under optimal food conditions at 20 °C had identical net reproduction rates. Dahril (1997) found that the number of eggs of an AF of *B. calyciflorus* was larger than that of an UMF when they were fed freshwater *Chlorella* sp. at a density of $6.0\times10^6$ cells $ml^{-1}$. However, we found that the reproductive output of UMF was higher than that of AF, and that of AF was higher than or equal to that of the FMF, also depending on the type of diet. It seems possible that the relationship among the number of eggs produced, respectively, by AF, UMF and FMF depended on experimental temperature, tested food type and concentration, and rotifer strains.

## Acknowledgements

We thank Dr Norbert Walz, Prof. Dr Jian-Kang Liu and two anonymous reviewers for their comments on the manuscript. This research was supported by the National Natural Science Foundation of China (Grant No. 39870158).

## References

Dahril, T., 1997. A study of the freshwater rotifer *Brachionus calyciflorus* in Pekanbaru, Riau, Indonesia. Hydrobiologia 358: 211–215.

Galindo, M. D. & C. Guisande, 1993. The reproductive biology of mictic females in *Brachionus calyciflorus* Pallas. J. Plankton Res. 15: 803–808.

Gilbert, J. J., 1963. Mictic female production in rotifer *Brachionus calyciflorus*. J. exp. Zool. 153: 113–124.

King, C. E., 1970. Comparative survivorship and fecundity of mictic and amictic female rotifers. Physiol. Zool. 43: 206–212.

Korstad, J. & O. Vadstein, 1989. Life history characteristics of *Brachionus plicatilis* (Rotifera) fed different algae. Hydrobiologia 186/187: 43–50.

Li, S. H., H. Zhu, Y. Z. Xia, M. J. Yu, K. S. Liu, Z. Y. Ye & Y. X. Chen, 1959. The mass culture of unicellular green algae. Acta hydrobiol. Sinica 4: 462–472.

Mitchell, S. A. & J. H. B. Joubert, 1986. The effect of elevated pH on the survival and reproduction of *Brachionus calyciflorus*. Aquaculture 55: 215–220.

Pilarska, J., 1972. The dynamics of growth of experimental populations of the rotifer *Brachionus rubens* Ehrenbg. Pol. Arch. Hydrobiol. 19: 265–277.

Pourriot, R., 1973. Recherches sur la biologie des rotiferes III- Fécondité et durée de vie comparées chez les femelles amictiques et quelques espèces. Ann. Limnol. 9: 241–258.

Pourriot, R. & C. Rougier, 1991. Importance volumétrique des oeufs chez les Rotifères planctoniques. Ann. Limnol. 27: 15–24.

Ruttner-Kolisko, A., 1984. Der Einflus von Quantitat und Qualitat des Futters auf Lebensparameters, Klonwachstum und Korpermase einiger planktischer Rotatorienarten. Jber. Biol. Stn. Lunz. 7: 181–191.

Snell, T. W., 1986. Effect of temperature, salinity and food level on sexual and asexual reproduction in *Brachionus plicatilis* (Rotifera). Mar. Biol. 92: 157–162.

Xi, Y. -L. & X. -F. Huang, 1999. The effect of food supply in both food quality and quantity on the population dynamics of *Brachionus urceolaris*. Acta hydrobiol. Sinica 23: 227–234. (Chinese with English abstract).

Zhang, Z. -S., 1991. Method of determining the weight of various phytoplankta. In: Zhang & Huang (eds), Methods for Study on Freshwater Plankton. Science Press, Beijing: 340–344.

Hydrobiologia **446/447:** 99–105, 2001.
L. Sanoamuang, H. Segers, R.J. Shiel & R.D. Gulati (eds), Rotifera IX.



# Why do rotifer populations present a typical sigmoid growth curve?

Tatsuki Yoshinaga[1], Atsushi Hagiwara[2] & Katsumi Tsukamoto[1]
[1]*Ocean Research Institute, The University of Tokyo, 1-15-1 Minamidai, Nakano, Tokyo 164-8639, Japan*
*E-mail: yosinaga@ori.u-tokyo.ac.jp*
[2]*Faculty of Fisheries, Nagasaki University, Bunkyo 1-14, Nagasaki 852-8521, Japan*

*Key words: Brachionus plicatilis*, population dynamics, rotifera, sigmoid growth curve

## Abstract

To determine the underlying processes to population growth in the rotifer *Brachionus plicatilis*, we conducted an experiment using 1.5 ml cultures for 70 days. All individuals were transferred daily to culture media containing algae, and the number of individuals, clutch sizes and number of deaths were counted. The population dynamics showed a typical sigmoid curve. The population density increased exponentially from 10 to 682 individuals during the first 7 days (exponential growth phase), and gradually up to about 1500 individuals during the next 30 days (post-exponential growth phase). The population density then remained at a constant level with small fluctuations during the rest of the experimental period (stationary phase). Mortalities appeared from the post-exponential growth phase and were almost constant at about 2% throughout the experimental period. The clutch size decreased from 5 to 1 during the first 5 days, and afterwards females laid only one egg each. The proportion of non-reproductive females increased from 30% (exponential growth phase) to 80% (post-exponential growth phase) to 90% (stationary phase). These results suggest that the exponential growth phase resulted from the imbalance between a high birth rate and a low death rate, while the stationary phase was maintained by the compensation between low birth and death rates.

## Introduction

Animal populations live in a diversity of environments. Therefore, their population dynamics are regulated by a complex mixture of environmental factors. Since early in the last century, studies have been conducted on laboratory population dynamics in order to understand population growth in a controlled environment.

Many kind of organisms in various taxa, e.g. protozoa (Gause, 1934 in *Paramecium aurelia*), zooplankton (Terao & Tanaka, 1928 in the daphnid *Moina macrocopa*; King, 1967 in the rotifer *Euchlanis dilatata* ) and insects (Pearl, 1927 in the fruit fly *Drosophila melanogaster*), have been used in the laboratory, and sigmoid growth curves have been found as the commonest behavioral dynamic regardless of the differences in the species. However, unanswered questions remain regarding their mechanism. First, carrying capacity is a simple concept in the determination of the population size, although its actual role is unknown. One classical experimental study on laboratory population dynamics was conducted by Terao & Tanaka (1928) in the water-flea *Moina macrocopa*. They cultured *M. macrocopa* populations at 20, 25 and 34 °C, and found that the maximum growth rate ($r_{max}$ at 34 °C) did not yield the maximum population density (at 25 °C), i.e. the carrying capacity. Similar observations have been reported in other daphnid populations (Slobodkin, 1954; Smith, 1963), and no constant relationship between $r_{max}$ and carrying capacity is thought to exist (Ito et al., 1980). Second, there are two possibilities which would yield a sigmoid growth curve: (1) When there occurs a density-dependent suppression of the birth rate and a density-independent death rate. In this case, birth rate will be suppressed with the increase in population density, and dead individuals will appear randomly. (2) When there occurs a density-independent birth rate and a density-dependent increase in the death rate. If a higher population density yields a higher mortality, the number of dead individuals will increase with population density. Even though birth rate is constant at the maximum level over time, population growth rate will drop with the increase in population density. Furthermore, while the sigmoid growth curve is the commonest population phenomenon in the laboratory, it has never been observed in natural conditions except for the ant *Atta sexdens* in the tropical rain forest (Allee et al., 1949). The sigmoid growth curve may be only an artifact caused by specific laboratory culture conditions. Thus, the results of a detailed examination of the time-courses of birth and death rates would be relevant to the understanding of the population growth mechanism.

In this study, we investigated the underlying processes to the sigmoid growth curve of monogonont *Brachionus plicatilis* populations. We monitored the number of females, clutch size and number of deaths during long-term culture in order to evaluate the density dependency of birth and death rates.

## Materials and methods

### Materials

For the experiment, we used the rotifer *Brachionus plicatilis* Ishikawa strain and the dietary microalgae *Tetraselmis tetrathele* described in Yoshinaga et al. (1999). *B. plicatilis* Ishikawa strain reproduces only by parthenogenesis. Both the rotifers and the algae were cultured in Brujewicz artificial seawater (Subow, 1931) sterilized by 0.45 $\mu$m filtration before use at 25 °C and with 33 ppt salinity. The algal culture medium was enriched with additional nutrients (Yoshinaga et al., 1999). The algal cells were suspended in the rotifer culture medium at $0.5 \times 10^6$ cells ml$^{-1}$. Stock and experimental rotifer cultures were kept under total darkness except during observations.

### Culture

Thirty newborns (age <2 h) were randomly divided into three groups ($N$ = 10; populations A, B and C), and cultured in 1.5 ml of culture medium containing algae. To prevent evaporation of cultures, petri dishes (diameter, 3 cm) were placed in a 6 well culture plate filled with distilled water in the other wells. Every day, all individuals were transferred to newly prepared culture medium containing algae, and the number of females, clutch sizes and numbers of deaths were counted. Experimental periods were of 70, 42 and 46 days in the populations A, B and C, respectively.

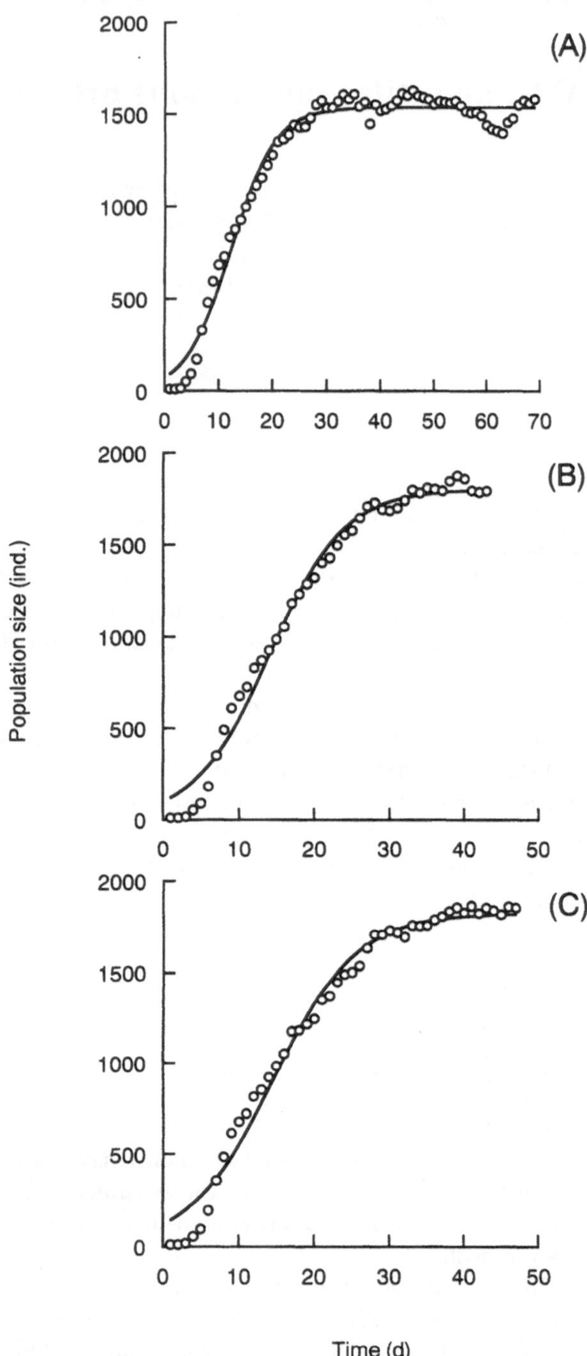


*Figure 1.* Population dynamics (circle) and logistic models (solid line) of the rotifer *Brachionus plicatilis* in triplicate cultures (populations A, B and C).

### Estimation of births

We calculated birth rate using the following procedures. Females without eggs included three kinds

*Table 1.* Parameters describing population dynamics and logistic models of the rotifer *B. plicatilis* in triplicate cultures

| Population | Time (days) | Number of individuals | | | Logistic model[a] | | |
|---|---|---|---|---|---|---|---|
| | | Initial ($t$ = 0) | Maximum | Mean ± SE | $K$ | $c$ | $r$ |
| A | 70 | 10 | 1628 | 1263.6 ± 57.3 | 1536.2 | 20.0 | 0.24 |
| B | 42 | 10 | 1876 | 1213.6 ± 96.0 | 1804.0 | 17.1 | 0.20 |
| C | 46 | 10 | 1865 | 1252.1 ± 90.8 | 1825.3 | 14.3 | 0.18 |

[a]Logistic model: $N(t) = K/\{1+c^* \exp(-r^*t)\}$; where $N$ is population size and $t$ is time (d), according to Aoki (1999).

of females: immature females, reproductive females without eggs and post-reproductive females. In order to classify them, first we assumed that there were no females in post-reproductive ages because we considered that few females survive beyond reproduction. Next, we estimated the number of newborns (immature females) based on the number of eggs in the population. We did this on the basis that (1) the duration of embryonic development is about 1 day, and (2) first amictic egg production age is at day 1, at 25 °C (Yoshinaga et al., 1999). Consequently, the number of eggs in the population on day $t$ will be the number of newborns on day $t$+1 as shown in following equation:

$$(\text{Eggs})_t = (\text{Estimated newborns})_{t+1}$$

In a laboratory population, changes in the number of individuals can be expressed by the following equation:

$$(\text{Females})_t + 1 = (\text{Females})_t + (\text{Newborns})_{t+1} - (\text{Dead})_{t+1}$$

In the above equation, (Dead)$t$+1 is the number of deaths from day $t$ to $t$ +1. To test the validity of the above estimation of the number of newborns, we calculated the estimated total number of females using the following equation:

$$(\text{Estimated females})_{t+1} = (\text{Females})_t + (\text{Estimated newborns})_t - (\text{Dead})_t$$

In the above equation, $(\text{Females})_t$ and $(\text{Dead})_t$ are known parameters. If $(\text{Estimated females})_{t+1}$ coincide with the counted number of females, $(\text{Females})_{t+1}$, then $(\text{Estimated newborns})_{t+1}$ will be ascertained. After the test, we calculated the number of non-reproductive females by subtracting the number of newborns from the number of females without eggs.

## Results

### *Population dynamics*

All the population dynamics showed typical sigmoid curves, each fitting well with the logistic model (Fig. 1; Table 1). During the first 2 days of culture, the number of females did not increase (lag phase). From day 3 to day 9, the number of females increased exponentially ($r$> 0.5) from 10 individuals to 682, 675 and 676 individuals in populations A, B and C, respectively (exponential growth phase). During the exponential growth phase, the reproductive females laid a maximum of five eggs each (Fig. 2). In the next 30–37 days, the number of females gradually increased to maximums of 1628 individuals (day 45), 1876 individuals (day 38) and 1865 individuals (day 40) in the populations A, B and C, respectively (post-exponential growth phase). In contrast to during the exponential growth phase, reproductive females laid only one egg each during the post-exponential growth phase (Fig. 2). After reaching the maximum, the number of females stayed at around 1500 individuals with small fluctuations during the rest of the experimental period (stationary phase).

### *Estimated newborns*

Changes in the estimated number of females were almost the same as the counted number of females in the three populations. Mean errors of estimation, (counted number − estimated number) / counted number, were 0.86, 0.51 and 0.64% in populations A, B and C, respectively. Based on these results, estimation of newborns from the number of eggs was ascertained as explained in the methods section.

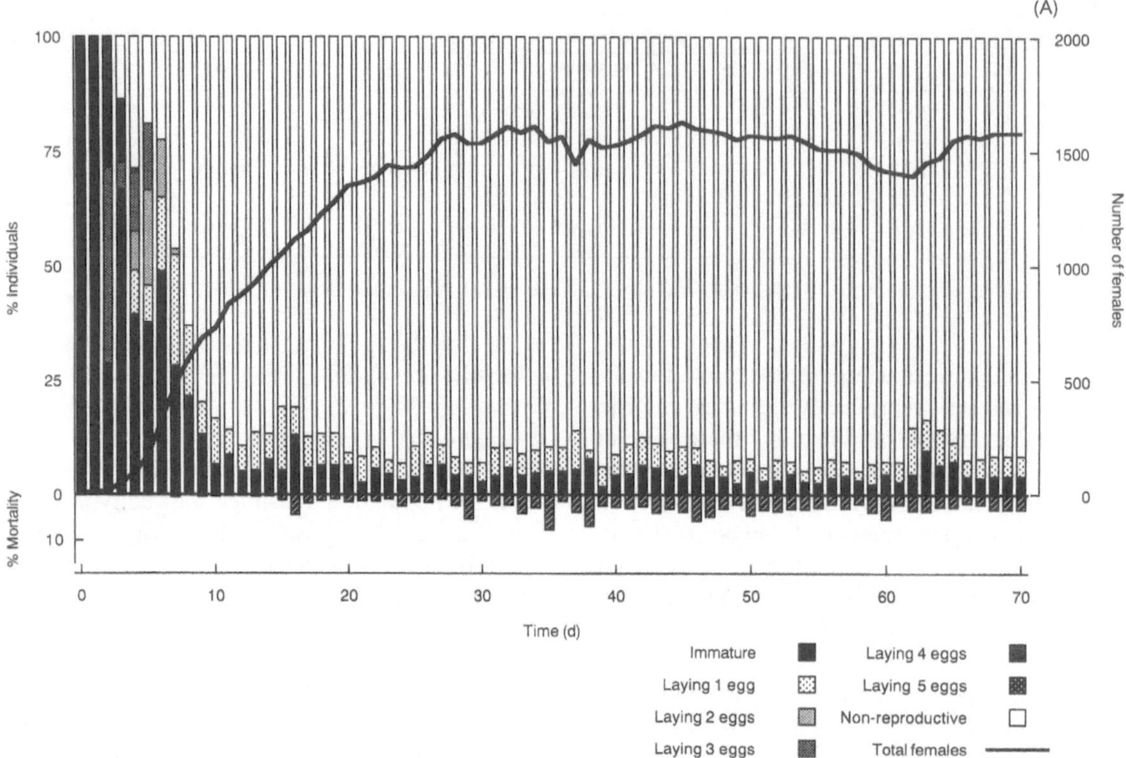


*Figure 2a.* Changes in % composition of females at various reproductive stages (columns on left vertical axis, upper), daily mortalities (lower) and population dynamics (solid lines on right vertical axis, upper) of the rotifer *B. plicatilis* in populations A, B and C.

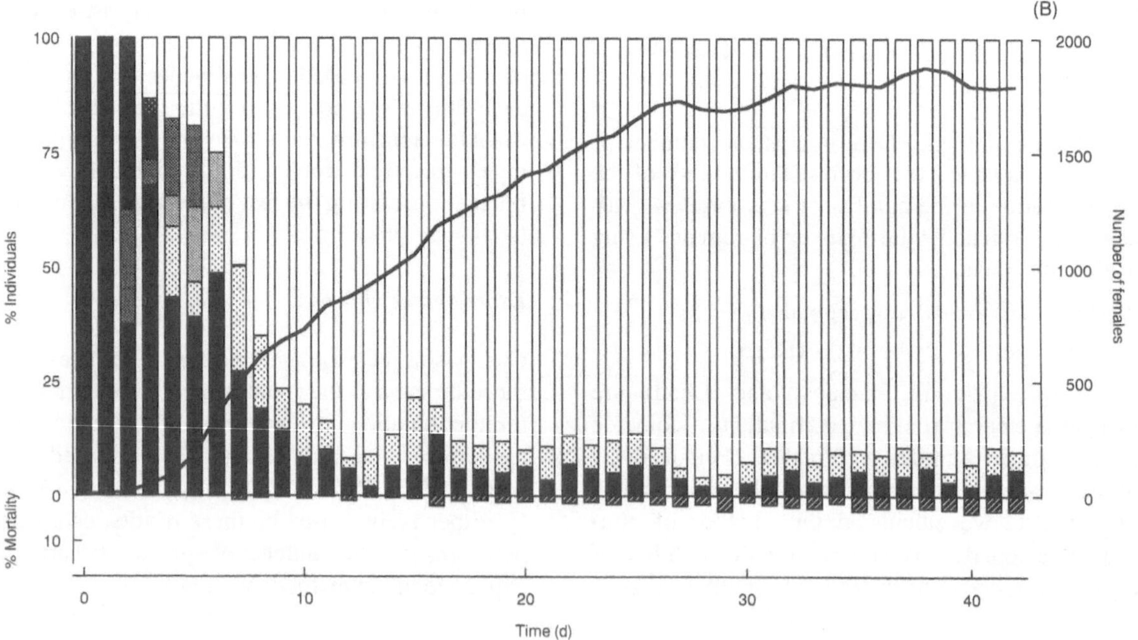

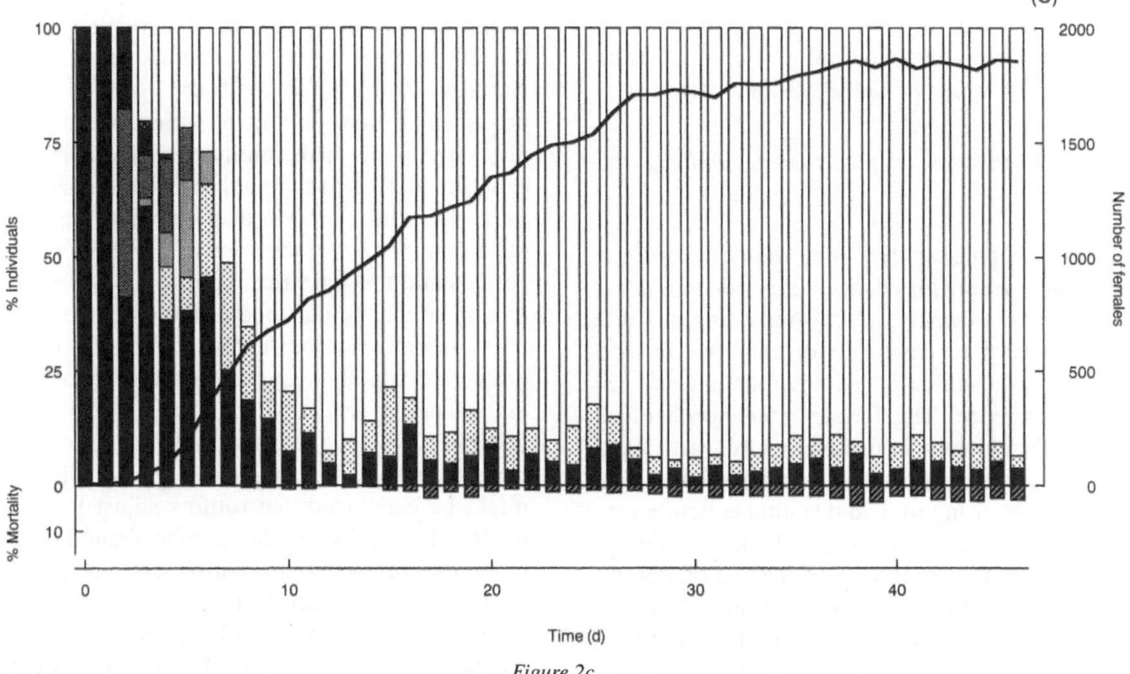


*Figure 2c.*

*Changes in proportion*

In the lag phase, populations consisted of immature females (Fig. 2a, b, c). Females with eggs and immature females dominated the exponential growth phase. The proportion of non-reproductive females increased from 30% (exponential growth phase) to 80% (post-exponential growth phase) to 90% (stationary phase).

Dead individuals first appeared on days 7, 7 and 8, and mean death rates (dead/total females) of the whole experimental period were 2.3, 1.5 and 1.5% in populations A, B and C, respectively.

## Discussion

*How do rotifer populations present a sigmoid growth curve?*

The growth curve observed in this study was divided into 4 phases: lag, exponential-growth post-exponential growth and stationary. The lag phase is a delay until the inoculated population starts reproducing (Yúfera & Navarro, 1995). In this study, the lag phase was of two days, as we inoculated newborns at day 0. In the exponential growth phase, females produced 3–5 offspring per day, as estimated by on the number of eggs each laid (Fig. 2). Therefore, the populations in this phase showed the highest growth rate of the entire experimental period. In the post-exponential growth phase, the number of females increased gradually, although the proportion of females laying eggs decreased markedly, and they had only one egg each (Fig. 2). Deaths started occurring in the post-exponential growth phase, although mortalities did not increase during the experimental period and stayed at a constant low level (Fig. 2). From this evidence, it is clear that birth rate was the key density-dependent factor which regulated the population growth rate, and that death rate was the density-independent factor in the laboratory rotifer populations. Therefore, the sigmoid growth curve is caused by changes in birth rate which decreases with increasing population density. On the basis of these results, we conclude that the exponential growth phase is caused by the imbalance between a high birth rate and a low death rate, while the stationary phase is maintained by the compensation between low birth and death rates. A similar pattern was reported in the bdelloid rotifer *Philodina roseola* (see Gatto et al., 1992).

According to Yúfera & Navarro (1995), *B. plicatilis* laboratory population growth has 4 phases: lag, exponential growth, post-exponential growth and decline. In this study, a stationary phase was observed instead of the decline phase. This difference may be explained by the different culture methods used: Yúfera & Navarro (1995) did not renew the medium in their cultures, while we did so medium daily.

*Why do rotifer populations present a sigmoid growth curve?*

In laboratory population dynamics, the environmental factors which arise with changes in population density are known as 'density effect.' According to Utida (1941), density effect can be divided into two factors: (1) biological conditioning of environment and (2) mutual interference. The former refers to chemical changes in the culture medium caused by waste accumulation. In a previous study, we examined the effect of this factor by using individual cultures with a precultured medium ('conditioned medium'; Yoshinaga et al., 1999). With the increase in population density during preculture, the conditioned medium showed a positive effect on population growth rates. However, conditioned media have been reported to have toxin-like negative effects in the freshwater rotifer *Synchaeta pectinata* (Kirk, 1998) and in the cladoceran *Daphnia* (Seitz, 1984; Goser & Rattle, 1994; Burns, 1995). In laboratory *S. pectinata* populations, the autotoxin is considered to cause density-dependent mortality and to stabilize population fluctuations (Kirk, 1998). Even though the biological significance of the conditioned medium has not yet been clarified, it might not act as density-dependence factor in the present study, as the accumulation of wastes was eliminated by daily medium renewal. Furthermore, it was found that mortalities were kept at very low levels regardless of population density over the entire experimental periods (Fig. 2).

On the other hand, the simplest pattern of mutual interference is the competition for limited food resources among individuals. In natural populations, the abundance of food is the major factor which controls rotifer population dynamics (Nelson & Edmondson, 1955). In this study, we fed algal cells at constant concentrations throughout the experimental periods. The rotifer populations in the lag and the exponential growth phases fed continuously on algal cells, while in the post-exponential and stationary phases, they were allowed to feed for only 3 and 1 h daily, respectively. Based on these observations, we suggest that the competition for limited food resources is the essential factor which limited birth rate in the post-exponential and stationary phases. In a previous study (Yoshinaga et al., 2000), we experimentally examined the effect of periodical starvation on the life history parameters of *B. plicatilis* using individual cultures. Rotifers fed for both 1 and 3 h per day produced less than half the offspring, but survived significantly longer than continuously fed animals. This experimental evidence suggests that females in the post-exponential growth and stationary phase reproduced at much lower rates and had longer reproductive periods and lifespans than animals in the exponential growth phase. In the food-rich environment, rotifers reproduce at maximum levels and therefore population grows very rapidly (exponential growth phase). On the other hand, when population densities are saturated and thus create limited food resources, rotifers suppress their fecundity and achieve longer lifespans. It might be concluded that rotifers adjust to high population density by regulating their fecundity in spite of their reproductive potential, and such alteration in their individual life history parameters would stabilize the population. Rotifer populations present a typical sigmoid growth curve because individuals flexibly change their life history parameters to adapt to variations in environmental conditions.

## Acknowledgements

We thank two anonymous reviewers for reading and improving our manuscript.

## References

Allee, W. C., A. C. Emerson, O. Park, T. Park & K. P. Schmidt, 1949. Principles of Animal Ecology. Saunders, Philadelphia.

Aoki, S., 1999. Black-Box – data analysis on the WWW -. Logistic regression. available at http://aoki2.si.gunma-u.ac.jp/JavaScript/fit-logistic.html (in Japanese).

Burns, C. W., 1995. Effects of crowding and different food levels on growth and reproductive investments of *Daphnia*. Oecologia 101: 234–244.

Gatto, M., C. Ricci & M. Loga, 1992. Assessing the response of demographic parameters to density in a rotifer population. Ecol. Model. 62: 209–232.

Gause, G. F., 1934. The Struggle for Existence. Williams & Wilkins, Baltimore.

Gosser, B. & H. T. Rattle, 1994. Experimental evidence of negative interference in *Daphnia magna*. Oecologia 98: 354–361.

Ito, Y., N. Norihashi & K. Fujisaki, 1980. Animal Populations and Communities. University of Tokai Press, Tokyo. pp. 79–90 (in Japanese).

King, C. E., 1967. Food, age and the dynamics of a laboratory population of rotifers. Ecology 48: 111–127.

Kirk, K. L., 1998. Enrichment can stabilize population dynamics: autotoxins and density dependence. Ecology 79: 2456–2462.

Nelson, P. R. & W. T. Edmondson, 1955. Limnological effects of fertilizing Bare Lake, Alaska. U.S. Fish and Wildlife Ser. Fish Bull. 102: 414–436.

Pearl, R., 1927. The growth of populations. Q. Rev. Biol. 2: 532–548.

Seitz, A., 1984. Are there allelopathic interactions in zooplankton? Laboratory experiment with *Daphnia*. Oecologia 62: 94–96.

Slobodkin, L. B., 1954. Population dynamics in *Daphnia obtusa* Kurz. Ecol. Monogr. 24: 69–88.

Smith, F. E., 1963. Population dynamics in *Daphnia magna* and a new model for population growth. Ecology 44: 651–663.

Subow, N. N., 1931. Oceanographical tables. U.S.S.R., Oceanogr. Institute, Hydro-meteorol. com., Moscow. 208 pp.

Terao, A. & T. Tanaka, 1928. Population growth of the water-flea, *Moina macrocopa Strauss*. Proc. Imp. Acad. 4: 550–552.

Utida, S., 1941. Studies on experimental population of the azuki weevil, *Callosobruchus chinensis*. I. The effect of population density on the progeny population. Mem. Coll. Agr. Kyoto Imp. Univ. 48: 1–30.

Yoshinaga, T., A. Hagiwara & K. Tsukamoto, 1999. Effect of conditioned media on the asexual reproduction of the monogonont rotifer *Brachionus plicatilis* O.F. Müller. Hydrobiologia 412: 103–110.

Yoshinaga, T., A. Hagiwara & K. Tsukamoto, 2000. Effect of periodical starvation on the life history of *Brachionus plicatilis* O.F. Müller (Rotifera): a possible strategy for population stability. J. exp. mar. Biol. Ecol. 253: 253–260.

Yúfera, M. & N. Navarro, 1995. Population growth dynamics of the rotifer *Brachionus plicatilis* cultured in non-limiting food condition. Hydrobiologia 313/314: 399–405.

Hydrobiologia 446/447: 107–114, 2001.
L. Sanoamuang, H. Segers, R.J. Shiel & R.D. Gulati (eds), Rotifera IX.




# Grazing by a dominant rotifer *Conochilus unicornis* Rousselet in a mountain lake: in situ measurements with synthetic microspheres

X. Armengol[1], L. Boronat[1], A. Camacho[1] & W. A. Wurtsbaugh[2]
[1]*Departament de Microbiologia i Ecologia, Facultat de Biologia, Universitat de València, E-46100 Burjassot, València, Spain*
*E-mail: javier.armengol@uv.es*
[2]*Department of Fisheries and Wildlife/Ecology Center. Utah State University, Logan, UT 84322-5210, U.S.A.*

*Key words:* rotifer, grazing, distribution, zooplankton, temperature, cladocerans

## Abstract

Grazing rates of zooplankton were analysed in the summer of 1999 in Yellow Belly Lake, an oligotrophic system in the Sawtooth Mountains of Idaho (U.S.A.). The colonial rotifer *Conochilus unicornis* was a dominant species in the epilimnion, with densities reaching 20 colonies $l^{-1}$ (ca. 400 ind. $l^{-1}$). Clearance rates were measured with an in situ Haney Grazing chamber and synthetic microspheres 5, 9 and 23$\mu$m in diameter. At epilimnetic temperatures of around 14 °C, mean clearance rates for 5$\mu$m particles ranged from 30 to 65 $\mu$l ind.$^{-1}$ h $^{-1}$. Clearance rates were 2–9 times higher on the 5$\mu$m spheres than on the 9 $\mu$m spheres, and *C. unicornis* almost never fed on the 23 $\mu$m spheres. Grazing rates did not change over the diel cycle. Clearance rates declined more than 10-fold as temperatures declined from 14 °C in the epilimnion to 7 °C in the metalimnion. In the epilimnion, grazing by *C. unicornis* was more important than grazing by crustaceans in the community, at least on particles $\leq$9$\mu$m. The results show the importance of grazing by rotifers in lakes, and the significance of spatial variations that influence grazing rates.

## Introduction

Considerable attention has been placed on grazing rates of zooplankton (Peters & Downing, 1984) to help understand trophic transfer in lentic systems. Most of these analyses, however, have focused on copepods and cladocerans. Rotifers have been excluded from most grazing studies, presumably because of their small size and sometimes limited biomass. Smaller organisms like rotifers, however, have greater weight-specific feeding rate (Peters, 1983) than larger crustaceans, and when densities are high rotifers can play an important role in the trophic web of freshwater lakes (Pace & Orcutt, 1981; Bogdan & Gilbert, 1982; Weisse et al., 1990; Hansson et al., 1998). One of the methodologies commonly employed to study zooplankton feeding is the use of latex microspheres (Wilson, 1973; DeMott, 1986, 1988). Microspheres have also been used to calculate grazing of rotifers in ecotoxicological tests (Juchelka & Snell, 1994), to estimate the effect of colony size on clearance rates in the colonial rotifer *Synantherina socialis* Linnaeus (Wallace, 1987) and to study grazing in different rotifer species (Rothhaupt, 1990; Ronneberger, 1998).

During our study (summer 1999), a colonial rotifer, *Conochilus unicornis* Rousselet was a dominant member of the epilimnetic zooplankton community in Yellow Belly Lake, an oligotrophic mountain lake. In order to evaluate the grazing capacity of this rotifer and their contribution to the total community grazing, we designed several experiments using *in situ* incubations of latex microspheres in a Haney grazing chamber. Specifically for *Conochilus*, we determined: (1) how different concentration of beads affected ingestion and clearance rates; (2) if feeding changed over a diel cycle; (3) if feeding varied with depth, and; (4) if particles of different sizes were grazed at different rates. Finally, we compared grazing rates of this rotifer and crustaceans to show the relative importance of rotifers in this lake.

## Study site

Our in situ analyses were performed in the summer of 1999 in Yellow Belly Lake (44 °N, 115 °W), a 0.73 km$^2$ system located at an elevation of 2157 m in the granitic Sawtooth Mountains of central Idaho, a subsystem of the Rocky Mountains. The lake has mean and maximum depths of 14 and 26 m, respectively. It is oligotrophic, with epilimnetic chlorophyll concentrations near 0.5 $\mu$g l$^{-1}$ and Secchi disc transparency depths of 12–15 m for most of the summer. The phytoplankton is dominated by small taxa, with 31% of the chlorophyll in size fractions <1 $\mu$m, and 95% <30 $\mu$m (Budy et al., 1995). The lake has a pronounced deep chlorophyll layer in the metalimnion. Wurtsbaugh et al. (1997) provide additional limnological information. There are relatively few fish in the lake, and consequently the zooplankton is usually dominated by the crustaceans *Daphnia rosea* Sars, *Holopedium gibberum* Zaddach, *Aglodiaptomus* sp. and *Leptodiaptomus* sp. However, during our study in summer 1999, the colonial rotifer *Conochilus unicornis* was dominant in the epilimnion, reaching densities of 20 colonies l$^{-1}$, with an average of 20 ind. per colony. Other rotifers were present but in lower densities.

## Methods

In most experiments, we used short-term in situ incubations employing two 3-l Haney Chambers (Haney, 1971). Normally, two replicates were deployed at each time or depth. A Haney Chamber was lowered to the specified depth and then triggered to enclose the zoo- and phytoplankton and to simultaneously release a mixture of latex microspheres (Bangs Laboratories, Inc.) into the container. Microspheres of three different sizes and colours were used: 5$\mu$m (blue), 9$\mu$m (red) and 23$\mu$m (white). After 8 min of incubation at depth, the chamber was raised up and the contents filtered through 45$\mu$m Nitex netting. The animals were then anaesthetised with $CO_2$-saturated water, and fixed with 4% formaldehyde (final concentration). In the laboratory, the rotifers and crustacean zooplankton were enumerated in counting chambers. *Conochilus* colonies were then placed on slides in 10% glycerine and sealed with nail varnish for subsequent counting of the beads. The beads were enumerated under a microscope at 100–400 magnification. In most cases, *Conochilus* colonies remained intact during collection and processing, and beads were counted in the entire colony. We also processed the small number of free *Conochilus* individuals, assuming that they were detached from the colonies. Estimates of ingestion and clearance rates were based on all animals analysed, whether or not they had beads in their guts. The mean number of individuals analysed for a sample was 400, ranging from 14 to 1300.

A summary of the different experiments performed is presented in Table 1. The first two experiments tested methodological procedures. In the first experiment, we measured grazing rates on beads using water without phytoplankton to determine if *Conochilus* was selectively eating beads, or if they were just accidentally ingested when phytoplankton were being consumed. In this experiment, we gently filtered *Conochilus* onto 45$\mu$m Nitex, and then washed them into a 1.25 l Erlenmeyer flask containing 0.45$\mu$m filtered lake water without phytoplankton. The *Conochilus* were left in the water for 15 min to clear their guts and then filtered again through Nitex-and resuspended in 0.45$\mu$m filtered water. It is likely that this water contained algal exudates, but lacked phytoplankton >0.45 $\mu$m. A mixture of 5, 9 and 23 $\mu$m beads was then introduced at concentrations of ca. 1400, 1100 and 200 beads ml$^{-1}$, respectively. Three replicate flasks were incubated for 8 min at 3 m and another three at 9 m. Temperatures at these depths were 14 and 8 °C. The short incubation time used in our study were always below the gut evacuation time of 20–30 min, as known for other rotifers (Starkweather & Gilbert, 1977), and consequently the results should represent actual grazing rates. In the second experiment, we studied the effect on IR and CR of different concentrations of beads. For this, we measured grazing rates of *Conochilus* at five different concentrations of beads. We started with a concentrated solution 'C' (ca. 1200, 800 and 200 ml$^{-1}$for the 5, 9 and 23$\mu$m beads, respectively). This solution was diluted to provide additional concentrations of C/2, C/4, C/8 and C/16.

In the last two experiments, we measured temporal and spatial variations of grazing in the lake. In Experiment 3, we analysed grazing rates at 3 m, 12 m and 18 m during four periods of the day (see Table 1; daylight, dusk, night and dawn). During this experiment, however, *Conochilus* were only encountered in epilimnetic samples (3 m). Bead concentrations in this experiment were approximately 1200, 800 and 200 ml$^{-1}$for the 5, 9 and 23$\mu$m beads, respectively. In Experiment 4, we measured vertical differences in grazing rates dur-

*Table 1.* Main characteristics and differences among the four 'in situ' experiments done in Yellow Belly Lake. 'C' represents a concentration of ca. 1200, 800 and 200 beads ml$^{-1}$for the 5, 9 and 23 $\mu$m beads, respectively

| Experiment number | Objective | Depth (temperature) | 'Food' | Beads concentration | Date | Starting Time |
|---|---|---|---|---|---|---|
| 1 | Method verification | 3 and 9 m (8–14 °C) | Only beads | C | 14-Jul-1999 | 20h to 21h |
| 2 | Method verification | 9 m (9 °C) | Beads and phytoplankton | C, C/2, C/4, C/8 and C/16 | 7-Aug-1999 | 11h to 12h |
| 3 | Diel cycle | 3–12–18 m (12–6–4 °C) | Beads and phytoplankton | C | 1 and 2-Aug-1999 | 15h, 20h, 02h, 08h |
| 4 | Depth variations | 2–4–6–8–10 m (14–14–13–10–7 °C) | Beads and phytoplankton | C/2 | 14-Jul-1999 | 19h to 20h |

ing the day-time: in epilimnion at three depths (2, 4 and 6 m) and in the metalimnion at two depths (8 and 10 m). Bead concentrations used were approximately 600, 400 and 100 ml$^{-1}$for the 5, 9 and 23$\mu$m beads, respectively. Vertical profiles of temperature and oxygen were measured with a Yellow Springs Instrument Model 58 Dissolved Oxygen meter.

Ingestion rates (IR; beads ind.$^{-1}$ h$^{-1}$) were calculated as:

IR = Total number of beads observed / no. of individuals examined

The clearance rate (CR; $\mu$l ind.$^{-1}$ h$^{-1}$) was estimated according to the equation of Bogdan and Gilbert (1984):

CR = IR / TC

where TC is the concentration of particles (no. $\mu$l$^{-1}$) in the experimental vessel. This concentration was measured in some samples for each experiment by taking 100 ml of sample and counting the beads after formaldehyde fixation and sedimentation in Utermöhl chambers. Counts were made using an inverted microscope at 400 magnification.

## Results

### *Methodological experiments*

In Experiment 1, we observed *C. unicornis* consuming beads even in the absence of phytoplankton cells (Table 2). CR in this experiment were, however, lower than in the other experiments using the Haney Chambers (see below). It is possible that this reduction in clearance was related to the considerable manipulation of the rotifers prior to the grazing rate measurement.

*Table 2.* Ingestion rates (IR) and clearance rates (CR) of *Conochilus unicornis* feeding on beads of different sizes in media from which phytoplankton had been removed. Flasks were incubated *in situ* at depths of 3 m (14 °C) and 9 m (8 °C). Standard deviations (sd) are for within treatment flask replicates

| Depth (m) | IR (beads ind.$^{-1}$ h$^{-1}$) ± sd | | CR ($\mu$l ind.$^{-1}$ h$^{-1}$) ± sd | |
|---|---|---|---|---|
| | 9 $\mu$m | 5 $\mu$m | 9 $\mu$m | 5 $\mu$m |
| 3 | 4.5±0.7 | 19.9 ±7.0 | 3.9±0.6 | 14.9±5.3 |
| 9 | 0.6±0.3 | 5.1 ±2.6 | 0.5±0.2 | 4.1±2.1 |

Ingestion rates were a function of bead density and size of the beads (Fig. 1a). In Experiment 2, ingestion rates of 5$\mu$m beads varied from a low of 0.7 beads ind.$^{-1}$ h$^{-1}$ at the lowest bead concentration to approximately 14 beads ind.$^{-1}$ h$^{-1}$ at concentration of 600 or 1200 beads ml$^{-1}$. Ingestion rates of 9-$\mu$m beads were much lower, ranging from 0.02 at the lowest concentration to 0.7 beads ind.$^{-1}$ h$^{-1}$ at a concentration of 400 beads ml$^{-1}$. Ingestion rates of 23$\mu$m beads were negligible (data not shown).

In contrast to IR, estimated clearance rates were relatively constant at all bead densities, but were much higher for 5-$\mu$m than for 9-$\mu$m beads. Mean CR for 5$\mu$ beads was 13 $\mu$l ind.$^{-1}$ h$^{-1}$, whereas on 9$\mu$m

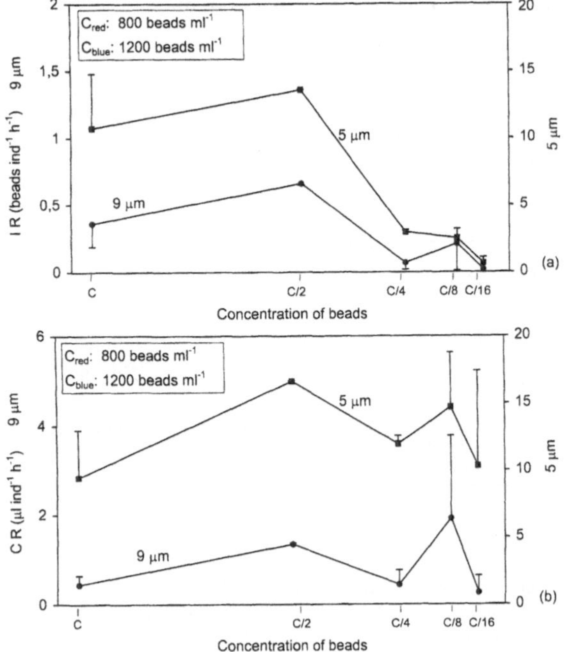


*Figure 1.* (a) Ingestion rates (beads ind $^{-1}$ h$^{-1}$) and (b) Clearance rates ($\mu$l ind.$^{-1}$ h$^{-1}$) of *Conochilus unicornis* measured *in situ* with different concentrations of latex beads of two sizes (5 $\mu$m and 9 $\mu$m). Error bars show standard deviations of 2 replicates.

beads, the mean rate was only 0.9 $\mu$l ind.$^{-1}$ h$^{-1}$ (Fig. 1b).

## *Spatial-temporal variation in grazing*

Clearance rates of individual *C. unicornis* were nearly constant throughout the diel cycle at 3 m depth (Fig. 2). As in the previous two experiments, mean clearance rates were much higher for 5 $\mu$m beads than for 9$\mu$m beads (56 vs. 7 $\mu$l ind.$^{-1}$ h$^{-1}$). Although grazing appeared quite constant throughout all the period, the results from the night and dawn samples were obtained from counting only 14 and 25 individuals, respectively, as the number of colonies in the lake declined precipitously in these samples (Fig. 2b).

Clearance rates were much higher in the warm epilimnion than in the cooler metalimnion (Fig. 3). Estimated clearance rates on 5$\mu$m beads in the epilimnion at temperature near 14 °C were near 40 $\mu$l ind.$^{-1}$ h$^{-1}$, whereas at 7 °C at 10 m in the metalimnion, the rate was only 3 $\mu$l ind.$^{-1}$ h$^{-1}$ (Fig. 3a). The estimated rates at 8 m were unexplainably variable. Clearance rates on 9 $\mu$m beads were much lower than on the smaller beads, but followed the same pattern with depth. The number of *C. unicornis* colonies de-

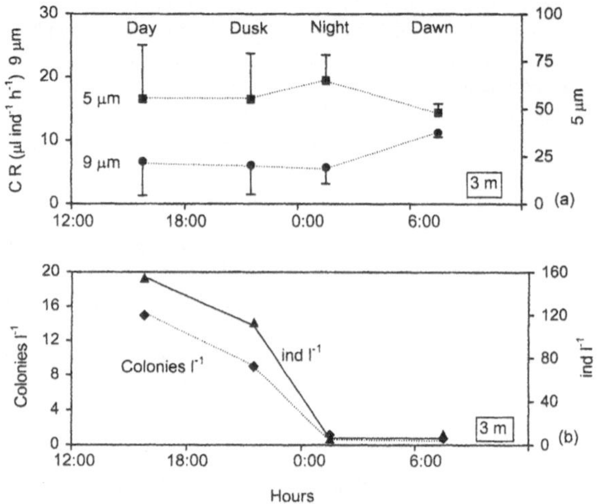


*Figure 2.* Changes in the (a) Clearance rate ($\mu$l ind.$^{-1}$ h$^{-1}$) of *Conochilus* individuals during four different periods of the day, on two bead sizes (5 $\mu$m and 9 $\mu$m). The density of *Conochilus* colonies and the number of individuals found in the samples are shown in the lower panel (b). Error bars show standard deviations within samples.

clined significantly in the metalimnion (Fig. 3c); in the epilimnion, there were about 18 colonies l$^{-1}$ and 450 ind. l$^{-1}$, declining at 10 m to 4 colonies l$^{-1}$ and 135 ind. l$^{-1}$, respectively.

Clearance rates estimated by individual CR times the density of individuals indicated that epilimnetic grazing rates were high for *C. unicornis* but declined precipitously with depth (Fig. 3b). Clearance rates by the *Conochilus* population reached 18.9 ml h$^{-1}$ (453 ml d$^{-1}$) in the epilimnion, but were only 0.4 ml h$^{-1}$ at 10 m. Ingestion rates were clearly related to depth (Fig. 4) and temperature (Fig. 5). Although the data varied considerably, the correlation between temperature and the ingestion rates was significant for the both 5 and 9 $\mu$m beads ($p < 0.001$). At the cooler temperatures in the metalimnion, the rotifers cleared a smaller proportion of large beads (Fig. 4). That is true also when we estimate the IR of a colony (Fig. 5). The coefficient of variation (%) of CR for the colonies was large, ca. 65%, irrespective of the depth and water temperature (Fig. 5).

Other rotifers, including *Ascomorpha ecaudis* Perty, *Gastropus stilyfer* Imhof, *Kellicotia longispina* Kellicott, *Keratella quadrata* Müller, *Polyarthra* sp. and *Synchaeta pectinata* Ehrenberg, were found in our samples, but showing low densities. In contrast to *C. unicornis*, beads were only infrequently found in these other species, and the numbers were too low to reliably calculate their clearance rates.

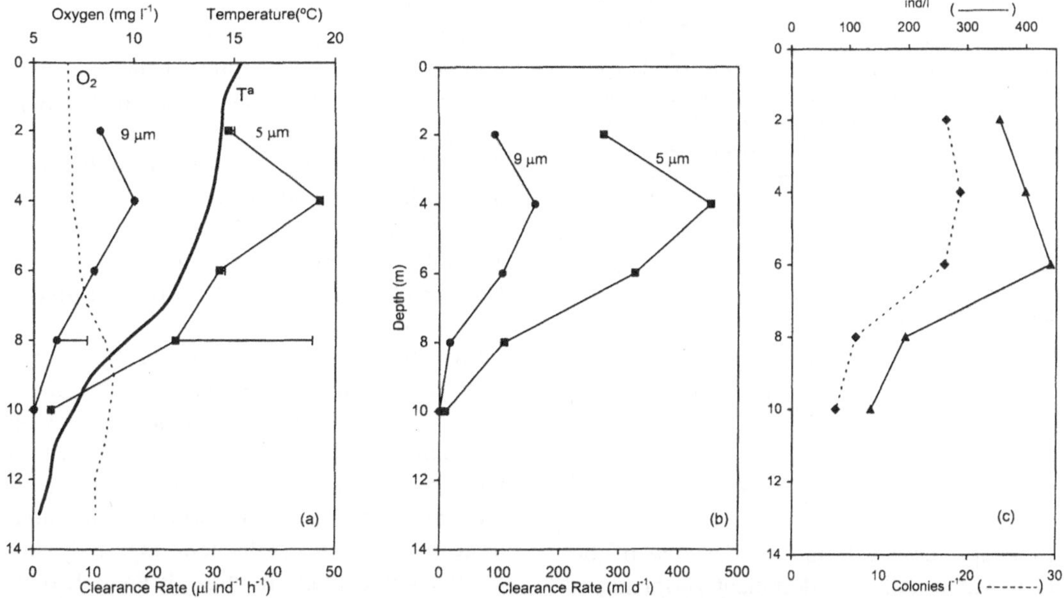


*Figure 3.* (a) Mean Clearance Rates ($\mu$l ind.$^{-1}$ h$^{-1}$) of *Conochilus* individuals calculated at different depths in the epilimnion and metalimnion of Yellow Belly Lake on 14 July 1999. Feeding rates on two sizes of beads (5 $\mu$m and 9 $\mu$m) are shown here. (b) Overall grazing rates of *C. unicornis* estimated by multiplying individual clearance rates times the density of individuals (c) Densities of *Conochilus* colonies and individuals at different depths. Temperature and oxygen profiles are also plotted. Error bars show standard deviations between two replicates.

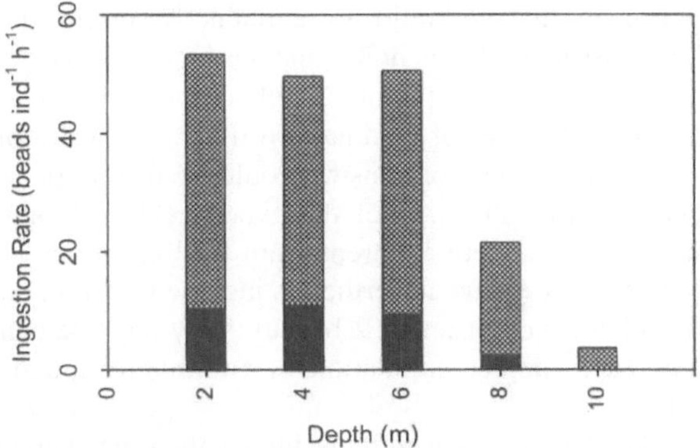


*Figure 4.* Accumulated ingestion rates (beads ind.$^{-1}$ h$^{-1}$) of *Conochilus unicornis* individuals on two sizes of latex beads (5 $\mu$m and 9 $\mu$m) at different depths in Yellow Belly Lake. These measurements were done *in situ* on 14 July 1999.

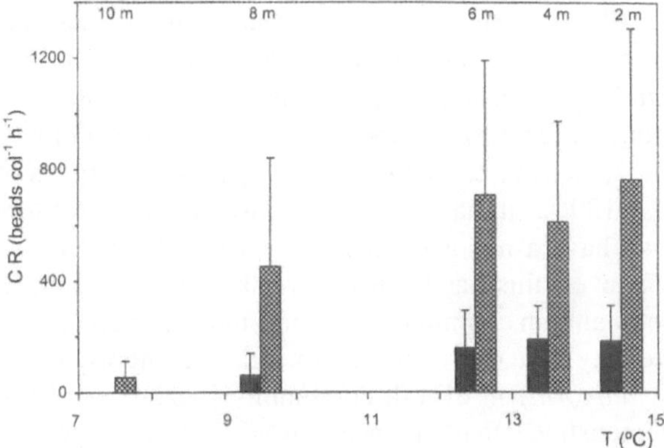


*Figure 5.* Ingestion rates of *Conochilus* colonies (beads colony$^{-1}$ h$^{-1}$) on two sizes of latex beads (5 $\mu$m, and 9 $\mu$m), plotted as a function of temperature. Error bars show standard deviations among colonies of the same depth (temperature).

Grazing by *C. unicornis* contributed significantly to the zooplankton community CR (The sum of the CR for Rotifera, Cladocera and Copepoda: Table 3). In the profile described above, we also measured the ingestion rates of the most abundant cladocera (*Daphnia* and *Holopedium*) and copepod (*Aglodiaptomus* and *Leptodiaptomus*). In the epilimnion, *C. unicornis* consumed 83% of the 5$\mu$m beads and 70% of the 9$\mu$m beads. In the metalimion, the proportion of beads ingested by this rotifer decreased. Nevertheless, *C. unicornis* still ate 62% of the 5$\mu$m beads consumed, even though rotifer densities declined considerably with depth. Summarising, these results showed an important decrease in community CR with depth as well as with the particle size.

## Discussion

*C. unicornis* is a common species in temperate lakes. In the lakes of Cascade Mountains of western North

*Table 3.* Percentage of the community grazing expressed as clearance rate (CR) calculated on beads of different sizes ingested by the rotifers, cladocera and copepoda in the epilimnion and metalimnion of Yellow Belly Lake on 14-7-99. On this date and depths, *Conochilus* was the only significant rotifer in the community. *Daphnia* and *Holopedium* were the only cladocerans, and the copepods were filter-feeding calanoids

| | Epilimnion (2, 4 &6 m) | | | Metalimnion (8 &10 m) | | |
|---|---|---|---|---|---|---|
| | 5 $\mu$m | 9 $\mu$m | 23 $\mu$m | 5 $\mu$m | 9 $\mu$m | 23 $\mu$m |
| Rotifera (%) | 83 | 70 | 0 | 62 | 27 | 0 |
| Cladocera (%) | 16 | 29 | 99 | 34 | 66 | 97 |
| Copepoda (%) | 1 | 1 | 1 | 4 | 7 | 3 |
| Community CR (ml d$^{-1}$ l$^{-1}$) | 625 | 272 | 44 | 131 | 49 | 25 |

America, this species was found in over 70% of the lakes, and it was usually the dominant zooplankton (Deimling et al., 1997). In coastal Lake Washington, *C. unicornis* was found only sporadically (Edmondson & Litt, 1987). In our study in Yellow Belly Lake, *Conochilus* was the dominant rotifer and the most abundant zooplankter in epilimnetic layers.

Also, it was a very important grazer in the epilimnion of the lake, consuming 70–80% of the 5-9 $\mu$m beads consumed by the entire zooplankton community (Table 3). In other studies (Bogdan & Gilbert, 1982; Scheda & Cowell, 1988; Quiblier-Llobéras et al., 1996), it has also been observed that rotifers may have a major grazing impact in different lakes. However, this has been rarely taken into account in zooplankton community grazing studies. Even though we may have overestimated the relative importance of *C. unicornis* in overall community grazing, we have demonstrated that this species was at least as important as cladocera for grazing epilimnetic phytoplankton. The high clearance rates of rotifers is expected, in part, because of their high weight-specific metabolic and feeding rates in relation to the larger crustaceans (Peters, 1983). Some caution is needed in generalising the importance of rotifers based on our study using of beads. We might have underestimated the grazing by calanoid copepods, because they are known to reject beads. DeMott (1988) compared grazing rates of crustaceans on microspheres and phytoplankton and found that calanoids preferred algae, whereas cladocerans did not discriminate between them.

The other rotifer species were inefficient in ingesting beads and there are several possible reasons to explain the difference between *C. unicornis* and the other rotifers. First, the particle sizes used in our experiment (5, 9 and 23 $\mu$m) may have been too large for the other species. Ronneberger (1998), using fluorescent latex beads, found that *Keratella quadrata, Polyarthra dolichoptera* Idelson and *S. pectinata* preferred particles having a diameter of 0.5–3 $\mu$m. Secondly, rotifer species differ markedly in their feeding mechanisms and dietary specialisation (Pourriot, 1977; Starkweather, 1980), and some of these species are not considered filter-feeders (e.g. *Ascomorpha, Gastropus, Synchaeta*). Thirdly, negative selection against microspheres may have occurred, although for rotifers this is unclear (Snell, 1998). Some species, however, can discriminate between flavoured and unflavoured beads whereas other cannot (DeMott, 1986). *Conochilus* feeds with a tube that crosses the gelatinous matrix of the colony, and the presence of this tube could interfere with particle tasting and make it more difficult to reject the items ingested. This feeding mode could explain the high IR and CR obtained for this species when feeding on beads.

We frequently found extremely high variances in the number of beads consumed by different colonies within the same sample. Similarly, Ronneberger (1990) encountered high variances in their experiments that he could not explain. Assuming that bead distribution was homogeneous inside the Haney Chambers, another possible explanation is that there are natural cycles of feeding in rotifers. In short-term incubations, some organisms would be resting or inactive during the time of the experiment. Although many rotifers were apparently not feeding in our experiments, we have nevertheless included them in our calculations of IR and CR because they must be considered to depict population or community grazing rates.

The CR we measured are higher than the average values obtained by other investigators, where few species exceed levels of 50 $\mu$l ind.$^{-1}$ h$^{-1}$ (Starkweather, 1980; Nogrady et al., 1993). However, Wallace & Starkweather (1983, 1985), using radioactively-labelled yeast as a food resource, found CR ranging from 19 to 240 $\mu$l ind.$^{-1}$ h$^{-1}$ in the colonial sessile rotifer *Floscularia conifera* hudson. Nevertheless, these authors found CR of only 1 $\mu$l ind.$^{-1}$ h$^{-1}$ for *C. unicornis*, far lower than the rates we measured.

Diel variations in rotifer feeding have also been described in previous studies (Nauwerck, 1959; Bogdan & Gilbert, 1982; Arndt & Heerkloss, 1989). Although we did not observe any significant diurnal variations in feeding (Fig. 2), one needs to be cautious because in the night and dawn samples, we encountered *Conochilus* in only one of the replicates and their densities were low, reducing our confidence in these results.

The decreased ingestion with increasing depth (Fig. 3) was likely due related to temperature (Fig. 4). The reduction in feeding activity with decreased temperature is expected for ectothermic organisms like rotifers (Cossins & Bowler, 1987), and has been found in studies with other rotifer species (Bogdan & Gilbert, 1982). Many studies on crustacean grazing, however, have found a minimal influence of temperature on grazing rates (Haney, 1973; Gulati, 1978; Lampert & Taylor, 1985); Peters & Downing (1984) argued that this was due to acclimatisation of the test animals. In our analysis, however, the rotifers should have been acclimated to their environment as the tests were done *in situ* on animals trapped in the Haney Chambers. Nevertheless, grazing rates at 14 °C in the epilimnion were 14-times higher than at 6 °C in the hypolimnion, suggesting that temperature had a very large influence on the feeding rate of *C. unicornis.* Although we did not check for other environmental factors such as algal availability that changed with depth, we also found lower grazing rates at colder temperatures when *Conochilus* was feeding only on microspheres (Exp.1), suggesting that temperature may have been the controlling variable. Nevertheless, we observed a significant decrease with decreasing temperature that was much larger than reported from laboratory studies (Hansen et al., 1997). So, we cannot dismiss the possibility that some other factor such as the physiological state or metabolic state of the rotifers, or the presence of algal metabolites at different depths may have influenced grazing rates.

In summary, our results suggest that grazing by the colonial rotifer *Conochilus unicornis* contributed substantially to overall community clearance rates, particularly on particles near 5 $\mu$m. Vertical variations in densities of *Conochilus*, and in grazing rate with temperature greatly determined the overall importance of this species in the community. Future studies on grazing rates in freshwater systems should take into account the potentially large contribution of the rotifer community.

## Acknowledgements

This work was funded by a grant from the U.S.-Spain Science & Technology Program to W. Wurtsbaugh and M. Miracle, and by U.S. National Science Foundation grant DEB9708927 to W. Wurtsbaugh and C. Luecke. X. Armengol and L. Boronat received a grant from Conselleria de Cultura, Educacion y Ciencia from Generalitat Valenciana, and from Spanish Ministerio de Educaciön y Ciencia, respectively. We also acknowledge the members of Utah State University Limnological Laboratory for their collaboration during this study.

## References

Arndt, H. & R. Heerkloss, 1989. Diurnal variation in feeding and assimilation rates of planktonic rotifers and its possible ecological significance. Int. Rev. ges. Hydrobiol. 74: 261–272.

Bogdan K. G. & J. J. Gilbert, 1982. Seasonal pattern of feeding by natural populations of *Keratella, Polyarthra* and *Bosmina*: clearance rates, selectivities and contributions to coommunity grazing. Limnol. Oceanogr. 27: 918–934.

Bogdan K. G. & J. J. Gilbert, 1984. Body size and food size in freshwater zooplankton. Proc. Nat. Acad. Sci. U.S.A. 81: 6427–6431.

Budy, P., C. Luecke, W. A. Wurtsbaugh, H. P. Gross & C. Gubala, 1995. Limnology of Sawtooth Valley Lakes with respect to potential growth of juvenile Snake River Sockeye Salmon. Northwest Science 69: 133–150.

Cossins, A. R. & K. Bowler, 1987. Temperature biology of animals. Chapman and Hall. London: 339 pp.

Deimling, E. A., W. J. Liss, G. L. Larson, R. L. Hoffman & G. A. Lomnicky, 1997. Rotifer abundance and distribution in the northern Cascade mountains, Washington, U.S.A. Arch. Hydrobiol. 138: 345-363.

DeMott, W. R., 1986. The role of taste in food selection by freshwater zooplankton. Oecologia 69: 334–340.

DeMott, W. R., 1988. Discrimination between algae and artificial particles by freshwater and marine copepods. Limnol. Oceanogr. 33: 397–408.

Edmondson, W. T. & A. H. Litt, 1987. *Conochilus* in Lake Washington. Hydrobiologia 147: 157–162.

Gulati, R. D., 1978. Vertical changes in the filtering, feeding and assimilation rates of dominant zooplankters in a stratified lake. Verh. int. Ver. Limnol. 20: 950–956.

Haney, J. F., 1971. An *in situ* method for the measurement of zooplankton grazing rates. Limnol. Oceanogr. 16: 970–977.

Haney, J. F., 1973. An *in situ* examination of the grazing activities of natural zooplankton communities. Arch. Hydrobiol. 72: 87–132.

Hansen, P. J., P. K. Bjørnsen & B. W. Hansen, 1997. Zooplankton grazing and growth: scaling within the 2–2000-$\mu$m body size range. Limnol. Oceanogr. 42: 687–704.

Hansson, L. A., E. Bergman & G. Cronberg, 1998. Size structure and succesion in phytoplankton communities: the impact of interactions between herbivory and predation. Oikos 81: 337–345.

Juchelka, C. M. & T. W. Snell, 1994. Rapid toxicity assessment using rotifer ingestion rate. Arch. Envir. Contam. Toxicol. 26: 549–554.

Lampert, W. & B. E. Taylor, 1985. Zooplankton grazing in a eutrophic lake: implications of diel vertical migration. Ecology 66: 68–82.

Nauwerck, A., 1959. Zur Bestimmung der Filtrierrate limnischer Planktontiere. Arch. Hydrobiol., Suppl. 25: 83–101.

Nogrady, T., R. L. Wallace & T. W. Snell, 1993. Rotifera 1: Biology, Ecology and Systematics. In Guides to the Identification of the Microinvertebrates of the Continental Water of the World 4. SPB Academic Publishing. The Hague, The Netherlands: 142 pp.

Pace, M. L. & J. D. Orcutt, Jr., 1981. The relative importance of protozoans, rotifers and crustaceans in a freshwater zooplankton community. Limnol. Oceanogr. 26: 822–830.

Peters, R. H., 1983. The Ecological Implications of Body Size. Cambridge University Press. New York: 329 pp.

Peters, R. H. & J. Downing, 1984. Empirical analysis of zooplankton filtering and feeding rates. Limnol. Oceanogr. 29:763–784.

Pourriot, R., 1977. Food and feeding habits of Rotifera. Arch. Hydrobiol. Beih. Ergeb. Limnol. 8: 243–260.

Quiblier-Llobéras, C. G. Bourdier, Ch. Amblard & D. Pepin, 1996. A qualitative study of zooplankton grazing in an oligo-mesotrophic lake using phytoplanktonic pigments as organic markers. Limnol. Ocanogr. 41: 1767–1779.

Ronneberger, D., 1998. Uptake of latex beads as size-model for food of planktonic rotifers. Hydrobiologia 387/388: 445–449.

Rothhaupt, K. O., 1990. Differences in particle size-dependent feeding efficiencies of closely related rotifer species. Limnol. Oceanogr. 35: 16–23.

Scheda, S. M. & B. C. Cowell, 1988. Rotifer grazers and phytoplankton: seasonal experiments on natural communities. Arch. Hydrobiol. 114: 31–44.

Snell, T. W. 1998. Chemical ecology of rotifers. Hydrobiologia 387/388: 267–276.

Starkweather, P. L., 1980. Aspects of the feeding behavior and trophic ecology of suspension-feeding rotifers. Hydrobiologia 73: 63–72.

Starkweather, P. L. & J. J. Gilbert, 1977. Radiotracer determination of feeding in *Brachionus calyciflorus*: the importance of gut passage times. Arch. Hydrobiol. Beih. 8: 261–263.

Wallace, R. L., 1987. Coloniality in the phylum Rotifera. Hydrobiologia 147: 141–155.

Wallace, R. L. & P. L. Starkweather, 1983. Clearance rates of sessile rotifers: *In situ* determinations. Hydrobiologia 104: 379–383.

Wallace, R. L. & P. L. Starkweather, 1985. Clearance rates of sessile rotifers: *in vitro* determinations. Hydrobiologia 121: 139–144.

Wilson, D. S., 1973. Food size selection among copepods. Ecology 54: 909–914.

Weisse, T., H. Müller, R. M. Pinto-Coelho, A. Schweizer, D. Springmann & G. Baldringer, 1990. Response of the microbial loop to the phytoplankton spring bloom in a large prealpine lake. Limnol. Oceanogr. 35: 781–794.

Wurtsbaugh, W. A., H. P. Gross, C. Luecke & P. Budy, 1977. Nutrient limitation of oligotrophic sockeye salmon lakes of Idaho (U.S.A.). Verh. int. Ver. Limnol. 26: 413–419.

Hydrobiologia **446/447:** 115–121, 2001.
L. Sanoamuang, H. Segers, R.J. Shiel & R.D. Gulati (eds), Rotifera IX.




# Observations of insect predation on rotifers

Stephanie E. Hampton & John J. Gilbert
*Dartmouth College, Department of Biological Sciences, 6044 Gilman, Hanover, NH 03755-3576, U.S.A.*
*E-mail: stephanie.e.hampton@dartmouth.edu john.j.gilbert@dartmouth.edu*

*Key words:* Rotifera, notonectids, backswimmers, Odonata, juvenile diet, aquatic insects

## Abstract

Interactions between rotifers and their insect predators have not received adequate attention, possibly due to the assumption that rotifers are too small for insects to eat. In laboratory experiments, we offered the rotifers *Hexarthra mira*, *Plationus patulus* and small and large *Synchaeta pectinata* to four common insect predators: the notonectids *Notonecta lunata* and *Buenoa macrotibialis*, the smaller hemipteran *Neoplea striola* and small (1.5 mm) aeschnid dragonfly larvae. Excepting *Plationus* offered to dragonflies, all rotifer preys were consumed to some degree. No size selectivity was apparent for predators that ate few rotifers, but small instar *Buenoa* ate significantly more large (420 $\mu$m) than small (300 $\mu$m) *Synchaeta*. Predator size appeared to be less important than predatory style and prey morphology in determining ingestion rates. *Neoplea* and dragonflies ate more *Hexarthra* than *Plationus*, while the pattern was reversed for *Buenoa*, possibly because *Buenoa* is able to manipulate the hard lorica of *Plationus* better. Insect predators are capable of direct suppression of rotifer populations, an interaction which may be particularly important in littoral zones and fishless ponds where macroinvertebrates are numerous.

## Introduction

Despite the common co-occurrence of insects and rotifers, trophic links between these groups are very little studied. This neglect is probably partly due to the assumption that rotifers are too small for most insects to attack. Many insects show positive size selection for prey (Cooper, 1983; Scott & Murdoch, 1983; Murdoch & Scott, 1984; Giller, 1986), and often have a variety of larger prey from which to choose in their littoral habitats. However, several common insect predators have been shown to be opportunistic rather than selective (Cooper et al., 1985), attacking any prey encountered. In addition, juvenile insects may be too small to handle the larger prey of adults and thus may be more likely to take rotifers and protozoans as prey (Kormondy, 1959; Lawton, 1970). We addressed two *a priori* hypotheses: (1) larger rotifers are eaten more than smaller rotifers, and (2) smaller insects feed more on rotifers than do larger insects. In addition, we directly observed notonectids feeding on several rotifers and briefly describe those interactions.

## Materials and methods

### Predators

We conducted trials with four insect predators: the notonectids (backswimmers) *Buenoa macrotiabilis* and *Notonecta lunata*, the smaller related hemipteran *Neoplea striola* (pleids or pygmy backswimmers) and larval aeschnid dragonflies. The aeschnids were too small (ca. 1.5 mm) for us to identify further.

The notonectids and pleids, as true bugs, feed by piercing a prey item with the rostrum, injecting digestive juices and sucking the liquefied contents from the prey. They are typically ambush predators, remaining still until prey are detected by mechanoreception or vision, although notonectids will also forage more actively (Gittelman, 1974; Cooper et al., 1985; Streams, 1987). Prey size typically increases with notonectid size (Streams, 1974; Fox, 1975; Giller, 1986). Development is direct through six distinct stages (juvenile I–V, adult VI), punctuated by a molt. *Notonecta lunata* is larger than *Buenoa macrotibialis*. *Neoplea striola* is relatively small, and more closely associated with vegetation.

Aeschnid larvae can use both ambush and stalking strategies (Cooper et al., 1985; Johansson, 1991). Prey size is usually positively correlated with predator size, but aeschnids are often highly opportunistic in prey capture (recently reviewed in Corbet, 1999). They capture prey with a uniquely prehensile labium, or underlip, which is normally held in a closed position but can be swung out quickly to grasp prey and bring it to the chewing mouthparts. Thus, while notonectids and pleids leave some sort of carcass after feeding, dragonflies may consume prey entirely.

*Prey*

All rotifer prey co-occurred with the predators in Johnson Pond, Vermont, U.S.A. *Hexarthra mira* (ca. 200 $\mu$m) is well known for its evasiveness; encounters with predators elicit a long jump (Iyer & Rao, 1996). *Plationus patulus* (ca. 200 $\mu$m) has a hard lorica with elaborate anterior and posterior spines. It secretes a strand of mucus that secures it to a substrate, and it will feed within the same small area for hours at a time in the laboratory (Stephanie E. Hampton, unpublished). This string can be quickly severed when *P. patulus* is disturbed, and we frequently find them in open water samples. *Synchaeta pectinata* (300–420 $\mu$m) is free-swimming and soft-bodied. It reacts to contact with predators by retracting the corona to form itself into a turgid ball, which may not be easily ingested by some invertebrates (Stemberger & Gilbert, 1987).

*Feeding trials*

We conducted feeding experiments from May 1998 to July 1999. Insect predators and *Hexarthra mira* were collected from Johnson Pond (Norwich, Vermont, U.S.A.) 2–4 h before experiments. *Plationus patulus* and *Synchaeta pectinata* were cultured in GF/F filtered lake water (Storr's Pond, Hanover, New Hampshire, U.S.A.) on *Cryptomonas erosa*. Cryptomonads were-grown in modified MBL medium (Stemberger, 1981). Replication varied among experiments (3 – 6 replicates in a treatment), dependent on availability of prey and predators.

In experiments comparing predation by different insects on different prey, 10 rotifer prey of a given species were pipetted into 15 ml of filtered lake water in a plastic Petri dish (60 × 15 mm), and one insect predator was introduced. Dish location was randomized in a 19 °C lighted chamber. After 6 h, we removed all insect predators and counted live rotifers. We attempted to visually identify prey remains in predator treatments.

We also compared insect ingestion of small (ca. 300 $\mu$m) and large (ca. 420$\mu$m) *S. pectinata*. Ten rotifers of each size class were pipetted into 15 ml of filtered lake water. Because we were initially uncertain about the length of time needed to obtain definite results, these experiments were run for 4–6 h.

Because most experiments were done on different days, subject to availability of predators and prey, we analyzed each experiment separately (JMP, SAS Institute). Experiments with two factors were analyzed with 2-way ANOVA. Single factor experiments were analyzed with $t$-tests or ANOVA if data were normally distributed and had equal variance. If the data did not meet these criteria, we analyzed the results with Welch's ANOVA. All prey individuals were recovered from control dishes (prey with no predators), so these data were omitted from the analyses to simplify comparisons.

## Results

*Rotifer body size*

Pleids did not consume large or small *Synchaeta* differentially, nor did 1st and 2nd instar pleids differ in total *Synchaeta* consumption (Table 1, Fig. 1). Small *Buenoa* consumed more *Synchaeta* than did *Buenoa* V and *Neoplea* adults. Among *Buenoa*, large *Synchaeta* were eaten more than were small *Synchaeta*.

*Insect predators*

All predators showed some ability to eat rotifers (Table 1, Fig. 2). Except for *Plationus* offered to aeschnids, all rotifers offered to predators were consumed to some degree. Consumption of *Hexarthra* did not differ between *Buenoa* and *Notonecta* (Table 1). Different instars of pleids ate similar numbers of *Hexarthra* and *Plationus*; both reduced *Hexarthra* more than *Plationus* (Table 1). Although comparisons of data for different days must be made with caution, it is notable that small *Buenoa,* particularly *Buenoa* II, ate more *Plationus* than *Hexarthra*, in contrast to aeschnids and pleids (Fig. 2).

Rotifer carcasses were frequently found in the notonectid treatments. *Hexarthra* and *Synchaeta* carcasses were difficult to identify, because their soft bodies were crumpled. However, the corona often could

*Table 1.* Descriptions of feeding trials carried out on different dates. All experiments on a given date were run concurrently under identical conditions. Differences in prey remaining among treatments were analyzed separately by date, with no comparisons among dates. Bold indicates significant effects ($P < 0.05$)

| Experiment date | Predator–prey combinations | Analysis and results (prey eaten) |
|---|---|---|
| 16 June 1998 | *Neoplea* I – large &small *Synchaeta*<br>*Neoplea* II – large &small *Synchaeta* | 2-way ANOVA<br>predator × prey: *P*=0.54<br>large vs. small: *P*=0.17<br>*Neoplea* I vs. *Neoplea* II: *P*=0.54 |
| 19 June 1998 | *Buenoa* I – large & small *Synchaeta*<br>*Buenoa* V – large &small *Synchaeta* | 2-way ANOVA<br>predator × prey: *P*=0.22<br>**small < large: *P*=0.02**<br>*Buenoa* **V** < *Buenoa* **I: *P* = 0.04** |
| 23 June 1998 | *Buenoa* II – large & small *Synchaeta*<br>*Buenoa* IV – large &small *Synchaeta*<br>*Neoplea* VI – large &small *Synchaeta* | 2-way ANOVA<br>predator × prey: *P*=0.32<br>**small < large: P < 0.01**<br>*Neoplea* **VI** < *Buenoa* **II, IV:** *P* < **0.01** |
| 30 May 1999 | Aeschnid – *Hexarthra* | Welch's ANOVA<br>**Aeschnid > control on** *Hexarthra*: ***P* = 0.02** |
| 31 May 1999 | Aeschnid – *Synchaeta* | Welch's ANOVA<br>**Aeschnid > control on** *Synchaeta*: **P < 0.01** |
| 24 June 1999 | *Buenoa* I – *Hexarthra*<br>*Notonecta* I – *Hexarthra* | *t*-test<br>*Buenoa* I vs. *Notonecta* I on *Hexarthra*: *P* = 0.22 |
| 29 June 1999 | *Buenoa* II – *Hexartha*<br>*Notonecta* II – *Hexarthra* | *t*-test<br>*Buenoa* II vs. *Notonecta* II on *Hexarthra*: *P* = 0.45 |
| 6 July 1999 | *Neoplea* I – *Plationus*<br>*Neoplea* I – *Hexarthra*<br>*Neoplea* II – *Plationus*<br>*Neoplea* II – *Hexarthra*<br>Aeschnid – *Plationus* | 2-way ANOVA<br>predator × prey: *P* = 0.12<br>*Neoplea* I vs. *Neoplea* II: *P* = 0.21<br>*Plationus* < *Hexarthra*: *P* < **0.01**<br><br>ANOVA<br>**Aeschnid <** *Neoplea* **I, II on** *Plationus*: ***P* = 0.02** |
| 9 July 1999 | *Buenoa* I – *Plationus* | Welch's ANOVA<br>*Buenoa* **I > control on** *Plationus*: *P* < **0.01** |
| 30 July 1999 | *Buenoa* II – *Plationus* | *t*-test<br>*Buenoa* **II > control on** *Plationus*: *P* < **0.01** |

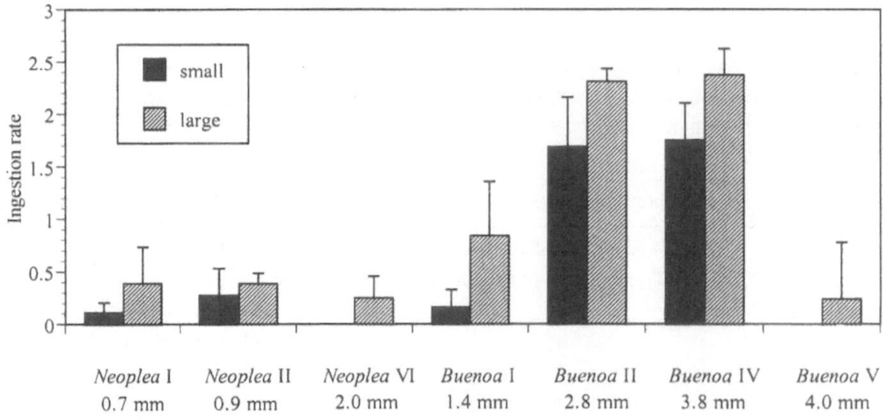


*Figure 1.* Predator ingestion rate (rotifers $h^{-1}$) of small (300 $\mu$m) and large (420 $\mu$m) *Synchaeta pectinata*. Predator body length is shown.

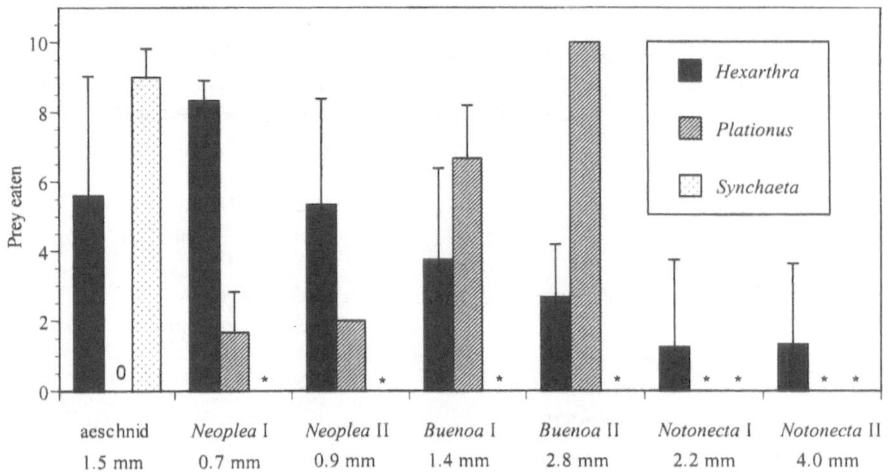


*Figure 2.* Prey eaten in predator treatments after 6 h exposure. Experiments were begun with 10 prey rotifers in each dish. Predator body length is shown. * indicates predator-prey combinations that were not tested. 0 indicates that no prey were eaten in any of the predator treatment vessels.

be discerned once a carcass was located. *Plationus* carcasses were conspicuous, as the lorica generally kept its shape after a predation event (Fig. 3). Trophi were always identifiable within *Plationus* loricae after predation events (Fig. 3).

## Discussion

No size selection among *Synchaeta* was discernible among the predators that consumed relatively few individuals of this rotifer (*Neoplea* I, II, VI, and *Buenoa* V). However, the smaller instars of *Buenoa* selected larger *Synchaeta* (Fig. 1). This observation is consistent with those in previous studies, which report that notonectids typically eat the largest available prey that they can ingest (Cooper, 1983; Murdoch & Scott, 1984; Gilbert & Burns, 1999; Hampton et al., 2000).

Contrary to our second hypothesis that smaller insects feed more on rotifer preys, size could not explain differences among the predator taxa. There were no significant differences in rotifer ingestion between juvenile *Notonecta* and *Buenoa* (Table 1). In addition, adult pleids ate fewer *Synchaeta* than juvenile *Buenoa* (Table 1), despite *Neoplea's* smaller size (Fig. 1). This result is consistent with Gittelman's (1977) conclusion that the construction of *Neoplea*'s forelegs provides mechanical advantages in capturing relatively large prey, but is not well suited for capturing and subduing small, mobile prey.

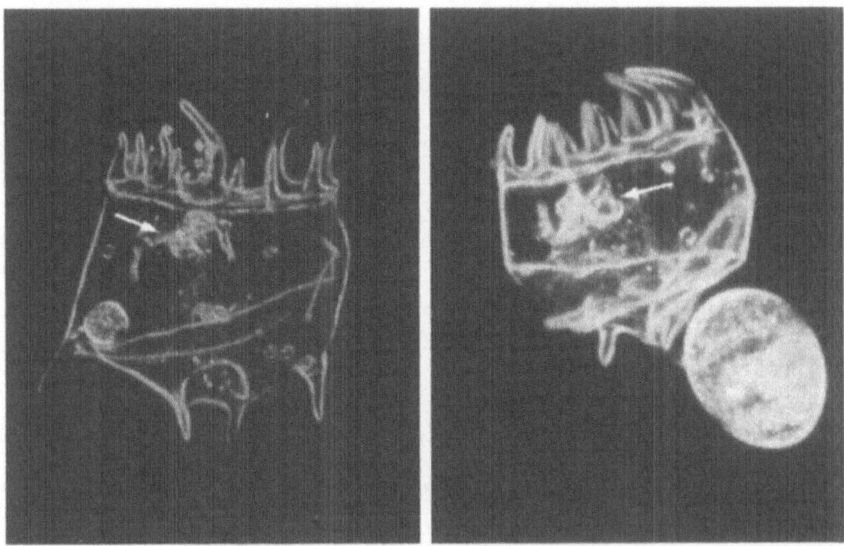

*Figure 3.* Carcasses of *Plationus patulus* after notonectid predation. Arrows indicate trophi.

However, predator size may explain ingestion rates in *Buenoa*, as 1st instar *Buenoa* ate more *Synchaeta* overall than did 5th instar *Buenoa* (Table 1). Ellis & Borden (1970) suggest that larger notonectids may be morphologically ill-equipped to handle small prey. Notonectids must manipulate prey toward the rostrum with the forelegs, and then pierce the prey and insert digestive enzyme through the rostrum. If the prey is small relative to the predator, these manipulations may be difficult. Of the notonectids, *Buenoa* and *Anisops* are probably best-equipped to handle small prey. Their forelegs are modified in such a way that long setae on the forelegs create a 'basket' in which small prey can be held (Truxal, 1953; Gittelman, 1977).

Despite the well known evasive maneuvers of *Hexarthra*, it was captured and eaten by all the insect predators. In contrast, *Plationus* was seemingly preferred over *Hexarthra* by the smaller instars of *Buenoa*, but eaten less by pleids and dragonflies. *Plationus* is far less evasive than *Hexarthra*, but it has a sturdy lorica, which may have discouraged some of the predators. Rowe (1994) reported that, unless the prey was oriented in a certain position, larval damselflies were unable to eat *Alona guttata* due to the hardness of the carapace. Pearlstone (1973) directly observed that the carapace of amphipods was too hard to be pierced by the labial hooks of dragonfly larvae, and thus these crustaceans were not easily captured.

We did not directly observe the capture of any rotifer, although we did witness the consumption of *Plationus* and *Synchaeta* by pleids. With *Plationus*, the pleid inserted its rostrum through the anterior end of the lorica, without puncturing the lorica itself. As previously mentioned, the loricae were never torn, suggesting that notonectids must manipulate the live rotifer until the rostrum can be inserted into the corona. We have witnessed this type of manipulation with ostracods and *Bosmina*, as the notonectids turn the ostracods until the ventral opening is found. In contrast, pleids may puncture the body wall of *Synchaeta* in many places. We observed no discernible pattern in the point of entrance.

With both rotifers, we noted that the pleid held its forelegs away from the prey once it was affixed to the rostrum. This behavior contradicts other reports of backswimmer feeding. Researchers have repeatedly noted that this characteristic feeding pose, the forelegs positioned near the rostrum, can be used to discern when a notonectid is feeding (Streams, 1974; Fox, 1975). This position may well be highly correlated with predation on larger prey items that are more unwieldy, but if it is common for notonectids to open the forelegs while feeding on small prey, then small prey are likely underestimated during direct observations when using this method.

This complication is one of several in describing notonectid diet. Because notonectids suck the contents of their prey in liquid form, no recognizable prey items can be seen in the gut. Thus, direct observations in the field are the primary method of determining the natural diet. Small prey items are likely to be missed not only for the potential reason discussed above, but also because the time in which the predator is feeding is likely to be shorter than it would be with larger prey (Streams, 1974). Molecular techniques have been used to identify gut contents (Giller, 1986), but this requires building a 'library' of the molecular fingerprints of potential prey against which gut contents can be matched. The discovery that trophi are consistently left intact in *Plationus* (Fig. 3) diminishes hope that trophi might be small enough to be sucked into the notonectid gut and later identified. Consequently, molecular techniques are still the most reliable methods for gut content analysis of notonectids.

Apart from the many studies describing the predatory behavior of *Chaoborus*, very little information exists about insect predation on rotifers. Damselfly larvae fed on *Euchlanis dilatata* in laboratory experiments (Walsh, 1995). Several odonate researchers have noted that protozoans and protozoan-sized metazoans are probably important prey for the first free-swimming odonate instar (Kormondy, 1959; Lawton, 1970) or state that rotifers comprised some part of odonate diet (Pearlstone, 1973). These reports are vague in the description of the rotifer prey. Lawton (1970) reported that "a rotifer, possibly *Diurella* {*Trichocerca*}" was disproportionately consumed by damselfly larvae in a mixture of protozoans and rotifers. Pearlstone's (1973) reference to rotifer remains in damselfly larvae fecal pellets suggests that the rotifer prey were loricate rather than soft-bodied.

Here, we have demonstrated that dragonfly larvae, pleids, and two notonectid genera are capable of feeding on rotifers. These data support the conclusions of Hampton et al. (2000) that 2nd and 4th instar *Buenoa* reduce *Hexarthra* populations in field enclosures through direct predation. In those enclosures with *Buenoa*, *Hexarthra* and the copepod *Tropocyclops* were greatly reduced. Populations of smaller rotifers, most notably *Polyarthra*, increased dramatically in the presence of 4th instar *Buenoa*, suggesting release from predation, competition or both. Such indirect effects of insect predation are probably quite common, since many insects select larger prey and thus probably the most influential competitors or predators. Direct predation by insects on small zooplankters such as rotifers may also be common, but has been very little studied. Macroinvertebrate predation may be particularly important in fishless ponds, where insects can proliferate, and especially where macrophyte stands harbor high abundance of insect predators.

## Acknowledgements

We greatly appreciate the generosity of Mr W.C. Johnson in allowing access to his property, statistical advice from T.S. Sillett and L.R. Nagy, helpful comments on the manuscript from T.R. Rao and C.E. King, and the financial support of NSF Dissertation Improvement Grant #9902177 and a Sondra and Charles Gilman Graduate Research Fellowship.

## References

Cooper, S. D., 1983. Selective predation on cladocerans by common pond insects. Can. J. Zool. 61: 879–886.

Corbet, P. S., 1999. Dragonflies: Behavior and Ecology of Odonata. Cornell University Press, Ithaca, New York, U.S.A. pp

Ellis, R. A. & J. H. Borden, 1970. Predation by *Notonecta undulata* (Heteroptera: Notonectidae) on larvae of the yellow-fever mosquito. Ann. ent. Soc. am. 63: 963–973.

Fox, L. R., 1985. Some demographic consequences of food shortage for the predator, *Notonecta hoffmani*. Ecology 56: 868–880.

Gilbert, J. J. & C. W. Burns, 1999. Some observations on the diet of the backswimmer, *Anisops wakefieldi* (Hemiptera: Notonectidae). Hydrobiologia, 412: 111–118.

Giller, P. S., 1986. The natural diet of the Notonectidae: field trials using electrophoresis. Ecol. Ent. 11: 163–172.

Gittelman, S. H., 1974. Locomotion and predatory strategy in backswimmers (Hemiptera: Notonectidae). Am. Midl. Nat. 92: 496–500.

Gittelman, S. H., 1977. Leg segment proportions, predatory strategy and growth in backswimmers (Hemiptera: Pleidae, Notonectidae). Kans. ent. Soc. 50: 161–171.

Hampton, S. E., J. J. Gilbert & C. W. Burns, 2000. Direct and indirect effects of juvenile *Buenoa macrotibialis* (Hemiptera: Notonectidae) on the zooplankton of a shallow pond. Limnol. Oceanogr. 45: 1006–1012.

Iyer, N. & T. R. Rao, 1996. Responses of the predatory rotifer *Asplanchna intermedia* to prey species differing in vulnerability: laboratory and field studies. Freshwat. Biol. 36: 521–533.

Johansson, F., 1991. Foraging modes in an assemblage of odonate larvae – effects of prey and interference. Hydrobiologia 209: 79–87.

Kormondy, E. J., 1959. The systematics of *Tetragoneuria*, based on ecological, life history and morphological evidence (Odonata: Corduliidae). Misc. Publs. Mus. Zool. Univ. Mich. 107: 1–79.

Lawton, J. H., 1970. Feeding and food energy assimilation in larvae of the damselfly *Pyrrhosoma nymphula* (Sulz.) (Odonata: Zygoptera). J. anim. Ecol. 39: 669–689.,

Murdoch, W. W. & M. A. Scott, 1984. Stability and extinction of laboratory populations of zooplankton preyed on by the backswimmer *Notonecta*. Ecology 65: 1231–1248.

Pearlstone, P. S. M., 1973. The food of damselfly larvae in Marion Lake, British Columbia. Syesis 6: 33–39.

Rowe, R. J., 1994. Predatory behaviour and predatory versatility in young larvae of the dragonfly *Xanthocnemis zealandica* (Odonata, Coenagrionidae). N. Z. J. Zool. 21: 151–166.

Scott, M. A. & W. W. Murdoch., 1983. Selective predation by the backswimmer, *Notonecta*. Limnol. Oceanogr. 28: 352–366.

Stemberger, R. S., 1981. A general approach to the culture of planktonic rotifers. Can. J. Fish. aquat. Sci. 38: 721–724.

Stemberger, R. S. & J. J. Gilbert., 1987. Defenses of planktonic rotifers against predators, p. 227–239. In Kerfoot, W. C. & A. Sih (eds), Predation: Direct and Indirect Impacts on Aquatic Communities. University Press of New England, Hanover (NH): 227–239.

Streams, F. A., 1974. Size and competition in Connecticut *Notonecta*. In Beard, R. L. (ed.), 25th Anniversary Memoirs, Connecticut Entomological Society. Connecticut Entomological Society, New Haven (CT): 215–225.

Streams, F. A., 1987. Foraging behavior in a notonectid assemblage. Am. Midl. Nat. 117: 353–361.

Truxal, F. S., 1953. A revision of the genus *Buenoa* (Hemiptera Notonectidae). Univ. Kans. sci. Bull. 35: 1351–1523.

Walsh, E. J., 1995. Habitat-specific predation susceptibilities of a littoral rotifer to two invertebrate predators. Hydrobiologia 313/314: 205–211.

Hydrobiologia **446/447**: 123–127, 2001.
L. Sanoamuang, H. Segers, R.J. Shiel & R.D. Gulati (eds), Rotifera IX.




# The effect of bacteria on interspecific relationships between the euryhaline rotifer *Brachionus rotundiformis* and the harpacticoid copepod *Tigriopus japonicus*

Min-Min Jung[1] & Atsushi Hagiwara[2]
[1]*National Fisheries Research and Development Institute (NFRDI), South Sea Fisheries Research Institute (SSFRI), 347, Anpo-ri Hwayang-myun Yosu-si Chullanam-do, 556-820, Korea*
*E-mail: jungminmin@nfrda.re.kr jungminmin@hanmail.net*
[2]*Faculty of Fisheries, Nagasaki University, Bunkyo, Nagasaki 852-8521, Japan*
*E-mail: hagiwara@net.nagasaki-u.ac.jp*

*Key words:* axenic culture, *Brachionus rotundiformis*, copepod, co-existing bacteria, interspecific relation, rotifer, synxenic culture, *Tigriopus japonicus*

## Abstract

Inconsistent results have been obtained on the population growth of *Brachionus rotundiformis* and *Tigriopus japonicus*, when results from single-species and two-species mixed cultures are compared. Bacteria growth was not regulated in these experiments, which could be the cause for this. In order to test this possibility, we conducted similar experiments under axenic and synxenic (with presence of one species of bacteria) conditions. The population growth of *B. rotundiformis* was suppressed by the presence of *T. japonicus* in axenic cultures. *T. japonicus* could not persist in axenic cultures, but its population increased when grown in synxenic cultures. *T. japonicus* used RT bacteria strain as a food source, while these bacteria were toxic to *B. rotundiformis*. These results suggest that bacteria can modify the interspecific relationship between *B. rotundiformis* and *T. japonicus*.

## Introduction

Several studies are available on the interspecific relations between zooplankton (Pace & Orcutt, 1981; Gilbert & Stemberger, 1985; Burns & Gilbert, 1986a, b; Gilbert, 1988, 1989; Gilbert & Jack, 1993; Hagiwara et al., 1995; Jung et al., 1997). Controversial results were reported by Hagiwara et al. (1995) and Jung et al. (1997) on the interspecific relations between rotifers (*B. rotundiformis*) and copepods (*T. japonicus*), notwithstanding that the experiments were conducted under similar circumstances. Hagiwara et al. (1995) reported that the copepod *T. japonicus* did not affect rotifer growth, while *T. japonicus* grew better with rotifers than when cultured alone. On the other hand, Jung et al. (1997) reported that contamination of rotifer cultures by *T. japonicus* strongly suppressed *B. rotundiformis* population growth when compared to *B. rotundiformis* single species cultures. At the same time, the growth of *T. japonicus* was better when cultured alone than when the two species were kept in mixed cultures.

Jung et al. (1997) suggested that the interspecific relations between rotifers and copepods could be influenced by co-existing aquatic bacteria occurring in the cultures. Here, we test this hypothesis, by investigating the effect of co-existing aquatic bacteria on the interspecific relationship of the rotifer *B. rotundiformis* and the copepod *T. japonicus* in axenic and synxenic cultures.

## Materials and methods

### *Test organisms*

The tested *B. rotundiformis* was collected and isolated from the brackish Kai-ike Lake, Koshiki Island, Kagoshima prefecture, Japan. This strain has been cultured in the laboratory for several years and is known as the Koshiki strain. It is characterized by low occurrence of mixis. *T. japonicus* was collected from brackish Hamana Lake, Japan and was maintained in the experiments as monospecies culture originating from one ovisac-carrying female. *Nannochloropsis oculata* used in the laboratory was obtained from the phytoplankton bank of Nagasaki University, Japan.

*Axenic cultures of tested organisms*

Axenic *N. oculata* originated from monospecies stock cultures. An axenic culture of *N. oculata* was obtained by processing, management and repeated plating, collection and dilution of cultures in a bacteria free room. They were cultured using modified Erd-Schreiber medium (Hagiwara et al., 1994) in 200 ml flat bottom glass flasks. We used a modified antibiotic mixture solution AM9 (Sakamoto & Hirayama, 1983) for axenic conditioning of *B. rotundiformis* and *T. japonicus*.

*Experimental conditions*

The experimental design for testing the interspecific relation between *B. rotundiformis* and *T. japonicus* was based on the experimental design of Hagiwara et al. (1995) and Jung et al. (1997). Rotifers and copepods cultures were initiated with seven amictic females of *B. rotundiformis* and one ovisac-carrying female *T. japonicus*. They were cultured in seawater at a salinity of 22 ppt at 25 °C. The seawater utilized was GF/C- filtered, autoclaved and diluted with distilled water. Test organisms were fed axenic *N. oculata* at a concentration of $7\times10^5$cell ml$^{-1}$ every 2 days during the experimental period. Experimental culture volumes were 40 ml in 200 ml flat bottom glass flasks (PYREX, Japan) with three replicates each treatment. All of the experimental flasks were covered with a silico stopper. Cultures were kept in total darkness, except during feeding periods when they were transferred in bacteria free room. The experimental culture period was 16 days. Population density was determined on two occasions only, at the beginning and end of experiments in order to maintain axenic conditions.

*Isolation of bacteria from cultures*

Bacteria were collected from all synxenic experimental culture flasks on the last day of experiment. They were cultured on Zobell 2216E marine agar plate medium, and only one type of colony as defined by Claus (1978) was isolated. This colony was named RT

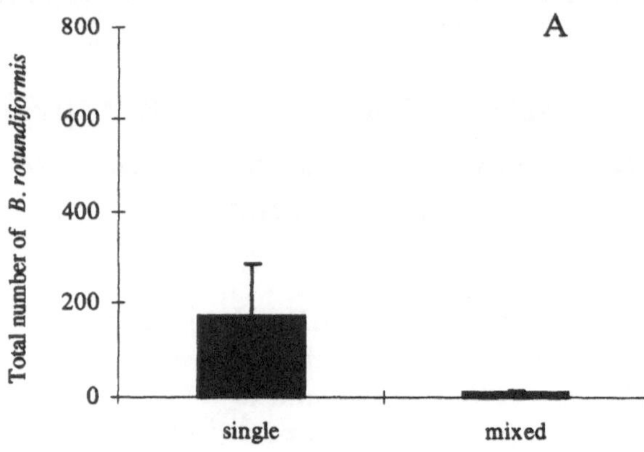


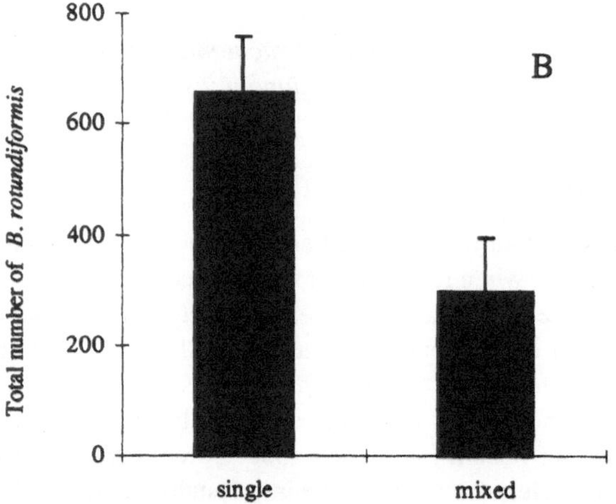


*Figure 1.* Comparative growth patterns of rotifer *B. rotundiformis* in synxenic (A) and axenic (B) conditions. *B. rotundiformis* was cultured alone or with *T. japonicus*.

strain and its effect on axenic cultures of *N. oculata*, *B. rotundiformis* and *T. japonicus* was investigated.

*Statistical analysis*

The difference in population growth between single culture vs. mixed culture, and population growth of synxenic culture vs. axenic culture conditions were compared by $t$-test at 95% confidence level.

## Results

Figure 1 shows that rotifer *B. rotundiformis* populations were suppressed by co-cultured *T. japonicus* in synxenic mixed cultures (Fig. 1A, $P<0.05$) and axenic (Fig. 1B, $P<0.05$) when compared with *B.*

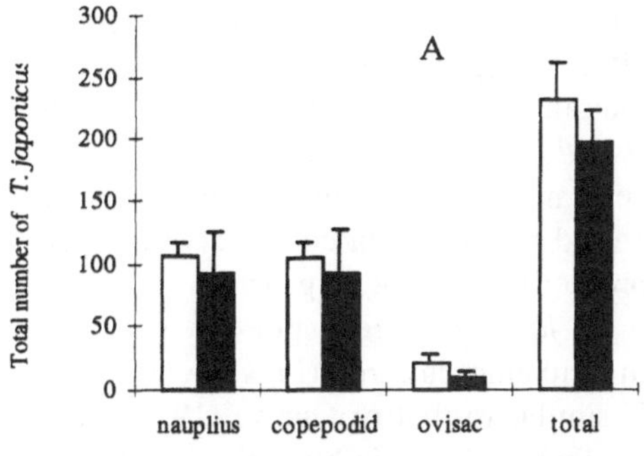


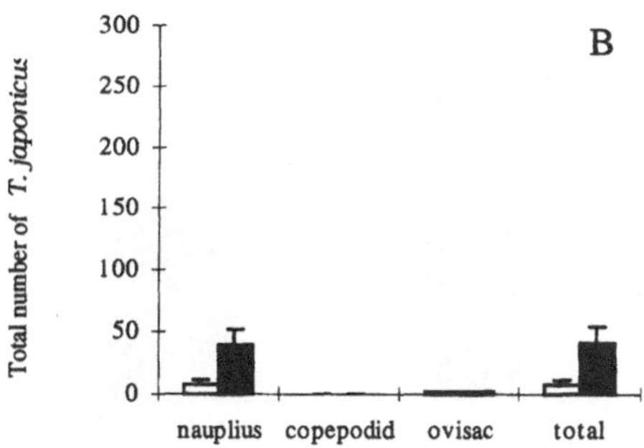


*Figure 2.* Comparative growth patterns of the copepod *T. japonicus* under synxenic (A) and axenic (B) conditions. *T. japonicus* was cultured alone (white bars) or with *B. rotundiformis* (black bars).

*rotundiformis* single culture under synxenic and axenic cultures conditions. The rotifer *B. rotundiformis* population almost disappeared when cultured with *T. japonicus* in the synxenic culture, compared with axenic mixed cultures (Fig. 1A, $P<0.05$).

The population of *T. japonicus* developed well in the synxenic culture condition regardless of the existence of *B. rotundiformis*. We observed appearance of all developmental stages of *T. japonicus* from nauplius stage to ovisac-carrying adults (Fig. 2A, $P<0.05$). We did not observe copepodites of *T. japonicus* in the axenic culture, indicating that the young nauplii did not continue to develop in these cultures, in both *T. japonicus* single species culture and in mixed axenic cultures with *B. rotundiformis* (Fig. 2B, $P<0.05$).

The remaining concentration of food *N. oculata* in all of culture flasks showed that less than 10% of the total number of cells remained in all synxenic cultures. However, single species axenic cultures of *T. japonicus* indicated that almost no algae were consumed as food. It is quite clear that *T. japonicus* did not consume *N. oculata* in the *B. rotundiformis* and *T. japonicus* mixed axenic culture (Fig. 3, $P<0.05$).

In the synxenic culture (Figs 1A and 2A), only one form of colony appeared on the 2216E marine agar medium. The isolated bacteria presented colony color white, colony size 1–2 mm, texture when probed dry, colony edge smooth, and surface appearance convex. The RT bacteria strain suppressed population growth of *N. oculata* (Fig. 4A, $P<0.001$), and completely destroyed axenically cultured *B. rotundiformis* population (Fig. 4B, $P<0.001$). On the other hand, presence of RT bacteria positively influenced population growth of axenically cultured *T. japonicus*, whereas these cultures did not persist without RT bacteria (Fig. 4C, $P<0.001$), and population growth of RT bacteria was greatly suppressed in mixed cultures with *T. japonicus*

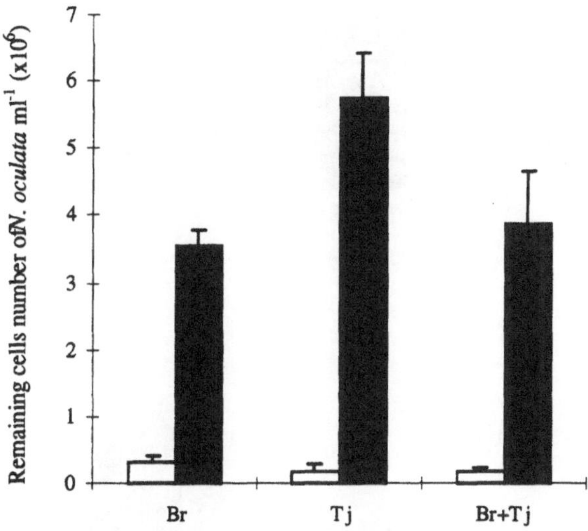


*Figure 3.* Leftover *Nannochloropsis oculata*, after the experiments in synxenic (white bars) and axenic (black bars) cultures. Br: *B. rotundiformis* in single culture, Tj: *T. japonicus* in single culture, Br+Tj: *B. rotundiformis* and *T. japonicus* in mixed culture.

(Fig. 4D, $P<0.05$). RT bacteria population growth increased when cultured with *N. oculata*.

## Discussion

One bacterial colony of a unique type (RT strain) was found to suppress population growth of *B. rotundiformis* (Figs. 1A and 4B) and *N. oculata* (Fig. 4A). RT bacteria thus appear to be harmful to *B. rotundiformis* and *N. oculata*, with an impact exceeding that of *T. japonicus*. Yu et al. (1990) showed a negative effect of bacteria *Vibrio alginolyticus* isolated from 150 $m^3$ rotifer mass culture tank on rotifer cultures. On the other hand, Yu et al. (1988) isolated eight strains, six of which were of the genus *Pseudomo-*

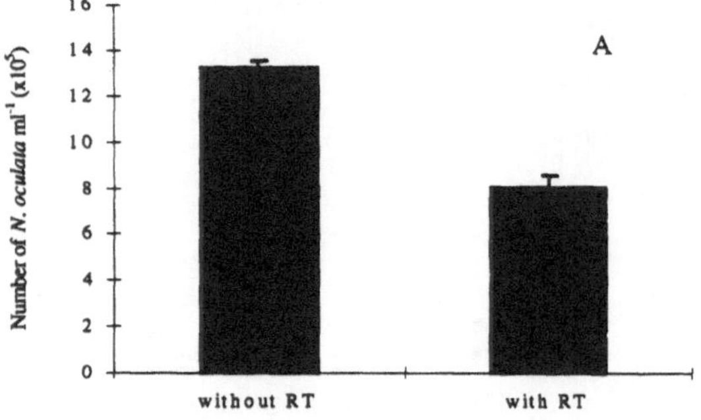


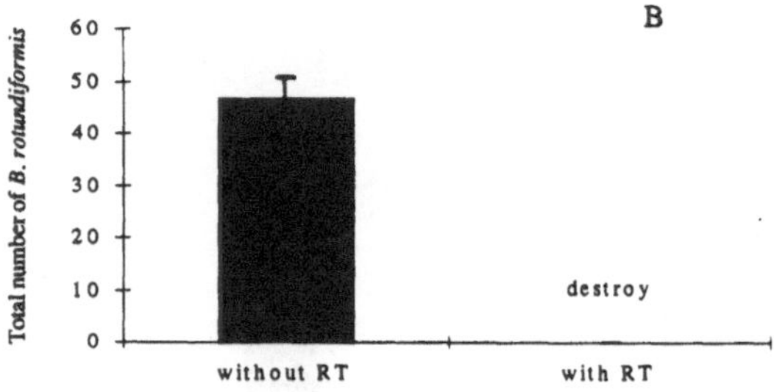


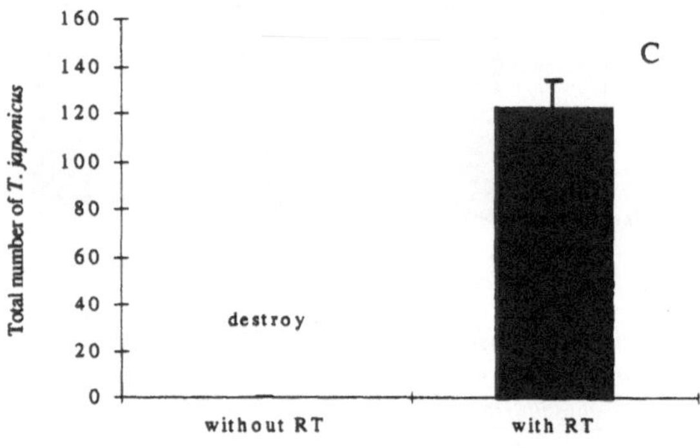


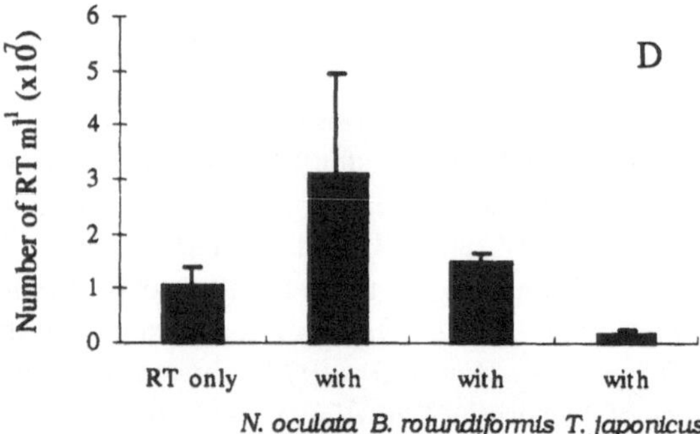


*Figure 4.* Relationship between RT bacteria strain and each tested organism in axenic cultures. A: *N. oculata,* B: *B. rotundiformis*, C: *T. japonicus*, D: RT bacteria strain, when cultured in isolation and with each of the testing organisms.

*nas*, of vitamin $B_{12}$ producing bacteria that positively affected population growth of rotifers. The rotifer population increased rapidly when vitamin $B_{12}$ producing *Pseudomonas* were supplied in a range of $10^7$–$10^{11}$ cells $ml^{-1}$. Moreover, Yasuda & Taga (1980) suggested that bacteria could serve as food for rotifers, promoting population growth.

*T. japonicus* grew smoothly when fed RT bacteria in synxenic cultures (Figs 2A and 4C). This result is similar to that of Ogawa (1977), who reported on feeding on bacteria by *Calanus*. *T. japonicus* ceased developing when cultured in axenic conditions and died at the nauplius stage, both in single or mixed cultures, although abundant *N. oculata* and rotifer feces remained present (Figs 2B and 3). This suggests that at least in the early nauplius stage of *T. japonicus* there is no competition for food (*N. oculata*) with *B. rotundiformis*. This contrasts with the report that *Calanus helgolandicus* could feed on living and dead phytoplankton cells, ocean detritus and it's own fecal pellets (Paffenhofer & Strickland, 1970). In the present study, nauplii of *T. japonicus* utilized RT bacteria rather then *N. oculata*, and growth of RT bacteria was greatly suppressed by *T. japonicus* (Fig. 4D). These results indicate that the aquatic bacteria serve as food for *T. japonicus*. *T. japonicus* fed on bacteria, directly (feeding to RT bacteria, Fig. 4C) and, indirectly, on feces, through bacterial growth on the latter. Furthermore, this study showed that RT bacteria had a beneficial effect from *N. oculata* and *B. rotundiformis* (Fig. 4D).

RT bacteria caused a decline in population growth of axenic *N. oculata* in this study (Figs 3 and 4A). Sukoso & Sakata (1996) maintained synxenic *Chattonella marina* cells for 30 days with several co-existing bacteria, whereas, axenically cultured *C. marina* populations decreased on day 10 and were exterminated by day 30 under bacteria-free culture conditions.

Various zooplankton species utilize aquatic bacteria or fecal bacteria as a nutrient source (Ogawa, 1977). In *T. japonicus*, the nauplii are apparently required to feed on bacteria, without which they are unable to develop into the copepodite stage. This may be a consequence of the primitive filter feeding apparatus of young *T. japonicus* nauplii (Fraser, 1936). The importance of aquatic bacteria as food for *T. japonicus* nauplii is best illustrated in Figs 2A and 4C, showing that young nauplii of *T. japonicus* only survive and grow to the copepodite stage in the presence of co-existing bacteria.

In the present study, we demonstrate that the presence or absence of bacteria in the aquatic environment can alter the relationship between species. Co-existing RT bacteria had a pivotal impact on the interspecific relation between *B. rotundiformis* and *T. japonicus* in experimental conditions. This indicates that co-existing bacteria alter the relations between different zooplankton organisms, and of these organisms with their environment.

## References


Burns, C. W. & J. J. Gilbert, 1986a. Effects of daphnid size and density on interference between *Daphnia* and *Keratella cochlearis*. Limnol. Oceanogr. 31: 848–858.

Burns, C. W. & J. J. Gilbert, 1986b. Direct observations of the mechanisms of interference between *Daphnia* and *Keratella cochlearis*. Limnol. Oceanogr. 31: 859–866.

Claus, G. W., 1978. Understanding microbes. W. H. Freeman and Company, New York, U.S.A. 547 pp.

Fraser, J. H., 1936. The occurrence, ecology and life history of *Tigriopus fulvus* (Fischer). J. mar. biol. Ass. U. K. 20: 523–536.

Gilbert, J. J. & R. S. Stemberger, 1985. Control of *Keratella* populations by interference competition from *Daphnia*. Limnol. Oceanogr. 30: 180–188.

Gilbert, J. J., 1988. Susceptibilities of ten rotifer species to interference from *Daphnia pulex*. Ecology 69: 1826–1838.

Gilbert, J. J., 1989. Competitive interactions between the rotifer *Synchaeta oblonga* and the cladoceran *Scapholeberis kingi* Sars. Hydrobiologia 186/187: 75–80.

Gilbert, J. J. & J. D. Jack, 1993. Rotifers as predators on small ciliates. Hydrobiologia 255/256: 247–25.

Hagiwara, A., K. Hamada, S. Hori & K. Hirayama, 1994. Increased sexual reproduction in *Brachionus plicatilis* (Rotifera) with the addition of bacteria and rotifer extracts. J. exp. mar. Biol. Ecol. 181: 1–8.

Hagiwara, A., M.-M. Jung, T. Sato & K. Hirayama, 1995. Interspecific relation between marine rotifer *Brachionus rotundiformis* and zooplankton species contaminating in the rotifer mass culture tank. Fish. Sci. 61: 623–627.

Jung, M.-M., A. Hagiwara & K. Hirayama, 1997. Interspecific interactions in the marine rotifer microcosm. Hydrobiologia 358: 121–126.

Ogawa, K., 1977. The role of bacteria floc as food for zooplankton in the sea. Bull. Japan Soc. Sci. Fish. 43: 395–407.

Pace, M. L. & J. D. J. Orcutt, 1981. The relative importance of protozoans, rotifers and crustaceans in a freshwater zooplankton community. Limnol. Oceanogr. 26: 822–830.

Paffenhofer, G. A. & J. D. H. Strickland, 1970. A note on the feeding of *Calanus helgolandicus* on detritus. Mar. Biol. 5: 97–99.

Sakamoto, H. & K. Hirayama, 1983. Dietary effect of *Thiocapsa roseopersicina* (Photosynthetic bacteria) on the rotifer *Brachionus plicatilis*. Bull. Fac. Fish. Nagasaki. Univ. 54: 13–20.

Sukoso & T. Sakata, 1996. Effect of co-existent bacteria on the growth of *Chattonella marina* in non-axenic culture. Fish. Sci. 62: 210–214.

Yasuda, K. & N. Taga, 1980. Culture of *Brachionus plicatilis* Muller using bacteria as food. Bull. Japan Soc. Sci. Fish. 46: 933–939.

Yu, J.-P., A. Hino, R. Hirano & K. Hirayama, 1988. Vitamin $B_{12}$-producing bacteria as a nutritive complement for a culture of the rotifer *Brachionus plicatilis*. Nippon Suisan Gakkaishi 54: 1873–1880.

Yu, J.-P., A. Hino, T. Noguchi & H. Wakabayashi, 1990. Toxicity of *Vibrio alginolyticus* on the survival of the rotifer *Brachionus plicatilis*. Nippon Suisan Gakkaishi 56: 1455–1460.

*Hydrobiologia* **446/447:** 129–137, 2001.
*L. Sanoamuang, H. Segers, R.J. Shiel & R.D. Gulati (eds), Rotifera IX.*



# Bed and Breakfast: the parasitic life of *Proales werneckii* (Ploimida: Proalidae) within the alga *Vaucheria* (Xanthophyceae: Vaucheriales)

Robert L. Wallace[1], Donald W. Ott[2], Sheri L. Stiles[2] & Carla K. Oldham-Ott[3]
[1]*Department of Biology, Ripon College, Ripon, WI 54971-0248, U.S.A.*
*E-mail: WallaceR@Mail.Ripon.EDU*
[2]*Department of Biology, The University of Akron, Akron, OH 44325-3908, U.S.A.*
[3]*Department of Biological Sciences, Kent State University, Kent, OH 44242-0001, U.S.A.*

*Key words:* Chrysophyta, gall formation, life history, Monogononta, Rotifera, SEM, TEM

## Abstract

The unusual parasitic association between *Proales werneckii* (Ehrenberg, 1834) (Ploimida: Proalidae) and the psychrophilic, coenocytic, filamentous alga, *Vaucheria* De Candolle, 1801 (Xanthophyceae: Vaucheriales), is documented using light and electron microscopy. A young female rotifer colonizes a *Vaucheria* filament (ca. 80 $\mu$m × 10 cm) by gaining entrance to the cell at a growing region where the wall has not yet matured. After achieving access to the cell, it disrupts development of either a gametophore or an apical tip by inducing cell hypertrophy and formation of an excrescent gall (ca. 80–120 × 140–1500 $\mu$m). Remaining within the vacuole of the gall for the rest of her life, the female deposits her eggs after feeding on the rich food supply furnished by the alga's organelles which continue to translocate within the cytoplasm. This abundant nourishment provides the adult rotifer with energy sufficient to permit an enormous reproductive potential (e.g. galls with >80 subitaneous eggs have been seen). Upon hatching, some offspring emigrate to other immature gametophores or growing tips within the same filament, while others may leave to colonize new filaments. Here, we present the life history of this parasitic relationship, highlighting previously unknown aspects of the association as well as noting differences among the accounts of previous authors. To our knowledge, this is the first description of this parasitic relationship to include electron photomicrographs.

## Introduction

Although no census has been taken of the some 2000 species of rotifers, relatively few are considered to be parasitic (May 1989). Of those parasitic associations that have been explored, the relationship between *Proales werneckii* (Ehrenberg, 1834) and species of the chrysophyte genus *Vaucheria* De Candolle, 1801 (Xanthophyceae: Vaucheriaceae) is well suited for additional study for a number of reasons. (1) Parasitized *Vaucheria* filaments are easily recognized by the presence of galls: short, excrescent, lateral branches, resembling flasks or cylinders. Unparasitized filaments lack these structures. (2) The *Proales–Vaucheria* association is wide-spread, being reported in both the northern and southern hemispheres (Verb et al., 1999). (3) *Vaucheria* is cultured with relative ease without the need for intensive maintenance schedules (Davis & Gworek, 1973; Stiles, 1978; Del Grosso, 1988; Verb et al., 1999). (4) Populations of *Proales* live within *Vaucheria* cultures at levels sufficient for detailed observational and experimental work. (5) *Proales werneckii* parasitizes both fresh and brackish water *Vaucheria* (Christensen, 1987) and *Dichotomosiphon*, a coenocytic chlorophyte (Del Grosso, 1988; cf. Davis & Gworek, 1973). (6) Another *Proales* species (*P. decipiens*) has been reported to parasitize the filaments of *Vaucheria* in a manner different from *P. werneckii* (Thompson, 1892).

Here, we document details of the life history of the *Proales–Vaucheria* association, highlight formerly unknown aspects, discuss differences among previously published accounts, and enumerate areas in need of additional research. To our knowledge, this is the first published report of this parasitism to include electron photomicrographs.

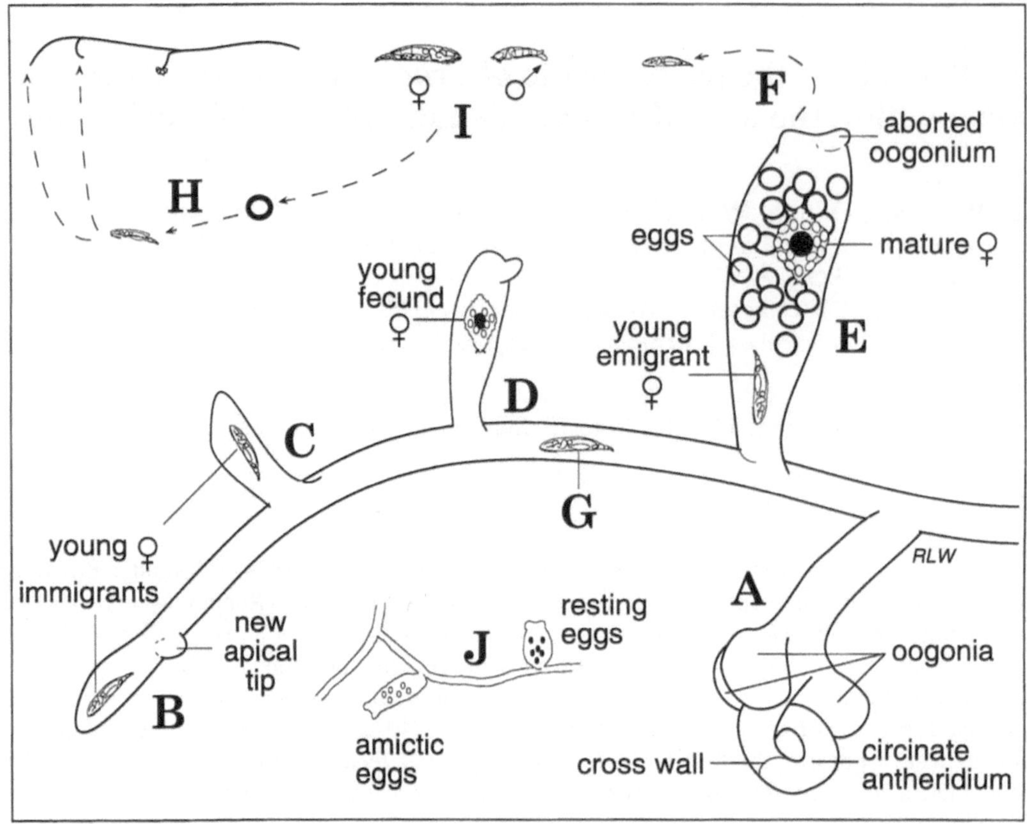


*Figure 1.* A schematic representation of the life history of the host-parasite relationship of *Vaucheria* (Chrysophyta: Xanthophyceae: Vaucheriaceae) and *Proales* (Proalidae). The elements in this representation are not drawn to scale. Letters in bold print refer to specific events in the life history of the parasite and its host (consult the text for a discussion of those events).

## Materials and methods

Specimens of *Vaucheria* infected with *Proales werneckii* (hereafter *Proales*) were collected from three habitats: (1) a bog (Firestone Park, Summit County; Ohio), (2) a shallow ditch (Ponchatoula, Tangipahoa Parrish; Louisiana), and (3) a small pond (Bath Nature Preserve; Summit County; Ohio). In the laboratory, algae were separated into small mats which were placed in culture dishes containing pond water. Small amounts of distilled water were added periodically to compensate for evaporation. For EM work we maintained our cultures at 10 °C with fluorescent lighting under a 12:12, light:dark photoperiod. Materials to be used for light microscopy work were maintained in the same way or were kept at ambient room conditions.

Galls of various ages were isolated and examined with the aid of a dissecting microscope. To better observe the rotifers, some galls were torn open with fine needles. All materials were either observed alive or were fixed with osmium tetroxide fumes before observation or before further processing for electron microscopy work (Stiles, 1978). Light microscopical observations were done using Zeiss Universal microscopes equipped with bright and dark field and Nomarski differential interference contrast.

## Biology of the parasitism

The following account records the life history of the *Proales-Vaucheria* association (Figs 1–15). In this narrative we use capital letters in bold print (**A–J**) to refer to specific events depicted in Figure 1. As *Proales* parasitizes several species of *Vaucheria* (Verb et al., 1999), our presentation is a generalization of the relationship.

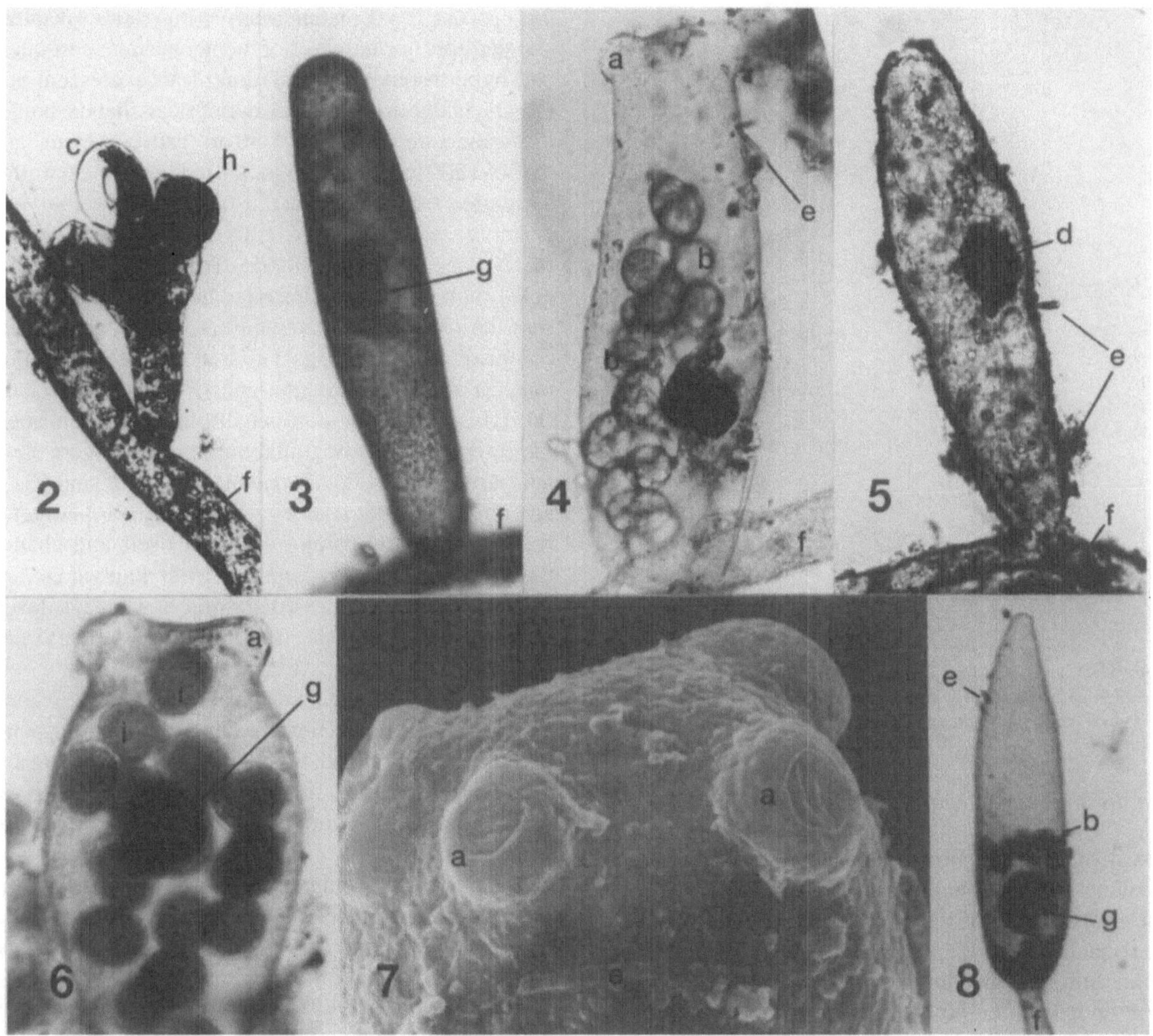


*Figures 2–8. Vaucheria* galls. **Figure 2**. Mature gametophore of *Vaucheria* with gametangia. (gametophore length ≈350 μm.) **Figure 3.** Young gall with a female, but no eggs are visible. (gall length ≈650 μm.) **Figure 4.** Old gall with mature *Proales* and ≈20 subitaneous embryos. (gall length ≈600 μm.) **Figure 5**. Very old, transparent, thick-walled gall containing few choroplasts. (gall length ≈1000 μm.) **Figure 6.** Gall with mature *Proales* and ≈14 smooth-walled resting eggs. (gall length ≈450 μm.) **Figure 7**. Scanning electron photomicrograph of the apex of a gall. (gall width ≈100 μm.) **Figure 8**. Gall developed at an apical tip with adult *Proales* and several embryos. (gall length ≈500 μm.) Symbols = (a) aborted oogonia; (b) amictic eggs; (c) circinate antheridium; (d) dead female; (e) epiphytes; (f) filament; (g) live female; (h) oogonia; (i) resting eggs.

## Vaucheria

*Vaucheria* is a psychrophilic, coenocytic, filamentous alga that lives in mats either submerged or on mud surfaces which are often flooded (Rieth, 1980; Bold & Wynne, 1985; Entwisle, 1988; Verb et al. 1999). Depending on the species, *Vaucheria* filaments range in size up to 80 μm in diameter and may be greater than 10 cm long. Filaments of uninfected specimens periodically bud transverse branches. If the alga produces reproductive structures, some of these branches may differentiate into gametophores, which if not colonized by *Proales*, will undergo normal maturation and, in due course, produce gametangia. Gametophore confirmation is a critical character in *Vaucheria* taxonomy. The mature gametophore illustrated in Figure 1 (**A**) possesses three oogonia and a circinate antheridium (see Fig. 2).

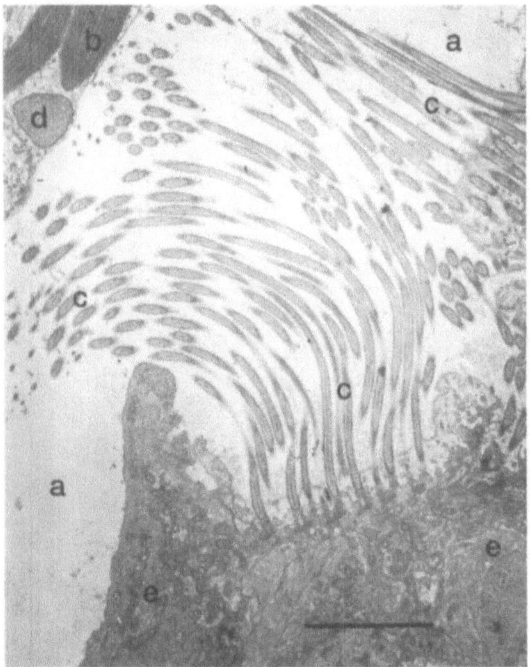


*Figure 9.* Anterior end of a *Proales* female in *Vaucheria* vacuole. (Bar=5 $\mu$m.) Symbols = (a) algal vacuole; (b) chloroplast; (c) cilia of *Proales*' corona; (d) mitochondrion; (e) rotifer corona.

## *Thallus invasion and gall formation*

Parasitization by *Proales* begins when a young female encounters a *Vaucheria* filament and locates a vulnerable spot whereby it gains entrance. To date, no one has explained how *Proales* locate the host filament. Some authors state that entrance is gained through open antheridia or other extant openings (Balbiani, 1878; Del Grosso, 1988), but others disagree (e.g. Rothert, 1896a,b) or at least remain neutral in their assessment (Davis & Gworek, 1973). It is most likely that entrance is secured at a growing region of the thallus where the wall, comprised of cellulosic microfibrils (Parker et al., 1963), has not yet matured (Weidner, 1952). This may be at an apical growing tip (**B**) or at the tip of a young gametophore (**C**). Apparently the rotifer eventually punctures the soft wall with her sharp unci after repeatedly projecting the trophi out of the mouth (Rothert, 1896a). The process of wound healing at this entry point has not been investigated nor has anyone explained how entrance into the alga is accomplished without permanently rupturing the plasma membrane.

Once a young *Proales* has invaded the thallus, it emigrates through the alga's vacuole to its final home, either young gametophore or an apical tip. Although the process is not understood, colonization of either site disrupts normal development, causing subsequent cell hypertrophy and formation of an excrescent gall (**B–E**). Galls are flask- to cylindrical-shaped bulges that range in size from $\approx$80 to 120 $\mu$m wide and $\approx$140–1500 $\mu$m long (Figs 3–8). These thick-walled structures (see below) are usually solitary, but two or more, closely-spaced galls may be confluent at their bases (Rothert, 1896a; Cribb, 1964). Mature galls, such as the one illustrated here (**E**), usually possess the founding female and her offspring; the later comprise numerous eggs and/or newborn emigrant female(s) about to leave the gall (Fig. 4). Some galls (**D**, **E**) have one or more small bumps at their apex; these represent the oogonial initials which were aborted early in their development (Figs 4, 6 and 7). If an emigrant rotifer reaches a mature gametophore, the rotifer apparently cannot establish itself and change the gametophore into a gall. To our knowledge, no research has been done to determine at what age developing gametophores are no longer susceptible to gall formation.

Galls that formed at an apical tip (**B**) lack the aborted oogonial initials (Fig. 8). Once an apical growing tip has been colonized, the filament stops growing and a new apical region may form at right angles to the previous growth. Vegetative growth will then continue from that point.

In our studies, the ratio of gall formation from aborted gametophores (**C–E**) to vegetative growing tips (**B**) was about 1.5–2.3:1, but we have not examined this aspect of the relationship closely (see also Davis & Gworek, 1973). Although *Proales* parasitizes both immature gametophores and the vegetative growing tips of *Vaucheria*, we have never seen colonization of zoosporangia.

## *Female* Proales

*Proales* females ($\leq$200 $\mu$m) live within the vacuole of *Vaucheria*, repeatedly invading the tonoplast with their oral apparatus to consume organelles and lipid droplets from the cytoplasm (Fig. 9). Exactly how this attack is accomplished without permanently rupturing this membrane is not understood. Figures 10–13 document, in part, digestion of plant organelles within *Proales*' stomach. While these figures show a progressive degradation of chloroplast integrity, digestion does not appear to be complete; as feeding continues, the female's gut grows much larger eventually form-

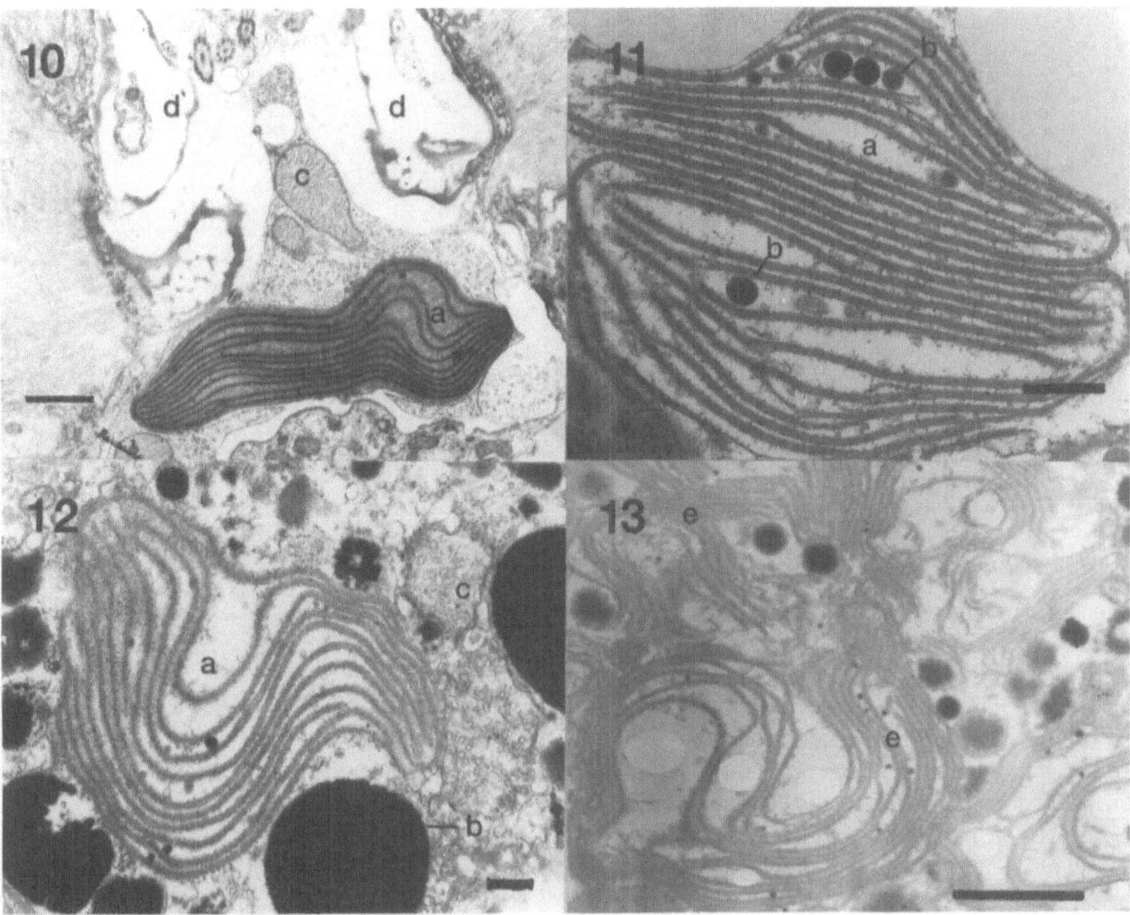


*Figures 10–13.* Digestion of plant organelles within the *Proales* gut. **Figure 10**. Intact chloroplast and mitochondrion within *Proales* pharynx. (Bar=2.5 $\mu$m.) **Figure 11**. Delamination of chloroplast thylakoid membranes. (Bar=1 $\mu$m.) **Figure 12**. Degradation of chloroplast and mitochondrial structure. (Bar=1 $\mu$m.) **Figure 13**. Disarticulated thylakoid membranes of several chloroplasts. (Bar=1 $\mu$m.) Symbols = (a) chloroplast; (b) lipid droplet; (c) mitochondrion (*Vaucheria*); (d) mastax; (e) thylakoid membranes.

ing a distended dark mass filled with partially digested plant organelles (Davis & Gworek, 1973).

Being illoricate, *Proales* possesses a flexible integument, and so adult shape is constantly changing as she squeezes past her eggs which are often tightly packed within the gall's vacuole. This flexibility also permits the female to enlarge to several times her original size as her gut becomes distended with algae materials. Some authors state that the females do not discharge fecal materials into the gall (e.g. Rousselet, 1897; Harring & Myers, 1924), but Davis & Gworek (1973) report that this does occur.

In most of our observations of galls, we have seen only one large, fecund female per gall (Figs 3–6 and 8); in very old galls, a dead female or her remains may be present along with assorted protists (Fig. 5). However, we also have seen up to six newborn emigrants within a gall along with numerous subitaneous eggs (**E**). Emigrating rotifers also may be seen moving slowly through the vacuole of the vegetative filament (see below; Brain, 1894; Rothert, 1896a). According to some authors (Rousselet, 1897), emigrants may leave a gall (**F**), but we have not witnessed this action. Once outside the filament, young females are reported to feed on tiny unicellular algae and debris (Davis & Gworek, 1973). To our knowledge, no one has confirmed their observation.

### *Rotifer movements*

Like Brain (1894), we have seen young females emigrating within the filaments (**G**), and like their counterparts in the galls, they feed on the organelles within the filament thereby developing the characteristic dark

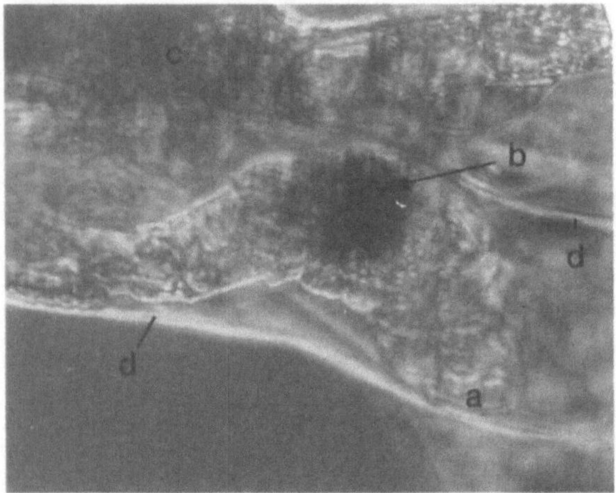


*Figure 14.* Emigrant rotifer within a filament. Symbols = (a) corona; (b) dark gut; (c) debris from another filament; (d) filament wall; rotifer length, ca. 175 $\mu$m.

gut (Fig. 14). While we have not seen these emigrants walled off within the filament, Davis & Gworek (1973) report that this can occur. Immigrants forming galls in a new section of the filament may have hatched from a subitaneous egg laid by the founding female in an older gall, immigrated from a different filament (**F**), or hatched from a resting egg (**H**). Bacterial decomposition of the aborted oogonial initials (Weidner, 1952) may provide egress for emigrants.

*Rotifer reproduction*

The reproductive potential of *Proales* in established galls (**E**) seems to be enormous. We routinely have seen galls with clutches of 20 or more subitaneous eggs (Fig. 4) and, while not common, we also have seen galls with more than 80 subitaneous eggs. In some cases, the eggs are so numerous that they are displaced from the gall into the filament itself (Davis & Gworek, 1973; Christensen, 1987). However, reports differ as to whether such large clutches represent the reproductive effort of a single female (Rothert, 1896a) or the collective efforts of several females who have sequentially deposited their eggs into the gall (Rousselet, 1897). It may well be possible that large clutches are deposited by a single female. We posit this because (1) it is very rare for more than one female to occupy a gall (Budde, 1925; Weidner, 1952; Figs 3–6 and 8) and (2) dead, mature females do not appear to decompose very rapidly within their galls (Fig. 5). In certain cases, females may deposit their eggs in the vacuole of the vegetative filament, but this appears to be rare as we have seen it only in very old, senescent cultures.

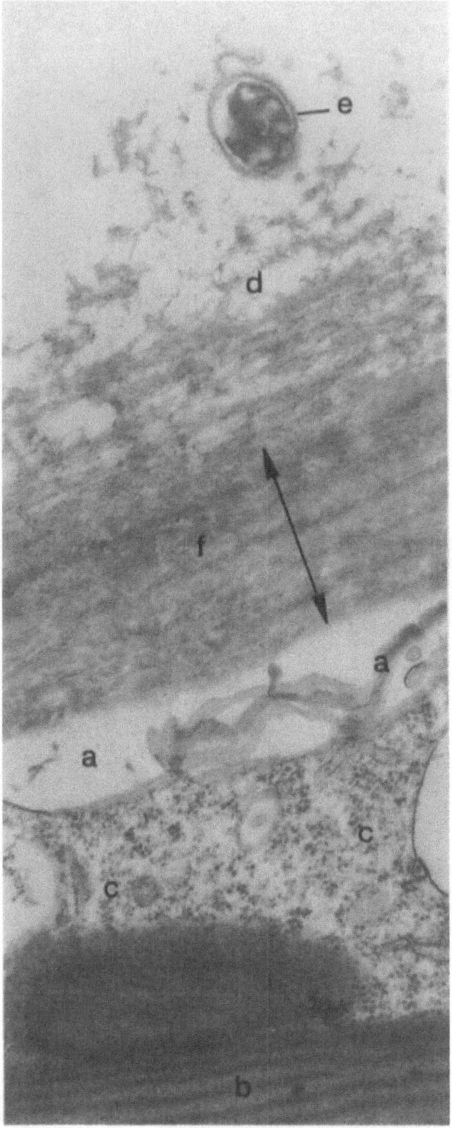


*Figure 15.* Cell wall of an old *Vaucheria* gall with about 7 additional walls. Symbols = (a) artifact in preparation (folded membrane, space); (b) chloroplast's thylakoid membranes; (c) cytoplasm; (d) epiphytic gelatinous material; (e) epiphyte; (f) supernumerary cellulosic walls; wall width as shown by arrow ≈1.25 $\mu$m).

Rotifer reproduction appears keyed to the growth of this psychrophilic alga, with much of the asexual reproduction being accomplished in the colder months followed by resting egg production (Rothert, 1896b). Unfortunately, the location where mating takes place remains unclear. Mixis has been reported to occur outside of the alga (**I**) with the resting eggs spending diapause within the sediments (Del Grosso, 1988; Koste & Shiel, 1990). However, several authors re-

*Table 1.* Questions regarding the *Proales-Vaucheria* association requiring additional research

| Questions needing additional research | Selected references |
|---|---|
| **Colonization of the alga** | |
| How do newborn or emigrant females gain entrance into the alga: is it through (i) the soft immature walls of growing tips or of immature gametophores (**H**), (ii) the aborted oogonial initials of galls (Figs 4, 6 and 7), or (iii) accidentally through existing damaged regions of the alga? | Balbiani (1878); Debray (1890); Rothert (1896a,b); Rousselett (1897); Gabriel (1922a); Budde (1925); Davis & Gworek (1973); Del Grosso (1988). |
| How do emigrant females emerge from the galls (**F**): is this due to action of the rotifer's trophi or is it through the aborted oogonial initials (Figs 4, 6 and 7)? | Lister (1884); Debray (1890); Brain (1894); Rothert (1896a,b); Rousselett (1897); Davis & Gworek (1973); this study. |
| What is the ratio of newborn females that leave the parental filament (**F**) to those that remaining within it (**G**)? | Lister (1884); Brain (1894); Thompson (1892). |
| Does *Proales* have a preference for immature gametophores (**C**) over apical tips (**B**)? | This study. |
| Does *Proales* have a preference for certain species of *Vaucheria*? | Lister (1884); this study. |
| **Rotifer feeding** | |
| Do the young females, that are not within *Vaucheria*, feed? How efficient is the digestion of the algal organelles? | Rothert (1896b); Rousselett (1897); Budde (1925); Weidner, (1952); Davis & Gworek (1973). |
| Does the anus of the female function or not? Do males feed? | Rousselett (1897); Wesenberg-Lund (1923); Weidner (1952); Davis & Gworek (1973). |
| **Rotifer biology** | |
| How does the rotifer withstand osmotic differences between freshwater and the alga's vacuole? Are life history events in the rotifer keyed to the alga's physiology? | To our knowledge no one. |
| **Rotifer sexuality** | |
| When does mixis occur within the life history of the host? What controls mixis? Does mixis take place externally (**I**) and/or within the galls (**J**)? | Balbiani (1878); Rothert (1896a,b); Rousselett, (1897); Gabriel (1922b); Budde, (1925); Weidner (1952); Del Grosso (1988); this study. |
| Are the spinous resting eggs which are occasionally present those of a different species? Is *Proales* amphoteric or do the resting eggs within the galls have a pseudosexual etiology (**J**)? | Debray (1890); Rieth (1980); this study. |
| **Plant responses** | |
| Does this parasitism (i) significantly affect algal growth or reproduction, (ii) increase the probability of parasitic attack on *Vaucheria* by other organisms, or (iii) cause *Vaucheria* to initiate a defense against the rotifer? | Wollny (1879); Cribb (1964); Davis & Gworek, (1973). |
| How is gall formation induced by *Proales* and does it differ in gametophores (**C**) and apical growing tips (**B**)? | Rothert, (1896a); Gabriel (1922a,b); Davis & Gworek (1973); Ott (1977); this study. |
| How does *Proales* induce deposition of supernumerary cell wall material in the gall (Fig. 15)? | Davis & Gworek (1973); Ott (1977); this study. |

Questions posed here have not been completely resolved because they have (i) not yet been thoroughly investigated, (ii) various authors strongly contradict one another's observations, or (iii) the research has not be reported. (Letters in boldface type are keyed to Fig. 1; see the text for discussion.)

port galls with both subitaneous and resting eggs or subitaneous and male eggs (e.g. Rothert, 1896a; Rousselet, 1897; Budde, 1925; Weidner, 1952). In one instance, we found a gall with subitaneous eggs in close proximity (ca. 1300 μm) to another possessing resting eggs (**J**; Figs 4 and 6, respectively). The reports of spined resting eggs may indicate parasitization by another species or development of pseudosexual eggs (De Smet, 1996).

It is generally assumed that the resting eggs are freed when the gall decays and the tip falls off (Harring & Myers, 1924; Koste & Shiel, 1990; De Smet, 1996). However, Budde (1925: Abb. 5) provides a photomicrograph reportedly showing the egg cases of resting eggs that have hatched within a gall. In any case, it follows that when a resting egg hatches (**H**) the young *Proales* finds a suitable host filament and establishes a new infection.

*Damage to the alga*

The most striking aspect of the parasitism is gall formation, but even so, no one has attempted to quantify the impact this has on the algae. As the galls age, they eventually lose their chloroplasts becoming so transparent that the inhabitants (i.e. adult, juveniles, embryos, protists) may be easily seen within the gall (compare Figs 3–5). Davis & Gworek (1973) assert that the gradual loss of chloroplasts is due to the intense feeding by *Proales*. We disagree and suggest that at least some of this loss of chloroplasts occurs through the normal algal processes of continuously translocating its organelles from older regions toward the growing tips.

Normal gametophores deposit a cell wall comprising a single layer about 200 nm thick, but galls usually lay down supernumerary, cellulosic walls with a small space between each layer (Fig. 15). The result of this is that total wall thickness increases as a function of gall age (Weidner, 1952; Ott, 1977; cf. Christensen, 1987). We have found galls with as many as 25 additional layers of cell wall material comprising a total wall thickness of approximately 7 μm. The result of these over-thickened walls is that a coverglass will break before the gall is crushed should sufficient pressure be applied to a microscopic preparation (Ott, pers. obs.). We speculate that *Proales* feeding stimulates the abundant dictyosomes and their associated mitochondria to lay down additional cell wall material. While septa are often visible in filaments that have been damaged by physical injury or through invasion by fungal parasites, we have not seen such cross walls in *Vaucheria* that have been parasitized by *Proales* (cf. Davis & Gworek, 1973; Del Grosso, 1988).

When a *Proales* infection begins in a mat of *Vaucheria* it can spread rapidly (Debray, 1890) and last the entire growing season of the alga (Verb et al., 1999; this study). *Vaucheria* cultures and perhaps natural populations can be subject to extirpation should the *Proales* infestation become severe (Davis & Gworek, 1973; Ott, pers. obs.).

*Unresolved issues*

This account stresses that knowledge regarding the *Proales-Vaucheria* association is far from complete. Besides the controversial points discussed in the narrative, there are several other areas that have yet to be thoroughly explored. These issues are summarized in Table 1. Continued study of the *Proales-Vaucheria* association will clarify other aspects of rotifer biology and biogeography, as well as provide new insights into the disciplines of parasitology, phycology and cell biology. In future contributions on this topic, we expect to elucidate the details of the subjects we have introduced here. We hope that this contribution will encourage others to join our efforts in examining the life history of this amazing relationship.

## Acknowledgements

We thank our respective institutions for the support they provided us in the preparation of this work. Elements of the research discussed here were originally described in the graduate thesis submitted by Sheri Lynn Stiles in partial fulfillment of the M.Sc. degree from the University of Akron. Douglas Light, Margaret Stevens, Russ Shiel and an anonymous reviewer read and improved this manuscript. We also thank Giulio Melone for pointing out an important reference that we had overlooked and Ramesh Gulati for his suggestion to change our original title.

## References

Balbiani, E. G., 1878. Observations sur le Notommate de Werneck et sur son parasitisme dans les tubes des Vauchéries. Ann. Sci. Nat.; Zool. Biol. Anim. Paris (ser.6) 7: 1–40.

Bold, H. C. & M. J. Wynne, 1985. Introduction to the Algae. 2nd edn. Prentice-Hall, Englewood Cliffs, NJ.

Brain, J. L., 1894. An inhabitant of *Vaucheria*. Science-Gossip 1: 201–202.

Budde, E., 1925. Die parasitischen Rädertiere mit besonder Berücksichtigung der in der Umgegend von minder I. W. beobachteten Arten. Z. Morph. Ökol. Tiere 3: 706–784.

Christensen, T., 1987. Some collections of *Vaucheria* (Tribophyceae) from south-eastern Australia. Aust. J. Bot. 35: 617–630.

Cribb, A. B., 1964. An occurrence of rotifers within *Vaucheria*. Qld. Nat. 17: 74.

Davis, J. S. & W. F. Gworek, 1973. A rotifer parasitizing *Vaucheria* in a Florida spring. Trans. am. Microsc. Soc. 92: 135–140.

De Smet, W. H., 1996. Rotifera Volume 4: The Proalidae (Monogononta). In Dumont, H. J. (ed.), Guides to the Identification of the Microinvertebrates of the Continental Waters of the World. SPB Academic Publishers, The Hague, The Netherlands: 102 pp.

Del Grosso, F., 1988. On the formation of galls in *Dichotomosiphon tuberosus* (A. Br.) Ernst (Chlorophyceae, Dichotomosiphonales), produced by *Proales werneckii* (Ehrenberg) Hudson et Gosse. Nova Hedwigia 46: 157–164.

Debray, F., 1890. Sur *Notommata werneckii* Ehrenberg, Parasite des Vaucheriees. Bull. Biol. Fr. Belg. 22: 222–242.

Entwisle, T. J., 1988. A monograph of *Vaucheria* (Vaucheriaceae, Chrysophyta) in south-eastern mainland Australia. Aust. Syst. Bot. 1: 1–77.

Gabriel, C., 1922a. Cécidies de *Vaucheria aversa* produites par *Notommata werneckii*. Compt. Rend Séanc. Soc. Biol. (Paris) 86: 453–455.

Gabriel, C., 1922b. La ponte de *Notommata werneckii* dans les galles de *Vaucheria aversa*. Compt. Rend Séanc. Soc. Biol. (Paris) 86: 696–698.

Harring, H. K. & F. J. Myers. 1924. The rotifer fauna of Wisconsin II: revision of the Notommatid rotifers, exclusive of the Dicranophorinae. Trans. Wisc. Acad. Sci. Arts Lett. 21: 415–49.

Koste, W. & R. J. Shiel, 1990. Rotifera from Australian inland waters. VI. Proalidae, Lindiidae (Rotifera: Monogononta). Trans. R. Soc. S. Aust. 114: 129–143.

Lister, A., 1884. On the parasitism of rotifers on *Vaucheria*. Proc. Essex Nat. Field Club 3: 45–48.

May, L., 1989. Epizoic and parasitic rotifers. Hydrobiologia 186/187: 59–67.

Ott, D. W., 1977. Ultrastructural observations on parasitism of *Vaucheria prona* by *Proales werneckii*. J. Phycol. 13 (suppl.): 51.

Parker, B. C., R. E. Preston & G. E. Fogg, 1963. Studies of the structure and chemical composition of the walls of Vaucheriaceae and Saprolegniaceae. Proc. R. Soc., B. 158: 435–445.

Rieth, A., 1980. Xanthophyceae. Part 2. In Ettl, H., J. Gerloff & H. Heynig (eds), Süsswasserflora Von Mitteleuropa. Vol. 4. Stuttgart, Gustav Fisher Verlag: 147 pp.

Rothert, W., 1896a. Über die Gallen der Rotatorie *Notommata werneckii* auf *Vaucheria walzi* n. sp. Jahrb. Wiss. Bot. 29: 525–594.

Rothert, W., 1896b. Zur Kenntniss der in *Vaucheria*-Arten parasitirenden Rotatorie *Notommata werneckii* Ehrenberg. Zool. Jahrb., Abt. Syst. 9: 672–713.

Rousselet, C. F., 1897. On the male of *Proales wernecki*. J. Quek. Micr. Cl. Ser. 2, 6: 415–418.

Stiles, S.L., 1978. Light and electron microscopy of the parasitic rotifer, *Proales werneckii* (Ehrenberg). M.Sc. Thesis. University of Akron, Akron, Ohio.

Thompson, P. G., 1892. Notes on the parasitic tendency of rotifers of the genus *Proales*; with an account on a new species. Hardwicke's Science-Gossip 28: 219–221.

Verb, R. G., M. L. Vis, D. W. Ott, & R. L. Wallace, 1999. New records of *Vaucheria* species (Xanthophyceae) with associated *Proales werneckii* (Rotifera) from North America. Cryptogamie Algol. 20: 67–73.

Weidner, H., 1952. *Proales wernecki*, ein in *Vaucheria* parasitierendes Rotator. Nachricht Mus. Stadt Aschaffenburg 35: 39–46.

Wesenberg-Lund, C., 1923. Contributions to the biology of the Rotifera. I. The males of the Rotifera. Kgl. Dansk. Vidensk. Selsk. Skrift., Naturvid. Afd. 8, Reakke IV(3): 191-345.

Wollny, R., 1879. Parasitism of *Notommata* on *Vaucheria*. J. R. Microsc. Soc. 2: 291.

Hydrobiologia **446/447**: 139–148, 2001.
L. Sanoamuang, H. Segers, R.J. Shiel & R.D. Gulati (eds), Rotifera IX.




# The ecology of periphytic rotifers

Ian C. Duggan
*Department of Biological Sciences, The University of Waikato, Private Bag 3105, Hamilton, New Zealand*

*Key words:* rotifers, periphyton, littoral, macrophytes, substrate

## Abstract

The ecology of rotifer assemblages in the periphyton has received little attention relative to that of pelagic rotifers. This paper reviews the ecology of periphytic rotifers, with particular emphasis on the role of macrophytes in the structuring of rotifer assemblages spatially and temporally, and compares these aspects with the dynamics of better known pelagic rotifer communities. Littoral rotifer periphyton communities are typically diverse in lakes, and have composition dissimilar to that of the open water. In rivers, diversity and composition in the pelagic and littoral appear to be similar. Rotifers show preference for macrophyte species they associate with, probably through differences in physical structure or complexity, food concentration or composition, chemical factors, macrophyte age, and differences in the degree of protection from predation provided by macrophytes. These mechanisms are in general not well investigated in rotifers. Factors affecting the seasonal dynamics of periphytic communities appear to be similar to pelagic communities, with seasonal dynamics of substrates and disturbance by flooding or drying also being important.

## Introduction

The periphyton is a complex community of microbiota (bacteria, fungi, algae and animals) that is found on, or in association with, inorganic or organic substrata (Wetzel, 1983). Rotifers have long been known to have diverse and abundant communities on aquatic macrophytes and other substrates, and the majority of known species are found in these habitats. Although some are sessile as adults (e.g. Wallace, 1980), most are usually not firmly attached but instead move over the surfaces of the substrate. Despite the diversity and abundance of these assemblages, their ecology has received little attention relative to that of planktonic rotifers, possibly because they are more difficult to sample quantitatively. Rotifers may be functionally important in periphytic communities, however, as conduits of energy transfer to higher consumers in food webs. Rotifer grazing also may be important for increasing macrophyte growth by suppression of periphyton, as by larger grazers (e.g. molluscs) (e.g. Rogers & Breen, 1983; Jones et al., 1999). In turn, aquatic macrophytes are likely to play an important role in the structuring of rotifer communities, acting as substrates, providing food sources in the form of various components of the periphyton, and by influencing interactions between predators and prey. This paper reviews the ecology of periphytic rotifers, with particular emphasis on the role of macrophytes in affecting the diversity and structuring of rotifer assemblages spatially and temporally, and compares these aspects with the dynamics of the comparatively better known pelagic communities. The focus is mainly on the more mobile epiphytic rotifers that move over the surface of, or live in close association with, macrophytes. Sessile rotifers, which differ from the other periphytic forms in their lack of mobility, and generally also in their modes of feeding, were reviewed by Wallace (1980) and Hutchinson (1993) and so will not be examined in detail. Communities associated with moss are not treated here because this generally is a separate, specific habitat. By way of introduction, the communities of vegetated and non-vegetated habitats will be compared. The variation within vegetated habitats will then be examined.

## Diversity and composition of littoral rotifers

The littoral region is often the most diverse part of

*Table 1.* Diversity of rotifers (number of taxa recorded) in the vegetated littoral and plankton of lakes

| Lake | Littoral | Plankton | |
|---|---|---|---|
| Deggs Lake | 19 | 9 | Pennak (1966) |
| Frentus Lake | 11 | 11 | " |
| Highland Lake | 14 | 7 | " |
| Little Swede No. 1. Lake | 11 | 3 | " |
| Parvin Lake | 7 | 4 | " |
| Standley Lake | 8 | 7 | " |
| Hayden's Lake | 11 | 7 | " |
| Bedrock Lake | 7 | 4 | " |
| Triangle Lake | 10 | 5 | Havens (1991) |
| Lake Rotomanuka | 58 | ~40 | Duggan et al. (1998) |

lake or river ecosystems, commonly supporting a variety of macrophytes, their associated microflora and a large number of animal species. The presence of macrophytes has been found to affect the diversity and composition of rotifer communities. Diversity is often high in the vegetated habitats compared with the pelagic (Table 1). Pennak (1966), examining the rotifers in the plankton and macrophyte zones of a number of Colorado lakes, U.S.A., distinguished an equal or greater diversity of taxa in the littoral than plankton. He notes, however, that his sampling technique, a littoral sampling tube (see Pennak, 1962), was less effective in removing rotifers from macrophytes than were nets, so that he is likely to have underestimated both littoral diversity and the difference between littoral and limnetic habitats. In Triangle Lake, Ohio, U.S.A., during summer, Havens (1991) obtained similar results. These results represent a limited number of samples spatially and temporally. Duggan et al. (1998) found 58 species in the littoral of Lake Rotomanuka, New Zealand, across a transect sampled over a period of 10 months. This diversity exceeds that found in the plankton of well studied New Zealand lakes, which commonly have an $\alpha$ diversity of around 40 species (Sanoamuang, 1992; author's unpublished data). It also exceeds diversities recorded in planktonic communities elsewhere, which commonly have fewer than 25 species over a year (e.g. Elliott, 1977; Berner-Fankhauser, 1983; Vasconcelos, 1990; Andrew & Fitszimmons, 1992). This greater richness in the littoral than the pelagic reflects the more heterogeneous nature of the environment and hence greater niche diversity (e.g. Havens, 1991; Duggan et al., 1998).

There are also compositional differences between communities of vegetated and non-vegetated habitats. Species have long been recognised as having preference for one or other habitat (e.g. Hudson & Gosse, 1889). The diversity and density of submerged vegetation were among the most important factors found to affect the distribution of rotifers between lakes and other waterbodies by Morales-Baquero et al. (1989) and Jersabek (1995), even when lakes had similar morphometric characteristics. Both found a decreasing importance of typically benthic and periphytic species with decreasing development of vegetation. Similarly, Green (1972) and Bonecker et al. (1998) found correlations between rotifer composition and macrophyte development in small lakes and rivers of the Mato Grosso, Brazil and Langley et al. (1995) found the percentage of macrophyte cover to be important in determining rotifer distribution in English ponds. On the roots of floating vegetation in a Brazilian lagoon, Koste (1974) recorded many species that had not been recorded previously from the lower Amazon basin, and considered this to be due to lack of sampling from periphytic habitats.

Within lakes, rotifer community composition is also commonly markedly different between adjacent littoral and pelagic zones. Based on his findings in Colorado lakes, Pennak (1966) divided rotifers into three groups of differing habitat affinities: (1) Limnetic species that are never abundant in vegetation, (2) species common in both vegetation and open water, and (3) species essentially restricted to vegetation, seldom found in open water (Table 2). He listed only three species as chiefly limnetic, and 21 species essentially restricted to the vegetation, including five *Lecane* and seven *Trichocerca*. In Triangle Lake, Havens (1991) also found differences in the composition and dominance of species between the littoral and limnetic, with *Kellicottia bostoniensis, Synchaeta oblonga* and *Polyarthra vulgaris* dominating the zooplankton in the limnetic, but rare in the littoral. Of the 58 species recorded in the Lake Rotomanuka littoral by Duggan et al. (1998), no *Asplanchna, Filinia* or *Conochilus* species were found, although *Asplanchna priodonta* and two species of both *Conochilus* and *Filinia* are present in the plankton (author's unpublished data). Of 58 species found in the littoral, 10 were *Trichocerca*, eight *Lecane* and five were *Cephalodella*.

Much of the difference in species composition between vegetated and non-vegetated regions is likely to come about because of the morphological adaptations required for life in these habitats (Ruttner-Kolisko, 1974). For example, periphytic species lack

*Table 2.* Pennak's (1966) habitat affinities of rotifers

| Essentially restricted to vegetation; seldom in open water | Common in both vegetation and open water | Chiefly limnetic; never abundant in vegetation |
|---|---|---|
| *Colurella colurus* | *Asplanchna brightwelli* | *Filinia longiseta* |
| *Euchlanis dilatata* | *A. priodonta* | *Hexarthra mira* |
| *Lecane bulla* | *Brachionus angularis* | *Kellicottia longispina* |
| *L. mira* | *B. calyciflorus* | |
| *L. ohioensis* | *Conochilus unicornis* | |
| *L. papuana* | *Keratella cochlearis* | |
| *L. quadridentata* | *K. quadrata* | |
| *Lepadella patella* | *L. luna* | |
| *Mytilina mucronata* | *L. lunaris* | |
| *Plationus patulus* | *Polyarthra minor* | |
| *Platyais quadricornis* | *P. vulgaris* | |
| *Scaridium longicaudum* | *Pompholyx complanata* | |
| *Testudinella patina* | | |
| *Trichocerca elongata* | | |
| *T. gracilis* | | |
| *T. intermedia* | | |
| *T. longiseta* | | |
| *T. rattus* | | |
| *T. rosea* | | |
| *T. similis* | | |
| *Trichotria tetractis* | | |

the structures that increase buoyancy in planktonic species, such as spines or setae, they are known to weigh relatively more, and in general do not carry their eggs (Ruttner-Kolisko, 1974; Dumont et al., 1975). Periphytic species also have particularly well developed foot regions and adhesive organs, while in planktonic species, the foot is commonly lost (De Beauchamp, 1909; Pejler, 1962; Ruttner-Kolisko, 1974). Behavioural adaptations have also developed to ensure that species remain in their preferred habitat, with planktonic species found to actively avoid the shore. Preissler (1977a, b), in experiments carried out in a circular perspex arena in the shore region of different lakes, found that the planktonic species *Asplanchna priodonta* and *Synchaeta pectinata* swam away from the shore, and Saunders-Davies (1989) found that *Keratella cochlearis* and *Polyarthra vulgaris* densities increased with distance from the shore in a small eutrophic lake. Preissler (1980) provided evidence that avoidance of the shore was a result of optical orientation, with planktonic rotifers avoiding shading. Preissler was unable to test for optical orientation in the periphytic forms, although Clément (1977) has shown the periphytic *Notommata copeus* to be positively phototactic when hungry, and for the proportion observed swimming or crawling to be dependent on wavelength. Positive phototaxis may, however, reduce the chance of these animals remaining in the littoral. No other behavioural adaptations of periphytic rotifers are known that may aid in maintaining their position in the littoral. Donner (1964) notes, however, that some periphytic species actively cling to substrates and most contract when disturbed, behaviours which are both likely to prevent species entering the pelagic.

It therefore appears usual for lake margins with vegetation to have higher diversity and different species composition of rotifers than the limnetic zone, although there are exceptions to this. For instance, Goddard & McDiffett (1983) found that in Montandon Marsh, U.S.A., diversity was greater in the plankton than the vegetated littoral. In sparsely or non-vegetated margins, limnetic species dominate, and few typically littoral taxa (e.g. *Lecane, Trichocerca* and *Cephalodella*) are found (Pennak, 1966; Lemley & Dimmick, 1982; Sollberger & Paulson, 1992). Also, diversity and composition of the periphyton and plankton of rivers appears to be similar (e.g. Kofoid, 1908; Meuche, 1939; Green, 1960; Pejler, 1962; Donner, 1970; Modennuti & Claps, 1988). However, in contrast to unvegetated lake littorals, periphytic forms are most common in the potamoplankton, commonly including many *Lecane* and/or *Trichocerca* species (e.g. Ferrari et al., 1989; Onwudinjo & Egborge, 1994; Kobayashi et al., 1998, Modenutti 1998). These species appear to be commonly derived from communities from shore impoundments and marginal vegetation that have been washed into the river (Kofoid, 1903; Green 1960).

## Dynamics in vegetated habitats

### *Substrate specificity*

Pejler (1962) divided rotifers into six ecological groups based on their dependence to substrate: (1) planktonic forms, (2) benthic-periphytic forms that swim about more or less freely, which have no great dependence on the substrate (e.g. *Testudinella parva, Trichotria pocillum, Lepadella patella*), (3) Mainly periphytic forms that are more intimately connected with their substrate, often attached by means of their toes or anchoring filaments (e.g. *Euchlanis dilatata, E. meneta, Trichocerca rattus*), (4) sessile forms, tied to the vegetation and occasionally dependent on a special substrate, (5) psammonic forms, growing in or moving through sand, and (6) loose sediment forms. For those in close association with macrophytes, Pennak (1966) distinguished similar groups: (1) limnetic species that invade the littoral but do not browse there, (2) chiefly limnetic species that may occasionally revert to browsing (e.g. *Brachionus calyciflorus*), (3) browsing species that leave the substrate frequently and cruise about in the spaces between vegetation (e.g. *Lecane lunaris*), (4) browser species that rarely leave the substrate (e.g. *Colurella colurus*), and (5) sessile forms. Except for the most mobile (planktonic) forms and the sessiles, these categories are not sharply defined from one another. Many species are unlikely to fit exactly into the intermediate categories, with these likely to be all intergrades in the way of relative browsing, with less mobile forms spending relatively more time browsing and less time filter feeding than those more mobile (Pejler, 1962; Pennak, 1966). To find instances of dependence on substrate, Pejler (1962) suggested that the less mobile creeping (e.g. bdelloids, *Euchlanis* species) or sessile forms be investigated, with more freely swimming forms (e.g. *Testudinella parva, Trichotria pocillum, Lecane lunaris*) expected to be less dependent on their substrate. Therefore, the more mobile a species, the better it can tolerate different kinds of substrate. Of the rotifers in the periphyton, the sessiles appear to be the most restricted to their substrate. For example, Wallace (1977a, b) and Wallace & Edmondson (1986) note that many (but not all) sessile species are confined to particular or few macrophyte species. *Cupelopagis vorax*, for example, is predominantly distributed on broad flat macrophyte surfaces, *Stephanoceros fimbriatus* on plants with dissected leaves (e.g. *Utricularia*), and *Collotheca gracilipes* has preference for *Elodea*. Because sessile rotifers are immobile as adults, their reproductive success is dependent on the suitability of the substrate (or the environment surrounding the substrate) chosen at the larval stage (Wallace, 1977a). Habitat selection is thus important in many sessile species. Unlike the sessile rotifers, if a habitat of a motile organism deteriorates, it is able to emigrate to a more favourable locality. The dependence of the mobile periphytic rotifers on their substrate is, therefore, likely to be less absolute. Pejler (1962) found many species that are periphytic in habit to be also found in the benthos, although generally restricted to firmer substrates, with absolute restriction to one or other habitat apparently rare. However, he did note that some species, e.g. *Euchlanis dilatata, E. meneta, E. alata, E. deflexa, Colurella colurus, Proales theodora, Trichocerca uncinata, T. longiseta* and *T. rattus*, had a more or less marked preference for macrophytes over the benthos (or alternately an aversion to the benthos). Pejler & Berzins (1993e) also noted that trichocercids generally prefer periphytic environments to the benthos.

Pejler & Berzins (1989, 1993a–e, 1994) published a number of papers on habitat or substrate specificity of many of the mobile periphytic and benthic rotifer families and genera. Comparison was made of the percentage occurrences of rotifer species on different substrates. In the groups examined (Brachionidae, *Cephalodella*, Colurellidae, Dicranophoridae, *Lecane*, Trichocercidae, and also the bdelloids; Pejler & Berzins, 1989, 1993b–e, 1994), there was in general an overall low preference for macrophyte species, morphology or substrate type by species. However, using this approach, Pejler & Berzins (1993a) also found that sessile rotifers showed no real preference for substrate, except that a higher number of species was found on *Utricularia*. Based on these findings, Pejler & Berzins (1995) suggested that most periphytic species do not show any obvious selectivity for the macrophytes they colonize. They did note that some macrophytes house a higher abundance of species and individuals, especially the fine leaved plants (e.g. *Utricularia, Myriophyllum*), and that *Alopecurus,* a grass which grows close to the shore which may be dry at low water levels, is not colonized well. It is likely that the methods used, although possibly identifying species restrictions if they existed, were not suitable to indicate species preferences for macrophyte species. Quantitative methods, comparing species abundance between macrophytes, are likely to better indicate substrate preference.

## Abundance on macrophytes

Although non-sessile periphytic rotifers appear not to be confined to certain macrophytes, quantitative studies show they do have preference, or at least differences in abundance, between macrophyte species and growth forms. Pontin & Shiel (1995) found total rotifer abundance to differ between macrophyte species of different architecture, with abundance greatest under *Azolla pinnata* fronds, lower on *Myriophyllum crispatum*, and least within *Pseudoraphis spinescens*. There were also differences in the densities of dominant rotifers (*Mytilina acanthophora, Lecane bulla, L. hamata* and *Trichocerca similis*) amongst plants. In experimental ponds, Irvine et al. (1990) found an increase in the total abundance, and changes in the dominance of rotifer species (increases in *Keratella* and *Polyarthra* species), associated with changes in macrophyte dominance from a predominantly submerged to emergent community over a 3-year period. Duggan et al. (submitted) found significant differences in the abundances of all periphytic rotifer species (e.g. *Euchlanis dilatata, Lecane closterocerca*) between *Egeria densa, Eleocharis sphacelata* and *Myriophyllum propinquum*, except for those that are more planktonic in form (*Ascomorpha saltans, Keratella cochlearis* and *Synchaeta oblonga*). These differences in rotifer abundance between macrophytes were found even when different macrophyte species grew in close proximity to one another. *Eleocharis* had lower abundances of most rotifer species than adjacent growing *Egeria*, except the more mobile forms. The degree of dependency that rotifers have for a substrate (as shown in the lists of Pejler, 1962 and Pennak, 1966), therefore, does appear to have an effect on substrate choice, with more mobile littoral species less dependent on substrate type than less mobile species.

Rotifer diversity has also been found to differ between macrophyte species. Pontin & Shiel (1995) found significant differences in diversity between macrophytes in Ryans 3 Billabong, Australia, with diversity highest under *Azolla pinnata*, lower under *Myriophyllum crispatum*, and lowest in *Pseudoraphis spinescens*. Duggan et al. (submitted) found greater diversities on *Myriophyllum* than *Egeria*, which were greater than on *Eleocharis*. Thus, macrophyte species composition appears to have a great influence on the structuring of periphytic rotifer communities within vegetated areas. The mechanisms for substrate preference, however, are generally not well known and can only be inferred. Preference may be due to (a) physical structure or complexity, (b) differences in food concentration or composition, (c) chemical factors, (d) macrophyte age, or (e) differences in the degree of protection given from predation (e.g. Duggan et al. submitted).

### *Physical structure and complexity*

Different macrophyte species have different architectural designs and levels of complexity, in particular the fineness and density of the leaves and the heterogeneity of surfaces. The studies of Pontin & Shiel (1995) and Duggan et al. (submitted) show a general increase in abundance and diversity of rotifers with macrophyte architectural complexity. Studies of macroinvertebrates (e.g. Dudley, 1988) have found similar relationships, and have suggested that increased macrophyte complexity provides greater variety, sizes and shapes of habitat, so that a greater variety, size and shape of animals can be packed in. More complex shapes provide more small scale habitat, supporting disproportionately more small fauna (Jeffries, 1993). Similarly, higher rotifer diversity within the more complex macrophytes is also likely to be largely due to an increased diversity of microhabitats (e.g. a greater variety of physical structures leading to greater habitat heterogeneity).

### *Food composition and concentration*

Food concentration and composition is known to play an important role in structuring plankton communities, particularly where rotifers have specialized diets (e.g. Gilbert & Bogdan, 1981; Bogdan & Gilbert, 1987). Similarly, differences in food concentration and composition may be predominant causes of variation

*Table 3.* Feeding types and dietary items of rotifers from submerged macrophytes (translated from Donner 1964)

| | |
|---|---|
| **1. Herbivores** | |
| Green algae | *Philodinavus paradoxus, Encentrum mustela, Habrotrocha tridens* |
| Blue-green algae | Same as above |
| Diatoms | *Brachionus quadridentatus brevispinus, Cephalodella apocolea, C. forficata, Euchlanis deflexa, E. oropha, Notholca squamala, Proales theodora, Ptygura beauchampi, Embata laticeps, Philodina flaviceps, Philodina acuticornis odiosa* |
| Bacteria | Not enough known, most likely Bdelloidea |
| **2. Carnivores** | |
| Protozoans | *Dicranophorus uncinatus* (including testate amobae) |
| Predators, feeders on rotifers | *Cephalodella gibba, Dicranophorus secretus* |
| Saprophages | *Proales fallaciosa, Encentrum mustela* |
| **3. Detrivores** | Most rotifers, particularly bdelloids |

in rotifer species composition between macrophytes. Pourriot (1977) noted that feeding types in periphytic rotifers appear to be varied, with food preferences more or less strict. Donner (1964) observed most species in the River Danube to be detrivores, with a strict mode of feeding rare (Table 3). Pourriot (1977), however, felt that even species with wide preferences were likely to show some choice. Many periphytic rotifers are likely to feed directly on the periphyton to greater or lesser degrees. The composition of algal periphyton can vary among different macrophyte hosts, with architecture alone shown to affect the growth of attached algae (Eminson & Moss, 1980; Blindow, 1987; Jones et al., 1999). Modenutti & Claps (1988) and Duggan et al. (submitted) both found lower rotifer densities associated with emergent macrophytes (*Schoenoplectus* and *Eleocharis*, respectively), and these forms are known to have poor algal periphyton communities, because of shading (e.g. Grimshaw et al., 1997). Macrophyte surface microtopography may also influence rotifer communities, with differing macrophyte surface architectures providing selective and unique microhabitats for periphytic algal cells (Burkholder, 1996). The high degree of hydrophobic cuticle development on some macrophyte species (e.g. *Scirpus* culms) is also known to retard the development of algal periphyton (Goldsborough & Hickman, 1991).

*Chemicals*

It has been suggested that aquatic macrophytes are capable of influencing their algal periphyton through allelopathy (e.g. Fitzgerald, 1969; Wium-Anderson et al., 1982), which would, therefore, be likely to affect the distribution of rotifers indirectly by influencing food composition and concentration. However, the release of allelopathic chemicals by macrophytes at biologically relevant concentrations is contested (Forsberg et al., 1990), and it has been suggested that they instead act as deterrents to macrophyte grazing (Ostrofsky & Zettler, 1986). By contrast, Bronmark (1985) found that snails, attracted by dissolved organic matter excreted by some (healthy) macrophytes, remove the epiphytic cover, lengthening that macrophyte's life. The grazing impacts of rotifers in the periphyton are unknown, although based on their high abundance and probable high production in the littoral, it is possible they may remove substantial quantities of periphyton. The attraction of grazers such as rotifers to macrophytes should, therefore, be favoured by the macrophytes. Wallace & Edmondson (1986) found that reduced levels of ambient $Ca^{+2}$ are important in controlling sessile rotifer larval settlement, with larval attachment limited if concentrations of $Ca^{+2}$ are above a certain level. This possibly assures that larvae choose only an actively photosynthesizing plant. It is not known if there are similar chemical mechanisms that attract more mobile species.

*Macrophyte age*

Macrophyte age has been suggested to be important

by Pontin & Shiel (1995), who found that older established beds of *Myriophyllum crispatum* had greater rotifer abundance than new growths in Ryans 3 Billabong. Duggan et al. (submitted) found low abundances of rotifers on *Myriophyllum propinquum* compared with other macrophytes, and thought this may be because most of its biomass was new growth. Older macrophytes generally have greater growths of epiphytic algae than younger ones (Millie & Lowe, 1983; Roos, 1983; Burkholder & Wetzel, 1989), and macrophyte age is, therefore, likely to affect rotifers through variations in the colonization, and subsequent composition and concentration of algal periphyton on which most rotifers are likely to feed.

*Predation*

In the plankton, rotifers are often the target of intense predation by invertebrates and small fish (e.g. Moore & Gilbert, 1987; Hewitt & George, 1987). The structural complexity of macrophyte communities is likely to provide a wide variety of potential refugia from predators, with the presence of vegetation decreasing predation susceptibility by increasing predator search and pursuit times (e.g. Goddard & McDiffett, 1983). Walsh (1995) proposed that predators might be one of the principal determinants of habitat choice in periphytic rotifers, with plants of greater architectural complexity offering more protection from predators. She investigated the habitat-specific predation susceptibilities of *Euchlanis dilatata* on three aquatic macrophytes, *Myriophyllum exalbescens, Elodea canadensis* and *Ceratophyllum demersem*, to two predators, damselfly nymphs *Enallagma carunculata* and the cnidarian *Hydra* sp. Of the macrophytes, rotifer survival was greatest on *M. exalbescens* in the presence of both predators. The presence of macrophyte species (cf. their absence) increased the susceptibility to predation by damselfly nymphs, by increasing the nymphs ability to forage. Increasing plant structural complexity, however, resulted in increased rotifer survival. The susceptibility of rotifers to different predators, for example copepods and fish, within macrophytes of different architecture is not known and needs examination.

*Oviposition behavior*

Walsh (1989) found that the choice of egg deposition sites by *Euchlanis dilatata* was determined in part by substrate type. Eggs were laid preferentially on *Myriophyllum* when there was a choice between it, *Elodea* and *Ceratophyllum*. Behavioural adaptations of this sort are likely to provide benefits including exploitation of favorable physical conditions for embryonic development and subsequent resource utilization by neonates.

Nothing is yet known of the microspatial distribution of the periphytic rotifers on macrophytes. Wallace & Edmondson (1986) found preference of the sessile *Collotheca gracilipes* for undersurfaces of *Elodea canadensis*. Based on the variability between macrophytes, preferences by more mobile forms may be expected to be not as strong as those of the sessiles, but for their abundance to be greatest where food concentration and composition are greatest.

As well as variability of rotifers between macrophytes caused by plant specific factors, macrophytes also cause variability in physical and chemical conditions, as a result of the differential effects of photosynthesis, shading, decomposition and by macrophytes forming barriers to water movement. Duggan et al. (1998), however, found greatest spatial variability in rotifer composition across a vegetated littoral when physical and chemical variability was least, indicating that factors relating to the macrophytes themselves are the most important in structuring the littoral rotifers.

## Seasonality

Study of the seasonality of periphytic rotifers is difficult because accurate assessment of rotifer abundance often requires methods destructive to the habitat. Koste (1965) gives some basic information on the seasonal distribution of rotifers over a 2-year period from Engelbergs Moor, Germany, by taking scrapings of water lily surfaces, noting annual fluctuations in species composition and species abundance. Succession in species dominance and abundance has also been observed in the rotifers between macrophytes in a number of Colorado lakes (Pennak, 1966), and in rotifers collected in artificial substrates in the littoral of Lake Rotomanuka, New Zealand (Duggan et al., 1998). Duggan et al. (1998) examined seasonality over a period of 10 months, and using multivariate techniques found that seasonal community changes coincided with a rise in water level, and a change from marked spatial gradients in physical and chemical variables in summer and autumn to more homogeneous conditions in winter and spring. Donner (1964), in the River Danube periphyton, did not examine seasonality specifically, but noted different species in winter (e.g. *Cephalodella delicata, C. ventripes*) and summer (e.g. *Colurella colurus, Euchlanis deflexa*). He also observed compositional changes in rotifer species after disturbance by flooding or drying. Seasonal distribution of species was found by Duggan et al. (1998) to be associated mainly with temperature and pH. However it was suggested that, as in the plankton (e.g. Herzig, 1987), distribution is also likely affected by competition for food, predation and parasitism. The seasonal growth cycles of macrophytes will also affect rotifer seasonal dynamics (for example, see '*Macrophyte age*' section, above). In rivers, however, Modenutti & Claps (1988) found little seasonal variation in community composition of the periphytic rotifers, and much less so than in the plankton where marked changes occurred. These authors found seasonal changes in abundance to relate to changes in water level, with low abundances of species occurring during high water. Similarly, Donner (1964) and Erben (1987) found that the composition or abundance of periphytic rotifers in rivers was related to current speeds, with low rotifer numbers being found in fast flowing water, and highest species diversity occurring at times of low flow.

## Conclusion

The study of rotifers in the periphyton needs far greater attention. As feeders on components of the periphyton, they are likely to extend the life of macrophytes, and so contribute to the maintenance of a vegetated littoral zone. However, nothing is known of their grazing impacts on the periphyton. Periphytic rotifers are known to vary in abundance between macrophytes (or other substrates), which is likely to be a result of a number of factors, including differences in macrophyte morphology, macrophyte age, epiphytic algal growths and the differential effects of predation between macrophytes. Other than predation (Walsh, 1995), the mechanisms for this variability have not been investigated, and the relative importance of these factors is unknown.

Another major problem in studies of periphytic rotifers to date has been the lack of uniformity of methods employed, for example, top to bottom sampling using nets or tubes (Pennak, 1966), removal of macrophytes (Goddard & McDiffet, 1983; Duggan et al., submitted), removal from the surfaces using a syringe (Pontin & Shiel, 1995), and artificial substrates (Duggan et al., 1998). Such variation in sampling techniques makes valid comparisons between studies difficult or impossible. The development of standardised methods will be worthwhile for future studies.

## Acknowledgements

I thank Kareen Schnabel and Meredith Pearson for aiding in the translation of the German texts. I thank J. D. Green and the anonymous referees for improving the manuscript, and Michelle White for help with manuscript formatting.

## References

Andrew, T. E. & A. G. Fitzsimons, 1992. Seasonality, population dynamics and production of planktonic rotifers in Lough Neagh, Northern Ireland. Hydrobiologia 246: 147–164.

Berner-Fankhauser, H., 1983. Abundance, dynamics and succession of planktonic rotifers in Lake Biel, Switzerland. Hydrobiologia 104: 349–352.

Blindow, I., 1987. The composition and density of epiphyton on several species of submerged macrophytes – the neutral substrate hypothesis tested. Aquat. Bot. 29: 157–168.

Bogdan, K. G. & J. J. Gilbert, 1987. Quantitative comparison of food niches in some freshwater zooplankton. Oecologia 72: 331–340.

Bonecker, C. C., F. A. Lansac-Tôha & L. M. Bini, 1998. Composition of zooplankton in different environments of the Mato Grosso Pantanal, Mato Grosso, Brazil. Anais do VIII seminario regional de ecologia 3: 1123–1135.

Bronmark, C., 1985. Interactions between macrophytes, epiphytes and herbivores: an experimental approach. Oikos 45: 26–30.

Burkholder, J. M., 1996. Interactions of benthic algae with substrata. In. Stevenson, R. J., M. L. Bothwell & R. L. Lowe (eds), Algal Ecology: Freshwater Benthic Ecosystems. Academic Press. 253–297.

Burkholder, J. M. & R. G. Wetzel, 1989. Epiphytic microalgae on a natural substratum in a phosphorus-limited hardwater lake: seasonal dynamics of community structure, biomass and ATP content. Arch. Hydrobiol. Suppl. 83: 1–56.

Clément, P., 1977. Phototaxis in rotifers (action spectra). Arch. Hydrobiol. Beih. 8: 67–70.

De Beauchamp, P. M., 1909. Recherches sur les Rotiferes: les formations tegumentaires et l'appareil digestif. Arch. Zool. Exp. Gen. 4: 1–410.

Donner, J., 1964. Die Rotatorien-Synusien submerser Makrophyten der Donau bei Wien und mehrerer alpenbache. Arch. Hydrobiol. Suppl. 27: 227–324.

Donner, J., 1970. Die Rädertierbestande submerser Moose der Salzach und anderer Wasser-Biotope des flussgebietes. Arch. Hydrobiol. Suppl. 36: 109–254.

Dudley, T. L., 1988. The roles of plant complexity and epiphyton in colonization of macrophytes by stream insects. Verh. Int. Ver. Theor. Agnew. Limnol. 23: 1153–1158.

Duggan, I. C., J. D. Green, K. Thompson & R. J. Shiel, 1998. Rotifers in relation to littoral ecotone structure in Lake Rotomanuka, North Island, New Zealand. Hydrobiologia 387/388: 179–197.

Duggan, I. C., J. D. Green, K. Thompson & R. J. Shiel, submitted. The influence of macrophytes on the spatial distribution of littoral rotifers. Freshwat. Biol.

Dumont, H. J., I. V. De Velde & S. Dumont, 1975. The dry weight estimate of biomass in a selection of Cladocera, Copepoda and Rotifera from the plankton, periphyton and benthos of continental waters. Oecologia 19: 75–97.

Elliott, J. I., 1977. Seasonal changes in the abundance and distribution of planktonic rotifers in Grasmere (English Lake District). Freshwat. Biol. 7: 147–166.

Erben, R., 1987. Rotifer fauna in the periphyton of Karst rivers in Croatia, Yugoslavia. Hydrobiologia 147: 103–105.

Eminson, D. F. & B. Moss, 1980. The composition and ecology of periphyton communities in freshwaters. I. The influence of host type and external environment on community composition. Br. Phycol. J. 15: 429–446.

Ferrari, I, A. Farabegoli & R. Mazzoni, 1989. Abundance and diversity of planktonic rotifers in the Po River. Hydrobiologia 186/187: 201–208.

Fitzgerald, G. P., 1969. Some factors in the competition or antagonism among bacteria, algae and aquatic weeds. J. Phycol. 5: 351–359.

Forsberg, C., S. Kleiven & T. Willen, 1990. Absence of allelopathic effects of *Chara* on phytoplankton *in situ*. Aquat. Bot. 38: 289–294.

Gilbert, J. J. & K. G. Bogdan, 1981. Selectivity of *Polyarthra* and *Keratella* for flagellate and aflagellate cells. Verh. int. Ver. Limnol. 21: 1515–1521.

Goddard, K. A. & W. F. McDiffett, 1983. Rotifer distribution, abundance, and community structure in four habitats of a freshwater marsh. J. Freshwat. Ecol. 2: 199–211.

Goldsborough, L. G. & M. Hickman, 1991. A comparison of periphytic algal biomass and community structure on *Scirpus validus* and on a morphologically similar artificial substratum. J. Phycol. 27: 196–206.

Green, J., 1960. Zooplankton of the River Sokoto. The Rotifera. Proc. zool. Soc. Lond. 135: 491–523.

Green, J., 1972. Freshwater ecology in the Mato Grosso, Central Brazil. III. Associations of Rotifera in meander lakes of the Rio Suiá Miss-u. J. nat. Hist. 6: 229–241.

Grimshaw, H. J., R. G. Wetzel, M. Brandenburg, K. Segerblom, L. J. Wenkert, G. A. Marsh, W. Charnetzky, J. E. Haky & C. Carraher, 1997. Shading of periphyton communities by wetland emergent macrophytes: decoupling of algal photosynthesis from microbial nutrient retention. Arch. Hydrobiol. 139: 17–27.

Havens, K. E., 1991. Summer zooplankton dynamics in the limnetic and littoral zones of a humic acid lake. Hydrobiologia 215: 21–29.

Herzig, A., 1987. The analysis of planktonic rotifer populations: a plea for long term investigations. Hydrobiologia 147: 163–180.

Hewitt, D. P. & D. G. George, 1987. The population dynamics of *Keratella cochlearis* in a hypereutrophic tarn and the possible impact of predation by young roach. Hydrobiologia 147: 221–227.

Hudson, C. T. & P. H. Gosse, 1889. The Rotifera or Wheel Animalcules, both British and Foreign. Longmans, Green & Co., London. 2 vols.

Hutchinson, G. E., 1993. A Treatise on Limnology, Volume 4: The Zoobenthos. John Wiley & Sons, New York: 944 pp.

Irvine, K., H. Balls & B. Moss, 1990. The entomostracan and rotifer communities associated with submerged plants in the Norfolk Broadland – effects of plant biomass and species composition. Int. Rev. ges. Hydrobiol. 75: 121–141.

Jeffries, M., 1993. Invertebrate colonisation of artificial pondweeds of differeing fractal dimension. Oikos 67: 142–148.

Jersabek, C. D., 1995. Distribution and ecology of rotifer communities in high-altitude alpine sites – a multivariate approach. Hydrobiologia 313/314: 75–89.

Jones, J. I., J. O. Young, G. M. Haynes, B. Moss, J. W. Eaton & K. J. Hardwick, 1999. Do submerged aquatic plants influence their periphyton to enhance the growth and reproduction of invertebrate mutualists? Oecologia 120: 463–474.

Kobayashi, T., R. J. Shiel, P. Gibbs & P. I. Dixon, 1998. Freshwater zooplankton in the Hawkesbury-Nepean River: comparison of community structure with other rivers. Hydrobiologia 377: 133–145.

Kofoid, C. A., 1908. The plankton of the Illinois River, 1889–1899. Part II. Constituent organisms and their seasonal distribution. Bull. Ill. State Lab. Nat. Hist. 8: 1–361.

Koste, W., 1965. Die Rotatorien des Naturdenkmals 'Engelbergs Moor' in Druchhorn, Kreis Bersenbrück. Veröff. Naturw. Ver. Osnabrück 31: 49–82.

Koste, W., 1974. Zur Kenntnis der Rotatorienfauna der 'schwimmenden Wiese' einer Uferlagune in der Várzea Amazoniens, Brasilien. Amazoniana 5: 25–59.

Langley, J. M., S. Kett, R. S. Al-Khalili & C. J. Humphrey, 1995. The conservation value of English urban ponds in terms of their rotifer fauna. Hydrobiologia 313/314: 259–266.

Lemley, A. D. & J. F. Dimmick, 1982. Structure and dynamics of zooplankton communities in the littoral zone of some North Carolina lakes. Hydrobiologia 88: 299–307.

Meuche, A., 1939. Die fauna im algenbewuchs. Nach untersuchungen im litoral östholsteinischer Seen. Arch. Hydrobiol. 34: 349–520.

Millie, D. L. & R. L. Lowe, 1983. Studies on Lake Erie's littoral algae: host specificity and temporal periodicity of epiphytic diatoms. Hydrobiologia 99: 7–18.

Modenutti, B. E., 1998. Planktonic rotifers of Samborombon River Basin Argentina. Hydrobiologia 387/388: 259–265.

Modenutti, B. E. & M. C. Claps, 1988. Monogonota rotifers from plankton and periphyton of pampasic lotic environments Argentina. Limnol. 19: 167–175.

Moore, M. V. & J. J. Gilbert, 1987. Age specific *Chaoborus* predation on rotifer prey, Freshwat. Biol. 17: 223–236.

Morales-Baquero, R., L. Cruz-Pizarro & P. Carrillo, 1989. Patterns in the composition of the rotifer communities from high mountain lakes and pools in Sierra Nevada Spain. Hydrobiologia 186/187: 215–221.

Onwudinjo, C. C. & A. B. M. Egborge, 1994. Rotifers of Benin River, Nigeria. Hydrobiologia 272: 87–94.

Ostrofsky, M. L. & E. R. Zettler, 1986. Chemical defences in aquatic plants. J. Ecol. 74: 279–287.

Pejler, B., 1962. On the taxonomy and ecology of benthic and periphytic Rotatoria. Investigations in northern Swedish Lapland. Zool. Bidr. Upps. 33: 327–422.

Pejler, B. & B. Berzins, 1989. On choice of substrate and habitat in brachionid rotifers. Hydrobiologia 186/187: 137–144.

Pejler, B. & B. Berzins, 1993a. On relation to substrate in sessile rotifers. Hydrobiologia 259: 121–124.

Pejler, B. & B. Berzins, 1993b. On the ecology of *Cephalodella*. Hydrobiologia 259: 125–128.

Pejler, B. & B. Berzins, 1993c. On the ecology of Colurellidae (Rotifera). Hydrobiologia 263: 61–64.

Pejler, B. & B. Berzins, 1993d. On the ecology of Dicranophoridae (Rotifera). Hydrobiologia 259: 129–131.

Pejler, B. & B. Berzins, 1993e. On the ecology of Trichocercidae (Rotifera). Hydrobiologia 263: 55–59.

Pejler, B. & B. Berzins, 1994. On the ecology of *Lecane* (Rotifera). Hydrobiologia 273: 77–80.

Pejler, B. & B. Berzins, 1995. Relation to habitat in rotifers. Hydrobiologia 313/314: 267–278.

Pennak, R. W., 1962. Quantitative zooplankton sampling in littoral areas. Limnol. Oceanogr. 7: 487–489.

Pennak, R. W., 1966. Structure of zooplankton populations in the littoral macrophyte zone of some Colorado lakes. Trans. am. Microsc. Soc. 85: 329–349.

Pontin, R. M. & R. J. Shiel, 1995. Periphytic rotifer communities of an Australian seasonal floodplain pool. Hydrobiologia 313/314: 63–67.

Pourriot, R., 1977. Food and feeding habits of Rotifera. Arch. Hydrobiol. Beih. 8: 243–260.

Preissler, K., 1977a. Do rotifers show 'Avoidance of the shore'? Oecologia 27: 253–260.

Preissler, K., 1977b. Horizontal distribution and 'avoidance of shore' by rotifers. Arch. Hydrobiol. Beih. 8: 43–46.

Preissler, K., 1980. Field experiments on the optical orientation of pelagic rotifers. Hydrobiologia 73: 199–203.

Rodgers, K. H. & C. M. Breen, 1983. An investigation of macrophyte, epiphyte and grazer interactions. In Wetzel, R. G. (ed.), Periphyton of Freshwater Ecosystems. Dr W. Junk Publishers, The Hague: 217–226.

Roos, P. J., 1983. Dynamics of periphytic communities. In Wetzel, R. G. (ed.), Periphyton of Freshwater Ecosystems. Developments in Hydrobiology 17: 5–10.

Ruttner-Kolisko, A., 1974. Planktonic rotifers: biology and taxonomy. Die Binnengewasser Supplement. 26: 1–146.

Sanoamuang, L., 1992. The ecology of mountain lake rotifers in Canterbury, with particular reference to Lake Grasmere and the genus *Filinia* Bory de St. Vincent. Unpublished PhD thesis, The University of Canterbury, Christchurch, New Zealand.

Saunders-Davies, A. P., 1989. Horizontal distribution of plankton rotifers *Keratella cochlearis* (Bory de St. Vincent) and *Polyarthra vulgaris* (Carlin) in a small eutrophic lake. Hydrobiologia 186/187: 153–156.

Sollberger, P. J. & L. J. Paulson, 1992. Littoral and limnetic zooplankton communities in Lake Mead, Nevada-Arizona, U.S.A. Hydrobiologia 237: 175–184.

Vasconcelos, V., 1990. Seasonal fluctuation in the zooplankton community of Albizo reservoir (Portugal). Hydrobiologia 196: 183–191.

Wallace, R. L., 1977a. Adaptive advantages of substrate selection by sessile rotifers. Arch. Hydrobiol. Beih. 8: 53–55.

Wallace, R. L., 1977b. Distribution of sessile rotifers in an acid bog pond. Arch. Hydrobiol. 79: 478–505.

Wallace, R. L., 1980. Ecology of sessile rotifers. Hydrobiologia 73: 181–193.

Wallace, R. L. & W. T. Edmondson, 1986. Mechanism and adaptive significance of substrate selection by a sessile rotifer. Ecology 67: 314–323.

Walsh, E. J., 1989. Oviposition behavior of the littoral rotifer *Euchlanis dilatata*. Hydrobiologia 186/187: 157–161.

Walsh, E. J., 1995. Habitat-specific predation susceptibilities of a littoral rotifer to two invertebrate predators. Hydrobiologia 313/314: 205–211.

Wetzel, R. G., 1983. Opening Remarks. In Wetzel, R. G. (ed.), Periphyton of Freshwater Ecosystems. Dr W. Junk Publishers, The Hague: 3–4.

Wium-Anderson, S., U. Anthoni, C. Christophersen & G. Hoen, 1982. Allelopathic effects on phytoplankton by substances isolated from aquatic macrophytes (Charales). Oikos 39: 187–190.

*Hydrobiologia* **446/447:** 149–153, 2001.
*L. Sanoamuang, H. Segers, R.J. Shiel & R.D. Gulati (eds), Rotifera IX.*




# The psammic rotifer structure in three Lobelian Polish lakes differing in pH

Irena Bielańska-Grajner
*Department of Ecology, University of Silesia, Bankowa 9, 40-007 Katowice, Poland*
*E-mail: igrajner@us.edu.pl*

*Key words:* psammic rotifers, pH, Lobelian lakes

## Abstract

Thirty five species of rotifers were found in the psammon of three lakes of differing pH in open water. The largest number of rotifer species and their highest abundance was found in lake Rekowskie which had a slightly acid to neutral water pH. The lowest quantity and the number of rotifers species were observed in the lake with lowest pH. The results suggest that pH is one of the most important factors affecting the number and diversity of psammic rotifers in these lakes.

## Introduction

Since the basic works of Wiszniewski (1932, 1934a,b, 1936,1937), Pennak (1951) and Ruttner-Kolisko (1953,1954,1965) only a few authors have paid attention to the ecology of psammic rotifers of marine beaches (Tzschaschel, 1983; Turner, 1990, 1993). Some others (Evans, 1984; Schmid-Araya, 1993, 1994, 1995; Turner, 1996) carried out interesting investigations on interstitial rotifers of lotic waters.

The results obtained hitherto show a strong correlation between the species composition and number of rotifers and the type of waterbody. This is an indirect correlation, because the type of lake has an impact on the quality of the beach. It has been suggested that the most important factor influencing psammic rotifer communities is pH of the lake water (Wiszniewski, 1936). However, the qualitative and quantitative composition of the community also depends on the structure and characters of the shore, including grain size and detritus content. These factors depend also on the limnological type of the waterbody (Wiszniewski, 1932, 1934b).

The aim of this study was to determine whether differences exist between the rotifer communities living in three Lobelian lakes of differing pH.

*Table 1.* Salient features of the study lakes

| | Jeleń | Rekowskie | Pomysko |
|---|---|---|---|
| Surface area [ha] | 84 | 10 | 10.5 |
| Max. depth [m] | 33 | 33 | 6 |
| Catchment basin forestage [%] | 40 | 80 | 100 |
| Visibility [m] | 6.1 | 3.1 | 5.4 |
| Epilimnion pH | 7–8 | 6–7 | 5.0 |
| Interestitial water: | | | |
| Temperature °C | 17–19 | 14.5–17 | 14.5–17 |
| PH | 7.5–8 | 6–7.9 | 6–7.9 |
| Oxygen concentration [mg $O_2$/l] | 2–4.5 | 3–4.2 | 3–4.2 |
| Hardness [mg Ca $CO_3$/ l] | 7–55 | 53–70 | 53–70 |
| Sand grain size [mm] | 0.3–1.6 | 0.1–2.5 | 0.1–1.5 |

## Study area

The studies were carried out in three Lobelian lakes of similar trophy. The Lobelian lakes are located along the frontal moraine belt of West Pomerania in Poland. These lakes characterised by low primary production, small amount of biogens and high water transparency. The submerged vegetation of these lakes is on the wane and of the boreal-Atlantic origin (Kraska at al.,

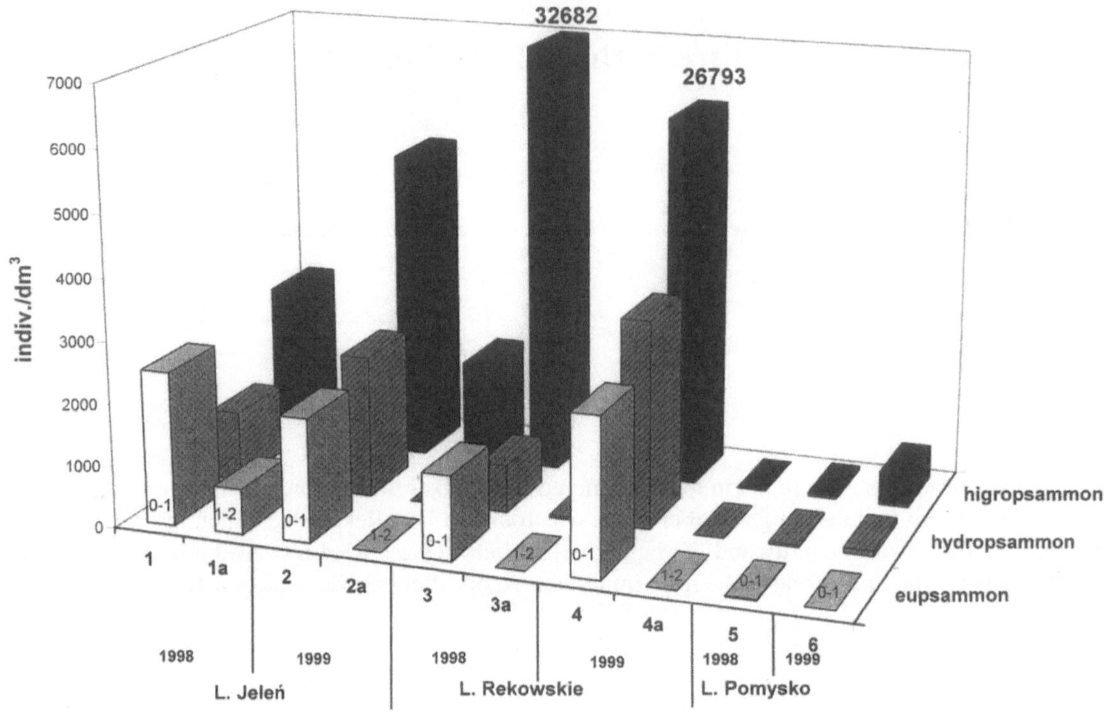


*Figure 1.* Distribution of psammic rotifers in three studies sites: Lake Jeleń: 1998, 0–1 cm - 1, 1–2 cm 1a; 1999, 0–1 cm -2, 1–2 cm - 2a. Lake Rekowskie: 1998, 0–1 cm - 3, 1–2 cm - 3a; 1999 0–1 cm - 4, 1–2 cm - 4a. Lake Pomysko: 1998, 0–1 cm - 5; 1999 0–1 cm - 6.

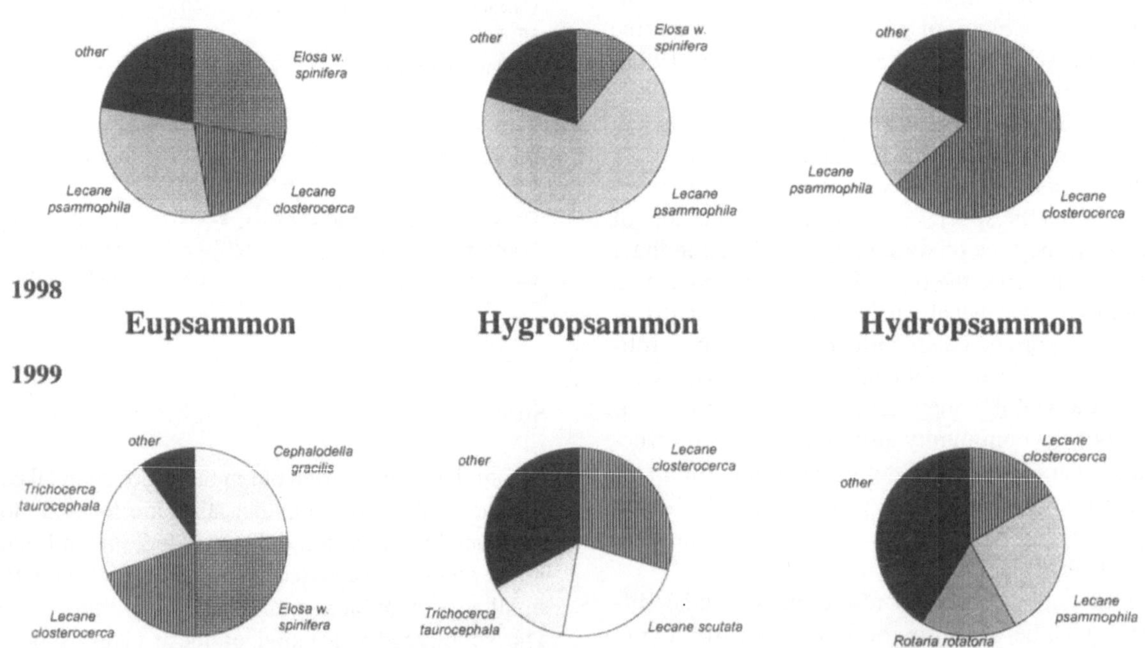


*Figure 2.* Percentage share of dominant species in the total abundance in eupsammon, higropsammon and hydropsammon in Lake Jeleń.

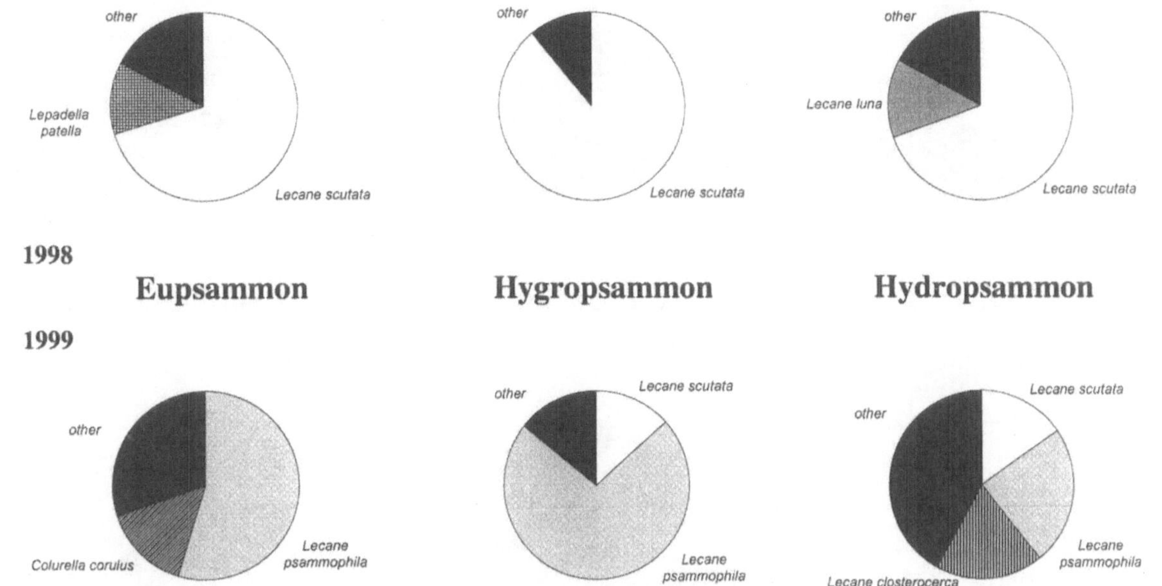


*Figure 3.* Percentage share of dominant species in the total abundance in eupsammon, higropsammon and hydropsammon in Lake Rekowskie.

1994). The lakes included in the analysis are described in Table 1.

Species diversity was calculated according to the Shannon–Weaver index (Pielou, 1975) and similarity coefficient of Sörensen (Krebs, 1989).

## Materials and methods

Rotifers were collected during the summer (July and August 1998, 1999). The methods of collection were based on those described by Evans (1984) and Wiszniewski (1934b). Samples of psammon were collected at 3 sites located at the water's edge and at 1 m above non-submerged interstitial and below submerged interstitial. All samples were collected in duplicate with a plastic corer 3.5 cm in diameter and 10 cm in length. Only two fractions, 0–1 cm and 1–2 cm sample depth, were analysed. Ten samples were collected at each site, five for qualitative analysis and the remaining five for quantitative analysis. Quantitative samples were preserved in a mixture of formaldehyde and glycerol (ratio 3:1). The qualitative analysis was conducted on live rotifers species.

Rotifers were counted in a Kolkwitz chamber (Schwoerbel, 1970) and the number of rotifers was expressed per 1 $dm^3$of sand.

Sand water for chemical analysis was sampled from the waters edge and 1-m above and below the waterline. Water samples were filtered through a mesh net (bolting silk).

## Results and discussion

Thirty five rotifer species were found in the lakes investigated (Table 1). The largest number of species (26) was observed in Lake Rekowskie in which the pH was slightly acid to neutral and interstitial water in sand -neutral to slightly alkaline . The smallest number of rotifers species (10) was found in Lake Pomysko, in which water had the lowest pH, even in the interstitial water (pH< 7.0). Oxygen concentration in the interstitial water was about half of that in the lake water, although the hardness was higher and remaining chemical parameters were similar in all three tested lakes. The most pronounced differences were those in the water pH. So, it seems that pH is one of the most important factors influencing the number and diversity of psammic rotifer species in Lobelian lakes. These results agree with those of Wiszniewski (1936, 1937) who also observed species poor rotifer communities in lakes with low pH of water.

Significant differences in the number of rotifer species were found among all of the lakes studied. In the lake with the lowest water pH, the number of rotifers was markedly low compared with the remaining

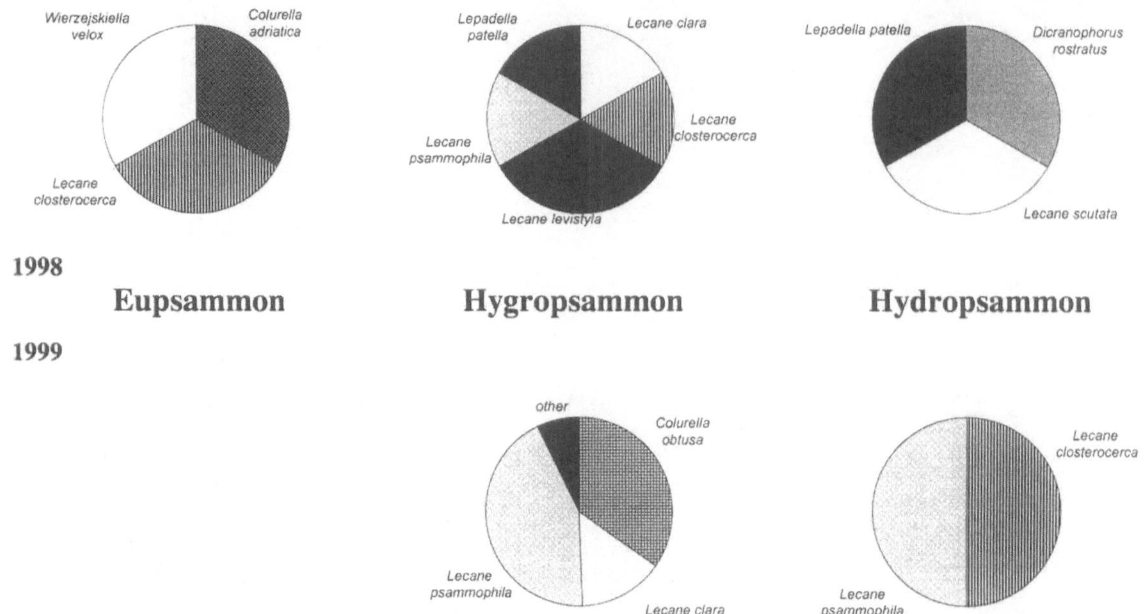


*Figure 4.* Percentage share of dominant species in the total abundance in eupsammon, higropsammon and hydropsammon in Lake Pomysko.

lakes (Fig. 1). In the first year of observation (1998), the highest number of rotifers was found in the hygropsammon and the lowest – in hydropsammon. In the following year (1999), the same distribution of rotifers was found only in the lake Pomysko, whereas in two others (Jeleń and Rekowskie), the highest number of rotifers was located in eupsammon. It is worth noting that Wiszniewski (1934a, b) and Ejsmont-Karabin (1998) found much higher numbers of organisms in the rotifer communities of wet beach sand of other Polish lakes.

Most psammic rotifers occurred in the superficial layer (0–1 cm), but in Lake Pomysko, no rotifer were found in the depth of 1–2 cm (Fig. 1). The same observation was made previously in lakes of Polesie Lubelskie (Radwan & Bielańska-Grajner, 2001). In contrast, Pennak (1951) reported that in the beach of Wisconsin Lake and in the psammon of the Ninnescah river most rotifers were found in the deeper layers (1–3 cm in depth).

In each of the lakes investigated, the same species dominated all of the zones of the psammolitoral examined. These were *Elosa woralli f. spinifera, Lecane closterocerca* and *L. psammophila* in Lake Jeleń, *Lecane scutata* and *L. psammophila* in Lake Rekowskie and in Lake Pomysko. Only *Rotaria rotatoria* preferred the hydropsammon, and *Trichocerca taurocephala* the eupsammon (Figs 2, 3 and 4).

As for the distribution and composition of rotifer communities living in different zones of the psammon of Lobelian lakes, it should be stressed that these differed from those found in lakes located in other regions of Poland and investigated by Wiszniewski (1932, 1934a), Ejsmont-Karabin (1998) and (Radwan & Bielańska-Grajner, 2001).

The Shannnon–Weaver diversity index showed that the community structure in each of the lakes studied was different. In Lake Jeleń, the highest diversity index was observed in the higropsammon (2.13), lower in the eupsammon (1.9) and hydropsammon (1.97). In two other lakes, a higher diversity index was observed in hydropsammon (Lake Rekowskie – 2.19; Lake Pomysko – 1.39), in eupsammon (1.92 and 1.10, respectively) and hygropsammon (1.24 and 1.38, respectively). In general, the Shannon–Weaver index was low which is considered to be characteristic for a highly variable environment. Also, Ejsmont-Karabin (1998) found the highest diversity of species living in hydropsammon of Lake Kuc (Mazury, Poland).

The Sorensen index showed community structure in Lake Jeleń and Lake Rakowskie was similar (0.76) but not in Jeleń and Pomysko lakes (0.35) and Rekowskie and Pomysko lakes (0.27).

In conclusion, the communities of psammic rotifers living in three selected Lobelian lakes in Western Pomerania had their own specific structure. This dif-

*Table 2.* Rotifer species in the psammon of lakes Jeleń, Rekowskie, Pomysko, Poland

| | Jeleń | Rekowskie | Pomysko |
|---|---|---|---|
| *Brycella tenella* (Bryce) | | + | |
| *Cephalodella catellina* (Müll.) | + | + | |
| *C. gibba* (Ehrb.) | + | + | |
| *C. gracilis* (Ehrb.) | + | + | |
| *C. ventripes* (Dix.-Nutt.) | | + | |
| *Colurella adriatica* Ehreb. | | | + |
| *C. colurus* (Ehrb.) | + | + | |
| *C. hindenburgi* Stein. | | + | |
| *C. obtusa* (Gosse) | + | | + |
| *C. uncinata* (Müll.) | | + | |
| *Dicranophorus hercules* Wiszn. | + | + | |
| *D. rostratus* (Dix.-Nutt. & Free.) | | | + |
| *Elosa woralli f. spinifera* Wiszn. | + | | |
| *Lecane bulla* (Gosse) | + | | |
| *L. clara* (Bryce) | | | + |
| *L. closterocerca* (Schm.) | + | + | + |
| *L. levistyla* (Olof.) | + | + | + |
| *L. luna* (Müll.) | + | + | |
| *L. lunaris* (Ehrb.) | + | + | |
| *L. psammophila* Wiszn. | + | + | + |
| *L. pyriformis* (Dad.) | + | | |
| *L. scutata* (Harr. & Myers) | + | + | + |
| *Lepadella acuminata* (Ehrb.) | + | | |
| *L. patella* (Müll.) | + | + | + |
| *Monommata astia* Myers | + | + | |
| *Notommata cyrtopus* Gosse | + | + | |
| *Proales minima* (Montet.) | + | + | |
| *Rotaria rotatoria* Pall. | + | + | |
| *Trichocerca iernis* (Gosse) | | + | |
| *T. insignis* (Herrick) | | + | |
| *T. intermedia* Sten | + | + | |
| *T. taurocephala* Hauer | + | + | |
| *T. tenuior* (Gosse) | + | + | |
| *T. rousseleti* Voigt | | + | |
| *Wierzejskiella velox* Wiszn. | | | + |
| | 24 | 26 | 10 |

ferentiated them from the psammic rotifers identified in other Polish lakes so far. However, no new species were found.

## Acknowledgements

I thank Prof. Dr Barbara Kłapcińska and Dr Zofia Piotrowska-Seget for help in preparing the English text.

## References

Ejsmont-Karabin, J., 1998. Is sandy beach in lakes an ecotone? (Psammon *Rotifera* in Lake Kuc) In Radwan, S. (ed.), Freshwater Ecotones. Structure, Types and Function. UMCS Publish., Lublin: 43–49 (English summary).

Evans, W., 1984. Seasonal abundances of the psammic rotifers of a physically controlled stream. Hydrobiologia 108: 105–114.

Kraska, M., H. Szyper & H. Romanowicz, 1994. Water trophy of 37 Lobelian lakes of the Tuchola Forest and of the Bytów Lake District. In Kraska, M. (eds), Lobelian Lake. Characteristic, Functioning and Protection. Idee Ekologiczne 6: 135–147 (English summary).

Krebs, J. C., 1989. Ecological Methodology. New York: 654 pp.

Pennak, R., 1951. Comparative ecology of the interstitial fauna of fresh-water and marine beaches. Année Biol. 27: 449–480

Pielou, E. C., 1975. Ecological Diversity. John Wiley & Sons, New York.

Radwan, S. & I. Bielańska-Grajner, 2001. Ecological structure of psammic rotifers in the ecotonal zone of Lake Piaseczno (eastern Poland). Hydrobiologia 446/447 (Dev. Hydrobiol. 153): 221–227.

Ruttner-Kolisko, A., 1953. Psammonstudien I. Das Psammon des Tronetträsk in Schwedisch – Lappland. Österr. Akad. Wiss. Math.-Nat. Kl. Sitzungsber. I, 162: 129–161.

Ruttner-Kolisko, A., 1954. Psammonstudien II. Das Psammon des Erken in Mittelschweden. Õsterr. Akad. Wiss. Math.-Nat. Kl. Sitzungsber. I, 163: 301–324.

Ruttner-Kolisko, A., 1965. Psammonstudien III. Das psammon des Lago Maggiore in Oberitalien. Mem. Ist. ital. Idrobiol. 9: 356–402.

Schmid-Araya, J. M., 1993. Benthic rotifera inhabiting the bed sediments of a mountain Jber. Biol. Stn Lunz 14: 75–101.

Schmid-Araya, J. M., 1994. Spatial and temporal distribution of micro-meiofaunal groups in an alpine gravel stream. Verh. int. Ver. Limnol. 25: 1649–1655.

Schmid-Araya, J. M., 1995. Disturbance and population dynamics of rotifers in bed sediments. Hydrobiologia 313/314: 279–290.

Schwoerbel, J., 1970. Methods of Hydrobiology (Freshwater biology). Pergamon Press Ltd: 200 pp.

Turner, P. N., 1990. Some Interstitial Rotifera from a Florida, USA, Beach. Trans. am. Microsc. Soc. 109: 417–421.

Turner, P. N., 1993. Distribution of rotifers in a Floridian saltwater beach, with a note on rotifer dispersal. Hydrobiologia 255/256: 435–439.

Turner, P. N., 1996. Preliminary data on rotifers in the interstitial of the Ninnescah river, U.S.A. Hydrobiologia 319: 179–184.

Tzschaschel, G., 1983. Seasonal abundance of psammon rotifers. Hydrobiologia 104: 275–278.

Wiszniewski, J., 1932. Les rotifères des rives sablonneuses du Lac Wigry. Arch. Hydrobiol. (Rybactwa Suwalki) 6: 86–99.

Wiszniewski, J., 1934a. Les rotifères psammiques. Ann. Mus. Zool. Pol. 10: 339–399.

Wiszniewski, J., 1934b. Recherches ècologiques sur le psammon. Arch. Hydrobiol. (Rybactwa Suwalki) 8: 149–271.

Wiszniewski, J., 1936. Notes sur le psammon III. Deux tourbières aux environs de Varsovie. Arch. Hydrobiol. (Rybactwa Suwalki) 10: 173–187.

Wiszniewski, J., 1937. Diffèrenciation ècologiques des rotifères dans le psammon d'eaux Douces. Ann. Mus. Zool. Pol. 12: 221–238.

Hydrobiologia **446/447:** 155–164, 2001.
L. Sanoamuang, H. Segers, R.J. Shiel & R.D. Gulati (eds), Rotifera IX.




# Distribution of rotifers in North Island, New Zealand, and their potential use as bioindicators of lake trophic state

I.C. Duggan[1], J.D. Green[1] & R.J. Shiel[2]
[1]*The Department of Biological Sciences, The University of Waikato, Private Bag 3105, Hamilton, New Zealand*
[2]*Murray-Darling Freshwater Research Centre, P.O. Box 921, Albury N.S.W. 2640, Australia*

*Key words:* rotifers, trophic state, bioindicators

## Abstract

The distribution and ecology of planktonic rotifers was investigated in 33 lakes in the North Island, New Zealand, between 1997 and 1999. A total of 79 species of monogonont rotifer were identified, with an average of 21 species per lake, a diversity which is high in comparison with many previous New Zealand studies. Most species recorded were cosmopolitan taxa, and were widespread in their distribution over the North Island. Multivariate analyses (Multi-Dimensional Scaling and Canonical Correspondence Analysis) did not distinguish distinct lake groupings based on rotifer communities, but rather gradients in assemblages, which were most highly associated with lake trophic state. Based on these responses, the development of potential rotifer bioindicator schemes for lake trophic state is described and discussed.

## Introduction

Diverse rotifer assemblages are known to exist in New Zealand, with approximately 400 species recorded to date (Shiel & Green, 1996). Many earlier studies focused on documenting new records of species from the country, rather than recording and exploring the distribution of species within and between individual lakes (see Shiel & Green, 1996, and references therein). Sanoamuang & Stout (1993) surveyed the distribution of species from 35 lakes in the South Island of New Zealand, the most widespread survey to date, although no attempt was made to infer the factors involved in determining the distribution of species between lake habitats. In the only study of factors affecting New Zealand rotifer community composition, Duggan et al. (in press) found the distribution of rotifers in 10 Rotorua lakes to be associated with lake trophic state. Based on this finding, Duggan et al. (in press) suggested rotifers may provide useful bioindicators of lake trophic state. This study was, however, based on single samples from each lake, and it is, therefore, not known if trophic state determines distribution when a wider range of seasonal variability, or a wider geographical area, is considered.

Trophic state has commonly been found to be important in determining distribution of rotifer communities elsewhere (e.g. Siegfried et al., 1989; Kaushik & Saksena, 1995; Ejsmont-Karabin, 1995). Several studies have provided lists of rotifer species indicative of different trophic states (e.g. Gannon & Stemberger, 1978; Mäemets, 1983; Berzins & Pejler, 1989a; Matveeva, 1991), although no quantitative community index has yet been developed based on these responses (however see Sládecek, 1983). This paper will assess the potential of rotifer communities as bioindicators of lake trophic status using a set of lakes spanning a wider geographical and temporal range than that used by Duggan et al. (in press), and develop a preliminary index that may potentially be utilised in the assessment of lake trophic state.

## Methods

Sampling localities are shown in Figure 1. Rotifer samples were collected by vertical hauls using a standardised 40 $\mu$m mesh net with a reducing cone. Sampling was carried out between 1997 and 1999 either by the author or by Regional Councils and

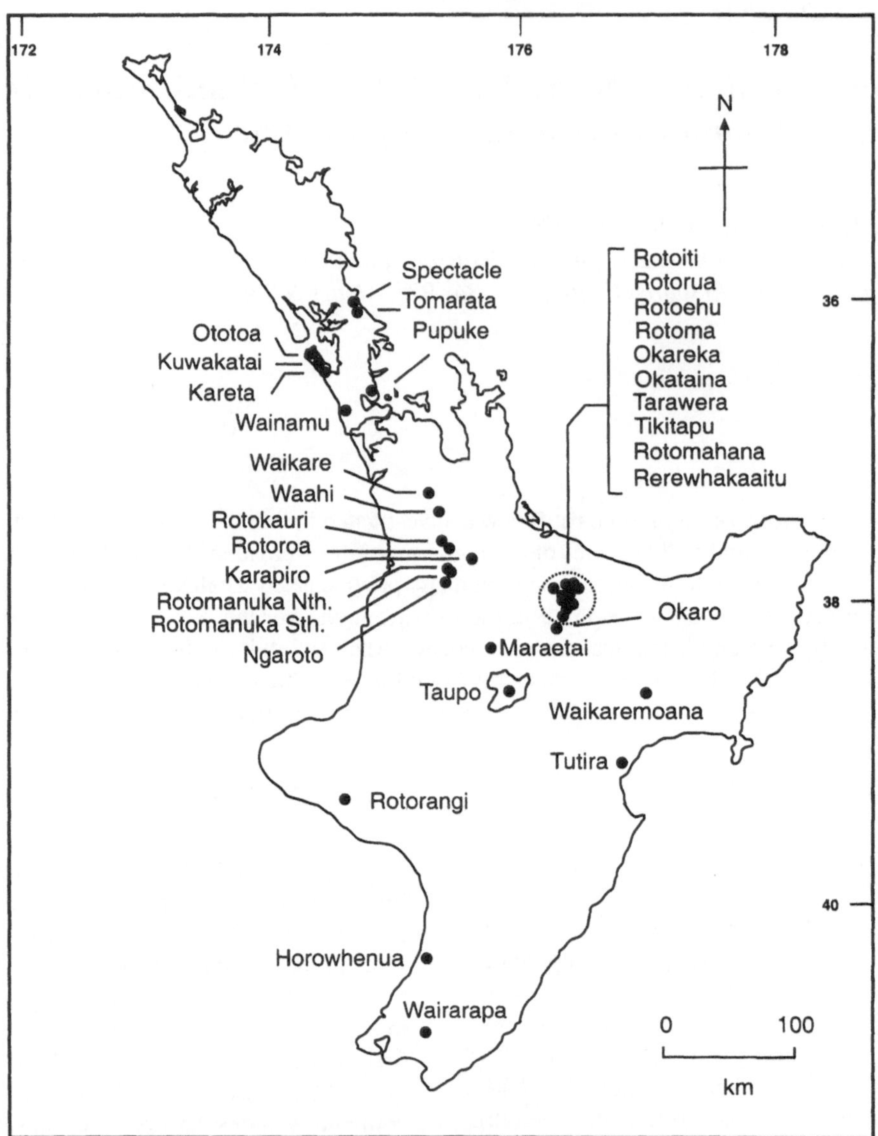


*Figure 1.* Sampled lakes in the North Island, New Zealand.

the National Institute of Water and Atmospheric Research (NIWA) during regular monitoring of lakes. Where possible, samples were taken once each season through the entire water column. Samples were preserved in 4% formalin. Temperature, dissolved oxygen (DO), pH, total phosphorus (TP), chlorophyll *a* and Secchi depth were determined at the time of sampling. In general, samples were taken from single central sites in each lake, except Lakes Rotorangi (2 sites) and Wairarapa (4 sites), and are initially assessed as separate entities.

For counting, samples were made up to a known volume, and enumerated until at least 1000 individuals were encountered, or until the entire sample was counted. Unknown animals were identified using Koste (1978) and Shiel (1995).

Initial analyses included only lakes for which there were more than two samples from different seasons. All data from each lake were averaged. Thirty five samples from 31 lakes were examined, with 27 species comprising greater than 4% in any sample from two lakes before averaging. As there were only single samples from Lakes Rerewhakaaitu and Rotomahana, these lakes were not included in these analyses. Cluster Analysis, Multidimensional Scaling (MDS) and Canonical Correspondence Analysis (CCA) were used to detect and classify groupings of species and to detect important physical, chemical or geographical variables associated with underlying trends. Cluster Analysis was based on the Bray-Curtis similarity coefficient calculated on the fourth root of the average relative abundances (percentages) of abundant rotifer species (comprising >4% in any two samples from different lakes). Species data were transformed to downweight dominant species. MDS was performed on the ranked Bray-Curtis similarity matrix produced by Cluster Analysis. CCA was performed using the average species relative abundance data (%), log ($x$+1) transformed to downweight dominant species. Environmental data was log ($x$+1) transformed to remove skew and subsequently standardised to zero mean and unit variance to remove the influence of differing scales of measurement. Missing temperature, DO and pH values were replaced with average values from all lakes. Standardised latitude, longitude and (log) mean lake depth data were included in the analysis, generally taken from data published by Irwin (1975), Livingston et al. (1986) and Lowe & Green (1987).

Bioindicator schemes are developed in relation to a single environmental gradient. The OECD scheme relies on separate assessments from TP, chlorophyll *a* and Secchi transparency measurements. The Trophic Lake Index (TLI) (Burns & Rutherford, 1998), developed for New Zealand conditions to enable easy comparison of trophic level of different lakes, provides an absolute indicator of lake trophic level by taking into account all of these three factors, plus total nitrogen (TN), to give a single numerical value of trophic state. On average, equivalent trophic levels are assigned using each indicator based on regression equations used in development, with the TLI value a single number that is an average of the four indicators. In the current study, TLI values were calculated only with the available data (generally four samples), without TN. The inference of lake trophic state using relative abundances of rotifer species in a sample requires a good statistical relation between the variable of interest (e.g. TP, TLI) and the rotifer taxa. In order to develop rotifer inference models for lake trophic state, the relative strengths of different trophic state indicators were estimated by running a series of Canonical Correspondence Analyses (CCA) constrained to single environmental variables (i.e. TLI, TP, chlorophyll *a* and Secchi transparency). In this type of analysis, there is one constrained axis related to each variable, and a series of unconstrained axes. The ratio of the first constrained eigenvalue ($\lambda_1$) to the second unconstrained eigenvalue ($\lambda_2$) indicates the relative importance of that environmental variable in explaining the species data (Ter Braak, 1988), and variables with high $\lambda_1/\lambda_2$ ratios are potential candidates for developing predictive models. Dixit et al. (1991) suggest using only those variables with $\lambda_1/\lambda_2$ ratios greater than 0.5 to develop inference models, with variables with lower ratios likely to be less robust. In the constrained CCAs and in the development of the bioindicator scheme, 121 samples (non-deseasonalised data) were used. In Lakes Wairarapa and Rotorangi, only samples from central sites were used so that additional samples did not to have over-representation in the construction of the bioindicator scheme. Analyses were performed using untransformed relative abundance data (%) of abundant species and raw (TLI) or log transformed (others) environmental data, averaged for each lake. Forty four species, comprising >1% of any one sample and present in five or more samples, were used to ensure a large initial species data set.

Tolerance downweighted Weighted Averaging regression (WA-tol) was used to construct an index for assessing lake trophic state from rotifer community composition, based on the trophic indicator with the strongest relationships with rotifer communities. This method is commonly used in paleolimnological studies for inferring past environmental conditions from the responses of modern taxa (Ter Braak & Van Dam, 1989; Birks et al., 1990), and involves both regression (Ter Braak & Looman, 1987) and calibration (Ter Braak, 1987) steps. WA regression is used to estimate species indicator values, which are equivalent to the optima of a species unimodal response curve to the variable of interest. A weighted average is taken of the environmental data, in which values are weighted proportional to the species relative abundance, i.e.:

$$u^* = (y_1x_1 + y_2x_2 + \ldots + y_nx_n)/(y_1 + y_2 + \ldots y_n)$$

where $u^*$ is the weighted average (= the species indicator value), $y_1, y_2, \ldots, y_n$ are the relative abundances of species at sites 1, 2 … $n$ and $x_1, x_2, \ldots, x_n$ are values of environmental variables at sites 1, 2 … $n$.

In the calibration step, the estimates of optima of taxa are used to infer environmental conditions based on the taxonomic composition of the sample. A WA estimate for a sample is the average of the optima of taxa in that sample, weighted with respect to their

*Table 1.* Total cumulative $\alpha$ species diversity from lakes with two or more samples. The data is ordered from highest to lowest diversity

| Lake | $\alpha$ | Lake | $\alpha$ | Lake | $\alpha$ |
|---|---|---|---|---|---|
| Maraetai | 35 | Okareka | 22 | Waikaremoana | 18 |
| Karapiro | 34 | Ototoa | 22 | Okaro | 17 |
| Rotorangi | 28 | Waahi | 22 | Rotoehu | 17 |
| Rotokauri | 26 | Tarawera | 21 | Tikitapu | 17 |
| Rotoiti | 26 | Rotomanuka South | 21 | Tomarata | 17 |
| Rotomanuka | 26 | Rotoroa | 21 | Wainamu | 17 |
| Kareta | 24 | Pupuke | 20 | Rotoma | 16 |
| Tutira | 24 | Waikare | 20 | Wairarapa | 10 |
| Kuwakatai | 23 | Rotorua | 19 | Taupo | 9 |
| Ngaroto | 23 | Spectacle | 19 | | |
| Horowhenua | 22 | Okataina | 18 | Average | 21.1 |

*Table 2.* Key to sample numbers for MDS and CCA ordinations of lake rotifer communities

| $n$ | Lake | $n$ | Lake | $n$ | Lake |
|---|---|---|---|---|---|
| 1 | Horowhenua | 13 | Rotoiti | 25 | Tikitapu |
| 2 | Karapiro | 14 | Rotokauri | 26 | Tomarata |
| 3 | Kareta | 15 | Rotoma | 27 | Tutira |
| 4 | Kuwakatai | 16 | Rotomanuka | 28 | Waahi |
| 5 | Maraetai | 17 | Rotomanuka South | 29 | Waikare |
| 6 | Ngaroto | 18 | Rotorangi (L2) | 30 | Waikaremoana |
| 7 | Okareka | 19 | Rotorangi (L3) | 31 | Wainamu |
| 8 | Okaro | 20 | Rotoroa | 32 | Wairarapa 1 |
| 9 | Okataina | 21 | Rotorua | 33 | Wairarapa 2 |
| 10 | Ototoa | 22 | Spectacle | 34 | Wairarapa 3 |
| 11 | Pupuke | 23 | Tarawera | 35 | Wairarapa 4 |
| 12 | Rotoehu | 24 | Taupo | | |

relative abundances, i.e.:

$$x_o = (y_1u_1 + y_2u_2 + ... + y_nu_n)/(y_1 + y_2 + ...y_n)$$

where $x_o$ is the assessed environmental variable, $y_1$, $y_2, \ldots, y_n$ are the responses (relative abundances) of species ($1, 2 \ldots n$) at the sites, $u_1, u_2, \ldots, u_n$ are the indicator values of species ($1, 2 \ldots n$) (i.e., the weighted average values, $u^*$, determined above).

In WA-tol, each taxon is downweighted by its respective variance.

$$x_o = [(y_1u_1/t_1^2) + (y_2u_2/t_2^2) + \ldots + (y_nu_n/t_n^2)] \\ /[(y_1/t_1^2) + (y_2/t_2^2) + \ldots (y_n/t_n^2)]$$

In WA-tol reconstructions, averages are taken twice, once in the development of indicator values and again in the inference of environmental conditions. The resulting 'shrinkage' of the inferred environmental variables is corrected for using a deshrinking regression equation (Birks et al., 1990). Classical deshrinking was used as it provided more reliable estimates than inverse deshrinking. Regression analyses were performed on deseasonalised observed *versus* inferred TLI values. Deseasonalised inferred WA-tol values were obtained as an average of each lakes inferred WA-tol values from different seasons.

Cluster analysis and MDS were performed using the Plymouth Routines in Multivariate Research statistical package (PRIMER, 1994; Clarke & Warwick, 1994), and CCA using CANOCO 4.0 (Ter Braak & Smilauer, 1998). WA-tol regression and calibration and calculation of deshrinking equations were performed using CALIBRATE v. 0.8.2 (Juggins & Ter Braak, 1998).

## Results

Seventy nine species of monogonont rotifer were found during this study. Mean lake species diversity, estimated from lakes with two or more samples, was 21.1 species per lake (Table 1). Lakes Taupo and Wairarapa contained the lowest diversities with nine and ten species, respectively, and Lakes Karapiro and Maraetai the greatest with 34 and 35 species.

Cluster analysis (Fig. 2) of the rotifer communities averaged seasonally for each lake revealed eight groupings of lakes at a 58% similarity level. The majority of samples were found in two clusters only (Clusters B and C). The remaining samples generally did not form distinct cluster groupings, indicating a gradation in community structure. Rotifer distribution appears to broadly relate to lake trophic state. Clusters A and B comprised mainly oligotrophic to mesotrophic lakes, Cluster C mesotrophic to eutrophic lakes, and Clusters D, E, F and G eutrophic to hypereutrophic lakes. Samples taken from different positions of the same water body were generally closely related to one another, e.g. the samples from the two sites in Lake Rotorangi, the four sites in Lake Wairarapa and the connected hydroelectric reservoirs, Lakes Karapiro and Maraetai. There appeared to be some geographical basis to the distribution of communities, with many of the Waikato riverine lakes (e.g. Lakes Ngaroto, Rotokauri and Rotomanuka South) or Rotorua volcanic lakes (e.g. Lakes Rotoma, Okataina and Okareka) being closely related to one another, although these lakes generally had similar trophic states.

The MDS of these lakes shows a general gradation in species composition based on trophic state (Fig. 3,

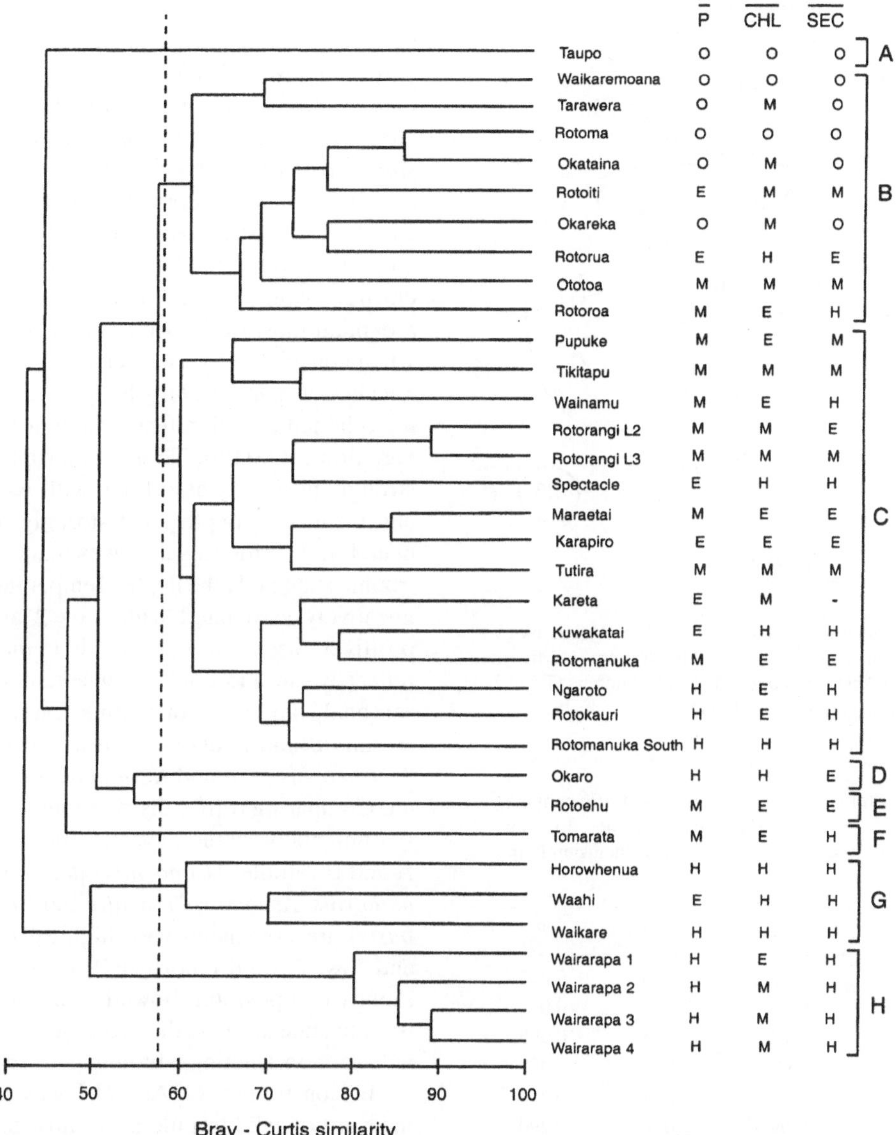


*Figure 2.* Cluster analysis of lakes based on mean percentage rotifer species composition, and inferred trophic states, using mean total phosphorus, mean chlorophyll *a* concentration, and mean Secchi transparency, based on the OECD (1982) fixed boundary system. Dashed line indicates 58% level of similarity.

Table 2). Lakes with relatively low trophic state appear at the bottom of the ordination, medium trophic state lakes central, and more highly trophic state lakes near the top and top left of the ordination. This distribution was reflected within clusters also, with the more oligotrophic lakes in Cluster B (e.g. Waikaremoana and Rotoma) distributed nearer to Cluster A, and more mesotrophic lakes (e.g. Rotoiti, Rotoroa and Rotorua) closer to Cluster C. The higher trophic state lakes in Cluster C (e.g. Ngaroto, Rotokauri and Rotomanuka South) distributed nearer to Clusters D, E, H and G. Lake Tomarata (Cluster F) was found to separate from the other clusters, with distribution perhaps determined by an environmental forcing factor other than trophic state.

CCA (Fig. 4, Table 2) revealed a distribution of lakes that was similar to those shown by cluster analysis and MDS. Lower trophic state lakes (Clusters A and B) were negatively associated with Axis 1, and mesotrophic to eutrophic samples (Cluster C)

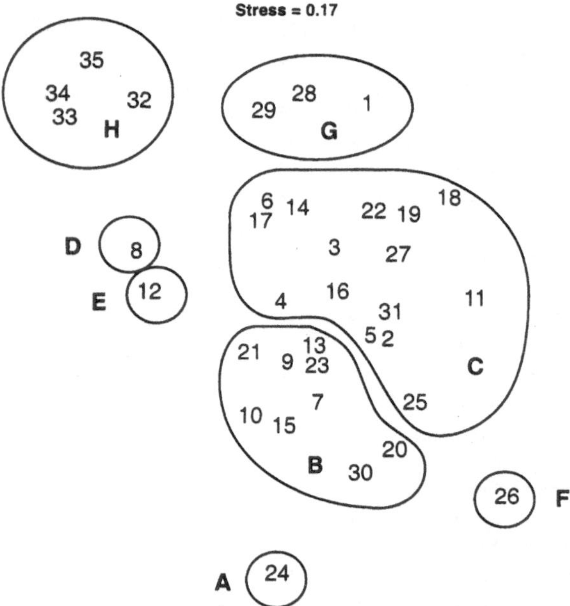


*Figure 3.* Non metric multi-dimensional scaling (MDS) plot of lakes based on mean% rotifer species composition (fourth root transformation). Numbers correspond to sample numbers (Table 2). The MDS is overlaid with groupings found in cluster analysis.

*Table 3.* Results of forward selection and Monte Carlo permutation tests from CCA of North Island rotifer species. Environmental variables are listed by the order of their inclusion in the model (lambda-A)

| | Lambda-1 | Lambda-A | *P* |
|---|---|---|---|
| Secchi | 0.21 | 0.21 | 0.005 |
| Temperature | 0.10 | 0.11 | 0.005 |
| Chlorophyll *a* | 0.12 | 0.09 | 0.005 |
| DO | 0.09 | 0.06 | 0.080 |
| Mean lake depth | 0.16 | 0.04 | 0.200 |
| Latitude | 0.09 | 0.04 | 0.290 |
| TP | 0.20 | 0.04 | 0.420 |
| Longitude | 0.08 | 0.03 | 0.375 |
| pH | 0.05 | 0.05 | 0.215 |

samples were generally found central in the ordination. Clusters D, E, G and H samples were positively associated with Axis 1, with Cluster E (Lake Rotoehu) strongly positively associating with Axis 2. *Conochilus dossuarius, Conochilus unicornis, Ascomorpha ovalis* and *Conochilus coenobasis* strongly negatively associated with Axis 1, and *Keratella tropica, Keratella tecta, Brachionus budapestinensis, Synchaeta oblonga* and *Hexarthra intermedia* were strongly positively associated. *Filinia terminalis, Trichocerca pusilla, Trichocerca longiseta, H. intermedia* and *C. dossuarius* were strongly positively associated with Axis 2, and *Anuraeopsis fissa, Keratella slacki, B. budapestinensis* and *K. tecta* strongly negatively associated. Forward selection and associated Monte Carlo permutation tests of the significance of environmental variables (Table 3) indicate that Secchi transparency, TP, average depth and chlorophyll *a* explain most of the variation in species distribution when considered by themselves. After the addition of Secchi transparency, only temperature and chlorophyll *a* explained any significant amount of the remaining variation ($P<0.05$). TP and chlorophyll *a* were most strongly positively associated with Axis 1, and Secchi and mean lake depth most strongly negatively associated, indicating this axis was mainly related to lake trophic state and lake depth. Temperature was strongly negatively associated with Axis 2 and DO strongly positively associated. This axis is therefore likely to reflect warm and cold water assemblages caused by seasonal, latitudinal or altitude variation, e.g. *Filinia terminalis* and *Trichocerca pusilla* in colder water and *Anuraeopsis fissa* and *Keratella slacki* in warmer water. Comparing rotifer distributions with measured environmental variables, species associated with Cluster A and B samples (*Conochilus dossuarius, Conochilus unicornis, Ascomorpha ovalis* and *Conochilus coenobasis*) are associated with high Secchi transparency, and low TP and chlorophyll *a* levels. Rotifer community composition showed strong relationships with trophic state so was, therefore, investigated as a variable for constructing a bioindicator scheme.

In constrained CCAs, TLI was identified as the measure of lake trophic state most highly associated with rotifer distribution ($\lambda_1/\lambda_2$ ratio = 0.505), followed by TP (0.490), Secchi transparency (0.465 ) and chlorophyll *a* (0.427). TLI is the only variable whose first constrained eigenvalue can be considered large (>0.5) compared to the other unconstrained eigenvalues (cf. Dixit et al., 1991). The predictive ability of the inference model was assessed from the coefficient of determination ($R^2$) between the observed and the rotifer inferred TLI or RCI values, and the deviation between observed and inferred values of samples (Fig. 5, Table 4). Only Lake Waahi was assessed as having TLI levels greater than 2 TLI units different from the observed. The WA optima and tolerance of species is given in Table 5. The WA-tol classical deshrinking

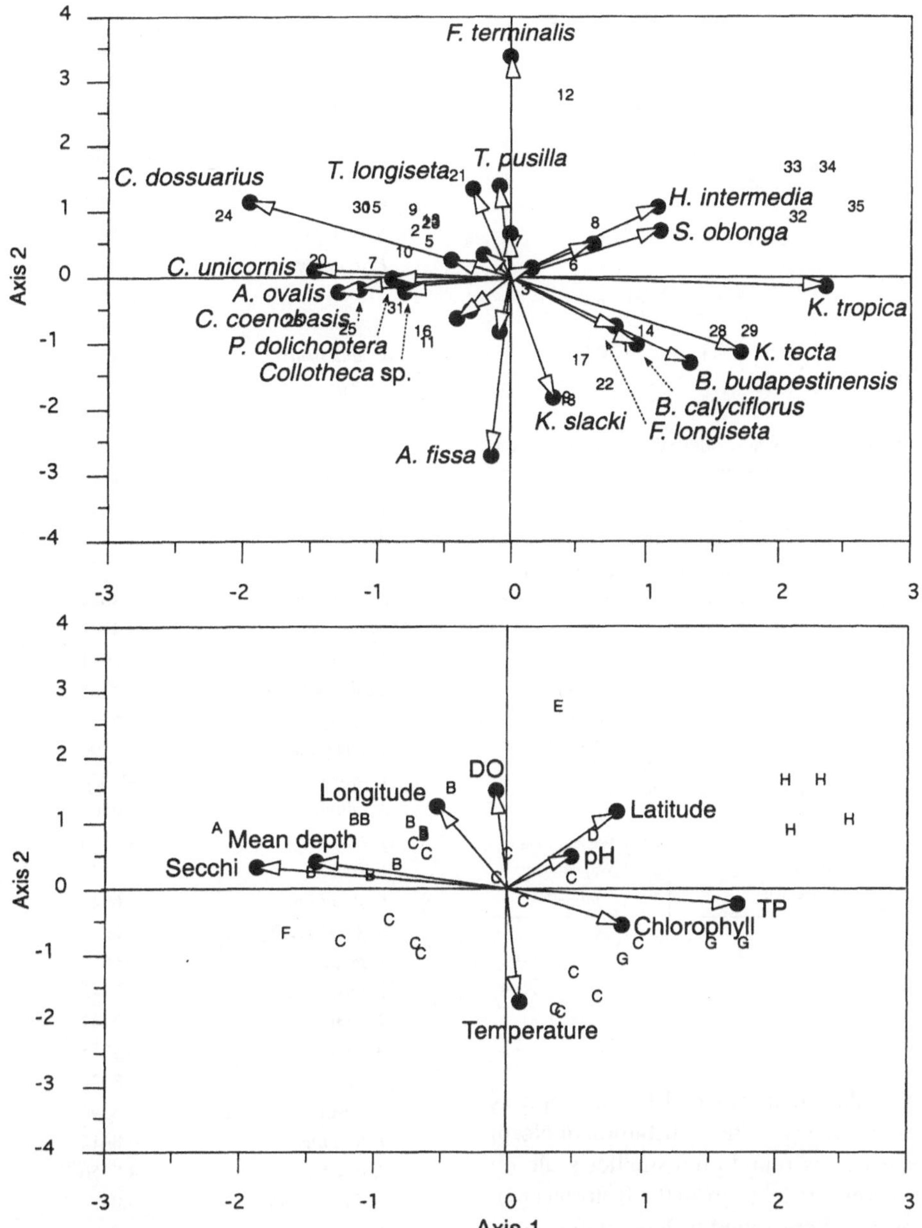


*Figure 4.* Ordination diagram based on canonical correspondence analysis (CCA) of North Island rotifer species with respect to environmental variables. Eigenvalues for the first two axes = 0.243 and 0.125. Scores of rotifer species and environmental variables were scaled to fit the sample ordination. Numbers correspond to sample numbers (Table 2). Species weakly associating with both Axes 1 and 2 have been omitted from the ordination for clarity.

equation is:

$$\mathrm{TLI_{final}} = (\mathrm{TLI_{initial}} - 2.932)/0.371.$$

Estimates of the average yearly TP and chlorophyll *a* can be calculated from the TLI value using the following equations derived from Burns & Rutherford (1998):

$$\log(\mathrm{TP}) = (\mathrm{TLI} - 0.218)/2.92$$

$$\log(\mathrm{chlorophyll}\,a) = (\mathrm{TLI} - 2.22)/2.54$$

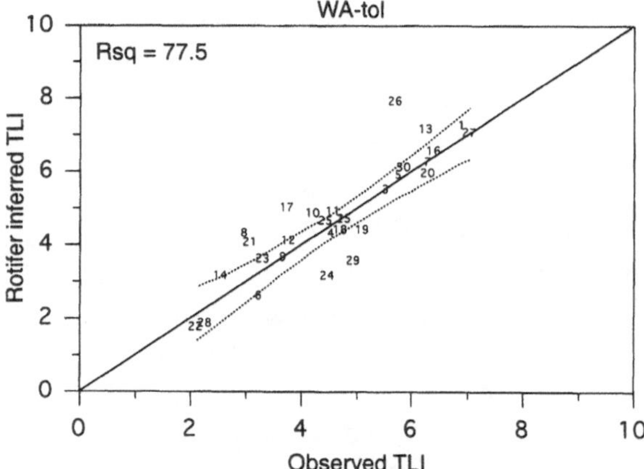


*Figure 5.* Plot of observed TLI against inferred TLI, derived from WA-tol. Solid line indicates regression line, and dashed lines indicate 95% confidence limits.

*Table 4.* Key to sample numbers for inferred *versus* observed TLI

| *n* | Lake | *n* | Lake | *n* | Lake |
|---|---|---|---|---|---|
| 1 | Horowhenua | 11 | Rotoehu | 21 | Tarawera |
| 2 | Karapiro | 12 | Rotoiti | 22 | Taupo |
| 3 | Kuwakatai | 13 | Rotokauri | 23 | Tikitapu |
| 4 | Maraetai | 14 | Rotoma | 24 | Tomarata |
| 5 | Ngaroto | 15 | Rotomanuka | 25 | Tutira |
| 6 | Okareka | 16 | Rotomanuka South | 26 | Waahi |
| 7 | Okaro | 17 | Rotorangi | 27 | Waikare |
| 8 | Okataina | 18 | Rotoroa | 28 | Waikaremoana |
| 9 | Ototoa | 19 | Rotorua | 29 | Wainamu |
| 10 | Pupuke | 20 | Spectacle | 30 | Wairarapa |

## Discussion

Trophic state was the environmental factor that was most strongly associated with the distribution of North Island rotifer species, as found on a smaller scale by Duggan et al. (in press) in lakes from the Rotorua area. Many rotifer species were found to have preferences for either particular ranges or extremes in trophic state. For example, *Ascomorpha ovalis, Conochilus coenobasis, Conochilus dossuarius, Conochilus unicornis, Polyarthra dolichoptera* and *Synchaeta longipes* were found associated with low trophic state, and *Brachionus budapestinensis, Brachionus calyciflorus, Filinia longiseta, Keratella slacki, Keratella tecta* and *Keratella tropica* with high trophic state. Species found with poor statistical associations with trophic state in the current study may, however, also have distributions more indicative of trophic state that will become apparent once a larger data set is obtained. Species

*Table 5.* Weighted average (WA) optima and tolerance data for TLI for abundant North Island rotifer species. Species are ordered by TLI optima

| Species | TLI optimum | TLI tolerance |
|---|---|---|
| *C. dossuarius* | 3.0989 | 0.9545 |
| *S. longipes* | 3.3232 | 0.3942 |
| *P. dolichoptera* | 3.4396 | 1.3568 |
| *T. stylata* | 3.7553 | 1.2586 |
| *G. minor* | 3.7921 | 0.3662 |
| *C. unicornis* | 3.8036 | 1.1211 |
| *G. hyptopus* | 3.8248 | 1.0634 |
| *C. coenobasis* | 3.9056 | 0.6608 |
| *A. ovalis* | 3.9558 | 0.8746 |
| *L. closterocerca* | 4.1376 | 0.5969 |
| *L. bulla* | 4.1650 | 0.7413 |
| *T. patina* | 4.3055 | 1.1250 |
| *S. oblonga* | 4.3875 | 1.2897 |
| *A. priodonta* | 4.4042 | 1.3888 |
| *A. navicula* | 4.4189 | 0.5391 |
| *S. pectinata* | 4.5011 | 0.9830 |
| *Collotheca* sp. | 4.5186 | 1.6649 |
| *F.* cf.*pejleri* | 4.5193 | 0.9545 |
| *F. terminalis* | 4.5290 | 0.1414 |
| *H. mira* | 4.6060 | 0.7787 |
| *E. dilatata* | 4.6508 | 0.5745 |
| *T. tetractis* | 4.6885 | 0.1556 |
| *S. stylata* | 4.6926 | 0.5304 |
| *C. catellina* | 4.6947 | 0.4421 |
| *A. brightwelli* | 4.6949 | 0.9757 |
| *T. tenuior* | 4.6982 | 0.1236 |
| *T. porcellus* | 4.7448 | 0.3732 |
| *T. similis* | 4.7747 | 0.9012 |
| *A. fissa* | 4.8205 | 1.1542 |
| *K. cochlearis* | 4.8324 | 1.1914 |
| *F. novaezealandiae* | 4.8392 | 1.4754 |
| *T. longipes* | 4.8412 | 1.0207 |
| *T. pusilla* | 4.8556 | 0.7896 |
| *H. intermedia* | 5.0850 | 1.4825 |
| *K. procurva* | 5.2296 | 1.1108 |
| *P. complanata* | 5.2315 | 1.1958 |
| *A. sieboldi* | 5.6245 | 1.3073 |
| *K. tropica* | 5.8483 | 1.0890 |
| *B. quadridentatus* | 5.9200 | 0.9669 |
| *K. slacki* | 5.9414 | 0.9749 |
| *K. tecta* | 6.0166 | 1.1050 |
| *B. calyciflorus* | 6.1631 | 0.4242 |
| *F. longiseta* | 6.3957 | 0.7238 |
| *B. budapestinensis* | 6.5324 | 0.4540 |

found to be indicative of high and low trophic state in general correspond with those found in studies elsewhere, e.g. Mäemets (1983), Sládecek (1983), Pejler (1983) and Berzins & Pejler (1989a). Pejler (1983) attributed species preferences along a trophic gradient to the size and nature of the particulate food present, with ultraoligotrophic rotifer species, e.g. *Conochilus*, predominantly feeders of minute algal particles, hypereutrophic species, e.g. *Brachionus*, predominantly feeders of bacteria, and those between these ranges feeders on coarser particles. However, as with phytoplankton communities, distribution of rotifer species along the trophic gradient is likely to be caused not by competition for a single factor, e.g. food type (or nutrients in the case of phytoplankton), but by the consequential impacts of a variety of factors along the gradient (Reynolds, 1998). Examples of this are likely to include the appearance of toxic cyanobacteria (Snell, 1980; Starkweather & Kellar, 1987), the increased importance of predation on rotifers by *Asplanchna* (Ejsmont-Karabin, 1974; Pejler, 1983) and the degree of oxygen depletion (e.g. Berzins & Pejler, 1989b; Mikschi, 1989) with increasing trophic state. These factors will weigh in favour of the growth and survival of particular rotifer species, with the realised niches of species the outcome of several dimensions of variability along a common gradient.

Based on the relationships between rotifer species composition and trophic state, rotifers have previously been suggested as bioindicators. Trophic state inference in these studies has, however, generally been fairly coarse, relying on the simple comparison of rotifer composition with species lists of indicative taxa (e.g. Mäemets, 1983; Matveeva 1991). Sládecek (1983), suggested a more quantitative method that used the ratio of the number of species of *Brachionus* to *Trichocerca*, based on his findings that *Brachionus* are associated with eutrophic waters and *Trichocerca* with oligotrophic waters. However, because this relationship will generally be based on a limited number of species, it is likely to provide only a very coarse measure. Also, *Trichocerca* species were in general not found to be indicative of oligotrophic conditions in the current study, and *Brachionus* species are known to be limited in their distribution globally (e.g. Dumont, 1983), likely affecting this relationship from place to place.

Biotic score systems, based on community composition, have generally been preferred over indices based on either a single indicator taxon or diversity measures in freshwater biomonitoring (e.g. Armitage et al., 1983; De Pauw & Vanhooren, 1983; Stark, 1985). Score systems are based on the responses of a number of species, with inferred conditions based on those present in a sample, but not affected by species absence. Because of the general cosmopolitan nature of rotifer species, and the similarity between species–trophic state relations shown in this and other studies, the development of a score system for assessing lake trophic state based on North Island rotifer species may be relevant not only in New Zealand, but also possibly elsewhere with little or no modification. This provides advantages over bioindicator schemes using macroinvertebrates as these species, and therefore indices, are generally regionally specific (cf. Armitage et al., 1983; Stark, 1985).

WA-tol gave a high coefficient of correlation (77.5) and a low spread around the regression line because species with narrow tolerances are given a greater weighting than those with wider tolerances, without the removal of rare species. Ideally, indicator species need to have narrow tolerances to the variable of interest (Rosenberg & Resh, 1993), and WA-tol therefore gives preference to these species. This scheme also results in a numerical value of use to freshwater managers because it provides inferences of commonly used measures of trophic state such as TP and TLI. This, therefore, appears to be a useable method for inferring trophic state from rotifer community composition. The close relationship between this rotifer index and trophic state variables indicates that rotifers have potential as bioindicators of lake trophic state in the North Island of New Zealand. The advantage of rotifer indices over the TLI and traditional OECD assessment is that they require the measure of a single factor, so saving time and cost.

## Acknowledgements

We thank Lee Laboyrie and Gavin Reynolds for aid with sampling, and Michelle White for help with manuscript formatting. We thank also NIWA and the following Regional Councils for collecting samples and providing data: Auckland Regional Council; Manawatu-Wanganui Regional Council; Wellington Regional Council; Environment BOP; Environment Waikato; Taranaki Regional Council; Northland Regional Council. We are grateful to George Payne of NIWA for allowing use of the NIWA spectrofluorometer. Financial support was provided by the Hillary Jolly Memorial Scholarship, and the project was funded by an Environment Waikato Research grant.

## References

Armitage, P. D., D. Moss, J. F. Wright & M. T. Furse, 1983. The performance of a new biological water quality score system based on macroinvertebrates over a wide range of unpolluted running-water sites. Wat. Res. 17: 333–347.

Berzins, B. & B. Pejler, 1989a. Rotifer occurrence and trophic degree. Hydrobiologia 182: 171–180.

Berzins, B. & B. Pejler, 1989b. Rotifer occurrence in relation to oxygen content. Hydrobiologia 183: 165–172.

Birks, H. J. B., J. M. Line, S. Juggins, A. C. Stevenson & C. J. F. Ter Braak, 1990. Diatoms and pH reconstruction. Phil. Trans. r. Soc. Lond., series B 327: 263–278.

Burns, N. M. & J. C. Rutherford, 1998. Results of Monitoring New Zealand Lakes 1992–1996, Vol. 2 – Commentary on Results. NIWA Client Report: MFE80216: 125 pp.

Clarke, K. R. & R. M. Warwick, 1994. Change in marine communities: an approach to statistical analysis and interpretation. National Environment Research Council, U.K.: 144 pp.

De Pauw, N. & G. Vanhooren, 1983. Method for biological quality assessment of watercourses in Belgium. Hydrobiologia 100: 153–168.

Dixit, S. S., A. S. Dixit & J. P. Smol, 1991. Multivariate environmental inferences based on diatom assemblages from Sudbury (Canada) lakes. Freshwat. Biol. 26: 251–266.

Duggan, I. C., J. D. Green & K. Thomasson, in press. Do rotifers have potential as bioindicators of lake trophic state? Verh. int. Ver. für Theor. Angewan. Limnol. 27.

Dumont, H. J., 1983. Biogeography of rotifers. Hydrobiologia 104: 19–30.

Ejsmont-Karabin, J., 1974. Research on the feeding of planktonic polyphage *Asplanchna priodonta* Gosse (Rotatoria). Ekol. Pol. seria A 22: 311–317.

Ejsmont-Karabin, J., 1995. Rotifer occurrence in relation to age, depth and trophic state of quarry lakes. Hydrobiologia 313/314: 21–28.

Gannon, J. E. & R. S. Stemberger, 1978. Zooplankton (especially crustaceans and rotifers) as indicators of water quality. Trans. am. Microsc. Soc. 97: 16–35.

Irwin, J., 1975. Checklist of New Zealand lakes. Vol. 74. New Zealand Oceanographic Institute. Wellington.

Juggins, S., & C. J. F. Ter Braak 1998. CALIBRATE version 0.8.2. A C++ Program for analysing and visualising species environmental relationships and for predicting environmental values from species assemblages.

Kaushik, S. & D. N. Saksena, 1995. Trophic status and rotifer fauna of certain water bodies in Central India. J. envir. Biol. 16: 283–291.

Koste, W., 1978. Rotatoria. Die Rädertiere Mitteleuropas. 2 vols, Gebrüder Borntraeger, Berlin, Stuttgart, West Germany: 673 pp, 234 pp.

Livingston, M. E., B. J. Biggs & J. S. Gifford, 1986. Inventory of New Zealand lakes: Part I, North Island. Water and Soil Miscellaneous Publication 80, Wellington, New Zealand: 199 pp.

Lowe, D. J. & J. D. Green, 1987. Origins and development of the lakes. In Viner, A. B. (ed.), Inland Waters of New Zealand. New Zealand Department of Scientific and Industrial Research bulletin 241: 1–64.

Mäemets, A., 1983. Rotifers as indicators of lake types in Estonia. Hydrobiologia 104: 357–361.

Matveeva, L. K., 1991. Planktonic rotifers as indicators of trophic state. Bulletin of the Moscow Naturalists' Society, Biology Section 96: 54–62.

Mikschi, E., 1989. Rotifer distribution in relation to temperature and oxygen content. Hydrobiologia 186/187: 209–214.

OECD (Organization for Economic Co-operation and Development), 1982. Eutrophication of waters: monitoring assessment and control. OECD, Paris: 154 pp.

Pejler, B., 1983. Zooplanktic indicators of trophy and their food. Hydrobiologia 101: 111–114.

PRIMER (Plymouth routines in multivariate ecological research), 1994. PRIMER v. 4.0. Plymouth Marine Laboratory, Plymouth.

Reynolds, C. S., 1998. What factors influence the species composition of phytoplankton in lakes of different trophic status? Hydrobiologia 369/370: 11–26.

Rosenberg, D. M. & V. H. Resh, 1993. Introduction to Freshwater Biomonitoring and Benthic Macroinvertebrates. In Rosenberg, D. M. & V. H. Resh (eds), Freshwater Biomonitoring and Benthic Macroinvertebrates. Chapman and Hall, New York and London: 1–9.

Sanoamuang, L. & V. M. Stout, 1993. New records of rotifers from the South Island lakes, New Zealand. Hydrobiologia 255/256: 481–490.

Shiel, R. J., 1995. A guide to the identification of rotifers, cladocerans and copepods from Australian inland waters. Albury, N.S.W., Co-operative Research Centre for Freshwater Ecology, Murray-Darling Freshwater Research Centre: 144 pp.

Shiel, R. J. & J. D. Green, 1996. Rotifera recorded from New Zealand, 1859–1995, with comments on zoogeography. New Zealand J. Zool. 23: 193–209.

Siegfried, C. A., J. A. Blomfield & J. W. Sutherland, 1989. Planktonic rotifer community structure in Adirondack, New York, U.S.A. lakes in relation to acidity, trophic status and related water quality characteristics. Hydrobiologia 175: 33–48.

Sládecek, V., 1983. Rotifers as indicators of water quality. Hydrobiologia 100: 169–201.

Snell, T. W., 1980. Blue-green algae and selection in rotifer populations. Oecologia 46: 343–346.

Stark, J. D., 1985. A Macroinvertebrate Community Index of Water Quality for Stony Streams. Water and Soil Miscellaneous Publication No. 87. Wellington, NZ: 53 pp.

Starkweather, P. L. & P. E. Kellar, 1987. Combined influences of particulate and dissolved factors in the toxicity of *Microcyctis aeruginosa* (NRC-SS-17) to the rotifer *Brachionus calyciflorus*. Hydrobiologia 147: 375–378.

Ter Braak, C. J. F., 1987. Calibration. In Jongman, R. H. G., C. J. F. Ter Braak & O. F. R. Van Tongeren (eds). Data Analysis in Community and Landscape Ecology. Pudoc, Wageningen: 78–90.

Ter Braak, C. J. F., 1988. Partial canonical correspondence analyses. In Bock, H. H. (ed.), Classification and Related Methods of Data Analysis. North Holland, Amsterdam: 551–558.

Ter Braak, C. J. F. & C. W. N. Looman, 1987. Regression. In Jongman, R. H. G., C. J. F. Ter Braak & O. F. R. van Tongeren (eds), Data Analysis in Community and Landscape Ecology. Pudoc, Wageningen. The Netherlands: 29–77.

Ter Braak, C. J. F. & P. Smilauer, 1998. Canoco for Windows Version 4.02. Wageningen. The Netherlands.

Ter Braak, C. J. F. & H. Van Dam, 1989. Inferring pH from diatoms: a comparison of old and new calibration methods. Hydrobiologia 178: 209–223.

Hydrobiologia **446/447:** 165–171, 2001.
L. Sanoamuang, H. Segers, R.J. Shiel & R.D. Gulati (eds), Rotifera IX.




# Urban rotifers: structure and densities of rotifer communities in water bodies of the Poznań agglomeration area (western Poland)

Jolanta Ejsmont-Karabin[1] & Natalia Kuczyńska-Kippen[2]
[1]*Institute of Ecology, Polish Academy of Sciences, Hydrobiological Station, Lesna 13, 11-730 Mikolajki, Poland*
[2]*Department of Hydrobiology, Adam Mickiewicz University, Marcelinska 4, 60-801 Poznań, Poland*

*Key words:* Poznań agglomeration, urban habitats, Rotifera, species diversity

## Abstract

The rotifer fauna of 19 mostly small water bodies (natural and artificial ponds, clay-pits and pools) in Poznań was studied on four occasions during 1996–98 to determine the suitability of urban areas for rotifer habitats. Rotifers were present in all the water bodies studied, with 114 species in 39 genera found, representing ca. 25% of all rotifers recorded from Poland. Mean diversity was 10 spp (range 1–36). Most common were: *Brachionus angularis* and *Keratella cochlearis* (spring), *Colurella uncinata, Lecane closterocerca* and *Lepadella patella* (summer) and *L. closterocerca* and *K. cochlearis* (autumn). Rotifer densities (1-1503 ind $l^{-1}$), Shannon's diversity (H′ 0.00-3.71) and dominant species differed in different water-bodies. The index of percentage similarity of community showed strong differences in qualitative structure of rotifer assemblages. The different types of water habitats, both the existing or the newly created in towns, may explain the relatively high diversity of rotifer communities observed in the urban areas studied.

## Introduction

Urban development is thought to affect negatively the richness of natural assemblages of animals like birds (Nichols et al., 1998; Blair, 1999), mammals, amphibians and invertebrates (Delis et al., 1996; Mazgajska, 1996). In the process of urbanisation natural aquatic environments – streams, ponds, wetlands – are replaced by mostly artificial small water bodies, ponds, clay-pits, pools and reservoirs. Although species richness of these sites has been found to be similar to relatively undamaged ones (Langley et al., 1995) rotifer communities inhabiting urban water bodies are poorly known.

Urban water-bodies usually receive large amounts of anthropogenic pollutants and become highly degraded environments. These inputs of nutrients are variable in time due to effects of weather (Carpenter et al., 1998). Stressed in this way, ecosystems are characterized by a 'distress syndrome' which is usually indicated by reduced biodiversity, increased dominance of exotic species and increased dominance of smaller opportunistic species (Rapport & Whitford, 1999). According to the intermediate disturbance hypothesis (Connell, 1978), environments under great stress are inhabited by very few species, whereas communities living in moderately stressed ecosystems achieve the highest diversity.

According to Jenkins & Buikema (1998), structurally different zooplankton communities may inhabit even newly constructed ponds of similar environmental 'quality'. The Poznań agglomeration is rich in small water bodies representing very differentiated habitats for invertebrates. It may be then expected that rotifer communities in the water bodies are different in their species richness and total density.

The aim of this study was to test the above hypothesis and to examine if urban areas may offer suitable habitats for rotifer fauna.

## Study area

Poznań is the fifth largest city in Poland, of ca. 261 $km^3$ area and ca. $600 \times 10^3$ inhabitants. Nineteen mostly small water bodies, including natural and artificial ponds, clay-pits, pools (Table 1), were sampled in the Poznań area in spring 1996, summer 1997 and

*Table 1.* List and general description of localities sampled during this study and summer values of total phosphorus (TP) and nitrogen (TN). Explanation: A = area in $m^2$; I, II, III – visits to the water-bodies in spring 1996, summer 1997 and autumn 1998, respectively

| | |
|---|---|
| 1 | Non-existing at present pool (A = 5000) in Zawady; concrete walls; devoid of macrophytes; high concentration of suspended clay material; I; |
| 2 | Clay-pit (A = 30 000) at Głogowska St. – a large and polluted with domestic sewage reservoir with reed belt all round; *Typha* sp., *Potamogeton* spp., *Ceratophyllum* sp., *Myriophyllum* sp., *Ranunculus circinatus* Sibth; TP- 32 $\mu g\ l^{-1}$, TN – 130 $\mu g\ l^{-1}$; I, II, III + summer 1998; |
| 3 | Pond (A = 4000) in the Bajka Estate; surrounded with houses and gardens; dense stands of submerged and submersed vegetation (*Typha latifolia* L., *T. angustifolia* L., *Sparganium* sp., *Polygonum amphibium* L., *Sagittaria* sp., *Ranunculus* sp., *Ceratophyllum* sp., frogbit, sedges); introduced fish; TP- 232 $\mu g\ l^{-1}$, TN – 1880 $\mu g\ l^{-1}$; I, II + summer 1998; |
| 4 | Fountain (A = 2200) in front of the opera-house, reservoir in a centre of the city; drained for winter and refilled in late spring; submerged vegetation lacking; dense populations of insect larvae and big crustaceans; TP- 142 $\mu g\ l^{-1}$, TN – 880 $\mu g\ l^{-1}$; I, II, III + summer 1998; |
| 5<br>6 | Sołacki Pond with two basins (upper – No 5 and lower – No 6; A = 25 000) connected with a flood-gate; the upper basin with stone edge; both surrounded with trees and bushes of a park; single specimens of *Phragmites* sp. and *Typha* spp.; **5**: TP- 402 $\mu g\ l^{-1}$, TN – 3150 $\mu g\ l^{-1}$; **6**: TP- 218 $\mu g\ l^{-1}$, TN – 2100 $\mu g\ l^{-1}$; I, II, III; |
| 7 | Pond (A = 3200) with fountain at the Academy of Economy in the centre of the city in a park neighbouring with very crowded streets; concrete reservoir with *Lemna* sp. and *Ceratophyllum* sp.; TP- 119 $\mu g\ l^{-1}$, TN – 930 $\mu g\ l^{-1}$; I, II + summer 1998; |
| 8 | Pond (A = 3000) in the Wodziczko's Park; small and overgrown with reed; non-existing at present; I, II; |
| 9 | Natural pond (A = 3500) in a park on the River Warta bank; dense stands of *Phragmites* sp., *Typha* sp.and *Ceratophyllum* sp.; TP- 160 $\mu g\ l^{-1}$, TN – 280 $\mu g\ l^{-1}$; I, II, III; |
| 10 | Pond (A = 30 000) at Głogowska St. – close to a very busy road and surrounded with public edifices, factories and houses; clouds of *Daphnia magna* Straus; some macrophytes like *Typha* sp., *Ceratophyllum* sp. and from time to time thick layer of *Lemna* sp.; TP- 125 $\mu g\ l^{-1}$, TN - 2100$\mu g\ l^{-1}$; I, II; |
| 11 | Small concrete pool (A = 2000) in the Citadel (Rosary); no macrophytes; frogs abundant; I, II; |
| 12 | Large, natural pond (A = 11 500) in St. George's Park; sand beaches; stands of *Phragmites* sp. and *Typha* sp.; quite abundant submerged vegetation – *Myriophyllum* sp., *Ceratophyllum* sp.; TP- 553 $\mu g\ l^{-1}$, TN – 230 $\mu g\ l^{-1}$; I, II, III; |
| 13 | Temporary pool (A = 400) in Lech's Estate; submerged mosses, sedges, rushes, *Butomus umbellatus* L.; stands of *Phragmites* sp. and *Typha* sp.; I, III; |
| 14 | Pond (A = 4000) in Golęcin; surrounded with pine-forest; littoral zone with *Phragmites* sp. and *Elodea canadensis* Michx.; used for recreation though littered with garbage; TP- 80 $\mu g\ l^{-1}$, TN – 1000 $\mu g\ l^{-1}$; I, II, III + summer 1998; |

*Continued on p. 167*

*Table 1.* contd.

| | |
|---|---|
| 15 | Very shallow pond (A = 1400) situated in a park surrounded with high buildings; covered with dense stands of macrophytes (*Chara* sp., *Phragmites* sp., *Typha* sp., *B. umbellatus*, *Potamogeton* spp., sedges, mosses and *Ceratophyllum* sp.); TP- 348 $\mu$g $l^{-1}$, TN – 2500 $\mu$g $l^{-1}$; II, III; |
| 16 | Pond (A = 950) in the Dąbrowski's Park; extremely polluted mostly due to occupation by very numerous ducks; single stand of *Typha* sp.; from time to time the surface covered with *Lemna* sp.;TP- 1267 $\mu$g $l^{-1}$, TN – 8650 $\mu$g $l^{-1}$; II, III; |
| 17 | Large concrete pool (A = 1000) in zoological garden; polluted due to presence of birds; no macrophytes (for the same reason); TP- 226 $\mu$g $l^{-1}$, TN – 1780 $\mu$g $l^{-1}$; II, III; |
| 18 | Small concrete pool (A = 300) in zoological garden; no macrophytes; TP- 796 $\mu$g $l^{-1}$, TN – 4330 $\mu$g $l^{-1}$; III; |
| 19 | Mankol – pond (4500) surrounded with houses, factories, shops; wide belt of reeds at the margin; some stands of *Ceratophyllum* sp.; fish and small invertebrates abundant; TP- 1322 $\mu$g $l^{-1}$, TN – 1450 $\mu$g $l^{-1}$; II, III + summer 1998; |

1998, and autumn 1998. Some of the water bodies disappeared during the study and some were included in summer and autumn 1998 (Table 1), for a total of 49 samples.

## Material and methods

Zooplankton was sampled from an open water zone, 2 m from the shoreline. Samples were taken at 0.5 m intervals from the surface to the bottom of water-bodies with a 1-l water sampler. Five litres were pooled to get a composite sample, filtered through a plankton net of 30 $\mu$m mesh size and fixed with 4% formalin. In the laboratory, rotifers were transferred to sedimentation chambers and counted using a Nikon Optiphot-2PH microscope at 100× magnification. Subsamples were counted equal to 1 l of lake water. In summer 1997, the concentration of total nitrogen was determined by digestion in sulphuric acid (Kjeldahl method) and total phosphorus, by digestion in perchloric acid (Golterman, 1969).

The index of percentage similarity of community (PSC) (Whittaker & Fairbanks, 1958) was used:

$$PSC = 100 - 0.5 \cdot (a\text{-}b) = \cdot \min.(a, b),$$

where $a$ and $b$ are percentages of individuals of each species in total numbers of the communities of lakes A and B, compared in pairs. The index takes into account the quantitative relations between different pairs of species.

Another index used in this paper was the species diversity index according to Shannon and Weaver (Margalef, 1957).

## Results

Rotifers were encountered in all the sampled water bodies. In total, 39 rotifer genera and 114 species were found (Table 2), i.e. ca. 25% of the 448 Rotifera recorded from Poland (Ejsmont-Karabin & Karabin, 1999).

We found 1–36 species per sample (mean=10). The most frequent species were *Brachionus angularis* and *Keratella cochlearis* in spring, *Colurella uncinata*, *Lecane closterocerca* and *Lepadella patella* in summer, and *L. closterocerca* and *K. cochlearis* in autumn. As many as 48 species were found only once, 3 – in spring, 17 – in winter and as many as 28 in autumn. The single records were noted in 15 water-bodies, mostly in natural ponds with more or less abundant submerged vegetation (Table 1 – pond nr. 9, 10, 12 and 14), in the concrete pond nr. 7, and the clay-pit nr. 2. *Keratella cochlearis* was found in 27 of 49 samples and *Lecane closterocerca* in 24 samples (Table 2).

Both rotifer densities and diversity differed in different water-bodies. Rotifer numbers were relatively low and ranged from 1-546 ind. $l^{-1}$ in spring, 1-1503 ind. $l^{-1}$ in summer and from 13-1375 ind. $l^{-1}$ in autumn. Values of Shannon's diversity index ranged

*Table 2.* List of rotifer species (Monogononta) and the number of records of particular species in samples collected on four occasions in water bodies of Poznań

| Species | Number of records in | | |
|---|---|---|---|
| | Spring | Summer | Autumn |
| Number of studied water-bodies—> | 14 | 22 | 13 |
| *Anuraeopsis fissa* (Gosse) | | 5 | |
| *Ascomorpha ecaudis* Perty | | 1 | |
| *Ascomorpha ovalis* (Bergendal) | | 1 | |
| *Ascomorpha saltans* Bartsch | | 2 | |
| *Asplanchna brightwelli* (Gosse) | | 2 | |
| *Asplanchna priodonta* Gosse | 1 | 3 | |
| *Brachionus angularis* Gosse | 7 | 4 | 3 |
| *Brachionus budapestinenis* Daday | | 1 | |
| *Brachionus calyciflorus* Pallas | 3 | 2 | 1 |
| *Brachionus diversicornis* (Daday) | 2 | 2 | |
| *Brachionus quadridentatus* Hermann | 1 | 5 | |
| *Brachionus rubens* Ehrenberg | 1 | 1 | 1 |
| *Brachionus urceolaris* (Müller) | 4 | 4 | 2 |
| *Cephalodella auriculata* (Müller) | | 2 | |
| *Cephalodella catellina* (Müller) | 1 | 2 | |
| *Cephalodella dentata* Wulfert | | | 1 |
| *Cephalodella eva* (Gosse) | 1 | | |
| *Cephalodella gibba* (Ehrenberg) | | 4 | 2 |
| *Cephalodella gibboides* Wulfert | | 1 | |
| *Cephalodella gracilis* (Ehrenberg) | | 1 | |
| *Cephalodella hoodi* (Gosse) | | | 1 |
| *Cephalodella hyalina* Myers | | 1 | 1 |
| *Cephalodella sterea* (Gosse) | | 1 | 3 |
| *Cephalodella tenuior* (Gosse) | | | 1 |
| *Cephalodella tenuiseta* (Burn) | | | 1 |
| *Cephalodella ventripes* Dixon-Nuttall | | 2 | 5 |
| *Collotheca mutabilis* (Hudson) | | 1 | |
| *Collotheca ornata* (Ehrenberg) | | | 1 |
| *Collotheca pelagica* (Rousselet) | | 2 | |
| *Colurella adriatica* Ehrenberg | 1 | 1 | 5 |
| *Colurella colurus* (Ehrenberg) | | | 4 |
| *Colurella obtusa* (Gosse) | 1 | 1 | 3 |
| *Colurella uncinata* (Müller) | 1 | 8 | 4 |
| *Conochiloides dossuarius* (Hudson) | | 1 | |
| *Conochilus unicornis* Rousselet | 1 | 1 | |
| *Dicranophorus forcipatus* (Müller) | | 1 | |
| *Dicranophorus hercules* Wiszniewski | | 1 | |
| *Euchlanis deflexa* Gosse | | 1 | 1 |
| *Euchlanis dilatata* Ehrenberg | 2 | 4 | 2 |
| *Euchlanis incisa* Carlin | | | 1 |
| *Filinia longiseta* (Ehrenberg) | | 3 | |
| *Filinia terminalis* (Plate) | 1 | | |
| *Gastropus stylifer* Imhof | | 2 | |
| *Hexarthra mira* (Hudson) | 1 | 1 | |
| *Itura aurita* (Ehrenberg) | | 1 | |

*Table 2.* contd.

| Species | Number of records in | | |
|---|---|---|---|
| | Spring | Summer | Autumn |
| Number of studied water-bodies—> | 14 | 22 | 13 |
| *Itura viridis* (Stenroos) | | 1 | |
| *Kellicottia longispina* (Kellicott) | | 2 | |
| *Keratella cochlearis* (Gosse) | 7 | 11 | 9 |
| *Keratella quadrata* (Müller) | 6 | 6 | 7 |
| *Keratella testudo* (Ehrenberg) | | 2 | 2 |
| *Lecane arcuata* (Bryce) | | 2 | |
| *Lecane bulla* (Gosse) | 1 | 8 | |
| *Lecane closterocerca* (Schmarda) | 3 | 11 | 10 |
| *Lecane flexilis* (Gosse) | | | 1 |
| *Lecane furcata* (Murray) | | 2 | |
| *Lecane hamata* (Stokes) | 1 | 5 | 2 |
| *Lecane intrasinuata* (Olofsson) | | 1 | |
| *Lecane luna* (Müller) | | 7 | 2 |
| *Lecane lunaris* (Ehrenberg) | 1 | 1 | 3 |
| *Lecane nana* (Murray) | | 2 | |
| *Lecane stenroosi* (Meissner) | 1 | | |
| *Lecane tenuiseta* Harring | | | 1 |
| *Lepadella acuminata* (Ehrenberg) | | 1 | |
| *Lepadella ovalis* (Müller) | | 7 | 1 |
| *Lepadella patella* (Müller) | 2 | 9 | 6 |
| *Lepadella quadricarinata* (Stenroos) | | 1 | 1 |
| *Lepadella rhomboides* (Gosse) | | 1 | |
| *Lepadella triptera* Ehrenberg | | | 3 |
| *Lophocharis oxysternon* (Gosse) | | 4 | 2 |
| *Lophocharis salpina* (Ehrenberg) | | 1 | |
| *Mytilina bisulcata* (Lucks) | | | 1 |
| *Mytilina mucronata* (Müller) | | 5 | 1 |
| *Mytilina trigona* (Gosse) | | 1 | |
| *Mytilina ventralis* (Ehrenberg) | | 6 | |
| *Monommata grandis* Tessin | | | 1 |
| *Monommata longiseta* (Müller) | | | 2 |
| *Notholca acuminata* (Ehrenberg) | 4 | | |
| *Notholca squamula* (Müller) | 3 | | |
| *Notommata cyrtopus* Gosse | | | 1 |
| *Notommata tripus* Ehrenberg | | | 1 |
| *Platyas quadricornis* (Ehrenberg) | | 2 | 1 |
| *Plationus patulus* (Müller) | | | 1 |
| *Pleurotrocha petromyzon* Ehrenberg | | 1 | 1 |
| *Polyarthra dolichoptera* Idelson | 2 | 2 | |
| *Polyarthra euryptera* Wierzejski | | 1 | |
| *Polyarthra longiremis* Carlin | | 4 | 1 |
| *Polyarthra remata* Skorikov | | 3 | 5 |
| *Polyarthra vulgaris* Carlin | | 3 | 3 |
| *Pompholyx complanata* Gosse | | 1 | |
| *Pompholyx sulcata* Hudson | 3 | 3 | 2 |
| *Proales sigmoidea* (Skorikov) | | | 1 |

*Continued on p. 169*

*Table 2.* contd.

| Species | Number of records in | | |
|---|---|---|---|
| | Spring | Summer | Autumn |
| Number of studied water-bodies—> | 14 | 22 | 13 |
| *Proalides tentaculatus* Beauchamp | | 1 | |
| *Resticula melandocus* (Gosse) | | | 1 |
| *Scaridium longicaudum* (Müller) | | 1 | |
| *Squatinella mutica* (Ehrenberg) | | 1 | |
| *Stephanoceros fimbriatus* (Goldfuss) | 1 | | |
| *Synchaeta kitina* Rousselet | 1 | | 3 |
| *Synchaeta oblonga* Ehrenberg | 3 | | 5 |
| *Synchaeta pectinata* Ehrenberg | 6 | 2 | 5 |
| *Synchaeta tremula* (Müller) | 2 | | 1 |
| *Taphrocampa selenura* Gosse | | 1 | |
| *Testudinella emarginula* (Stenroos) | | 1 | |
| *Testudinella mucronata* (Gosse) | | 1 | |
| *Testudinella patina* (Hermann) | 1 | 9 | 1 |
| *Trichocerca bicristata* (Gosse) | | 1 | |
| *Trichocerca capucina* Wierz. et Zach. | 2 | | |
| *Trichocerca cylindrica* (Imhof) | | 1 | |
| *Trichocerca pusilla* (Lauterborn) | | 1 | 3 |
| *Trichocerca rattus* (Müller) | | 2 | 2 |
| *Trichocerca relicta* Donner | | | 1 |
| *Trichocerca similis* (Wierzejski) | | 3 | 1 |
| *Trichocerca weberi* (Jennings) | | | 1 |
| *Trichotria pocillum* (Müller) | 1 | 1 | 3 |
| *Trichotria truncata* (Whitelegge) | | 1 | |

from 0.00 to 2.93 with a rather low mean value of 1.63 (± 0.84) in spring. In summer and autumn, they were markedly higher, ranging from 1.00 to 3.18 (mean 2.22±0.79) and from 0.53 to 3.71 (mean 2.32±0.96), respectively.

The studied water-bodies were very different in their species dominants (>20% of rotifer numbers) (Table 3). 21 species dominated in 49 samples. The most frequent dominants were *Keratella cochlearis* and *K. quadrata*, which are eurytopic species common in all types of lakes, and littoral *Colurella uncinata*. They occurred as dominants in as few as 14%, 8% and 8% (respectively) of samples. Although samples were taken from an open water zone, the list of dominants included littoral and benthic species such as *Lecane bulla, L. closterocerca, Colurella adriatica, C. uncinata,* and *Mytilina ventralis.*

The strong differences in qualitative structure of rotifer assemblages in the studied water-bodies may be well illustrated by the PSC index (Whittaker & Fairbanks, 1958). In spring, values of PSC index

*Table 3.* List and dominants, number of sites with their dominance (in parentheses) and the species share in rotifer numbers (in %)

| | Dominant species | Spring | | Summer | | Autumn | |
|---|---|---|---|---|---|---|---|
| 1 | *Anuraeopsis fissa* | | | (1) | 49 | | |
| 2 | *Brachionus angularis* | (1) | 77 | | | | |
| 3 | *Brachionus calyciflorus* | (1) | 37 | | | | |
| 4 | *Brachionus diversicornis* | (2) | 78–95 | | | | |
| 5 | *Colurella adriatica* | | | | | (2) | 30–31 |
| 6 | Colurella uncinata | | | (4) | 31–55 | | |
| 7 | *Conochiloides dossuarius* | | | (1) | 38 | | |
| 8 | *Itura viridis* | | | (1) | 36 | | |
| 9 | *Keratella cochlearis* | (1) | 84 | (4) | 23–55 | (2) | 41–60 |
| 10 | *Keratella quadrata* | (1) | 59 | (1) | 53 | (2) | 57–60 |
| 11 | Lecane bulla | | | (1) | 33 | | |
| 12 | *Lecane closterocerca* | | | | | (1) | 33 |
| 13 | *Lepadella patella* | (1) | 44 | (1) | 56 | | |
| 14 | *Lophocharis oxysternoon* | | | (1) | 52 | | |
| 15 | *Mytilina ventralis* | | | (1) | 25 | | |
| 16 | Notholca squamula | (1) | 41 | | | | |
| 17 | *Polyarthra longiremis* | | | (1) | 53 | (1) | 92 |
| 18 | *Pompholyx sulcata* | (1) | 69 | | | | |
| 19 | Synchaeta oblonga | (1) | 52 | | | | |
| 20 | *Synchaeta pectinata* | (1) | 64 | | | (1) | 20 |
| 21 | *Trichotria pocillum* | (1) | 24 | | | | |

(Fig. 1) ranged from 0.0–95.0%, with extremely low mean value of 7.7% (SD=13.6%, $n$=91) for all compared pairs. In summer, the mean PSC value was almost identical (7.6%, SD=12.0, $n$=66), though the range of values differed widely (0–51.7%). As many as 73% of pairs had PSC values lower than 10%. In autumn, PSC values ranged from 0.0-63.8% (mean 14.1%, SD=17.1, $n$=78), and 63% pairs had less than 10% similarity of their rotifer communities. In spring, 42% pairs had no species in common. In summer, the percentage of such pairs was much lower (20%), and the lowest was in autumn (12%).

Generally, only 20 compared pairs (i.e. 8.5% of pairs summed for the three periods) had PSC values higher than 30%. The highest similarities of rotifer communities were found for pairs natural ponds versus concrete reservoirs or pools. They constituted 60% of all pairs having PSC values above 30%.

## Discussion

The recorded 114 species in the water bodies of Poznań is comparable to totals from some natural ecosystems. Chengalath & Koste (1987) reported 97 species of rotifers belonging to 32 genera – from

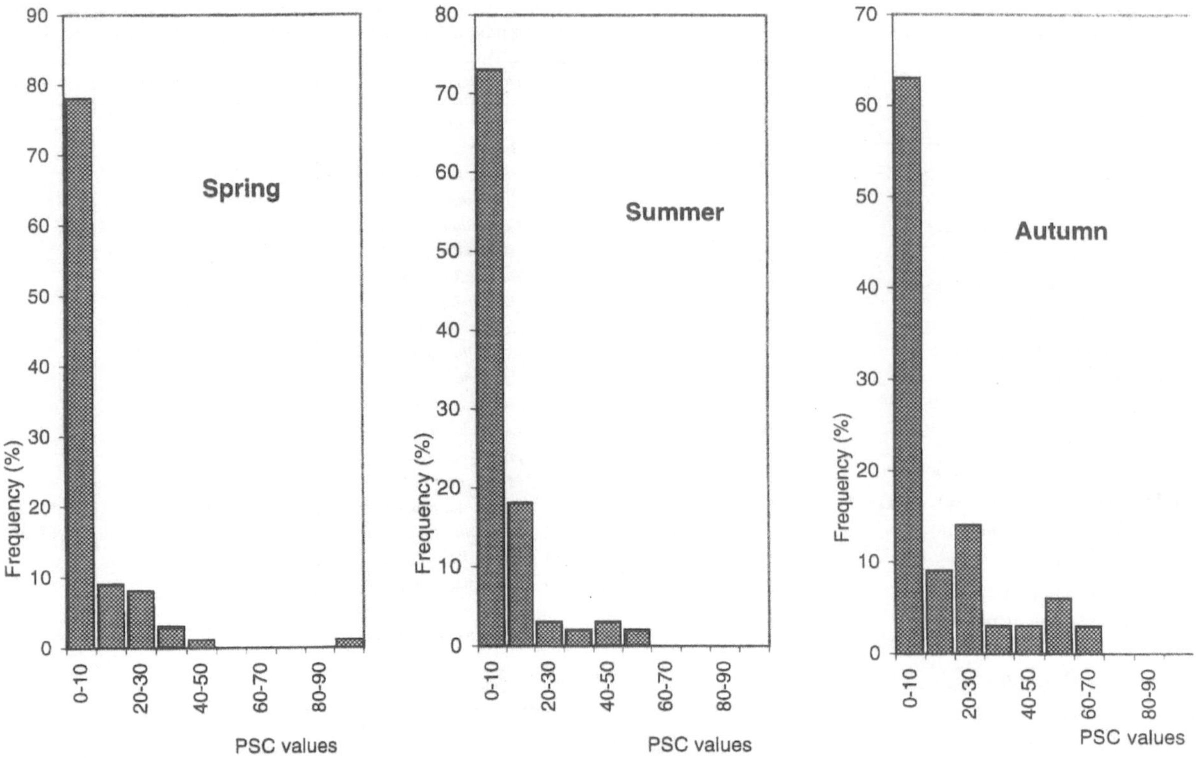


*Figure 1.* Frequency of 10 classes of values of percentage similarity of community (PSC) indices calculated for all possible pairs of Poznań water-bodies.

British Columbia and the Yukon Territory of Canada though, similar to this study, their sampling was carried out in very different ecosystems. However, in studies covering larger areas, or longer time frames the recorded number of rotifer species may be markedly higher. For example. 234 rotifer species were identified over a 14-year period from 30 sites in Florida (Turner & Taylor, 1998) and 250 taxa were recorded from diverse aquatic ecosystems of Tasmania, (Shiel et al., 1989).

Rotifer communities of small urban water-bodies should rather be compared with small, natural reservoirs. Klimowicz (1970) in studies of 16 astatic water-bodies in the vicinity of Mikołajki (Masurian Lake District, Poland) monthly in the years 1956/1957, found 123 species of monogonont rotifers. Compared to urban water-bodies, species diversity of rotifer communities in natural small water bodies was lower in spring (mean 1.25±0.93) and higher in summer (mean 2.54±0.59) (calculated from Klimowicz, 1970).

Rotifer communities of the astatic water-bodies were dominated by *Polyarthra vulgaris, P. remata, Anuraeopsis fissa, Keratella testudo, K. valga, Plationus patulus, Synchaeta pectinata, Lepadella ovalis* (Klimowicz, 1970). Except for *A. fissa* and *S. pectinata*, the species listed above were rare or absent in water-bodies of Poznań.

The differences between rotifer communities of urban water-bodies seem to be much greater than those observed in lakes of different trophy and mixis by Ejsmont-Karabin et al. (1980). They found the mean PSC value 32±25% ($n$ = 6) for four lakes: shallow dystrophic and polytrophic and deep dystrophic and eutrophic. Rotifer communities of the more similar 17 lakes of the River Krutynia watershed (Karabin & Ejsmont-Karabin, 1996) had PSC values ranging from 9 to 83% with their mean ca. 43±15% ($n$=136). Percentage similarity of rotifer communities (PSC) of temporary water bodies in the Masurian Lake District (calculated from Klimowicz, 1970) was similar to urban sites in spring, with the mean value of 5.5±13.2% ($n$=78), but markedly higher in summer (mean 15.6±12.8, $n$=78).

Thus, urban water habitats and their rotifer communities seem to be more differentiated than lake systems and similar in this respect to temporary water bodies. These differences make the rotifer assemblage of the Poznań agglomeration rich in species, although species diversity of rotifer communities of particular water-bodies may be low. Our studies did not support the hypothesis of poverty of urban communities of Rotifera. If heavy pollution of urban water sites influences the qualitative structure of their rotifer communities, it does not seem to impoverish them.

## Acknowledgements

The authors are deeply indebted to Dr Ramesh D. Gulati and an anonymous referee for valuable comments that greatly improved the manuscript.

## References

Blair, R. B., 1999. Birds and butterflies along an urban gradient: surrogate taxa for assessing biodiversity? Ecol. Appl. 9: 164–170.

Carpenter, S. R., N. F. Caraco, D. L. Correll, R. W. Howarth, A. N. Sharpley & V. H. Smith, 1998. Nonpoint pollution of surface waters with phosphorus and nitrogen. Ecol. Appl. 8: 559–568.

Chengalath, R. & W. Koste, 1987. Rotifera from Northwestern Canada. Hydrobiologia 147: 49–56.

Connell, J. H., 1978. Diversity in tropical rainforests and coral reefs. Science 109: 1304–1310.

Delis, P. R., H. R. Mushinsky & E. D. McCoy, 1996. Decline of some west-central Florida anuran populations in response to habitat degradation. Biodivers. Conserv. 5: 1579–1595.

Ejsmont-Karabin, J., L. Bownik-Dylińska, T. Wêgleńska & A. Karabin, 1980. The effect of mineral fertilization on lake zooplankton. Ekol. pol. 28(1): 3–44.

Ejsmont-Karabin, J. & A. Karabin, 1999. Will faunal studies come into fashion? Some fairly pessimistic reflections suggested by the work on guides, monographs or expertise. Acta. Hydrobiol. 41: 55–60.

Golterman, H. L., 1969. Methods for Chemical Analysis of Fresh Waters, IBP Handbook no. 8, Blackwell Scientific Publ., Oxford, Edinburgh. 166 pp.

Jenkins, D. G. & A. L. Buikema Jr., 1998. Do similar communities develop in similar sites? A test with zooplankton structure and function. Ecol. Monogr. 68: 421–443.

Karabin, A. & J. Ejsmont-Karabin, 1996. Structure, abundance and differentiation of the zooplankton of the Krutynia river lakes (Mazurian Lakeland). In Hillbricht-Ilkowska A. & R. Wiœniewski (eds), The Functioning of River-lake System in a Lakeland Landscape: River Krutynia (Mazurian Lakeland, Poland), Oficyna Wydawnicza IE PAN, Zesz. Nauk. Kom.Nauk. 'Człowiek i Środowisko' 13: 155–172 (in Polish).

Klimowicz, H., 1970. Rotifers (Rotatoria) of astatic waters. Zesz. Nauk. Inst. Gosp. Komunalnej, Warszawa, 30: 253 pp (in Polish).

Langley, J. M., S. Kett, R. S. Al.-Khalili & C. J. Humphrey, 1995. The conservation value of English urban ponds in terms of their rotifer fauna. Hydrobiologia 313/314: 259–266.

Margalef, R., 1957. Information theory in ecology. Gen. Sys. 3: 36–71.

Mazgajska, J., 1996. Distribution of amphibians in urban water bodies (Warsaw agglomeration, Poland). Ekol. pol. 44: 245–257.

Nichols, J. D., T. Boulinier, J. E. Hines, K. H. Pollock & J. R. Sauer, 1998. Inference methods for spatial variation in species richness and community composition when not all species are detected. Conserv. Biol. 12: 1390–1398.

Rapport, D. J. & W. G. Whitford, 1999. How ecosystems respond to stress. Bioscence 49: 193–203.

Shiel, R. J., W. Koste & L. W. Tan, 1989. Tasmania revisited: rotifer communities and habitat heterogeneity. Hydrobiologia 186/187: 239–245.

Turner, P. N. & H. L. Taylor, 1998. Rotifers new to Florida, U.S.A. Hydrobiologia 387/388: 55–62.

Whittaker, R. H. & C. W. Fairbanks, 1968. A study of plankton copepod communities in the Columbia Basin, Southeastern Washington. Ecology 39: 46–65.

Hydrobiologia **446/447:** 173–177, 2001.
L. Sanoamuang, H. Segers, R.J. Shiel & R.D. Gulati (eds), Rotifera IX.




# Rotifer distributions in the coastal waters of the northeast Pacific Ocean

Steven C. Fradkin
*University of Oregon, Oregon Institute of Marine Biology, P.O. Box 5389, Charleston, OR 97420, U.S.A.*
*Present address: National Park Service, Olympic National Park, Coastal Resources Office, Forks, WA 98331, U.S.A.*
*Tel: [+1]-360-374-1222. Fax: [+1]-360-374-1250. E-mail: Steven_Fradkin@nps.gov*

*Key words:* marine rotifers, coastal upwelling, estuarine export *Synchaeta*, *Trichocerca*

**Abstract**

Absolute abundance of rotifers was assessed from 5 to 80 km across the continental shelf off of the southern Oregon coast (U.S.A.) in the northeast Pacific Ocean. A total of 97 vertically stratified water samples were collected at 49 stations from two depths, 3 and 30 m. Coastal upwelling conditions were indicated, with decreased temperature, increased salinity and higher chlorophyll-*a* concentrations closer to shore. Two rotifer genera, *Synchaeta* and *Trichocerca* occurred within 16 km of shore with densities increasing closer to shore. *Synchaeta* reached densities of 64 inds $l^{-1}$ while *Trichocerca* was sparse (<1 inds. $l^{-1}$). Rotifers were most abundant at 3 m and the densest aggregations appear to be associated with estuary outlets, suggesting that estuaries may be important in exporting rotifers to nearshore coastal waters.

## Introduction

Little information exists on the distribution and abundance of marine rotifers. Existing studies are few, and take place in estuarine and enclosed sea environments (e.g. Hollowday, 1949; Johansson, 1982; Henroth, 1983; Dolan & Gallegos, 1992). These studies suggest that rotifers can be seasonally abundant (>1000 $l^{-1}$), though temporally variable, and can play an important role in energy transfer in marine food webs (Capriulo & Carpenter, 1980; Gifford, 1988; Mallin & Paerl, 1994). In the open ocean, however, rotifers are thought to be sparse and of little significance to food web dynamics, possibly due to resource limitation. Open ocean data, however, are essentially non-existent.

In the coastal northeast Pacific during summer months, north winds drive warm coastal surface waters offshore causing a coastal sea level drop that draws dense, cool, nutrient rich water from depth to the surface (Csanady, 1981). Such upwelled water is responsible for increased primary and secondary productivity in nearshore regions, relative to nutrient limited offshore waters. In the present study, the distribution and abundance of marine rotifers was examined in the northeastern Pacific Ocean during coastal upwelling conditions, from the nearshore waters of the northwestern United States (Oregon coast) to across the continental shelf. Quantitative sampling of rotifer distributions was used to test the hypothesis that rotifer abundance would be highest in the upwelling zone, and would decrease with distance from shore.

## Study site

From 26 May to 1 June 1998, the NOAA Ship *RV/McArthur* sampled physical and biological variables on a grid of 49 stations (Fig. 1) comprising 8 transects across the continental shelf off of the coast of southern Oregon, U.S.A. coast (Fig. 1). The four southern-most transects were composed of eight sample stations, 5, 8, 16, 24, 32, 48, 64 and 80 km from shore, while the four northern transects were composed of five sample stations, 5, 8, 16, 24 and 32 km from shore. The continental shelf begins approximately 18 km offshore in this area.

## Materials and methods

Sampling during the cruise occurred around the clock.

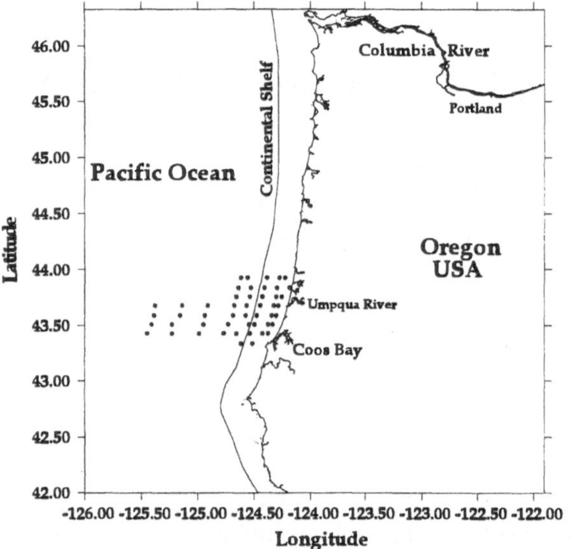


*Figure 1.* Vicinity map of cross-shelf transect sampling stations off of the northwestern U.S.A. coast.

At each sampling station, stationary casts of a CTD (Seabird 911 Plus) equipped with a fluorometer (Wetstar 100) measured depth profiles of temperature, salinity and chlorophyll-*a* concentration.

Water samples were collected at the surface (~3 m) and above the thermocline (~30 m) using a rosette of eight 2.5 l niskin bottles. At each depth, four bottles were triggered remotely, collecting a total volume of 10 l. Samples were concentrated using a 25 $\mu$m filter and preserved in EtOH (~25%). In the laboratory, samples were stained with acid lugol's and all rotifers were counted using a dissecting microscope at 25× magnification. Rotifers were identified to the genus level.

Distributional graphs of physical variables and rotifer densities (Figs. 2 and 3) were created with Surfer (Version 6.0) using a kreiging interpolation algorithm.

## Results

Upwelling conditions occurred during the cruise, as evidenced by trends in temperature, salinity and chlorophyll-*a* (Figs 2 and 3). Nearshore temperatures were substantially cooler compared to offshore at the surface (3 m), with a differential of approximately 3° (Fig. 2E). At 30 m, the differential was less at approximately 1.5° (Fig. 3E). Chlorophyll-*a* concentrations were higher near shore at both depths (Figs 2D and 3D).

Salinity was also higher closer to shore, with an approximate 2 ppt differential between nearshore and cross-shelf waters at the surface. An exception was the distinct surface plume of lower salinity water extending approximately 10 km offshore from the mouth of the Umpqua River (Fig. 2C). This plume represents input of freshwater-influenced estuarine water. At depths below the surface plume (i.e. 30 m), salinity was higher than at the surface, and slightly higher (~1) nearshore (Fig. 3C).

Two rotifer genera were encountered, *Synchaeta* and *Trichocerca*. *Synchaeta* was the most abundant (maximum density of 64 inds $l^{-1}$), while *Trichocerca* was sparse (<1 inds $l^{-1}$). Both genera were found over the continental shelf, and no rotifers were found further than 16 km from shore (Figs 2A, B, and 3A, B). Rotifer densities increased with proximity to the coast. The highest concentrations of both genera occurred in the plume of lower salinity water at the mouth of the Umpqua River. *Synchaeta* did not occur in the more saline water below the Umpqua plume; however, it did occur in slightly lower salinity water at that depth further south by Coos Bay (Fig. 3A, C). Rotifer densities appeared to be associated with the mouth of Coos Bay at both depths, however sample coverage of this area is limited. *Synchaeta* was relatively abundant at both 3 and 30 m at the mouth of Coos Bay (Fig. 3A, C), with maximum densities of approximately 32 inds $l^{-1}$.

## Discussion

Marine rotifer communities are typically less diverse than freshwater communities, with marine constituents that are different from rotifers species common in inland saline lakes (Wallace & Snell, 1991). Marine rotifer diversity is likely a function of the evolutionary radiation of Eurotatoria from freshwater environments. The genus *Synchaeta* has been particularly successful in invading marine environments, with some 20 of 32 species (Nogrady, 1982) described as marine. Studies from estuarine and enclosed ocean environments in the Atlantic Ocean (Henroth, 1983; Johansson, 1983; Brownell, 1988; Ambrogi et al., 1989; Arndt et al., 1990; Dolan & Gallegos, 1991) have shown *Synchaeta* to be the dominant rotifer community constituent.

In the present study, *Synchaeta* was the dominant rotifer off of the Oregon coast, with *Trichocerca* occurring at very low densities. The lack of rotifers in the open ocean past the continental shelf suggests

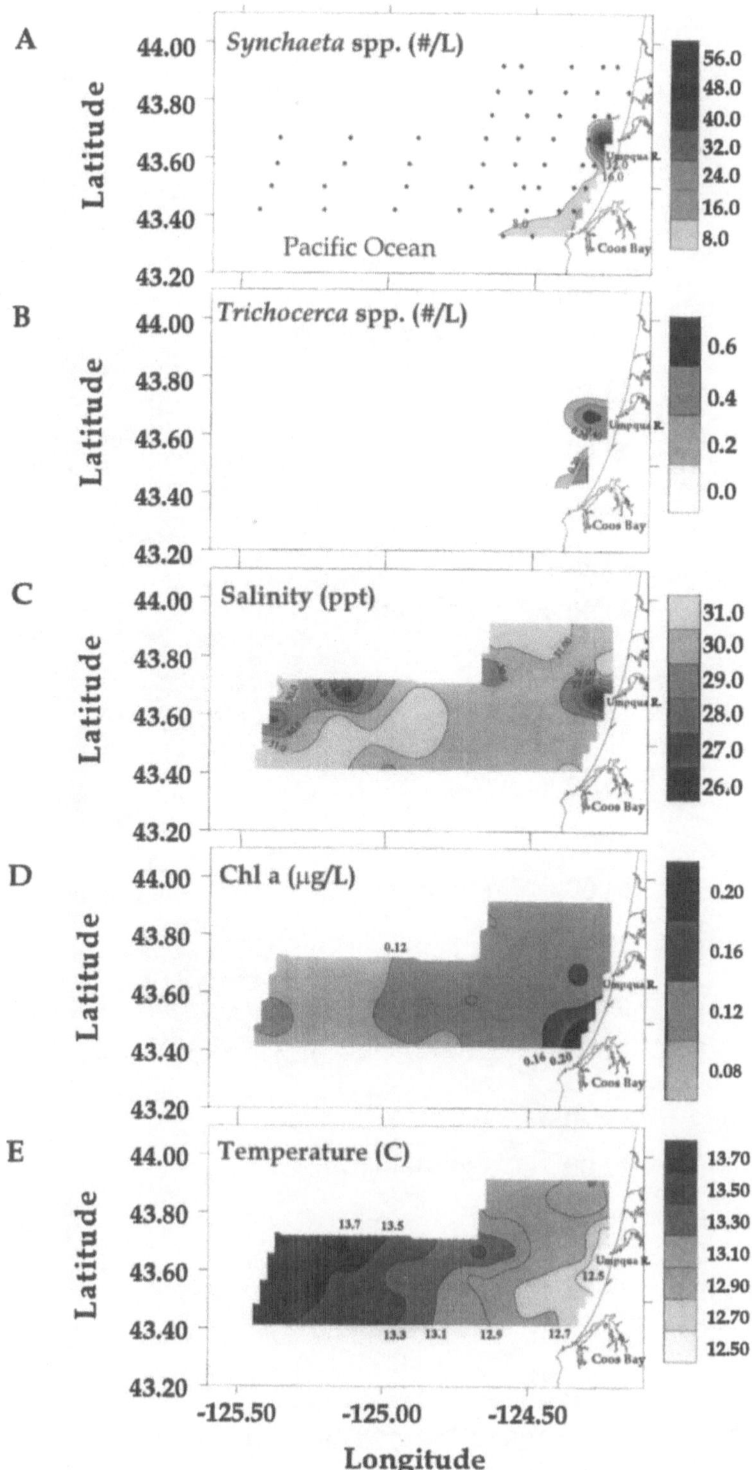


*Figure 2.* Cross-shelf surface (~3 m) isopleth distribution patterns of physical and biological variables. (A) *Synchaeta* spp. density. Dots denote sampling stations. (B) *Trichocerca* spp. density. (C) Salinity. (D) Chlorophyll-*a* concentration. (E) Temperature.

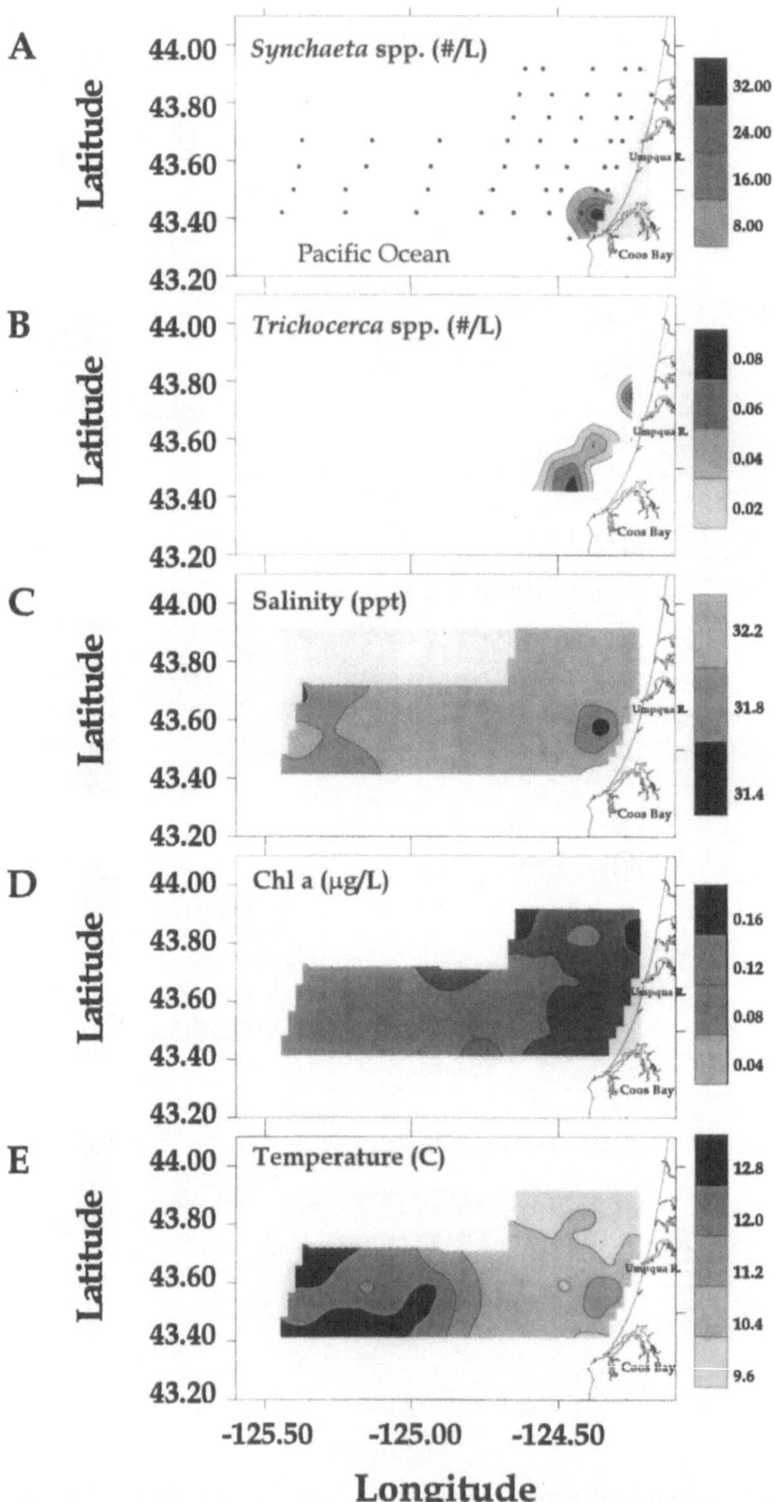


*Figure 3.* Cross-shelf 30 m isopleth distribution patterns of physical and biological variables. (A) *Synchaeta* spp. density. Dots denote sampling stations. (B) *Trichocerca* spp. density. (C) Salinity. (D) Chlorophyll-*a* concentration. (E) Temperature.

that conditions, either predation, resource or physical, were not suitable to sustain populations. No data were available on the distribution and abundance of potential predators. Temperature and salinity appear unlikely to be limiting factors on rotifer distributions. In the nearshore area, *Synchaeta* was found across a large portion of the observed temperature and salinity range (Figs 2 and 3).

The highest rotifer densities were associated with areas of estuarine water input (Figs 2A and 3A). *Synchaeta* is temporally common and numerically abundant in both the lower Umpqua and Coos Bay estuaries (Fradkin, unpublished data). The estuarine plume from the Umpqua River was much more pronounced in comparison to Coos Bay (Fig. 2), reflecting its greater flushing rate (C. Roegner, pers. comm.).

In the present study, rotifers were not identified to the species level, thus enabling confirmation of their euryhaline, estuarine-derived species status. The presence of obligate marine species would suggest that estuaries improve nearshore habitat conditions. More likely, the occurrence of high rotifer densities in estuarine-influenced water suggests that estuaries play an important role in exporting rotifers into the nearshore ocean. Ambrogi et al. (1989) similarly concluded that an estuarine lagoon plays an important role in exporting plankton biomass, particularly *Synchaeta*, into Mediterranean coastal surface waters.

Persistence of rotifers in nearshore waters off the Oregon coast may be further facilitated by upwelling conditions. Coastal upwelling was evident during the study, with cooler, more saline waters containing higher chlorophyll-*a* concentrations occurring closer to shore. During pronounced upwelling events, such conditions may produce high resource availability for rotifers, thereby facilitating their persistence.

## Acknowledgements

I would like to thank the crew of the *RV/McArthur* and S. Rumrill for the enabling my participation in the cruise. C. Roegner provided much needed help and advice on data analysis and presentation. R. Emlet kindly provided funding and lab facilities at OIMB. Two anonymous reviewers provided comments that improved the manuscript.

## References


Ambrogi, R., I. Ferrari & S. Geraci, 1989. Biotic exchange between river, lagoon and sea: The case of zooplankton in the Po Delta. Topics in Marine Biology. Sci. Mar. 53: 2–3.

Arndt, H., C. Schroder & W. Schnese, 1990. Rotifers of the genus *Synchaeta*- an important component of the zooplankton in the coastal waters of the southern Baltic. Limnologica 21: 233–235.

Brownell, C. L., 1988. A new pelagic marine rotifer from the southern Benguela, *Synchaeta hutchingsi*, n. sp. with notes on its temperature and salinity tolerance and methods of culture. Hydrobiologia 162: 225–233.

Capriulo, G. M. & E. J. Carpenter, 1980. Grazing by 35–202 $\mu$m micro-zooplankton in Long Island Sound. Mar. Biol. 56: 319–326.

Csanady, G. T., 1981. Circulation in the coastal ocean. Adv. Geophys. 23: 101–183.

Dolan, J. & C. L. Gallegos, 1992. Trophic role of planktonic rotifers in the Rhode River Estuary, spring–summer 1991. Mar. Ecol. Prog. Ser. 85: 187–199.

Gifford, D. J., 1988. Impact of grazing by microzooplankton in the northwest arm of Halifax harbor, Nova Scotia. Mar. Ecol. Prog. Ser. 47: 249–258.

Henroth, L., 1983. Marine pelagic rotifers and tintinnids- important trophic links in the spring plankton community of the Gullmar Fjord, Sweden. J. Plankton Res. 5: 835–846.

Hollowday, E. D., 1949. Preliminary report on the Plymouth marine and brackish-water rotifera. J. mar. biol. Ass. U.K. 28: 239–253.

Johansson, S., 1983. Annual dynamics and production of rotifers in a eutrophication gradient in the Baltic Sea. Hydrobiologia 104: 335–340.

Mallin, M. A. & H. W. Paerl, 1994. Planktonic trophic transfer in an estuary: seasonal, diel and community structure effects. Ecology 75: 2168–2184.

Nogrady, T., 1982. Rotifera. In Parker, S. P. (ed.), Synopsis and Classification of Living Organisms. McGraw Hill, New York: 865–872.

Wallace, R. L. & T. W. Snell, 1991. Rotifers. In Thorp, J. H. & A. P. Covich (eds), Ecology and Classification of North American Freshwater Invertebrates. Academic Press: 173–248.

Hydrobiologia **446/447:** 179–185, 2001.
L. Sanoamuang, H. Segers, R.J. Shiel & R.D. Gulati (eds), Rotifera IX.




# Species diversity and dominance in the planktonic rotifer community of the Pripyat River in the Chernobyl region (1988–1996)

Galina A. Galkovskaya & Dmitry V. Molotkov
*Institute of Zoology, National Academy of Sciences of Belarus, Academycheskaya st., 27, Minsk 220072 Belarus*

*Key words:* rotifers, species richness, radioactive pollution, recovery


## Abstract

Here, we report the results of monitoring the rotifer community in the Pripyat River within the 30-km evacuation zone of the Chernobyl Nuclear Power Plant over the period 1988–1996. While radionuclide concentration in water did not exceed 4.07 Bq $l^{-1}$, the radioactivity in the bottom sediment was quite high, varying irregularly between 113 and 824 kBq $m^2$. Radionuclide concentration in the seston also ranged widely: riverbed = 659–2491; backwater = 168–32 832 Bq $kg^{-1}$. The rotifer density varied in the range of 65–17 970 individuals $l^{-1}$. Sixty-seven rotifer species were identified in the Pripyat, with nine species being previously unknown to this river. Species richness (jackknife estimate) in both the riverbed and the backwater stations was similar and was characterized by a very great variability: riverbed = 66.1 (df=20, SD=39.50); back-water = 66.2 (df=20, SD=42.17). Correlation between the heterogeneity of rotifer community ($H'$) and the number of species and relative density of the dominant species was evident. The degree of statistical interrelation between $H'$ and relative density of the dominant species was especially high in the riverbed station ($r^2$= 0.74, $p$= 0.00001). However, no significant correlation between radionuclide concentration and rotifer biodiversity was found.


## Introduction

On April 26, 1986, the Chernobyl Nuclear Power Plant (CNPP), located in the Kiev region of northern Ukraine, 12 km from the Belarussian border, exploded. As a result of the accident 1.95 $10^{18}$ Bq of radioactive material, the largest amount ever reported, was released into the environment (Kuzmenko, 1998). A radionuclide content exceeding 37 kBq $m^{-2}$ resulted in contamination of over 46 500 $km^2$of surface soil in Belarus. This area amounts to 23% of the total territory of the republic. Within this contaminated region a 30-km zone around the CNPP was so heavily contaminated that obligatory resettlement was inforced; much of the Pripyat River was included in this zone.

During the first years following the accident, a decrease in radionuclide concentration of aquatic sediments was observed. For example, the average level in Cs-137 concentration in the Pripyat River sediment decreased 8 fold over the period from 1987 to 1991 (Savchenko, 1995). However, at present, the level of contamination of the aquatic ecosystems in this region is driven by secondary processes, including transfer of radionuclides from bottom sediments and surface transfer from soils in the surrounding watershed (i.e. via floods and rainwater).

Investigation of the long-term impact of radionuclides on the zoocoenoses in the contaminated region is of interest because a precise description of the changes in the faunal complexes will permit a better understanding of the nature and rate of recovery. As part of a larger study of the impact of the CNPP event on the zoocoenoses of the Pripyat River and surrounding ecosystems, we undertook a study of the rotifer fauna. Here, we describe the results of our 9-year monitoring program (1988–1996).

## Materials and methods

### *Study area and field methods*

The Pripyat River is the largest tributary (>500 km long) of the Dnieper. Its basin is characterized by the presence of numerous swamps, a slow current within a

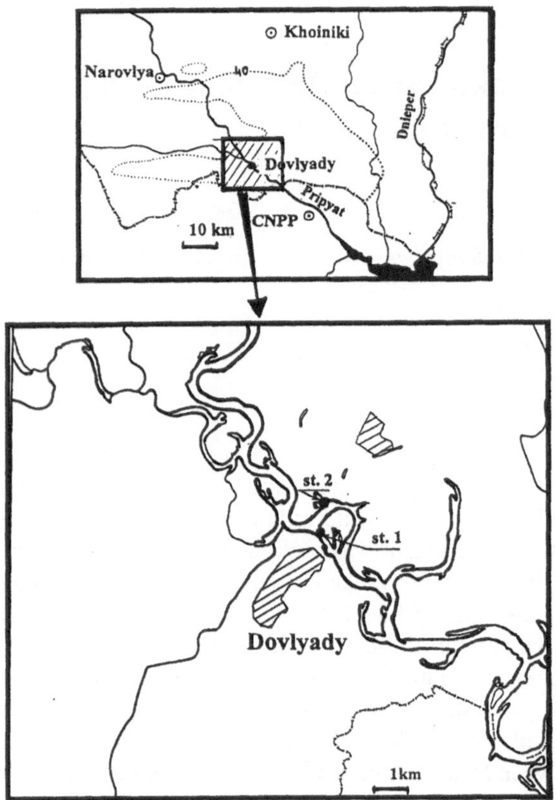


*Figure 1.* Map-showing the location of the sampling stations.

meandering channel, and banks boardered by weakly pronounced valleys.

Samples were collected in the evacuation zone near the village of Dovlyady in the center of the riverbed (station 1) and in backwater (station 2) (Fig. 1). During sample collection, the current did not exceed 0.75 m $s^{-1}$in the riverbed station. There was no noticeable current in the backwater station. The water level at the sampling stations varied in the range of 0.5–3m.

Because of safety restrictions on visiting this region, samples were taken only 2–3 times a year during the months of April–September. The samples were collected using a Ruttner sampler (1L), specimens were narcotized by $CO_2$-saturated water, and preserved with formalin (final concentration = 2%). In the laboratory, the samples were allowed to settle. After 3 days, of settling the supernatant was siphoned off so that the final sample volume was about 80–100 ml. All the rotifers in this volume were counted. Addition samples were collected to enumerate species with concentration of less than 1 individuals $l^{-1}$. This was done by filtering 50 l of water through a plankton net (45 $\mu$m mesh sieve).

Radionuclide concentration in water, bottom sediments and sestonic samples also were measured. We collected at least 100 g wet weight of seston by towing a plankton net (45 $\mu$m sieve, 45 cm diameter). At the same time, the temperature was measured and samples were taken for the determination of chlorophyll *a* concentrations.

### *Laboratory procedures*

To determine the radionuclide concentration, we used a 6-crystal, scintillation, $\gamma$-spectrometer (Laboratory of the Institute of Physics, National Academy of Sciences, Belarus). Chlorophyll concentration was determined by the extraction method (90% acetone) using a 2-beam spectrophotometer (Specord UV-VIS). Rotifer samples were counted using a MBS-10 microscope (100×). Species identifications were made using a Amplival compound microscope (500×) (Kutikova, 1970).

### *Statistical methods*

The jackknife estimate was used for species richness measures (Krebs, 1999). The Shannon–Wiener index was used as heterogeneity measure. Differences in the values of parameters were checked with the *t*-test (Rokitsky, 1973).

## Results

### *Radionuclide concentration*

Radionuclide concentration in river water did not exceed 4.07 Bq $l^{-1}$over the period of investigations. The radioactivity of the bottom sediments was high, varying irregularly within the limits of 113–824 kBq $m^2$. Radionuclide concentration in seston also varied irregularly and over a wide range: riverbed = 659 (1992) – 2491 (1994); backwater = 168 (1991) – 32832 (1993) Bq $kg^{-1}$. As seen in Figure 2, the maximum radionuclide concentration in the backwater is an order of magnitude higher than in the riverbed.

### *Species richness, heterogeneity and dominance in rotifer communities*

Sixty-seven rotifer species were identified in the riverbed and backwater at the stations near Dovlyady

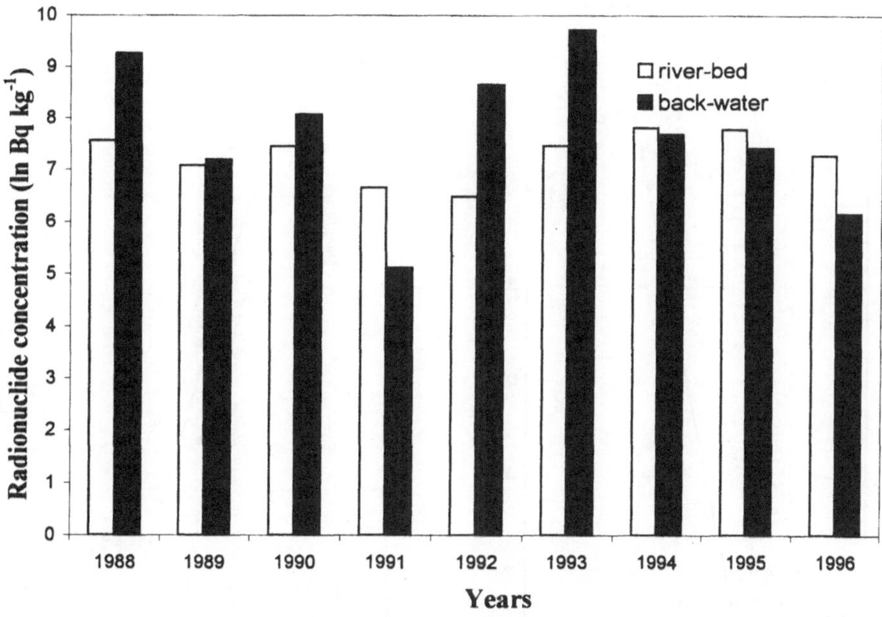


*Figure 2.* Radionuclide concentration in seston.

village (Table 1). Of these, nine species are new records for the Pripyat River. One of these, *Brachionus nilsoni*, was detected in one sample of 1988 and is considered to be an inhabitant of polluted water. The other eight species were detected in samples of 1991–1996 and, according to saprobical classification, are attributed an x-o saprobic degree.

As seen in Table 1, 36 species are common to the riverbed (1) and the backwater (2) in the evacuation zone, 13 species were detected only in the riverbed and 18 were found only in the backwater. The similarity coefficient is 0.69. Species richness, both in the riverbed and the backwater was similar and characterized by a great variability: riverbed = 66.1 (df=20, SD=39.50); backwater = 66.2 (df=20, SD=42.17). The Shannon–Wiener index of the species diversity varied irregularly during the subsequent years but exhibited some seasonal causality, being higher in summer than in spring and autumn.

The rotifer density ranged from 65 to 10 575 ind. $l^{-1}$ in the riverbed to 117 to 17 970 ind. $l^{-1}$ in the backwater. In the riverbed station, *Trichocerca pusilla* dominated in all summer samples (from 48 to 70% of the total rotifer density), while *Synchaeta oblonga* dominated in most spring and autumn samples (35–80%). In the backwater station, *Keratella cochlearis* dominated (28–70%) in most spring and autumn samples, however, *Filinia longiseta* (41%), *Brachionus angularis* (37%), *Polyarthra vulgaris* (39%) and *K. cochlearis* (15%) dominated in the summer samples. In the April samples, the dominance curves (i.e. the rank order of the natural logarithm of species density) were straight, a characteristic generally interpreted as indicating a severe environment (Odum, 1986). Possibly, this occurred because of low food levels as indicated by the low chlorophyll *a* concentration (<1 $\mu$g $l^{-1}$). In samples collected from May to September, the dominance curves were S-shaped. During that time, the chlorophyll concentration varied between 6.3–88.9 $\mu$g $l^{-1}$, so we assume that food availability was not limiting.

We could not observe the seasonal succession of rotifer plankton because the frequency of sampling was too infrequent as a result of personal safety considerations. Consequently, rotifer diversity was explored by comparing the following parameters: (1) natural logarithm of the rotifer plankton density, (2) species diversity, (3) relative density of the most abundant species, (4) temperature, (5) natural logarithm of radionuclide concentration, and (6) Shannon–Wiener index ($H'$). A significant tendency towards increasing rotifer density with temperature was seen with a temperature effect being particularly pronounced in the riverbed station (Fig. 3a). Correlations between heterogeneity of the rotifer community ($H'$) and both number of species (Figs 3d and 4d) and the relative density of the dominant species (Figs 3c and 4c) are particularly evident. The degree of statistic in-

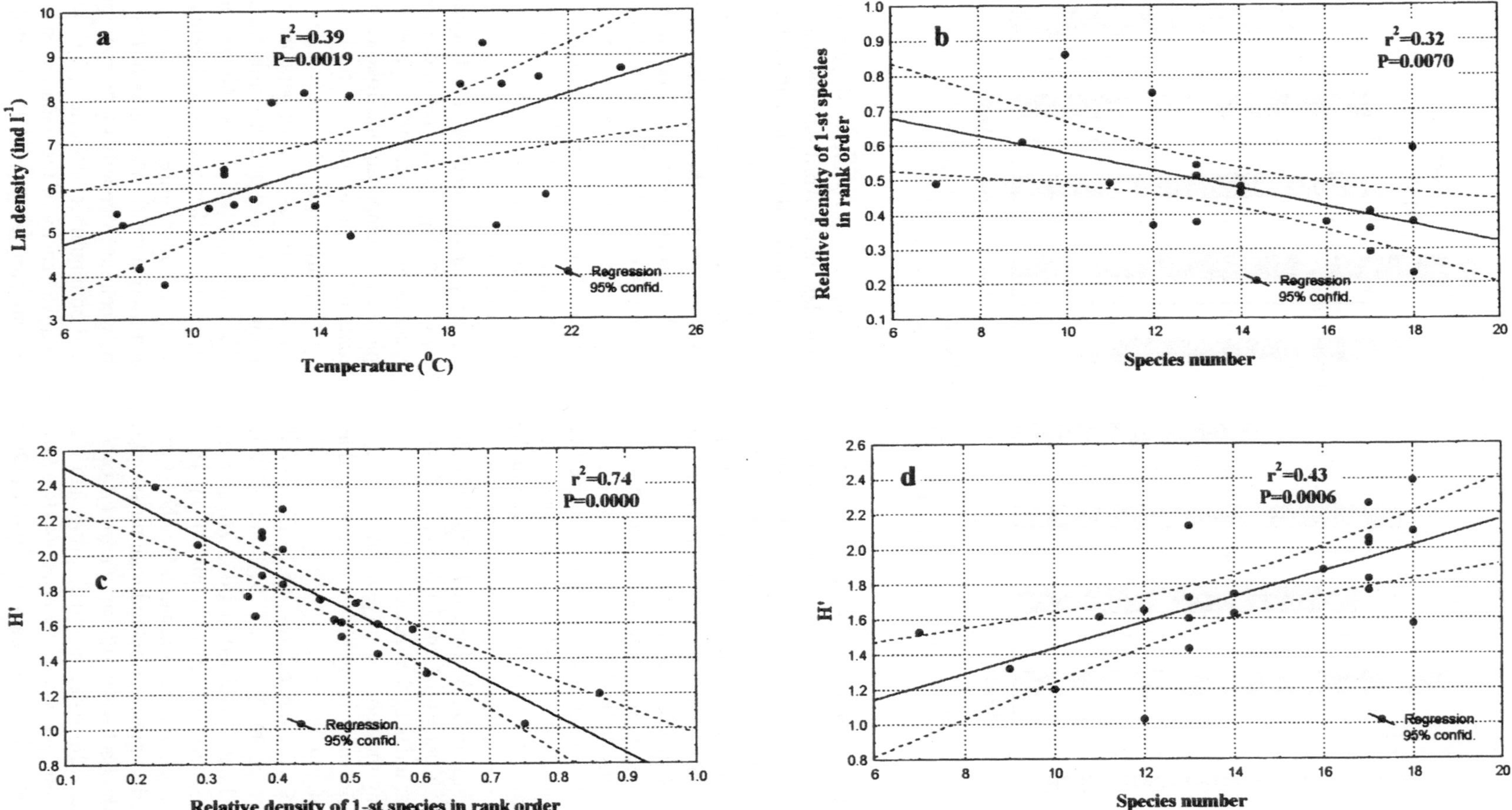


*Figure 3.* Regression analysis of biodiversity parameters in the riverbed station.

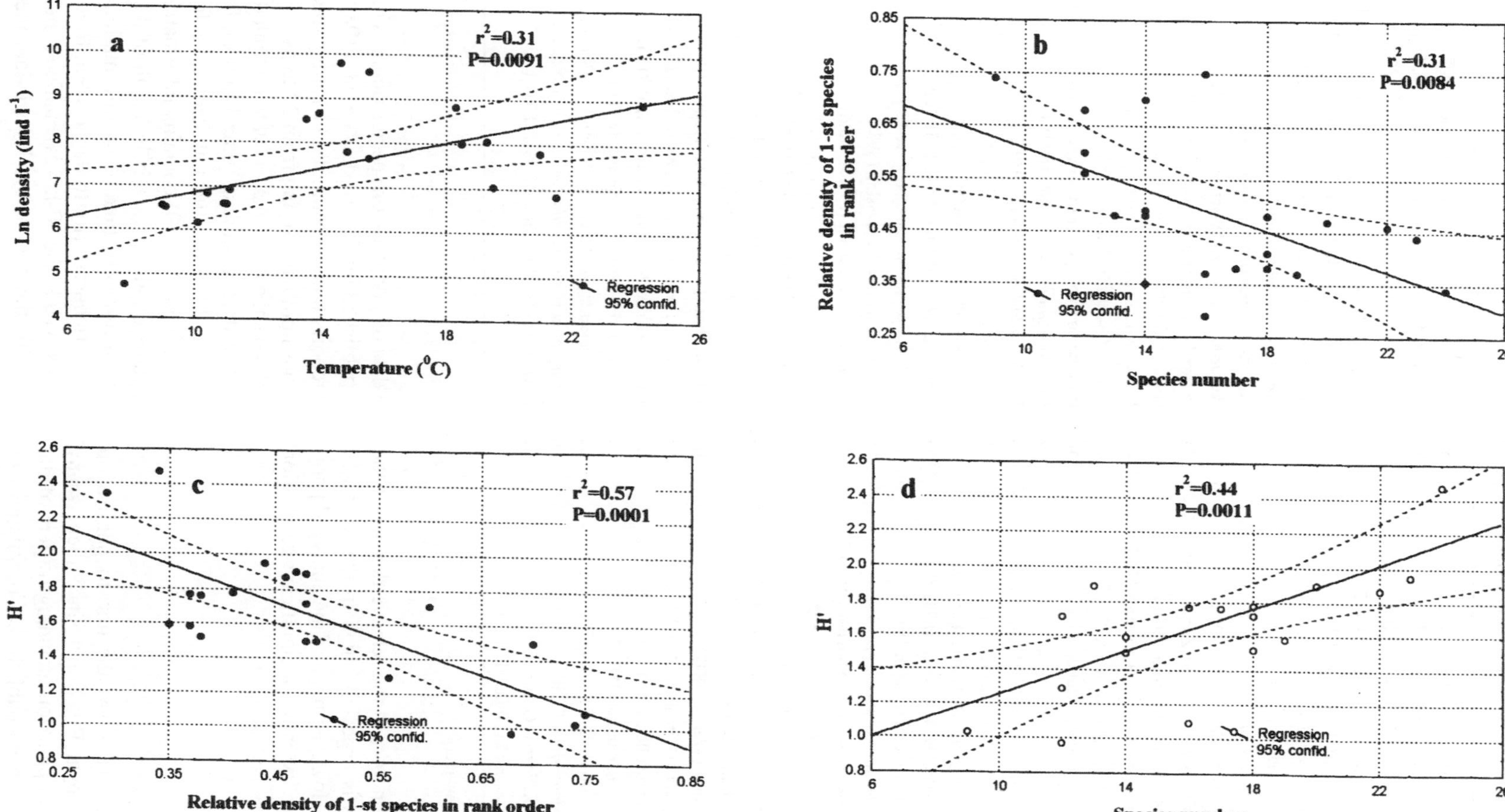


*Figure 4.* Regression analysis of biodiversity parameters in the backwater station.

*Table 1.* List of Rotifera found in the riverbed (1) and backwater (2) stations

| | | | |
|---|---|---|---|
| 1. | *Anuraeopsis fissa* (Gosse, 1851) 1, 2 | 35. | *Keratella valga monospina* (Klausener, 1908) 2 |
| 2. | *Ascomorpha ecaudis* Perty, 1850 2 | 36. | *Lecane (Monostyla) bulla bulla* (Gosse, 1832) 1 |
| 3. | *Asplanchna priodonta* Gosse, 1850 1, 2 | 37. | *Lecane (Monostyla) lunaris* (Ehrenberg, 1832) 2 |
| 4. | *Brachionus angularis* Gosse, 1851 1, 2 | 38. | *Lecane (s.str.) verecunda* Harring et Myers, 1926 1 |
| 5. | *Brachionus bennini* Leissling, 1924 1 | 39. | **Lepadella (s.str.) imbricata Harring, 1914 1 *** |
| 6. | *Brachionus budapestinensis* Daday, 1885 1, 2 | 40. | *Notholca acuminata* (Ehrenberg, 1832) 1, 2 |
| 7. | *Brachionus calyciflorus* Pallas, 1766 1, 2 | 41. | *Notholca squamula* (Muller, 1786) 1, 2 |
| 8. | *Brachionus diversicornis* (Daday, 1883) 1, 2 | 42. | **Philodina acuticornis odiosa Milne, 1916 1, 2*** |
| 9. | *Brachionus falcatus* Zacharias,1898 2 | 43. | *Polyarthra euryptera* Wierzejski, 1891 1, 2 |
| 10. | *Brachionus leydigii tridentatus* Zernov, 1901 1, 2 | 44. | *Polyarthra major* Burckhardt, 1900 1, 2 |
| 11. | **Brachionus nilsoni Ahlstrom, 1940 2*** | 45. | *Polyarthra vulgaris* Carlin, 1934 1, 2 |
| 12. | *Brachionus quadridentatus* Hermann,1783 2 | 46. | *Pompholyx complanata* Gosse, 1851 1 |
| 13. | *Brachionus urceus* (Linnaeus,1758) 1, 2 | 47. | *Pompholyx sulcata* Hudson, 1885 1, 2 |
| 14. | *Cephalodella catellina* (Muller, 1786) 1, 2 | 48. | **Postclausa hyptopus (Ehrenberg, 1838) 2*** |
| 15. | *Cephalodella gibba* (Ehrenberg, 1832) 2 | 49. | *Postclausa minor* (Rousselet, 1892) 1 |
| 16. | *Cephalodella forficata* (Ehrenberg, 1832) 1 | 50. | *Proales fallaciosa* Wulfert, 1939 2 |
| 17. | *Chromogaster ovalis* (Bergendal, 1892) 1, 2 | 51. | *Rotaria rotatoria* (Pallas, 1766) 2 |
| 18. | *Collotheca mutabilis* (Hudson, 1885) 1, 2 | 52. | **Synchaeta kitina Rousselet, 1902 1, 2*** |
| 19. | *Collotheca pelagica* (Rousselet, 1893) 1, 2 | 53. | *Synchaeta oblonga* Ehrenberg, 1831 1, 2 |
| 20. | *Colurella adriatica* Ehrenberg, 1831 2 | 54. | *Synchaeta pectinata* Ehrenberg, 1832 1, 2 |
| 21. | *Colurella uncinata* (Muller, 1773) 1, 2 | 55. | *Synchaeta stylata* Wierzejski, 1893 2 |
| 22. | **Conochiloides coenobasis Skorikov, 1914 1, 2*** | 56. | *Testudinella mucronata* (Gosse, 1886) 2 |
| 23. | *Conochilus unicornis* Rousselet,1892 2 | 57. | *Testudinella patina* (Hermann, 1783) 1 |
| 24. | *Dicranophorus forcipatus* (Muller, 1786) 2 | 58. | *Trichocerca (D.) dixon-nuttalli* (Jennings, 1903) 1, 2 |
| 25. | *Euchlanis dilatata lucksiana* Hauer, 1830 2 | 59. | *Trichocerca (D.) inermis*(Linder, 1904) 1, 2 |
| 26. | **Filinia brachiata (Rousselet, 1901) 1, 2*** | 60. | *Trichocerca (D.) similis* (Wierzejski, 1893) 1, 2 |
| 27. | **Filinia cornuta (Weisse, 1847) 1*** | 61. | **Trichocerca (D.) taurocephala (Hauer, 1931) 1 *** |
| 28. | *Filinia longiseta* (Ehrenberg, 1834) 1, 2 | 62. | *Trichocerca (s.str.) capucina* (Wierzejski et Zacharias, 1893) 1, 2 |
| 29. | *Filinia passa* (Muller, 1786) 1, 2 | 63. | *Trichocerca (s.str.) cylindrica* (Imhof, 1891) 1, 2 |
| 30. | *Gastropus stylifer* Imhof, 1891 2 | 64. | *Trichocerca (s.str.) iernis* (Gosse, 1887) 1 |
| 31. | *Hexarthra mira* (Hudson, 1871) 2 | 65. | *Trichocerca (s.str.) longiseta* (Schrank, 1802) |
| 32. | *Kellicottia longispina* (Kellicott, 1879) 1 | 66. | *Trichocerca (s.str.) pusilla* Lauterborn, 1898) 1, 2 |
| 33. | *Keratella cochlearis* (Gosse,1851) 1, 2 | 67. | *Trichocerca (s.str.) stylata* (Gosse, 1851) 1, 2 |
| 34. | *Keratella quadrata* (Muller, 1786) 1, 2 | | |

*Asterisks indicate new records.

terrelation between $H'$ and the relative density of the dominant species is especially high in the river bed station (Fig. 3c). However, none of the analyzed parameters of biodiversity shows a significant correlation with radionuclide concentration.

## Discussion

Voronkov (1907, 1915) was the first to publish on rotifers in the Pripyat River, but since then, many other papers on zooplankton including rotifers of the Pripyat River have been published. Compilations of data on zooplankton species richness of the river were published by Galkovskaya et al. (1983, 1985). These reports noted a total of 111 rotifer species, with species numbers varying from 22 to 59 in any given year. Rassashko & Savitsky (1989) compiled a complete list of rotifers, including some 145 taxa, from the Pripyat River, but only the publication of Radzimovsky & Polishchuk (1970) contains information about rotifer plankton along the whole length of the Pripyat River. Unfortunately, there is no information on rotifer species richness prior to the accident for the river section involved in the 30-km evacuation zone.

To the best of our knowledge, there are no published reports on the effect of radionuclides on rotifers. Although Shekanova (1983) reported that radionuclide concentration of 1 $10^{-6}$ Cu $l^{-1}$ (37 kBq $l^{-1}$) was harmless to plankton crustaceans, those findings were obtained under short-term experimental conditions and should not be extrapolated to field conditions, with chronic radiation contamination.

Monitoring the terrestrial communities in the evacuation zone has already shown that cessation of human activity has been a key factor in the recolonization by wildlife (Plenin & Rury, 1996). We have not discovered any significant relationship between rotifer biodiversity and radionuclide concentration. However, the cessation of human disturbance necessitated by the contamination in the CNPP region represents an excellent opportunity for scientific investigation of a variety of topics of interest to managers and ecologists.

## Acknowledgements

We thank Dr Zarubov and Mr Evdokimov for their help with sampling, and Dr R. L. Wallace for help in revising our manuscript.

## References

Galkovskaya, G. A., D. V. Molotkov & I. A. Smirnova, 1983. The changes of species composition of zooplankton of the Pripyat river during the century. In Sushchenya, L. M. (ed.), The Biological Bases of Assimilation, Reconstruction and Protection of Byelorussian Fauna. Minsk: 6–7 (in Russian).

Galkovskaya, G. A., D. V. Molotkov, A. I. Zarubov & I. A. Smirnova, 1985. Species composition and density of zooplankton of the Pripyat river along the length from Lemeshevichi to Narovlya. Vesti AN BSSR, ser.biol.3: 92–97 (in Byelorussian).

Krebs, C. J., 1999. Ecological Methodology. 2nd edn. Benjamin Cummings, Addison Wesley Longman, CA. 620 pp.

Kutikova, L. A., 1970. Rotifera Fauna of the USSR. Subclass Eurotatoria. Nauka, Leningrad. 744 pp (in Russian).

Kuzmenko, M. I., 1998. Radioecological problems in Ukrainian water reservoirs. Gidrobiol. Zh. 34(6): 95–119 (in Russian).

Odum, E. P., 1986. Ecology. V.2. Mir, Moskva. 376 pp (in Russian).

Plenin, A. E. & P. M. Rury, 1996. Evacuation zone changes in Belarussian wildlife populations following the Chernobyl nuclear accident. 1996 SETAG Annual Meeting, Washington, D.C.: 1–22.

Radzimovsky, D. O. & V. V. Polishchuk, 1970. The Plankton of the Pripyat River. Naukowa dumka, Kiev. 211 pp (in Ukrainian).

Rassashko, I. F. & B. P. Savitski, 1989. The Zooplankton of Reservoirs and Runwaters of the Byelorussian Polessye (the data bank from 1888 to 1985). Dep.N 1178–B89. VINITI, Lubertzy: 125 pp (in Russian).

Rokitsky, P. F., 1973. Biological Statistics. Minsk, University Press. 319 pp (in Russian).

Savchenko, V. K., 1995. The ecology of the Chernobyl catastrophe. Scientific outlines of an International programme of collaborative research. In Jeffers, J. N. R. (ed.), Man and the Biosphere Series. Vol.16. UNESCO and Parthenon Publishing Group, Paris: 200 pp.

Shekhanova, I. A., 1983. Radioecology of Fishes. Moscow. 208 pp (in Russian).

Voronkov, N. V., 1907. Rotifers collected by the expedition of ichthyological section in the West. Tran. Russ. Soc. Pl. An. Accl. Ichthyol. Sect. 6: 147–215 (in Russian).

Voronkov, N. V., 1915. The rotifers of the Dnieper river and the bayou lakes of the Trukhanov island. Trans. Dnieper biol. stat. 2: 1–89 (in Russian).

Hydrobiologia **446/447:** 187–194, 2001.
L. Sanoamuang, H. Segers, R.J. Shiel & R.D. Gulati (eds), Rotifera IX.




# Variability and instability of planktonic rotifer associations in Lesotho, southern Africa

Jim Green
*17 King Edwards Grove, Teddington, Middx. TW11 9LY, England*
*E-mail: jimgreen17keg@btinternet.com*

*Key words:* climate, instability, altitude, rotifers, *Lovenula*, *Paradiaptomus*, copepods, *Xenopus* tadpoles, predation

## Abstract

Seventeen localities over a 1500 m altitudinal range were sampled in January 1991 and October 1992 in the subtropical highlands of Lesotho, southern Africa. Spatial and temporal variability in rotifer assemblages are described briefly, and factors driving community heterogeneity are considered. Predation was identified as one significant factor; some sites with low rotifer diversity had dense populations of large calanoid copepods (*Lovenula falcifera, Paradiaptomus warreni*), and in other localities, tadpoles of the clawed toad, *Xenopus laevis.* Gut content analysis demonstrated that both calanoids and tadpoles were consuming rotifers.

## Introduction

Variation in rotifer biodiversity at different latitudes on the various continents has received more attention than has altitudinal variation in assemblages (e.g. Green, 1993). It could be predicted that an inverse relationship between altitude and species diversity would apply, with limiting factors such as temperature extremes and availability of free water. To confound this prediction, evidence is emerging that habitat ephemerality may in fact increase species diversity – Australian temporary waters, for example, were considered to be important evolutionary loci (Williams, 1988), a contention supported for the rotifers of southern Australia in a survey of >100 ephemeral habitats (Shiel et al., 1998).

A sample series collected from a heterogeneous series of waters in the highlands of Lesotho during two visits in 1991–92 provided an opportunity to contrast the rotifer assemblages of an altitudinal range of habitats which also differed in their permanence, with lowland assemblages. Furthermore, despite only two visits to most of the sites, clear evidence was obtained of at least some of the forces driving community succession. I report here the variations in rotifer associations between the Lesotho sites, the seasonal instability apparent in them, and speculate on some of the selective forces operating, in particular the evidence for structuring of the rotifer communities by predation.

## Study area

Lesotho is a landlocked kingdom in southern Africa, between latitudes 28 and 31° S. It has an area of 30 355 sq. km, and the lowest point is over 1000 m asl. Much of the country is over 2000 m asl. (Fig.1), and the highest point, in the Drakensberg, rises to 3482 m asl. There is a considerable effect of altitude on temperature – the mean monthly maximum at Maseru (alt. 1600 m) ranges up to 29.5 °C in December, while the mean minimum falls to 0 °C in July (Staples & Hudson, 1938). In the high mountains, at 3000 m, the mean monthly maximum reaches 15.5 °C in January and the mean monthly minimum falls to −11.7 °C in July.

Rainfall is very variable from year to year. Over much of Lesotho about 750 mm falls per year. Much of this water runs out of the country via an extensive network of streams and rivers. There are very few natural standing waters. The largest natural lake in Lesotho, Tsa Kholo, averages about 1 sq. km in area, but shows such wide fluctuations that it was not possible to sample it. On the first visit it was the centre of a raging torrent, and on the second visit it was dry. All

*Table 1.* Localities sampled, arranged in order of increasing altitude

| | Locality | Altitude m. asl | Area sq.m | Surface Jan 1991 | T °C Oct 1992 | Cond. $\mu$S | Notes |
|---|---|---|---|---|---|---|---|
| 1. | Leribe pool | 1501 | 14 000 | 24 | 23.2 | 164 | Well vegetated |
| 2. | Bolaea Tau Dam | 1550 | 18 750 | 22.9 | dry | 65 | Very silty |
| 3. | Linyane Pool | 1578 | 39 275 | 20 | dry | 115 | Clear water, *Elodea* abundant |
| 4. | University Pond | 1615 | 10 000 | 25.1 | 24.2 | 130 | Well vegetated |
| 5. | Tso Litiama | 1661 | 200 000 | 25.9 | 23.7 | 140 | Very silty |
| 6. | Collett Dam | 1722 | 462 000 | 25.7 | 20 | 175 | Vegetation sparse |
| 7. | Semonkong Dam | 2210 | 15 000 | 23.8 | 15.4 | NR | *Xenopus* larvae abundant in Jan. |
| 8. | Sehlabathebe (Ngoangoana Gate) | 2301 | 23 565 | 18.6 | 22.7 | 18 | Shallow, may dry |
| 9. | Letseng la Letsie | 2380 | 562 500 | 22.4 | NR | 121 | Adult *Xenopus* present |
| 10. | Sehlabathebe (*Potamogeton* Pool) | 2362 | 2500 | 27 | NR | 24 | Well vegetated |
| 11. | Sehlabathebe Pool 1, N. of Lodge | 2362 | 4000 | 21.4 | 10 | 12 | Appears to have dried before Oct. |
| 12. | Sehlabathebe Pool 2, N. of Lodge | 2368 | 3000 | 20.1 | 10.5 | 9 | .. |
| 13. | Sehlabathebe Pool 3, N. of Lodge | 2425 | 5000 | NR | 12.5 | NR | More permanent than pools 1 & 2 Adult *Xenopus* |
| 14. | Letseng la Terae | 2947 | 60 000 | 17.6 | 10.5 | 425 | A diamond mine tailing dam |
| 15. | Khalong la Lithunya *Lagarosiphon* pool | 3072 | 7 | 18.2 | NR | 60 | Only 40 cm deep |
| 16. | Khalong la Lithunya *Crassula* pool | 3079 | 4 | 18.3 | 11.5 | 39 | Only 35 cm deep |
| 17. | Khalong la Lithunya *Aponogeton* pool | 3082 | 1 | 12.4 | NR | 45 | Only 5 cm deep |

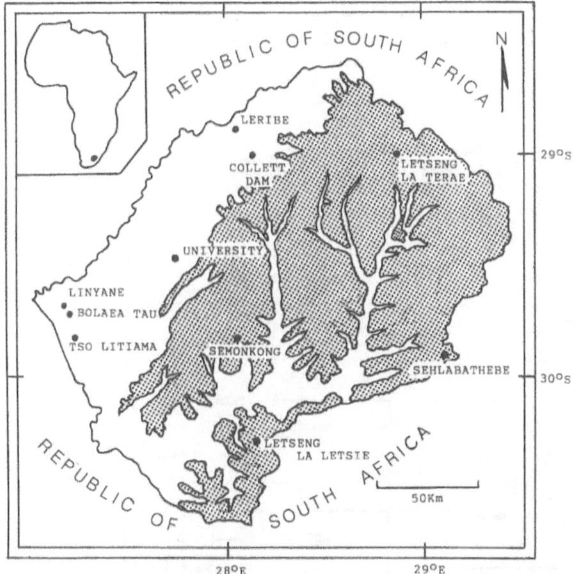


*Figure 1.* Map of Lesotho, showing land over 2000 m asl., and the main sampling localities. Khalong la Lithunya ( loc. 15, 16, 17) lies in the same region as Letseng la Terae, but at a higher altitude (cf. Table 1).

the standing water bodies sampled in this study have areas less than 1 sq. km.

## Methods

Waterbodies were located using 1:50 000 maps produced by the U.K. Directorate of Overseas Surveys, and available at the Department of Lands, Survey and Physical Planning in Maseru. Altitudes and areas estimated from these maps can only be regarded as approximate. Conductivity was measured with a Schott Gerate Konductometer, and temperature with an electronic probe calibrated against a mercury thermometer.

Net tows were taken from the banks of the water bodies using Hydrobios 55 $\mu$m-mesh nets, and were preserved in 5% formaldehyde. From some of the very small water bodies samples were taken by pouring water through the nets. Coarser nets (250 $\mu$m mesh) were used to collect copepods and *Xenopus* tadpoles. Details of the counting technique are given in Green (1993), where the limitations of momentary samples are also discussed. Dominance was calculated simply as the percentage of the total rotifers made up by the most numerous species.

## Results

### *Spatial variability*

Although 84 rotifer taxa were identified across the range of sites (Appendix), within-site diversity was low; only two sites of the 25 samples which contained sufficient rotifers to estimate% composition had >15 species present. The remainder had 3–14 species (Fig. 2). Twenty-six species (ca. 31%) were recorded only once, with another 20 taxa (24%) recorded from only two sites.

There was no obvious relationship between increasing altitude and species diversity, but clearly obvious in Figure 2 is the change in species dominants between sites. A few species were dominant at widely separated altitudes. For example, *Keratella reducta* was dominant in locality 2 (1550 m) in January, and in locality 9 (2380 m) in October. Both these localities lacked vegetation, and both had silty water. *Plationus patulus* dominated localities 9 and 11 in January, and was also relatively abundant, although not dominant in locality 1 (1501 m) in October. *Keratella cochlearis* also was dominant at different altitudes, dominating locality 5 (1661 m) in the second October sample, and locality 13 (2425 m) in October.

In sites with different macrophyte dominants, the rotifer communities also differed in their dominants. For instance, the three highest localities (15, 16, 17) are very small water bodies. In Locality 15, the dominant macrophyte was *Lagarosiphon*, and the dominant rotifer was *Lepadella patella.* In the next locality (16), the dominant plant was *Crassula natans* and the most abundant rotifer was *Trichocerca bicristata.* In the highest, smallest locality (17), the dominant plant was *Aponogeton ranunculiflorus,* and the rotifers were dominated by *Scaridium longicaudum.*

Assemblages also differed appreciably in the proportional composition of taxa, ranging from equitable distributions (dominance <30%) (sites 4, 5, 10 (January)) to >80% dominance by single species (sites 7, 13, 14, 16 & 17 (January) and 4, 5 & 13 (October)). Extreme dominance appears to be more frequent than equitable distribution in these samples from Lesotho. Extreme dominance was associated with low species numbers. Eight samples with extreme dominance had a mean species number of 5.6 (s.e. 0.9), while the nine samples with a dominance of less than 50% had a mean species number of 12.5 (s.e. 0.9 ).

### *Temporal variability*

As shown in Figure 2, there was little overlap in spe-

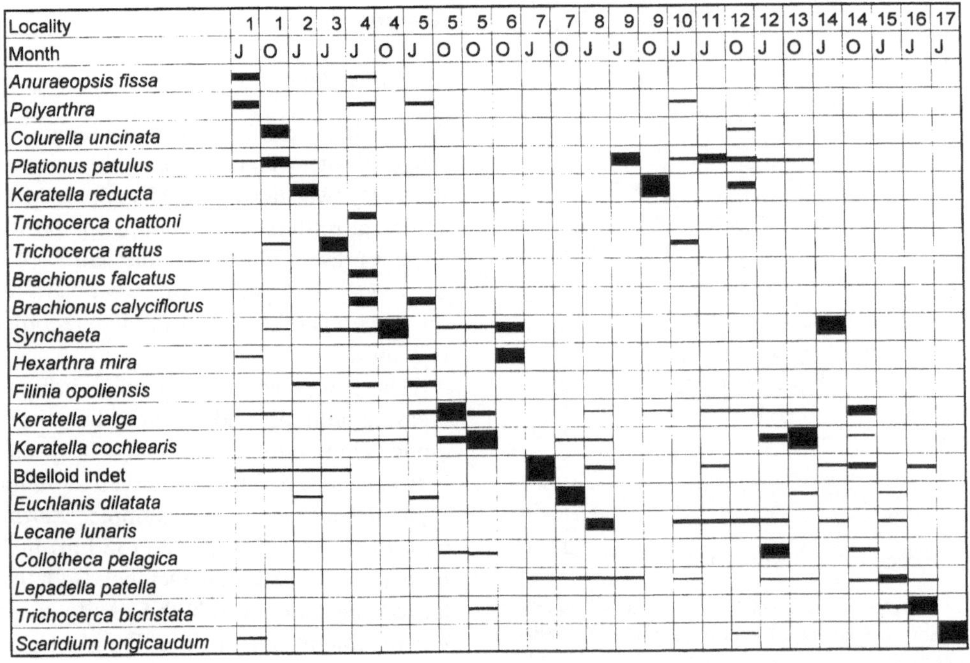


*Figure 2.* Dominant species of Rotifera in 17 localities in Lesotho, arranged in order of increasing altitude. The extent to which each square is filled is proportional to the percentage of that species in the sample. J – January 1991; O – October 1992.

*Table 2.* Sorensen similarities of rotifer associations within Lesotho, January 1991/October 1992

| | |
|---|---|
| Semonkong Dam | 62 |
| University Pond | 40 |
| Crassula Pool | 55 |
| Leribe Pool | 30 |
| Tso Litiama (1) | 14 |
| " (2) | 24 |
| Sehlabathebe, pool 1 | 14 |
| " pool 2 | 23 |
| Ngoangoana | 0 |
| Letseng la Terae | 0 |
| Mean | 23.8 |
| S.E. | 6.8 |

cies composition at sites sampled in the successive years. It was possible to obtain samples from both January and October from 10 localities which gave good estimates of the percentage composition of rotifer species. Table 2 shows the 10 sites compared by means of the Sorensen Index, which is based on the number of species common to each of two samples. Semonkong Dam showed the greatest similarity of associations in the two seasons, but the overall mean similarity was only about 24%, and in three localities there was no overlap of the species found in the two seasons.

Successional events are evident from the dominants shown in Figure 2. In Tso Litiama (site 5), for example, of nine species recorded in January 1991, only one, *Keratella valga*, was collected again in October 1992. In Letseng la Terae (site 14), not one of the four species collected in the first visit was collected again on the second (12 taxa recorded). These two localities (5 and 14) are permanent, but still show large differences in their rotifer associations at different seasons.

*Impact of predators*

*Calanoid copepods*

In two samples, rotifers were sparse when large calanoid copepods were abundant. The copepods present in both localities were *Lovenula falcifera* (Loven), *Paradiaptomus warreni* Rayner, and the much smaller *Metadiaptomus meridianus* (van Douwe).

1. Sehlabathebe, Ngoangoana Gate (locality 8) 18 January 1991. Although sparse, 11 taxa of rotifers were identified (dominants shown in Figure 2). Most of the rotifers in the sample were not strictly planktonic, and appear to have been swept

*Table 3.* Comparison of plankton composition and gut contents of the copepods *Lovenula falcifera* and *Paradiaptomus warreni*, from Sehlabathebe, Pool 3

| Taxa identified | Plankton % comp. | No. *L. falcifera* containing | No. *P. warreni* containing |
|---|---|---|---|
| Rotifera | | | |
| *Euchlanis dilatata* | 1 | – | – |
| *Keratella cochlearis* | 95 | 16 (88)[a] | 4 (8) |
| *Lepadella patella* | 1 | – | – |
| *Mytilina ventralis* | – | | 1 (1) |
| *Plationus patulus* | 1 | – | – |
| *Testudinella* sp. | | – | 1 (1) |
| *Trichocerca capucina* | 1 | – | – |
| *Trichocerca weberi* | | 1 (1) | – |
| *Cephalodella* sp. | 1 | 1 (1) | – |
| Cladocera | | | |
| *Chydorus* sp. | | 2 (8) | – |
| Copepoda | | | |
| Nauplius larva | | 2 (2) | – |
| *Metadiaptomus meridianus* | | 4 (4) | – |
| *Microcyclops* sp. | | 4 (4) | – |
| Insecta | | | |
| Chironomid larvae | | 5 (5) | – |
| Total animal records | | 35 (113) | 6 (10) |
| Total algal records | | 11 | 69 |

[a]Numbers in brackets give total individuals found.

off the vegetation on the floor of the pool. When the same pool was sampled in October 1992 no rotifers were found.

2. Sehlabathebe, Pool 3 (locality 13) 11 October 1992: Rotifers were again sparse, but enough were found to give a percentage count (Table 3). *Keratella cochlearis* was overwhelmingly dominant.

The gut contents of 20 specimens of each of the two large copepods were examined to see if they were eating rotifers (Table 3). It is very clear that *Lovenula falcifera* consumed much more animal material than *Paradiaptomus warreni*. Conversely, algae are more abundant in the latter species. Both species consume rotifers, but *L. falcifera* is more voracious.

*Table 4.* Net tow composition vs. foregut contents of large tadpoles of *Xenopus laevis* from Semonkong Dam, January 1991

| Tadpole No. | 1 | 2 | 3 | 4 | 5 | Total | Net tow % comp. |
|---|---|---|---|---|---|---|---|
| Crustacea | | | | | | | |
| *Chydorus* | 25 | 22 | 10 | 2 | 20 | 79 | |
| *Pleuroxus* | 2 | 5 | 1 | – | 12 | 20 | |
| *Alonella + Alona* | 11 | 4 | 1 | – | 1 | 17 | |
| *Ceriodaphnia* | – | 1 | 3 | – | 9 | 13 | |
| *Bosmina* | – | – | 4 | – | – | 4 | |
| *Cyclops* | 1 | 1 | – | 1 | 1 | 4 | |
| Ostracoda | 5 | 3 | 2 | – | 3 | 13 | |
| | | | | | | 150 | |
| Rotifera | | | | | | | |
| Bdelloids | 1 | – | – | 3 | 3 | 7 | 93 |
| *Euchlanis dilatata* | – | – | – | – | – | – | 1 |
| *Keratella cochlearis* | – | – | 1 | – | – | 1 | |
| *Lecane bulla* | – | – | – | – | – | – | 1 |
| *Lecane closterocerca* | 1 | 3 | – | 2 | 5 | 11 | 1 |
| *Lepadella patella* | – | – | – | – | – | – | 2 |
| *Mytilina ventralis* | – | – | – | – | 1 | 1 | 2 |
| *Trichocerca* sp. | – | – | – | – | 1 | 1 | |
| | | | | | | 21 | |
| Chironomid larvae | 1 | – | – | – | 1 | 2 | |
| | | | | | | 173 | |

### Xenopus *tadpole populations*

Tadpoles of *Xenopus* feed by an elaborate pumping mechanism forcing water past two mucous strings, which trap particles of various sizes (Wassersug, 1996). Rotifers were difficult to find in samples where the tadpoles were abundant. To determine if rotifers were entrapped by the feeding mechanism, the gut contents of tadpoles were examined.

Two samples were rich in *Xenopus* tadpoles and poor in rotifers:

1. Semonkong Dam (locality 7) 13 January 1991: Large (ca. 60 mm total length) tadpoles were abundant. Rotifers were sparse, and dominated by an unidentified, contracted bdelloid. The foreguts of five tadpoles were examined (Table 4), and revealed that most of the animals ingested were crustaceans, but some rotifers were entrapped, including some species not found in the net samples.
2. Khalong la Lithunya, *Crassula* Pool (locality 16) 20 January 1991: Small (ca. 20 mm total length) tadpoles were abundant. Rotifers were very sparse

*Table 5.* Net tow composition vs. gut contents of *Xenopus laevis* tadpoles from Khalong la Lithunya *Crassula* Pool, January 1991

| Food | No./20 with | Net tow % comp. |
|---|---|---|
| Algae | 20 | |
| Rhizopoda | 12 | |
| Nematoda | 2 | |
| Rotifera: *Cephalodella* sp. | 2 | 1 |
| *Colurella* sp. | 1 | |
| *Lecane hamata* | 8 | |
| *L. signifera* | 1 | 1 |
| *L. ungulata* | 1 | |
| *Lepadella patella* | 3 | 1 |
| *L. rhomboides* | 1 | |
| *Mytilina ventralis* | 11 | 6 |
| *Trichocerca bicristata* | 6 | 87 |
| *T. capucina* | 1 | 1 |
| *T.* (*Diurella* sp.) | 1 | |
| *T. weberi* | – | 2 |
| Indet. | – | 1 |
| Cladocera: Chydoridae | 10 | |
| Copepoda: nauplii | 9 | |
| copepodids | 6 | |
| Ostracoda | 1 | |

and dominated by *Trichocerca bicristata* (Table 5). Examination of the gut contents of 20 small tadpoles (Table 5) showed that the majority had been feeding on algae, but some had ingested rotifers, and more had ingested *Mytilina ventralis* than any other rotifer. Although the numbers of rotifers ingested by each tadpole were relatively small, the large population of tadpoles could exert a restraining influence on the development of rotifer populations.

## Discussion

With more than half of the identified taxa recorded from only one or two sites, the rotifers of the Lesotho Highlands demonstrate similar patchiness to that reported from ephemeral habitats elsewhere (cf. Shiel et al., 1998). The variability of rotifer assemblages in Lesotho is shown well in Figure 2. Of the 21 species in the diagram, only five are dominant in more than one locality. This is in marked contrast to the localities in the Nyanga hills of Zimbabwe, studied by Green (1990), where seven of the 10 localities were dominated by the same species. Figure 2 also shows that there is an altitudinal component in the local variation within Lesotho, but it appears to be operating at the species level rather than the community level.

Other factors likely to cause variation are the size and depth of the water body, and its vegetation. The development of vegetation increases structural complexity, so providing more niches for rotifers. In a large water body with a complex littoral zone, the number of rotifer species can reach over 200 (Segers & Dumont, 1995; Dumont & Segers, 1996). Conversely, in a very small locality, dominated by a single species of macrophyte, as in localities 16 and 17, conditions may favour a single species to the extent that it may show extreme dominance, with few other species present.

Seasonal variation is clearly driven by climate. In the extreme case, a locality may dry out. This, the ultimate form of instability, happened in several of the localities in Lesotho. But even localities with permanent water showed instability, revealed by marked changes in the specific composition of the rotifer associations between January 1991 and October 1992.

An additional source of variability is the presence or absence of predators. Copepods are well known as predators of rotifers (cf. Williamson, 1983, 1987), and Green et al. (1999) recently reported selective predation by large Australian freshwater calanoids. Table 3 indicates that both *Lovenula falcifera* and *Paradiaptomus warreni* found some rotifers that I did not find in the samples. An important effect of the presence of these large copepods was the reduction in the abundance of rotifers. It was very difficult to find sufficient rotifers to make a percentage count when the copepods were abundant.

The tadpoles of *Xenopus* also take in rotifers along with many other small particles. Table 4 shows that large tadpoles take in more crustaceans than rotifers, but Table 5 shows that small tadpoles take a higher proportion of rotifers. These predators also found rotifers that I did not.

The overall variability of rotifer associations in Lesotho is a result of the combined effects of climate, altitude , small size of water bodies and selective predation by copepods and *Xenopus* tadpoles.

## Acknowledgements

This work was done on two visits sponsored by the British Council through a link between the Biology

Departments of Queen Mary and Westfield College, London and the National University of Lesotho. I am most grateful to Kevin Broadfoot, of the British Council in Maseru, and to Dr Joseph, Head of Biology at the University, for the provision of facilities, and general smoothing of our path. I also thank Prof. J.G. Ducket and Dr R. Ligroni for their companionship and help in the field.

## References

Dumont, H. J. & H. Segers, 1996. Estimating lacustrine zooplankton species richness and complementarity. Hydrobiologia 341: 125–132.

Green, J., 1990. Zooplankton associations in Zimbabwe. J. Zool. Lond. 222: 259–283.

Green, J., 1993. Diversity and dominance in planktonic rotifers. Hydrobiologia 255/256: 345–352.

Green, J. D., R. J. Shiel & R. A. Littler, 1999. *Boeckella major* (Copepoda: Calanoida): a predator in Australian ephemeral pools. Arch. Hydrobiol. 145: 181–196.

Segers, H. & H. J. Dumont, 1995. 102+ rotifer species (Rotifera: Monogononta) in Broa Reservoir (S.P. Brazil) on 26 August 1994, with a description of three new species. Hydrobiologia 316: 183–197.

Shiel, R. J., J. D. Green & D. L. Nielsen, 1998. Floodplain biodiversity: why are there so many species? Hydrobiologia 387/388: 39–46.

Staples, R. R. & W. K. Hudson, 1938. An ecological survey of the mountain area of Basutoland. Crown Agents for the Colonies. London. 68 pp.

Wassersug, R., 1996. The biology of *Xenopus* tadpoles. In Tinsley, R. C. & H. R. Kobel (eds), The Biology of *Xenopus*. Clarendon Press: Oxford: 195–211.

Williams, W. D., 1988. Limnological imbalances: an Antipodean viewpoint. Freshwat. Biol. 20: 406–420.

Williamson, C. E., 1983. Invertebrate predation on planktonic rotifers. Hydrobiologia 104: 385–396.

Williamson, C. E., 1987. Predator–prey interaction between omnivorous diaptomid copepods and rotifers: the role of prey morphology and behaviour. Limnol. Oceanogr. 32: 167–177.

*Appendix 1.* List of rotifers found in samples from Lesotho in January 1991 and October 1992. The numbers following each name refer to the localities listed in Table 1.

*Anuraeopsis fissa* (Gosse): 1, 4;
*Asplanchna brightwelli* (Gosse): 4, 5, 6, 7;
*A. sieboldi* (Leydig): 4;
*Ascomorpha ecaudis* (Perty): 10, 11;
*A. saltans* Bartsch: 4, 5, 6, 7;
*Brachionus angularis* (Gosse): 4, 5;
*B. bennini* (Leissling): 5:
*B. bidentatus* Anderson: 9;
*B. calyciflorus* Pallas: 4, 5;
*B. falcatus* Zacharias: 1, 4, 5;
*B. forficula* Wierzejski: 9;
*B.quadridentatus* Hermann: 2, 9;
*Cephalodella forficula* (Ehrenberg): 1, 5,7,14;
C. sp.: 8, 11, 13, 16;
*Collotheca pelagica* (Rousselet): 5,11,12,14;
*Colurella obtusa* (Gosse): 1;
*C. tesselata* (Glasscott): 12;
*C. uncinata* (Müller): 1, 12;
*Conochilus* sp.: 1, 2, 11, 12;
*Euchlanis dilatata* Ehrenberg: 2, 5, 13, 15;
*E.* sp.: 1, 4, 7;
*Filinia longiseta* (Ehrenberg): 1, 4, 5, 9;
*Filinia opoliensis* (Zacharias): 2, 4, 5;
*F. terminalis* (Plate): 12, 14;
*Gastropus hyptopus* (Ehrenberg):4,5,14;
*Hexarthra mira* (Hudson): 1, 5, 6;
*Keratella cochlearis* (Gosse): 4, 5, 7, 8, 12, 13, 14;
*K. quadrata* (Müller): 1;
*K. reducta* Huber-Pestalozzi: 2, 9, 12;
*K. tropica* (Apstein): 4;
*K. valga* (Ehrenberg): 1, 5, 8, 9, 11, 12, 13, 14;
*Lecane bulla* (Gosse): 1, 2, 4, 5, 7, 10;
*L. clara* (Bryce): 12;
*L. closterocerca* (Schmarda): 1, 7;
*L. curvicornis* (Murray): 1, 14;
*L. furcata* (Murray): 1, 8:
*L. hamata* (Stokes): 15;
*L. ludwigii* (Eckstein): 1;
*L. luna* (Müller): 1, 4, 5, 10, 15, 17;
*L. lunaris* (Ehrenberg): 8, 10, 11, 12, 14, 15;
*L. mira* (Murray): 10, 12;
*L. papuana* (Murray): 1, 4;
*L. signifera* (Jennings): 8, 10, 12, 16, 17;
*L. stichaea* Harring: 4, 11, 12;
*Lecane tabida* Harring & Myers: 11;
*L. tenuiseta* Harring: 12, 14;
*L. undulata* Hauer: 4;
*L. unguitata* (Fadeev): 5;
*L. ungulata* (Gosse): 10, 12;
*Lepadella biloba* (Hauer): 1;
*L. cristata* (Rousselet): 11, 12;
*L. ovalis* (Müller): 8;
*L. patella* (Müller): 1, 7, 8, 9, 10, 12, 13, 14, 15, 16;
*L. quadricarinata* (Stenroos): 14;
*Lophocharis oxysternon* (Gosse): 1, 4;
*Monommata* sp.: 8, 12;
*Mytilina mucronata* (Müller): 15;
*M. ventralis* (Ehrenberg): 1, 7, 15, 16;
*Notholca squamula* (Müller): 14;
*Plationus patulus* (Müller): 1, 2, 9, 10, 11, 12, 13;
*Platyias quadricornis* (Ehrenberg): 4;
*Polyarthra dolichoptera* Idelson: 4, 5;
*P. vulgaris* Carlin: 1, 4, 10;
*Pompholyx sulcata* (Hudson): 4;
*Rotaria neptunia* Ehrenberg: 4;
*Scaridium longicaudum* (Müller): 1, 12, 17;
*Synchaeta* sp.: 1, 3, 4, 5, 6, 12, 14;
*Testudinella patina* Hermann: 1, 3, 9, 14;
*T. trilobata* Shephard: 1;
*Trichocerca bicristata* (Gosse): 5, 15, 16;
*T. bidens* (Lucks): 10, 11, 12, 13;
*T. brachyura* (Gosse): 1;
*T. braziliensis* (Murray): 10;
*T. capucina* Wierzejski & Zacharias: :13, 16;
*T. chattoni* (De Beauchamp): 2, 4, 5, 14;
*T. elongata* (Gosse): 3, 10, 15;
*T. iernis* (Gosse): 10;
*T. longiseta* (Schrank): 10;
*T. rattus carinata* (Ehrenberg): 1, 3, 10;
*T. weberi* Jennings: 15, 16;
*T.* sp.: 3, 11, 12;
*Trichotria tetractis* (Ehrenberg): 8, 15;
*Tripleuchlanis plicata* Levander: 1;
*Wolga spinifera* (Western): 5;

Hydrobiologia **446/447:** 195–201, 2001.
L. Sanoamuang, H. Segers, R.J. Shiel & R.D. Gulati (eds), Rotifera IX.




# Diurnal vertical distribution of rotifers (Rotifera) in the *Chara* zone of Budzyńskie Lake, Poland

Natalia Kuczyńska-Kippen
*Department of Hydrobiology, Adam Mickiewicz University, Marcelinska 4, 60-801 Poznań, Poland*

*Key words:* rotifers, diurnal vertical distribution, *Chara*, shallow lake

## Abstract

Diurnal vertical distribution of rotifers was investigated in the *Chara* bed and the water immediately above it in the shallow region (ca. 1 m depth) of Budzyńskie Lake (Wielkopolski National Park, Poland) in early September 1998. Eighty one rotifer species were identified – 71 among *Chara* and 59 in the open water. Significant differences in rotifer densities were observed in the *Chara*, with highest numbers during the day (2316 ind. $l^{-1}$) and lowest numbers early morning (521 ind. $l^{-1}$) and at dusk (610 ind. $l^{-1}$). Above the *Chara*, the numbers of rotifers did not change significantly (615–956 ind. $l^{-1}$). Littoral- or limnetic-forms differed in their diel vertical distribution between both zones. One group of littoral species was characterized by increased densities in the *Chara* in the daytime, while a second group increased in density during the night. The densities of limnetic species, which were much higher in open water, decreased in the morning or daytime in this zone. These differences in the diel behaviour of particular groups of rotifers may be dependent on microhabitat and may also be related to different kinds of predation, the exploitative competition for shared food resources between rotifers and crustaceans, as well as typical adaptation to littoral or limnetic life.

## Introduction

Diel vertical migration of zooplankton is well documented in the pelagic zone of freshwater lakes. A combination of factors such as temperature, light, oxygen as well as food and invertebrate and fish predation are responsible for vertical distribution of species (Stich & Lampert, 1981; Wright & Shapiro, 1990; Lampert, 1993; Loose & Dawidowicz, 1994).

The potential for vertical migration in shallow lakes, in contrast to the situation in deep lakes, is limited due to the absence of stratification (Lauridsen & Buenk, 1996). Only the low light intensity near the bottom, or macrophytes covering only the bottom part of a lake, may offer zooplankton refuge. In shallow lakes, the large macrophyte-covered areas may provide zooplankton with a spatial refuge and thus be a factor strongly associated with diurnal horizontal migrations (Gliwicz & Rybak, 1976; Schriver et al., 1995; Kairesalo et al., 1998). Although diurnal horizontal migrations between the pelagic and littoral zones and the migrations of zooplankton within the littoral have been the subject of much research (Kairesalo, 1980; Paterson, 1993; Lauridsen et al., 1994), our knowledge about them is still limited. In particular, we know little about the migration distances covered by zooplankton. It is known that the kind of migration is dependent on the kind of predation, but there is no clear answer with regard to the zooplankton migration pattern when they are influenced by both fish and invertebrate predators inhabiting spatially segregated areas. There should also be more research done on the interactions between plant-associated zooplankton and organisms which stay temporarily or permanently in the water within the plant stand (Jeppesen et al., 1998). Moreover, information concerning the factors that are responsible for zooplankton to undergo the vertical or horizontal movement is still poor.

The factors influencing rotifer community structure among macrophytes, apart from food availability and predator avoidance, might also be related to different habitat conditions present, providing differences in the suitability for typically littoral and pelagic species

(Preissler, 1977; Jose-de-Paggi, 1993) and the morphological adaptations of rotifers inhabiting particular zones of the lake (Preissler, 1983).

The distribution of rotifers in the shallow lake might also be connected with the relative distances that rotifers have to cover between particular macrophyte stands, as they are not very good swimmers and do not usually travel more than 3–5 m (Nogrady et al., 1993). In this case, vertical migration seems to be advantageous as the distance between the macrophyte bed and the open water just above it is not too far.

The aim of this study was to determine whether in a shallow lake with well-developed macrophyte cover, for which diurnal horizontal migrations of zooplankton are characteristic (Kuczyńska-Kippen, 1997; Kuczyńska-Kippen & Cerbin, 1998), vertical migrations of rotifers can take place. Thus, this work was undertaken in order to:

(1) compare the distribution of rotifers in the *Chara* bed and the zone of water immediately above the vegetation,
(2) compare the diurnal vertical dynamics between typically littoral and pelagic rotifer species.

## Study area

Lake Budzyńskie is a shallow lake situated in the southern part of the Wielkopolski National Park in the catchment area of the river Warta. It has an area of 17.4 ha, maximum depth of 2.7 m and a mean depth of 1.4 m (Jańczak et al., 1996).

The lake consists of two basins – the south-east which is deeper and only partially covered by submerged vegetation (only along the bank) and the north-west basin which is fully overgrown. A belt of emergent macrophytes (95% of the shoreline) with *Typha angustifolia* L. and *Phragmites australis* (Cav.) Steud. surrounds the whole basin of the lake. Nearly 75% of the basin is covered by submerged macrophytes, predominately *Chara tomentosa* L. and *Myriophyllum verticillatum* L. The *Chara* beds grow in the shallower margins.

## Methods

Zooplankton was sampled four times within 24 h (6 September, 1998). Samples were collected from the north-west basin of the lake from two sites: (a) from among the *Chara* bed and (b) from the zone of water immediately above the vegetation. Samples were taken in triplicate at each site using a plexiglass core sampler (Ø50 mm). The 10 l samples were concentrated using a 45-$\mu$m plankton net and were fixed immediately with 4% formalin.

The *t*-test was used for statistical analysis in order to evaluate the differences in density of zooplankton between particular habitats ($N$=12).

The ANOVA test was used in order to evaluate diurnal differences in rotifer densities at particular stations ($N$=12). The *posteriori* Tukey Test was applied in order to find where the differences in time occurred.

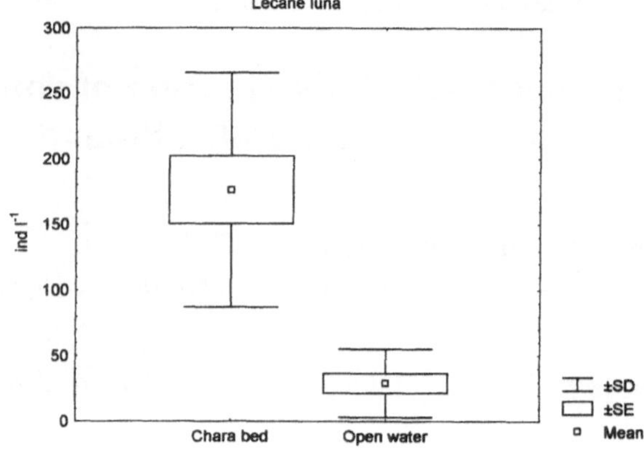


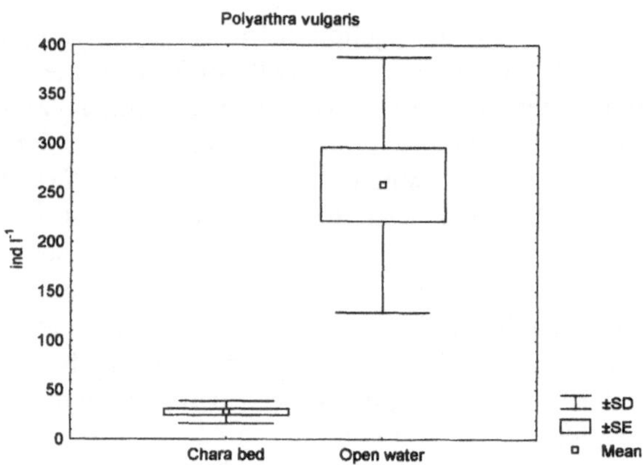


*Figure 1.* Habitat preferences of particular species in the *Chara* bed/open water zone above the vegetation stand of Budzyńskie Lake.

## Results

Eighty one rotifer species were identified; 71 among *Chara* and 59 in the open water.

*Table 1.* Statistically significant habitat preferences of particular species in the *Chara* bed and open water (*t*-test)

| Species | Chara bed | | Open water | | *t* | df | sig. |
|---|---|---|---|---|---|---|---|
| | M | SD | M | SD | | | |
| *Ascomorpha ecaudis* | 3.792 | 3.799 | 10.333 | 9.633 | −2.188 | 14.341 | 0.040 |
| *Ascomorpha saltans* | 0.625 | 0.980 | 9.750 | 10.524 | −2.991 | 11.191 | 0.012 |
| *Collotheca edentata* | 49.167 | 30.963 | 0.333 | 0.888 | 5.461 | 11.018 | 0.000 |
| *Collotheca mutabilis* | 4.250 | 5.627 | 42.083 | 21.284 | −5.953 | 12.530 | 0.000 |
| *Conochilus unicornis* | 424.583 | 586.852 | 22.917 | 21.056 | 2.369 | 11.028 | 0.037 |
| *Dicranophorus* sp. | 10.208 | 6.727 | 0.208 | 0.397 | 5.141 | 11.076 | 0.000 |
| *Euchlanis dilatata* | 5.250 | 4.575 | 0.542 | 1.157 | 3.456 | 12.402 | 0.005 |
| *Gastropus stylifer* | 1.208 | 1.725 | 5.000 | 3.790 | −3.154 | 15.370 | 0.006 |
| *Keratella cochlearis* | 90.583 | 61.007 | 168.417 | 116.479 | −3.516 | 20.551 | 0.002 |
| *Lecane closterocerca* | 3.792 | 3.702 | 0.833 | 1.155 | 2.643 | 13.121 | 0.020 |
| *Lecane luna* | 176.500 | 89.334 | 29.417 | 26.082 | 5.475 | 12.862 | 0.000 |
| *Monommata* sp. | 41.333 | 31.454 | 11.167 | 7.518 | 3.231 | 12.253 | 0.007 |
| *Polyarthra vulgaris* | 27.583 | 11.147 | 258.5000 | 129.636 | −6.148 | 11.163 | 0.000 |
| *Ptygura melicerta* | 15.333 | 17.732 | 0.000 | 0.000 | 2.995 | 11.000 | 0.012 |
| *Testudinella parva* | 4.083 | 4.852 | 0.208 | 0.582 | 2.747 | 11.317 | 0.019 |
| *Trichocerca similis* | 18.083 | 7.038 | 107.083 | 78.189 | −3.927 | 11.178 | 0.002 |

*Collotheca edentata* (Collins), *Conochilus unicornis* (Rousselet), *Dicranophorus* sp., *Euchlanis dilatata* (Ehrenberg), *Lecane closterocerca* (Schmarda), *L. luna* (Müller), *Monommata* sp., *Ptygura melicerta* (Ehrenberg) and *Testudinella parva* (Terentz) were found to have significantly greater numbers in the *Chara* zone, regardless of the sampling period. However, *Ascomorpha ecaudis* (Perty), *A. saltans* (Bartsch), *Collotheca mutabilis* (Hudson), *Gastropus stylifer* (Imhof), *Keratella cochlearis* Gosse, *Polyarthra vulgaris* Carlin and *Trichocerca similis* Wierzejski were significantly correlated with the zone of open water above the macrophyte bed (Table 1, Fig. 1).

The significant differences in rotifer densities in the *Chara* bed were observed in the diurnal cycle ($F$=6.227; df=3; $p$=0.017). The highest numbers of rotifers were noticed during the daytime (2316 ind. $l^{-1}$), while the lowest were observed between early morning (521. ind $l^{-1}$) and at dusk (610 ind. $l^{-1}$). The numbers of rotifers in the *Chara* zone differed significantly between the day-morning and day-dusk samplings ($T$-Tukey, $p$<0.05). In the open water above the *Chara*, rotifer numbers did not change significantly ($F$=0.625; df=3; $p$=0.619) between sampling periods (615–956 ind. $l^{-1}$).

A similar pattern of diurnal distribution, with an abundance increase within the macrophyte stand during the daytime examination, was observed for *Colurella uncinata* (Müller), *C. edentata*, *C. unicornis*, *L. luna*, *Monommata* sp. and *T. parva*. (Fig. 2).

Another group of rotifers (*Heterolepadella ehrenbergii* (Perty), *L. closterocerca*, *Scaridium longicaudum* (Müller)) was characterised by a different pattern of diurnal distribution. The numbers of these species decreased during the daytime sampling within the macrophyte stand (Fig. 2).

The densities of the following species: *A. saltans*, *A. ecaudis*, *C. mutabilis*, *G. stylifer*, *P. vulgaris* and *T. similis* were much higher in the zone of open water, decreasing in the morning or daytime in this zone. In the *Chara* bed, they were usually present only in very small numbers within the whole 24-h cycle (Fig. 2).

## Discussion

The littoral zone is typically characterised by rich and diverse zooplankton communities (Gliwicz & Rybak, 1976; Havens, 1991). In Budzyńskie Lake, the *Chara* bed was characterised by a more diverse rotifer community than the open water zone above it. However, the open water zone also possessed a diverse rotifer fauna, likely because of the small distance between both of the examined zones, which resulted in specimens of many littoral species occurring in this area.

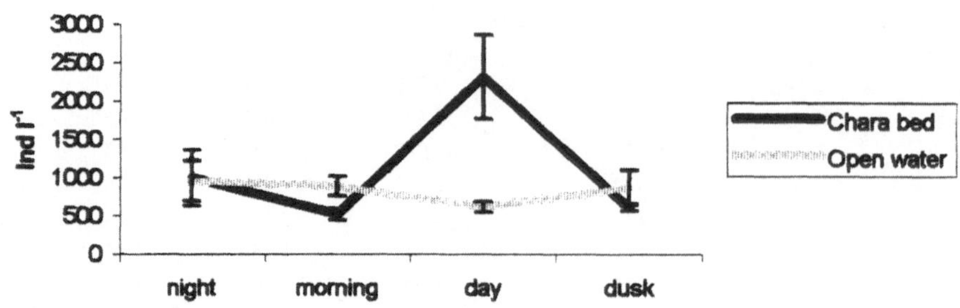


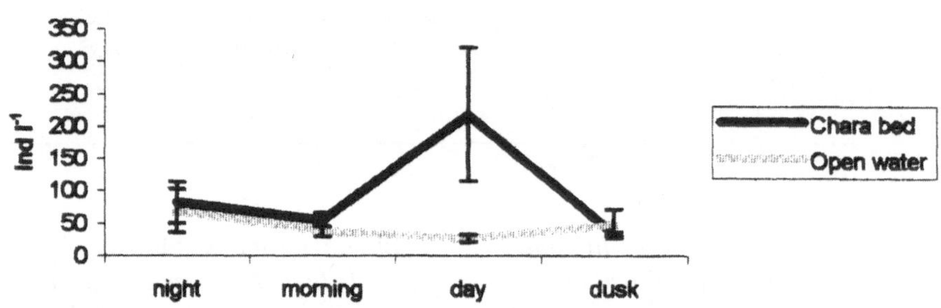


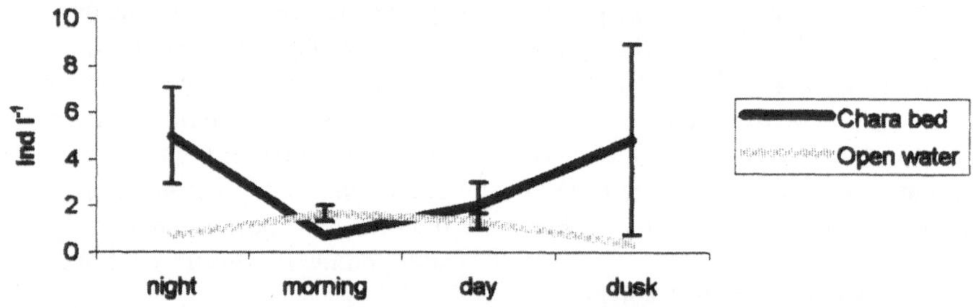


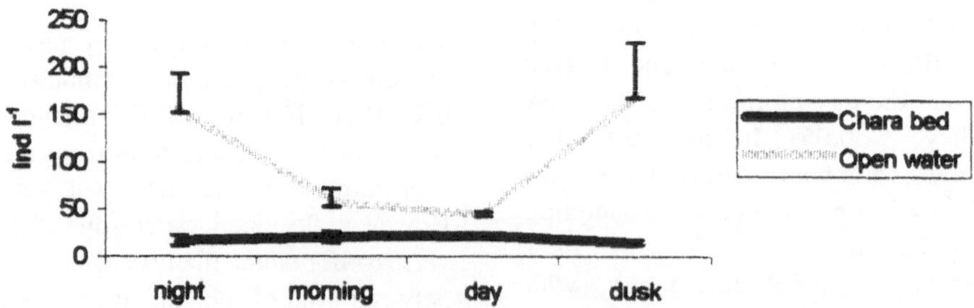


*Figure 2.* Types of diurnal vertical distribution of rotifers in Budzyńskie Lake.

The total numbers of rotifers (as in Fig. 1 top) in the zone of *Chara* and the open water zone in the diurnal cycle showed a distribution which resembles a model for zooplankton in a shallow lake with fish predation present, where the highest numbers of all individuals are found among the thick macrophyte beds during the day as described by, e.g., Timms & Moss, (1984). In the case of Budzyńskie Lake, where horizontal migrations took place, it was expected that vertical migrations which would involve much shorter distances would be preferable. Unfortunately, as the result of this study, a clear model of vertical migration was not observed due to the absence of a drop in the open water above the *Chara* bed during the daytime. The significant increase in rotifer abundance in the *Chara* during the day together with their slight but insignificant increase in the open water during the night might suggest that young fish predation may be a factor influencing their vertical distribution. Rotifers are preyed upon by many fish, especially during fish young stages (Williamson, 1987; Lair et al., 1996). Telesh (1993) found rotifers made up the majority of food items in fish fry guts. In Budzyńskie Lake, numerous shoals of fish (especially of perch, bream and roach), swimming between the vegetation beds, were frequently observed during the study. This kind of zooplankton behaviour suggests that the *Chara* bed is a refuge from predation. Timms & Moss (1984) and Moss et al. (1998) have shown that macrophytes are effective refuges for zooplankton from fish predation.

These results are not contrary to those of Jeppesen et al. (1998) who found that in the littoral zone of eutrophic Lake Stigsholm large-bodied predation-vulnerable zooplankton such as *Daphnia* spp. and adult cyclopoid copepods exhibited diurnal vertical migration, whereas smaller forms as nauplii and rotifers did not.

However, such a model of diurnal distribution in Budzyńskie Lake was also characteristic for littoral species *C. uncinata*, *C. edentata*, *C. unicornis*, *L. luna*, *Monommata* sp. and *T. parva*. This group of rotifers, which was littoral-associated, remained mostly in the *Chara* bed within the whole period of examination. Their densities in the zone of open water were much smaller and the number increase at night was not as high as that in the *Chara* bed in the daytime. It is possible that these species were hidden near the bottom of the *Chara* bed during the day, which was not examined due to the sampling method. Animals may have been hidden in the surface of the sediment during the daytime and at night they had a small-scale migration, or else they also migrated horizontally. Kairesalo (1980) also observed a similar behaviour of the *Bosmina*-swarms in oligotrophic Lake Pääjärvi. Another alternative for finding rotifers in much greater abundance within the *Chara* bed in the daytime might be the autecology of numerous littoral species. Many rotifers are generally closely associated with the macrophytes themselves, becoming more free swimming during the day and hence sampled at greater densities.

In the Budzyńskie Lake *Chara* zone, some of the littoral species also revealed changes in diurnal vertical distribution, although at a small scale. In this case the migrating distances were small (approx. 0.5 m) and for some species they might have been preferable to horizontal migrations, which may involve much longer distances.

There was also a group of rotifers that revealed slight discrepancies in their diurnal behaviour which might have been a result of their also undergoing intense invertebrate predation apart from the predatory fish. In Budzyńskie Lake, the invertebrate predators (mostly the dragonfly and *Chaoborus* sp. larvae (Michalkiewicz, 1990) as well as numerous representatives of Cyclopoida (Cerbin, unpubl. data)) occurred numerously. For these species (*H. ehrenbergii, L. closterocerca, S. longicaudum*), the higher importance of invertebrate predation was suggested in stimulating their diel behaviour. Smiley & Tessier (1998) described a similar pattern of diel behaviour of *Ceriodaphnia* sp., caused by *Chaoborus* predation. This kind of rotifer behaviour might also be connected with the exploitative competition for shared food resources between rotifers and crustaceans, which manifests itself by the suppression of rotifers (Wickham & Gilbert, 1990). It is difficult to distinguish between such competition and invertebrate predation, so a reverse pattern of migration might tend to reduce the exploitative competition (Lair et al., 1996).

In the case of species that were classified as open water-associated, the vertical changes of rotifer numbers were not pronounced in such a clear way, although a decrease in density was observed in the zone of water above the *Chara* bed during the daytime (*A. saltans, T. similis, G. stylifer, P. vulgaris*) or in the morning (*A. ecaudis, C. mutabilis*). These species did not undergo typical vertical migration, remaining mostly in the zone of water above the *Chara* stand. Most of these species are characterized as limnetic species (Pennak, 1966; Jose-De-Paggi, 1993; Pejler, 1995). Limnetic species may have evolved special morphological features which could reduce the probability of predator success, such as long, bristle-like spines, which makes the capture of prey more difficult (e.g. *K. cochlearis*), quick escape reactions (e.g. *P. vulgaris*) or gelatinous sheaths (e.g. *C. mutabilis*) which give them greater protection by increasing their body size and by appearing less distinct (Williamson, 1987; Lampert & Sommer, 1993; Pejler, 1995).

The diel behaviour of rotifers within the *Chara* bed and the open water above the macrophyte stand seems to be very complex, dependent on microhabitat and may be possibly related to young fish and invertebrate predation, the exploitative competition for shared food resources between rotifers and crustaceans, as well as typical adaptation to littoral or limnetic life. Smiley & Tessier (1998) stated that a food source plays an important role in determining the distribution of zooplankton and may be a factor here also. They suggested that typically littoral species of zooplankton are able to survive feeding on the pelagic as well as the littoral seston, while pelagic species do not feed well on the food which is present in the littoral.

## Acknowledgements

I would like to thank Jolanta Ejsmont-Karabin for her support and encouragement during my investigations. Furthermore, I wish to thank two anonymous reviewers for their valuable suggestions and comments to the manuscript.

## References

Gliwicz, Z. M. & J. I. Rybak, 1976. Zooplankton. In Pieczyńska, E. (ed.), Selected Problems of Lake Littoral Ecology. Wyd. Uniwersytetu Warszawskiego, Warszawa: 69–96.

Havens, K. E., 1991. Summer zooplankton dynamics in the limnetic and littoral zones of a humic acid lake. Hydrobiologia 215(1): 21–29.

Jańczak, J., B. Brodzińska, A. Kowalik & R. Sziwa, 1996. Atlas of Lakes of Poland. T. I, Bogucki Wydawnictwo Naukowe. Poznań.

Jeppesen, E, Ma. Søndergaard, Mo. Søndergaard, K. Christoffersen, J. Theil-Nielsen, K. Jürgens, S. Bosselmann & L. Schlüter, 1998. Cascading trophic interactions in the littoral zone of a shallow lake. In E. Jeppesen Doctor's Dissertation – Ecology of Shallow Lakes – Trophic Interactions in the Pelagial. Ministry of Environment and Energy. National Environmental Research Institute. Technical Report No 247: 207–220.

Jeppesen, E., T. L. Lauridsen, T. Kairesalo & M. R. Perrow, 1998. Impact of submerged macrophytes on fish-zooplankton interactions in lakes. In Jeppesen, E., Ma. Søndergaard, Mo. Søndergaard & K. Christoffersen (eds), The structuring Role of Submerged Macrophytes in Lakes. Ecological Studies Series. Springer, New York: 131: 91–114.

Jose-De-Paggi, S., 1993. Composition and seasonality of planktonic rotifers in limnetic and littoral regions of a floodplain lake (Parana River system), Rev.-Hydrobiol.-Trop. 26(1): 53–63.

Kairesalo, T., 1980. Diurnal fluctuations within a littoral plankton community in oligotrophic Lake Pääjärvi, southern Finland. Freshwat. Biol. 10: 533–537.

Kairesalo, T., I. Tátrai & E. Luokkanen, 1998. Impacts of waterweed (*Elodea canadensis* Michx) on fish-plankton interactions in the lake littoral. Verh. int. Ver. Limnol. 26: 1846–1851.

Kuczyńska-Kippen, N., 1997. The role of refuge in horizontal distribution of Rotatoria of Budzyńskie Lake, Wielkopolski National Park, Poland. In proceedings of Conference: Aquatic Life Cycle Strategies, Plymouth, U.K. 14–17 April.

Kuczyńska-Kippen, N & S. Cerbin, 1998. Diurnal changes in horizontal distribution of rotifers and crustaceans of a polymictic lake. In proceedings of Conference: Shallow Lakes. Trophic Interactions in Shallow Freshwater and Brackish Lakes. Berlin.

Lair, N., H. Taleb & P. Reyes-Marchant, 1996. Horizontal distribution of the rotifer plankton of Lake Aydat (France). Aquat.-Sci. 58 (3): 253–268.

Lampert, W., 1993. Ultimate causes of diel vertical migration of zooplankton: new evidence for the predator-avoidance hypothesis. Archiv für Hydrobiol. Beihefte Ergenisse der Limnol. 39: 79–88.

Lampert, W. & U. Sommer, 1993. Limnoökologie. Georg Thieme Verlag Stuttgart, New York.

Lauridsen, T. L. & I. Buenk, 1996. Diel changes in the horizontal distribution of zooplankton in the littoral zone of two shallow eutrophic lakes. Arch. Hydrobiol. 137(2): 161–176.

Lauridsen, T. L., L. J. Pedersen, E. Jeppensen & M. Søndegaard, 1994. Horizontal migration of zooplankton in the littoral zone in some shallow Danish lakes. Lake Reservoir Manage. 9, no. 2.

Loose, C. J. & P. Dawidowicz, 1994. Trade-offs in dial vertical migration by zooplankton: the costs of predator avoidance. Ecology 75 (8): 2255–2263.

Michalkiewicz, M., 1990. Ecological basis of macrozoobentos occurrence. The dynamics of numbers and biomass. In Functioning of Water Ecosystems. Protection and Recultivation. Part II, Ecology of Lakes, Their Protection And Recultivation. Experiments on ecosystems. No. 50, SGGW-AR, Warszawa.

Moss, B., M. Beklioglu & R. Kornijów, 1998. Differential effectiveness of nymphaeids and submerged macrophytes as refuges against fish predation for herbivorous Cladocera. Verh. int. Ver. Limnol. 26: 1863.

Nogrady, T., R. Wallace & T. Snell, 1993. Rotifera. Vol. 1: Biology, Ecology and Systematics. SPB Academic Publishing.

Paterson, M., 1993. The distribution of microcrustacea in the littoral zone of a freshwater lake. Hydrobiologia 263: 173–183.

Pejler, B., 1995: Relation to habitat in rotifers. Hydrobiologia 313/314: 267–278.

Pennak, R. W., 1966. Structure of zooplankton populations in the littoral macrophyte zone of some Colorado lakes. Trans. am. Micr. Soc. 85: 329–349.

Preissler, K., 1977: Do rotifers show 'avoidance of the shore'. Oecologia 27(3): 253–260.

Preissler, K., 1983. Adaptations in anatomy and orientation behaviour of rotifers and crustaceans of the littoral and pelagic region. Verh.-Ges.-Oekol. 10: 575–582.

Schriver, P. J., Bøgestrand, E. Jeppesen & M. Søndergaard, 1995. Impact of submerged macrophytes on fish-zooplankton-phytoplankton interactions: large scale enclosure experiments in a shallow eutrophic lake. Freshwat. Biol. 33: 255–270.

Smiley, E. A. & A. J. Tessier, 1998. Environmental gradients and the horizontal distribution of microcrustaceans in lakes. Freshwat. Biol. 39: 397–409.

Stich, H. B. & W. Lampert, 1981. Predator evasion as an explanation of diurnal vertical migration by zooplankton. Nature 293: 396–398.

Telesh, I. V., 1993. The effect of fish on planktonic rotifers. Hydrobiologia 255/256: 289–296.

Timms, R. M. & B. Moss, 1984. Prevention of growth of potentially dense phytoplankton populations by zooplankton grazing, in the presence of zooplanktivorous fish, in a shallow wetland ecosystem. Limnol. Oceanogr 29: 472–486.

Wickham, S. A. & J. J. Gilbert, 1980. Relative vulnerability of natural rotifer and ciliate communities to cladocerans: laboratory and field experiments. Freshwat. Biol. 26: 77–86.

Williamson, C. E., 1987. Predator–prey interactions between omnivorous diaptomid copepods and rotifers: the role of prey morphology and behaviour. Limnol. Oceanogr. 31: 393–402.

Wright, D. & J. Shapiro, 1990. Refuge availability: a key to understanding the summer disappearance of *Daphnia*. Freshwat. Biol. 24: 43–62.

Hydrobiologia **446/447:** 203–211, 2001.
L. Sanoamuang, H. Segers, R.J. Shiel & R.D. Gulati (eds), Rotifera IX.



# Hatching from the sediment egg-bank, or aerial dispersing? – the use of mesocosms in assessing rotifer biodiversity

J.M. Langley[1], R.J. Shiel[2], D.L. Nielsen[2] & J.D. Green[3]
[1]*Biological and Environmental Sciences, Middlesex University, Trent Park, Bramley Road, London N14 4YZ, U.K. E-mail: john@rotifer.demon.co.uk*
[2]*Murray-Darling Freshwater Research Centre, P.O. Box 921 Albury, NSW 2640, Australia*
[3]*Centre for Biodiversity and Ecology Research, Biological Sciences, University of Waikato, P.O. Box 3105, Hamilton, New Zealand*

*Key words:* aerial colonisation, emergence, mesocosms, passive dispersal, resting eggs, rotifers, sediment egg bank

## Abstract

Rotifer emergence from dry billabong sediments was studied from 3 sites on the River Murray floodplain, near Wodonga, northern Victoria, Australia. The sites had different flood histories, ranging from annual to approximately 25-year flooding intervals. Half of each sediment type was sterilized by $\gamma$-radiation to contrast the contribution of recruitment from the egg bank with recruitment from passive dispersal. A series of mesocosms was employed to assess differences between treatments, i.e. sediment sterilization and flood history. Analysis by Canonical Correspondence Analysis and Similarity Percentages suggested that some species were passively dispersed. Of the 54 species colonizing after 35 days, four were undescribed, one was a new record to Australia, three species had previously been recorded only from Tasmania and five more were new to the study area. Overall, 22% of species were previously unrecorded from the study area. This suggests that habitat poor mesocosms, may be more successful in locating passively dispersed taxa than examination of natural temporary waters.

## Introduction

The billabongs (oxbows) of the River Murray floodplain on Ryan's property, near Wodonga, Victoria are probably the most intensively studied in Australia. During 1976–1995, ca. 250 monogonont rotifer species and subspecies were recorded from Ryan's 1 (cf. 570 Monogononta presently known from Australia). Prior to our study, Ryan's 1 had been sampled weekly over 2 years, and rotifer community mosaics examined in relation to microhabitats (Shiel, unpublished). Daily sampling quantified pre- and post-flooding rotifer communities (Tan & Shiel, 1993), with 63 rotifer taxa recorded from 24 visits, at densities ranging from 1076 to 13 264 ind. $l^{-1}$. Similar species diversity was recorded from nearby Ryan's 3, a smaller ephemeral habitat by Pontin & Shiel (1995), who contrasted open water and periphytic rotifer species, recording 51 taxa over a 5 week study period. >200 monogonont taxa have been recorded from this site (to December 1998), with >100 taxa recorded from a single net tow on one occasiôn (Shiel et al., 1998). High biodiversity is characteristic of these floodplain habitats (cf. Segers et al., 1993).

Over the course of microfaunal studies in the Ryan's sites, it had become apparent that rotifer assemblages rapidly recolonized the smaller ephemeral waters on re-wetting, and also responded rapidly to flooding events in the larger more permanent waters. In the flooding study by Tan & Shiel (1993), some taxa, e.g. *Filinia*, had large and rapid population pulses which could only be attributed to triggering of resting egg hatching in response to the flood event. A preliminary examination of dry sediments from three of the Ryan's sites established the presence of a large resting egg bank, with >1000 ind. $ml^{-1}$ of surficial sediment at the Ryan's 3 site (Shiel et al., 2000), however the viability of this egg bank was not determined in the initial study. To provide further evidence of the contribution of the store of resting eggs to the resident rotifer community we designed an *in situ* mesocosm experiment which would reduce some of the variability implicit in field studies in a flood-drought environment.

Aquatic mesocosms have been used for a variety of research purposes, but primarily to evaluate the effects of xenobiotic chemicals on aquatic communities (Boyle & Fairchild, 1997). For rotifers, Snell & Janssen (1995) reported their use to determine EC50 and LC50 concentrations in response to agricultural pesticides. Other uses have included: assessment of rotifer population dynamics in response to the release of 'genetically defined' bacteria into the environment (Hofmann & Hofle, 1993); the level of taxonomic resolution required to determine the impact of a flood event on rotifer community structure (Nielsen et al., 1998) and determination of density-dependent effects of crustacea planktivory, on zooplankton communities, e.g. Whalstrom & Westman (1999). Laboratory studies of rotifer emergence from sediments have been conducted by e.g. May (1986) and Colville (1995), both using a 12:12 light, dark cycle under artificial light. The use of field based mesocosms to support existing billabong biodiversity studies is new.

For this study, 30 mesocosms were constructed to compare the contributions of recruitment from the egg bank (from dry sediment), with passive dispersal (sterilized dry sediment). The mesocosms were used to simulate a period of flooding after a rainfill event. It had been shown from laboratory studies, that age of the sediment may influence the resulting community (Boulton & Lloyd, 1992). To accommodate this, previously visited sites of known flood history were included in this study: Ryan's 1, Ryan's 3 and an infrequently flooded part of the Murray floodplain at the crest of the bank of Ryan's 2.

Only the rotifer responses are reported here. The taxonomic component, and non-rotifer responses, will be reported elsewhere (Shiel unpubl. and the authors unpubl., respectively).

## Study area

The field study was carried out in March–April, 1999 (late summer–autumn), within the grounds of the Michael Ryan Ecology Laboratory, located on the River Murray floodplain, near Wodonga, Victoria, Australia (36° 08′ S/146° 58′ E) (Fig. 1). The laboratory grounds are fenced to prevent cattle incursions. The primary land use of the Ryan's property is cattle grazing. The surrounding vegetation is dominated by grassland with scattered River Red Gum trees (*Eucalyptus camaldulensis*) and dense patches of exotic thistle. The climate is temperate, with maximum daytime temperatures during the study period of 19–33 °C. Rainfall during the study period was 89.5 mm. Two large billabongs and the smaller ephemeral pool are within 100 m of the study site. At the beginning of the study, all billabongs on Ryan's property were dry, following two of the driest years in Australian records, however the previous flood history of the sites prior to the 1998–99 drought was: annually flooded (Ryan's 1, about 100 m north of the study site); flooded every 2–3 year (Ryan's 3, 100 m east of the study site), and flooded every 25–30 year (banktop of Ryan's 2, i.e. the floodplain proper, 30 m to the west of the study site). The floodplain was last flooded to >1 m above the present surface in 1974 and before that in 1956.

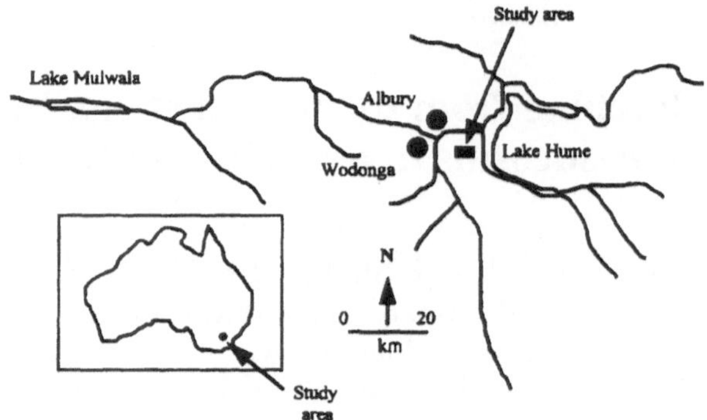


*Figure 1.* Location of the study site, northern Victoria, Australia.

## Methods

Surfacial sediment samples (4–5 kg) were collected during December 1998, by hand (to a depth of 10 cm) from the central areas of the dry sites to ensure a precise flood history. For instance, although Ryan's 1 is annually flooded, it does not flood to the same depth every year, therefore marginal areas may be flooded less frequently. The dry sediment samples were analysed for moisture content, organic content (loss on ignition), soluble reactive phosphorus and $NO_x$(by a Lachat QuickChem 8000 Flow Injector Analyser system, after filtration through a 0.45 $\mu$m cellulose acetate membrane filter). Half of each sediment sample was then sterilized by exposure to $\gamma$-radiation (25 kG for 5.5 h, repeated three times), to create 'sterile' and 'non-sterile' inocula.

The mesocosms, each measuring 1.6 m × 1.2 m, were filled to a depth of 20 cm using rainwater col-

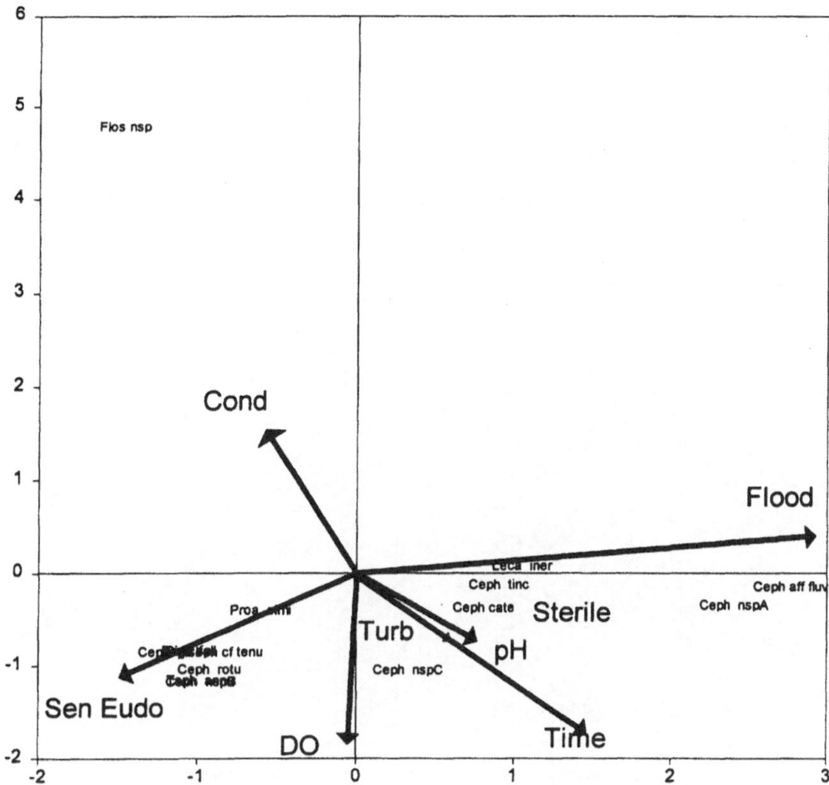


*Figure 2.* CCA ordination of selected taxa and environmental variables.

lected in tanks from the roof of the laboratory, after filtration through a 10 $\mu$m mesh. The mesocosms were filled over one 14 h period (mostly at night). The effectiveness of the 10 $\mu$m mesh in excluding microinvertebrates and larger phytoplankton was tested by sampling and examining one mesocosm (prior to adding sediment). The mesocosms were arranged in a grid separated by 1 m, to ensure that each was equally available to aerial colonisation. The treatments 'flood history' (i.e. 1:1, 1:3 or 1:30) and 'sterile' (S or NS) were randomly allocated to the mesocosms. Two kilograms of sediment (+/- 1 g) were added to each mesocosm.

Microfaunal sampling was quantitative, using a modified (4-l) Haney-Patalas perspex sampler. Six trap samples were taken from randomly determined locations within each mesocosm to provide a 24-l water sample. Water samples were then passed through a 35 $\mu$m mesh net and the organisms preserved in 70% ethanol. To avoid cross contamination between mesocosms, nets and samplers were washed in detergent (5% DECON 90 solution), with a pre- and post-rinse of rainwater. The mesocosms were sampled 11 times over 5 weeks (days 1, 2, 3, 4, 6, 8, 10, 14, 21, 28 and 35). Physicochemical parameters (pH, conductivity, temperature, dissolved oxygen and turbidity) were measured on site using an Horiba U10 water quality meter. After each sample visit, the water level was topped up to 20 cm. This was repeated between sample visits, during periods of low rainfall. Algal samples (125 ml, preserved in Lugol's $I_2KI$) also were taken from each mesocosm for later analysis.

Quantitative samples were examined for microcrustacea, nematodes, macroinvertebrates, larger algae, as well as rotifers. Rotifers were identified to species using an Olympus BH2 microscope, and the keys of Koste (1978), Nogrady et al. (1995), Segers (1995), Shiel (1995) and De Smet (1996, 1997). Rotifer abundance was determined by enumeration of the whole sample, with densities converted to counts $l^{-1}$ of the sampled volume.

The rotifer and environmental results were analysed using canonical correspondence (direct gradient) analysis (CCA), with log abundance values and downweighting for rare species, by the package CANOCO 4.0 (Ter Braak & Smilauer, 1998). Similarity Per-

*Figure 3.* Placement of the mesocosms at the Ryan's field station.

centages (SIMPER) were also calculated (for each rotifer species) between treatments using the package PRIMER.

## Results and discussion

During this study, 54 monogonont taxa were identified from the $\gamma$-irradiated (sterile) and non-sterile, treatments (Table 1). These taxa included four new species: three *Cephalodella* spp., one *Floscularia* sp. and in addition one new record to Australia, *Cephalodella rotunda* Wulfert, 1937, a common species in this study. Three taxa: *Cephalodella tinca* Wulfert, 1937; *Proales fallaciosa*, Wulfert, 1937 and *Proales similis* De Beauchamp, 1907, were previously only recorded from Tasmania. Another 5 species were previously unrecorded from the Ryan's billabongs: *Cephalodella catellina* (Müller, 1786); *Lecane inermis* (Bryce, 1892); *Cephalodella* cf.*tenuior* (Gosse, 1886), *Sinantherina* sp. and a *Cephalodella* aff. *fluviatilis* (Zavadovsky). The latter two are yet to be confirmed as new species. Hence, 22% of the taxa from this study were previously unknown from this site.

CCA was used to investigate the relative contributions of the environmental variables to the taxa assemblages, using the full data set. Figure 3 shows the CCA biplot of taxa against environmental factors. For purposes of presentation, the 54 rotifer taxa have been reduced to 16 taxa, focusing on the new taxa to this study area. The environmental variables tested are shown by arrows, except for 'sterile' which is a nominal value and therefore represented by its centroid (Jongman et. al., 1995). The resulting ordination diagram expresses not only the pattern of variation between taxa, but also the main associations between the taxa and each of the environmental variables. Environmental variables with arrows close to the canonical axes may well explain a large portion of the variation accounted for by that axis. The longer the arrow, the more variation that factor may explain. The statistical relevance of the first canonical axis was tested by a global permutation test (Ter Braak & Smilauer, 1998) shown in Table 2. This indicated that the first canonical axis was statistically meaningful, ($p<0.01$). This table also showed that based on the sum of all the canonical eigenvalues (the trace),

*Table 1.* Taxa found during the study period March to April, 1999

| | Day 1 | Day 2 | Day 3 | Day 4 | Day 6 | Day 8 | Day 10 | Day 14 | Day 21 | Day 28 | Day 35 |
|---|---|---|---|---|---|---|---|---|---|---|---|
| Bdelloidea | | + | | + | + | + | + | + | + | + | + |
| *Keratella procurva* | | | | + | + | + | + | + | + | + | + |
| *Floscularia* nov. sp. | | | | + | + | + | | | | | |
| *Filinia passa* | | | | + | | + | | | | | |
| *Filinia opoliensis* | | | | + | | + | | | | | |
| *Brachionus quadridentatus* | | | | + | + | + | + | + | + | + | + |
| *Lecane hamata* | | | | + | + | + | + | + | + | + | + |
| *Lecane closterocerca* | | | | + | + | + | + | + | + | + | |
| *Polyarthra dolichoptera* | | | | + | | | | | | | |
| *Lecane bulla* | | | | | + | + | + | + | + | | |
| *Brachionus angularis* | | | | | + | | | | + | + | + |
| *Lecane pyriformis* | | | | | + | + | | | | | |
| *Lecane inermis* | | | | | | + | + | + | + | + | + |
| *Cephalodella intuta* | | | | | | + | + | + | + | + | + |
| *Cephalodella catellina* | | | | | | | + | + | + | + | + |
| *Cephalodella* cf *tenuior* | | | | | | | + | + | + | + | |
| *Taphrocampa annulosa* | | | | | | | + | + | + | + | + |
| *Cephalodella ventripes* | | | | | | | + | | | | + |
| *Cephalodella gisleni* | | | | | | | + | + | | | |
| *Lecane quadridentata* | | | | | | | + | | | | |
| *Monommata dentata* | | | | | | | + | + | + | + | + |
| *Trichocerca bidens* | | | | | | | + | + | + | + | + |
| *Cephalodella rotunda*[A] | | | | | | | | + | + | + | + |
| *Cephalodella forficata* | | | | | | | | + | | | + |
| *Brachionus calyciflorus* | | | | | | | | + | + | + | + |
| *Cephalodella* nov.sp.A | | | | | | | | + | + | + | + |
| *Cephalodella gibba* | | | | | | | | + | + | + | |
| *Encentrum uncinatum* | | | | | | | | + | + | + | + |
| *Lecane luna* | | | | | | | | + | + | + | + |
| *Trichocerca weberi* | | | | | | | | + | + | + | |
| *Scaridium longicaudum* | | | | | | | | | + | + | + |
| *Taphrocampa selenura* | | | | | | | | | + | | |
| *Notommata tripus* | | | | | | | | | | + | + |
| *Anuraeopsis fissa* | | | | | | | | | | + | + |
| *Trichocerca similis* | | | | | | | | | | + | |
| *Cephalodella* nov.sp.B | | | | | | | | | | + | + |
| *Collotheca* cf *pelagica* | | | | | | | | | | + | |
| *Cephalodella* aff *fluviatilis* sp. nov.? | | | | | | | | | | + | |
| *Cephalodella forficula* | | | | | | | | | | + | |
| *Cephalodella tinca* [B] | | | | | | | | | | + | + |
| *Cephalodella* nov.sp.C | | | | | | | | | | + | + |
| *Colurella adriatica* | | | | | | | | | | + | |
| *Lecane ludwigii* | | | | | | | | | | + | |
| *Heterolepadella ehrenbergi* | | | | | | | | | | + | |
| *Lepadella ovalis* | | | | | | | | | | + | + |
| *Proalides tentaculatus* | | | | | | | | | | + | + |
| *Brachionus urceolaris* | | | | | | | | | | | + |
| *Proales similis*[B] | | | | | | | | | | | + |
| *Trichocerca pusilla* | | | | | | | | | | | + |
| *Sinantherina* sp. | | | | | | | | | | | + |
| *Dicranophorus forcipatus* | | | | | | | | | | | + |
| *Dicranophorus lutkeni* | | | | | | | | | | | + |
| *Proales fallaciosa*[B] | | | | | | | | | | | + |
| *Trichocerca* cf *intermedia* | | | | | | | | | | | + |

[A] Indicates a new record to Australia.

[B] Indicates that the species was previously only known from Tasmania.

*Table 2.* Summary of significance of canonical axes using Global (Monte Carlo) Permutation Test

| Test of significance of first axis: | Eigenvalue | 0.529 |
|---|---|---|
| | $F$-ratio | 24.643 |
| | $p$-value | 0.002 |
| Test of significance of all axes | Trace | 1.087 |
| | $F$-ratio | 6.843 |
| | $p$-value | 0.002 |

*Table 4.* Summary of significance of environmental variables using Global permutation test

| Variable | Lambda A | $F$-value | $p$-value |
|---|---|---|---|
| Flood | 0.40 | 18.73 | 0.002 |
| Sterile | 0.17 | 8.38 | 0.002 |
| Conductivity | 0.12 | 6.35 | 0.008 |
| Senescent *Eudorina* | 0.11 | 5.54 | 0.008 |
| Time | 0.11 | 6.46 | 0.008 |
| Dissolved Oxygen | 0.03 | 1.90 | 0.042 |

*Table 3.* Summary of the Canonical Correspondence Analysis

| Axes | 1 | 2 | 3 | 4 | Total inertia |
|---|---|---|---|---|---|
| Eigenvalues | 0.529 | 0.204 | 0.132 | 0.077 | 3.682 |
| Taxa-environment correlations | 0.939 | 0.802 | 0.755 | 0.609 | |
| Cumulative% variance: | | | | | |
| of species | 14.4 | 19.9 | 23.5 | 25.6 | |
| of taxa-environment relation | 48.6 | 67.4 | 79.5 | 86.6 | |

there was a significant ($p$<0.01) association between the taxa and the environmental variables.

The CCA summary (Table 3) indicated that the first canonical axis explained 14.4% of the variance (inertia) in the taxa results. The first two canonical axes together explained almost 20% of the variance in the taxa results. Apparently low percentage variance values are not unusual in field experiments, e.g. the first canonical axis of the model 'Dune Meadow data' (Batterink & Wijffels, 1983, cited in Ter Braak & Smilauer, 1998), accounted for 21.8% of the variance in taxa results. The relative importance of the environmental variables in influencing the species associations (Table 4), showed that the treatments flood history and sterile had the greatest influence on the taxa assemblages, both these parameters were significantly associated with canonical axis 1 ($p$<0.01). Conductivity appeared most closely associated with canonical axis 2 ($p$<0.01).

Further analysis was undertaken to distinguish between the treatments flood history and sterility using Similarity Percentages with PRIMER (Tables 5a–c). Only taxa with a minimum 5% dissimilarity between treatments are displayed. The SIMPER analysis indicated that some taxa showed differential success dependent upon the sterile/non-sterile treatment. *Lecane inermis* consistently had higher abundance in the sterile treatments (Table 5a–c), with dissimilarity percentages ranging from 8.03 to 12.84. The results for *Cephalodella catellina* conflicted, reaching slightly higher abundance in the non-sterile annually flooded sediments, whereas in the sediments flooded approximately every 25–30 years, the abundance was considerably greater in the sterile treatment. Passively dispersed taxa would be expected to colonise all mesocosms, but pioneer species would be expected to reach highest abundance in early successional environments. *Cephalodella* sp. nov. A reached highest abundance in sterile sediments (Table 5c), the first indication of its ecology (passively dispersed, pioneer species).

Conversely, *Lecane hamata* (Stokes, 1896), *Keratella procurva* (Thorpe, 1891) and *Taphrocampa annulosa* Gosse, 1851 (Table 5a) were restricted to non-sterile treatments. Bdelloidea were also most abundant in the non-sterile sediments, although later reached moderate numbers in trent-annually flooded sterile mesocosms. The above taxa were all found early in the colonisation sequence and thus are expected to have been recruited from the egg bank.

The CCA ordination diagram (Fig. 2), showed not only the variation between species, but also the main relations between the species and each of the environmental variables. Species associated with the head of the arrow have their abundance closely related to that variable. Hence, Figure 2 showed that *Lecane inermis*, *Cephalodella catellina* and *Cephalodella tinca* were associated with the variable 'sterile'. *Cephalodella* sp. nov. A lies between 'sterile' and flood history, supporting recorded capture at the least frequently flooded, sterile sites. *Cephalodella* aff. *fluviatilis* and *Floscularia* sp. nov. were rarely encountered in this study (1 and 6 specimens, respectively), therefore their locations reflected this lack of abundance and possibly represent outliers to this data set.

*Table 5a.* Similarity percentages for sterile treatment (annually flooded sediments)

| Taxa | Annually flooded, non-sterile, average abundance | Annually flooded, sterile, average abundance | Percent |
|---|---|---|---|
| Leca iner | 1.23 | 26.77 | 10.78 |
| Leca hama | 57.03 | 0.00 | 7.79 |
| Ceph cate | 17.58 | 6.18 | 6.26 |
| Kera proc | 3.18 | 0.00 | 6.05 |
| Taph annu | 38.9 | 0.02 | 5.55 |

*Table 5b.* Similarity percentages for sterile treatment (tri-annually flooded sediments)

| Taxa | Tri-annually flooded, non-sterile, average abundance | Tri-annually flooded, sterile, average abundance | Percent |
|---|---|---|---|
| Bdel loid | 1.65 | 0.00 | 13.93 |
| Leca iner | 1.36 | 3.58 | 12.84 |

*Table 5c.* Similarity percentages for sterile treatment (trent-annually flooded sediments)

| Taxa | Trent-annually flooded, non-sterile, average abundance | Trent-annually flooded, sterile, average abundance | Percent |
|---|---|---|---|
| Bdel loid | 1.27 | 0.44 | 10.95 |
| Leca iner | 1.36 | 11.92 | 8.03 |
| Ceph cate | 2.09 | 51.98 | 6.27 |
| Ceph nspA | 0.28 | 219.11 | 5.25 |

Larger algae were noted contemporaneously as rotifers were studied. The environmental variable 'senescent *Eudorina*', was included because the development of these mesocosms was characterised by population blooms and crashes of this species. Many taxa are associated with this variable, indeed species overlie one another. Some taxa were included in Figure 2, because their ecology was known e.g. *Taphrocampa annulosa*, a creeping detritovore (Nogrady et al., 1995), which was located close to the head of the 'senescent *Eudorina*' variable. This overlies *Cephalodella* sp. nov. B, indicating a similar ecology. Above these taxa lies *Cephalodella rotunda*, originally described from a 'ditch near Merseberg Germany' (Nogrady et al., 1995). This reinforced the validity of this variable. *Sinantherina* sp., *Proales fallaciosa* and *Cephalodella gisleni* Berzins, 1953 overlie each other. *Proales fallaciosa* is a detritovore, consuming decomposing microcrustaceans, macroinvertebrates, as well as bacteria and algae (De Smet, 1996). *Sinantherina* sp. is likely to be a bacteriovore, given the body shape and small trophi. Although the above taxa were associated with 'senescent *Eudorina*', this does not imply direct feeding on *Eudorina*, merely the conditions under which the taxa were most abundant.

The Ryan's billabongs have been studied for >20 years, and of these, most intensively the billabong Ryan's 1. Without these accumulated records it would not be possible to contrast the value of mesocosm studies in terms of biodiversity. During this 5 week study, the 'Ryan's list' was increased by 12 taxa, and at least 4 were new. It is not suggested that these mesocosms emulate the complex mosaics of microhabitats found within billabongs, indeed during this study, Ryan's 1 filled with rainwater and showed a much more diverse macrophyte structure. Cursory examination of water samples collected from Ryan's 1 showed a much higher abundance of microcrustacea. Another contrast between the mesocosms and billabongs was the time of rainfill. When the mesocosms were filled, the billabongs of Ryan's property were dry, yet within 20 min of filling, some were colonised by predatory Notonectidae (primarily *Anisops deanii* and *Enthires* sp.). The relationship between Australian Notonectidae and microcrustacea has been studied in farm dams without fish. Large populations of *Anisops deanii* were found to decimate the zooplankton, virtually eliminating some daphnids and reducing populations of *Asplanchna* spp. (Geddes, 1986). In this study, similar trends were observed, despite initial strong emergence, few microcrustacea were found after week 3 and no *Asplanchna* were found (despite their presence in a similar mesocosm in the grounds of the MDFRC, Albury, 20 km away). This suggests that in the absence of refugia, Notonectidae consumed major rotifer predators, permitting lower rotifer predation in the mesocosms than in billabongs.

Rotifer communities were dominated by *Cephalodella* spp. (14 taxa) and *Lecane* spp, the second richest genus (8 taxa). Laboratory based microcosm experiments which investigated emergence of resting eggs from dry Ryan's 3 sediment (Colville, 1995), found that bdelloids and *Lecane* spp. dominated the community. Boulton & Lloyd (1992), investigated laboratory emergence of rotifers from sediments of different ages from the Chowilla floodplain downstream of our study site. These rotifers were dominated by the genera *Brachionus*, *Keratella*, *Asplanchna* and *Filinia*. Nogrady et al. (1995) stated that "*Cephalodella* species never occur in as large numbers as planktonic taxa do". Therefore, the prevalence of *Cephalodella* spp. seems anomalous. Assuming that this is not because of preferential survivorship in the egg bank, and the profile of the mesocosms (flat bottoms with vertical sides) does not favour non-planktonic species, then relative lack of predation or passive dispersal, or both, may account for this.

The recruitment from the egg bank in this study was much greater than from passive dispersal. This is the traditional viewpoint; in the Northern Hemisphere, resting egg densities range between 1 and 40 million/$m^3$ of sediment (Hairston, 1996). May (1986) compiled a complete species list for Loch Leven (of open water taxa) from sediment emergence. This success for stable permanent lakes was not reproducible using similar techniques in the macrophyte rich, temporary billabong, Ryan's 3 (Colville, 1995).

The mesocosms were based on sediments from 2 billabongs (Ryan's 1 and Ryan's 3), plus an infrequently flooded part of the Murray flood plain. This study recorded 54 taxa, which is well below the number of taxa associated with either billabong. Koste 1990 (reported in Shiel et al., 1998), acquired more than 100 species from a single visit to Ryan's 3. Despite the low number of taxa observed, these mesocosms have yielded a range of previously unknown taxa from this site (irrespective of the intensity of previous study), uncommon taxa (the capture of *Cephalodella gisleni* from the non-sterile sediment of Ryan's 1 was the 3rd record for Australia), and at least 4 new species. As a comparison, Shiel et al. (1998) visited 112 temporary rain filled floodplain pools once, which yielded 252 taxa and 3 new species. Thus, mesocosms may provide a less labour intensive way to find uncommon taxa. Moreover, they may be utilized independently of rainfall. In this manner mesocosms are suggested as a useful adjunct to biodiversity assessment. By using a sterile sediment treatment, this study has been able to indicate that *Lecane inermis*, *Cephalodella catellina*, *Cephalodella tinca* and *Cephalodella* sp. nov. A are passively distributed (aerially or by phoresis) and been able to contrast species which are strongly associated with colonisation from the egg bank e.g. *Lecane hamata*, *Keratella procurva*, *Taphrocampa annulosa*. A further factor, low predation environments, with initially harsh conditions (especially in the sterile treatments) permit eurytopic pioneer species to reach high densities, which could otherwise be missed in assessing the biodiversity of a mature zooplankton community. Despite the statistical significance attached to the sterile and non-sterile treatments, the experimental design could have been improved by including two controls: a set of sterile treatments sealed in the field (or with fine mesh), to exclude passive colonisers, providing a further test of the sterility of the sediments; and one set of sterile and non-sterile treatments covered with 2 mm mesh to exclude Notonectidae, to elucidate whether reduced predation caused the dominance of *Cephalodella* spp. Further work on the temporal sequence of rotifer emergence from these sediments (in relation to environmental factors) and comparison of mesocosm communities with recently inundated billabong communities, would provide a greater understanding of rotifer emergence using field mesocosms.

## Acknowledgements

The authors would like to thank the Murray-Darling Freshwater Research Centre, Albury, and in particularly the Director, Terry Hillman, for the excellent research environment provided. JML would like to thank his colleagues at Middlesex for making this sabbatical possible.

## References

Boulton, A. J. & L. N. Lloyd, 1992. Mean flood recurrence frequency and invertebrate emergence from dry sediments of the Chowilla floodplain, River Murray, Australia. Regulated Rivers: Res. Manage. 7: 137–151.

Boyle, T. P. & J. F. Fairchild, 1997. The role of mesocosm studies in ecological risk analysis. Ecol. Appl. 7: 1099–1102.

Colville, V., 1995. Effect of temperature and other chemical factors on emergence of resting eggs and encysted organisms from Ryan's 3 billabong. Undergraduate thesis, MDFRC.

De Smet, W. H., 1996. Rotifera. Vol. 4. The Proalidae (Monogononta). Guides to the Identification of the Microinvertebrates of the Continental Waters of the World 9. SPB Academic Publishing: 102 pp.

De Smet, W. H., 1997. Rotifera. Vol. 5. The Dicranophoridae (Monogononta). Guides to the Identification of the Microinvertebrates of the Continental Waters of the World 12. SPB Academic Publishing: 325 pp.

Geddes, M. C., 1986. Understanding zooplankton communities in farm dams: the importance of predation. In DeDeckker, P. & W. D. Williams (eds), Limnology in Australia. Dr W. Junk Publishers, Dordrecht: 387–400.

Hairston, N. G., 1996. Zooplankton egg banks as biotic reservoirs in changing environments. Limnol. Oceanogr. 41: 1087–1092.

Hoffmann, W. & M. G. Hofle, 1993. Rotifer population dynamics in response to increased bacterial biomass and nutrients: a mesocosm experiment. Hydrobiologia 255/256: 171–176.

Jongman, R. H. G., C. J. F. Ter Braak & O. F. R. van Tongeren (eds), 1995. Data Analysis in Community and Landscape Ecology. Cambridge University Press: 299 pp.

Koste, W., 1978. Rotatoria. Die Radertiere Mitteleuropas. 2 Vols. Gebruder Borntraeger, Stuttgart, West Germany. May, L., 1986. Rotifer sampling – a complete species list from one visit? Hydrobiologia 134: 117–120.

Nielsen, D. L., R. J. Shiel & F. J. Smith, 1998. Ecology *versus* taxonomy: is there a middle ground? Hydrobiologia 387: 451–457.

Nogrady, T., R. Pourriot & H. Segers, 1995. Rotifera. Vol. 3. The Notommatidae and the Scaridiidae (Monogononta). Guides to the Identification of the Microinvertebrates of the Continental Waters of the World 8. SPB Academic Publishing: 248 pp.

Pontin, R. M., & R. J. Shiel, 1995. Periphytic rotifer communities of an Australian seasonal floodplain pool. Hydrobiologia 313: 63–67.

Segers, H., 1995. Rotifera. Vol. 2. The Lecanidae (Monogononta). Guides to the Identification of the Microinvertebrates of the Continental Waters of the World 6. SPB Academic Publishing: 226 pp.

Segers, H., C. S. Nwadiaro & H. J. Dumont, 1993. Rotifera in some lakes in the floodplain of the River Niger (Imo State. Nigeria). II. Faunal composition and diversity. Hydrobiologia 250: 63–71.

Shiel, R. J., 1995. A Guide to Identification of Rotifers, Cladocerans and Copepods from Australian inland waters. CRCFE Identification Guide 3. MDFRC, Albury: 144 pp.

Shiel, R. J., J. D. Green & D. L. Nielsen, 1998. Floodplain biodiversity: why are there so many species? Hydrobiologia 387/388: 39–46.

Shiel, R. J., J. D. Green & L.-W. Tan, 2000. Microfaunal and resting stage heterogeneity in ephemeral floodplain pools, upper River Murray floodplain, Australia. Verh. int. Ver. Limnol. 27, in press.

Snell, T. W. & C. R. Janssen, 1995. Rotifers in ecotoxicology: a review. Hydrobiologia 313/314: 231–247.

Tan, L.-W. & R. J. Shiel, 1993. Responses of rotifer billabong communities to inundation. Hydrobiologia 255/256: 361–370.

Ter Braak, C. J. F. & P. Smilauer, 1998. CANOCO Reference manual and User's Guide to CANOCO for Windows: Software for Canonical Community Ordination (version 4). Ithaca, NY, U.S.A.: 352 pp.

Wahlstrom, E. & E. Westman, 1999. Planktivory by the predacious cladoceran *Bythrotrephes longimanus*: effects on zooplankton size structure and abundance. Can. J. Fish. aquat. Sci. 56: 1865–1872.

Hydrobiologia **446/447:** 213–220, 2001.
L. Sanoamuang, H. Segers, R.J. Shiel & R.D. Gulati (eds), Rotifera IX.




# Increase of rotifer diversity after sewage diversion in the hypertrophic lagoon, Albufera of Valencia, Spain

Rafael Oltra[1], M. Teresa Alfonso[1], Maria Sahuquillo[2] & Maria Rosa Miracle[1*]
[1]*Departament de Microbiologia i Ecologia, Facultat de Biologia, Universitat de València, 46100 Burjassot, Spain*
*E-mail: rosa.miracle@uv.es*
[2]*Conselleria de Medi Ambient de la Generalitat Valenciana, Gregorio Gea 27, 46009 València, Spain*
*(*Author for correspondence)*

*Key words:* eutrophication, shallow lake, turbid state, water management, *Brachionus* spp., *Brachionus angularis*

## Abstract

The Albufera of Valencia is a large oligohaline hypertrophic lagoon, regulated by sluice gates according to the needs of the surrounding rice field cultivation. It is in a turbid state with permanent cyanobacterial blooms. A slight improvement was detected after diversion in the 1990s of part of the sewage flowing into it. After sewage diversion, we found that: (1) Chlorophyll concentration and rotifer densities decreased; (2) Rotifer proportions declined, due mainly to a relative increase in cladocerans; (3) Rotifer diversity increased. The two dominants of the 1980s, *Polyarthra* spp. in the colder period and *Brachionus angularis* in the warmer one, reverted after sewage diversion to a more diverse assemblage reminiscent of the 1970s, with a higher number of dominant species. In the summer of 1998, both *Brachionus calyciflorus* and its predator *Asplanchna brightwelli*, dominant in 1973, became abundant again. In 1998, an increase in the number of dominant species was also observed during water renewal periods, some of these species were new or seldomly found before in the lagoon (*Proalides tentaculatus-digitus, Trichocerca pusilla* at the end of rice culture, *Brachionus variabilis* at the end of winter flooding). Another change that indicates an improvement of water conditions is a more distinct and longer clear water phase, which occurs in the water renewal period at the end of winter and involves a *Daphnia magna* peak. The increased importance of this phase, promoted the flourishment of *Brachionus variabilis*, a facultative *Daphnia* epibiont never found before in the lake.

## Introduction

Coastal lagoons have suffered great human impacts, leading to a severe deterioration of their water quality. Many large lagoons have become eutrophic and in developed countries some management measures have been undertaken to correct the situation. Only a few publications exist on changes of rotifer abundance and composition after restoration measures without obtaining clear results on general trends (Gulati, 1990; Van Tongeren et al., 1992; Ronneberger et al., 1993). The Albufera of Valencia, a large lagoon in the Spanish Mediterranean coast, is a good example of recent eutrophication. At the beginning of 1960, the water of the Albufera of Valencia was still clear to the bottom and magnificent subaquatic prairies (mainly charophytes) could be seen from the surface, but eutrophication advanced in this decade and in the early 1970s the degradation of submerged macrophytes was widespread over most of the lake (Pardo, 1942; Gil & Martinez, 1972; Docavo, 1979).

The great increase thereafter of urban and industrial development led to a further eutrophication of the lagoon that reached its greatest peak in the 1980s (Vicente & Miracle, 1992). Macrophytes totally disappeared (Carretero & Boira, 1989) because of the extraordinary growth of plankton and the lagoon became an hypertrophic system dominated by permanent cyanobacterial blooms (Miracle et al., 1984a, b; Romo & Miracle, 1993). At the end of the 1990s, a partial diversion of sewage waters to treatment plants was undertaken and a slight improvement in water quality was detected. The hydrology of the lagoon traditionally has been manipulated by man and since 1953,

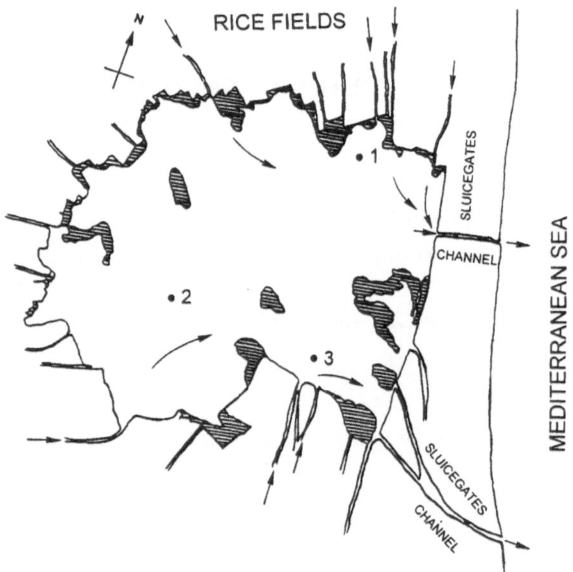


*Figure 1.* Outline of Albufera lake indicating the sampling points (1 North, 2 Center, 3 South).

when the last hydraulic works for drainage of the lagoon were undertaken, the same seasonally regulated flow, according to rice cultivation needs, has been applied.

In this paper, rotifer population dynamics are described for the years 1997–98, i.e. after partial diversion of sewage. Rotifer data from these years are compared with those from several annual cycles in the 1980s (studied also by some of us; Oltra & Miracle, 1984, 1992; Alfonso & Miracle, 1990; Oltra, 1993; Alfonso, 1996), when sewage and industrial discharges went directly into the brooks and irrigation channels and subsequently into the lagoon. The aim of this paper is to describe main changes in rotifer populations after partial diversion of the sewage load in the Albufera of Valencia lagoon. We further attempt to assess resilience in rotifer species composition by comparing the seasonal assemblages in recent years with those from the first quantitative zooplankton studies that exist from this lagoon, that are from the years 1972 to 1973 (Blanco, 1976).

## Material and methods

### *Study area*

The Albufera of Valencia is a shallow lagoon (mean depth of 1 m) situated 12 km south of Valencia, Spain. The lagoon occupies an area of 23.2 km$^2$ (Fig. 1) and is surrounded by 223 km$^2$ of rice fields (Fig. 1). This area includes more than 30 towns with more that 400 000 inhabitants and more than 4000 industries of different types. The Albufera has been the great purifier of most of the sewage from the catchment area. Water entering the Albufera has high levels of nutrients but on leaving the lagoon is nutrient free, although laden with cyanobacterial biomass (Vicente & Miracle, 1992). At the beginning of 1970, phytoplankton was still dominated by diatoms, although some cyanobacterial populations developed in summer (Blanco, 1976), but since 1980 phytoplankton was dominated all year around by cyanobacteria (Miracle et al., 1984a; Romo & Miracle, 1993).

The hydrology of the lagoon has long been manipulated, in medieval times mainly for fishing and subsequently more and more regulated for rice cultivation. Water inflow in the lagoon takes place through some streams and a series of channels dug for rice field irrigation. The Albufera is also freshwater-fed by springs located either within the lagoon or in the surrounding marshland. The lagoon outflows through three main channels to the sea, all of them artificially excavated, because the natural water outlets got clogged. The whole lagoon acts as a reservoir to regulate rice fields seasonal inundations. The flow is controlled by a system of sluice gates. This manipulation has been more or less the same during the period studied in this paper (1972–98), it influences the lagoon water level and its hydrological annual cycle, in which 5 phases could be distinguished, two with an important water flow and the rest in a steady state, (Fig. 2): (1) November–January, the sluice gates are closed, the rice fields remain flooded and the water level is about 20 cm above the usual; (2) February–April, water flows to the sea, the level goes down and the rice fields remain dry; (3) May–June, by partially closing the sluice gates the level of the water rises until the rice fields are flooded and the rice cultivation begins; (4) July–August, after a brief period of a few days of drought of the rice fields at the end of June–beginning of July for weed control, rice cultivation proceeds with a slow water flow; and (5) September–October, sluice gates are open to empty rice fields for harvest and water flow is again important; the fields remain dry throughout the period, unless it rains. The conductivity of the water is also influenced by this manipulation. It increases through summer and decreases in the water renewal periods of February–April and September–October (ranging from 1 to 2–3 mS cm$^{-1}$, Vicente & Miracle, 1992).

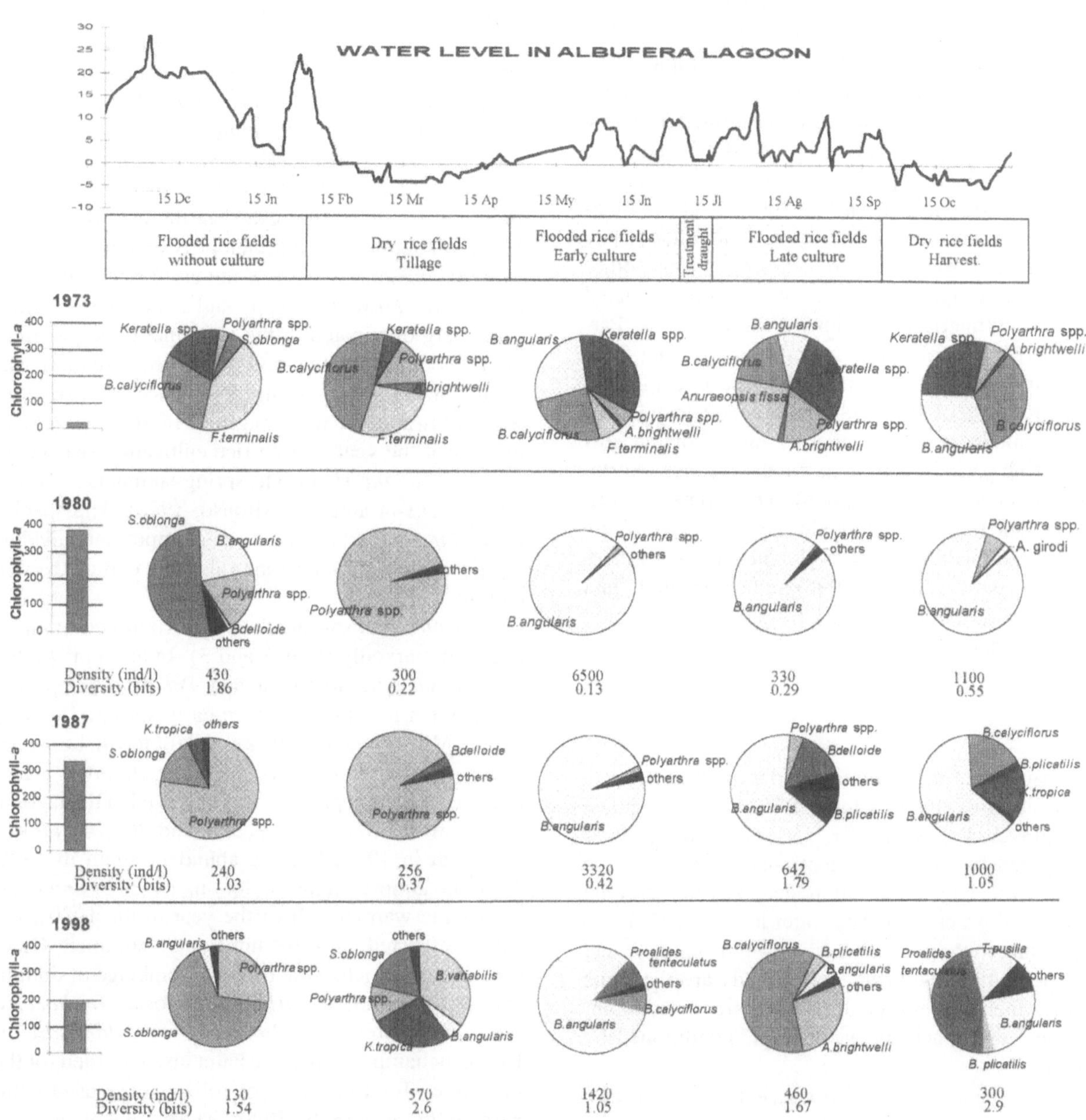


*Figure 2.* Interannual variation of rotifer assemblages in Albufera of Valencia lagoon. Circles show the proportions of the main rotifer species for each seasonal period (top) in the different years indicated at the left. The corresponding values of rotifer density and diversity are shown below. All the data are monthly means during the indicated seasonal phase of averaged values from samples corresponding to three main Albufera zones (North, South and Center). Data for 1973 and 1980 has been re-elaborated from Blanco (1976) and Oltra & Miracle (1992). Top: Water level of Albufera lagoon in 1998 with indication of the 5 phases in which the annual cycle is divided, according to sluice gate regulation for the needs of rice farming. The five periods are well indicated by the water level of the lagoon with respect to that of rice fields (0 point), when water level is well below that point the rice fields are being emptied and the water flow is higher. Left: mean annual concentration of chlorophyll *a* in $\mu g\ l^{-1}$ (mean is calculated from monthly averaged values corresponding to several stations in main Albufera zones (North, South and Center).

*Methods*

Samples were taken monthly at 3 points, North, center and South of the lagoon (Fig. 1) from November 1997 to November 1998. Zooplankton samples (3 l) were collected from the surface (25 cm) with a 1 l plastic bottle. Zooplankton was concentrated *in situ* by filtering through a 35 $\mu$m mesh nytal filter and fixed in 4% formalin. Zooplankton species were counted using a inverted microscope at 100 and 200× magnifications (Oltra & Miracle, 1992). Simultaneously water samples were taken for chlorophyll and other analysis. Chlorophyll *a* concentrations were evaluated spectrophotometrically, following Strickland & Parsons (1972). Zooplankton and chlorophyll data from the years 1987 were computed to have equivalent means to those of 1997–98 for Figures 2 and 4; that is to obtain averages of monthly means of the corresponding periods of time in the three zones (North, South and center of the lagoon). These data was obtained with the same methodology by some of the authors (zooplankton data is detailed in Alfonso & Miracle (1990), Alfonso (1996) and chlorophyll data is taken from Soria (1997)). Diversity was estimated using the Shannon-Wiener index.

## Results

The changes during the three decades in rotifer species composition of the Albufera of Valencia are shown in Figure 2. For each year, species composition is represented for the above mentioned 5 phases of the annual cycle, which are well indicated by the seasonal variation of water level. The water level variation corresponding to the year 1998 is shown in Figure 2, and since it is man regulated, main trends are the same for the other study years, although there are some interannual differences in water level, conditioned by rainfall.

Mean annual chlorophyll concentrations are also represented in Figure 2 which indicates the drastic increase of the rate of eutrophication from 1973, when mean concentration was 19 $\mu g\ l^{-1}$ (Blanco, 1976) to reach concentrations over 400 $\mu g\ l^{-1}$ in the 1980s (Vicente & Miracle, 1992). Due to some sewage diversion and purification of part of the waters inflowing into the Albufera of Valencia, chlorophyll concentration has decreased, as evident from the mean of 180 $\mu g\ l^{-1}$ in 1998.

Rotifer composition has changed because of changes in the trophic state of the lagoon (Fig. 2). *Filinia terminalis* Plate, which was one of the most important winter spring species in 1973, practically disappeared from the plankton since 1980. *Anuraeopsis fissa* Gosse, a estival species which had great abundance in July 1973 was almost nonexistent in the samples of 1980 and those taken after. The other dominant species in 1973, *Brachionus calyciflorus* Pallas and *Keratella* species *(Keratella cochlearis* Gosse, *Keratella quadrata* Müller and *Keratella tropica* Apstein), were much diminished in the next decade. In 1980, plankton was completely dominated by *Brachionus angularis* Gosse in the early rice cultivation period and *Polyarthra* spp. and *Synchaeta oblonga* Ehrenberg dominated in the autumnal period of ricefield flooding. By the mid 1980s the situation was similar; there was a strong dominance of *Polyarthra* spp. from autumn to spring and of *B. angularis* in the rest of the year, during rice cultivation season (*B. angularis angularis* in late spring-summer, *B. angularis bidens* in autumn, Alfonso, 1996). The species *B. calyciflorus* and *K. tropica* were important only during the period of water renewal at the end of the rice culture.

In 1998, after sewage diversion, rotifer community changed markedly (Figs 2 and 3). In autumn–winter, *S. oblonga* was dominant and *Polyarthra* spp. became much less important. In early spring, *Daphnia magna* Müller had high peaks, and for the first time the occurrence of *Brachionus variabilis* Hempel, a facultative *Daphnia* epibiont, was observed in the lagoon. Furthermore, *Keratella* species and *B. calyciflorus*, dominant in 1973, became abundant again in 1998. Then the almost exclusive dominance of *B. angularis* during the warmer half of the year in the 1980s, became restricted in 1998 until early summer. Other species that enriched the rotifer assemblage of dominants were *Asplanchna brightwelli* Gosse, *Trichocerca pusilla* Lauterborn and *Proalides tentaculatus-digitus* De Beauchamp-Donner, the latter also recorded for the first time. Total densities of rotifers decreased compared with those in the 1980s. The annual cycle, with a main peak in spring and a secondary one in autumn, was changed in 1998, maintaining only the main peak in late spring due to *B. angularis*, but with lower numbers. The autumn maximum made up mainly of *B. angularis bidens* did not occur (Fig. 3). A small peak of *B. angularis* was accompanied by those of *P. tentaculatus-digitus* and *T. pusilla*.

The structure of rotifer community changed also from the 1980s to 1998. Diversity indices were much higher in 1998, especially in the periods of wa-

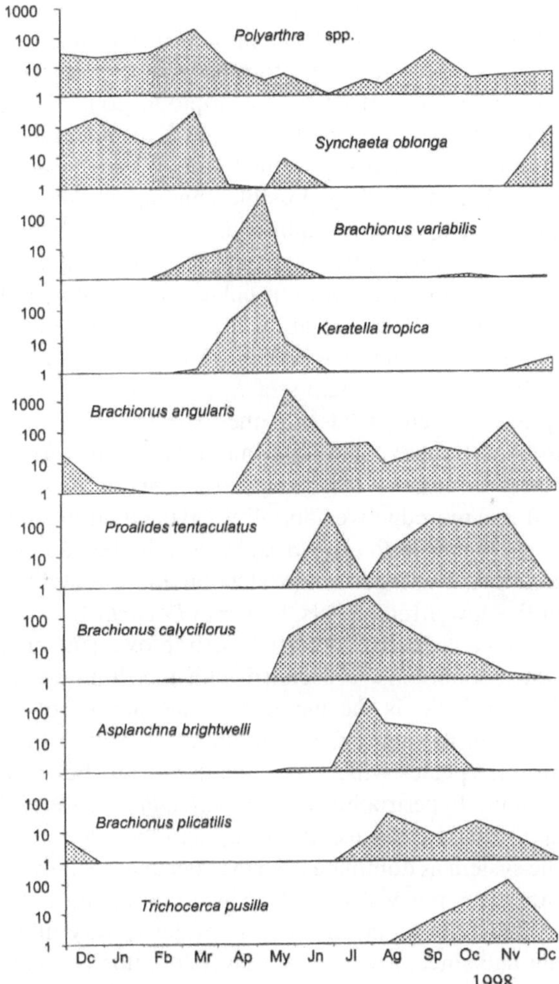


*Figure 3.* Population dynamics of main rotifer species during 1998. Densities in ind $l^{-1}$ are the averages of the three sampling points.

ter renewal, in which the species composition was completely different from that of the former decade. Density and diversity in 1998 was closer to the values of the 1970s than to the ones of the 1980s. In 1998, the species composition tended to revert to that in 1973, especially in summer, mainly because of the dominance of *B. calyciflorus* and its predator *A. brightwelli*.

Other than rotifers, the zooplankton has been mainly made up of the copepod *Acanthocyclops robustus* during the three studied decades (1972–1998). Its density, like rotifer densities, also decreased in 1998 with respect to 1987. On the contrary, cladocerans, which have been relatively not important in this lagoon especially during the 1980s, have been favoured after sewage diversion, showing higher abundance in 1998 than in 1987. However, their presence has been always restricted to brief periods, *D. magna* in February–March and *Moina micrura* Kurz in August.

## Discussion

The eutrophication of the Albufera lagoon caused marked changes in the rotifer community. Already in 1972–73, when the first quantitative studies of zooplankton were conducted (Blanco, 1976), the lake was highly eutrophic. Phytoplankton dominated and macrophytes had by then almost disappeared. The chlorophyll content, however, was still relatively low. Thereafter, in 1980 when the eutrophication drastically increased, the chlorophyll concentrations increased by an order of magnitude. The turbid state reached its maximum in the mid-1980s (Secchi disk depth annual means <0.2 m in most of the lake) and rotifer composition shifted to two main dominants: *Brachionus angularis* from late spring to early autumn and *Polyarthra longiremis* Carlin and other species of this genus, during the rest of the year. In the period of highest eutrophication levels, *F. terminalis* disappeared. The other major species in 1973, *B. calyciflorus* and several *Keratella* spp. *(K. cochlearis, K. quadrata, K. tropica*) were not dominant anymore and *A. fissa* has become a very rare species.

The increased and extended dominance of *B. angularis* during the warmer half of the year is the most outstanding change linked to the increased eutrophication process. In the first existing report of Albufera planktonic rotifers, Arevalo (1918) recorded *B. angularis* as a rare species, scarce in May. Then we have no data until the study of zooplankton dynamics during the years 1972–73 (Blanco, 1976) in which the maximum recorded density of *B. angularis* was 500 ind $l^{-1}$in May 1973. The next year with zooplankton data from this lagoon is 1980 and *B. angularis* reached densities over 10 000 ind $l^{-1}$, maintaining a high dominance thereafter (data from the authors of this paper, partially published in Oltra & Miracle, 1984, 1992 and Alfonso & Miracle, 1990). However, after sewage load has been decreased, *B. angularis* dominance was restricted again to late spring and the maximum density recorded in 1998 was 2300 ind $l^{-1}$. Gulati et al. (1992) indicate that the important factors to be examined for changes in zooplankton composition and abundance are zooplankton food and predation. We do not know to what extend predation could change in these years, fish in the lagoon are largely benthivorous or littoral and eutrophication favoured mullets. Fish biomass has been constituted through all studied years (Blanco, 1976; Docavo 1979; archives from "Comunidad de pescadores del Palmar"), mainly by mullets (*Mugil cephalus* L. the most abundant but other *Liza* species are also abundant) and some percentage of carp (*Cyprinus carpio* L.) and mosquitofish (*Gambusia holbrooki* Giraud) (the latter restricted to the littoral). No remarkable changes have been made in the manipulation of sluice gates and fisheries in these years. Rice cultivation was changed at the end of the 1960s to direct seeding, thus increasing and extending the herbicide treatments since then and nothing has been done as far as we know to diminish them, except replacements of the chemicals used.

We will centre the discussion on the shift of rotifer populations in relation to the differences in phytoplankton, which is the most evident change in the limnology of the lagoon and it is well documented, since phytoplankton samples were taken simultaneously with zooplankton. Differences in feeding can account for the displacement of *B. calyciflorus* by *B. angularis* when phytoplankton became completely dominated by Oscillatoriaceae in the 1980s. At the beginning of the 1970s phytoplankton was still dominated by green algae, although some cyanobacterial populations developed in summer (Blanco, 1976), but in the 1980s, phytoplankton was completely dominated by thin filamentous oscillatoriaceae in summer *(Pseudoanabaena, Geitlerinema)* and *Planktothrix agardhii* Anagn. Kom. the rest of the year (Romo & Miracle, 1993, 1994). Although *B. angularis* eats algal cells, it is an important consumer of detritus (Pourriot, 1965, 1977). Furthermore, according to Erman (1962), this species consumes detritus originating from the decomposition of cyanobacterial blooms (of *Microcystis* in that study). In the Albufera, it may eat fragments detached from filaments of *Oscillatoria* and their associated bacteria. On the other hand, *B. calyciflorus* eats preferentially algae of a larger size than *B. angularis* and it has a complex mechanism for selecting food (Starkweather, 1980, 1995). *B. calyciflorus* is able to feed on filamentous cyanobacteria such as *Anabaena* sp. (Starkweather & Kellar, 1983) or *P. agardhii* (Weithoff & Walz, 1995). The rotifer seems to benefit from these filaments only if they are supplemented by more edible algae, therefore it is in disadvantage with respect to *B. angularis* in a turbid system with important amounts of detritus and completely dominated by *Oscillatoria* filaments. Recent works on feeding using different kinds of tracer particles (Ooms-Wilms et al., 1993; Ooms-Wilms 1997) showed that *B. angularis* behaves quite differently to other species considered also to be bacterivorous and triptophagous such as *Filinia longiseta* Ehrb, *A. fissa* and *K. cochlearis*. For instance, *B. angularis* seems to prefer bigger particle sizes than *A. fissa*, a possible competitor due to its similar temperature requirements that was abundant also in summer 1973 and is one of the main species in similar shallow, eutrophic lakes such as Lake Loosdrecht (Gulati et al., 1992). Moreover, experiments on population dynamics (Walz, 1987) indicate that demographic factors of *B. angularis* correspond to those of an r-strategist, the expected life strategy in hypertrophic conditions and that they are very different from those of *K. cochlearis*. They do not only differ in reproductive, mortality and population growth rates, higher in *B. angularis*, but also in the efficiency to allocate energy to reproduction and the ability to utilize lower food levels, higher in *K. cochearis*.

A change that indicated an improvement of the water conditions, associated with the diminution of *B. angularis*, is the increase in the number of species contributing to the rotifer biomass. Some of them are the species which were the dominants before the extreme hypertrophication (*B. calyciflorus*, *Keratella* spp.) that are discussed above, but others are new to the system as dominants and they occur especially during water renewal periods. Since the lagoon is very shallow, the replacement of its water is very important and since, after sewage diversion, this is done by cleaner inflows, the shifts to other algal and rotifer communities should be markedly apparent in those periods. The absolute dominance of a small form of *B. angularis bidens* in the early autumn period was replaced by the new dominants *P. tentaculatus-digitus* and *T. pusilla*. In the period of water renewal at the end of winter, the appearance of *B. variabilis* is the most remarkable change. This species is an epibiont on *Daphnia magna*, and its occurrence is associated with the peaks of *D. magna* in winter–early-spring. The presence of *D. magna* is an indication of improvement of water conditions, because *D. magna* growth is accompanied by a clear water phase, which occurs more frequently and is more apparent, since the diversion of sewage. The clear water phase is determined by the water renewal at the end of winter, when rice fields which have been without cultivation are emptied and their water circulates through the Albufera to the sea. A cleaner water inflow, after sewage removal, favours a shift in phytoplankton populations, densities of

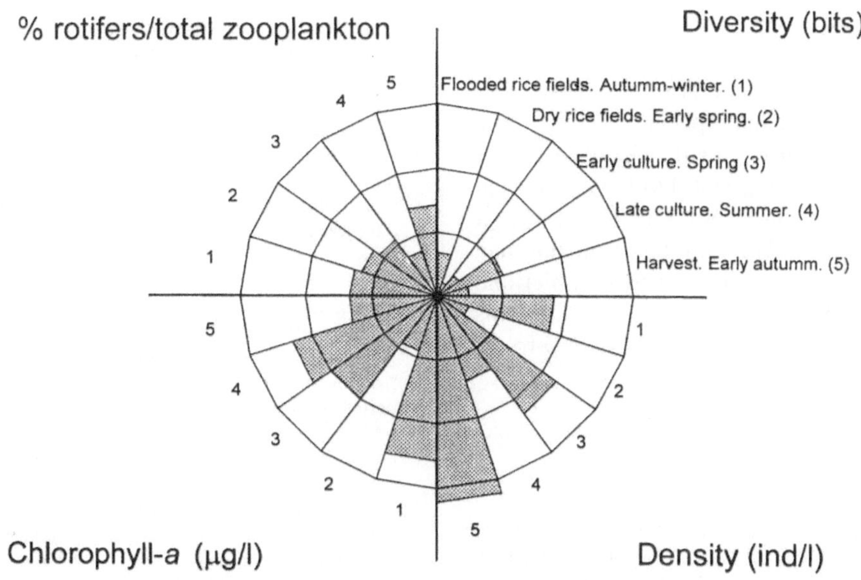


*Figure 4.* Comparison between the years 1998 and 1987 of several characteristics of Albufera of Valencia indicating an improvement of this environment. Rotifer density, diversity and proportion with respect to total zooplankton (rotifers+crustaceans), as well as chlorophyll concentration is compared for each seasonal period. The smallest inner circle corresponds to the state of the lake in 1998 and it is used as a reference frame. The shaded part represents the state in 1987 (with respect to that of 1998). The magnitude of the deviations represented by the two additional circles are two and three times the values of 1998.

cyanobacterial filaments are diluted, underwater light is higher, therefore they are further reduced and the proportions of centric diatoms, cryptophytes and green algae increase. Thus, *D. magna* is favoured and makes an explosive growth with its associated *B. variabilis*. Both these species are causal in the further decrease of phytoplanktonic biomass and the water becomes clear to the bottom for a few weeks. *D. magna* drops soon thereafter, due mainly to the production of resting eggs induced by overpopulation, although mortality due to agricultural treatments or fish predation should be also involved in its decline. However, *B. variabilis*, that begins its development a little later than *D. magna*, remains then abundant during longer time.

Improvement in conditions since sewage diversion, are highlighted in Figure 4. A comparison of the main macroscopic variables in the different seasonal phases, taking 1998 as a reference state and over it the deviations corresponding to 1987 measurements. That year 1987 corresponds to the last study of the zooplankton annual cycle, before main sewage diversion was undertaken. We can observe that after sewage diversion:

(1) Chlorophyll contents have been reduced by more than half in the steady phases, but not in the periods of water renewal (late winter-early spring, late summer–early autumn); this indicates that fluctuations are much smoother in 1998 than in 1987.

(2) The rotifer densities have also been reduced, except in the period of water renewal of late winter–early spring, in which the density is high in 1998 due to *B. variabilis* favoured by *D. magna* pulse. The reduction is more noticeable, to half or one third, in late spring-summer and early autumn, corresponding to the wane of *B. angularis*.

(3) On the other hand, rotifer diversity has increased, especially in late winter–early spring water renewal period, corresponding to the potential clear water phase.

(4) The percentage of rotifers with respect to total zooplankton was lower in 1998, as expected from the improvement of water quality, due to an increased importance of cladocerans (*D. magna* and *Moina micrura*).

From these results, evident changes corresponding to slight improvement of water quality are detected, however these changes are relatively small and the lake is at present still in a extreme hypertrophic state. Nonetheless, the study reflects how shallow lakes can respond to any small nutrient and organic matter reduction, towards a somewhat better and more diversified state.

## Acknowledgements

Our sincere thanks to Conselleria de Medi Ambient for 1998 chlorophyll data and especially to the people from the Natural Park office for their help in all the sampling trips. We are also grateful to J. M. Benavent and the "Oficina Técnica Devesa-Albufera" for their interest and assistance. The helpful comments and language corrections by R. Gulati and R. Shiel and an anonymous rewiewer are greatly appreciated. This study was partially supported by a SWALE, EC project (contract n° ENV4-CT97-0420).

## References

Alfonso, M. T., 1996. Estudio de las comunidades zooplanctónicas de los ecosistemas acuáticos del Parque Natural de la Albufera de Valencia. Ph. D. Thesis, Universidad de Valencia, Valencia: 310 pp.

Alfonso, M. T. & M. R. Miracle, 1990. Distribución espacial de las comunidades zooplanctónicas de la Albufera de Valencia. Scientia gerundensis 16/2: 11–25.

Arevalo, C., 1918. Algunos rotíferos planctónicos de la Albufera de Valencia. Anal. Ins. Gral. Tec. Valencia 8: 1–74.

Blanco, C., 1976. Estudio de la contaminación de la Albufera de Valencia y de los efectos de dicha contaminación sobre la fauna y flora del lago. Ph. D. Thesis, Universidad de Valencia, Valencia: 193 pp.

Carretero, J. L. & H. Boira, 1989. Flora y Vegetación de la Albufera. Bases para su recuperación. València, Edicions Alfons el Magnànim: 83 pp.

Docavo, I., 1979. La Albufera de Valencia, sus peces y sus aves (Ictiofauna y avifauna). València, Institució Alfons el Magnanim, 240 pp.

Erman, L. A., 1962. On the quantitative aspects of feeding and selectivity of food in the plankton rotifer *Brachionus calyciflorus* Pall. Zool. Zh. 41: 34–38.

Gil, M. & R. M. Martínez, 1972. La Albufera de Valencia. I. Flora. Rev. Agroq. Tecn. Aliment. (CSIC) 12 (4): 562–568.

Gulati, R. D., 1990. Zooplankton structure in the Loosdrecht lakes in relation to trophic status and recent restoration measures. Hydrobiologia 191: 173–188.

Gulati, R. D., A. L. Ooms-Wilms, O. F. R. Van Tongeren, G. Postema & K. Siewertsen, 1992. The dynamics and role of limnetic zooplankton in the Loosdrecht lakes (The Netherlands). Hydrobiologia 233: 69–86.

Miracle, M. R., M. P. García & E. Vicente, 1984a. Heterogeneidad espacial de las comunidades fitoplanctónicas de la Albufera de Valencia. Limnetica 1: 20–31.

Miracle, M. R., E. Vicente & E. Garay, 1984b. L'Albufera de València i la problemàtica de la contaminació de les aigües continentals costaneres. XII Congrés de Metges i Biòlegs de LLengua Catalana: 153–166.

Oltra, R., 1993. Estudio del zooplancton de dos lagunas litorales mediterráneas: el Estany de Cullera y la Albufera de Valencia. Ph. D. Thesis, Universidad de Valencia, Valencia: 437 pp.

Oltra, R. & M. R. Miracle, 1984. Comunidades zooplanctónicas de la Albufera de Valencia. Limnetica 1: 51–61.

Oltra, R. & M. R. Miracle, 1992. Seasonal succession of zooplankton populations in the hypertrophic lagoon Albufera of Valencia (Spain). Arch. Hydrobiol. 124: 187–204.

Ooms-Wilms, A. L., 1997. Are bacteria an important food source for rotifers in eutrophic lakes? J. Plankton Res. 19: 1124–1141.

Ooms-Wilms, A. L., G. Postema & R. D. Gulati, 1993. Clearance rates of bacteria by the rotifer *Filinia longiseta* (Ehrb.) measured using three tracers. Hydrobiologia 255/256: 255–260.

Pardo, L., 1942. La Albufera de Valencia. Estudio Limnográfico, biológico, económico y antropológico. Madrid, Instituto Forestal de Investigaciones y Experiencias, no. 24: 263 pp.

Pouriot, R., 1965. Recherches sur l'Ècologie des Rotifères. Vie et Milieu, suppl. 21: 224 pp.

Pouriot, R., 1977. Food and feeding habits of Rotifera. Arch. Hydrobiol. Beih. Ergebn. Limnol. 8: 243–260.

Romo, S. & M. R. Miracle, 1993. Long-term periodicity of *Planktothrix agardhii, Pseudoanabaena galeata* and *Geitlerinema* sp. in a shallow hyertrophic lagoon, the Albufera of Valencia (Spain). Arch. Hydrobiol. 126: 469–486.

Romo, S. & M. R. Miracle, 1994. Population dynamics and ecology of subdominant phytoplankton species in a shallow hyertrophic lake (Albufera of Valencia, Spain). Hydrobiologia 273: 37–56.

Ronneberger, D., P. Kasprzak & L. Krienitz, 1993. Long-term changes in the rotifer fauna after biomanipulation in Haussee (Feldberg, Germany, Mecklenburg-Vorpommern) and its relationship to the crustacean and phytoplankton communities. Hydrobiologia 255/256: 297–304.

Soria, J., 1997. Estudio limnológico de los ecosistemas acuáticos del "Parc Natural de l'Albufera" de Valencia. Ph.D. Thesis, Universidad de Valencia, Valencia: 289 pp.

Starkweather, P. L., 1980. Aspects of feeding behavior and trophic ecology of suspension feeding. Hydrobiologia 73: 63–72.

Starkweather, P. L., 1995. Near-coronal fluid flow patterns and food cell manipulation in the rotifer *Brachionus calyciflorus*. Hydrobiologia 313/314: 191–195.

Starkweather, P. L. & P. E. Kellar, 1983. Utilization of cyanobacteria by *Brachionus calyciflorus*: *Anabaena flos-aquae* (NRC-44-1) as a sole or complementary food source. Hydrobiologia 104: 373–377.

Strickland, J. R. & T. R. Parsons, 1972. A practical handbook of sea water analysis. J. Fish. Res. Bd Can: 311 pp.

Van Tongeren, O. F. R., L. Van Liere, R. D. Gulati, G. Postema & P. J. Boesewinkel-De Bruijn, 1992. Multivariate analysis of the plankton communities in the Loosdrecht lakes: relationship with the chemical and physical environment. Hydrobiologia 233: 105–117.

Vicente, E. & M. R. Miracle, 1992. The coastal lagoon Albufera de Valencia: an ecosystem under stress. Limnetica 8: 87–100.

Walz, N., 1987. Comparative population dynamics of the rotifers *Brachionus angularis* and *Keratella cochlearis*. Hydrobiologia 147: 209–213.

Weithoff, G. & N. Walz, 1995. Influence of the filamentous cyanobacterium *Planktothrix agardhii* on population growth and reproductive pattern of the rotifer *Brachionus calyciflorus*. Hydrobiologia 313/314: 381–386.

Hydrobiologia **446/447:** 221–228, 2001.
L. Sanoamuang, H. Segers, R.J. Shiel & R.D. Gulati (eds), Rotifera IX.



# Ecological structure of psammic rotifers in the ecotonal zone of Lake Piaseczno (eastern Poland)

Stanisław Radwan[1] & Irena Bielańska-Grajner[2]
[1]*Department of Hydrobiology and Ichthyobiology, University of Agriculture, Akademicka 13, 20-950 Lublin, Poland*
[2]*Department of Ecology, Silesian University, Bankowa 9, 40-007 Katowice, Poland*
*E-mail: Radwan@ursus.ar.lublin.pl*

*Key words:* benthos, interstitial, lentic, sediment

## Abstract

The structure of the psammic rotifer community in three ecotonal zones (arenal, interface, eulittoral) of a mesotrophic lake (Lake Piaseczno; Leczna – Wlodawa Lakeland, eastern Poland) was studied in June and September of 1996 and 1997. A total number of 48 taxa, belonging to three different groups of psammon (eupsammon, hygropsammon and hydropsammon), were identified: 6 psammobionts, 16 psammophiles and 26 psammoxens. In each ecotonal zone, rotifers were clearly differentiated both qualitatively and quantitatively, depending upon the season and chemical factors. For instance, species diversity indicates a positive correlation with three chemical parameters: pH, chemical oxidation demand and $PO_4^{3-}$ concentration.

## Introduction

Studies on the biology and ecology of freshwater psammic rotifers are still quite rare. Within this rather limited literature, descriptive work on the rotifers of rivers and streams prevails (e.g. Neiswestnova-Shadina, 1935; Urban, 1971; Evans, 1984; Pennak, 1988; Turner, 1996). In contrast, there are few publications that focus on the biology and ecology of psammic rotifers in lakes. Among these are the pioneering papers of Wiszniewski (1934, 1937) on Lake Wigry in North Poland and those of Myers (1936), Pennak (1940) and Evans (1982) in the U.S.A. Recently, some papers on the qualitative and quantitative structure of the psammic rotifer communities living in mesotrophic lakes have been published (Ejsmond-Karabin, 1998; Radwan et al., 1998).

According to Wiszniewski (1934a,b, 1937) the ecotonal zone (shore) comprises three regions (arenal, interface zone and eulittoral), each of which are inhabited by a unique assemblage of minute invertebrates. The eupsammon live in the arenal or sandy beach area, a region which is flooded only occasionally. The hygropsammon occur in a relatively narrow interface where the sandy beach of the land meets the water. The hydropsammon are found in the sandy eulittoral, a zone of constant inundation. Here, we offer a contribution to the understanding of the processes that shape the biocenotic structure of the rotifer community in the ecotonal zone of Lake Piaseczno (eastern Poland).

## Materials and methods

This study was carried out in Lake Piaseczno, a mesotrophic lake situated in Leczna–Wlodawa Lakeland (eastern Poland) (Fig. 1). Lake Piaseczno has a surface area of 83.8 ha, a maximum depth of 38.8 m, and an ecotone with a well developed arenal region occupying about 60% of the lake shore. This area is characterised by a broad (ca. 5–10 m) sandy beach (sediment size ≈0.1–1.5 mm), lacking emerged macrophytes in the eulittoral, but with a rich submerged macrophyte community in the sublittoral and deep littoral.

Samples (collected as microtransects) were taken in spring (June) and autumn (September) of 1996 and 1997 from three regions: (1) arenal (beach), 1 m from the land-water interface; (2) interface zone; (3) littoral, ≈1 m from the interface contact zone at a depth of between 0.5 and 0.7 m. All samples were collected

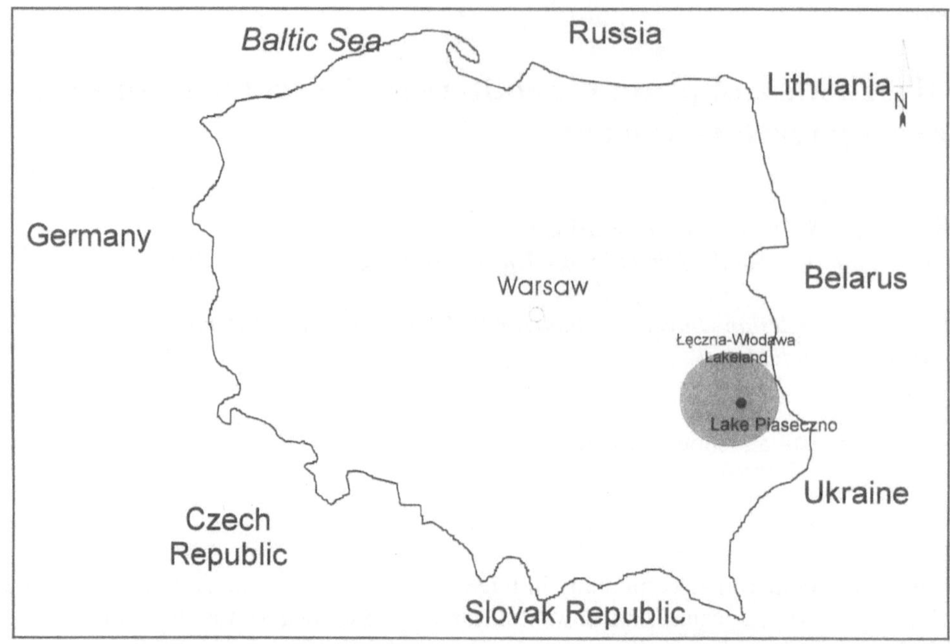


*Figure 1.* Location of Lake Piaseczno in Poland.

*Table 1.* Physical and chemical parameters of the water in ecotonal zones of the Piaseczno Lake

| Year | 1996 | | | | | | 1997 | | | | | |
|---|---|---|---|---|---|---|---|---|---|---|---|---|
| Parameter Season | VI | | | IX | | | VI | | | IX | | |
| Zone | A | I | E | A | I | E | A | I | E | A | I | E |
| Temperature | 17.0 | 21.6 | – | 11.0 | 14.7 | 15.1 | 17.0 | 21.0 | 21.7 | 13.0 | 18.3 | 18.0 |
| pH | 5.77 | 6.85 | 6.90 | 6.04 | 8.42 | 7.80 | 5.47 | 6.98 | 7.90 | 4.98 | 7.60 | 6.99 |
| Temp. ($C^0$) | – | 21.6 | – | – | 14.7 | 15.1 | – | 21.0 | 21.7 | – | 18.3 | 18.0 |
| COD - $mgO_2$/l | 13.20 | 5.70 | 4.60 | 14.14 | 6.70 | 5.20 | 18.50 | 5.70 | 5.50 | 18.60 | 6.70 | 6.00 |
| N-$NH_4^+$ - mg/l | 0.84 | 0.14 | 0.15 | 1.70 | 0.19 | 0.12 | 0.27 | 0.10 | 0.09 | 0.30 | 0.21 | 0.10 |
| N-$NO_3^-$ - mg/l | 0.07 | 0.05 | 0.02 | 0.19 | 0.07 | 0.09 | 0.04 | 0.05 | 0.0 | 0.16 | 0.23 | 0.29 |
| $PO_4^{-3}$ - mg/l | 0.12 | 0.00 | 0.00 | 0.17 | 0.02 | 0.02 | 0.09 | 0.00 | 0.00 | 0.12 | 0.00 | 0.06 |
| Conductivity - ?S?$cm^{-1}$ | 0.49 | 85.00 | 101.00 | 0.90 | 93.00 | 98.00 | 0.40 | 85.00 | 96.00 | 1.19 | 93.00 | 85.00 |

Ecotonal zone: A –arenal. I– interface zone. E – eulittoral.

using a plastic cylinder 10 cm long and 3.5 cm in diameter with a sharpend leading edge to aid cutting into the sediment (Evans, 1984). On each sampling date, two samples were taken, one for qualitative and the other for quantitative analyses. Each sample consisted of five replicates, each 2 cm in depth. The material in the first sample was kept alive, while the second was preserved with Lugol's and 4% formalin solutions. Each quantitative sample was sliced into two layers (0–1 cm and 1–2 cm). Rotifer density was expressed as number of individuals per $cm^3$ of sand and their biodiversity was estimated using the Shannon–Wiener index (Krebs, 1994).

At the same time that the sediment cores were taken, 500 samples of piesometer water and lake water were collected and placed into plastic bottles. These water samples were later analysed for pH, organic matter expressed as chemical oxidation demand (COD), [N-$NH_4^+$], [$PO_4^{3-}$], and electrolytic conductivity. Electrolytic conductivity and pH were measured *in situ* by means of Hydrolab $4^{r®}$ apparatus. Biogenic compounds were analysed according to the methods by Golterman (1972).

## Results and discussion

### *Physical and chemical factors*

The physical and chemical properties of the water did not show large seasonal changes. Average concentrations of N-ammonium ranged from 0.9 to 1.7 mg $NH_4^+ l^{-1}$, N-nitrate from 0.00 to 0.29 mg $NO_3^- l^{-1}$, and phosphate from 0.00 to 0.17 mg $PO_4^{-3} l^{-1}$. Mean electrolytic conductivity varied from 0.4 to 101 $\mu$S $cm^{-1}$ (Table 1). COD in the water samples ranged from 18.6 to 4.6 mg $O_2 l^{-1}$. All these chemical factors exhibited a clear spatial pattern. Organic matter, [N-$NH_4^+$], and [$PO_4^{3-}$] reached their highest level in the arenal and the lowest in the eulittoral or in the contact interface zone, while [N-$NO_3^-$], pH and electrolytic conductivity showed a different pattern. The highest values occurred at the contact interface zone or in eulittoral, with generally lower levels appearing in the arenal (Table 1).

### *Qualitative structure of psammic rotifer community*

The psammic rotifer community in Lake Piaseczno consisted of 48 taxa (Table 2). In both study years, the highest number of taxa was observed in September, when the number of taxa varied from 5 in the hydropsammon (living in eulittoral) to 19–21 in the eupsammon (living in arenal). In June, there were 3–9 and 8–19 taxa in these two groups, respectively. In September, the eupsammon was the richest fauna, while in June the hygropsammon (living in interface zone) was the richest. The lowest number of taxa was found in June in the hydropsammon and eupsammon (Fig. 2a). Shannon–Wiener index changed from 2.47–2.17 to 0.91–0.69, respectively.

According to Wiszniewski's (1937) classification, the psammic rotifer communities consisted of psammophiles (6), psammobionts (16) and psammoxens (26) species. In general, species richness of the psammon is the same in lótic and lentic systems (Urban, 1971; Evans, 1984). An exception to this generalisation is the high number of species (ca. 80) reported from one river in the U.S.A. (Turner, 1996).

### *Quantitative structure*

Rotifer abundance in each zone varied considerably during the year, ranging from 9 inds. $cm^{-3}$ in the eulittoral in June 1996 to as many as 6587 inds. $cm^{-3}$ in the interface zone in September 1997. The majority of individuals occurred in the upper 0–1 cm layer of sand, while only few individuals were recorded in the lower sediment layer (1–2 cm).

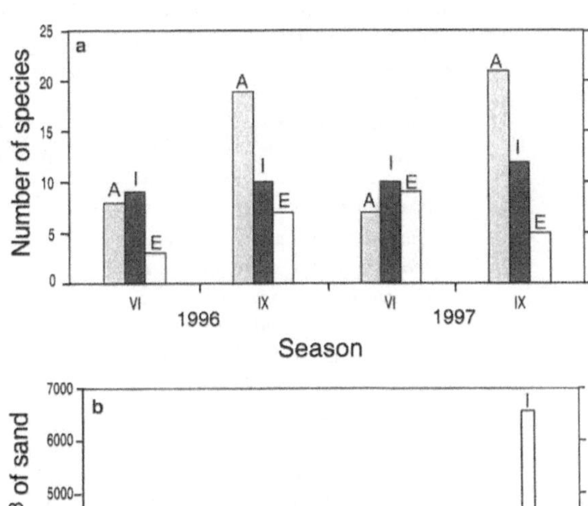


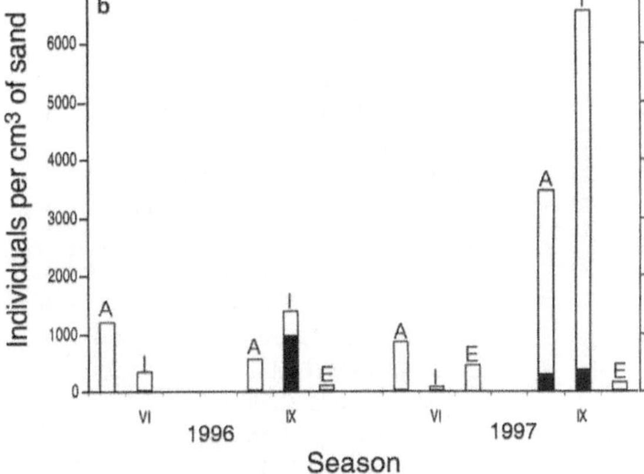


*Figure 2.* Number of species and abundance of psammic rotifers in Lake Piaseczno. Ecotonal zones: A – arenal; I – interface; E –eulittoral. (a) Psammon assemblage: shaded bars, eupsammon; solid bars, hygropsammon; open bars, hydropsammon. (b) Shaded bars, 0–1 cm; solid bars, 1–2 cm.

The total density of rotifers underwent seasonal changes. In June, the highest density occurred in the eupsammon (arenal), while in September the highest density occurred in the hygropsammon (interface zone). In turn, in both seasons rotifers had their lowest density in the hydropsammon (eulittoral) (Fig. 2b). In both years in all these assemblages, either psammobionts or psammophiles dominated. The only exception was June and September of 1996 when psammoxens dominated the hydropsammon and hygropsammon, respectively (Fig. 3).

Abundance of each assemblage of rotifers changed both seasonally and yearly. Among the dominant species, many are psammobionts and psammophiles: *Elosa worralli*, *Elosa worralli f. spinifera*, *Lecane psammophila*, *Lecane closterocerca*, *Lepadella patella*, *Cephalodella gibba*, *Cephalodella misgurnus* and *Trichocerca taurocephala*. In June 1996, the

*Table 2.* Psammic rotifers in the ecotonal zones of Lake Piaseczno (Ecotonal zone: A - - arenal, I – interface zone, E – eulittoral)

| Year | 1996 | | | | | | 1997 | | | | | |
|---|---|---|---|---|---|---|---|---|---|---|---|---|
| Parameter Season | VI | | | IX | | | VI | | | IX | | |
| Zone | A | I | E | A | I | E | A | I | E | A | I | E |
| Psammobionts | | | | | | | | | | | | |
| *Dicranophorus hercules* Wiszn. | + | + | + | + | | | | | | | | |
| *Elosa worallii* f. *spinifera* Wiszn. | + | + | + | | | | | | | | | |
| *Encentrum diglandula* (Zawad) | + | + | | | | | | | | | | |
| *Lecane psammophila* Wiszn. | | + | + | + | + | + | + | + | + | + | + | + |
| *Trichocerca taurocephala* Hauer | + | + | + | + | + | + | | | | | | |
| *Wierzejskiella velox* Wiszn. | | + | + | | | | | | | | | |
| Psammophilous | | | | | | | | | | | | |
| *Bryceella tenella* (Bryce) | | + | | | | | | | | | | |
| *Cephalodella auriculata* (Mull.) | + | | | | | | | | | | | |
| *Cephalodella catellina* (Mull.) | + | + | + | | | | | | | | | |
| *Cephalodella gibba* (Ehrb.) | | + | + | | | | | | | | | |
| *Cephalodella gracilis* (Ehrb.) | | + | | | | | | | | | | |
| *Cephalodella ventripes* (Dix.-Nutt.) | + | | | | | | | | | | | |
| *Colurella colorus* (Ehrb.) | | + | + | | | | | | | | | |
| *Colurella obtusa* (Gosse) | + | + | + | + | + | + | + | | | | | |
| *Lecane closterocerca* (Schm.) | + | + | + | + | + | + | + | + | | | | |
| *Lecane flexilis* (Gosse) | + | | + | | | | | | | | | |
| *Lecane lunaris* (Ehrb.) | | | + | + | + | + | + | | | | | |
| *Lepadella patella* (Mull.) | | + | + | + | + | + | + | + | | | | |
| *Proales minima* (Montet.) | | + | + | + | | | | | | | | |
| *Rotaria rotatoria* Pall. | + | | + | + | + | | | | | | | |
| *Trichocerca intermedia* Sten. | | + | + | + | + | | | | | | | |
| *Trichocerca tenuior* (Gosse) | | + | | | | | | | | | | |
| Psammoxens | | | | | | | | | | | | |
| Bdelloidea n. det. | | | | + | | + | | | | | | |
| *Cephalodella misgurnus* Wulf. | | + | + | | | | | | | | | |
| *Cephalodella sterea* (Gosse) | | + | | | | | | | | | | |
| *Conochilus unicornis* Rouss. | + | | | | | | | | | | | |
| *Dicranophorus hauerians* Wiszn. | + | | | | | | | | | | | |
| *Dicranophorus unicatus* (Milne) | + | | | | | | | | | | | |
| *Elosa worallii* Lord | | + | | | | | | | | | | |
| *Filinia limnetica* (Zach.) | | + | | | | | | | | | | |
| *Itura aurita* (Ehrb.) | | | + | | | | | | | | | |
| *Itura viridis* (Stenn.) | | | + | | | | | | | | | |
| *Keratella cochlearis* (Gosse) | | + | + | + | + | + | + | | | | | |
| *Keratella quadrata* (Mull.) | | + | + | + | | | | | | | | |
| *Kellicotia longispina* (Kell.) | | + | + | | | | | | | | | |
| *Lecane depressa* (Bryce) | | | + | + | | | | | | | | |
| *Lecene luna* (Mull.) | | | | + | + | + | + | | | | | |
| *Lecane scutata* (Harring i Myers) | | | | | | | | | | | | |
| *Lepadella acuminata* (Ehrb.) | | + | | | | | | | | | | |
| *Monommata* sp. | | + | | | + | | | | | | | |
| *Notholca labis* (Gosse) | | | | | | | | | | | | |
| *Notommata cyrtopus* (Gosse) | | + | + | | | | | | | | | |
| *Trichocerca capucina* (Wierz. i Zach.) | + | | | | | | | | | | | |
| *Trichocerca cylidrica* (Imh.) | | + | | | | | | | | | | |
| *Trichocerca dixonnuttalli* (Jenn.) | + | | | | | | | | | | | |
| *Trichocerca iernis* (Gosse) | | | | | | | | | | | | |
| *Trichocerca inermis* (Linder) | | + | | | | | | | | | | |
| *Trichocerca weberi* (Jenn.) | | + | | | | | | | | | | |
| Total species | 8 | 9 | 2 | 19 | 10 | 7 | 7 | 10 | 9 | 21 | 12 | 5 |

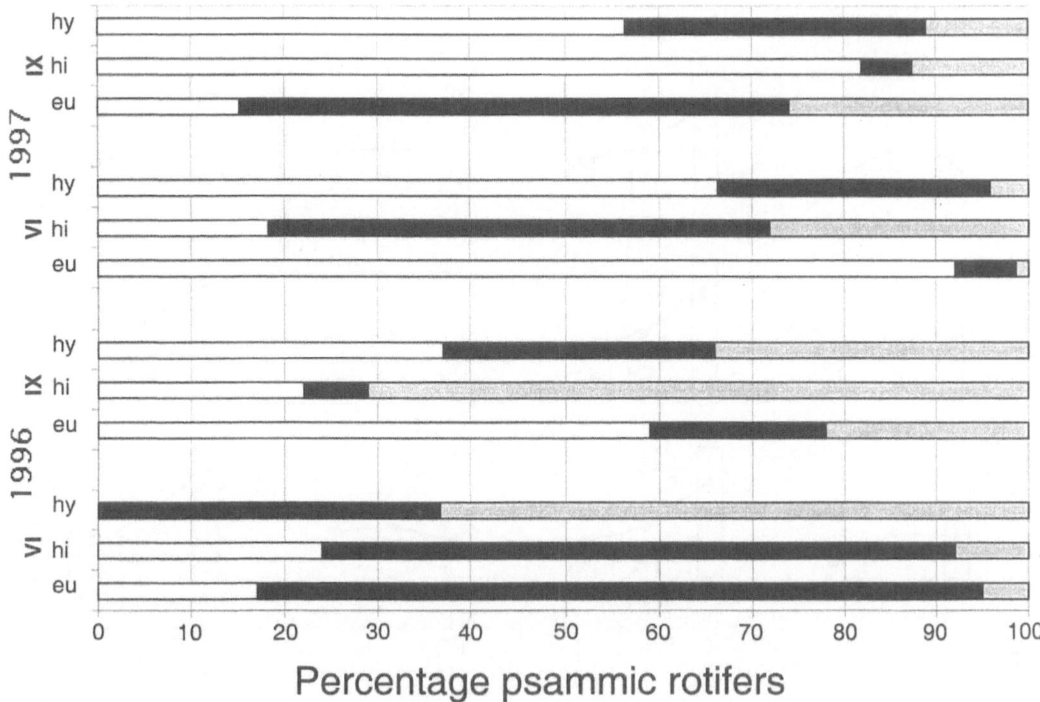


*Figure 3.* Percentage share of psammobionts, psammophilous and psammoxens in the psammic rotifer community of Lake Piaseczno. Ecological group: open bars, psammobionts; solid bars, psammophilous; shaded bars, psammonxens. Psammon assemblage: eu – eupsammon; hi – hygropsammon; hy – hydropsammon.

highest abundance of psammoxens was achieved by *Keratella cochlearis* f.*tecta*. Similar trends in frequency changes were observed in Lake Kuc in the Masurian Lakeland of Poland (Ejsmond-Karabin, 1998). The psammon of that lake also was composed of the following genera: *Lecane*, *Cephallodella* and *Trichocerca*. In rivers, however, psammic rotifers usually belong to the genera *Encentrum*, *Lecane*, *Colurella*, *Cephalodella* and *Trichocerca* (Evans, 1994; Turner, 1996).

*Relationship between environmental parameters and rotifers communities*

Rotifers inhabiting these ecotonal zones probably live in permanently stressed environments. It is likely that this stress is responsible for the observed changes in species diversity and numbers (Wiszniewski, 1934a,b). Some authors point to the fact that the biocenosis of habitats undergoing a moderate stress are highly differentiated (Fibiszewski, 1995; Ejsmond-Karabin 1998).

In the ecotone of Lake Piaseczno, the highest diversity of psammic rotifers occurred in the arenal (inhabited by eupsammon), and the lowest in the eulittoral (inhabited by hydropsammon). In Lake Kuc (northern Poland), the highest diversity was noted in the eulittoral. The highest species diversity of eupsammon in Lake Piaseczno was probably connected with high content of organic matter in the arenal zone. This can be seen by the positive relationship between number of species and [$PO_4^{3-}$], pH and chemical oxidation demand (Fig. 4). The remaining physical and chemical parameters did not appear to influence species diversity. A similar correlation between chemical oxidation demand, [$PO_4^{3-}$] and diversity of plankton rotifers was found in the pelagial region of Leczna-Wlodawa lakes in Eastern Poland (Radwan, 1984). We found that changes in numbers of psammic rotifers were more related to seasonality than with physical and chemical parameters in the ecotonal zones of Lake Piaseczno (Fig. 5). However, other workers offer data showing a positive correlation between density of psammic rotifers and alkalinity, and a negative correl-

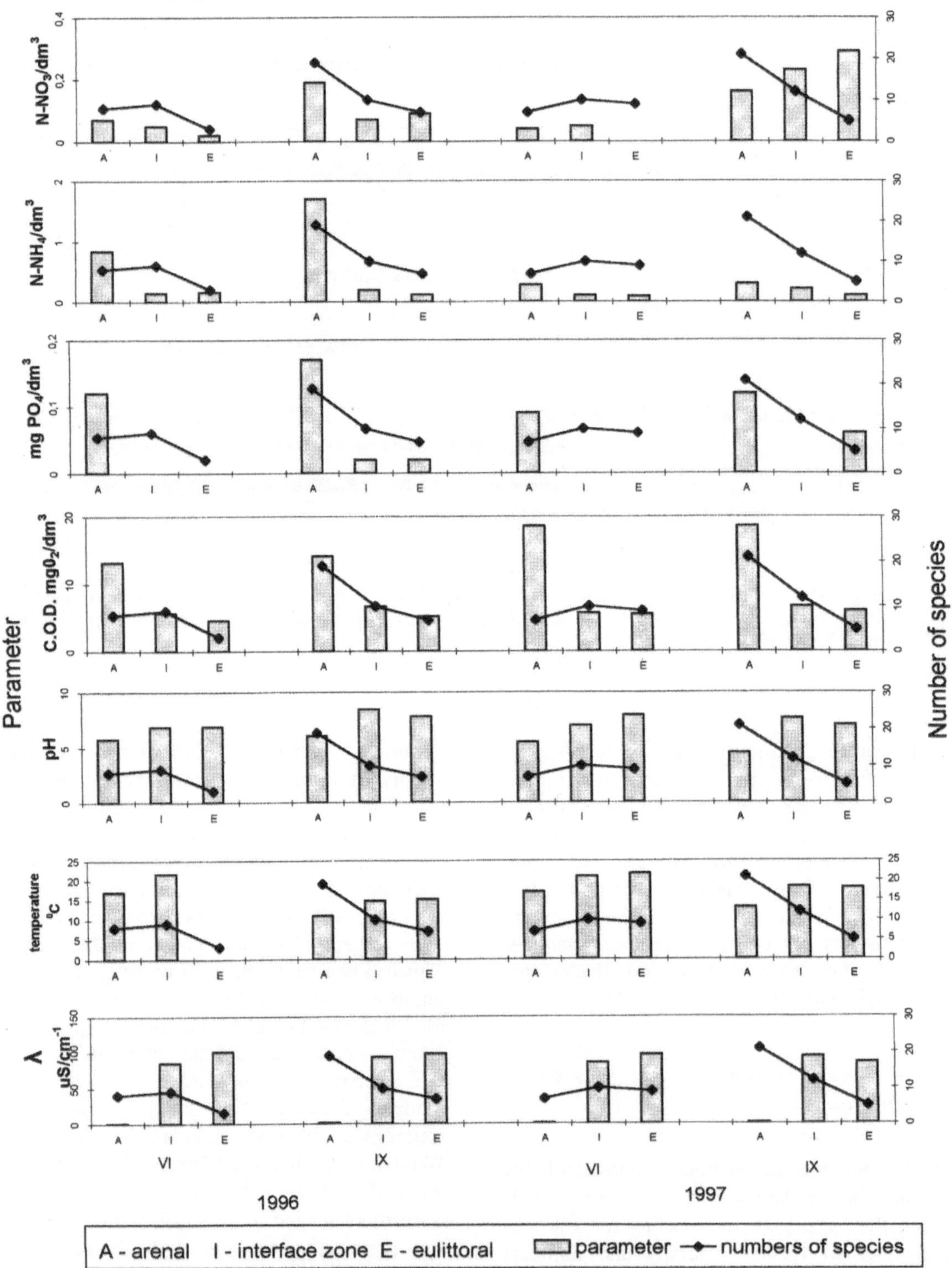


*Figure 4.* Relationships between number of species and physical and chemical parameters of psammic rotifers in Lake Piaseczno. Ectonal zones: A – arenal; I – interface; E – eulittoral. Bars, physical parameter; lines, numbers of species.

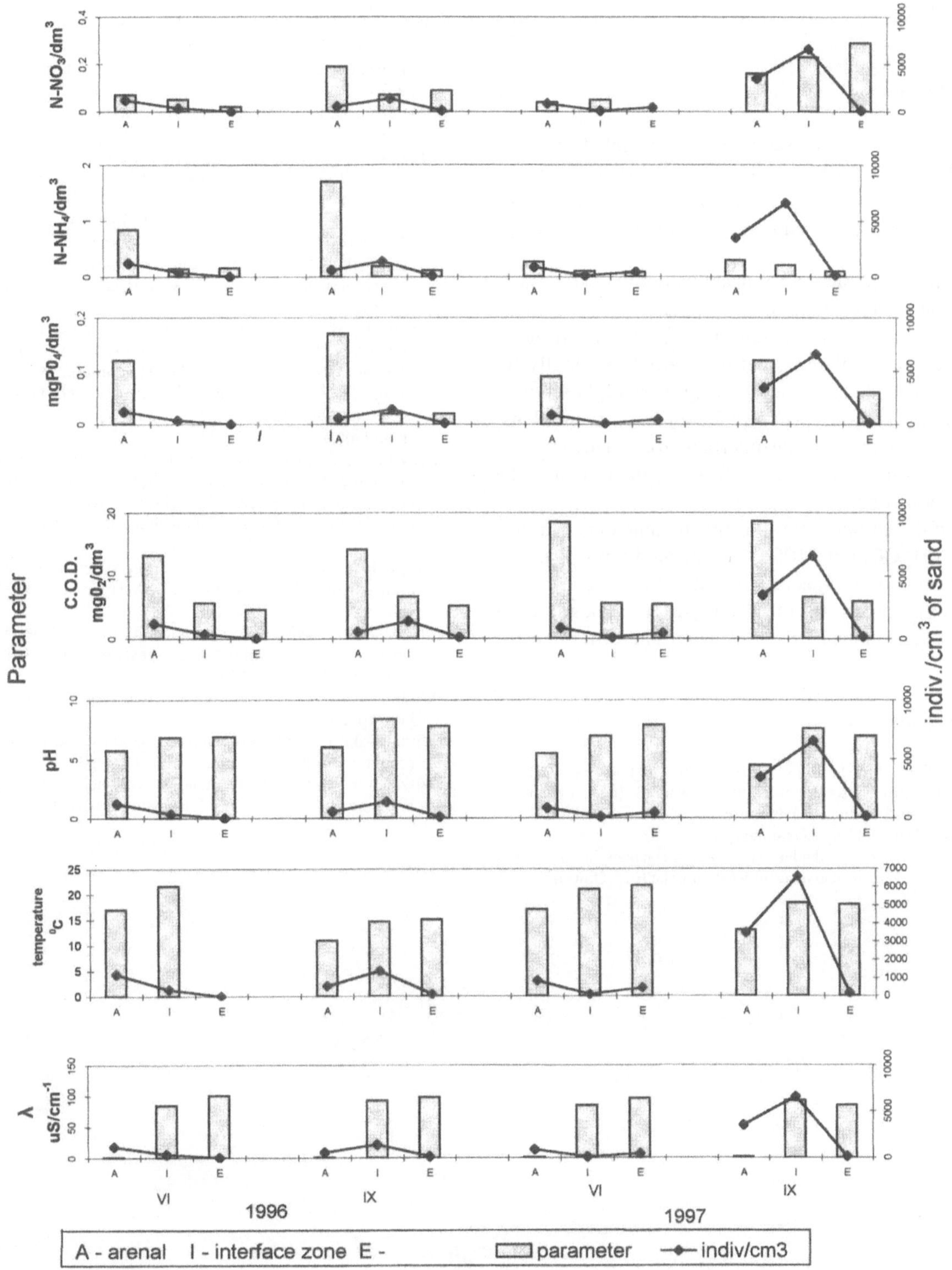


*Figure 5.* Relationships between abundance and some physical and chemical parameters of psammic rotifers in Lake Piaseczno. Ectonal zones: A – arenal; I – interface; E – eulittoral. Bars, physical parameters; lines, density (inds.cm$^{-3}$).

ation between density and magnesium concentration (Evans, 1984).

## Conclusion

The community of psammic rotifers inhabiting the ecotonal zones of Lake Piaseczno was found to be rather rich, totalling 48 taxa belonging to three different group (eupsammon, hygropsammon and hydropsammon).

All communities were usually dominated by psammobionts and psammophilous species.

Numbers of species and rotifers density differed in each ecotonal zone, but this varied seasonally. In September, diversity and density were highest found in the arenal zone, while in June diversity and density were highest in the interface zone. However, in both seasons, density of rotifers was the lowest in the eulittoral zone.

Of the abiotic environmental parameters examined only pH, COD and [$PO_4^{3-}$] indicated a positive correlation with the species diversity of psammic rotifers in the ecotonal zones of Lake Piaseczno. These factors did not have much effect on changes in rotifer density.

## References


Ejsmond-Karabin, J., 1998. Is the lake beach an ecotone? (Psammon Rotifera of Lake Kuc). In Radwan, S. (ed.), Freshwater Ecotones-Structure, Types and Function. UMCS Lublin: 43–49 (in Polish with English summary).

Evans, W. A., 1982. Abudances of micrometazoans in three sandy beaches in the island areas of western Lake Erie. Ohio J. Sci. 82: 246–251.

Evans, W. A., 1984. Seasonal abundances of the psammonic rotifers of a physically controlled stream. Hydrobiologia 108: 105–114.

Fibiszewski, J., 1995. Species diversity among of the plants. In Andrzejewski, R. & R. J. Wisniewski- (eds), Problems of the Biological Diversity. I. E. PAS Publish. Dziekanów Lesny: 29–36.

Golterman, H. L., 1971. Methods for Chemical Analysis of Fresh Waters – IBP Handbook 3: 1–166.

Krebs, C. J., 1994. Ecology. Harper Collins Publishers: 1–735.

Myers, F., 1936. Psammolittoral rotifers of Lenape and Union Lakes, New Jersey. Ann. Mus. Nov. 830: 1–22.

Neiswestnowa-Shadina, K., 1935. Zur Kenntnis des rheophilen Microbenthos. Arch. Hydrobiol. 28: 555–582.

Pennak, R. W., 1940. Ecology of the microscopic metazoa inhabiting the sandy beach of some Wisconsin lakes. Ecol. Monogr. 10: 539–615.

Pennak, R. W., 1988. Ecology of the fresh–water meiofauna. In Higgins, R. P. & H. Thiely (eds), Introduction to the Study of Meiofauna. Smithsonian Press Washington DC: 38–60.

Radwan, S., 1984. The influence of some abiotic factors on the occurrence of rotifers of Leczna and Wlodawa Lake District. Hydrobiologia 112: 117–124.

Radwan, S., T. J. Chmielewski & T. Ozimek, 1998. Structure and function of land/water ecotones in different trophically lakes of Polesie Lubelskie Region. In Radwan, S. (ed.), Freshwater Ecotones. Structure, Types and Function. UMCS Publish., Lublin: 17–41 (in Polish with English summary).

Turner, P. N., 1996. Preliminary data on rotifers in the interstitial of the Ninnescah river, Kansas, U.S.A. Hydrobiologia 319: 179–184.

Urban, R. D., 1971. The psammon of bars and beach in two small northwestern Minnesota streams. Ph.D. Dissertation, Univ. North Dakota, Fargo 154 pp.

Wiszniewski, J., 1934a. Les Rotiferes psammiques. Ann. Mus. Zool. Pol., 10: 339–396.

Wiszniewski, J,. 1934b. Recherches ecologiques sur les rotiferes du psammon. Arch. Hydrobiol. Ryb. 8: 149–399.

Wiszniewski, J., 1937. Differenciation ecologique des rotiferes dans le psammon d'eaux douces. Ann. Mus. Zool. Pol. 13: 1–13.

Hydrobiologia 446/447: 229–232, 2001.
L. Sanoamuang, H. Segers, R.J. Shiel & R.D. Gulati (eds), Rotifera IX.




# Lake characteristics influence recovery of microplankton in arctic LTER lakes following experimental fertilization

Parke A. Rublee[1] & Neil D. Bettez[2]
[1]*Biology Department, University of North Carolina at Greensboro, Greensboro, NC 27402, U.S.A.*
*E-mail: rublee@uncg.edu*
[2]*Ecosystems Center, Marine Biological Laboratory, Woods Hole, MA 02543, U.S.A.*

*Key words:* microplankton, arctic lakes, eutrophication, rotifers

### Abstract

Lakes N-1 and N-2 at the Arctic Long Term Ecological Research site at Toolik Lake, Alaska, U.S.A. were fertilized with nitrogen and phosphorus for 5 and 6 years, respectively. The response and recovery of the microplankton community (protozoans, rotifers and crustacean nauplii) differed in the two lakes. Microplankton biomass in Lake N-1 increased five-fold while that in Lake-N-2 only doubled, despite larger nutrient additions to N-2. Microplankton community structure in Lake N-1 shifted toward dominance by few taxa, while the community in Lake N-2 maintained diversity. Finally, the recovery of Lake N-1 to near prefertilization microplankton biomass levels was rapid, while Lake N-2 showed at least a 1-year lag in recovery. These differences appear to be related to differences in the structure of lake sediments.

## Introduction

Alaskan arctic aquatic ecosystems are of interest because the arctic region is expected to undergo significant global climate change (e.g. Lachenbruch & Marshall, 1986; Abelson, 1989; Schindler et al., 1990). The impact of increased temperature and altered rainfall is expected to lead to changes in the physical, chemical, and biological character of the relatively pristine lakes and streams in this region (McDonald et al., 1996; Rouse et al., 1997; Hobbie et al. In Press). Since microbial communities are characterized by rapid metabolism and short life cycles, they include organisms that could demonstrate rapid response to environmental change.

Previously, we have reported on the increased biomass and species changes of the microplankton community in response to increased nutrient (N and P) loadings in artificially divided Lake N-2 (Rublee, 1992), and in Lake N-1 (Rublee, 1998; Rublee & Bettez, 1995). In this paper, we report observations that demonstrate subtle differences in the recovery of the microplankton community after fertilization of the lakes stopped.

## Site description

The arctic LTER site lies in the northern foothills of the Brooks Mountain range of Alaska (68° N, 149° W). The area is a region of arctic tundra, underlain by permafrost, with an average annual temperature of about −9 °C and low rainfall. Lakes are covered by up to 2 m of ice from late September to mid-June. The site has been under study for 25 years, although microplankton populations have been monitored only since 1989. Lakes are of glacial origin and generally highly oligotrophic. More complete descriptions of Toolik and other LTER lakes can be found in O'Brien (1992) and O'Brien et al. (1997).

## Methods

We sampled from three lakes: two of these Lake N-1 and Lake N-2 were subject to experimental nutrient loading and the third one, Toolik Lake, served as a reference lake (Table 1). Water samples (2-l) collected weekly from a depth of 1 m by a Van Dorn bottle from June to August were concentrated to 60 ml by reverse

*Table 1.* Description of lakes sampled

| Name | Area (ha) | Depth (m) | Description of lake/manipulation |
|---|---|---|---|
| Toolik | 150 | 25 | Ultra-oligotrophic reference lake. |
| N - 2 | 2 | 7 | Divided lake: treatment side fertilized a 5× natural loading, 1985–1990. Highly oligotrophic prior to fertilization. |
| N - 1 | 4 | 12 | Whole lake fertilized at 4× natural loading, 1990–1994. Highly oligotrophic prior to fertilization. |

flow filtration through a 20 $\mu$m mesh net. Glutaraldehyde was added to a final concentration of 1% for fixation. For enumeration, 5–20 ml of the concentrated sample was stained with rose bengal for 10–15 min (Baldock, 1986), followed by observation under a light microscope at 100–400× magnification for identification and enumeration. Biomass (as $\mu$g carbon $l^{-1}$) was derived from literature values (cf. Rublee, 1992)

## Results

Microplankton biomass in Toolik Lake varied from 3 to 9 $\mu$g C $l^{-1}$ from 1989 to 1999 (Fig. 1). Rotifers averaged 15% (range 7–22%) of the biomass of the microplankton over the 11 year sampling period. Both Lake N-1 and the fertilized side of Lake N-2 demonstrated a clear response to the nutrient amendment. In Lake N-1, microplankton mean biomass increased five-fold during fertilization from a mean value of 11 $\mu$g C $l^{-1}$ to 50 $\mu$g C $l^{-1}$(Fig. 1). Microplankton populations shifted from a mixed assemblage to one dominated by the rotifer *Conochilus unicornis* during the first- year of fertilization, to peritrich protozoans during the second year of fertilization, the rotifers *Synchaeta* and *Polyarthra* during the third year of fertilization, and finally to a dominance of *Trichocerca* species in the fourth and fifth years (for details see Rublee & Bettez, 1995). The biomass of microplankton returned to near prefertilization levels in the first year following fertilization. Although biomass of rotifers was higher during years of fertilization, the proportion of microplankton biomass comprised by rotifers was similar during fertilized (average 30%, range 3–71%) vs non-fertilized years (average 26%, range 3–65%).

The microplankton community in Lake N-2, a divided lake (reference and treatment sides) was followed during 1989 and 1990, the last 2 years of fertilization, and for 9 additional years (Fig. 1). The average biomass of the microplankton in Lake N-2 during the years of fertilization was 20 $\mu$g C $l^{-1}$, about twice the average value found in the control side of Lake N-2. The microplankton community was a mixed assemblage of ciliate and peritrich protozoans and rotifers. No individual taxon comprised more than about 20% of the biomass of the community, and the rotifer assemblage was similar to that in unfertilized control lakes, mostly composed of *Keratella cochlearis, K. quadrata, Kellicotia longispina* and *Polyarthra* and *Conochilus* species. After fertilization ceased, biomass in the treatment side of Lake N-2 returned to prefertilization levels after a one to two year lag (Fig. 1).

## Discussion

Despite annual variability shown in microplankton values in Toolik Lake, and in the reference side of Lake N-2, there were pronounced responses of the microplankton community to nutrient additions. Differences in the response and recovery of Lake N-1 and Lake N-2 included: (1) a much greater increase in biomass in Lake N-1 as compared with Lake N-2, despite a heavier nutrient loading regime in Lake N-2; (2) decrease in the diversity of microplankton in Lake N-1 to a community dominated by a single or few protozoan or rotifer taxa; and (3) slower return of Lake N-2 to prefertilization levels of microplankton biomass.

The differences in response of Lake N-1 and N-2 to fertilization are likely the result of the character of the bottom sediments. Soils in this region are characterized by high iron content, which strongly adsorbs phosphorus (Prentki et al., 1980; Kipphut, 1988;

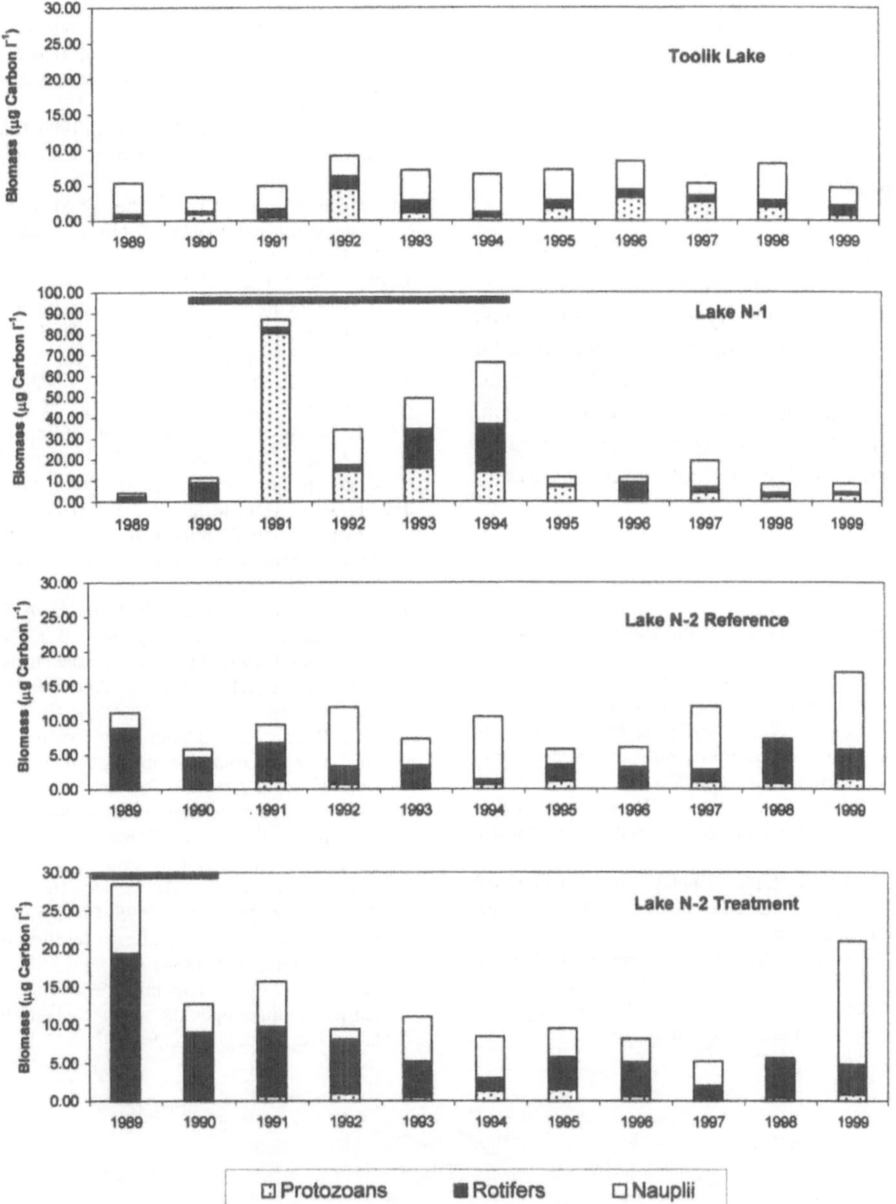


*Figure 1.* Microplankton biomass at 1 m depth in Arctic LTER lakes, 1989–1999. Horizontal bars indicate periods of experimental fertilization of lakes N-1 and N-2.

Cornwell & Kipphut, 1992; Sugai & Kipphut, 1992). Thus, in Lake N-2, which has a relatively thick layer of sediment, phosphorus was adsorbed onto sediments during the earlier years of fertilization, effectively reducing the loadings to the pelagic. However, by 1989 and for several years after fertilization ceased, there was a net flux of phosphorus out of sediments (cf. Sugai & Kipphut, 1992; Hobbie et al., In Press) which appears to have had the effect of extending the term of nutrient enrichment in this lake. This effect was minimal in Lake N-1, which has a rocky bottom and a very thin layer of sediment. Microplankton are affected indirectly by the supply of nutrients through phytoplankton growth, which provides food for microplankton.

In summary, relatively subtle differences in lake structure may contribute to differential responses of aquatic organisms to environmental change. Thus, designing studies to determine response to global warming or other perturbations must take into account any unique characteristics of study sites.

## Acknowledgements

This work was supported by NSF grants: OPP9096154, The Microbial Loop in Alaskan Arctic Aquatic Systems: Response to Whole System Manipulations; BSR-8702328, An LTER Program for the Alaskan Arctic; NSF-DEB-9211775, The Arctic LTER Project: Terrestrial and Freshwater Research on Ecological Controls. We also thank the Arctic LTER research community for sharing data, insight and discussion.

## References

Abelson, P., 1989. The arctic: a key to world climate. Science 243: 873.

Baldock, B. M., 1986. A method for enumerating protozoa in a variety of freshwater habitats. Microbiol. Ecol. 12: 187–191.

Bettez, N. D., P. A. Rublee, W. J. O'Brien, & M. C. Miller., In prep. Changes in abundance, composition and controls within the microbial loop of a fertilized arctic lake.

Cornwell, J. C. & G. W. Kipphut, 1992. Biogeochemistry of manganese- and iron-rich sediments in Toolik Lake, Alaska. Hydrobiologia 240: 45–60.

Hobbie, J. E., M. Bahr & P. A. Rublee, 1999. Controls on microbial food webs in oligotrophic arctic lakes. Archiv. Hydrobiol. Spec. Issues Advanc. Limnol. 54: 61–76.

Hobbie, J. E. B. J. Peterson, N. Bettez, L. Deegan, W. J. O'Brien, G. W. Kling, G. W. Kipphut, A. E. Hershey & B. Bowden, 1999. Impact of global change on the biogeochemistry and ecology of Arctic freshwater system. Polar Research 18: 207–214.

Kipphut, G. W., 1988. Sediments and organic carbon cycling in an arctic lake. In Degens, E. T. et al. (eds), Transport of Carbon and Minerals in Major World Rivers, Lakes and Estuaries. Part 5. Hamburg, SCOPE/UNEP: 129–135.

Lachenbruch, A. H. & B. V. Marshall, 1986. Changing climate: geothermal evidence from permafrost in the Alaskan arctic. Science 234: 689–696.

McDonald, M. E., A. E. Hershey & M. C. Miller, 1996. Global warming impacts on lake trout in arctic lakes. Limnol. Oceanogr. 41: 1102–1108.

O'Brien, W. J. (ed.), 1992. Toolik Lake - Ecology of an Aquatic Ecosystem in Arctic Alaska. Developments in Hydrobiology 78. Kluwer Academic Publishers, Dordrecht: 276 pp. Reprinted from Hydrobiologia 240.

O'Brien, W. J., M. Bahr, A. Hershey, J.Hobbie, G. W. Kipphut, G. W. Kling, H. Kling, M. McDonald, M. C. Miller, P. Rublee & J. R. Vestal, 1997. The limnology of Toolik Lake. In Milner, A. M. & M. W. Oswood (eds), Freshwaters of Alaska: Ecological Synthesis. Springer-Verlag, New York: 61–106.

Prentki, R. T., M. C. Miller, R. J. Barsdate, V. Alexander, J. Kelly & P. Coyne, 1980. Chemistry. In, Hobbie, J. E. (ed.), Limnology of Tundra Ponds. Dowden, Hutchinson & Ross, Stroudsburg, PA, U.S.A.: 76–178.

Rouse, W. R., M. V. Douglas, R. E. Hecky, A. E. Herhsey, G. Kling, L. Lesack, P. Marsh, M. McDonald, B. J. Nicholson, H. T. Roulet & J. P. Smol, 1997. Effects of climate change on the freshwaters of arctic and subarctic North America. Hydrol. Processes 11: 873–902.

Rublee, P. A., 1992. Community structure and bottom-up regulation of heterotrophic microplankton in arctic LTER lakes. Hydrobiologia 240: 133–141.

Rublee, P. A., 1998. Rotifers in arctic North America with particular reference to their role in microplankton community structure and response to ecosystem perturbations in Alaskan Arctic LTER lakes. Hydrobiologia 387/388: 153–160.

Rublee, P. A. & N. Bettez, 1995. Change of microplankton community structure in response to fertilization of an arctic lake. Hydrobiologia 312: 183–190.

Sugai, S. F. & G. W. Kipphut, 1992. The influence of light and nutrient addition upon the sediment chemistry of iron in an arctic lake. Hydrobiologia 240: 91–101.

*Hydrobiologia* **446/447**: 233–246, 2001.
L. Sanoamuang, H. Segers, R.J. Shiel & R.D. Gulati (eds), Rotifera IX.



# Zoogeography of the Southeast Asian Rotifera

Hendrik Segers
*Laboratory of Animal Ecology, Zoogeography and Nature Conservation, Dept. of Biology, Ghent University, K.L. Ledeganckstraat 35, B-9000 Gent, Belgium*
*E-mail: Hendrik.Segers@rug.ac.be*

*Key words:* Rotifera, zoogeography, diversity, Southeast Asia

## Abstract

The distribution and taxonomic composition of Rotifera in Southeast Asia is reviewed. For some countries, records are poor: Brunei, Cambodia and Laos are almost *terra incognita* for rotifers (<10 taxa recorded), while the Thai rotifer fauna is the best documented (ca. 310 taxa on record). However, analysis of the available data is impeded by fuzzy taxonomy and the questionable reliability of many records. Most studies focus on the pelagic or littoral of freshwater habitats. Other habitats are largely ignored. Similarly, few studies deal comprehensively with illoricate Monogononta, sessile Flosculariacea and Collothecacea and, especially, Bdelloidea. The genera *Lecane, Brachionus* and *Trichocerca* are the best represented, with littoral taxa predominant. Fisheries-related studies dealing with highly productive pelagic environments tend to over report the contribution of *Brachionus*. Most taxa are thermophilic character, exemplified by the dominance of tropic-centred *Lecane* and *Brachionus.* Some cold-water taxa have been recorded, but the relative climatological homogeneity of the region and low number of studies on high-altitude environments prevent the discrimination of clear latitudinal or altitudinal variation in the distribution of rotifers within Southeast Asia. The majority of Southeast Asian rotifers are widely distributed, including true cosmopolites and thermophilic taxa. There are several local or Oriental endemic Rotifera, mostly *Lecane.* The American *Brachionus havanaensis* and *Keratella americana* appear to have been introduced to the region. The taxonomy of some Rotifera described from the region is commented upon; *Brachionus murphyi* Sudzuki is recognised as senior synonym of *B. niwati* Sanoamuang et al. (syn. nov.). Some cases of geographical and/or ecological vicariant species-pairs are suggested. The Southeast Asian rotifer fauna contains a sizeable fraction of taxa occurring in the tropical regions of the Old World, most of which also occur in tropical Australia or the Austro-Malayan region. A tropical Australasian faunal component is present, but consists of few taxa only. Hence, affinities between the rotifer fauna of the Ethiopian, Oriental and tropical Australian and Austro-Malayan regions are supported, rather than an affinity between the Indo-Asian or Indo-Malaysian and tropical Australian fauna.

## Introduction

To date, little is known of the zoogeography of Southeast Asian Rotifera. There are quite a few publications reporting rotifers, but the accumulated data still constitute only a fragmentary record. The first reports included sketchy reports on collections from the Philippines (Semper, 1875: description of *Trochosphaera aequatorialis*), and Java (Indonesia: Weber, 1907; Van Oye, 1922a–c, 1924; Slonimski, 1925). An interesting early contribution is by Heinis (1928), who reported on the recovery of the moss-dwelling microfauna from Krakatau after the massive volcanic explosion of 1883. The publication of Hauer's (1937, 1938) "Die Rotatorien von Sumatra, Java und Bali" constituted a milestone in the development of the scientific record of Southeast Asian Rotifera.

The biogeography of Southeast Asia is of special interest because of the particularly complex geological and biological history of the region. Mainland Southeast Asia joined with Eurasia in the Triassic, but the movements and collisions of its lithospheric plates, and the northwardly movement of the Australian continent, result in ongoing processes of mountain formation. The southern islands of Southeast Asia have a diverse geological origin, which can be illustrated by the case of Sulawesi (Celebes): its northern and western parts were part of Borneo until they rifted away in the Eocene. In contrast, its eastern part was originally a fragment of Australasia, and joined the rest of the island only in the early Miocene (Moss & Wilson, 1998).

During the Pleistocene glaciations, the continental shelf of Southeast Asia was repeatedly exposed and submerged due to sea level fluctuations, with amplitudes of up to ca. 120 m (Fairbanks, 1989; Gallup et al., 1994). Consequently, the area of different habitat types has gone through important fluctuations, with possible fragmentation and subsequent reunification (Fig. 1; Brown & Lomolino, 1998; Cox & Moore, 1999). The present-day climate of the region is predominantly tropical humid (Köppen system). A short dry winter season occurs in the extreme south (part of Java, Lesser Sunda Islands) and the north (north, and part of south Thailand). The northern and eastern mountainous regions of Laos and northern Vietnam have a warm humid, subtropical climate. The coolest part is the central mountain range of Borneo.

A puzzling aspect of the region's biogeography relates to the transitional flora and fauna of the series of islands between the limits of the Southeast Asian (more or less corresponding with Wallace's line, see Fig. 1) and Australian (idem, Lydekker's line) continental shelves, a region sometimes named 'Wallacea'. It is not surprising that biogeographers devoted much effort to study the extent to which different organisms have succeeded in expanding their range across the islands in either direction. Hauer (1941) was the first rotifer researcher to acknowledge this, by publishing on the Rotifera of the "Zwischengebiet Wallacea".

Taking advantage of the IXth International Rotifer Symposium, the first to be held in a Southeast Asian country, the distribution of Rotifera in Southeast Asia is reviewed. The objectives of this review are to give an account of the available information on the distribution of rotifers in the region, to formulate general characteristics of the Southeast Asian rotifer fauna, if possible, and to identify the major gaps in our knowledge. For practical reasons, the territories of Brunei, Cambodia, Indonesia, Laos, Malaysia, Myanmar, Singapore, Thailand and Vietnam will be considered.

## The Rotifera of Southeast Asia: the state of the art

### *Restrictions of the data set*

The credibility of any application or analysis of data collected from different literature sources is inevitably restricted by the potential or actual introduction of biased, dated or simply erroneous entries. This is particularly relevant in the context of rotifer zoogeography, as such errors are inferred to be at least partly responsible for the alleged potential cosmopolitanism of all Rotifera (see Ruttner-Kolisko In Pejler, 1977b; Dumont, 1980; Dumont, 1983; Koste & Shiel, 1989; Segers, 1996; 1999). The identity of many records is questionable due to taxonomic controversies or changes in perception of the names used. This is especially so for names included in older works. Examples are *Brachionus plicatilis* (Müller), *Filinia terminalis* (Plate) and *Keratella valga* (Ehrenberg), which are frequently recorded from the region (see De Ridder & Segers, 1997). These taxa were previously considered to refer to eurytopic, cosmopolitan taxa, a view that was negated recently (Berzins, 1955; Pejler, 1977b; Fu et al., 1991a, b; Gómez et al., 1995; Segers, 1995b; Segers et al., 1996). It is unlikely that the cold-water nominate species are common here, and many of the records probably concern their respective warm-water relatives *B. rotundiformis* Tschugunoff, *F. novaezaelandiae* Shiel & Sanoamuang and *K. tropica* (Apstein), or related taxa. Judging from the relevant drawing, this is the case for Shirota's (1966) record of *B. plicatilis*. Such misidentified illustrated records are not uncommon in literature, as has been shown for Lecanidae (Segers, 1995a, 1996). This leads to doubt of the reliability of non-illustrated records. Fernando & Zankai (1981) mention the occurrence in their samples of several probably undescribed taxa, and such undoubtedly exist. In addition, intraspecific morphological variants of Brachionidae (*Brachionus, Keratella*) are sometimes treated as taxonomically significant entities, thereby confusing and artificially increasing the record. The situation is even worse for Bdelloidea, for which the requirement to examine active specimens, the absence of English identification keys and taxonomic specialists present major objections to the study of the organisms. Apart from this, the usual *caveat*, that present-day taxonomy may not have sufficient resolution to distinguish between morphologically similar yet distinct species of Rotifera, applies (e.g. Pejler, 1977a, b; Dumont, 1980; Koste & Shiel, 1989; Snell, 1989; Serra et al., 1997).

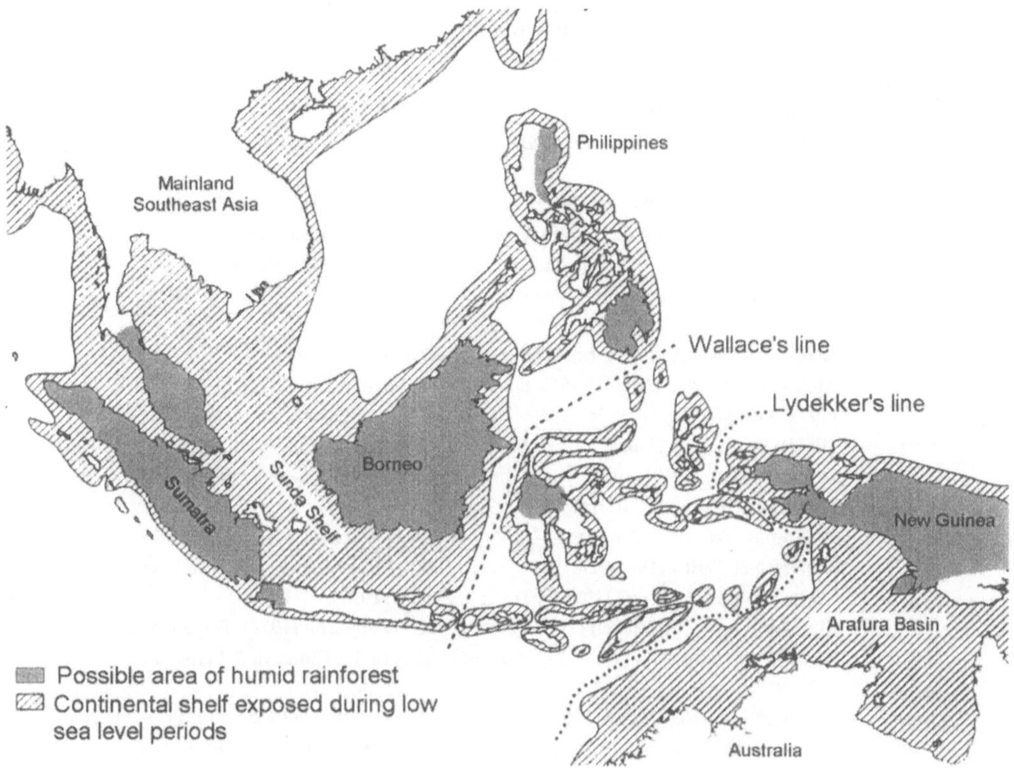


*Figure 1.* Change in sea level and possible area of humid rainforest during the last glacial maximum (ca. 20 000 years B.P.). Wallace's line separates regions with Australian and Asian fauna, Lydekker's line is the western boundary of strictly Australian fauna. (After Tallis, 1991; Brown & Lomolino, 1998; Cox & Moore, 1999).

The flaws of the available taxonomic apparatus, and low reliability of rotifer records are generally recognised as major stumbling blocks to the interpretation and comparison of published data on rotifers (e.g. Pejler, 1977a, b; Dumont, 1983; Ruttner-Kolisko, 1989, 1991; Segers, 1996; Serra et al., 1997). They make repeated pleas for more, and better, taxonomic research and education (Dumont, 1980; Dumont & Tundisi, 1984; Koste & Shiel, 1989; Nogrady et al., 1993; Segers, 1999). As a consequence, literature data need to be interpreted with great caution, similar to the findings for other freshwater micro-organisms and Crustacea (e.g. algae: Round, 1981; Anomopoda: Frey, 1986, 1987).

*Monogononta*

A summary of the literature dealing with monogonont rotifers from Southeast Asia is given in Table 1. This review remains incomplete, as records included in formally unpublished works (theses, reports) or local journals with limited distribution are known to exist, but could not be traced.

A striking aspect of the record of monogononts of several Southeast Asian countries or regions is its poverty. Less than 10 taxa are recorded from Brunei, Cambodia and Laos, which only stresses the dearth of information. The record from most other political or geographical units relies on one or a few major contributions only. This holds for Java and Sumatra (Indonesia: Hauer, 1937, 1938), Borneo (Indonesia: Koste, 1988), Myanmar (Burma: Koste, 1990), the Philippines (Mamaril & Fernando, 1978) and Vietnam (Shirota, 1966). Only the record from Malaysia, Singapore and Thailand rests on a richer literature, and reach totals of up to 220 taxa. However, such records from a region endowed with numerous and diverse freshwater habitats can only be considered poor, taking into account that extensive single-lake inventories can yield 200+ to 250+ rotifer taxa (Dumont & Segers, 1996). It is, therefore, unlikely that the inventory of any political or geographical unit in Southeast Asia is anywhere near complete. Only the Thai rotifer record comes near that of the more intensively studied, but climatically and geographically more heterogeneous

*Table 1.* Publications on monogonont rotifers from Southeast Asia

**Political units:*

- **Brunei** (2): Segers (1994).
- **Cambodia** (<10): Berzins (1955), Berzins (1962), Mizuno & Mori (1970), Berzins (1973).
- **Indonesia** (ca. 200): Weber (1907), Van Oye (1922a–c, 1924), Slonimski (1925), Heinis (1928), Hauer (1937, 1938), Hauer (1941), Green & Oey Biauw Lau (1974), Green et al. (1976), Green et al. (1978), Berzins (1982), Koste (1988), Green & Kramadibrata (1988), Fu et al. (1991a, b), Kutikova & Fernando (1995), Segers (1995a), Segers (1997).
- **Laos** (8): Heckman (1974), Koste & Shiel (1987), Segers (1995a).
- **Malaysia** (ca. 220): Ueno (1966*), Dunn (1970), Mizuno & Mori (1970), Lee (1973*) De Maeseneer (1976), Lai & Chua (1976), Karunakaran & Johnson (1978), Fernando & Zankai (1981), Sudzuki (1989), Ali (1990), Fu et al. (1991a, b: *B. plicatilis*), Green (1995), Kutikova & Fernando (1995), Segers (1995a).
- **Myanmar** (Burma)(ca. 100): Koste (1990).
- **Philippines** (ca. 115): Semper (1875), Lewis (1973), Mamaril & Fernando (1978), Petersen & Carlos (1984), Davis & Green (1990), Fu et al. (1991a, b), Kutikova & Fernando (1995), Segers (1995a); Tuyor & Segers (1999).
- **Singapore** (ca. 170): Russell (1958), Karunakaran & Johnson (1978), Fernando & Zankai (1981), Sudzuki (1989, 1991), Fu et al. (1991a, b), Kutikova & Fernando (1995).
- **Thailand** (ca. 310): Weber (1907), Mizuno & Mori (1970), De Ridder (1971), Koste (1975), Heckman (1979), Boonsom (1984a, b), Fu et al. (1991a, b), Segers & Sanoamuang (1994), Kutikova & Fernando (1995), Sanoamuang et al. (1995), Segers (1995a), Sanoamuang (1996), Sanoamuang & Segers (1997), Segers & Pholpunthin (1997), Pholpunthin & Chittapun (1998), Sanoamuang (1998), Chittapun et al. (1999), Sanoamuang & Savatenalinton (1999, present volume).
- **Vietnam** (80): Weber (1907), Shirota (1966), Segers (1995a).

**Indonesian and Malaysian islands:*

- **Bali** (27): Hauer (1937, 1938), Green et al. (1978), Segers (1995a).
- **Celebes** (<10): Hauer (1941), Berzins (1982).
- **Borneo** (ca. 150): Koste (1988), (Fernando & Zankai, 1981[a]), Segers (1994).
- **Flores** (1): Green et al. (1988).
- **Java** (ca. 110): Weber (1907), Van Oye (1922a-c, 1924), Slonimski (1925), Heinis (1928), Hauer (1937, 1938), Green et al. (1976), Green & Oey Biauw Lau (1974), Segers (1995a), Segers (1997).
- **Sumatra** (ca. 110): Hauer (1937, 1938), Berzins (1982).

Between brackets: approximate number of taxa on record.
[a] Fernado & Zankai (1981) report on rotifers from West and East Malaysia, but make no distinction between records from the Malay peninsula and the Bornean provinces Sarawak and Sabah.

Indian subcontinent (330 taxa: see Sharma, 1998). Sudzuki (1989) considered the Thai rotifer fauna as one of the poorly documented, which testifies to recent progress.

Scrutiny of the available literature indicates that the majority of studies dealing with Southeast Asian rotifers focussed on the pelagic or littoral of lowland freshwater environments. Only Hauer (1938) and Green et al. (1978) published on high-altitude water bodies. Marine and inland or coastal brackish waters have been investigated for rotifers by few researchers only (Shirota, 1966; Green, 1995), although there are aquaculture-related studies that involve the brackish-water *Brachionus plicatilis* (e.g. Fu et al., 1991a, b). Interstitial rotifers have never been looked at. As for taxonomic groups, there are relatively few records of the speciose but difficult groups of illoricate Monogononta (e.g. Lindiidae, Notommatidae, Proalidae), and only Koste (1975, 1990) provides more than

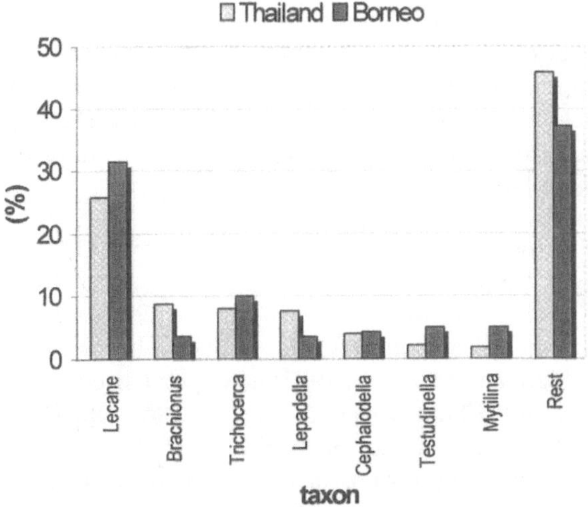


*Figure 2.* Procentual contribution of taxa to total species diversity, for Thailand (with short, dry winter season) and Borneo (humid tropical). Thailand: data from all sources (see Table 1), Borneo: data from Koste (1988) and Segers (1994).

an occasional record of sessile Flosculariacea and/or Collothecacea.

### *Bdelloidea*

Records of bdelloid rotifers from South East Asia are particularly rare (Table 2). The first author to record bdelloids from the region was Hauer (1937, 1938), who listed 9 taxa from Sumatra and 8 from Java, respectively. He also described five new species, including one new genus (*Synkentronia*, with three species). Only one of his taxa, *Henoceros caudatus* Hauer, 1937, is generally considered valid, the others were contracted hence unrecognisable animals. Bartoš (1963) published the only extensive study on Bdelloidea from Java, Indonesia. He recorded 20 taxa, four of which are named as new. However, only *Habrotrocha elliptica* is adequately described, his other new names are excluded from zoological nomenclature as they are based on the case inhabited by the bdelloids only, the animals themselves are unknown (see ICZN, 1999: art. 1.3.6). Additional reports on Bdelloidea from Southeast Asia are from Malaysia by Karunakaran & Johnson (1978: 6 taxa), Fernando & Zankai (1981: 2 taxa), Shirota (1966: 4 taxa) and Sudzuki (1989: 26 taxa). Koste (1975, 1988, 1990) lists 6, 5 and 5 species from Thailand, East Kalimantan (Indonesian Borneo) and Myanmar, respectively.

## *The distribution of Rotifera in Southeast Asia*

### *Diversity and ecology*

A comparison of the diversity of different rotifer genera, for two regions of which the available information is recent (Thailand and Borneo, see Fig. 2), reveals that *Lecane*, and to a lesser degree *Brachionus* and *Trichocerca*, contribute most to rotifer diversity in Southeast Asia. That these genera constitute the major component of the tropical freshwater rotifer fauna was noted by several authors (e.g. see Fernando & Zankai, 1981; Dussart et al., 1984; Segers & Dumont, 1995; Vazquez et al., 1998, Sharma & Sharma, present volume; to cite a few). Of the most diverse genera, all but one (*Brachionus*) are predominantly littoral-periphytonic taxa (*Lecane*, *Trichocerca, Lepadella, Cephalodella, Testudinella* and *Mytilina*). This illustrates the well-documented fact that such habitats are inhabited by a diverse rotifer taxocoenosis (e.g. Segers & De Meester, 1994; Sanoamuang et al., 1995), resulting from the higher habitat heterogeneity, and, consequently, higher microhabitat diversity of these environments (see Hassler & Jones, 1949; Pennak, 1966). Small differences in the relative contribution of different genera (*Lecane* relatively less, *Brachionus* relatively more represented in the Thai than in the Borneo rotifer fauna) may be a bias resulting from the relatively narrow range of habitats included in Koste's (1988) study (12 samples from 8 localities at short distance from one another). On the other hand, some authors (e.g. Fernando, 1980a, b) note a dominance of Brachionidae rather than Lecanidae. Several, not mutually exclusive factors connected to ecological differences between *Brachionus* and *Lecane* can be inferred to account for this. While most *Brachionus* inhabit the pelagic of productive, hard waters, *Lecane* is abundant in the littoral or benthos of soft, acid waters (Harring & Myers, 1928; Dussart et al., 1984). Inventories related to studies on freshwater resources for fisheries will tend to focus on the first-mentioned habitat type. In addition, pelagic habitats are more readily sampled and studied than littoral or benthic ones because of the easier methodology, less complex ecology and more accessible taxonomy of the rotifer fauna encountered.

### *Latitudinal and altitudinal variation*

Latitudinal variation in the distribution of rotifers, directly or indirectly induced by climatological factors, is well documented in Rotifera (Green, 1972, 1994; Pejler, 1977b; De Ridder, 1981; Dumont, 1983; Segers, 1996). Southeast Asia is a climatologically rather homogeneous tropical region, so it can be expected that a major component of its rotifer fauna consists of thermophiles. The predominance of the tropic-centred genera *Lecane* and *Brachionus*, and relative scarcity of representatives of speciose cold-water genera like *Cephalodella* and *Synchaeta*, confirms this. For *Lecane* and *Brachionus*, two taxa for which species' areas are relatively well-known and reliable (Pejler, 1977b; Dumont, 1983; Segers, 1995a, 1996), the relative importance of distributional groups is presented in Table 3. Apart from important fractions of geographically restricted taxa (see further), several groups of widespread animals can be recognised. Both cosmopolitan and tropicopolitan taxa are important, but the relative abundance of pantropical (including all recognised pantropical *Lecane*) taxa is notable. Reliable records of cold-water *Lecane* or *Brachionus* are absent, although these can occur in high-altitude environments in the region. Unfortunately, only few studies of such habitats are available to date (e.g. Hauer, 1938; Green et al., 1978). Shirota (1966) records and illustrates several cold-water taxa, including *Kellicottia bostoniensis* (Rousselet) and *K. longispina* (Kellicott), *Keratella hiemalis* Carlin, *K. quadrata* (Müller), *K. serrulata* (Ehrenberg), *K. valga* and *Notholca acuminata* (Ehrenberg) from lowland localities in South Vietnam. He also records and illustrates the fresh- and brackish water cold-stenotherm*S. tremula* (Müller) from marine plankton. While some of these records are clearly misidentified (Shirota's Figure of *K. bostoniensis* is *K. longispina*, that of *K. valga* is *K. tropica*), and others cannot be confirmed as the relevant illustrations appear redrawn from literature sources, some are taxa that can hardly be misidentified (e.g. *K. longispina*).

*Endemic and Oriental rotifers*

The Southeast Asian rotifer fauna contains several locally endemic taxa (Fig. 3), including a genus, *Architestudinella* Berzins. The biogeographic relevance of the taxa that have been recorded only at the occasion of their description) is evidently questionable (*Architestudinella mekongensis* Berzins: Fig. 3a; *Brachionus schwoerbeli* Koste: Fig. 3c; *Cephalodella songkhlaensis* Segers & Pholpunthin: Fig. 3e; *Lecane isanensis* Sanoamuang & Savatenalinton; L. *junki* Koste, 1975: Fig. 3j; *L. spiniventris* Segers: Fig. 3n; *Wulfertia kindensis* Koste & Tobias: Fig. 3r, and both bdelloid taxa, see Table 2). However, some are particularly noteworthy, like the *Architestudinella*, and the Myanmar *Wulfertia kindensis*, considering its possible vicariance with the East African *W. kivuensis* De Smet (see De Smet, 1996). Examples of endemic taxa are the following:

1. *Brachionus murphyi* Sudzuki, 1989 (Fig. 3a, stat.nov.). An Eastern Oriental taxon recorded from Singapore (Sudzuki, 1989), Thailand (Sanoamuang et al., 1995: sub. *B. niwati* Sanoamuang et al., 1995: new synonym), Hainan, South China (Koste & Zhuge, 1998: sub. *B. niwati*). A comparison of the description and relevant figures of *Brachionus angularis murphyi* Sudzuki, (Sudzuki, 1989, 1991) and *B. niwati* revealed the synonymy of both. A status as subspecies of *B. angularis* as suggested by Sudzuki (1989) is refuted by the weak morphological resemblance of *B. murphyi* with *B. angularis*, close relationship of the taxon with *B. budapestinensis* (see Sanoamuang et al., 1995), and absence of any indication of subspecificity. Consequently, the species name *murphyi* is hereby elevated to species rank.
2. *Colurella sanoamuangae* Chittapun et al. (Fig. 3d). A Southern Thailand endemic (Songkla and Phuket provinces: Chittapun et al., 1999).
3. *L. minuta* Segers (Fig. 3k). An Indo-Malayan element, recorded from Borneo and Southern Thailand (Segers & Pholpunthin, 1997).
4. *L. segersi* Sanoamuang (Fig. 3l). A Thai endemic, described from Northeast Thailand (Sanoamuang & Segers, 1997). The species has recently been found in abundance in the psammon of a peat swamp in Phuket (Southern Thailand: Segers, unpublished). This psammophilic species is a close relative and possible ecological vicariant of the tropicopolitan and littoral *L. papuana*. This case is similar to that of the eurytopic cosmopolitan *L. closterocerca* and its close relative *L. fadeevi*, a central and East European psammon species.
5. *L. shieli* Segers & Sanoamuang (Fig. 3m). A Thai endemic, recorded from Northeast and South Thailand (Sanoamuang & Segers, 1997; Chittapun et al., 1999)
6. *L. solfatara* (Hauer)(Fig. 3o). Endemic to Java and Sumatra (Segers, 1997). This taxon is morphologically similar to, and a possible vicariant of the cosmopolitan *L. bifurca* (Bryce).

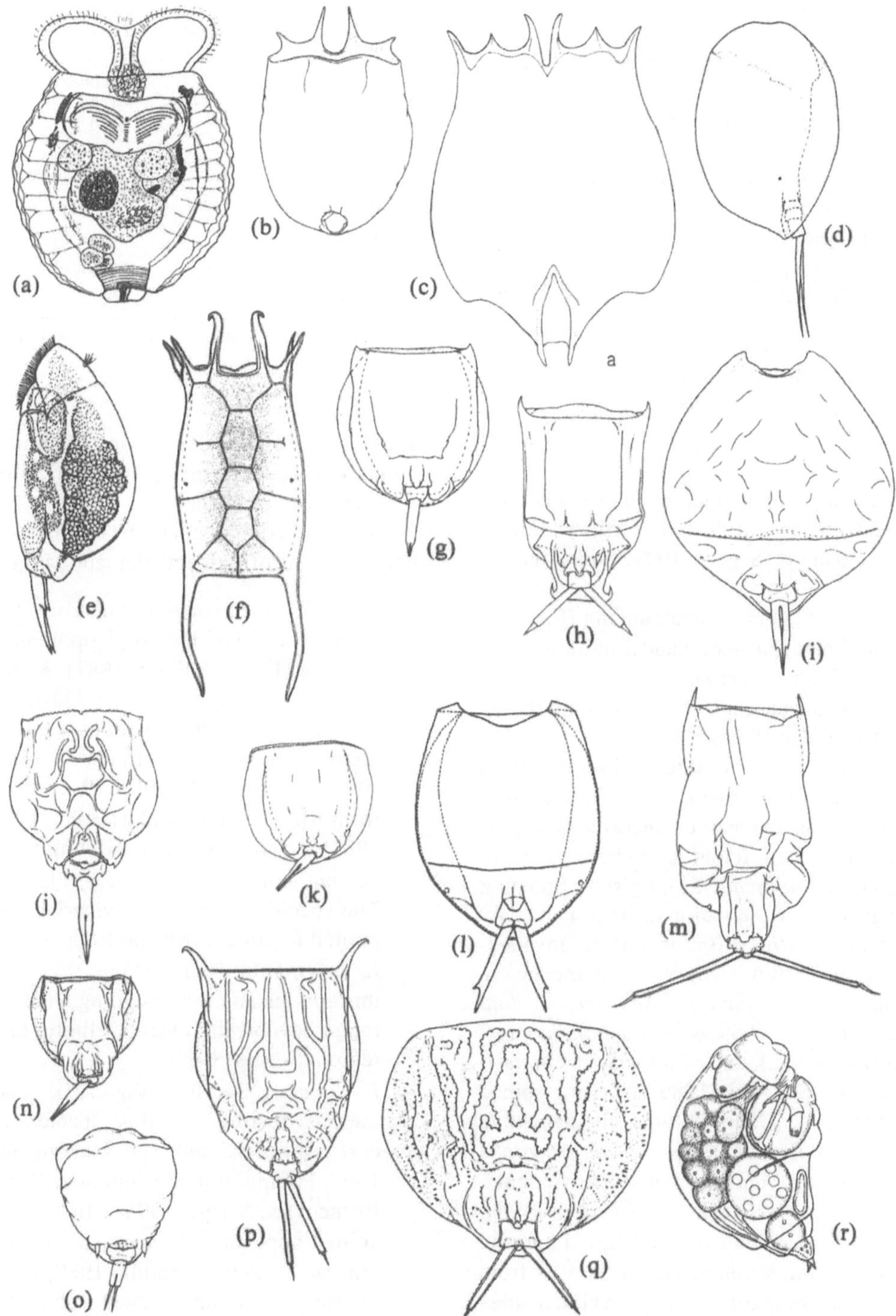


*Figure 3.* Local and Oriental endemic Southeast Asian Monogononta: (a): *Architestudinella mekongensis* Berzins, 1973; (b): *Brachionus murphyi* Sudzuki, 1989 (stat. nov.); (c): *B. schwoerbeli* Koste, 1988; (d): *Colurella sanoamuangae* Chittapun et al., 1999; (e): *Cephalodella songkhlaensis* Segers & Pholpunthin, 1997; (f) *Keratella edmondsoni* Ahlstrom, 1943; (g): *Lecane acanthinula* (Hauer, 1938); (h): *L. bifastigata* Hauer, 1938; (i): *L. blachei* Berzins, 1973; (j): *L. junki* Koste, 1975; (k): *L. minuta* Segers, 1994; (l): *L. segersi* Sanoamuang, 1996; (m): *L. shieli* Segers & Sanoamuang, 1994; (n): *L. spiniventris* Segers, 1994; (o): *L. solfatara* (Hauer, 1938); (p): *L. superaculeata* Sanoamuang & Segers, 1997; (q): *L. thailandensis* Segers & Sanoamuang, 1994; (r): *Wulfertia kindensis* Koste & Tobias, 1990. After Berzins (1973), Koste (1975, 1988), Sanoamuang (1996), Sanoamuang et al. (1995), Segers (1995a, 1997), Segers & Pholpunthin (1997), Segers & Sanoamuang (1994)(not to scale).

*Table 2.* Information on bdelloid rotifers from Southeast Asia

| * *Publications:* |
|---|
| - **Indonesia** (ca. 30): Hauer (1937, 1938), Bartoš (1963), Koste (1988) |
| - **Malaysia** (ca. 30): Karunakaran & Johnson (1978), Fernando & Zankai (1981), Sudzuki (1989) |
| - **Myanmar** (Burma: 5): Koste (1990) |
| - **Thailand** (6): Koste (1975) |
| - **Vietnam** (4): Shirota (1966) |
| * *Endemic taxa:* |
| *Henoceros caudatus* Hauer, 1937: Bali, Indonesia |
| *Habrotrocha elliptica* Bartoš, 1963: Java, Indonesia |

*Table 3.* Proportion of different distributional groups in the Southeast Asian rotifer fauna, for *Lecane* and *Brachionus*

| | *Lecane* | *Brachionus* |
|---|---|---|
| Oriental taxa[a] | 12 (18%) | 2 (8%) |
| Australasian taxa | 1 (1%) | 3 (12%) |
| Eastern hemisphere taxa[a] | 8 (12%) | 4 (15%) |
| Pantropical taxa[a] | 6 (9%) | 2 (8%) |
| Tropicopolitan taxa[a] | 23 (34%) | 5 (19%) |
| Cosmopolitan taxa | 18 (26%) | 10 (38%) |
| Total: | 68 | 26 |

[a]Groups considered to consist mainly or exclusively of thermophilic species.

7. *L. superaculeata* Sanoamuang & Segers (Fig. 3p). A local endemic in North and Northeast Thailand (Sanoamuang & Segers, 1997; Sanoamuang, 1998)
8. *L. thailandensis* Segers & Sanoamuang (Fig. 3q). Known from China, Northeast and Southern Thailand (Segers, 1995a; Sanoamuang & Segers, 1997). The species is a vicariant of the circumpolar *L. latissima* Yamamoto (Segers, 1996).

Some other taxa that were described rcently from Southeast Asia are not considered here, as their taxonomic identity is unclear or incorrectly reported, or they have been found elsewhere as well. The descriptions of *Macrochaetus longisetus* Sudzuki, 1991 (unrecognisable conservation artifact?) and *Mytilina mucronata sumatrana* Berzins, 1982 are inadequate and do not permit recognition of the taxon. Two more taxa are synonyms: *Anuraeopsis lata singapurensis* Sudzuki, 1989 is *A. coelata* (conservation artefact)(new syn.), and *Macrochaetus haueri-anus singapurensis* Sudzuki, 1989 is a long-spined *M. collinsi* (new syn.). The features distinguishing the subspecific taxa *Brachionus angularis orientalis* Sudzuki, 1989, *B. caudatus singapurensis* Sudzuki, 1989, *B. urceolaris semicircularis* Sudzuki, 1989, *Macrochaetus sericus coniunctus* Sudzuki, 1991 and *Testudinella tridentata insulana* Berzins, 1982 from the nominate taxon are taxonomically invalid, as they fit within the normal variability of the species (*Brachionus angularis orientalis, B. caudatus singapurensis, Testudinella tridentata insulana*), are based on a conservation artefact (*Brachionus urceolaris semicircularis*), or the misinterpretation of a feature and conservation artefacts (*Macrochaetus sericus coniunctus*). Finally, *Ptygura melicerta ctenoidea* Koste & Tobias, 1990, described from Myanmar, has been found in material from Australia and North Africa (Segers, unpublished).

A second component of the Southeast Asian rotifer fauna consists of Oriental elements (Figure 3):

1. *Keratella edmondsoni* Ahlstrom (Fig. 3f). A close relative of the tropicopolitan *K. procurva* (Thorpe)(Koste & Shiel, 1987), *K. edmondsoni* is known from India (Madras: Ahlstrom, 1943; Rajastan: Nayar, 1965; Northeast India: Sharma & Sharma, present volume) and Northeast Thailand (Sanoamuang et al., 1995). Records of this taxon from Africa and Australia (Pejler, 1977b; Dumont, 1983) are unverifiable and doubtful.
2. *Lecane acanthinula* (Hauer) (Fig. 3g): Oriental. The species is a possible vicariant of the cosmopolitan *L. furcata* (Murray)(Segers, 1996).
3. *L. bifastigata* Hauer (Fig. 3h): Oriental. Both this species and the foregoing have extended their range beyond the classical limits of the Oriental region (Segers, 1996).
4. *L. blachei* Berzins (Fig. 3I). There are numerous records of this species from Southeast Asia (Thailand: see Sanoamuang & Segers, 1997; Pholpunthin & Chittapun, 1998; Cambodia, Borneo: see Segers, 1995a). In India, however, the animal appears to be restricted to the Northeastern part (Calcutta: Sarma, 1988; Northeast India, Sharma & Sharma, present volume). Its distinctiveness makes it unlikely that the animal would pass unnoticed if encountered, so its absence from other parts of India may be factual. Hence, although the animal is Oriental, it is as yet unclear whether ecological or geographical factors account for its absence from the Western part of the Oriental region. *Lecane blachei* may be a vicariant of

the West African (Nigerian) *L. nwadiaroi* Segers (Segers, 1996).

The status of the Oriental *L. bulla* (Gosse) f. *diabolica* Hauer requires further study. Although commonly referred to as an Oriental endemic, *Brachionus donneri* Brehm (see Pejler, 1977; Dumont, 1983; Sharma, 1998) is pantropical (see De Ridder & Segers, 1997). The supposedly Oriental *Hexarthra insulana* Hauer, recently recorded by Green et al. (1978), Koste (1981) and Sudzuki (1989, 1991), was recognised as synonym of *H. intermedia* Wiszniewski by Hauer (1941).

Most of the Southeast Asian and Oriental endemic rotifers are *Lecane* species. The scarcity of endemic representatives of *Brachionus*, *Keratella* and *Lepadella* is remarkable, considering that these genera are notorious for their high rates of endemicity in Australia and South America (Dumont, 1983; Shiel & Koste, 1986). The areas of the taxa treated above illustrate a variety of distribution patterns, from local endemics to widespread Oriental taxa, and several species-pairs involving endemic taxa can be indicated in which geographical (*L. blachei-L. nwadiaroi*) or ecological vicariance (*L. papuana- L. segersi*), or both (*L. latissima -L. thailandensis*) can be hypothesised.

*Introduced species*

The literature on Southeast Asian rotifers contains several records of species which are known endemics of the New World. The majority of these are non-illustrated and may either represent misidentifications, sample contamination or true introductions. Reliable examples of introduced species appear to be the illustrated records of the American *Brachionus havanaensis* and *Keratella americana* (as *K. stipitata*) from Vietnam (Shirota, 1966). These introductions may result from the intense ship and aircraft traffic between South Vietnam and America during the 1960s. The first-mentioned species was also abundant in samples from Laguna de Bay and nearby localities in Luzon, the Philippines, in 1991 and 1993 (Segers & Dela Cruz Camacho, unpublished).

*Affinities of the Southeast Asian rotifer fauna*

A clear affinity between the tropical Australian (including Austro-Malayan) and Indo-Malayan rotifer fauna was postulated in a series of contributions on the Australian rotifer fauna (Koste, 1981; Shiel & Koste, 1986; Shiel & Williams, 1990; Shiel & Green, 1996). A contrasting hypothesis is that the Oriental rotifer fauna relates with the African and tropical Australian fauna (Segers, 1996). The tropical Australian and Austro-Malayan faunas are relatively poorly documented (Koste, 1981; Koste & Shiel, 1983; Tait et al., 1984; Shiel & Williams, 1990; Segers & De Meester, 1994), while more information is available on the Ethiopian fauna (Dussart et al., 1984; see De Ridder, 1986, 1987, 1991, 1994). Nevertheless, a comparison between the Australasian and Paleotropical components in the Southeast Asian rotifer fauna is revealing.

Seven taxa (Table 4) constitute the Australasian faunal component in the Southeast Asian rotifer fauna. Only three, probably four of the Australasian taxa (*Brachionus dichotomus* Shephard f. *reductus* Koste & Shiel, *Macrochaetus danneeli* Koste & Shiel, *Testudinella walkeri* Koste & Shiel and probably *Lecane batillifer* (Murray)) are Northern Australian, and can be considered representative elements of a tropical fauna element. These taxa present the only support for the affinity between the Indo-Asian or Indo-Malaysian and tropical Australian fauna. The geographic origin of the animals can not be traced at present. However, only the *reductus* variant of *B. dichotomus* occurs outside Australia. If the lower morphological variability reflects lower genetic variation in the Southeast Asian than in the Australian populations, then it can be hypothesized that these populations represent a recent expansion of the species' area, hinting at a possible Australian origin of the taxon.

Twenty-four taxa (Table 4) are restricted to the Eastern hemisphere, and all but two of these are warm-water animals. Some are even particularly common species (e.g. *Brachionus diversicornis* (Daday), *B. forficula* Wierzejski, *Keratella javana* Hauer, and *Lecane unguitata* (Fadeev)). Of the 22 (sub)tropical species, nine have not (yet) been recorded from Northern Australia or the Austro-Malayan region. Hence, a tropical Eastern hemisphere or Paleotropical component is notable. Support for the affinity between the rotifer fauna of the tropical regions of the Eastern hemisphere is stronger than for the affinity between the Southeast Asian and tropical Australian rotifer fauna. The Wallacea region appears to be no hindrance to their dispersal. It does, however, appear to be a barrier for the exchange of Southeast Asian and tropical Australian taxa (no endemic Austro-Malayan Rotifera are known to date), considering the limited number of extant Australasian taxa. So, although the small, resistant dormant stages and parthenogenetic

*Table 4.* List of Australian and Eastern Hemisphere rotifer species

• Australasian species:

*Brachionus dichotomus* Shephard, 1911 f. *reductus* Koste & Shiel, 1980: Australia, Northeast India, Laos, Thailand
*Brachionus kostei* Shiel, 1983: Australia, Papua New Guinea, Thailand, China
*Brachionus lyratus* Shephard, 1911: Australia, Thailand
*Lecane batillifer* (Murray, 1913): Australia, Northeast India, Thailand, The Philippines
*Macrochaetus danneeli* Koste & Shiel, 1983: Australia, South India, Thailand
*Testudinella walkeri* Koste & Shiel, 1980: Australia, Thailand
*Trichocerca orca* (Murray, 1913): New Zealand, Thailand

• Eastern hemisphere species:

*Asplanchna tropica* Koste & Tobias, 1987
*Brachionus africanus* Segers, 1994
†*Brachionus diversicornis* (Daday, 1883)
**Brachionus forficula* Wierzejski, 1891
**Dicranophorus halbachi* Koste, 1981
**Dipleuchlanis ornata* Segers, 1993
**Itura deridderae* Segers, 1993
**Itura symmetrica* Segers, Mbogo & Dumont, 1994
**Keratella javana* Hauer, 1937
*Lecane baimaii* Sanoamuang & Savatenalinton, 1999
**Lecane braumi* Koste, 1988
*Lecane eswari* Dhanapathi, 1976
**Lecane lateralis* Sharma, 1978
†*Lecane pumila* (Rousselet, 1906)
**Lecane serrata* (Hauer, 1933)
*Lecane simonneae* Segers, 1993
*Lecane stephensae* (Hutchinson, 1931)
**Lecane unguitata* (Fadeev, 1925)
**Lepadella discoidea* Segers, 1993
**Lepadella vandenbrandei* Gillard, 1952
*Scaridium grande* Segers, 1995
*Testudinella brevicaudata* Yamamoto, 1951
**Testudinella greeni* Koste, 1981
*Trichocerca tropis* Hauer, 1938

*(Sub)tropical species, included in the Northern Australian and/or Austro-Malayan fauna.
†Eurythermic species.

reproduction make rotifers superb passive dispersers (Gilbert, 1974; Pourriot & Snell, 1983; Ricci, 1987; Segers, 1996; but see Jenkins & Underwood, 1998), this ability appears not to be the same for all taxa (Dumont, 1983; Segers, 1996). The scarcity of data on rotifers from the Austro-Malayan region and North Australia impede a more profound analysis.

## Conclusions

Although fuzzy taxonomy and the questionable reliability of the available data impede the analysis of rotifer distributions in Southeast Asia, supporting the plea for more and better taxonomic research and education, some patterns emerge. Coverage of the region is heterogeneous, and generally incomplete. Countries like Brunei, Cambodia and Laos, or geographical entities like Sulawesi, are *terra incognita* as far as their rotifer fauna is concerned. Our knowledge on the Rotifera of those countries or geographical entities whose rotifer fauna has received some attention, is fragmentary at best. This even holds for the Thai rotifer fauna, which at present represents the best-documented rotifer fauna of Southeast Asia. Of habitat types, high-altitude water bodies, the psammon, athalassic or thalassic brackish waters and the marine environment are scarcely studied. Illoricate Monogononta, sessile Flosculariacea and Collothecacea and, especially, Bdelloidea are underreported.

The most diverse rotifer genera in the Southeast Asian fauna are *Lecane* and *Brachionus*, with several additional genera of littoral rotifer as sizeable components. The record is dominated by littoral taxa, a consequence of the littoral being a more fine-grained, diverse and complex environment than the pelagic. Occasional differences in taxon dominance reports are likely to result from differences in research focus: fisheries-related studies tend to focus on highly productive pelagic environments in which *Brachionus* abound, while biodiversity censuses and taxonomic studies are most rewarding on the highly diverse rotifer assemblages inhabiting acid, soft-water bodies.

The majority of rotifers recorded from Southeast Asia are widely distributed animals. Apart from true cosmopolitan taxa, the rotifer fauna mainly contains taxa with more or less pronounced thermophilic character, such as those belonging to tropic-centred genera like *Lecane* and *Brachionus*. The relative climatological homogeneity of the region and low number of studies on high-altitude environments prevent the discrimination of clear latitudinal or altitudinal variation in the distribution of rotifers, although some cold-water taxa have been recorded.

Many of the rotifers originally described from Southeast Asia have either been recognised as synonyms of widely distributed taxa, or found outside the region. However, there are several examples of local or Oriental endemic Rotifera, mostly belonging to the *Lecane*. A variety of distribution patterns is demonstrated by these taxa, some are geographical or ecological vicariants of sister taxa. The American *Brachionus havanaensis* and *Keratella americana* appear to have been introduced in the region.

The Southeast Asian rotifer fauna contains a relatively large number of taxa that occur in the tropical regions of the Old World and Notogaea. Eastern hemisphere taxa, most of which have also been recorded from tropical Australia or the Austro-Malayan region. A tropical Australasian fauna component is present, but consists of few taxa only. These observations indicate strong affinities between the rotifer fauna of the Ethiopian, Oriental and tropical Australian and Austro-Malayan regions, rather than affinity between the Indo-Asian or Indo-Malaysian and tropical Australian fauna, as far as rotifers are concerned.

## Acknowledgements

I wish to thank Prof. Ladda Wongrat for providing a copy of Shirota (1966), and the organizers of the VIIIth International rotifer symposium, for support to attend the IXth International rotifer symposium. Prof. J. Green, Dr L. Sanoamuang and Dr R.J. Shiel are thanked for reviewing the manuscript.

## References

Ahlstrom, E. H., 1943. A revision of the rotatorian genus *Keratella* with descriptions of three new species and five new varieties. Bull. am. Mus. Nat. Hist. 80: 411–457.

Ali, A. B., 1990. Seasonal dynamics of microcrustacean and rotifer communities in Malaysian ricefields used for fish farming. Hydrobiologia 206: 139–148.

Bartoš, E., 1963. Die Bdelloideen der Moosproben aus China und Java. Vestník csl. zool. spol. 27: 31–42.

Berzins, B., 1955. Taxonomie und Verbreitung von *Keratella valga* und verwandten Formen. Arkiv Zool. Ser. 2 Bd 8 (7): 549–559.

Berzins, B., 1962. Revision der Gattung *Anuraeopsis* Lauterborn (Rot.). K. fysiogr. Sällsk. Lund. Förh., 32: 33–47.

Berzins, B., 1973. Some rotifers from Cambodia. Hydrobiologia 41: 453–459.

Berzins, B., 1982. Short notes on Rotatoria. Inst. of Limnology, Univ. of Lund, Lund: 8 pp.

Boonsom, J., 1984a. Zooplankton feeding in the fish *Trichogaster pectoralis* Regan. Hydrobiologia 113: 223–229.

Boonsom, J., 1984b. The freshwater zooplankton of Thailand (Rotifera and Crustacea). Hydrobiologia 113: 223–229.

Brown, J. H. & M. V. Lomolino, 1998. Biogeography. 2nd edn. Sinauer Ass., Sunderland, MA: 691 pp.

Chittapun, C., P. Pholpunthin & H. Segers, 1999. Rotifera from peat-swamps in Phuket province, Thailand, with the description of a new *Colurella* Bory de St. Vincent Internat. Rev. Hydrobiol. 84: 587–593.

Cox, C. B. & P. D. Moore, 1999. Biogeography. An ecological and evolutionary approach. 6th edn. Blackwell Science Ltd., Oxford: 298 pp.

Davis, J. & J. Green, 1990. A preliminary study of Lake Manguao, Palawan, the Philippines. Asian Wetland Bureau, Publication No. 59: 1–39.

De Maeseneer, J., 1976. *Brachionus* and *Keratella* in fish ponds at Malacca. Meded. Fakult. Landbouw 410: 1–10.

De Ridder, M., 1971. Raderdieren uit het verre Oosten. Biol. Jb. Dodonaea 39: 361–391.

De Ridder, M., 1981. Some considerations on the geographical distribution of rotifers. Hydrobiologia 85: 209–225.

De Ridder, M., 1986. Annotated checklist of non-marine Rotifera from African inland waters. Zool. Doc., KMMA, Tervuren 21: 123 pp.

De Ridder, M., 1987. Distribution of rotifers in African fresh and inland saline waters. Hydrobiologia 147: 9–14.

De Ridder, M., 1991. Additions to the "Annotated checklist of non-marine rotifers from African inland waters". Rev. Hydrobiol. trop. 24(1): 25–46.

De Ridder, M., 1994. Additions II to the "Annotated checklist of non-marine rotifers from African inland waters". Biol. Jb. Dodonaea 61: 99–153.

De Ridder, M. & H. Segers, 1997. Rotifera Monogononta in six zoogeographical regions after publications between 1960 and 1992. In Van Goethem, J. (ed.), Studiedocumenten van het Koninklijk Belgisch Instituut voor Natuurwetenschappen: 481 pp.

De Smet, W. H., 1996. The Proalidae (Monogononta). Rotifera 4. In Nogrady, T. (ed.), Guides to the Identification of the Microinvertebrates of the Continental Waters of the World 9: 102 pp.

Dumont, H. J., 1980. Workshop on taxonomy and biogeography. Hydrobiologia 73: 205–206.

Dumont, H. J., 1983. Biogeography of rotifers. Hydrobiologia 104: 19–30.

Dumont, H. J. & J. G. Tundisi, 1984. Epilogue: the future of tropical zooplankton studies. Hydrobiologia 113: 331–333.

Dumont, H. J. & H. Segers, 1996. Estimating lacustrine zooplankton species richness and complementarity. Hydrobiologia 341: 125–132.

Dunn, I. G., 1970. Recovery of a tropical pond zooplankton community after destruction of algal bloom. Limnol. Oceanogr. 15: 373–379.

Dussart, B. H., C. H. Fernando, T. Matsumura-Tundisi & R. J. Shiel, 1984. A review of the systematics, distribution and ecology of tropical freshwater zooplankton. Hydrobiologia 113: 77–91.

Fairbanks, R. G., 1989. A 17 000-year glacio-eustatic sea level record: influence of glacial melting rates on the Younger Dryas event and deep-ocean circulation. Nature 342: 637–642.

Fernando, C. H., 1980a. The freshwater Zooplankton of Sri Lanka. Int. Rev. ges. Hydrobiol. 65: 85–125.

Fernando, C. H., 1980b. The Species and Size Composition of Tropical Freshwater Zooplankton with Special Reference to the Oriental Region (South East Asia). Int. Rev. ges. Hydrobiol. 65: 411–426.

Fernando, C. H. & N. P. Zankai, 1981. The Rotifera of Malaysia and Singapore with remarks on some species. Hydrobiologia 78: 205–219.

Frey, D. G., 1986. The non-cosmopolitanisms of chydorid Cladocera: Implications for biogeography and evolution. In Heck, K. L. & H. R. Gore (eds), Crustacean Issues 4. Crustacean Biogeography. Rotterdam: 237–256.

Frey, D. G., 1987. The taxonomy and biogeography of the Cladocera. Hydrobiologia 145: 5–17.

Fu, Y., K. Hirayama & Y. Natsukari, 1991a. Morphological differences between 2 types of the rotifer *Brachionus plicatilis* O.F. Müller. J. exp. mar. Biol. Ecol. 151: 29–41.

Fu, Y., K. Hirayama & Y. Natsukari, 1991b. Genetic divergence between S and L type strains of the rotifer *Brachionus plicatilis* O.F. Müller. J. exp. mar. Biol. Ecol. 151: 43–56.

Gallup, C. D., R. L. Edwards & R. G. Johnson, 1994. The timing of high sea levels over the past 200 000 years. Science 263: 796–800.

Gilbert, J. J., 1974. Dormancy in rotifers. Trans. am. Micr. Soc. 93: 490–413.

Gómez, A., M. Temprano & M. Serra, 1995. Ecological genetics of a cyclical parthenogen in temporary habitats. J. Evol. Biol. 8: 601–622.

Green, J., 1972. Latitudinal variation in associations of planktonic Rotifera. J. Zool. Lond. 167: 31–39.

Green, J., 1994. The temperate-tropical gradient of planktonic Protozoa and Rotifera. Hydrobiologia 272: 13–26.

Green, J., 1995. Associations of planktonic and periphytic rotifers in a Malaysian estuary and two nearby ponds. Hydrobiologia 313/314: 47–56.

Green, J. & H. Kramadibrata, 1988. A note on Lake Goang, an unusual acid lake in Flores, Indonesia. Freshwat. Biol. 20: 195–198.

Green, J. & Oey Biauw Lan, 1974. *Asplanchna* and the spines of *Brachionus calyciflorus* in two Javanese sewage ponds. Freshwat. Biol. 4: 223–226.

Green, J., S. A. Corbet, E. Watts & Oey Biauw Lan, 1976. Ecological studies on Indonesian lakes. Overturn and restratification of Ranu Lamongan. J. Zool. Lond. 180: 314–354.

Green, J., S. A. Corbet, E. Watts & Oey Biauw Lan, 1978. Ecological studies on Indonesian lakes. The montane lakes of Bali. J. Zool. Lond. 186: 15–38.

Hassler, A. D. & E. Jones, 1949. Demonstration of the antagonistic action of large aquatic plants on algae and rotifers. Ecology 30: 359–364.

Harring, H. K. & F. J. Myers, 1928. The rotifer fauna of Wisconsin IV. The Dicranophoridae. Trans. Wisc. Acad. Sci., Arts Letters 23: 667–808.

Hauer, J., 1937. Die Rotatorien von Sumatra, Java und Bali nach den Ergebnissen der Deutschen Limnologischen Sunda-Expedition. Teil I. Archiv für Hydrobiol., suppl. Bd. XV (2): 296–384.

Hauer, J., 1938. Die Rotatorien von Sumatra, Java und Bali nach den Ergebnissen der Deutschen Limnologischen Sunda-Expedition. Teil II. Archiv für Hydrobiol., suppl. Bd. XV (3): 507–602.

Hauer, J., 1941. Rotatorien aus dem 'Zwischengebiet Wallacea'. Int. Rev. ges. Hydrobiol. 41: 177–203.

Heckman, C., 1974. The seasonal succession of species in a rice paddy in Vientiane, Laos. Int. Rev. ges. Hydrobiol. 59: 489–507.

Heckman, C., 1979. Rice field ecology in Northeastern Thailand. Monographiae Biol., Den Haag 34: 228 pp.

Heinis, F., 1928. Die Moosfauna des Krakatau. Treubia. Batavia 10: 231–244.

International Commission on Zoological Nomenclature, 1999. International Code of Zoological Nomenclature. 4th edn. Int. Trust for Zool. Nomenclature, London: 306 pp.

Jenkins, D. J. & M. O. Underwood, 1998. Zooplankton may not disperse readily in wind, rain or waterfowl. Hydrobiologia 387/388: 15–21.

Karunakaran, L. & A. Johnson, 1978. A contribution to the rotifer fauna of Singapore and Malaysia. Malay. Nat. J. 32: 173–208.

Koste, W., 1975. Über den Rotatorienbestand einer Mikrobiozönose in einem tropischen aquatischen Saumbiotop, der *Eichernia-crassipes*-Zone im Litoral des Bung-Borapet, einem Stausee in Zentralthailand. Gewässer und Abwässer 57/58: 43–58.

Koste, W., 1981. Zur Morphologie, Systematik und Ökologie von neuen monogononten Rädertieren (Rotatoria) aus dem überschwemmungsgebiet des Magela Ck in der Alligator River Region, Australiens, N.T., Teil I. Osnabrücker naturwiss. Mitt. 8: 97–126.

Koste, W., 1988. Rotatorien aus Gewässern am Mittleren Sungai Mahakam, eimen Überschwemmungsgebiet in E-Kalimantan, Indonesian Borneo. Osnabrücker naturwiss. Mitt. 14: 91–136.

Koste, W., 1990. Zur Kenntnis der Rädertierfauna des Kinda-Stausees in Zentral-Burma (Aschelmintes: Rotatoria). Osnabrücker naturwiss. Mitt. 16: 83–110.

Koste, W. & R. J. Shiel, 1983. Morphology, systematics and ecology of new monogonont rotifers from the Alligator rivers Region, Northern Territory. Trans. r. Soc. S. Aust. 107: 109–121.

Koste, W. & R. J. Shiel, 1987. Rotifera from Australian Inland Waters. II. Epiphanidae and Brachionidae (Rotifera: Monogononta). Invertebr. Taxon. 7: 959–1021.

Koste, W. & R. J. Shiel, 1989. Classical taxonomy and modern methodology. Hydrobiologia 186/187: 279–284.

Koste, W. & W. Tobias, 1990. Zur Kenntnis der Rädertierfauna des Kinda-Stausees in Zentral-Burma (Aschelminthes: Rotatoria). Osnabrücker naturwiss. Mitt. 16: 83–110.

Koste, W. & Y. Zhuge, 1998. Zur Kenntnis der Rotatorienfauna (Rotifera) der Insel Hainan, China. Teil II. Osnabrücker Naturwissenschaftliche Mitteilungen 24: 183–222.

Kutikova, L. A. & C. H. Fernando, 1995. *Brachionus calyciflorus* Pallas (Rotatoria) in inland waters of tropical latitudes. Int. Rev. ges. Hydrobiol. 80: 429–441.

Lai, H. C. & T. E. Chua, 1976. Limnological features of Muda and Pedu reservoirs with an observation on their suitability for fish culture. Malaysian Agric. J. 50: 480–501.

Lee, C. L., 1973. The genus *Brachionus* (Rotifera: Monogononta, Brachionidae) from tropical fishponds. Mardi KDN publ. 707: 1–7 (cited by Fernando & Zankai, 1981).

Lewis, W. M., 1973: A limnological survey of Lake Mainit, Phillippines. Int. Rev. ges. Hydrobiol. 58: 801–818.

Mamaril, A. C. & C. H. Fernando, 1978. Freshwater zooplankton of the Philippines. (Rotifera, Cladocera & Copepoda). Bull. nat. Amml. Sci. Univ., Philppines 30: 109–183.

Moss, S. J. & M. E. J. Wilson, 1998. Biogeographic implications of the Tertiary paleogeographic evolution of Sulawesi and Borneo. In Hall, R. & J. D. Hollowday (eds), Biogeography and Geological Evolution of SE Asia. Leiden, Backhuys: 133–163.

Mizuno, T. & S. Mori, 1970. Preliminary hydrobiological survey of some Southeast Asian inland waters. Biol. J. linn. Soc. 2: 77–117.

Nayar, C. K. G., 1965. Taxonomic Notes on the Indian Species of *Keratella* (Rotifera). Hydrobiologia 26: 457–462.

Nogrady, T., R. L. Wallace & T. W. Snell, 1993. Rotifera 1: In Dumont, H. J. (ed.), Biology, Ecology and Systematics. Guides to the Identification of the Microinvertebrates of the Continental Waters of the World 4. SPB Academic Publishing, The Hague: 142 pp.

Pejler, B., 1977a. General problems of rotifer taxonomy and global distribution. Arch. Hydrobiol. Beih. Ergbn. Limnol. 8: 212–220.

Pejler, B., 1977b. On the global distribution of the family Brachionidae (Rotatoria). Arch. Hydrobiol. Suppl. 53: 255–306.

Pennak, R., 1966. Structure of zooplankton populations in the littoral macrophyte zone of some Colorado Lakes. Trans. Am. Microsc. Soc. 85: 329–349.

Petersen, F. & M. Carlos, 1984. A review of zooplankton in Philippine lakes. Fish. Res. j. Philipp. 9: 56–64.

Pholpunthin, P. & S. Chittapun, 1998. Freshwater Rotifera of the genus *Lecane* from Songkhla Province, southern Thailand. Hydrobiologia 387/388: 23–26.

Pourriot, R. & T. Snell, 1983. Resting eggs in rotifers. Hydrobiologia 104: 213–224.

Ricci, C. N., 1987. Ecology of bdelloids: how to be successful. Hydrobiologia 147: 117–127.

Round, F. E., 1981. The Ecology of Algae. Cambridge Univ. Press, Cambridge: 653 pp.

Russell, C. R., 1958. Some rotifers from Malaya. Trans. Soc. N. Zeal. 85: 433–437.

Ruttner-Kolisko, A., 1989. Problems in the taxonomy of rotifers, exemplified by the *Filinia longiseta-terminalis* complex. Hydrobiologia 186/187: 294–298.

Ruttner-Kolisko, A., 1991. Taxonomic problems with the species *Keratella hiemalis.* Hydrobiologia 255/256: 441–443.

Sanoamuang, L., 1996. *Lecane segersi* n.sp. (Rotifera, Lecanidae) from Thailand. Hydrobiologia 339: 23–25.

Sanoamuang, L., 1998. Rotifera of some freshwater habitats in the floodplain of the River Nan, northern Thailand. Hydrobiologia 387/388: 27–33.

Sanoamuang, L. & H. Segers, 1997. Additions to the *Lecane* fauna of Thailand. Int. Rev. ges. Hydrobiol. 82(4): 525–530.

Sanoamuang, L. & S. Savatenalinton, 1999. New records of rotifers from Nakhon Ratchasima province, northeast Thailand, with a description of *Lecane baimaii* n.sp. Hydrobiologia 412: 95–101.

Sanoamuang, L. & S. Savatenalinton, in press. The rotifer fauna of Lake Kud Thing, a shallow lake in Nong Khai province, northeast Thaliand. present volume.

Sanoamuang, L., H. Segers & H. J. Dumont, 1995. Additions to the rotifer fauna of South-East Asia: new and rare species from North-East Thailand. Hydrobiologia 313/314: 35–45.

Sarma, S. S. S., 1988. New records of freshwater rotifers (Rotifera) from Indian waters. Hydrobiologia 160: 263–269.

Segers, H., 1994. On four new tropical and subtropical *Lecane* (Lecanidae, Monogononta, Rotifera). Hydrobiologia 287: 243–249.

Segers, H., 1995a. World records of Lecanidae (Rotifera: Monogononta). In Van Goethem, J. (ed.), Studiedocumenten van het Koninklijk Belgisch Instituut voor Natuurwetenschappen 81: 114 pp (ISSN 0777-0111).

Segers, H., 1995b. Nomenclatural consequences of some recent studies on *Brachionus plicatilis* (Rotifera: Brachionidae). Hydrobiologia 313/314: 121–122.

Segers, H., 1996. The biogeography of littoral *Lecane* Rotifera. Hydrobiologia 323: 169–197.

Segers, H., 1997. Some Rotifera from the collection of the Academy of Natural Sciences of Philadelphia, including new species and new records.Proc. Acad. nat. Sci. Philad. 48: 147–156.

Segers, H., 1999. An analysis of taxonomic studies on Rotifera: a case study. Hydrobiologia 387/388: 9–14.

Segers, H. & L. De Meester, 1994. Rotifera of Papua New Guinea, with the description of a new *Scaridium* Ehrenberg, 1830. Arch. Hydrobiol. 131: 111–125.

Segers, H. & H. J. Dumont, 1995. 102+ rotifer species (Rotifera: Monogononta) in Broa reservoir (SP., Brazil) on 26 August 1994, with the description of three new species. Hydrobiologia 316: 183–197.

Segers, H. & P. Pholpunthin, 1997. New and rare Rotifera from Thale Noi Lake, Pattalang Province, Thailand, with a note on the taxonomy of *Cephalodella* (Notommatidae). Annls Limnol. 33(1): 13–21

Segers, H. & L. Sanoamuang, 1994. Two more new species of *Lecane* (Rotifera: Monogononta), from Thailand. Belg. J. Zool. 124: 39–46.

Segers, H., W. Koste & S. M. Yussuf, 1996: Contribution to the knowledge of the Monogonont Rotifera of Zanzibar, with a note on *Filinia novaezealandiae* Shiel & Sanoamuang, 1993. Int. Rev. ges. Hydrobiol. 81: 597–603.

Semper, C., 1875. IV. *Trochosphaera equatorialis*, a spherical rotifer found in the Philippine islands. – The Monthly Microscopical J. 237–245.

Serra, M., A. Galiana & A. Gomez, 1997. Speciation in monogonont rotifers. Hydrobiologia 358: 63–70.

Sharma, B. K., 1998. Rotifera. In Alfred, J. B. B, A. K. Das & A. K. Sanyal (eds), Faunal Diversity in India. Zoological Survey of India, Calcutta: 58–70.

Sharma, B. K. & S. Sharma, present volume. Biodiversity of Rotifera in some Tropical floodplain lakes of the Brahmaputra river basin, Assam (N. E. India). Hydrobiologia.

Shirota, A., 1966. The plankton of South Vietnam – fresh water and marine plankton. Overseas Technical Corporation Agency, Japan: 463 pp.

Shiel, R. J. & J. D. Green, 1996. Rotifera recorded from New Zealand, 1859–1995, with comments on zoogeography. New Zealand J. Zool. 23: 193–209.

Shiel, R. J. & W. Koste, 1986. Australian Rotifera: ecology and biogeography. In De Deckker, P. & W. D. Williams (eds), Limnology in Australia. CSIRO/Junk: 141–150.

Shiel, R. J. & W. D. Williams, 1990. Species richness in tropical fresh waters of Australia. Hydrobiologia 202: 175–183.

Slonimski, P., 1925. Sur la connaissance de *Brachionus caudatus* Barrois & Daday. C.R. des Seances de la Société de Biologie Paris 93: 948–951.

Snell, T. W., 1989. Systematics, reproductive isolation and species boundaries in monogonont rotifers. Hydrobiologia 186/187: 381–386.

Sudzuki, M., 1989. Rotifera from the Oriental Region and their characteristics. Special Issue celebrating the Centennial Anniversary of the Foundation of Nihon Daigaku University, Tokyo 3: 325–366.

Sudzuki, M., 1991. The Rotifera from Singapore and Taiwan. Proc. Japan. Soc. Syst. Zool. 43: 1–34, 1–23 pls.

Tait, R. D., R. J. Shiel & W. Koste, 1984. Structure and dynamics of zooplankton communities, Alligator Rivers Region, N.T. Australia. Hydrobiologia 113: 1–13.

Tallis, J. H., 1991. Plant Community History. Chapman & Hall, London.

Tuyor, J. B. & H. Segers, 1999. Contribution to the knowledge of the Philippine freshwater zooplankton: New records of monogonont Rotifera. Int. Rev. ges. Hydrobiol. 84: 175–180.

Ueno, M., 1966. Freshwater zooplankton of Southeast Asia. The Southeast Asian Studies 3: 94–103 (cited by Fernando & Zankai, 1981).

Van Oye, P., 1922a. Einteilung der Binnengewässer Javas. Internat. Rev. ges. Hydrobiol. 10: 7–22.

Van Oye, P., 1922b. Zur Biologie des Potamoplanktons auf Java. Internat. Rev. ges. Hydrobiol. 10: 362–393.

Van Oye, P., 1922c. Contribution à la connaissance de la flore et de la faune microscopiques des Indes Néerlandaises. Ann. Biol. Lac. 11: 130–151.

Van Oye, P., 1924. Zur Biologie des Potamoplanktons auf Java (II). Int. Rev. ges. Hydrobiol. 12: 48–59.

Vasquez, E., M. J. Pardo, E. Zoppi de Roa & C. Lopez, 1998. Rotifer fauna from Venezuela. Amazoniana 15: 11–24.

Weber, E. F., 1907. Rotateurs (Voyage du Dr. Walter Voltz). Zool. Jahrb. Abt. System. 24: 207–224.

Hydrobiologia **446/447**: 247–254, 2001.
L. Sanoamuang, H. Segers, R.J. Shiel & R.D. Gulati (eds), Rotifera IX.




# Rotifera from Burundi: the Lepadellidae (Rotifera: Monogononta)

Deo Baribwegure[1,2] & Hendrik Segers[1]
[1]*Laboratory of Animal Ecology, Zoogeography and Nature Conservation, Department Biology, University of Ghent, K.L. Ledeganckstraat 35, B-9000 Gent, Belgium*
*E-mail: deo.baribwegure@rug.ac.be Hendrik.Segers@rug.ac.be*
[2]*Département de Biologie, Faculté des Sciences, Université du Burundi, B.P. 2700 Bujumbura, Burundi*

*Key words:* Rotifera, Lepadellidae, Burundi

## Abstract

We studied the distribution of Lepadellidae (Rotifera) in freshwater habitats in the floodplain of the River Rusizi in northwest Burundi. Twenty-three species belonging to *Colurella* Bory de St. Vincent, 1824 (3 species), *Lepadella* Bory de St. Vincent, 1826 (18 species) and *Squatinella* Bory de St. Vincent, 1826 (2 species) are recorded, 22 of them are new to Burundi. One of the taxa encountered probably represents an unnamed species. *Lepadella arabica* Segers & Dumont, 1993 is recognised as junior subjective synonym of *Lepadella eurysterna* Myers, 1942 (syn. nov.). Most of the taxa recorded are cosmopolitan or tropicopolitan, two are restricted to the tropical regions of the Old World and Australia, and one, *Squatinella lunata* Segers, 1993 is an Ethiopian endemic.

## Introduction

To date, there are few publications dealing with the Burundian rotifer fauna (see De Ridder, 1987), although some contributions dealing with zooplankton from Lake Tanganyika exist (e.g. Coulter, 1991). Regarding Rotifera, Gillard (1957) recorded 62 rotifer taxa from the lake (listed by Coulter in De Ridder, 1991), but only one of his localities is from the Burundian part of the lake. Wulfert (1965), working on one of Gillard's samples, described two additional taxa from the surroundings of Lake Tanganyika: *Colurella collaris* Wulfert, 1965 (tentatively considered a synonym of *C. uncinata* (Müller, 1773) f. *bicuspidata* (Ehrenberg, 1832) by Koste, 1978) and *Lepadella minuta* (Weber & Montet, 1918) f. *africana* Wulfert (taxonomic validity rejected by Koste, 1978). From the information on the occurrence of species in the sample it can be inferred that Wulfert (1965) was studying a sample from Gillard's (1957) station 334, a swamp near Albertville (Kalemi), Congo. Recently, Segers & Baribwegure (1996) described an apparently endemic *Lecane* from swamps near Lake Tanganyika in Burundi.

In this contribution, we present results on the Lepadellidae found in samples from water bodies in the floodplain of River Rusizi in North-West Burundi. The family is one of the most important and speciose taxa of littoral-benthonic rotifers. To date, only a single representative, *Lepadella (Heterolepadella) ehrenbergi* (Perty, 1850), is known from Burundi (Gillard, 1957). A record consisting of a single Lepadellidae from a region naturally endowed with a wide variety of freshwater habitats is low when compared to surveys of other African waters. This dearth of records illustrates the need for a more extensive survey on Burundian rotifers.

## Materials and methods

Study samples were collected using a 35 $\mu$m-mesh plankton net and preserved in 4% formalin. For analysis, specimens were picked under a Wild M3 dissecting microscope, and drawn under a Medilux-12 microscope with drawing tube at 1000× magnification. For species identification, the keys by Koste (1978) and Koste & Shiel (1989) were used.

The floodplain of River Rusizi extends 80 km along the stream in west and northwest Burundi. It is located from the northern bay of Lake Tanganyika up to the confluence of River Rusizi with River Ruhwa at the border between Burundi and Rwanda. The study area is located between 2° 45′–3° 30′ S and 29° 10′–29° 15′ E at an altitude varying between 740 and 900 m. The climate is tropical with two rainy seasons (February–May and November–December).

The following localities were sampled (see Map 1):

1. Locality 1: Littoral zone of Lake Tanganyika between 'Musée Vivant' and the port of Bujumbura (one sample, leg. H.J. Dumont, 01 May 1983).
2. Localities 2 and 3: Two natural lakes in the delta of River Rusizi near Lake Tanganyika (Park of Rusizi) (Five and four samples, respectively, 5–11 July 1996).
3. Locality 4: Small water bodies and connecting ditch between localities 2 and 3, Park of Rusizi (3–9 July 1996).
4. Locality 5: Temporary pond in Rukoko (Reserve of Rusizi), North of Bujumbura airport (Three samples, leg. A. Caljon, 21 December 1981).
5. Locality 6: Pond near River Kajeke (One sample, leg. A. Caljon, 09 April 1990).
6. Localities 7 and 8: Two rice fields along a road to Rukaramu project near Bujumbura airport (Two samples each, 26 March 1996).
7. Locality 9: Abandoned fishpond in Rukaramu project (One sample, 26 March 1996).

## Results and discussion

The samples yielded a total of 23 Lepadellidae (Table 1). Three of the four genera of Lepadellidae are represented: *Colurella* Bory de St. Vincent, 1824 with three species, *Lepadella* Bory de St. Vincent, 1826 with eighteen species, and *Squatinella* Bory de St. Vincent, 1826 with two species. Two of the taxa found are taxonomically noteworthy. One, an apparent relative of *Lepadella acuminata* (Ehrenberg, 1834) (Figs 42–43) resembles *L. elongata* Koste, 1992 and *L. neglecta* Segers & Dumont, 1995, and has previously been reported as *L. cryphaea* Harring, 1916 by, amongst others, Segers et al. (1993a). We refrain from treating the taxon in full detail here, in view of the scarcity of specimens in the Burundian samples and the taxonomic confusion in the group.

The Burundian collection contained specimens of *Lepadella eurysterna* Myers, 1942 (Figs 35–37). A comparison of the present material with paratypes of *L. eurysterna* (Academy of Natural Sciences of Philadelphia rotifer collection nr. 964, Naomi Lake, Monroe Co., Pa, U.S.A. coll. 16 August 1939), conspecific specimens from Lenape Lake, Atlantic Co., NJ, U.S.A. (coll. 6 July 1996), and type specimens and topotypic material of *L. arabica* Segers & Dumont, 1993, revealed that *L. eurysterna* is a senior synonym of *L. arabica* (NEW SYNONYM). The lorica outline of the species is rather variable, but it can be recognised by its relatively flat lorica, as in *L. ovalis* (Müller), and convex rather than deeply concave posterior lorica margin (compare Figs 35–37 with Fig. 34). The original description of *L. eurysterna* by Myers (1942) is misleading by its brevity, and by the selection of animals depicted. This may explain why *L. eurysterna* has not been reported since its description. *Lepadella arabica* was described from northeast Saudi Arabia, and subsequently recorded from Nigeria (Segers et al., 1993a). The species has also been found in collections from two ponds in New Delhi, India (Dr S. Nandini, unpublished). Apparently, *L. eurysterna* is a cosmopolitan, and possibly warm-stenothermic species.

A number of species recorded were known from one or two African localities only. These are as follows:

1. *Lepadella apsicora* Myers, 1934 (Figs 15–16): The present record is the second of this species from the African continent, after a record from Nigeria (Segers et al., 1993a). It is also known from Madagascar (Berzins, 1982) and the Comoros Islands (Segers, 1992), and is tropicopolitan.
2. *Lepadella apsida* Harring, 1916 (Fig. 17): The only previous African record of this rare but cosmopolitan species is from Senegal (De Ridder, 1983). The animal can easily be confused with *L. lindaui* Koste, but has a smooth dorsal lorica.
3. *Lepadella discoidea* Segers, 1993 (Figs 30–31): To date, African records of this species are from Nigeria and Congo (Segers, 1993). The species is widespread in the tropical regions of the Old World and Australia.
4. *Lepadella lindaui* Koste, 1981 (Figs 27–28): The present record is the third African of this animal, after its description from Kenya (Koste, 1981), and a record from Nigeria (Segers et al., 1993a). In addition, the species has been recorded from North Australia (Koste, 1981), and Brazil (Segers et al., 1993b).
5. *Lepadella triba* Myers, 1934 (Figs 42–43): Our Burundian record is the third African of this cosmopolitan species, after reports from Senegal (De Ridder, 1983) and Nigeria (Segers et al., 1993a).

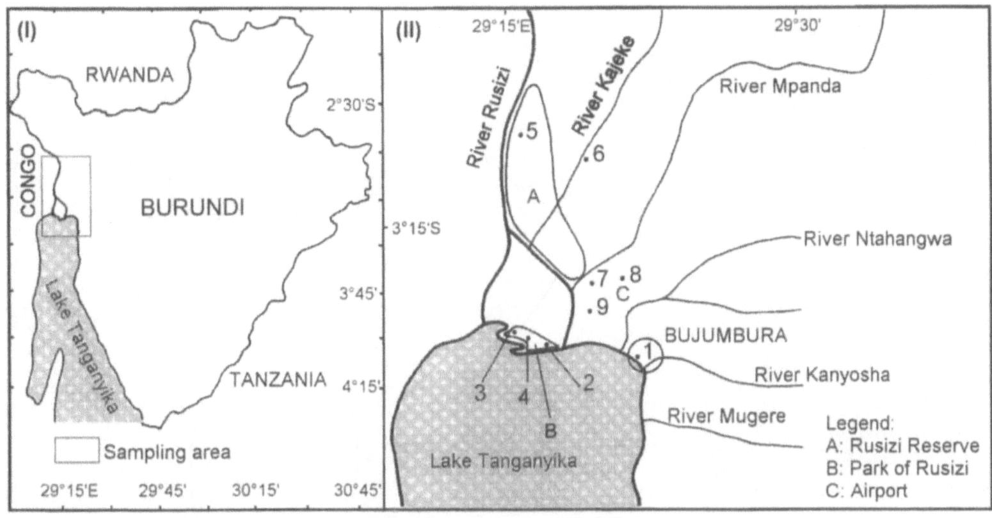


*Map 1.* (I) Map of Burundi, indicating the sampled region (inset), (II) sampled localities.

*Table 1.* Colurellidae recorded from Burundi

*Colurella adriatica* Ehrenberg, 1831 (Figs 1–3): Loc. 3 (c)
**Colurella obtusa* (Gosse, 1886) (Figs 4–6): Loc. 2, 4, 5, 8, 9 (c)
[1]***Colurella uncinata* (Müller, 1773) f. *bicuspidata* (Ehrenberg, 1832) (Figs 7–12): Loc. 2 (c)
*Lepadella amphitropis* Harring, 1916 (Figs 13–14): Loc. 2 (c)
*Lepadella apsicora* Myers, 1934 (Figs 15–16): Loc. 2, 4, 5 (t)
*Lepadella apsida* Harring, 1916 (Fig. 17): Loc. 2 (t)
*Lepadella biloba* Hauer, 1958 (Figs 18–19): Loc. 3, 4, 9 (c)
*Lepadella costatoides* Segers, 1992 (Figs 24–25): Loc. 2, 3, 4, 5, 7 (t)
*Lepadella discoidea* Segers, 1993 (Figs 30–31): Loc. 2, 3, 4, 5, 6, 7, 8, 9 (pal)
**Lepadella ehrenbergi* (Perty, 1850) (Figs 32–33): Loc. 3, 4 (c)
*Lepadella eurysterna* Myers, 1942 (Figs 35–37): Loc. 3, 4 (t)
**Lepadella heterostyla* (Murray, 1913) (Figs 38–39): Loc. 1, 4 (c)
*Lepadella latusinus* (Hilgendorf, 1899) (Figs 20–21): Loc. 1, 4, 8 (c)
*Lepadella lindaui* Koste, 1981 (Figs 26–27): Loc. 4 (pan)
**Lepadella ovalis* (Müller, 1786) (Fig. 34): Loc. 4, 5 (c)
**Lepadella patella* (Müller, 1786) (Figs 22–23): Loc. 2, 4, 7, 8 (c)
*Lepadella rhomboides* (Gosse, 1886) (Figs 40–41): Loc. 1, 5 (c)
*Lepadella* sp. near *acuminata* (Figs 44–45): Loc. 2, 4
*Lepadella triba* Myers, 1934 (Figs 42–43): Loc. 5 (c)
*Lepadella triptera* (Ehrenberg, 1830) (Figs 28–29): Loc. 3, 4, 8 (c)
**Lepadella vandenbrandei* Gillard, 1952 (Figs 46–48): Loc. 6 (pal)
*Squatinella lunata* Segers, 1993 (Fig. 49): Loc. 4 (e)
*Squatinella lamellaris* (Müller) f. *mutica* (Ehrenberg, 1832) (Figs 50–51): Loc. 2 (c)

(c): cosmopolitan; (e): endemic; (pal): tropical regions of the Old World and Australia; (pan): pantropical; (t): tropicopolitan.
*, **: species recorded by Gillard (1957) and Wulfert (1957), respectively, from Lake Tanganyika. Additional records from the lake are by Gillard (1957: *L. acuminata*, *L. cristata,* and *Paracolurella aemula*), and Wulfert (1957: *L. minuta* f. *africana*).
[1]As *C. collaris* Wulfert, 1965.

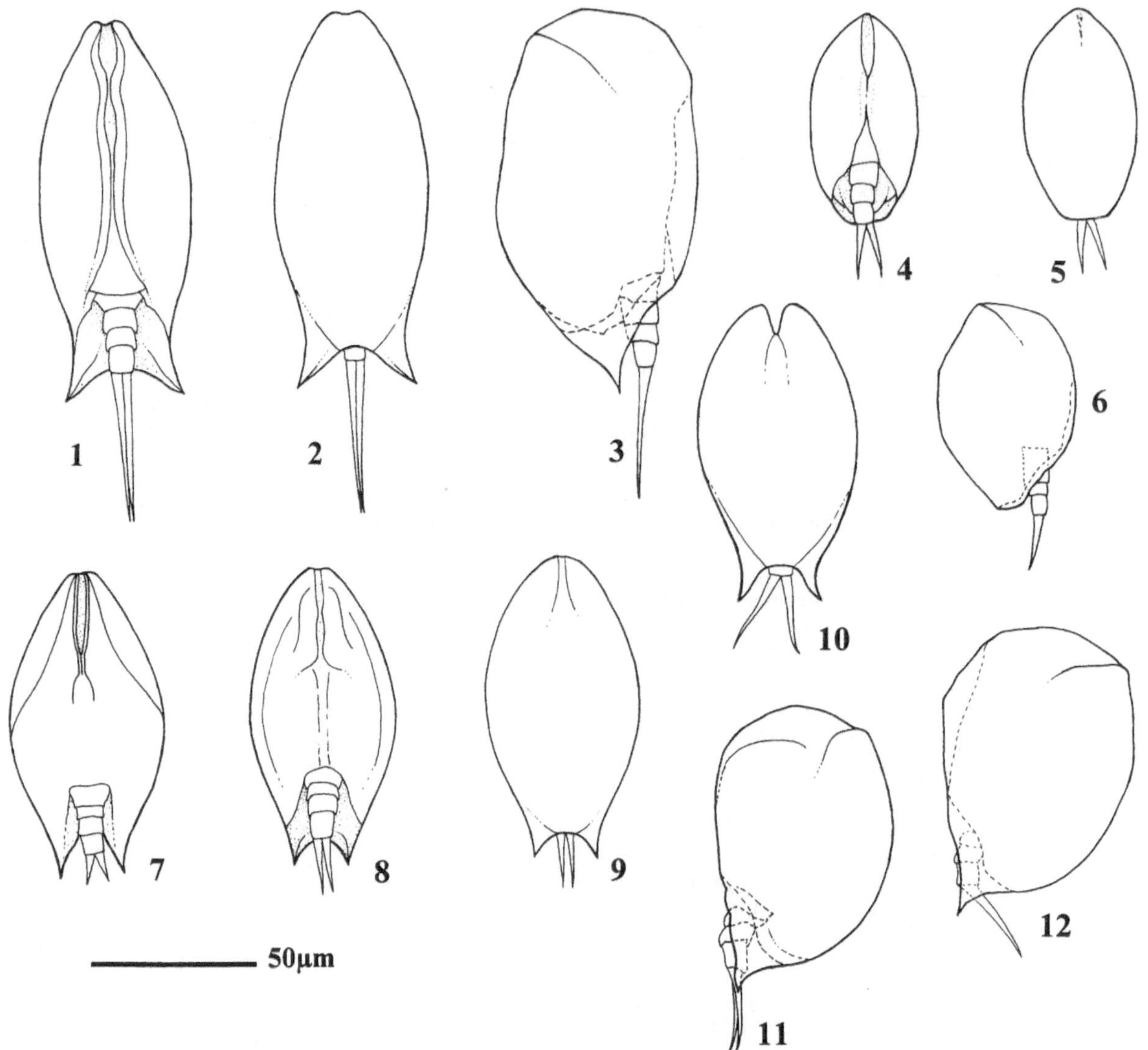


*Figures 1–12. Colurella* spp. 1–3: *C. adriatica*, 4–6: *C. obtusa*, 7–12: *C. uncinata* f. *bicuspidata*. 1, 4, 7, 8: ventral, 2, 5, 9, 10: dorsal, 3, 6, 11, 12: lateral.

6. *Squatinella lunata* Segers, 1993 (Fig. 49): This apparent African endemic was described from Nigeria (Segers, 1993), and is now also known from Burundi.

All taxa but *L. ehrenbergi* are new to the Burundian fauna, all had been recorded from Africa before. The majority of animals recorded are cosmopolitan (14 taxa, 61%), or widely distributed species with thermophilic character (tropicopolitans: 4 taxa, 17%) (see Table 1). One species is Pantropical (4%). Taxa of particular zoogeographic relevance are the Paleotropical *L. discoidea* and *L. vandenbrandei* Gillard, 1952 (Figs 46–48), and the Ethiopian endemic *S. lunata*. The diversity of Lepadellidae in the sampled localities appears relatively high, considering that De Ridder (1986, 1991, 1994) mentions a total of only 49 Lepadellidae from African freshwaters. A record of 23 Lepadellidae is high relative to the results by De Smet (1989, 1991), who lists eight and 13 Lepadellidae from water bodies on Mount Kilimanjaro, Tanzania, and from localities in the Bas-Zaïre, Republic of Congo, respectively. On the other hand, De Ridder (1981) records 22 Lepadellidae from East Congo, and Segers et al. (1993a) lists 33 species from various habitats in the lower delta of River Niger. It appears

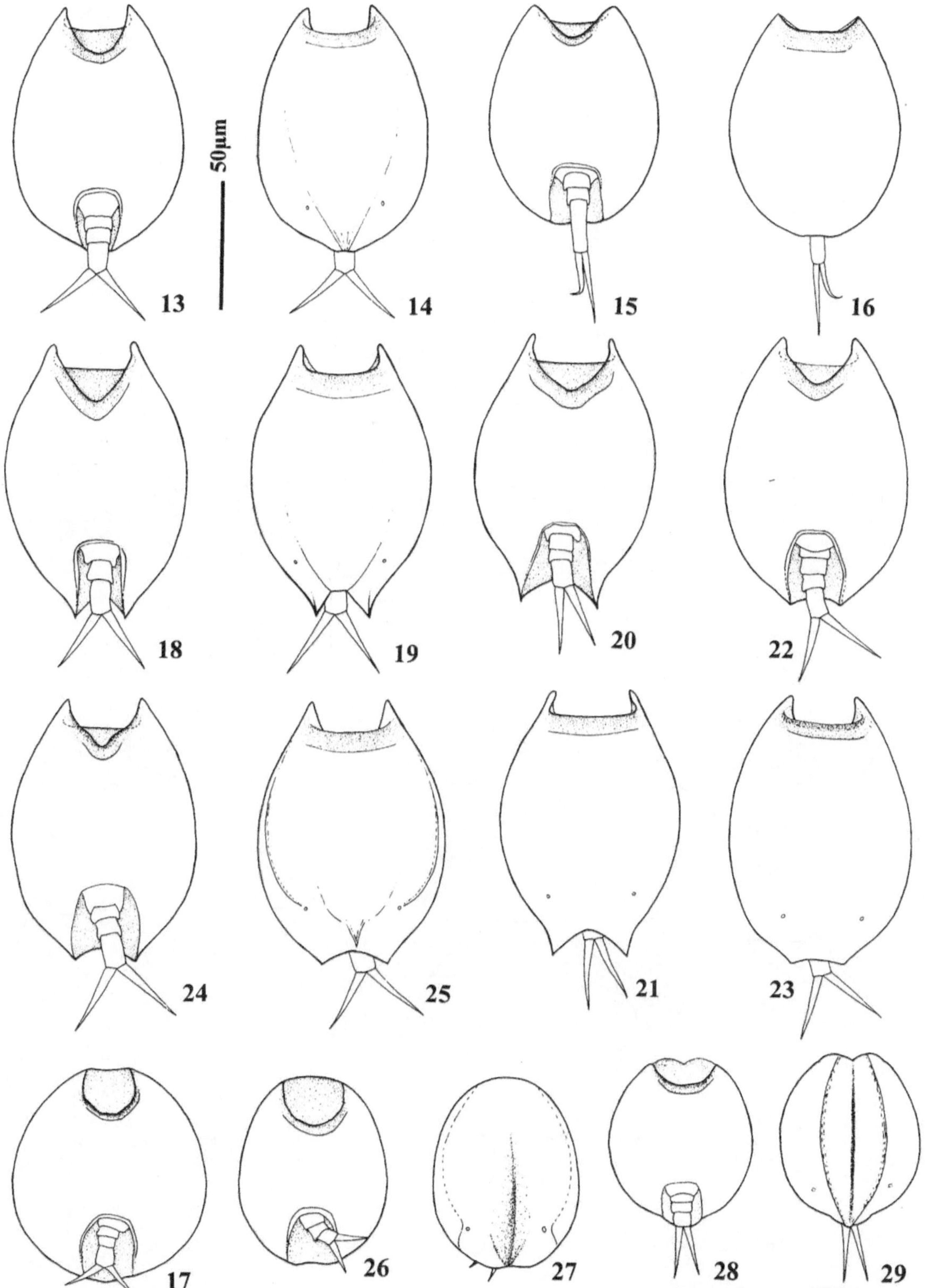


*Figures 13–29. Lepadella* spp. 13–14: *L. amphitropis*, 15–16: *L. apsicora*, 17: *L. apsida*, 18–19: *L. biloba*, 20–21: *L. latusinus*, 22–23: *L. patella*, 24–25: *L. costatoides*, 26–27: *L. lindaui*, 28–29: *L. triptera*. 13, 15, 17, 18, 20, 22, 24, 26, 28: ventral, 14, 16, 19, 21, 23, 25, 27, 29: dorsal.

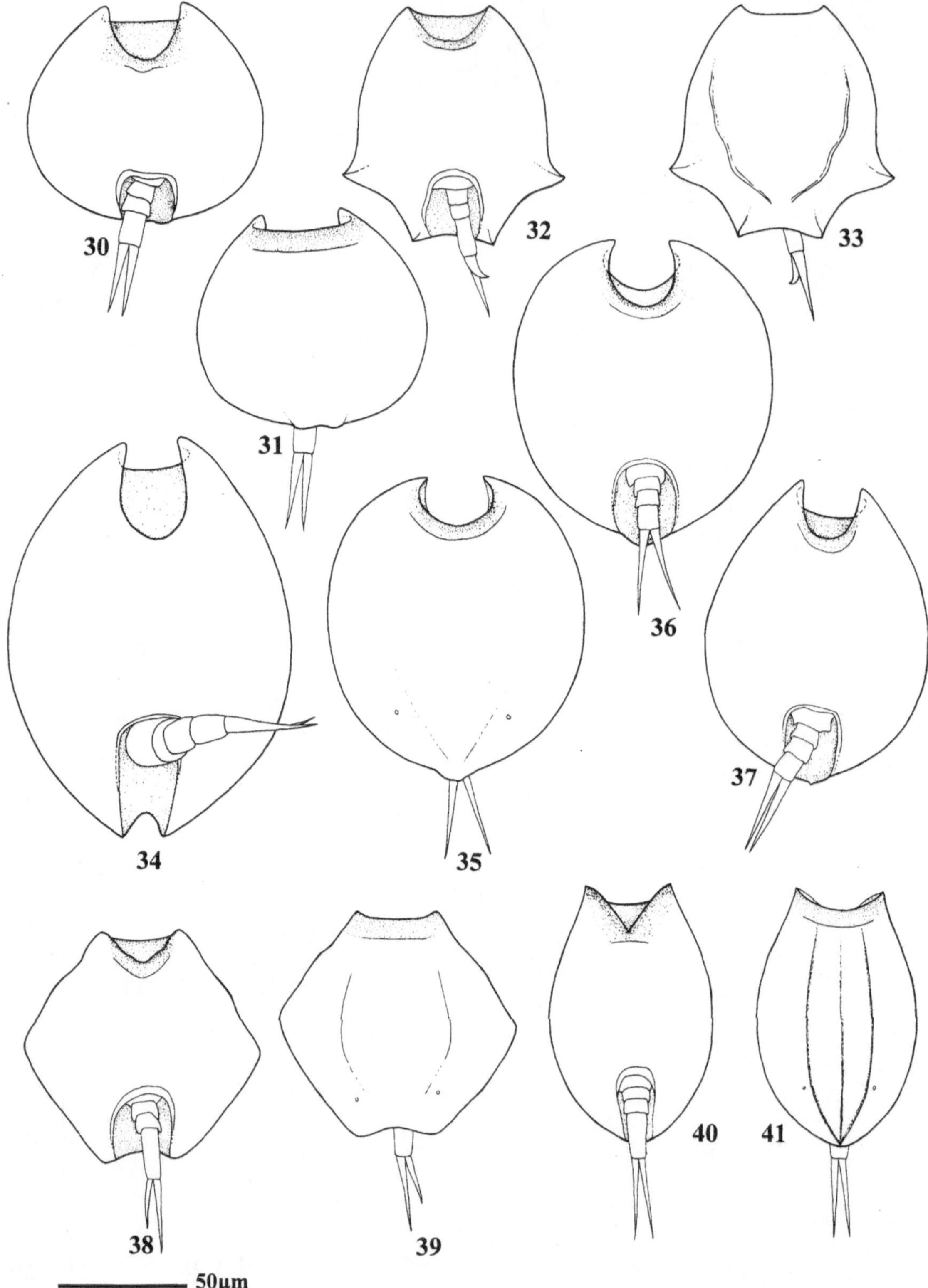


*Figures 30–41. Lepadella* spp. 30–31: *L. discoidea*, 32–33: *L. ehrenbergi*, 34: *L. ovalis*, 35–37: *L. eurysterna*, 38–39: *L. heterostyla*, 40–41: *L. rhomboides*. 30, 32, 34, 36–38, 40: ventral, 31, 33, 35, 39, 41: dorsal.

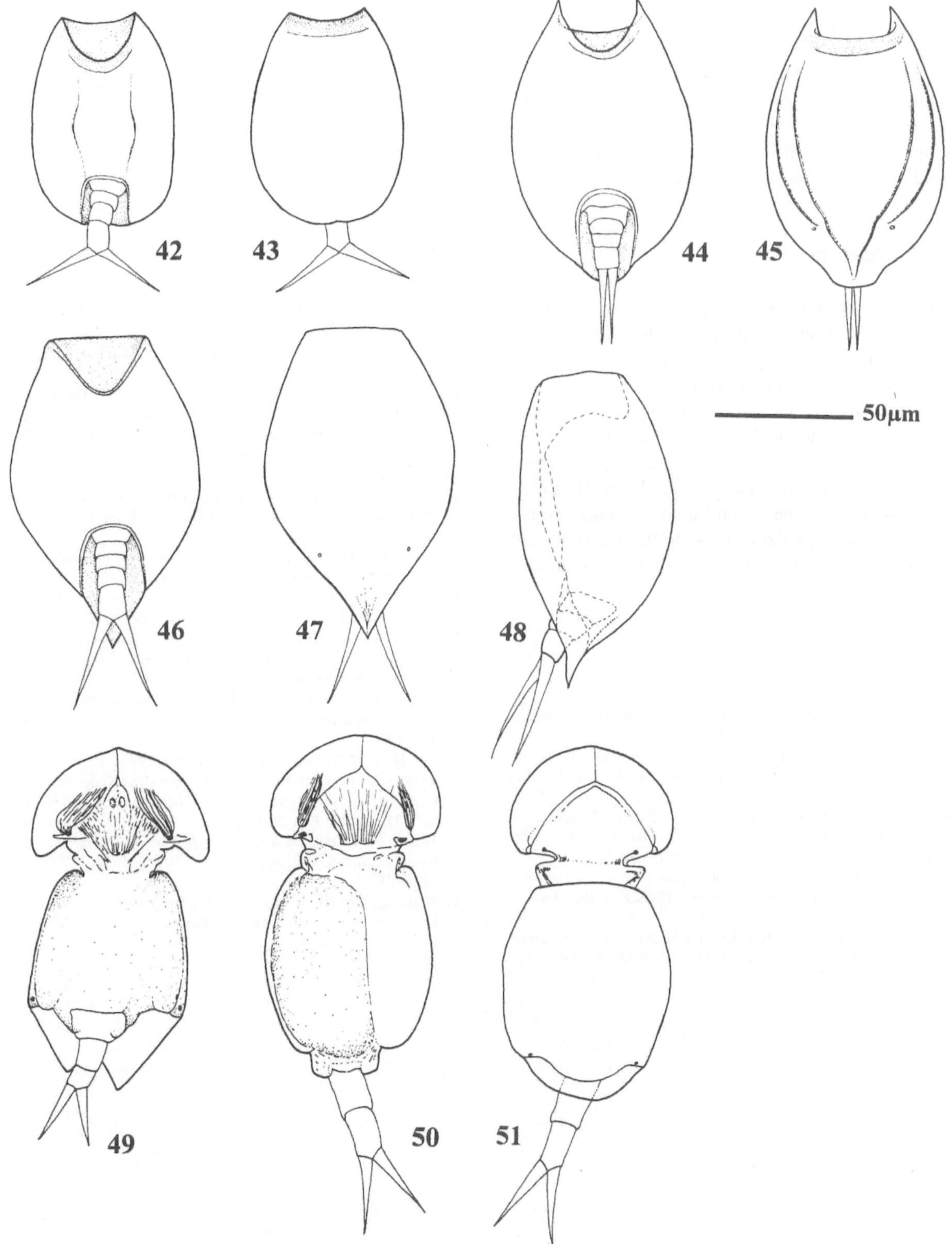


*Figures 42–51. Lepadella* spp. 42–43: *L. triba*, 44–45: *L.* sp. near *acuminata*, 46–48: *L. vandenbrandei*. *Squatinella* spp. 49: *S. lunata*, 50–51: *S. lamellaris* f. *mutica*. 42, 44, 46, 49, 50: ventral, 43, 45, 47, 51: dorsal, 48: lateral.

premature to draw any conclusions from these numbers, as differences in sampling intensity preclude comparisons. Considering the extent of the sampled area and sampling effort, the record illustrates a diverse rotifer taxocoenosis in Burundian freshwaters. Obviously, the present list is non-exhaustive, as most Burundian inland waters remain unstudied.

## Acknowledgements

The first author acknowledges a grant by the Belgian Administration for Development and Cooperation (B.A.D.C) to work on a Ph.D. dissertation. Samples from the littoral area of the Lake Tanganyika, from Kajeke and Rukoko were collected by Prof. H.J. Dumont and the late Dr A. Caljon. Sampling in Rusizi Natural Park was aided by logistic support from CRRHA (Centre Régional de Recherche en Hydrobiologie Appliquée). The second author acknowledges support from the organizers of the VIIIth International rotifer symposium, to attend the IXth International rotifer symposium.

## References

Berzins, B., 1982. Zur Kenntniss der Rotatorienfauna von Madagascar. Lund, AV centralen I Lund: 24 pp.

Coulter, G. W., 1991. Lake Tanganyika and its life. Natural History Museum publications. Oxford Univ. Press, Lond./Oxford, N.Y.: 354 pp.

De Ridder, M., 1981. Rotifera. In Symoens, J.-J. (ed.), Hydrobiological Survey of the Lake Bangweulu Luapula River Basin. Cercle Hydrobiologique de Bruxelles, Bruxelles: 191 pp.

De Ridder, M., 1983. Recherches écologiques et hydrobiologiques sur les Rotifères de la Basse-Casamance (Sénégal). Rev. Hydriobiol. trop. 16: 41–55.

De Ridder, M., 1986. Annotated checklist of non-marine Rotifera from African inland waters. Zool. Doc., kmMA, Tervuren 21: 123 pp.

De Ridder, M., 1987. Distribution of rotifers in African fresh and inland saline waters. Hydrobiologia 147: 9–14.

De Ridder, M., 1991. Additions to the "Annotated checklist of non-marine rotifers from African inland waters". Rev. Hydrobiol. trop. 24(1): 25–46.

De Ridder, M., 1994. Additions II to the "Annotated checklist of non-marine rotifers from African inland waters". Biol. Jb. Dodonaea 61: 99–153.

De Smet, W. H., 1989. Contributions to the rotifer fauna of the Bas-Zaïre. 1. The rotifers from some small ponds and a river. Biol. Jb. Dodonaea 56: 115–131.

De Smet, W. H., 1991. Rotifers from the Kilimanjaro. Biol. Jb. Dodonaea 58: 120–130.

Gillard, A., 1957. Exploration hydrobiologique du Lac Tanganyika (1946–1947). Inst. Roy. des Sc. Nat. de Belgique. Vol. III (fasc. 6). 3–26 pp.

Koste, W., 1978. Rotatoria. Die Rädertiere Mitteleuropas. Gebr. Borntraeger, Berlin, Stuttgart: 673 pp., 234 plates.

Koste, W., 1981. Zur Morphologie, Systematik und Ökologie von neuen monogononten Rädertieren (Rotatoria) aus dem Überschwemmungsbebiet des Magela Creek in der Alligator-River-Region Australiens, N.T. Teil i. Osnabrücker naturwiss. Mitt. 8: 97–126.

Koste, W. & R. J. Shiel, 1989. Rotifera from Australia inland waters IV. Colurellidae (Rotifera: Monogononta). Trans. r. Soc. S. Aust. 113 (3) 119–143.

Myers, F. J., 1942. The rotatorian fauna of the Pocono plateau and environs. Proc. Acad. nat. Sci. Philad. 44: 251–285.

Segers, H., 1992. Taxonomy and zoogeography of the Rotifer fauna of Madagascar and the Comores. J. Afr. Zool. 106: 351–361.

Segers, H., 1993. Rotifera of some lakes in the floodplain of the river Niger (Imo State, Nigeria). I. New species and other taxonomic considerations. Hydrobiologia 250: 39–61.

Segers, H. & D. Baribwegure, 1996. On *Lecane tanganyikae* new species (Rotifera: Monogononta, Lecanidae). Hydrobiologia 324: 179–182.

Segers, H., C. S. Nwadiaro & H. J. Dumont, 1993a. Rotifera of some lakes in the floodplain of the river Niger (Imo State: Nigeria). II. Faunal composition and diversity. Hydrobiologia 250: 63–71.

Segers, H., E. N. Dos Santos-Silva & A. L. De Oliveira-Neto, 1993b. New and rare species of *Lecane* and *Lepadella* (Rotifera: Lecanidae, Colurellidae) from Brazil. Belg. J. Zool.: 123. 113–121.

Wulfert, K., 1965. Rädertiere aus einigen afrikanischen Gewässern. Limnologica (Berlin) 3(3): 347–366.

Hydrobiologia 446/447: 255–259, 2001.
L. Sanoamuang, H. Segers, R.J. Shiel & R.D. Gulati (eds), Rotifera IX.




# The rotifer fauna of peat-swamps in southern Thailand

Supenya Chittapun & Pornsilp Pholpunthin
*Department of Biology, Faculty of Science, Prince of Songkhla University, Hat-Yai 90112, Thailand*
*E-mail: ppornsil@ratree.psu.ac.th*

*Key words:* Rotifera, peat-swamps, Thailand


## Abstract

The Rotifera from four peat-swamps in the provinces Suratthanee (Kra-Jood and Kun-Thu-Lee peat-swamps), Nakhonsri-thammarat (Khuan-Kreng peat-swamp) and Yala (Lan-Kway peat-swamp) in southern Thailand were examined by the study of qualitative samples collected on three occasions during July, October and November, 1998. A total of 96 species was identified, seventeen of which are new to the Thai rotifer fauna. The most diverse genera were *Lecane* (40.6%), followed by *Lepadella* (8.3%) and *Trichocerca* (7.3%). The most diverse rotifer fauna was found in Kra-Jood peat-swamp (61 species), followed by Kun-Thu-Lee (57 species), Lan-Kway (41 species) and Khuan-Kreng (24 species) peat-swamps, respectively.


## Introduction

In recent years, the number of studies on Thai Rotifera has increased steadily (Segers & Sanoamuang, 1994, 1997; Sanoamuang et al., 1995; Sanoamuang 1996, 1998; Pholpunthin, 1997; Segers & Pholpunthin, 1997; Sanoamuang & Segers, 1997; Sanoamuang & Savatenalinton, 1999). Most of the studies sampled canals, rivers, ponds, rice fields or reservoirs, but peat-swamps, an important wetland habitat in Thailand, have largely been ignored. The formation of these peat-swamps, by the accumulation of dead plant material and debris, has taken thousands of years. The vegetation of the peat-swamps depends on the peaty soil and on the fluctuation of the water levels. This water is acid and brownish (Phengklai et al., 1989). A previous study on Rotifera from Phuket peat-swamps reported 77 species, including 12 new records for Thailand, and a new species, *Colurella sanoamuangae* was described (Chittapun et al., 1999). This result illustrates that peat-swamps are interesting areas for the study of rotifer biodiversity. The purpose of the present study is to expand our knowledge of the rotifers inhabiting peat-swamps, by investigating four peat-swamps in southern Thailand.

## Materials and methods

Samples were collected qualitatively in four peat-swamps (Kra-Jood and Kun-Thu-Lee peat-swamps, Suratthanee province; Khuan-Kreng peat-swamp, Nakhornsri-thammarat province and Lan-Kway peat-swamp, Yala province, Southern Thailand (Fig. 1) on three occasions in July, October and November, 1998. Some physical and chemical parameters of the peat-swamps are as in Table 1. The samples consist of several horizontal hauls made using a 26-$\mu$m plankton net, which were immediately preserved in 4% formaldehyde solution. Specimens were sorted under an Olympus VM dissecting microscope. They were examined and drawn using an Olympus CH-2 compound microscope with camera lucida. Some specimens were prepared for scanning electron microscopy as described by Chittapun et al. (1999).

## Results and discussion

Ninety-six rotifer species, 17 of which are new to the Thai fauna, were identified from the four peat-swamps studied (Table 2). Of the 17 new records, two: *Lecane*

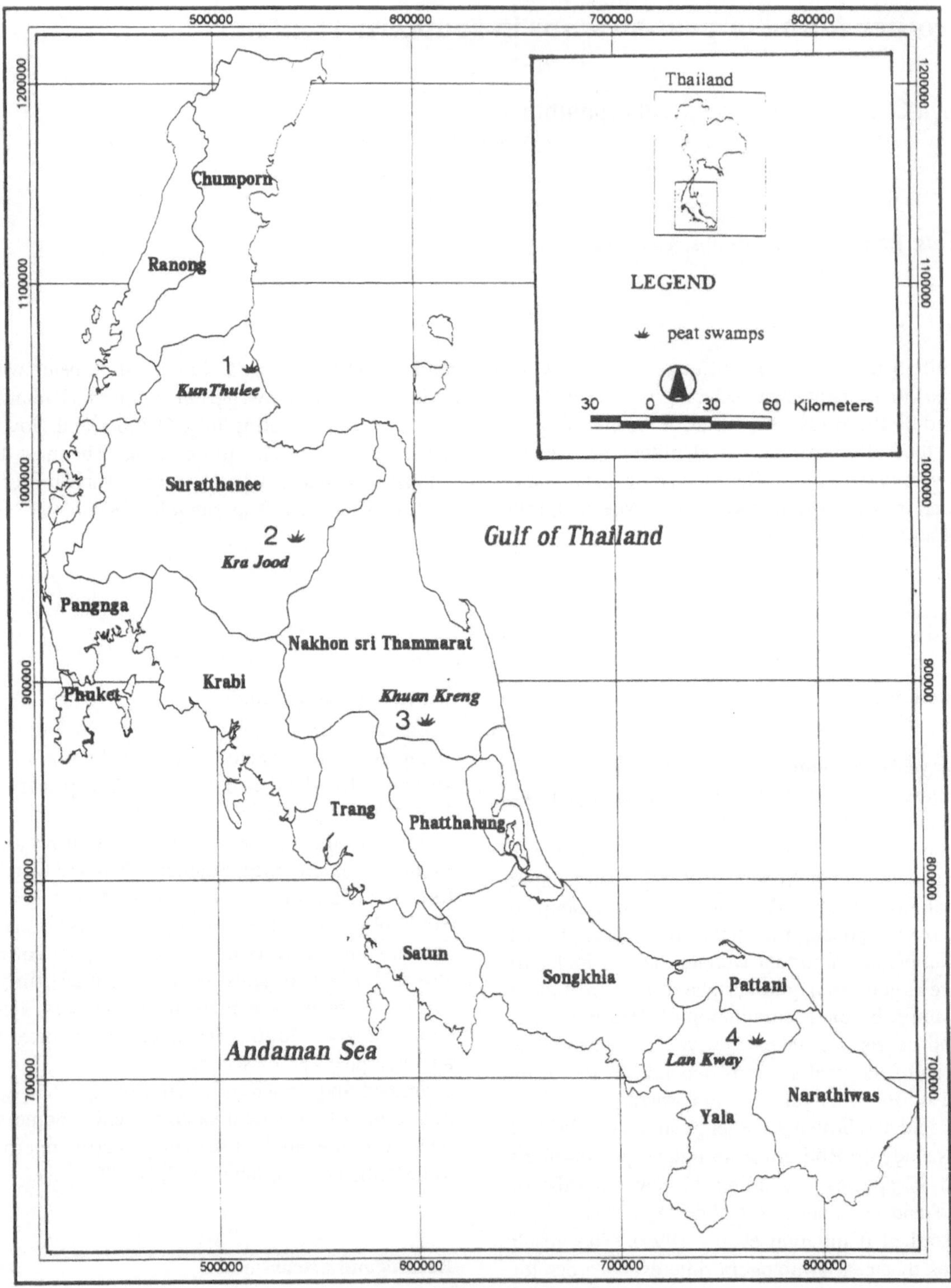


*Figure 1.* Map of Thailand showing the location of the southern part and the peat swamp areas.

*Table 1.* Physical and chemical parameters of the sampled peat-swamps

| Swamp | Water temp. (°C) | pH | Diss. $O_2$ (mg $l^{-1}$) | Conduct. (mS $cm^{-1}$) | Turbidity (NTU) |
|---|---|---|---|---|---|
| Kun Thulee | 25.7–26.6 | 4.93–5.37 | 2.27–4.30 | 0.03 | 11.0–12.0 |
| Kra Jood | 25.8–27.2 | 5.59–5.98 | 2.30–7.88 | 0.03 | 11.0–43.0 |
| Khuan Kreng | 27.2–30.1 | 5.20–7.14 | 5.60–7.40 | 0.04–0.07 | 11.0–69.0 |
| Lan Kway | 29.3–30.8 | 6.01–6.37 | 4.46–5.65 | 0.02 | 14.0–54.0 |

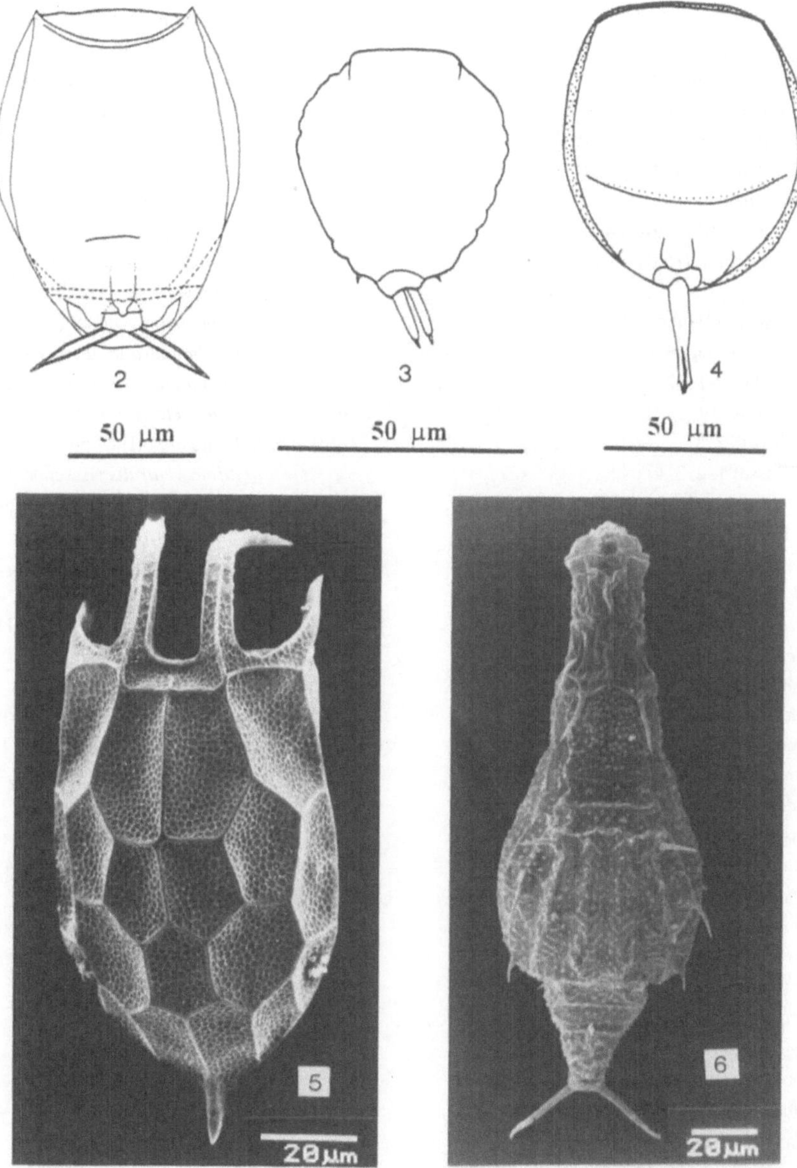


*Figures 2–6.* (2) *Lecane mitis* (Harring & Myers) – ventral view. (3) *Lecane palinacis* Harring & Myers – ventral view. (4) *Lecane syngenes* (Hauer) – ventral view. (5) *Keratella javana* (Hauer) SEM photograph – dorsal view. (6) *Dissotrocha aculeata* (Ehrenberg) SEM photograph – dorsal view.

*Table 2.* List of Rotifera from peat-swamps in Southern Thailand * New to Thailand; Numbers refer to localities as in Figure 1

*Anuraeopsis fissa* (Gosse) 1, 2, 3, 4
*Ascomorpha* sp. 4
*Brachionus quadridentatus* f. *mirabilis* (Daday) 2, 4
**Cephalodella mucronata* Myers 2
*Colurella adriatica* Ehrenberg 4
*C. colurus* (Ehrenberg) 3
*C. obtusa* (Gosse) 4
**C. sulcata* (Stenroos) 1, 2
**C. tesselata* (Glascott) 1, 2
*C. uncinata* (Müller) 1, 2, 3, 4
*Dicranophorus epicharis* Harring & Myers 1
*Dipleuchanis propatula* (Gosse) 1
*D. propatula macrodactyla* Hauer 2
**Dissotrocha aculeata* (Ehrenberg) 2, 3
*Euchlanis dilatata* Ehrenberg 1, 2
*E. incisa* Carlin 1
*Filinia opoliensis* (Zacharias) 4
*Floscularia conifera* (Hudson) 1
*Hexathra* sp. 4
**Keratella javana* Hauer 1, 2
**Lecane abanica* Segers 1
*L. aculeata* (Jakubski) 2
*L. arcuata* (Bryce) 1
*L. braumi* Koste 2, 3
*L. bulla* (Gosse) 1, 2, 3, 4
*L. calcaria* Harring & Myers 1
*L. clara* (Bryce) 1, 2
*L. closterocerca* (Schmarda) 1, 2
*L. crepida* Harring 4
*L. curvicornis* (Murray) 1, 2
*L. decipiens* (Murray) 3
*L. doryssa* Harring 1, 2, 4
*L. furcata* (Murray) 2, 4
*L. haliclysta* Harring & Myers 1, 2
*L. hamata* (Stokes) 1, 2, 3, 4
*L. hornemanni* (Ehrenberg) 1, 2
*L. inermis* (Bryce) 1, 3, 4
*L. inopinata* Harring &Myers 1, 2
*L. leontina* (Turner) 1, 2, 3, 4
*L. ludwigii* (Eckstein) 1, 2, 4
*L. lunaris* (Ehrenberg) 1, 2, 3, 4
**L. mitis* (Harring & Myers) 3
*L. monostyla* (Daday) 1, 2, 3
*L. obtusa* (Muray) 2, 3
**L. palinacis* Harring & Myers 1
*L. papauna* (Murray) 3
*L. pertica* Harring & Myers 1, 2, 3, 4
*L. pusilla* Harring 4
*L. pyriformis* (Daday) 1, 2, 4
*L. quadridentata* (Ehrenberg) 1, 2
*L. rhytida* Harring & Myers 1
*L. signifera* (Jennings) 1, 2, 3, 4
**L. simonneae* Segers 3
**L. syngenes* (Hauer) 1
*L. tenuiseta* Harring 2, 4
*L. thienemanni* Hauer 2
*L. undulata* Hauer 1
*L. unguitata* (Fadeev) 1, 2, 4
*L. ungulata* (Gosse) 1, 2, 3, 4
**Lepadella cristata* (Rousselet) 1, 2
*L. dactyliseta* (Stenroos) 4
*L. discoidea* Segers 1, 2
*L. ovalis* (Müller) 1, 2, 4
*L. patella* (Müller) 1, 2, 3
*L. quadricarinata* (Stenroos) 2
*L. rhomboides* (Gosse) 3
*L. vandenbrandei* Gillard 1, 2
*Macrochaetus collinsi* (Gosse) 1, 2, 4
**M. subquadratus* (Perty) 4
*Monommata grandis* Tessin 2
*M. longiseta* (Müller) 2
**M. maculata* Harring & Myers 1, 2
*Mytilina ventralis* (Ehrenberg) 1, 2
*Notommata pachyura* f. *spinosa* Koste 2, 4
**N. saccigera* Ehrenberg 1
*Plationus patulus* (Müller) 1, 2, 4
*Polyarthra vulgaris* Carlin 2, 4
*Scaridium elegans* Segers & De Meester 2
*S. grande* Segers 1, 2
*S. longicaudum* (Müller) 2, 4
*Squatinella mutica* (Ehrenberg) 4
*Testudinella amphora* Hauer 1, 4
*T. incisa ahlstromi* (Hauer) 1, 2, 4
**T. mucronata* (Gosse) 4
*T. parva* (Ternetz) 1, 2
*T. patina* (Hermann) 1, 2, 3, 4
*T. tridentata* Smirnov 1, 2
**Tetrasiphon hydrocora* Ehrenberg 4
*Trichocerca brasiliensis* (Murray) 4
**T. collaris* (Rousselet) 2
*T. flagellata* Hauer 2
*T. hollaerti* De Smet 1, 2
*T. insignis* (Herrick) 1
*T. similis grandis* Hauer 2, 3, 4
*T. tropis* Hauer 1
*Trichotria tetractis* (Ehrenberg) 1, 2, 3, 4

*mitis* (Harring & Myers) and *L. palinacis* Harring & Myers are new to the Oriental Region. The former species (Fig. 2) has been found in the Nearctic and Neotropical Regions, while the latter (Fig. 3) has been recorded from the U.S.A. and the Galapagos Islands (Segers, 1995; Segers & Dumont, 1995; De Ridder & Segers, 1997). Many of the new records (*Cephalodella mucronata* Myers, *Colurella sulcata* (Stenroos), *C. tesselata* (Glascott), *Keratella javana* Hauer, *Lecane syngenes* (Hauer), *Macrochaetus subquadratus* (Perty), *Notommata saccigera* Ehrenberg, *Tetrasiphon hydrocora* Ehrenberg and *Trichocerca collaris* (Rousselet)) are widely distributed in the Oriental Region, occurring in Indonesia, Malaysia, Singapore and Sri Lanka (De Ridder & Segers, 1997). Among them, *K. javana* (Fig. 5) and *N. saccigera* are cosmopolitan species, common in acid waters (Sudzuki, 1991; Nogrady & Pourriot, 1995), while *L. syngenes* (Fig. 4) is a rare, warm-stenothermic Pan(sub)tropical species (Segers, 1995). The rest of the new records, *Dissotrocha aculeata* (Ehrenberg), *Lecane abanica* Segers, *L. simonneae* Segers, *Lepadella cristata* (Rousselet), *Monommata maculata* Harring & Myers and *Testudinella mucronata* (Gosse), are new to Southeast Asia. However, *D. aculeata* (Fig. 6) is thought to be a cosmopolitan species (Koste & Shiel, 1986).

Of the 96 taxa, the most diverse genera were *Lecane* (40.6%), followed by *Lepadella* (8.3%) and *Trichocerca* (7.3%). These results, together with the most frequently encountered rotifer species in all peat-swamps agree well with the existing knowledge on the composition of the rotifer fauna of Phuket peat-swamps (Chittapun et al., 1999). Kra-Jood peat-swamp contained the most diverse rotifer taxocoenosis (61 species) followed by Kun-Thu-Lee (57 species), Lan-Kway (41 species) and Khuan-Kreng (24 species) peat-swamps, respectively.

## Acknowledgements

We are grateful to Dr Hendrik Segers for his valuable advice and critical reading this manuscript. We thank Ms Apinya Jantharangsri from Central Equipment Division, PSU for her asistance with SEM. This work was supported by TRF/BIOTEC Special Program for Biodiversity Research and Training grant BRT 541051 and partially supported by Royal Golden Jubilee Ph. D. Program No. 4.B.PS/42.

## References

Chittapun, S., P. Pholpunthin & H. Segers, 1999. Rotifera from peat-swamps in Phuket Province, Thailand, with the description of a new *Colurella* Bory de St. Vincent. Int. Rev. Hydrobiol. 84: 587–593.

De Ridder, M. & H. Segers, 1997. Monogonont Rotifera recorded in the World literature (except Africa) from 1960 to 1992. Studiedocumenten Van Het K.B.I.N., Documents de Travail De L'I.R.Sc.N.B., Belgium: 481 pp.

Koste, W. & J. Shiel, 1986. Rotifera from Australian inland waters. I. Bdelloidea (Rotifera : Digononta). Aust. J. Mar. Freshw. Res. 37: 765–792

Nogrady, T. & R. Pourriot, 1995. Rotifera. Vol 3 : The Notommatidae and The Scaridiidae. Guides to the Identification of the Microinvertebrates of the Continental Waters of the World 8. SPB Academic Publishing, The Hague, The Netherlands: 248 pp.

Phengklai, C., C. Niyomdham, A. Premrasmi, K. Chirathanakorn & R. Phuma, 1989. Peat swamp forest of Thailand. Thai For. Bull. 18: 1–42.

Pholpunthin, P., 1997. Freshwater zooplankton (Rotifera, Cladocera and Copepoda) from Thale-noi, south Thailand. J. Sci. Soc. Thailand 23: 23–34.

Sanoamuang, L., 1996. *Lecane segersi* n. sp. (Rotifera, Lecanidae) from Thailand. Hydrobiologia 339: 23–25.

Sanoamuang, L., 1998. Rotifera of some freshwater habitats in the floodplain of the River Nan, northern Thailand. Hydrobiologia 387/388: 27–33.

Sanoamuang, L. & S. Savatenalinton, 1999. New records of rotifers from Nakhon Ratchasima province, northeast Thailand, with a description of *Lecane baimaii* n. sp. Hydrobiologia 412: 95–101.

Sanoamuang, L. & H. Segers, 1997. Additions to the *Lecane* Fauna (Rotifera : Monogononta) of Thailand. Int. Rev. ges. Hydrobiol. 82: 525–530.

Sanoamuang, L., H. Segers & H. J. Dumont, 1995. Additions to the rotifer fauna of south-east Asia: new and rare species from north-east Thailand. Hydrobiologia 313/314: 35–45.

Segers, H., 1995. Rotifera. Vol. 2: The Lecanidae (Monogononta). Guides to the Identification of the Microinvertebrates of the Continental Waters of the World 6. SPB Academic Publishing, The Hague, The Netherlands: 226 pp.

Segers, H. & H. J. Dumont, 1995. 102+ rotifer species (Rotifera: Monogononta) in Broa reservoir (SP., Brazil) on 26 August 1994, with the description of three new species. Hydrobiologia 316: 183–197.

Segers, H. & P. Pholpunthin, 1997. New and rare Rotifera from Thale-Noi Lake, Pattalung province, Thailand, with a note on the taxonomy of *Cephalodella* (Notommatidae). Annals. Limnol. 33: 13–21.

Segers, H. & L. Sanoamuang, 1994. Two more new species of *Lecane* (Rotifera, Monogononta) from Thailand. Belg. J. Zool. 124: 39–46.

Sudzuki, M., 1991. The rotifera from Singapore and Taiwan. Proc. Jap. Soc. Syst. Zool. 43: 1–34.

Hydrobiologia **446/447:** 261–272, 2001.
L. Sanoamuang, H. Segers, R.J. Shiel & R.D. Gulati (eds), Rotifera IX.




# Freshwater Rotifera from plankton of the Kerguelen Islands (Subantarctica)

Willem H. De Smet
*Department of Biology, Laboratory of Polar Ecology, Limnology and Palaeobiology, University of Antwerp, R.U.C.A.-campus, Groenenborgerlaan 171, B-2020 Antwerpen, Belgium*
*E-mail: wides@ruca.ua.ac.be*

*Key words:* Rotifera, Subantarctica, Kerguelen Islands, zoogeography

## Abstract

During the austral summer 1997/98, a survey was made of a series of freshwater bodies on the Kerguelen Islands, Subantarctica (49° S–69° E), in the area of Port-aux- Français and Val Studer. The rotifer fauna from plankton of 17 lentic freshwaters is listed and commented upon. Species richness (39 taxa) and species diversity are fairly low. Only 6 taxa are euplanktic, the others are facultative plankters or littoral/benthic species. Three of the true plankters found are restricted to the Southern Hemisphere, whereas the rest of the rotifers (apart from two unidentified monogononts) shows a wide-spread or cosmopolitan distribution. Thirty-three taxa are new to the Kerguelen archipelago. A comparison is made with the rotifer assemblages observed on other Subantarctic islands. Taxonomic notes on little known species are added. The validity of *Cephalodella gibba microdactyla* as separate subspecies is rejected, and *Filinia terminalis kergueleniensis* is recognized as junior synonym of *F. pejleri*.

## Introduction

The Kerguelen archipelago (48° 58′–49° 73′ S, 68° 72′–70° 58′ E) is situated in the Southern Indian Ocean (Fig. 1), about halfway between South Africa (5000 km) and Australia (4800 km), and 2000 km from the nearest point of the Antarctic Continent. It consists of one major island, Grande Terre, and some 300 small islets, covering a total land area of about 7215 km$^2$. Lying at the northern limit of the Antarctic Convergence, the archipelago belongs to the cold temperate zone of the subantarctic region (Stonehouse, 1982). Due to the strong maritime influence, seasonal climatic variations are weak: the mean winter temperature is 3 °C and the mean summer temperature is 10 °C. The annual precipitation varies from 800 mm on the eastern coast to more than 3000 mm in the western part. Strong, mostly westerly winds blow almost continuously.

The rotifer fauna of the Kerguelen Islands is little known. Richters (1907) mentioned 6 unidentified bdelloids, belonging to the now obsolete genus *Callidina*, from aerophytic mosses. De Beauchamp (1940) reported *Encentrum permolle* (*sub Dicranophorus permollis* (Gosse)) and described *Philodina ? jeanneli* from unspecified habitats. In periphyton from stagnant and running waters, and among damp vegetation, Russell (1959) found *Cephalodella catellina* (Müller), *Colurella adriatica* Ehrenberg, *C. colurus* (Ehrenberg), *Filinia maior* (Colditz) (?syn. *F. terminalis*), *Keratella sancta* Russell, *Lecane mawsoni* n.sp. (considered spec. inq. by Segers, 1995), *Lepadella acuminata (*Ehrenberg), *L. patella* (Müller), *Lophocharis oxysternon* (Gosse), *Notommata cyrtopus* Hood, and *Adineta* spp. and *Philodina* spp. Lair & Koste (1984) described *Filinia terminalis kergueleniensis* and reported on *Keratella sancta* Russell and *Notholca* cf. *jugosa* (Gosse) from plankton of a lake. Besides these, there is a single report on marine rotifers by Zelinka (1927), who described *Trichocerca artmanni* (reported *sub Rattulus artmanni*) and *Synchaeta triophthalma* Lauterborn from Baie de l'Observatoire.

The present contribution deals with the rotifers found in plankton of stagnant freshwater bodies from the Kerguelen Islands. It is part of an ongoing study on the Rotifera from the French subantarctic islands Amsterdam, Crozet, Kerguelen and St. Paul.

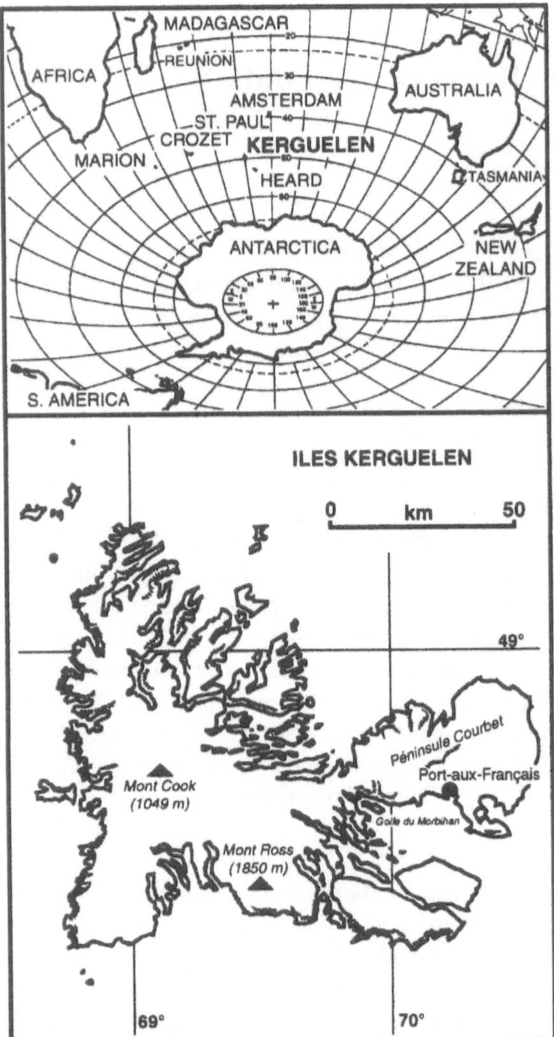


*Figure 1.* Upper: map showing Iles Kerguelen in relation to the continents and some other subantarctic islands. Lower: map of Iles Kerguelen.

## Study area

The region studied (Fig. 1) is located in the southern part of the Courbet peninsula (east of Grande Terre). It includes the area around the base Port-aux-Français at a distance of about 0.01–1 km from the sea and between 10 and 30 m a.s.l., and the area of Lac Supérieur (formerly called Lac Studer 1) in the Val Studer, 15 km WNW from the base, at an altitude of 70–80 m a.s.l. and distance to the sea of 7.4 km.

## Materials and methods

Plankton samples were taken from 17 permanent freshwater bodies (8 Lac Supérieur, 9 Port-aux-Français), ranging from a lake, lakelets, ponds and pools. They were collected by horizontal hauls, using a 40 $\mu$m mesh plankton net during the austral summer of 1997–1998 (02.02–08.02.1998). All samples were fixed on the spot with formalin up to a concentration of 3%. Basic physical and chemical characteristics of the water (temperature, pH, conductivity) were measured in the field at a depth of $\pm$15 cm from the surface with a WTW multiline P4. The total alkalimetric titer was determined in the field by titrating the water against 0.01 N HCl in the presence of methyl orange.

Rotifers were indentified using the keys in Koste (1978), Nogrady et al. (1995), Segers (1995), De Smet (1996) and De Smet & Pourriot (1997). Specimens were examined and drawn using a Leitz Orthoplan microscope with camera lucida. Preparation of rotifer trophi was done following De Smet (1998). Scanning electron microscopy (SEM) of trophi and critical-point dried specimens was performed with a Philips SEM 515, operated at 20 kV.

## Results and discussion

### *The habitats*

Basic information on the waters studied is summarized in Table 1. Water temperatures were fairly high due to the exceptionally warm weather at the moment of sampling. Extremely rapid thermal exchanges with the atmosphere were noted by Maire (1985). The waters are circum-neutral to slightly alkaline. There is a weak tendency for higher pH-values in the waters around Port-aux-Français (average pH 7.6 *versus* 7.0 for the area of Lac Supérieur). Alkalinity is low in both sampling areas with slightly higher values for the area of Port-aux-Français (average 0.24 *versus* 0.18 for the area of Lac Supérieur). On basis of the alkalinity all waters are to be classified as weakly buffered and very soft (0.1–0.5 mmol $l^{-1}$), and oligotrophic (alkalinity $<$0.3 mmol $l^{-1}$). Both sampling areas are clearly distinguished by the electrical conductivity. In the area around Lac Supérieur, conductivity varies from 40 to 60 $\mu$S cm$^{-1}$ (mean 46) which indicates a very low mineralization. The waters from Port-aux-Français show the highest conductivity values ranging between 100 and 340 $\mu$S cm$^{-1}$(mean 202), pointing

*Table 1.* List of waterbodies sampled for plankton, with some ecological information (Surf. = approximate surface $m^2$, Depth m, Temp. = water temperature °C, pH, Cond. = conductivity $\mu S\ cm^{-1}$, Alk. = total alkalinity titer mmol $l^{-1}$

| Site N°/name | Surf. | Depth | Temp. | pH | Cond. | Alk. |
|---|---|---|---|---|---|---|
| **Val. Studer** | | | | | | |
| 3. pool | | | | | | |
| 4. pool | 70 | 0.4 | 17 | 6.6 | 60 | 0.17 |
| 5. pond | 250 | – | 17 | 6.8 | 45 | 0.17 |
| 6. pond | 1.8 $10^3$ | 1+ | 16 | 6.9 | 40 | 0.16 |
| 7. Lac | 5 $10^3$ | 1+ | 15 | 7.2 | 40 | 0.23 |
| Supérieur | | | | | | |
| | 84.6 $10^4$ | 43 | 11 | 7.5 | 60 | 0.24 |
| 8. pond | | | | | | |
| 9. pool | 9 $10^3$ | 0.4 | 14 | 7.4 | 40 | 0.17 |
| 11. pool | 200 | 1 | 15 | 7.0 | 40 | 0.14 |
| | 50 | 0.4 | 17 | 6.7 | 40 | 0.14 |
| **Port-aux-Français** | | | | | | |
| 12. Lac des Sternes | 153 $10^3$ | 1+ | 14 | 7.4 | 340 | 0.21 |
| 14. Etang du Magnétisme | 80 $10^3$ | 1+ | 21 | 8.1 | 210 | 0.21 |
| 16. Etang de la Décharge | 37.5 $10^3$ | 1+ | 19 | 8.2 | 140 | 0.27 |
| 17. Etang des Collets | 11.8 $10^3$ | 1+ | 19 | 8.3 | 150 | 0.29 |
| 18. pond | 9.3 $10^3$ | 0.3+ | 18 | 7.7 | 100 | 0.24 |
| 20. pond | 4.3 $10^3$ | 0.4 | 18 | 6.7 | 300 | 0.05 |
| 27. pool | 24 | 0.5 | 13 | 7.6 | 150 | 0.24 |
| 28. pool | 21 | 0.4 | 15 | 6.7 | 240 | 0.23 |
| 30. pond | 1.9 $10^3$ | – | 12 | 7.7 | 190 | 0.45 |

towards low to medium mineralization. Undoubtedly, these higher conductivities found in the closest to the sea and low-lying sampling area, result from enrichment by sea salts. Limnological information on subantarctic islands, e.g. Marion Island (Grobbelaar, 1978), Macquarie Island (Tyler, 1972), Kerguelen Islands (Tyler, 1972; Gay, 1981, 1982; Maire, 1985), indicates that windborne seaspray contributes largely to the chemical composition of fresh waters, as shown by the ionic concentration and the frequencies of ionic dominance resembling that of seawater.

The phytoplankton was characterized by *Botryococcus braunii* Kütz., *Closterium kuetzingii* Bréb., *Mougeotia* sp., *Pediastrum boryanum* (Turp.) Menegh., *Pleurotaenium* sp. and *Spirogyra* sp., each of which could be bloom forming depending on the water-body. There usually was a well developed zooplankton consisting of cladocerans (*Daphniopsis studeri* Rühe, *Macrothrix* cf. *hirsuticornis* Norman & Brady, *Alona weinecki* Studer), copepods (*Pseudoboeckella volucris* Kiefer, *Acanthocyclops vernalis* (Fischer), *Paracyclops fimbriatus*-group) and rotifers.

### *Rotifer species composition*

An alphabetical list of the rotifer taxa found in the plankton samples is presented in Table 2. Bdelloidea were treated as a gross taxonomic group. In total, 39 taxa were found. Of the 38 Monogononta, 33 were identified to species level, 3 were assigned to genus rank only, and 2 could not be classified. The total number of genera recorded (the 2 species *incertae sedis* excepted) was 18, which is fairly high relative to the total number of species observed.

The number of monogonont species encountered at the different sites ranged from 3 to 23 (mean 9.5). Lac Supérieur (site 7) had the highest species richness, the lowest value was found at site 28. Species diversity (Shannon-Wiener) ranged from 0.18 to 2.68 (mean 1.86). Sites 12, 14 and 15 had a lower diversity (0.18–0.99) than the others (1.26–2.68). Equitability showed similar trends to species diversity, with lowest values at sites 12, 14, 15 (0.07–0.42) and higher ones at the other sites (0.49–0.83), indicating that for a similar species richness, species diversity was mainly determined by the uneven spread of the abundance among species.

No relationship was apparent between rotifer composition and morphometry or chemical and physical variables of the waters.

The most species-rich genera are *Cephalodella* (5 species), and *Lecane*, *Lepadella* and *Notholca* ( 4 species each). The most constantly occurring taxa in the samples are in decreasing order (number of samples with the taxon bracketed): *Lecane flexilis* and *Lepadella patella* (16), Bdelloidea and *Cephalodella gibba* (14), *Trichocerca bidens* (12), *Cephalodella forficula*, *Keratella sancta* and *Lecane closterocerca* (10). The predominance of non-planktonic species is confirmed by the presence of only 6 euplanktonic taxa in the species list. The scarceness of true planktonic species in antarctic and subantarctic locations is also evident from other studies (e.g. Dartnall & Hollowday, 1985; Dartnall, 1993, 1995a,b). A similar and yet unexplained phenomenon occurs with the planktonic

*Table 2.* Rotifer species recorded from plankton in waterbodies from Iles Kerguelen, during this study. Site numbers refer to Table 1. Abundance: r = individual specimens or rare, up to 10% of recorded rotifers, f = frequent, 11 – 20%, m = many, 20 – 50%, a = abundant, more than 50% of rotifers recorded; + = empty lorica; * = euplanktic species

Bdelloidea *indet*: 3a, 4a, 5m, 6m, 7r, 8a, 9f, 11m, 14r, 17r, 18r, 20a, 27m, 28f
*Cephalodella delicata* Wulfert: 30r
*C. forficula* (Ehrenberg): 3r, 4r, 5r, 6r, 7r, 8r, 9r, 11r, 17r, 27r
*C. gibba* (Ehrenberg): 3r, 4r, 6m, 7r, 8r, 9r, 11r, 14r, 15r, 17r, 18r, 20r, 27r, 30f
*C. stenroosi* Wulfert: 7r
*C. ventripes angustior* Donner: 6r
*Collotheca* sp.: 4r, 7r, 8r, 17r
*Colurella colurus* (Ehrenberg): 7r, 8r, 14r, 30r
*Dicranophorus forcipatus* (Müller): 7r
*Encentrum limicola* Otto: 7r
*E. mustela* (Milne): 7r
*E. uncinatum* (Milne): 8r
*Filinia pejleri** Hutchinson: 7f, 12r, 15r, 17m, 18r
*Keratella sancta** Russell: 7a, 8r, 9r, 11+, 12a, 14a, 15a, 17m, 18r, 27r
*Lecane closterocerca* (Schmarda): 3r, 4r, 5r, 7r, 8r, 9f, 11f, 14r, 17r, 18r
*L. flexilis* (Gosse): 3f, 4f, 5m, 6r, 7r, 8r, 9m, 11f, 14r, 15r, 17f, 18f, 20r, 27r, 28f, 30f
*L. hamata* (Stokes): 5r, 6r, 7r, 8r, 11r, 30r
*L. tenuiseta* Harring: 5r, 11r
*Lepadella acuminata* (Ehrenberg): 8+, 20r
*L. oblonga* (Ehrenberg): 3r, 11r, 17r, 18r
*L. patella* (Müller): 3r, 4r, 5f, 6r, 7r, 8r, 9m, 11f, 12r, 14r, 15r, 17r, 18m, 20m, 27r, 28a
*L. triptera* Ehrenberg: 4r, 9r, 14r, 17r
*Lindia torulosa* Dujardin: 7r, 8r, 11r, 14r, 17r, 20r
*Notholca hollowdayi** Dartnall: 3r, 4r, 6r, 17r, 18r, 27m, 28f
*N. labis** Gosse: 7r
*N. squamula** (Müller): 7r
*Notholca** sp.: 7r
*Notommata glyphura* Wulfert: 7r, 14r
*Proales minima* (Montet): 7r, 15r, 17r
*P. fallaciosa* Wulfert: 8r, 11r
*Ptygura* sp.: 7r, 14r, 17r
*Rhinoglena frontalis* (Ehrenberg): 30 m
*Scaridium longicaudum* (Müller): 11r
*Testudinella patina* (Hermann): 18r
*Trichocerca bidens* (Lucks): 3r, 6r, 7r, 8r, 9r, 11r, 12r, 14r, 17r, 18r, 20+, 27r
*T. rattus* (Müller): 17r
*T. relicta* Donner: 4r, 6r, 11r, 12r, 17r, 18r, 27f, 30f
*Incertae sedis* (2): 7r, 8r

freshwater diatoms which are absent in these regions (e.g. Jones, 1996). The preponderance of facultative plankters and true littoral/benthic forms is to be explained by (1) the shallowness of some ponds, (2) the relatively few number of euplankters, favouring the species less adapted to a planktonic way of life in the direct competition for food, and (3) wind mixing. Maire (1985) demonstrated that even the 43 m deep Lac Supérieur (site 7) is thoroughly mixed and never stratified, due to the continuing strong winds, and to a lesser extent by the small amplitude of the water temperature not allowing for a pronounced thermal stratification.

Only a few of the species found show a restricted distribution. *Keratella sancta* was described from New Zealand (Russell, 1944), and has also been reported from the Kerguelen Islands (Russell, 1959; Lair & Koste, 1984), Macquarie Island (Dartnall, 1993) and the Chatham Islands (R. Shiel, I. Duggan, pers. comm.). *Notholca hollowdayi* was hitherto only known from Heard Island (Dartnall, 1995a). Naturally, *Notholca* sp. and the two species *incertae sedis* are restricted to the Kerguelen Islands to date. Most other rotifers are amongst the most common, wide-spread or cosmopolitan species.

Of the 38 monogonont rotifers recorded, 31 of the taxa identified to genus or species level are new for the Kerguelen Islands. In the subantarctic, roughly defined as the zone between 45° and 65° S (Vekhov, 1993), about 55 species have been found (see Zelinka, 1927; De Beauchamp, 1940; Russell, 1959; José De Paggi, 1982; José De Paggi & Koste, 1984; Dartnall & Hollowday, 1985; Dartnall, 1993, 1995a; Hansson et al., 1996). Among the fully identified species recorded in this paper, 12, i.e. *Cephalodella stenroosi*, *C. ventripes angustior*, *Dicranophorus forcipatus*, *Encentrum limicola*, *Lecane flexilis*, *L. hamata*, *L. tenuiseta*, *Notholca labis*, *Proales minima*, *P. fallaciosa*, *Testudinella patina* and *Trichocerca relicta*, are new to the subantarctic.

*Comments on selected species*

*Cephalodella gibba* (Ehrenberg) Fig. 2A–D
Specimens of *Cephalodella gibba* sensu stricto (see Nogrady et al., 1995) and the short-toed *C. gibba microdactyla*, as well as transitions between the two were observed.

The subspecies *microdactyla* was erected by Koch-Althaus (1963) on basis of the shorter body and toes, the presence of a pleural rod laterally of each ramus, and a fan-shaped structure running to the base of the fulcrum on the right ramus. A long spur is depicted on the manubrium in the lateral view figure of Koch-Althaus (loc. cit., Fig. 7d), but should in fact be represented on the lateral end of the ramus according to Godske Eriksen (1969), Braioni & Gelmini (1983) and Turner & Palmer (1996). Examination by SEM of trophi of the *microdactyla* type (Fig. 2A,B) and *gibba* s. str. (Fig. 2C,D) reveals that the diagnostic pleural rods or spurs of *microdactyla*, are present in both types. It is the unilaterally fanned broom-like structures, called subunci by Wulfert (1937; also Koste, 1978; Nogrady et al., 1995), which have often been overlooked by light microscopy but prove present in most *Cephalodella* spp. examined by SEM to date (pers. obs.). The rays of the fan may lie close together as in *C. gibba*, or be spread as in *C. gigantea.* The ventral margin of the fan is connected to the antero-dorsal margin of the frontal plates of the rami. Likewise, the shaft can be connected to the rami dorso-laterally, or be free along the greater part of its length as in *C. gibba.* The shaft at least stands at a more or less right angle to the manubrium. It is clear that these structures are not associated with the unci, and instead of being called subunci, should be designated pleural rods. The nature of the U, V, Y-shaped structures which dissolve in hypochlorite solution, and are commonly called "pleural rods" in *Cephalodella* by, e.g. amongst others, Wulfert (1937), Koste (1978) and Nogrady et al., (1995) is unclear. They may turn out to be ligaments.

The subspecific status of *C. gibba microdactyla* is rejected, considering the transitions between *C. gibba* and *C. gibba microdactyla*, of the taxonomic irrelevance of features relative to body size and toe length , and the identical trophi morphology.

*Filinia pejleri* Hutchinson (Figs 3A–F and 5A,B)
A *Filinia* occurring in both sampling areas was identified as *F. pejleri* on basis of (1) the spindle-shaped body, (2) the greatly enlarged, in lateral view oblique base of the immoveable caudal seta (up to 40 $\mu$m wide) inserted terminally, (3) the size of the setae and body, and their respective ratios (Table 3), and (4) the number of unci teeth. SEM examination of trophi gave an asymmetrical teeth ratio of 16/17 and 17/18 (Fig. 5A, B).

Parthenogenetic (Fig. 3F), male (Fig. 3E) as well as resting eggs (Fig. 3C, D) were found in low numbers.

Lair & Koste (1984) described *F. terminalis kergueleniensis* from Lac Studer 2 (actually called Lac des Truites) at Kerguelen. This lake is near to, and partly fed by Lac Supérieur (present sampling site 7). As has already been noted by Sanoamuang (1993a), the external features of *F. terminalis kergueleniensis* are similar to those of *F. pejleri.* The taxon has a broad-based caudal seta, which is similar only to *F. pejleri*, *F.* cf. *pejleri* and *F. grandis* (Hutchinson, 1964; Koste, 1979; Shiel & Sanoamuang, 1993; Sanoamuang, 1993a,b). Moreover, compared to *F. terminalis*, the caudal seta is placed terminally, and not subdistally. The only resemblance with *F. terminalis* apparently concerns the symmetrical unci with 16/16 teeth, if one accepts the tooth formula 15–16/15–16 reported for the species by Lair & Koste (1984). However these counts were made using light microscopy,

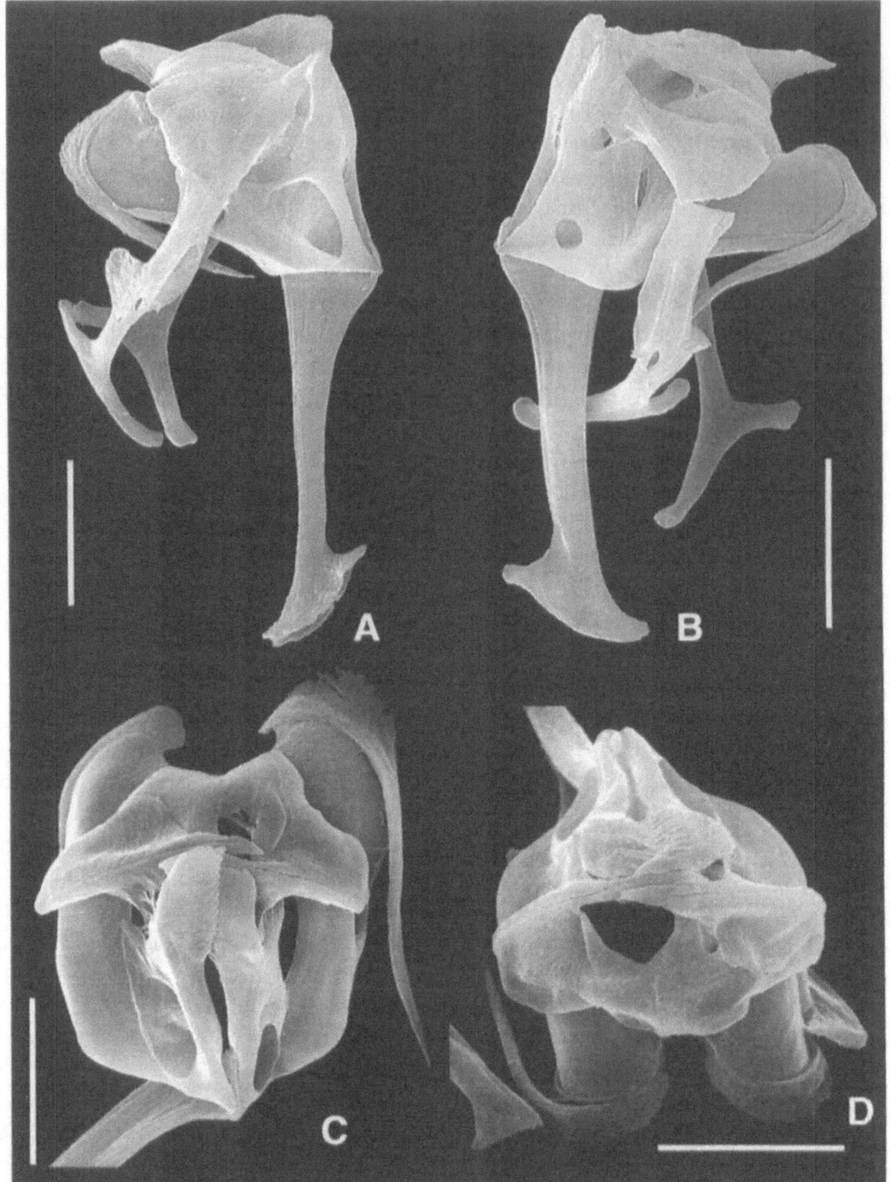


*Figure 2.* (A–D) *Cephalodella gibba* SEM of trophi. (A–B) "*microdactyla*", Iles Kerguelen, (A) lateral view, right; (B) lateral view, left. (C–D) *gibba* "sensu stricto", Antwerpen, Belgium. (C) detail rami, unci and pleural rods, ventral; (D) ibidem, frontal. Scale bar: 10 $\mu$m.

and have to be considered approximate. The tooth formula for *F. terminalis* from Europe and New Zealand, determined by SEM, is 14–15/16–17 (Sanoamuang, 1993a,b). Symmetrical unci, which has been reported often for other species as well (e.g. Koste, 1980), has never been confirmed with SEM (Sanoamuang, 1993a).

In the light of the foregoing discussion on the unci and the comparative measurements presented in Table 3, no distinguishing features are left between the taxa, and it therefore seems justified to synonymize *F. terminalis kergueleniensis* with *F. pejleri*.

Apart from the measurements by Lair & Koste (1984), all data from literature (Koste, 1979; Shiel & Sanoamuang, 1993; Sanoamuang, 1993a) on seta and body size of *F. pejleri* are invariably copied from Hutchinson (1964). Comparing the data summarized in Table 3, it is clear that there is some tendency for greater body and seta length in material from the Kerguelen Islands compared to the material of

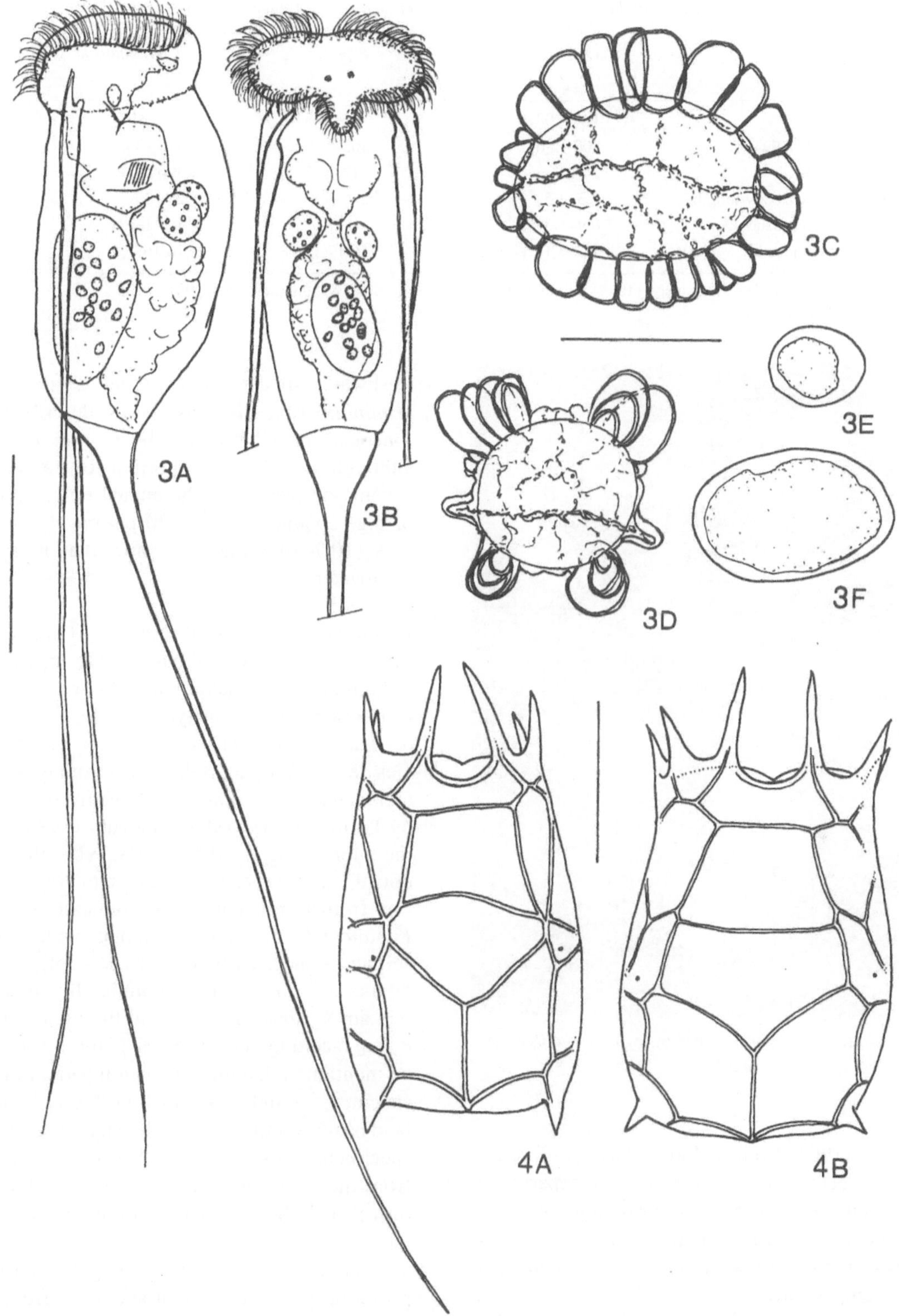


*Figures 3–4.* (3) (A–F) *Filinia pejleri.* (A) habitus, lateral view; (B) ibidem, ventral view; (C) resting egg, lateral view; (D) ibidem, top view; (E) male egg; (F) parthenogenetic egg. Scale bars: (A–B) 100 $\mu$m, (C–F) 50 $\mu$m. (4) (A–B) *Keratella sancta*, lorica dorsal view. Scale bar: 50 $\mu$m.

*Table 3.* Comparative measurements of *Filinia pejleri* and *F. terminalis kergueleniensis* (ranges in $\mu$m)

| | *F. pejleri* (1) | *F. pejleri* (2) | *F. terminalis kergueleniensis* (3) |
|---|---|---|---|
| Body length | 124–200 | 150–208 | 130–310 |
| Lateral setae length | 300–480 | 403–508 | 410–700 |
| Caudal seta length | 242–456 | 312–397 | 300–560 |
| Ratio lateral seta:body length | 2.3–3.5 | 2.4–3.7 | 2.1–3.6 |
| Ratio lateral seta:caudal seta | 0.9–1.3 | 1.1–1.5 | 1.1–1.4 |
| Number of unci teeth | 15/17,16/17,17/18 | 16/17,17/18 | 16/16 |

(1) Body and seta length after Hutchinson (1964), number of unci teeth after Shiel & Sanoamuang (1993).
(2) This study; (3) Lair & Koste (1984).

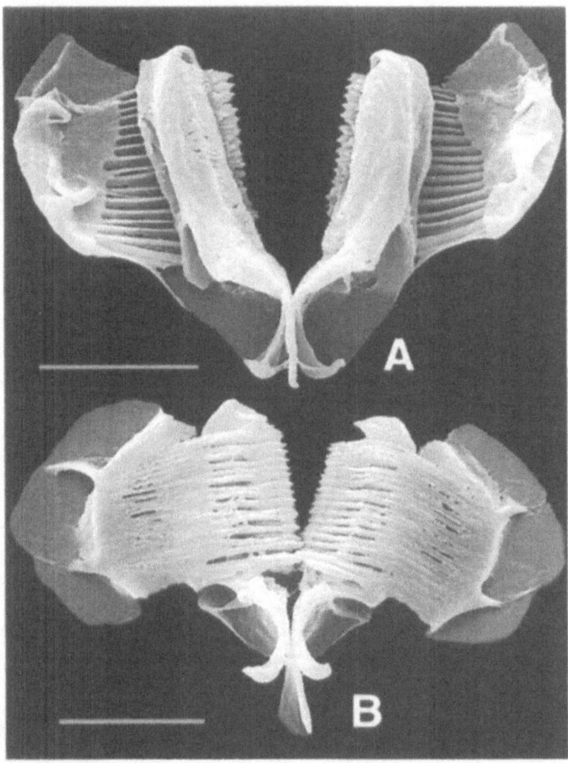


*Figures 5.* (A–B) *Filinia pejleri,* SEM of trophi. (A) dorsal view; (B) ventral view. Scale bar: 10 $\mu$m.

Hutchinson (1964) from S. India. These differences probably reflect the results of the culture experiments by Sanoamuang (1993b), demonstrating that body length, and lengths of lateral and caudal setae of rotifers grown at lower temperatures are longer than those at higher temperatures.

To date, *F. pejleri* was generally considered to be a warm stenotherm (e.g. Hutchinson, 1964; Lair & Koste, 1984), but a classification as eurytherm seems more appropriate. *F. pejleri* is an eurytopic freshwater species. Lac Studer 2 from which '*F. terminalis kergueleniensis*' was described, was erroneously called brackish by Lair & Koste (1984): although the relative proportions of the ions point to oceanic-derived salts, the annual ranges (Gay, 1981) of e.g. conductivity (51–70 $\mu$S cm$^{-1}$) and chloride ions (10.7–14.2 mg l$^{-1}$) show that it is definitely freshwater.

*Keratella sancta* Russell (Figs 4A, B and 6A–G)
*Keratella sancta* was all but the only species occurring in high numbers, making up 50% or more of the total rotifer population density.

There have been some doubts whether the species described by Russell (1944) from New Zealand is identical to the *K. sancta* recorded from Iles Kerguelen by Lair & Koste (1984). Sudzuki (1988) considered the morphological differences, viz. the pattern of dorsal facets, sufficiently great to treat the population from Kerguelen as taxonomically separate from *K. sancta.* Comparing my findings with those of Russell and Lair & Koste (loc. cit.), it follows that these doubts are due to some unreliability in the figures of Lair & Koste (loc.cit.) and the great variability in dorsal facetting of the species. In this respect, it should be mentioned that the foundation pattern of *K. sancta* specimens found by Dartnall (1993) at Macquarie Island is more similar to that shown for the New Zealand specimens (Russell, 1944) than that from specimens from the Kerguelen Islands figured by Lair & Koste (op. cit.). A short description of the specimens seen by me follows.

Anterior region of lorica slightly narrower than posterior, greatest width at about 1/3 from posterior; posterior margin almost straight. Ventral plate weakly pustulate, anterior margin sinuate with shallow median concavity. Dorsal plate more strongly pustulate.

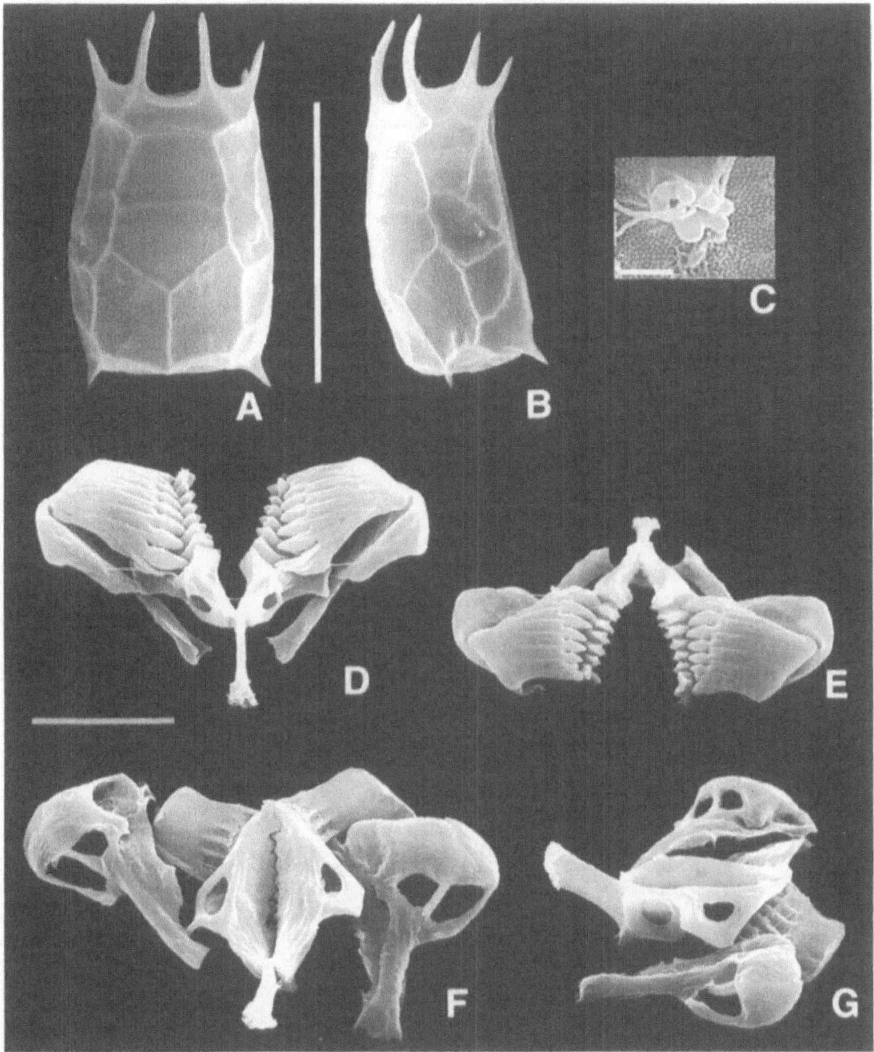


*Figures 6.* (A–G) *Keratella sancta*, SEM photographs. (A) lorica, dorsal view; (B) lorica, lateral view; (C) lateral antenna; (D) trophi, ventral view; (E) ibidem, frontal view; (F) ibidem, dorsal view; (G) ibidem, dorso-lateral view. Scale bars: (A–B) 50 $\mu$m, (C) 3 $\mu$m, (D–G) 10 $\mu$m.

Six stout anterior spines, median pair longest. Two short postero-lateral spines always present. Foundation pattern with two dorso-median facets. The ridge between the antero-median and meso-median facet may be absent, resulting in a single, large facet. Median frontal area broadly hexagonal. Antero-median facet large, wider than long, (1) hexagonal (Figs 4B and 6A), the postero-lateral ridges very short, or (2) quadrilateral by absence of postero-lateral ridges (Fig. 4B). Meso-median facet large, wider than long, (1) pentagonal (Figs 4B and 6A), when antero-median facet hexagonal, or (2) more or less hexagonal when antero-median facet quadrilateral; the antero-lateral and postero-lateral ridges of variable length (Fig. 4A, and Russell (1944: Fig. 1)). When the ridges separating the first two lateral facets originate at the junction of the antero-median facet and meso-median facet, the latter are quadrilateral and pentagonal, respectively. Two large, irregularly hexagonal antero-carinal facets, and two small postero-carinal facets. First two lateral facets irregularly hexagonal, closed; second lateral facet with lateral antenna; aperture of antennae bordered by 6–7 pustules (Fig. 6C). Third lateral facet triangular. Postero-lateral facet irregular, with posterior spine. A triangular marginal facet between first and second lateral facet.

*Trophi malleate* (Fig. 6D–G): Fulcrum fairly short, with weak basal plate. Rami triangular with weak

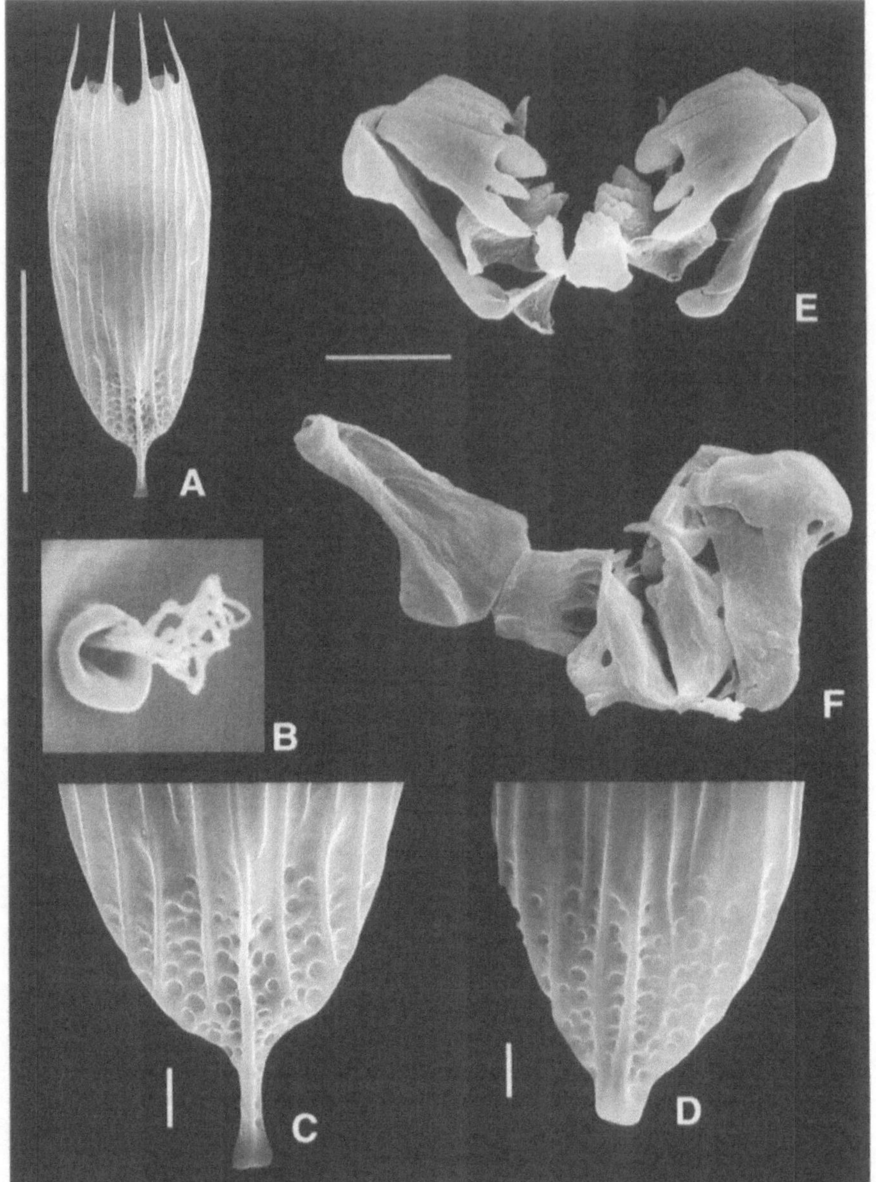


*Figures 7.* (A–F) *Notholca hollowdayi*, SEM photographs. (A) lorica, dorsal view; (B) lateral antenna; (C) posterior of lorica, dorsal view; (D) ibidem; (E) trophi, ventral view; (F) ibidem, dorsal view. Scale bars: (A) 50 μm, (C–F) 10 μm.

alulae; medullae with large projections. Unci with eight teeth, principal teeth with small anterior toothlet. Manubria with rod-shaped shaft; dorsal lamella with projection near connection with unci.

*Measurements:* Lorica length (exclusive of spines) 100–120 μm, lorica width 73–91 μm, antero-median spines 26–36 μm, antero-intermedian spines 13–23 μm, postero-lateral spines 8–13 μm, parthenogenetic egg 72–76×52–54 μm.

*Notholca hollowdayi* Dartnall (Fig. 7A–F)

*Notholca hollowdayi* was described by Dartnall (1995a) from subantarctic Heard Island. A more extensive description of the animals follows.

Dorsal plate with about 15 longitudinal ridges. Median two ridges fusing, between mid-length and posterior 1/4 of dorsal plate, into single ridge continuing on the posterior extension. Dorsal plate of all animals seen by me with posterior extension; the extension is absent in specimens from Heard Island. Posterior extension narrow and parallel-sided, or slightly expanded posteriorly (Fig. 7C); rarely short and broad (Fig. 7D). Grooves between longitudinal ridges of posterior region of dorsal plate dorsally with large circular fossettes of different diameter; fossettes also on ventral part of free end of dorsal plate, and on posterior extension. Ventral plate with 6 longitudinal ridges; posterior region with circular fossettes between the ridges (ventral plate reported undecorated by Dartnall (loc. cit.), but examination of holotype and additional material from type locality revealed the longitudinal ridges and posterior fossettes). Lateral antennae (Fig. 7B) near anterior 1/3, reckoning from notch of dorsal plate; aperture bordered by ring-shaped collar.

*Trophi of* Notholca-*type* (Fig. 7E, F): Fulcrum short, with weak basal plate. Rami with alula and rostellum; medullae with large denticulate projections. Unci with three differentiated main teeth and two vestigial ones. Manubria with straight shafts bearing broad dorsal membrane the whole of their length.

*Measurements* (specimens from Kerguelen): Dorsal plate length (from median notch till posterior end, exclusive extension) 151–178 $\mu$m, width 73–88 $\mu$m, anteromedian spine 30–45 $\mu$m, anterolateral spine 28–33 $\mu$m, posterior extension 7–32 $\mu$m.

*Notholca hollowdayi* is most probably related to *N. walterkostei* described from King George Island, South Shetland Islands (Antarctica) by José De Paggi (1982). Since then, specimens assigned to *N. walterkostei* have been reported from high-altitude lakes from Peru/Bolivia (see Hollowday & Hussey, 1989; Segers et al., 1994), Tierra del Fuego, Argentina (Battistoni, 1992), and several subantarctic and antarctic islands (Dartnall & Hollowday, 1985). However, among these there are noticeable differences concerning ridging of plates, arrangements of fossettes, etc. which in my and Segers's (pers. comm.) opinion, suggest that different species are involved.

## Acknowledgements

We gratefully acknowledge the "Institut Français pour la Recherche et la Technologie Polaires" for the logistic and financial support. The present study is part of the Terrestrial Ecology Program Biosol no. 136 (Dr Y. Frenot, Université de Rennes, Station Biologique Paimpont).

## References

Battistoni, P. A., 1992. Cinco especies del genero *Notholca* Gosse, 1886 (Rotatoria) de la Argentina, incluyendo *N. guidoi* sp. n. Iheringia, Sér. Zool., Porto Alegre 73: 35–45.

Braioni, M. G. & D. Gelmini, 1983. Rotiferi Monogononti (Rotatoria: Monogononta). Guide per il riconoscimento delle specie animali delle acque interne Italiane. 23. Consiglio Nazionale delle Ricerche. Verona: 180 pp.

Dartnall, H. J. G., 1993. The rotifers of Macquarie Island. ANARE Res. Notes 89: 41 pp.

Dartnall, H., 1995a. The rotifers of Heard Island: preliminary survey, with notes on other freshwater groups. Pap. Proc. R. Soc. Tasm. 129: 7–15.

Dartnall, H. J. G., 1995b. Rotifers, and other aquatic invertebrates, from the Larsemann Hills, Antarctica. Pap. Proc. R. Soc. Tasm. 129: 17–23.

Dartnall, H. J. G. & E. D. Hollowday, 1985. Antarctic rotifers. Brit. Antarct. Surv., Sci. Rep. 100: 46 pp.

De Beauchamp, P., 1940. Croisière du Bougainville aux îles australes françaises. XII. Turbellariés et rotifères. Mém. Mus. nat. Hist. Nat., n.s., 14: 313–327, pl. 9.

De Smet, W. H., 1996. Rotifera 4: The Proalidae (Monogononta). In Dumont, H. J. & T. Nogrady (eds), Guides to the Identification of the Microinvertebrates of the Continental Waters of the World 9. SPB Academic Publishing, The Hague, The Netherlands: 102 pp.

De Smet, W. H., 1998. Preparation of rotifer trophi for light and scanning electron microscopy. Hydrobiologia 387/388: 117–121.

De Smet, W. H. & R. Pourriot, 1997. Rotifera 5: The Dicranophoridae (Monogononta) and The Ituridae (Monogononta). In Dumont, H. J. & T. Nogrady (eds), Guides to the Identification of the Microinvertebrates of the Continental Waters of the World 12. SPB Academic Publishing, The Hague, The Netherlands: 344 pp.

Gay, C., 1981. Ecologie du zooplancton d'eau douce des îles Kerguelen. II. Le Lac Studer: physico-chimie, chlorophylle et biomasse zooplanctonique. C.N.F.R.A. 47: 59–74.

Gay, C., 1982. Les eaux douces des Iles Kerguelen et leur peuplement en entomostracés. C.N.F.R.A. 51: 93–99.

Godske Eriksen, B., 1969. Rotifers from two tarns in southern Finland, with a description of a new species, and a list of rotifers previously found in Finland. Acta Zool. Fenn. 125: 36 pp.

Grobbelaar, J. A., 1978. Mechanisms controlling the composition of fresh waters on the sub-antarctic island Marion. Arch. Hydrobiol. 83: 145–157.

Hansson, L.-A., H. J. G. Dartnall, J. C. Ellis-Evans, H. MacAlister & L. J. Tranvik, 1996. Variation in physical, chemical and biological components in the subantarctic lakes of South Georgia. Ecography 19: 393–403.

Hollowday, E. D. & C. G. Hussey, 1989. A re-appraisal of two members of the genus *Notholca* from the Andes, with a note on the fine structure of the lorica of *N. foliacea* (Ehrenberg). Hydrobiologia 186/187: 319–324.

Hutchinson, G. E., 1964. On *Filinia terminalis* (Plate) and *F. pejleri* sp. n. (Rotatoria: family Testudinellidae). Postilla 81: 1–8.

Jones, V. J., 1996. The diversity, distribution and ecology of diatoms from Antarctic inland waters. Biodiv. Conservat. 5: 1433–1449.

José De Paggi, S., 1982. *Notholca walterkostei* sp. nov. y otros rotiferos dulceacuicolas de la Peninsula Potter, Isla 25 de Mayo (Shetland del Sur, Antartida). Rev. Asor. Cienc. Nat. Litoral 13: 81–95.

José De Paggi, S. & W. Koste, 1984. Checklist of the rotifers recorded from Antarctic and Subantarctic areas. Senckenbergiana biol. 65: 169–178.

Koch-Althaus, B., 1963. Systematische und ökologische Studien an Rotatorien des Stechlinsees. Limnologica (Berlin) 1: 375–456.

Koste, W., 1978. Rotatoria. Die Rädertiere Mitteleuropas begründet von Max Voigt. Monogononta. $2^e$ Aufl. I. Textband. II. Tafelband, 234 Taf. Gebrüder Borntraeger, Berlin: 673 pp.

Koste, W., 1979. New Rotifera from the River Murray, Southeastern Australia, with a review of the Australian species of *Brachionus* and *Keratella*. Aust. J. mar. Freshwat. Res., 30: 237–253.

Koste, W., 1980. Ueber zwei Plankton-Rädertiertaxa *Filinia australiensis* n. sp. und *Filinia hofmanni* n. sp., mit Bemerkungen zur Taxonomie der *longiseta-terminalis*-Gruppe. Genus *Filinia* Bory de St. Vincent, 1824, Familie Filiniidae Bartos 1959, (Ueberordnung Monogononta). Arch. Hydrobiol. 90: 230–256.

Lair, N. & W. Koste, 1984. The rotifer fauna and population dynamics of Lake Studer 2 (Kerguelen Archipelago) with description of *Filinia terminalis kergueleniensis* n. ssp. and a new record of *Keratella sancta* Russell 1944. Hydrobiologia 108: 57–64.

Maire, P., 1985. Contribution à l'étude hydrobiologique du Lac Studer, écologie du phytoplancton et production primaire, Iles Kerguelen, Terres Australes et Antarctiques Françaises. Thèse. Université de Metz: 126 pp.

Nogrady, T., R. Póurriot & H. Segers, 1995. Rotifera 3: The Notommatidae and The Scaridiidae. In Dumont, H. J. & J. Nogrady (eds), Guides to the Identification of the Microinvertebrates of the Continental Waters of the World of 8. SPB Academic Publishing, The Hague, The Netherlands: 248 pp.

Richters, F., 1907. Die Fauna der Moosrasen des Gaussbergs und einiger südlicher Inseln. Deutsche Südpolar Expedition 1901–1903, Berlin 1907, Vol. 9, Zool. H9, 1: 258–302, Taf. 16–20.

Russell, C. R., 1944. A new rotifer from New Zealand. J. roy. microscop. Soc. 64: 121–123.

Russell, C. R., 1959. Rotifera. B.A.N.Z. Antarctic Research Expedition 1929–1931. Rep. Ser. B (Zool., Bot.), Vol. 8(3): 83–87.

Sanoamuang, L., 1993a. Comparative studies on scanning electron microscopy of trophi of the genus *Filinia* Bory de St. Vincent (Rotifera). Hydrobiologia 264: 115–128.

Sanoamuang, L., 1993b. The effect of temperature on morphology, life history and growth rate of *Filinia terminalis* (Plate) and *Filinia* cf. *pejleri* Hutchinson in culture. Freshwat. Biol. 30: 257–267.

Segers, H., 1995. Rotifera 2: The Lecanidae (Monogononta). In Dumont, H. J. & T. Nogrady (eds), Guides to the Identification of the Microinvertebrates of the Continental Waters of the World 6. SPB Academic Publishing, The Hague, The Netherlands: 226 pp.

Segers, H., L. Meneses & M. Del Castillo, 1994. Rotifera (Monogononta) from Lake Kothia, a high-altitude lake in the Bolivian Andes. Arch. Hydrobiol. 132: 227–236.

Shiel, R. J. & L. Sanoamuang, 1993. Trans-Tasman variation in Australasian *Filinia* populations. Hydrobiologia 255/256: 455–462.

Stonehouse, B., 1982. La zonation écologique sous les hautes latitudes australes. C.N.F.R.A. 51: 531–537.

Sudzuki, M., 1988. Comments on the antarctic Rotifera. Hydrobiologia 165: 89–96.

Turner, P. N. & M. A. Palmer, 1996. Notes on the species composition of the rotifer community inhabiting the interstitial sands of Goose Creek, Virginia with comments on habitat preferences. Quekett J. Microsc. 37: 552–565.

Tyler, P. A. , 1972. Reconnaissance limnology of sub-antarctic islands 1. Chemistry of lake waters from Macquarie Island and the îles Kerguelen. Int. Rev. ges. Hydrobiol. 57: 759–778.

Vekhov, N. N., 1993. Invertebrates (rotifers and crustaceans) of fresh and brackish waters of the Southern circumpolar district (the review of fauna and the results of investigations). Antarktika 32: 167–187. (in Russian).

Wulfert, K., 1937. Beiträge zur Kenntnis der Rädertierfauna Deutschlands. Teil III. Arch. Hydrobiol. 31: 592–635.

Zelinka, K., 1927. Die Rädertiere der deutschen Südpolar-Expedition 1901-1903. Berlin 1927, Bd. 19, Zool. 11: 422–446, Taf. 15–16.

Hydrobiologia **446/447:** 273–282, 2001.
L. Sanoamuang, H. Segers, R.J. Shiel & R.D. Gulati (eds), Rotifera IX.




# Rotifers in saline waters from Disko Island, West Greenland

Peter Funch[1] & Martin Vinther Sørensen[2]
[1]*Department of Zoology, Institute of Biological Sciences, University of Aarhus, Universitetsparken, Building 135, DK-8000 Århus C, Denmark*
*E-mail: peter.funch@biology.au.dk*
[2]*Invertebrate Department, Zoological Museum, University of Copenhagen, Universitetsparken 15, DK-2100 Copenhagen Ø, Denmark*

*Key words:* Arctic, marine relicts, morphology, SEM, taxonomy, trophi

## Abstract

Knowledge of rotifers from Greenland is poor. Those from marine and other saline waters have especially been poorly investigated. The authors have studied meiofauna from different saline localities at Disko Island, West Greenland on several occasions from 1990 to 1999. Samples from the intertidal zone, rock pools and a saline, radioactive spring contained 16 rotifer species. Six of these are new to Greenland. The radioactive salt spring has previously been suggested to contain marine relicts dating back to a warmer period with a higher sea level. The samples from the spring, however, did not yield any marine rotifers. It did contain the euryhaline *Cephalodella gracilis.* A description of its trophi adds to the known variability of this species. The trophi of *Notholca striata* and *N. psammarina* are studied in detail using light and scanning electron microscopy for the first time, and aspects of their taxonomy are discussed.

## Introduction

Disko Island is rich in homothermic springs, which maintain a fairly constant temperature during the year. The temperature of the springs varies between 0.5 and 19 °C and most springs have water with very low concentrations of electrolytes. However, one type of warm spring is radioactive and salt. Rotifers from such springs have not been studied in Greenland before and were, therefore, included in this study. Studies of other groups from these warm salt springs have revealed the occurrence of marine species (Reisinger & Steinböck, 1927; Røen, 1962; Kristensen, 1977; Nehring, 1988; Hansen et al., 1989). These have been regarded as relicts dating back to the hypsithermal period following the last glaciation. When the temperature rose and the glaciers retreated 9000 years ago the landmasses rose slowly. Springs, which today are situated up to 100 m a.s.l., were submerged in the warmer sea. Marine species, which typically are found in warmer climates, could in some cases survive in the warm salt springs. Several species of rotifers are euryhaline and capable of living under conditions of fluctuating salinity; such rotifers were well adapted to survive as relicts in homothermic springs as the climate cooled. However, euryhaline species are also well adapted to live in other Arctic habitats, in particular the tidal zone. Here the organisms have to cope with fluctuating salinities resulting from the influence of melt-water, rain at low tides and seawater at high tides. Given the fact that the warm salt springs of Disko are situated near the sea, it is also possible that rotifers living presently in the tidal zone could have migrated to the saline springs and become established there and may not necessarily therefore be marine relicts. In this case, we would expect the species of rotifers in the spring to be identical to some of the marine or brackish water species of Disko today.

The most comprehensive studies of the Greenlandic rotifer fauna are from freshwater habitats (Bergendal, 1892; De Smet et al., 1993). Studies of marine rotifers from Greenland are restricted to a few such as Vanhöffen (1897), Funch Andersen (1990) and Sørensen (1998). In this study, we investigated the rotifer fauna on Disko Island from different saline habitats. We included one salt spring and several marine and brackish localities from the tidal zone, lagoons, rock-pools and puddles. We also examined the possibility that the rotifer fauna of the salt spring contained any marine relicts.

## Materials and methods

### *Locality descriptions*

Disko (Qeqertarsuaq) is an island with an area of 8620 km$^2$. It is situated in Disko Bay on the west coast of Greenland (Fig. 1A, inset). The geology is varied but dominated by tertiary basaltic mountains rising to a height of up to 1.9 km above sea level. A ridge of Precambrian gneiss stretches N–S beneath the basalt. About 1/6 of the area is covered with glaciers. Permafrost is pronounced but many homothermic springs allows southern species to live. The tides of the sea surrounding Disko are mixed, with maximum amplitude around 3 m. Sea ice is typically present from November to May.

Sandy beaches dominate the eastern coast and two localities from there, Isunngua and Flakkerhuk (Fig. 1A), were included in this study. The southern and western coast is rocky with either gneiss or basalt. Rotifers were sampled from: a rock-pool in Fortuna Bay at the southern coast (Fig. 1A), three stations in the tidal zone at Nipisat (Fig. 1B), and a marine lagoon and an isolated puddle at Ungusivik (Fig. 1B). Finally, samples were taken from the special spring Tarajornitsoq in Disko Fjord (Fig. 1A). The spring water is homothermic, about 11 °C throughout the year (Feilberg, 1985), and contains elevated concentrations of $Na^+$, $Cl^-$, $SO_4{}^{2-}$ and $Ca^{2+}$ ions (Hansen et al., 1989). The gamma radiation is up to 1400 CPS, which is 35 times higher than the background and helium is present in the water, all of which indicate that a radioactive source is associated with the spring. The concentration of $F^-$ is also high and white fluoride rich gypsum is precipitating on the rocks and stones in the vicinity of the spring (Hansen et al., 1989). Data on the stations are given in Table 1.

### *Handling of material*

Samples taken from algae or large accumulations of bacteria and detritus were squeezed into a 30 $\mu$m mesh sieve. For sediment samples, the procedure outlined by Sørensen (1998) was followed. The plankton sample from Nipisat 2 was taken by pulling a 150 $\mu$m conical plankton mesh through the water for approximately 5 min.

The sorting and preparation of specimens and trophi for LM and SEM followed the procedures outlined by De Smet (1998) and Sørensen (1998). The specimens mounted on slides were examined and photographed using Nomarski differential interference contrast with an Olympus BX60 microscope and an Olympus DP10 digital camera. Drawings were made on a Leica DM-RXA Nomarski microscope using a camera lucida. Specimens for SEM were fixed in 4% buffered formaldehyde or trialdehyde (Kalt & Tandler, 1971) in 2/3 sea water, stained with 1% osmium tetroxide in cacodylate buffer, transferred through an acetone dehydration series culminating in clean acetone, dried in a critical point depression apparatus using carbon dioxide and mounted on a stub. All specimens for SEM were sputter coated with gold and examined with a JEOL JSM-840 microscope.

## Results

Sixteen species of rotifers (Table 2), distributed over seven genera and six families, were found in the saline waters from Disko Island. Six species are new to Greenland. One of these, *Notholca* cf. *marina* Focke, 1961, was represented by a single empty lorica only and this record therefore requires confirmation.

*Notholca angakkoq* Sørensen, 1998

*Notholca angakkoq* was found at four localities: Flakkerhuk, Nipisat 1, Ungusivik 1, and Ungusivik 2 (Fig. 1). This species was to date only known from its type locality, a drainage channel at Disko Island close to the lagoon at Flakkerhuk, which was investigated again during the present study.

The species is closely related to other *Notholca* spp. with movable lateral spines, the *Notholca striata*-group, which holds five species. The group appears to be circumpolar, but the distribution of this particular species is uncertain. Earlier, only one or two species, *N. striata* (Müller, 1786) and *N. bipalium* (Müller, 1786), were recognised and in some reports, these might have been confused with *N. angakkoq*, *N. ikaitophila* Sørensen & Kristensen, 2000, or *N. liepetterseni* Godske Björklund, 1972.

Until recently, the species in the *Notholca striata*-group were solely recognised by their spine and lorica dimensions (e.g. see Godske Björklund, 1972 and Koste, 1978), but these features alone are insufficient to characterise the species. Sørensen (1998) and

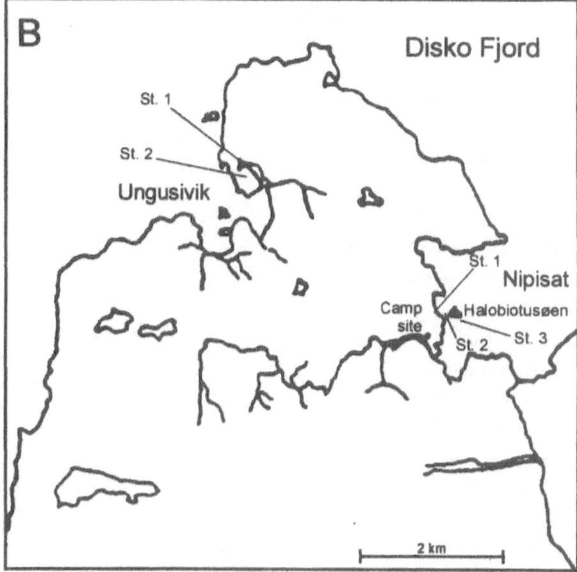


*Figure 1.* Maps showing the locations of sampling sites. (A) Disko Island; inset: Greenland; (B) Close up of sampling area in the south-western part of Disko Fjord (redrawn from Jørgensen & Schiøtt, 1996).

*Table 1.* Physical conditions at 9 localities at Disko Island. Salinity in rock pool at Fortuna Bay, marked with a *, is measured above the halocline

| | Isunngua | Flakkerhuk | Fortuna Bay | Nipisat 1 | Nipisat 2 | Nipisat 3 | Ungusivik | Tarajornitsoq |
|---|---|---|---|---|---|---|---|---|
| Collection date | 11 Aug. 1999 | 11 Aug. 1999 | 25 Aug. 1999 | 4 Aug. 1999 | 4 Aug. 1999 | 8 Aug 1990 | 4 Aug. 1999 | 2 Aug. 1999 |
| Position N | 69 °43.74' | 69 °36.97' | 69 °26.00' | 69 °25.93' | 69 °25.93' | 69 °25.93' | 69 °26.52' | 69 °33.40' |
| Position W | 051 °56.28' | 051 °53.64' | 053 °48.90' | 053 °10.87' | 053 °10.87' | 053 °10.87' | 054 °14.23' | 053 °36.40' |
| Height above sea level (m) | 0 | −0.5 | 10 | 0.5 | −0.5 | 0 | 0/0.5 | 30 |
| Short description | Tidal zone Low tide Exposed beach | Lagoon with fluctuating salinity | Spray zone Rock pool Exposed rocky coast | Tidal zone Large isolated puddle | Tidal zone Protected shallow bay | Tidal zone Protected shallow bay | Tidal zone Lagoon/ Small puddle | Near source of spring |
| Geology | Cretaceous sand | Cretaceous sand | Gneiss | Basalt | Basalt | Basalt | Basalt | Gneiss/basalt |
| Sampling media | Sediment | Sediment | Plankton | Mud/bacteria | Plankton | Algae | Mud/bacteria | Cyanobacteria |
| Sediment type | Coarse quartz sand | Sand/mud | Without sediment | Mud/bacteria | Rocks/gravel | Sand | Mud/bacteria | Rocks/mud |
| Remarks on Flora/fauna | – | Surface rich in bacteria and diatoms | No macroalgae | Numerous fish fry | Lots of *Fucus* and *Mytilus edulis* | Lots of *Fucus* and *Mytilus edulis* | Mats of filamentous algae | Mats of filamentous cyanobacteria |
| Temperature (°C) | – | – | 11 | 10 | 8 | 10 | 11 | 11 |
| Salinity (‰) | – | 23 | 16*/25 | 23 | 14 | 31 | 32/28 | 1 |

*Table 2.* List of rotifers from saline waters, Disko Island. Species marked with * are new to Greenland. [1]) Specimens represented by trophi in the stomach of *A. clydona*. [2]) One specimen represented by an empty lorica

| | Isunngua | Flakkerhuk | Fortuna Bay | Nipisat 1 | Nipisat 2 | Nipisat 3 | Ungusivik 1 | Ungusivik 2 | Tarajornitsoq |
|---|---|---|---|---|---|---|---|---|---|
| *Aspelta clydona** | | | | | + | + | | + | |
| *Cephalodella gracilis* | | | | | | | | | + |
| *Colurella colurus* | | | | + | + | + | + | + | |
| *C. uncinata* | | | | | | | | | + |
| *C. unicauda* | | | | | | + | + | | |
| *Encentrum algente** | | + | | + | | | | + | |
| *E.* cf. *tenuidigitatum** | + | | | | | | | | |
| *E. limicola* | | | | + | | | | | |
| *E. marinum* | | | + | + | +[1] | | + | + | |
| *Notholca angakkoq* | | + | | + | | | + | + | |
| *N. liepetterseni* | | | | | | + | | | |
| *N.* cf. *marina** | | +[2] | | | | | | | |
| *N. psammarina** | | + | | | | | | + | |
| *N. striata** | | | | + | | | | + | |
| *Proales reinhardti* | | | + | + | + | + | + | + | |
| *Synchaeta sp.* | | + | | | | | | | |

Sørensen & Kristensen (2000) give a comparison of diagnostic features in *N. angakkoq*, *N. ikaitophila* and *N. liepetterseni*. A comparison with *N. striata* is given below.

The investigated specimens were larger, with greater lorica widths than those recorded by Sørensen (1998). All specimens had well-developed anterior spines at the dorsal lorica plate and the median spines were always longer than the lateral. The ventral longitudinal fusion lines between the lorica plates were very distinct, especially in living animals. Furthermore, some preserved animals had a distinct transverse line at the posterior end of the ventral lorica plate. All living specimens had distinct longitudinal furrows on the dorsal lorica plate.

*Measurements* ($n$=15): Lorica length, 172–197 $\mu$m; lorica width, 88–121 $\mu$m, median anterior spines, 9–16 $\mu$m; lateral anterior spines, 8–12 $\mu$m; movable lateral spines, 18–26 $\mu$m; trophi, 28–35 $\mu$m; rami, 17–25 $\mu$m; fulcrum, 10–11 $\mu$m; unci, 18–25 $\mu$m; manubria, 22–32 $\mu$m.

*Notholca liepetterseni* Godske Björklund, 1972

*Notholca liepetterseni* dominated the meiofauna from Nipisat 3 (Fig. 1B). It has previously been recorded from rock pools along the Norway coast (Godske Björklund, 1972) and from Disko Fjord (Funch Andersen, 1990). The species belongs to the *Notholca*

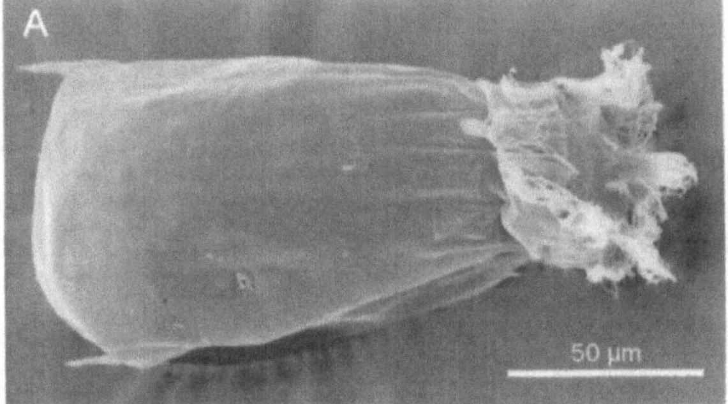


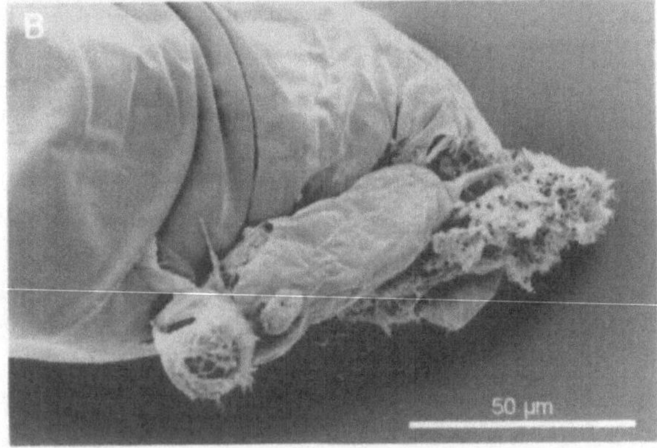


*Figure 2.* SEM pictures. (A) *Notholca liepetterseni*, dorsal view; (B) *Aspelta clydona*, female with attached dwarf male.

*striata*-group and is easily recognised by its smooth lorica and long movable lateral spines (Fig. 2A). The paired lateral antennae are multiciliate with an oval aperture. Only females were observed; some contained large eggs inside.

*Measurements* ($n$=4): Lorica length, 136–140 μm; lorica width, 69–81 μm; median anterior spines, 15–17 μm; lateral anterior spines, 15 μm; movable lateral spines, 22–23 μm; manubrium, 19–22 μm; rami, 11 μm; unci, 12–14 μm.

*Notholca psammarina* Buchholz & Rühmann, 1956

This species was found at two localities: Flakkerhuk and Ungusivik 2 (Fig. 1). The animal is considered Holarctic (De Ridder, 1981) and has been reported from Scandinavia (Godske Eriksen, 1968; Thane-Fenchel, 1968; Godske Björklund, 1972), North Germany (Buchholz & Rühmann, 1956; Focke, 1961) and Iceland (De Ridder, 1972). The species is often reported as rare, but at Ungusivik and especially Flakkerhuk it appeared in relatively high numbers. As in earlier reports (Thane-Fenchel, 1968; Godske Björklund, 1972), *N. psammarina* was found in bacteria and detritus rich sediments. The species is new to Greenland.

The morphology of the trophi was investigated with SEM (Fig. 3D–F). The tubercles, apically on the rami, are fused to three or four large knob-like teeth. Rostella are long and slender and bent distally, such that the inner margins appear semi-circular (Fig. 3 D–E). The margins of the subbasifenestrae (ventral fenestrae) are elevated and protrude caudally, such that they are visible even from a dorsal view (Fig. 3 D, F). The manubria have an opened anterior chamber, a closed median chamber forming a rod-shaped cauda and a closed posterior, which is the basic pattern in the genus. The fulcrum is characteristic in being very short and truncate (Fig. 3F).

*Measurements* ($n$=12): Lorica length: 103–126 μm; lorica width: 47–67 μm; median anterior spines: 6–7 μm; lateral anterior spines: 6–8 μm; caudal spine: 8–12 μm; trophi: 18–20 μm; rami: 10–11 μm; fulcrum: 2 μm; unci: 12–13 μm; manubria: 15–18 μm.

*Notholca striata* (Müller, 1786)

The species was found at two localities: Nipisat 1 and Ungusivik 2 (Fig. 1B). The species has been reported from several localities in the Palaearctic, Nearctic and Neotropic regions, i.e. the White Sea and the Barents Sea (Ghilarov, 1967), Spitsbergen (Amrén, 1964), Scandinavia (Godske Björklund, 1972), Iceland (De Ridder, 1972) and Alaska and Canada (Harring, 1921). As noted above, earlier reports on species belonging to the *Notholca striata*-group might be uncertain, according to the taxonomic and diagnostic confusions. The species is new to Greenland.

The investigated specimens were identified as *N. striata* because of their minute size and because their lateral anterior spines were longer or equal in length to their median anterior spines. It should be stressed that spine lengths only should be used with caution. Even though the median anterior spines often are shorter than the lateral, they can be highly variable (Focke, 1961). However, the character must be considered as useful, combined with other characters, as it may give a hint about the identity of the species.

The living specimens displayed the same variation in lorica shape as *N. angakkoq* and had similar lorica ornamentation. One individual contained a subitaneous egg, which refutes the idea that *N. striata* could be considered a newly hatched and immature *N. angakkoq*.

Trophi morphology was investigated using LM and SEM (Fig. 3A–C). The right uncus has three well-developed, finger-shaped teeth, while the left uncus has 4 teeth of varying size: Tooth 1 is large and finger-shaped; tooth 2 has the same shape as 1, but is smaller; tooth 3 is very large and flattened; tooth 4 is minute and fused with tooth 3. Both unci have well-developed subunci with sclerovilli at the internal surface (Fig. 3B). Both rami have a curved band of tubercles as described from *N. angakkoq*, but the tubercles can only be recognised close to the basal apophyses. Towards the rostella they are gradually reduced.

This study did not reveal any valuable diagnostic characters to distinguish *N. striata* from the other taxa in the *N. striata* group. A full biometrical analysis of the *N. stiata* group is strongly needed to clarify the taxonomy of this confusing group.

*Measurements* ($n$=6): Lorica length, 115–171 μm; lorica width, 77–112 μm; median anterior spines, 5–19 μm; lateral anterior spines, 7–12 μm; movable lateral spines, 15–19 μm; trophi, 19–28 μm; rami, 10–23 μm; fulcrum, 5–6 μm; unci, 13–14 μm; manubria, 16–26 μm; subitaneous egg, 92/69 μm.

*Aspelta clydona* Harring & Myers, 1928

This species was found at three localities: Nipisat 2, 3 and Ungusivik 2 (Fig. 1). Nipisat 2 and Ungusivik 2 were isolated saline puddles and only in contact with the sea during high tide or stormy weather. *A. clydona* is known to have a boreal distribution (De Smet, 1997)

*Figure 3.* SEM of trophi from *Notholca striata* (A–C) and *N. psammarina* (D–F). (A) Ventral view; (B) Ventrofrontal view; (C) Ventrolateral view; (D) Dorsal view; (E) Dorsal view, right uncus and manubrium loosened; (F) Dorsolateral view. Abbreviations: fu – fulcrum; kt – knob-like tooth; ro – rostellum; sc – sclerovilli; sf – subbasifenestra; tu – tubercles.

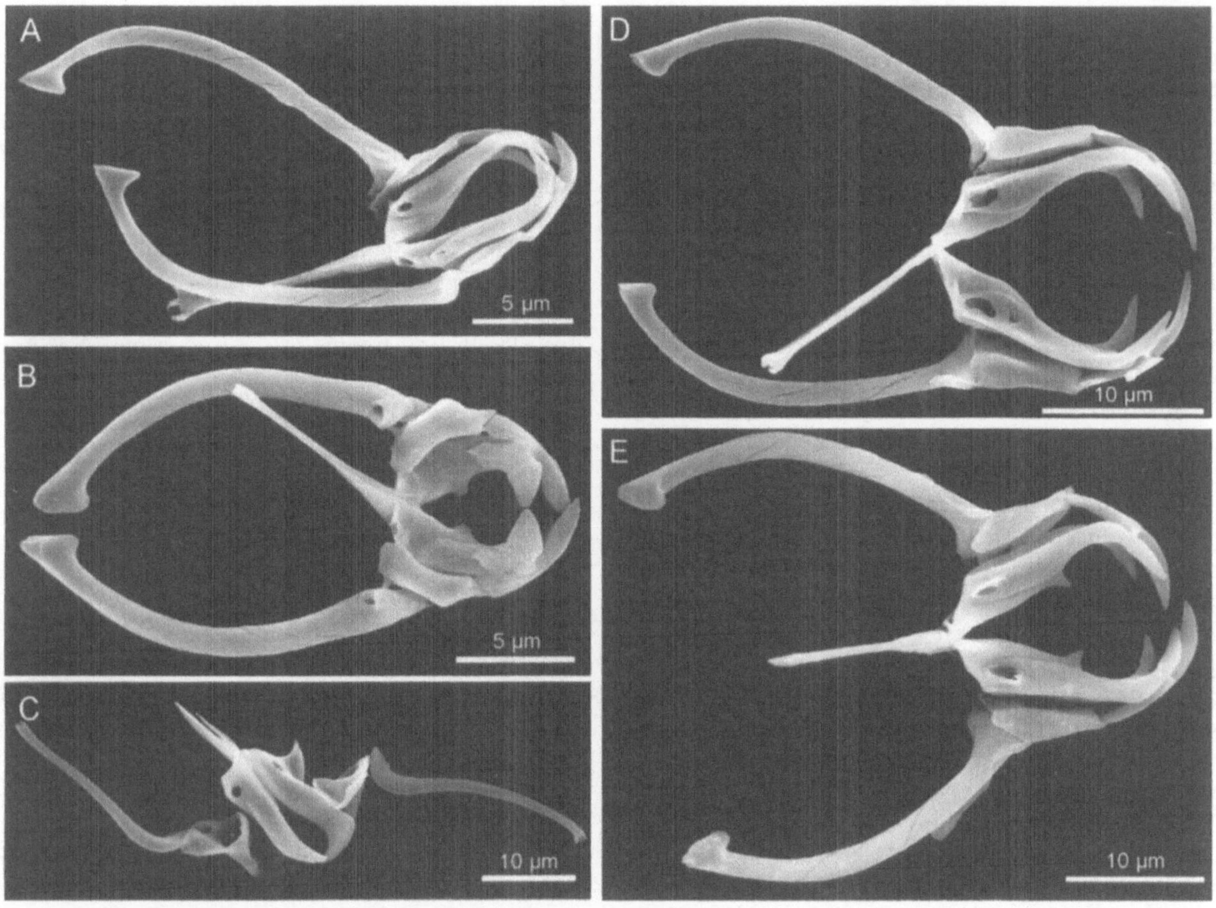


*Figure 4.* SEM of trophi from recorded dicranophorids. (A) *Encentrum marinum*, dorsal view; (B) *E.* cf. *tenuidigitatum*, ventral view; (C) *Aspelta clydona*, dorsal view; (D) *E. limicola*, dorsal view; (E) *E. algente*, dorsal view.

but has also been reported from Spitsbergen (De Smet, 1995). The species is new to Greenland.

Most species in this genus are carnivores (De Smet, 1997), and the strongly asymmetric trophi (Fig. 4C) might be an adaptation for manipulating food objects. *A. clydona* is known to feed on other rotifers such as *Encentrum* spp. (De Smet, 1997) and *Proales reinhardti* (see Jansson, 1967). The guts of some of the examined specimens contained trophi from ingested prey, thus it was easy to examine their food preferences. Trophi from three different species, *Colurella colurus*, *Encentrum marinum* and *Proales reinhardti*, were found in the gut. Trophi from the last two species occurred especially frequently. Males of *A. clydona* were present at Nipisat 3 (Fig. 2B).

*Encentrum* spp.

Four species of *Encentrum* were recorded in this study. Two of these, *E. limicola* Otto, 1936 (Fig. 4D) and *E. marinum* (Dujardin, 1841) (Fig. 4A), have earlier been recorded from Greenland. Sørensen & Kristensen (2000) recorded *Encentrum limicola* from the ikaite columns in Ikka Fjord. Bergendal (1892) recorded *E. marinum* from the Ilulissat/Aasiat area and Sørensen (1998) found it on the east coast of Disko Island.

*Encentrum* cf. *tenuidigitatum* De Smet, 2000. The sandy sediment from the tidal zone at Isunngua (Fig. 1A) contained only two rotifer specimens. In accordance to the appearance of their trophi (Fig. 4B), displaying stout single-toothed unci and rami with a pair of small triangular teeth at the inner margins, the species was identified as *E.* cf. *tenuidigitatum*. Until now the species has only been recorded at the type locality in Belgium (De Smet, 2000), thus it is new to Greenland.

*Encentrum algente* Harring, 1921

*Encentrum algente* (Fig. 4E) was found in the lagoon of Flakkerhuk and in the puddles of Nipisat 1 and Ungusivik 2 (Fig. 1). The species was originally described from a similar brackish lagoon in Alaska, but has also been recorded from Scandinavian saline waters and from Spitsbergen (De Smet, 1997). *E. algente* appears to be a cold stenothermal species and is new to Greenland.

So far, all known species of *Encentrum* are considered oviparous (De Smet, 1997), but some of the live specimens of *E. algente* appeared to contain one well-developed embryo inside. If this interpretation is correct then this would be the first record of viviparity or ovoviviparity in *Encentrum*.

*Measurements* ($n$=1): Length, 244 $\mu$m; toe, 17 $\mu$m; manubrium (adult), 31 $\mu$m; (embryo inside), 24 $\mu$m.

*Cephalodella gracilis* (Ehrenberg, 1832)

*Cephalodella gracilis* was found in the radioactive, saline spring Tarajornitsoq. It occurred in the thick assemblages of filamentous cyanobacteria, i.e. *Oscillatoria* sp. and *Rhizoclonium* sp. dominating the upper part of the spring. The species is a cosmopolitan and primarily a freshwater species. However, it has previously been recorded from saline waters, for instance the eastern Baltic Sea (Remane, 1929) and inland salt springs in east-central Germany (Althaus, 1957a, b). It has also been recorded from freshwater in northern Alaska (Chengalath & Koste, 1989) and in southern Greenland (De Smet et al., 1993).

The specimens in this study were identified as *C. gracilis* based on the presence of a vitellarium with only four nuclei, a frontal eyespot and a trophus type B (for explanation of trophi types see Koste, 1978 and Nogrady & Pourriot, 1995). However, the morphology of the species and the trophi are known to be highly diverse (Donner, 1950; Koste, 1978; Nogrady & Pourriot, 1995).

The body is short and gibbous, and the head is short with well-developed rostrum. The foot is conical and short. Toes are of medium length and almost straight or slightly ventrally bent, tapering gradually to tips. The eyespot is single and frontal, and the vitellarium has four nuclei (Fig. 5A).

Trophus is type B (Figs 5B–C and 6). The rami are asymmetrical, with small teeth on their inner margins (Figs 5B and 6B). Basifenestrae on both rami are large, and oval to elongate. The right basifenestra is slightly larger than the left. Subbasifenestrae are triangular and the edge of the left ramus protrudes laterocaudally into an alula. Right ramus has no alula. A pair of fringed pleural rods is attached to the rami dorsally (Fig. 6A, C). The unci are asymmetrical: right uncus is single toothed and expanded basally into a large apophysis; left uncus is larger, flattened, triangular, and with three small teeth (Figs 5C and 6C–D). The manubria are asymmetrical: the distal end of the right manubrium is slightly expanded laterally and inwardly bent; the distal end of the left is spatulate and only slightly bent. Both manubria have a minute, dorsal lamella medially. Fulcrum is slender and straight, slightly expanded distally (Figs 5B–C and 6B–C).

*Measurements* ($n$ = 4): Trunk length, 93–110 $\mu$m; toe, 25–26 $\mu$m; trophi, 21–23 $\mu$m; ramus, 6–7 $\mu$m; fulcrum, 15–16 $\mu$m; uncus, 3–4 $\mu$m; manubrium 13–15 $\mu$m.

## Discussion

The only rotifers found in the rock-pool were *Encentrum marinum* and *Proales reinhardti*. These two species are very common in marine waters in general and were present at most of the stations in this study (Table 2). At the locality Isunngua, only one species *E.* cf. *tenuidigitatum* was recorded.

Nipisat 1 and Ungusivik 2 are both isolated, saline puddles situated on Arctic seashores. At high tides, they are submerged. The two localities were sampled at the same day at low tide. The sediment of both puddles was mud with mats of bacteria. Three species of rotifers co-occurred at the two localities, namely *E. algente*, *N. angakkoq* and *N. striata*. Their occurrence in this study agrees with previous records (Ghilarov, 1967; Godske Björklund, 1972; De Smet, 1997; Sørensen, 1998).

It has been suggested that Tarajornitsoq and other saline, radioactive springs on Disko Island contain marine relicts from a time period with a higher sea level (see e.g. Kristensen, 1977; Nehring, 1988). However, nothing indicated that this was the case for the rotifer fauna in Tarajornitsoq (Table 2). *Colurella uncinata* is a known cosmopolitan. The other recorded species, *Cephalodella gracilis*, is a common euryhaline species, and this study added even more morphological variability to this already highly diverse species. None of the two species occurred in the marine and brackish water localities examined here (Table 2), and it is likely that both species have reached the spring through the transport of resting eggs.

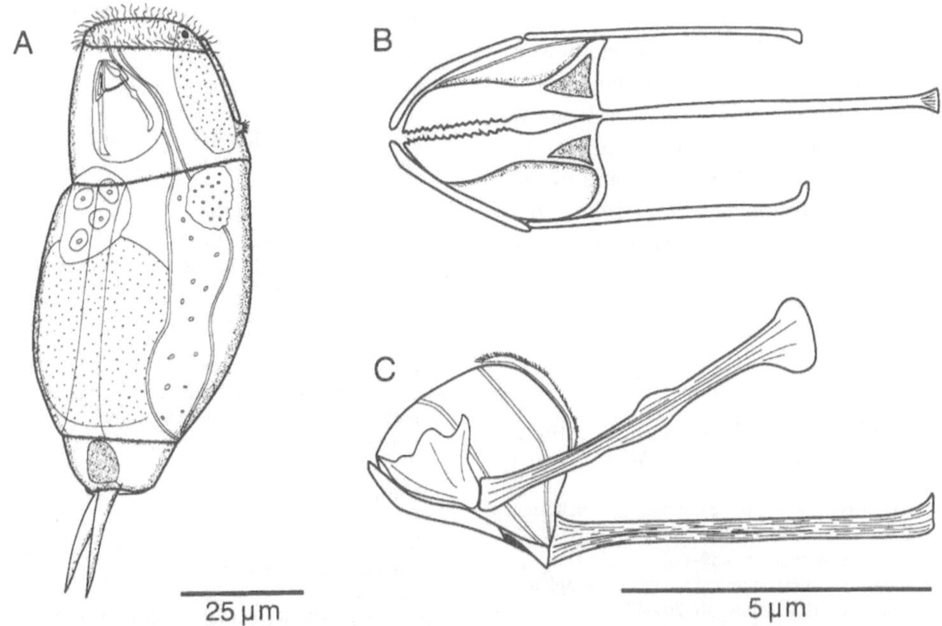


*Figure 5. Cephalodella gracilis.* (A) Female habitus, lateral view; (B) Trophi, ventral view; (C) Trophi, lateral view.

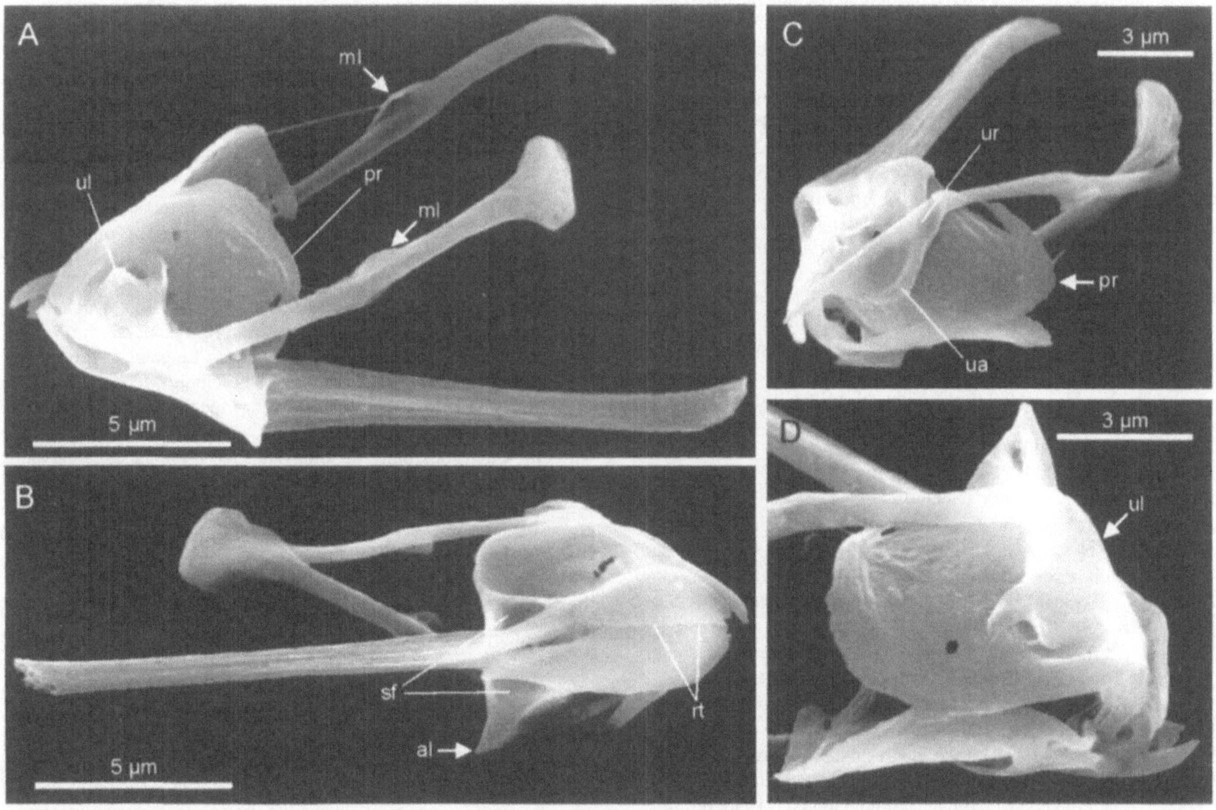

*Figure 6.* SEM of trophi from *Cephalodella gracilis.* (A) Lateral view; (B) Ventral view; (C) Laterofrontal view; (D) Rami, dorsal view. Abbreviations: al – alula; ml – manubrial lamella; pr – pleural rod; rt – ramal teeth; sf – subbasifenestra; ua – uncinal apophysis; ul – uncus, left; ur – uncus, right.

## Acknowledgements

We would like to thank the students Mette Mathiasen and Anne Dissing for their valuable assistance during the sampling and sorting processes, Dr Peter J. P. Croucher for the linguistic corrections, and Birgitte Rubæk for the line drawings. We are indebted to the staff and the board of the Danish Arctic Station. This work was financial enabled by a grant to P. Funch from the Danish Research Agency (no. 9801880).

## References

Althaus, B., 1957a. Faunistisch-ökologische Studien an Rotatorien salzhaltiger Gewässer Mitteldeutschlands. Wiss. Z. Martin Luther Univ. Halle-Wittenb. 6: 117–158.

Althaus, B., 1957b. Faunistisch-ökologische Studien an Rotatorien salzhaltiger Gewässer Mitteldeutschlands (Nachtrag). Wiss. Z. Martin Luther Univ. Halle-Wittenb. 6: 459–460.

Amrén, H., 1964. Ecological and taxonomical studies on zooplankton from Spitsbergen. Zool. Bidr. Upps. 36: 209–276.

Bergendal, D., 1892. Beitrage zur Fauna Grönlands. I. Zur Rotatorienfauna Grönlands. Acta Univers. Lund. 28: 1–180.

Buchholz, H. & D. Rühmann, 1956. *Notholca* und *Kellicottia*. Über einheimischer Vertreter zweier Rädertier-Gattungen. Mikrokosmos 45: 267–270.

Chengalath, R. & W. Koste, 1989. Composition and distributional patterns in arctic rotifers. Hydrobiologia 186/187: 191–200.

De Ridder, M., 1972. Rotatoria. The Zoology of Iceland. Vol. 2, part 13, Ejnar Munksgaard, Copenhagen and Reykjavik: 106 pp.

De Ridder, M., 1981. Some considerations on the geographical distribution of rotifers. Hydrobiologia 85: 209–225.

De Smet, W. H., 1995. Description of *Encentrum dieteri* sp. nov. (Rotifera, Dicranophoridae) from the High Arctic, with redescription of *E. bidentatum* (Lie-Pettersen, 1906) and *E. murrayi* Bryce, 1922. Belg. J. Zool. 125: 349–361.

De Smet, W. H., 1997. The Dicranophoridae (Monogononta). In Dumont, H. J. F. (ed.), Guides to the Identification of the Microinvertebrates of the Continental Waters of the World. Vol. 12, Rotifera. SPB Academic Publishing, Amsterdam: 1–325.

De Smet, W. H., 1998. Preparation of rotifer trophi for light and scanning electron microscopy. Hydrobiologia 387/388 (Dev. Hydrobiol. 134): 117–121.

De Smet, W. H., 2000. Three new species of the genus *Encentrum* (Rotifera, Monogononta, Dicranophoridae). Sarsia 85: 77–86.

De Smet, W. H., E. A. Van Rompu & L. Beyens, 1993. Contribution to the rotifer fauna of subarctic Greenland (Kangerlussuaq and Ammassalik area). Hydrobiologia 255/256: 463–466.

Donner, J., 1950. Rädertiere der Gattung *Cephalodella* aus Südmährens. Arch. Hydrobiol. 42: 304–328.

Feilberg, J., 1985. Grønlands varme kilder – naturens egne mistbænke. Forskning i Grønland 2: 10–22 (in Danish).

Focke, E., 1961. Die Rotatoriengattung *Notholca* und ihr Verhalten im Salzwasser. Kieler Meeresforsch. 17: 190–205.

Funch Andersen, P., 1990. Foreløbige resultater om marine Rotifera fra Disko Fjord, Grønland. In Funch Andersen, P., L. Düwel & O. S. Hansen. Feltkursus i Arktisk Biologi, Godhavn 1990, pp. 143–153, Zoological Museum, University of Copenhagen: 143–153 (in Danish, abstract in English).

Ghilarov, A. M., 1967. The zooplankton of arctic rock pools. Oikos 18: 82–95.

Godske Eriksen, B., 1968. Marine rotifers found in Norway, with descriptions of two new and one little known species. Sarsia 33: 23–34.

Godske Björklund, B., 1972. Taxonomic and ecological studies of species of *Notholca* (Rotatoria) found in sea- and brackish water, with a description of a new species. Sarsia 51: 25–66.

Hansen, L., K. E. Johansen, T. Lund & B. Simonsen, 1989. Homoterme kilder på Disko, Grønland. In Jørgensen, M., K. Johansen, T. Lund & B. Simonsen (eds), Feltkursus, Arktisk Zoological Museum, University of Copenhagen: Biologi. 143–205 (in Danish, abstract in English).

Harring, H. K., 1921. Rotatoria. Report of the Canadian Arctic Expedition 1913–18. Vol. 8: Mollusks, Echinoderms, Coelenterates, etc. F. A. Acland, Ottawa: 23 pp.

Jansson, A.-M., 1967. The food-web of the *Cladophora*-belt fauna. Helgol. wiss. Meeresunters. 15: 574–588.

Jørgensen, A. & M. Schiøtt, 1996. Kvantitativ undersøgelse af polychaeter i relation til sedimenttype. In Hansen, L.B., T. W. Kristensen, M. B. Christiansen & N. M. Schmidt (eds), Arktisk Biologisk Feltkursus, Qeqertarsuaq 1996. Zoological Museum, University of Copenhagen: 101–126 (in Danish, abstract in English).

Kalt, M. R. & B. Tandler, 1971 A study of fixation of early amphibian embryos for electron microscopy. J. Ultrastruct. Res. 36: 633–645.

Koste, W., 1978. Rotatoria, Die Rädertiere Mitteleuropas. – Textband + Tafelband. Gebrüder Borntraeger, Berlin, Stuttgard: VII+673 pp.

Kristensen, R. M., 1977 On the marine genus *Styraconyx* (Tardigrada, Heterotardigrada, Halechiniscidae) with description of a new species from a warm spring on Disko Island. Astarte 10: 87–91.

Nehring, S., 1988. Homothermische radioaktive Quellen auf Disko-Island (West-Grönland): Die Nematodenfauna – ein marines Relikt? In: Bericht über die Grönlandsexkursion des Institutes für Polarökologie vom 2–25 August 1987. Institut für Polarökologie (Kiel), Kiel: 186–201.

Nogrady, T. & R. Pourriot, 1995. The Notommatidae. In Dumont, H. J. F. (ed.), Guides to the Identification of the Microinvertebrates of the Continental Waters of the World. Vol. 8, Rotifera. SPB Academic Publishing, Amsterdam: 1–229.

Remane, A., 1929. Rotatoria. In Grimpe, G. & E. Wagler (eds), Tierwelt der Nord- und Ostsee. Vol. 7e, Akademische Verlagsgesellschaft Becker & Erler Kom.-Ges., Leipzig: 1–156.

Reisinger, E. & O. Steinböck, 1927. Foreløbig Meddelelse om vor zoologiske Rejse i Grønland 1926. Medd. om Grønl. 74: 33–42 (in Danish).

Røen, U., 1962. Salt- og brakvandsdyr i grønlandske farvande. Tidsskriftet Grønland 9: 335–341 (in Danish).

Sørensen, M. V., 1998. Marine Rotifera from a sandy beach at Disko Island, West Greenland, with the description of *Encentrum porsildi* n. sp. and *Notholca angakkoq* n. sp. Hydrobiologia 386: 153–165.

Sørensen, M. V. & R. M. Kristensen, 2000. Marine Rotifera from Ikka Fjord, SW Greenland, the description of a new species from the rare mineral ikaite. Meddr. Grønland, Bioscience 51: 1–46.

Thane-Fenchel, A., 1968. Distribution and ecology of non-planktonic brackish-water rotifers from Scandinavian waters. Ophelia 5: 273–297.

Vanhöffen, E., 1897. Die Fauna und Flora Grönlands, 1. Teil. In Von Drygalski, E. (ed.), Grönland-Expedition der Gesellschaft für Erdkunde zu Berlin 1891–93. W. H. Kühl, Berlin: 1–320.

*Hydrobiologia* **446/447:** 283–290, 2001.
*L. Sanoamuang, H. Segers, R.J. Shiel & R.D. Gulati (eds), Rotifera IX.*




# Reproductive isolation among geographically and temporally isolated marine *Brachionus* strains

T. Kotani[1], M. Ozaki[2], K. Matsuoka[2], T. W. Snell[3] & A. Hagiwara[2]
[1]*Graduate School of Marine Science & Engineering, Nagasaki University, Bunkyo 1-14, Nagasaki 852-8131, Japan*
*E-mail: f1054@cc.nagasaki-u.ac.jp*
[2]*Faculty of Fisheries, Nagasaki University, Bunkyo 1-14, Nagasaki 852-8521, Japan*
[3]*School of Biology, Georgia Institute of Technology, Atlanta, GA, 30332-0230, U.S.A.*

*Key words: Brachionus plicatilis, B. rotundiformis*, geographical isolation, temporal isolation, mating, copulation, mate recognition pheromone, sediments, resting eggs


## Abstract

Using a polyclonal antibody against the mate recognition pheromone (MRP) of *Brachionus rotundiformis* Koshiki strain, we investigated the behavioral reproductive isolation and the similarity of MRP among geographically and temporally isolated *B. rotundiformis* strains. Males of the Koshiki strain did not discriminate in mating attempts among females of the Koshiki strain and those of conspecific allopatric strains from Hamana, Fiji, Thailand and Spain. Likewise, Koshiki males attempted mating with statistically indistinguishable frequency with Koshiki females and *B. plicatilis* strains. However, copulation was not consummated between Koshiki males and *B. plicatilis* females. The amount of anti-MRP binding to three allopatric *B. rotundiformis* strains was similar to that of the Koshiki strain, but binding to Hamana and the *B. plicatilis* strain was significantly lower. Four temporally separated *B. rotundiformis* populations were hatched from resting eggs collected from 0, 5, 10 and 15 cm depth in the sediment of Kai-ike pond in Koshiki island, Japan. Sediment age was determined using the $^{210}$Pb method, allowing us to estimate that resting eggs from 15 cm depth were produced 65 years ago. Results of mating assays and anti-MRP binding showed that no behavioral reproductive isolation exists among the four temporally isolated Koshiki strains. *B. rotundiformis* appears to be reproductively isolated from *B. plicatilis*, but heterospecific matings are still attempted between *B. plicatilis* and *B. rotundiformis*, suggesting that the MRP remains sufficiently similar to elicit circling behavior.


## Introduction

In order to maintain the genetic cohesion of species, reproductive barriers preventing heterospecific gene exchange have evolved. A common mechanism of reproductive isolation is the geographic separation of one species from another. Geographical isolation is often regarded as a prerequisite for speciation and many studies of geographic variation suggest that allopatry is the initial stage of speciation (Mayr, 1970). Considerable genetic differentiation exists among geographically separated rotifer strains (Fu et al., 1991b). Much less is known, however, about genetic differentiation among temporally isolated strains and their degree of behavioral reproduction isolation. We have worked in a pond at Koshiki Island, Japan, where the sediments have been undisturbed for many years. An anoxic layer of water has prevented bioturbance of the sediments in Kai-ike Pond. A high density of photosynthetic bacteria exists above the anoxic layer, producing hydrogen sulfide ($H_2S$) and maintaining the anoxic conditions (Matsuyama, 1977). Genetic variation in allozymes was investigated among the rotifers hatched from resting eggs collected from different depths in the sediment of this pond (Fu, 1991). He found the variation in only one of six loci investigated. However, the reproductive isolation among these temporally isolated strains has not yet been studied.

In order to examine the behavioral reproductive isolation among geographically and temporally isolated strains, we investigated temporally isolated strains of *Brachionus rotundiformis* in Kai-ike pond and compared them with some conspecific geographically distinct population. Behavioral reproductive isolation between Koshiki males and *B. plicatilis* strain was also examined. We also investigated the similarity of the mate recognition pheromone (MRP) (Snell et al., 1995) using comparative binding studies of a polyclonal antibody against the MRP of *B. rotundiformis* from Kai-ike pond.

## Materials and methods

### *Geographically isolated strains*

Strains were collected in different geographic regions, including one *B. plicatilis* strain from Russia and five *B. rotundiformis* strains from Japan (Hamana lake and Kai-ike pond), Fiji, Thailand and Spain. The Russian strain (RUS) is that of Snell et al. (1995). The Hamana, Fiji and Thai strains were used in the study of Hagiwara et al. (1995) and the Spanish strain is that of Gómez & Serra (1995) known as SM1. The Koshiki strain was described in Fu et al. (1991a, 1993). For each strain, a clonal culture was established and used for experimentation.

### *Temporally isolated strains*

Kai-ike pond is a small isolated and stratified pond. The surface area is 0.15 $km^2$ and the maximum depth is 11.6 m (Matsuyama, 1977). Sediment samples from Kai-ike pond were collected with a 4 × 100 cm core-sampler in 1996 and 1997. Sampling was done around the deepest site. In six sediment layers (2–3, 5–6, 10–11, 20–21, 30–31, 40–41 cm) collected in 1996, the amount of $^{210}Pb$ was measured and the average rate of sediment accumulation was determined by Teledyne Brown Engineering Environmental Services, New Jersey, U.S.A.

The resting eggs used in the experiments were obtained from the sediments collected in 1997 at 0–1, 5–6, 10–11 and 15–16 cm depth with 1 × 30 cm core-sampler. One liter of seawater was used to suspend 1 g of the sediment and hatch the resting eggs. Salinity was 17 ppt, adjusted by dilution of seawater with deionized water. Temperature was 25 °C, and the light condition 3000–5000 lx, 24 L:0 D. At the third and fifth day of the culture period, the seawater was filtered through a 45 $\mu$m mesh plankton net and rotifers were harvested and re-cultured in fresh seawater. One rotifer was isolated from each culture and cloned. Culture temperature was 30 °C, salinity was 17 ppt, and diet was *Nannochloropsis oculata* added at a density of 7 million cells $ml^{-1}$ to stimulate sexual reproduction.

### *Mating assay*

The experimental procedure was as described in Hagiwara et al. (1995). The mating behavior of a brachionid rotifers has three main steps. First is head-on contact of a female by a male with his corona (contact). Next, the male makes at least 1 full circle around the female maintaining contact between his corona and her lorica (circling). Third is the insertion of his penis to the mouth or the base-of-foot of female and sperm transfer (copulation). Two males (<24 h old) and one female (<6 h old) collected from each strain were transferred into 15 $\mu$l of seawater on a plastic dish to maintain a spherical shape. This small test volume was necessary to promote frequent contact between males and the female.

Females from each geographic strain were tested with Koshiki males. In addition, all 16 combinations of males and females from each temporally isolated Koshiki strains were tested. The number of male circlings in 20 contacts with a female's body was counted, as well as the number of subsequent copulations. Each test was repeated 3 times with different individuals. Males and females were used only once. A chi square contingency test was employed to identify significant differences in how males reacted to females of their own strain and different strains.

### *Anti-MRP binding assay*

The procedure of MRP purification from the *B. rotundiformis* Koshiki strain was the same as described in Snell et al. (1995). Approximately 20 g wet weight of the Koshiki strain was homogenized in a tissue grinder. Serial lectin affinity chromatography was performed using two kinds of lectin (*Lens culinaris* and *Tetragonolobus purpureas*). The MRP (glycoprotein gp29) was purified from crude homogenate. A polyclonal antibody against the MRP (anti-*B. r.* (Koshiki) MRP) was produced by Cocalico Biologicals, Inc., Pennsylvania, U.S.A. using gp29 purified by lectin affinity chromatography. Approximately 130 $\mu$g IgG of this antibody was suspended in 100 $\mu$l of HMNK buffer (50 mM HEPES, 36 mM NaCl, 1 mM HCl, 0.1 mM $MgSO_4$, pH 7.4). The solution was added to 100 $\mu$l of DMSO containing 1 mg of biotin [succinimidyl

6-(biotinamido) hexanoate (biotinamidocaproate, N-hydroxysuccinimidyl ester), Molecular Probes Inc.]. The mixture was shaken at 150 rpm for 90 min to promote biotin binding to the antibody. To remove unbound biotin, the mixture was placed in a test tube with a 10 000 MW cut off filter (Ultrafree®-C, Millipore Inc.) and centrifuged at 14 000 rpm for 20 min. Biotinylated anti-MRP was washed off the filter with 100 $\mu$l of HMNK buffer.

Rotifer females carrying parthenogenetic eggs were collected from culture and placed in 500 $\mu$l of HMNK buffer along with *N. oculata* ($7 \times 10^6$ cells ml$^{-1}$). After 4–6 h, 20 neonate females were isolated and exposed to 50 $\mu$l biotinylated anti-MRP-HMNK solution for 1 h to allow binding of the antibody to the MRP on females. In order to remove unbound anti-MRP, rotifers were twice transferred to 500 $\mu$l of HMNK buffer. After washing, the females were exposed for 1 h to 550 $\mu$l of HMNK buffer containing $2 \times 10^9$ avidin-labeled fluorospheres (Molecular Probes Inc.). Avidin on the beads binds to biotin on the anti-MRP, fluorescently labeling MRP binding sites on females. Females were washed twice again by transferring them to 500 $\mu$l of HMNK buffer, then transferred to a slide for observation with an epifluorescent microscope (BH2-RFL, Olympus Inc.) at 100× using an excitation filter of 490 nm and an emission filter of 515 nm. Images were captured by a Macintosh Quadra 650 computer using an Olympus CCD camera (model FCD-725). Image data was processed using NIH image 1.56 software that permitted measurement of fluorescence intensity on a gray scale ranging from 1 to 256. A background measurement from the body of each rotifer was subtracted from the coronal measurement. One-way ANOVA to test the influence of strain on coronal fluorescence and Scheffé's *F* test to identify significant differences in pairwise comparisons among strains were employed.

*Anti-MRP inhibition of mating*

From each strain, 20 females were treated with anti-*B. r.* (Koshiki) MRP and washed twice with HMNK buffer. As a control, 20 neonate females were treated only with HMNK buffer. On the day before the experiment, females carrying unfertilized mictic eggs were isolated and transferred to 100 $\mu$l of buffer HMNK including *N. oculata* ($7 \times 10^6$ cells ml$^{-1}$). These ovigerous females provided newly born males that were used in the experiments. The mating assay was performed in the same way as described above, except that four females

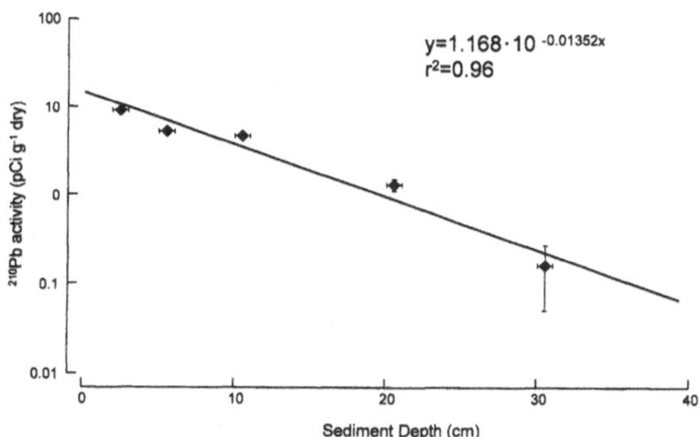


*Figure 1.* Activity of $^{210}$Pb (log scale) vs. sediment depth. The $^{210}$Pb activity measured near 0.5 pCi g$^{-1}$ dry at 40–41 cm depth was regarded as the background. Points correspond to the values subtracted from the background value. The horizontal error bar indicates the range of the sampled layer.

were used and five replicates were done. A *t*-test was performed comparing male mating attempts with females exposed to antibody and females exposed only to buffer. No males appeared in the Spanish strain so this strain was not used in the experiments.

## Results

Observation of the core samples showed no evidence of bioturbation. Figure 1 indicates the results of measurement of the amount of $^{210}$Pb. Assuming that the exponential decrease is due solely to radioactive decay of $^{210}$Pb, one can calculate a sediment accumulation rate from the expression:

$$s = -\lambda m^{-1}.$$

$s$ is the sediment accumulation rate (cm year$^{-1}$), $\lambda$ the multiplicative product of the decay constant of $^{210}$Pb (0.0311 yr$^{-1}$) and $\log_{10}e$ (0.4343), $m$ the slope of best-fit regression (−0.0598). This yields an average sediment accumulation in Kai-ike pond of 2.3 mm yr$^{-1}$. Consequently, it can be concluded that 0–1, 5–6, 10–11 and 15–16 cm depth sediments accumulated 0–4, 22–26, 43–47 and 65–69 years ago, respectively. The number of rotifer resting eggs that hatched from 0–1, 5–6, 10–11 and 15–16 cm depth layers was 72, 18, 3 and 1, respectively.

*Mating assay*

Table 1 shows % mating attempts (the percentage of circlings in 20 contacts by two males) and % copulation (the percentage of copulations in 20 contacts

*Table 1.* Summary of the mating attempts and copulation between *B. rotundiformis* male (Koshiki strain) and *Brachionus* females. Chi-square tests the hypothesis that self- and cross-circling and copulation occur with equal frequency

| Species | Females | Mating attempts | | | Copulation | | |
|---|---|---|---|---|---|---|---|
| | | % mating attempts | $\chi^{2a}$ value | *p* value | % copulation | $\chi^{2a}$ value | *p* value |
| *B. rotundiformis* | Koshiki | 14.1 | | | 4.4 | | |
| | Hamana | 1.7 | 3.1 | 0.077 | 0.0 | 0.7 | 0.414 |
| | Spain | 12.1 | 0.0 | 0.877 | 2.0 | 0.0 | 0.941 |
| | Thailand | 7.8 | 0.4 | 0.527 | 2.2 | 0.0 | 0.875 |
| | Fiji | 10.3 | 0.0 | 0.856 | 0.0 | 0.6 | 0.430 |
| *B. plicatilis* | Russia | 3.3 | 2.3 | 0.132 | 0.0 | 0.8 | 0.363 |

[a]2 × 2 Contingency table values.

by two males) of Koshiki males with females of geographically isolated strains. Koshiki males initiated circling with females from conspecific *B. rotundiformis* as well as *B. plicatilis* Russian strain. However, they completed penile attachment (copulation) only with conspecific females from Koshiki, Thailand and Spain. Although Koshiki males also failed to complete copulation with females from some conspecific strains like Hamana and Fiji, the Koshiki males' cross-mating with Hamana and Fiji females was not statistically different from the rate of copulation with Koshiki females.

Table 2 shows% mating attempts and% copulation among temporally isolated strains in Kai-ike pond. Koshiki males initiated mating irrespective of a female's temporal strain. In only one combination (5–6 cm male and 10–11 cm female), the frequency of mating attempts was significantly lower than intra-strain crosses. Copulations were observed in several inter-strain crosses and frequencies did not differ from intra-strain crosses.

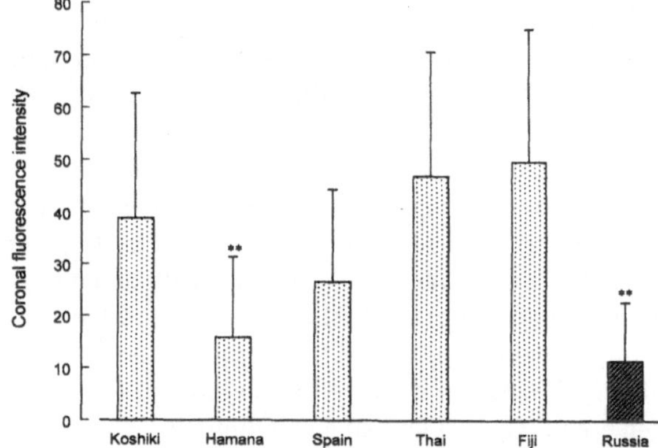


*Figure 2.* Antibody against *Brachionus rotundiformis* Koshiki strain mate recognition pheromone (anti-MRP) binding to females of geographically isolated strains. Coronal fluorescence is represented as dotted bars represent in *B. rotundiformis* strains and diagonal bars represent in *B. plicatilis* strains. Vertical bars indicate one standard deviation ($n$=20). Asterisks indicate anti-MRP binding significantly less than Koshiki control (**, $p<0.01$).

### *Anti-MRP binding assay*

One-way ANOVA showed that there were significant differences in anti-*B. r.* (Koshiki) MRP binding among geographically isolated strains (Fig. 2). The fluorescence intensity of Koshiki strain did not significantly differ (38.9±23.8, $n$=20) from that of the Fiji, Thai and Spanish strains (49.8±23.8, 46.9±23.8, 26.5±17.9, $n$=20, respectively). The Hamana strain had the lowest fluorescence among *B. rotundiformis* strains (15.8±15.6, $n$=20, $p<0.01$). The anti-*B. r.* MRP binding in *B. plicatilis* Russian strain was significantly lower (11.3±11.3, $n$=20, $p<0.05$) than in Koshiki strain.

There were no significant differences in coronal binding of the anti-*B. r.* MRP among temporally isolated strains in Kai-ike pond (Fig. 3).

### *Anti-MRP inhibition of mating*

In geographically isolated *B. rotundiformis* strains, the frequency of male mating attempts with females of their own strains exposed to anti-*B. r.* (Koshiki) MRP was significantly reduced compared with buffer controls (Fig. 4, $n$=5, $F$=8.0, $p<0.05$). However, anti-*B. r.* (Koshiki) MRP binding did not significantly reduce

*Table 2.* Summary of the circling and copulation among temporally isolated strains. cm refers to the depth of sediment collection in Kai-ike pond. Chi-square tests the hypothesis that self- and cross-mating attempts and copulation occur with equal frequency

| Males | Females | Mating attempts | | | Copulation | | |
|---|---|---|---|---|---|---|---|
| | | % mating attempts | $\chi^{2a}$ value | $p$ value | % copulation | $\chi^{2a}$ value | $p$ value |
| 0–1 cm | 0–1 cm | 11.9 | | | 0.0 | | |
| | 5–6 | 15.0 | 0.6 | 0.440 | 3.3 | 1.4 | 0.238 |
| | 10–11 | 6.8 | 0.3 | 0.589 | 0.0 | – | – |
| | 15–16 | 10.0 | 0.3 | 0.589 | 5.0 | 2.1 | 0.146 |
| 5–6 cm | 5–6 cm | 18.3 | | | 1.7 | | |
| | 0–1 | 8.3 | 2.6 | 0.107 | 1.7 | 0.0 | 1.000 |
| | 10–11 | 3.7 | 6.0 | 0.014 | 1.9 | 0.0 | 0.940 |
| | 15–16 | 13.3 | 0.6 | 0.453 | 6.7 | 1.8 | 0.171 |
| 10–11 cm | 10–11 cm | 11.7 | | | 1.7 | | |
| | 0–1 | 13.3 | 0.1 | 0.783 | 1.7 | 0.0 | 1.000 |
| | 5–6 | 8.3 | 0.4 | 0.543 | 0.0 | 1.0 | 0.315 |
| | 15–16 | 11.7 | 0.0 | 1.000 | 5.0 | 1.0 | 0.309 |
| 15–16 cm | 15–16 cm | 6.3 | | | 0.0 | | |
| | 0–1 | 15.0 | 2.1 | 0.150 | 3.3 | 1.6 | 0.202 |
| | 5–6 | 4.2 | 0.0 | 0.930 | 1.7 | 0.8 | 0.369 |
| | 10–11 | 6.3 | 0.2 | 0.646 | 0.0 | – | – |

[a]2 × 2 Contingency table values. –: Expected value is zero.

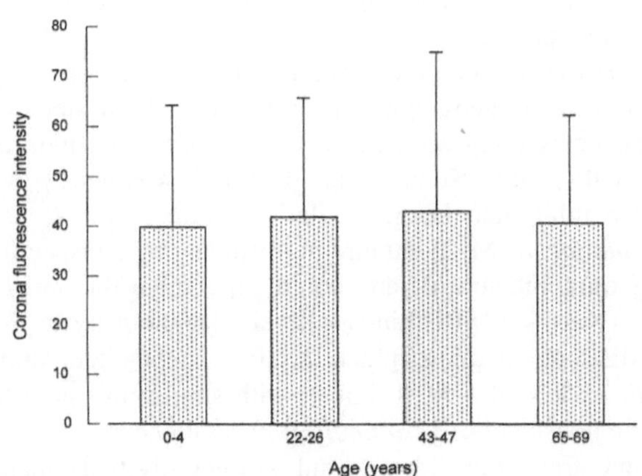


*Figure 3.* Anti-MRP binding to females of temporally isolated strains in Kai-ike pond. Vertical bars indicate one standard deviation ($n$=20).

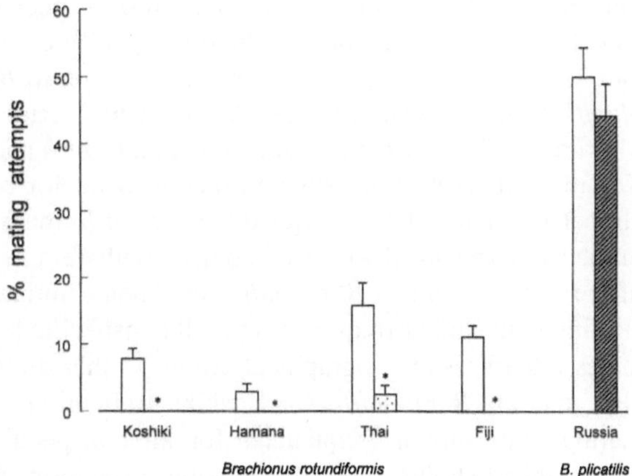


*Figure 4.* Anti-MRP binding to females reduces mating attempts by males of their own strain. Percent mating attempts refers to the proportion of intra-strain male-female encounters that resulted in an attempted mating. Clear bars represent controls where rotifer females were not exposed to anti-MRP. Diagonal bars represent *B. plicatilis* females exposed to anti-MRP, and dotted bars represent the same treatment for *B. rotundiformis* females. Vertical lines indicate one standard deviation ($n$=5). Asterisks indicate that anti-MRP binding significantly reduces male mating attempts ($p<0.05$).

conspecific male mating attempts for the *B. plicatilis* Russian strain ($n$=5, $F$=8.0, $p>0.05$).

In temporally isolated strains of Kai-ike Pond, the frequency of male mating attempts with females

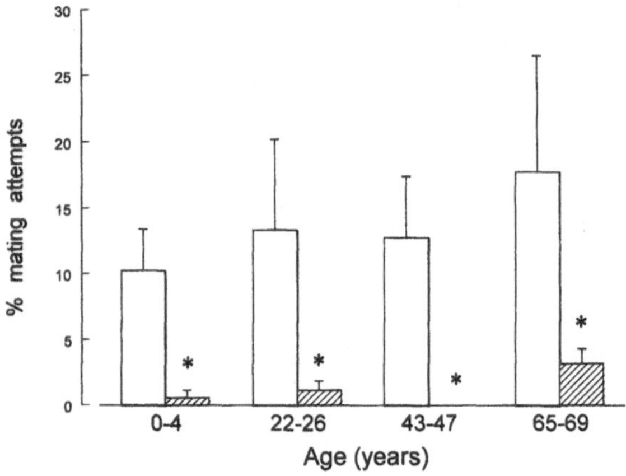


*Figure 5.* Anti-MRP binding to females of temporally isolated strains reduces male mating attempts. Percent mating attempts indicates the proportion of male-female encounters that resulted in an attempted mating. Clear bars represent controls where rotifer females were not exposed to anti-MRP. Dotted bars represent females exposed to anti-MRP. Vertical lines indicate one standard deviation ($n$=5). Asterisks indicate that anti-MRP binding significantly reduces male mating attempts ($p<0.05$).

from their own strain exposed to anti-*B. r.* (Koshiki) MRP was significantly reduced compared with buffer controls (Figure 5, $n$=5, $F$=8.0, $p<0.05$).

## Discussion

The finding of reproductive isolation among species and strains of the *Brachionus plicatilis* species complex was one factor causing its re-classification to *B. plicatilis* and *B. rotundiformis* (Segers, 1995). Rico-Martínez & Snell (1995a), Hagiwara et al. (1995) and Kotani et al. (1997) investigated mating behavior of allopatric strains. They reported that *B. plicatilis* males initiate circling in 50–70% of contact with females of the same strain and *B. rotundiformis* males initiate circling with similar frequencies for all conspecific females regardless of geographical origin. In this study also, males of *B. rotundiformis* Koshiki strain initiated mating with similar frequencies for all conspecific geographical strains, with the possible exception of Hamana (Table 1). Although mating was initiated with females from all other five geographic strains, Koshiki males did not complete copulation with females of Hamana and Fiji strains or females from *B. plicatilis* Russian strain (Table 1). This is similar to the results of Rico-Martínez & Snell (1995a), Hagiwara et al. (1995) and Kotani et al. (1997). Our study also investigated populations from the same pond but isolated in time, and found that frequencies of circlings and completed copulations among temporally isolated strains were similar in every combination, indicating no behavioral reproductive isolation (Table 2). Although we have demonstrated that there is little behavioral reproductive isolation among geographically separated *B. rotundiformis* strains worldwide, it is still possible that post-mating isolation exists.

Even though the *B. plicatilis* species complex was divided into *B. plicatilis* and *B. rotundiformis* (Segers, 1995), there is considerable morphological and genetic variation among allopatric and sympatric rotifer populations (King & Zhao, 1987; Fu et al., 1991a, b; Gómez et al., 1995, 1998). The probability of mating initiation varies with the combination of geographic strains, as a result of differences in the molecular structure of the MRP and its receptor on males (Rico-Martínez & Snell, 1995b, 1997; Kotani et al., 1997). Differences of MRP molecular structure may cause variation in mating frequencies among strains and may be a major contributor to the reproductive isolation among rotifer species. Kotani et al. (1997) compared binding of anti-MRP from the *B. plicatilis* Russian strain (anti-*B. p.* (Russia) MRP) to the same strains used in this study, except the Koshiki strain. They reported that anti-*B. p.* (Russia) MRP strongly bound to females of its own strain. This result was similar to that of Rico-Martínez & Snell (1997), and suggests that the MRP of *B. plicatilis* has a high species and strain-specificity. On the other hand, males of *B. rotundiformis* attempted mating with females of every conspecific geographic strain tested with similar frequencies (Hagiwara et al., 1995; Rico-Martínez & Snell, 1995b; Kotani et al., 1997). Consequently, we can infer that the molecular structure of the *B. rotundiformis* MRP is more uniform among conspecific geographic strains, and could expect that the anti-*B. r.* (Koshiki) MRP binds with similar intensity to the MRP of all geographical strains. In practice, anti-*B. r.* (Koshiki) MRP bound with similar intensity to the conspecific *B. rotundiformis* strains except Hamana strain, but did not bind so intensely to Hamana strain and *B. plicatilis* Russian strain (Fig. 2). Taking previous reports and the result of this study into consideration, we can conclude that the MRP of *B. plicatilis* Russian strain is different from that of *B. rotundiformis* Koshiki strain. On the other hand, Koshiki males did not distinguish females of all geographic strains, and never completed copulation with *B. plicatilis* and Hamana females to which anti-*B. r.* (Koshiki) MRP bound weakly, but neither with females of Fiji strain to which anti-*B. r.* (Koshiki) MRP bound intensely. Although the circling frequency of Koshiki males supported the hypothesis in relation to the similarity of the MRPs among *B. rotundiformis* strains, the copulation frequency and the binding of anti-*B. r.* (Koshiki) MRP did not support this hypothesis. In order to investigate the relationship between the MRP of *B. rotundiformis* and the mating behavior of those males, it is necessary to analyze the molecular details of MRP structure and compare it among geographic strains.

Since there is no oxygen in the lower layer of Kai-ike pond (Matsuyama, 1977), benthic organisms do not disturb the sediments and we can obtain a reliable time series of resting eggs of known age. Sediments in Kai-ike pond yielded resting eggs representing temporally isolated strains, but no reproductive isolation was found among them (Table 2). Moreover, the amount of anti-*B. r.* (Koshiki) MRP bound to temporally isolated strains is similar (Fig. 2) and inhibited male mating behavior of all strains (Fig. 4), while the anti-*B. r.* (Koshiki) MRP binding was not intensity in Hamana strain and *B. plicatilis* Russian strain. Fu (1991) hatched rotifers from resting eggs collected in Kai-ike pond up to 100 years old and investigated genetic variation among temporal strains with allozyme analysis. These are the oldest rotifer resting eggs ever to be hatched from natural sediments. He analyzed the variation of the six loci (*LDH*, *MDH*, *6PGD*, *SOD*, *PGM* and *GPI*), and found variation only in the *LDH* locus (ff, fh and hh), while the genotype of five other loci (*MDH*, *6PGD*, *SOD*, *PGM* and *GPI*) was same in each sample (bb, aa, cc, dd and hh, respectively). Annual samples of rotifers in the water column in the upper layer of the pond also revealed variation in *LDH* only. It can, therefore, be concluded that no allelic replacement has happened at these loci in the last 100 years. Therefore, we could expect that the anti-*B. r.* (Koshiki) MRP binding is similar among the temporally strains. Accordingly, we can conclude that the phenotypes of the six loci including five enzyme loci and the MRP locus have been maintained for at least 65 years, and that the rotifer in Kai-ike pond has been in stable genetically. Recently, DNA microsatellites have been described for marine *Brachionus* (Gómez et al., 1998; Boehm et al., 2000). Gómez et al. (1998) suggested that microsatellite loci would allow the analysis of long-term changes in strain structure by screening resting egg banks. In the future, by investigating variation in the MRP gene among temporally and geographically isolated strains, we hope to be able to clarify how behavioral reproductive isolation develops and how it contributes to speciation.

## Acknowledgements

This study was financially supported by the JSPS research fellowship. We thank two reviewers for improving the manuscript.

## References

Boehm, E. W. A., O. Gibson & E. Lubzens, 2000. Characterization of satellite DNA sequences from the commercially important marine rotifers *Brachionus rotundiformis* and *Brachionus plicatilis*. Mar. Biotechnol., 2: 38–48.

Fu, Y., 1991. Studies on genetic variations of the rotifer *Brachionus plicatilis* O.F. Müller. PhD Thesis, Nagasaki Univ: 144 pp.

Fu, Y., K. Hirayama & Y. Natsukari, 1991a. Morphological differences between two types of the rotifer *Brachionus plicatilis* O.F. Müller. J. exp. mar. Biol. Ecol. 151: 29–41.

Fu, Y., K. Hirayama & Y. Natsukari, 1991b. Genetic divergence between S and L type strains of the rotifer *Brachionus plicatilis* O.F. Müller. J. exp. mar. Biol. Ecol. 151: 43–56.

Fu, Y., A. Hagiwara & K. Hirayama, 1993. Crossing between seven strains of the rotifer *Brachionus plicatilis*. Nippon Suisan Gakkaishi 59: 2009–2016.

Gómez, A. & M. Serra, 1995. Behavioral reproductive isolation among sympatric strains of *Brachionus plicatilis* Müller, 1786: insights into the status of this taxonomical species. Hydrobiologia 313/314: 111–119

Gómez, A., M. Temprano & M. Serra, 1995. Ecological genetics of a cyclical parthenogen in temporary habitats. J. Evol. Biol. 8: 601–622.

Gómez, A., C. Clabby & G. R. Carvalho, 1998. Isolation and characterization of microsatellite loci in a cyclically parthenogenetic rotifer, *Brachionus plicatilis*. Mol. Ecol. 7: 1613–1621.

Hagiwara, A., T. Kotani, T. W. Snell, M. Assava-Aree & K. Hirayama, 1995. Morphology, genetics and mating behavior of small tropical marine *Brachionus* strains (Rotifera). J. exp. mar. Biol. Ecol. 194: 25–37.

King, C. E. & Y. Zhao, 1987. Coexistence of rotifer (*Brachionus plicatilis*) clones in Soda Lake, Nevada. Hydrobiologia 147: 57–64.

Kotani, T., A. Hagiwara & T. W. Snell, 1997. Genetic variation among marine *Brachionus* strains and function of mate recognition pheromone (MRP). Hydrobiologia 358: 105–112.

Matsuyama, M., 1977. Limnological features of Lake Kaiike, a small lake on Kamikoshiki Island, Kagoshima Prefecture, Japan. Jap. J. Limnol. 38: 9–18.

Mayr, E., 1970. Populations, Species, and Evolution. Belknap Press, Harvard Univ., Cambridge: 453 pp.

Rico-Martínez, R. & T. W. Snell, 1995a. Mating behavior and mate recognition pheromone blocking of male receptors in *Brachionus plicatilis* Müller (Rotifera) strains. Hydrobiologia 313/314: 105–110.

Rico-Martínez, R. & T. W. Snell, 1995b. Male discrimination of female *Brachionus plicatilis* Müller and *Brachionus rotundiformis* Tschugunoff (Rotifera). J. exp. mar. Biol. Ecol. 190: 39–49.

Rico-Martínez, R. & T. W. Snell, 1997. Comparative binding of an antibody to a mate recognition pheromone on female *Brachionus plicatilis* and *Brachionus rotundiformis*. Hydrobiologia 358: 71–76.

Segers, H., 1995. Nomenclature consequences of some recent studies on *Brachionus plicatilis* (Rotifera, Brachionidae). Hydrobiologia 313/314: 121–122.

Snell, T. W. & C. A. Hawkinson, 1983. Behavioral reproductive isolation among populations of the rotifer *Brachionus plicatilis*. Evolution 37: 1294–1305.

Snell, T. W., P. D. Morris & G. Cecchine, 1993. Localization of the mate-recognition pheromone in *Brachionus plicatilis* O.F. Müller (Rotifera) by fluorescent labeling with lectins. J. exp. mar. Biol. Ecol. 165: 225–235.

Snell, T. W., R. Rico-Martínez, L. S. Kelly & T. E. Battle, 1995. Identification of a sex pheromone from a rotifer. Mar. Biol. 123: 347–353.

Hydrobiologia **446/447**: 291–296, 2001.
L. Sanoamuang, H. Segers, R.J. Shiel & R.D. Gulati (eds), Rotifera IX.




# *Rhinoglena frontalis* (Rotifera, Monogononta): a scanning electron microscopic study

Giulio Melone
*Department of Biology, University of Milan, via Celoria 26, 20133 Milano, Italy*
*E-mail: Giulio.Melone@unimi.it*

*Key words:* morphology, female, male, trophi, SEM

## Abstract

Females and males of *Rhinoglena frontalis* (Monogononta, Epiphanidae) are observed by SEM and their external morphologies are compared. The two sexes differ in size and shape of the body. The female body is fusiform with a short, conical foot, while the male body is more slender and has a rather long foot. The rotatory apparatus (or corona) of both sexes is similar with only minor differences and consists of rows and tufts of cilia arranged around the mouth opening. The corona is made of two paired lobes lateral to the mouth and of a third prominent dorsal lobe, usually called proboscis. The three lobes are lined externally by dense rows of cilia, which constitute the cingulum, used for swimming. The central surface of the proboscis is covered with numerous longitudinal rows of cilia bent towards the mouth. The lateral lobes show, on their central surfaces, two concentric arcs of cirri (made of tightly packed cilia) bent towards the mouth. The similar organization of the rotatory apparatus of both sexes is related to the fact that the male, in this species, is able to feed and has a developed mastax and digestive system. The trophi of both sexes are illustrated and compared.

## Introduction

Usually, rotifers are being studied using light microscopy. During the last two centuries, rotifer students have been able to improve the accuracy of their descriptions and drawings. Methods of preparation of material and microscopic techniques have improved (e.g. Hudson & Gosse, 1886; De Beauchamp, 1909, 1965; Wesenberg-Lund, 1923, 1930; Remane, 1929–33). Nevertheless, observations by light microscope do not allow for understanding of the organization of the rotifer coronas in detail. Only recently, with the adoption of scanning electron microscopy (SEM), has the description of rotifer coronas improved (Clément & Wurdak, 1991; Melone & Ricci, 1995; Melone, 1998).

This paper presents a description of the morphology of female and male *Rhinoglena frontalis* Ehrenberg 1853 as revealed by SEM. The results are compared with descriptions present in the literature (e.g. Hudson & Gosse, 1886; Wesenberg-Lund, 1923, 1930; Ruttner-Kolisko, 1974).

## Materials and methods

Specimens of *Rhinoglena frontalis* used for this study originate from resting eggs provided by Thomas Schröder (Freie Universität, Berlin). Following the instructions of T. Schröder, hatching of *R. frontalis* was induced by exposing the resting eggs to room temperature (25 °C) for about 3 weeks and then transferring them to 5 °C. After this treatment, the resting eggs hatch about 12–14 days later. *R. frontalis* was cultured in a glass container of about 30 ml and fed *Chlorella minutissima.* The first males appeared in the culture after 3 weeks of parthenogenetic reproduction. Within 1 week, the population collapsed after the massive production of mictic females. These produce resting eggs which are not released but remain inside the body.

Females and males of *R. frontalis* were isolated from the culture, relaxed with bupivacaine, fixed and processed for observation by SEM following the method by Melone (1998). The specimens, dehydrated with ethanol and critical point dried with $CO_2$, were

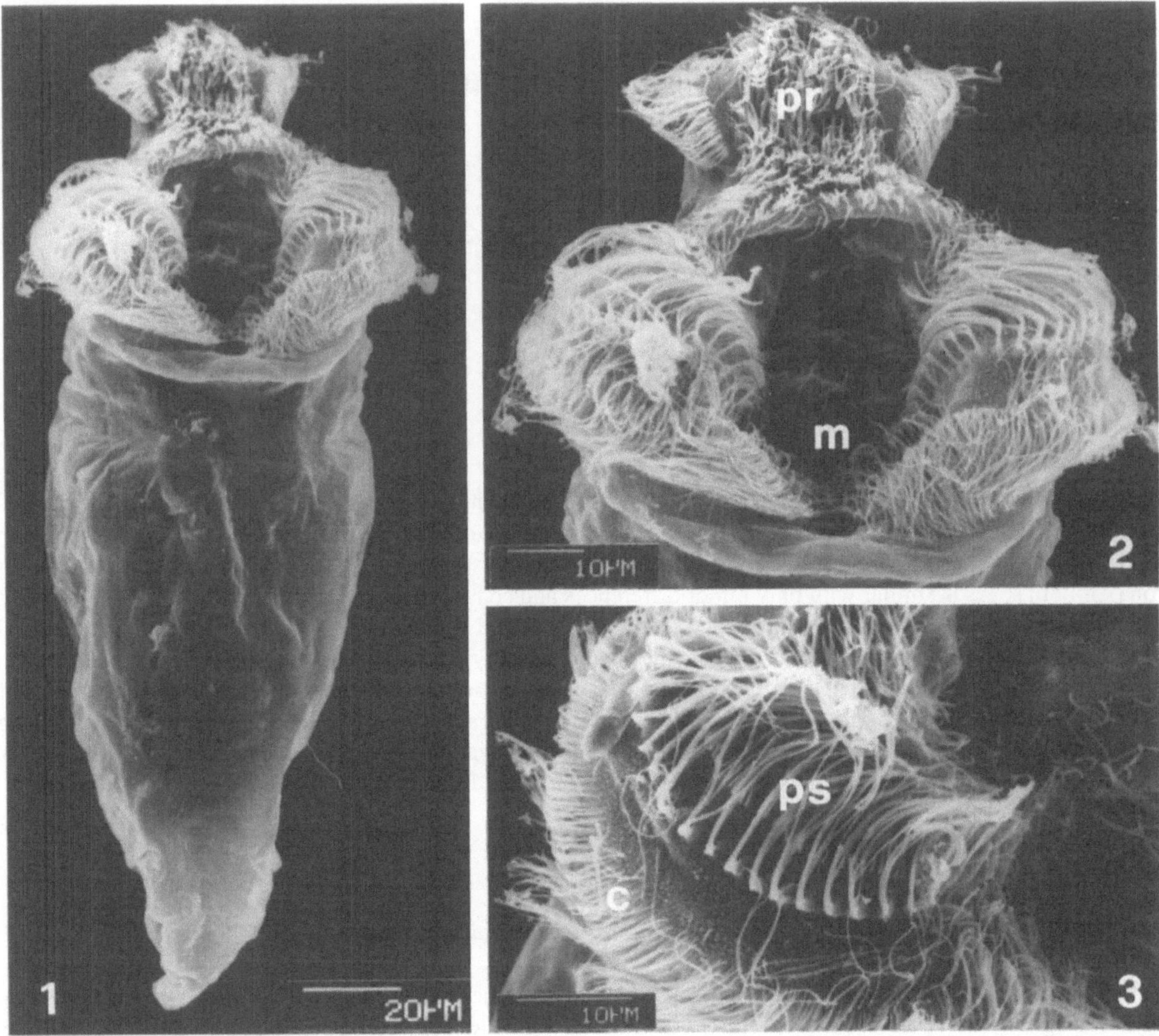


*Figures 1–3.* Female *Rhinoglena frontalis* by SEM. 1: General ventral view. 2: Head and corona in ventral view; m, mouth; pr, proboscis. 3: Right lobe of the corona; c, cingulum; ps, pseudotrochus.

mounted on a stub, coated with gold and observed using a Cambridge Stereoscan 250 Mk2.

## Results

### *Female*

Body conical, tapering to the foot, 200–250 $\mu$m long (Fig. 1). The wide head bears a complicated corona. The corona, consisting of pseudotrochus and cingulum, is arranged into three parts. Two of these are symmetrically located at both sides of the mouth, the third one is unpaired on a proboscis-like dorsal projection (Fig. 2).

The two symmetrical parts of the corona, here named auricles, are arched and present an external densely ciliated cingulum and internal pseudotrochus (Fig. 3). The cingulum consists of C-shaped ciliated band bordering each auricle. The pseudotrochus is made of two concentric arched rows of cirri, made of a group of adnate cilia. All cirri are curved toward the mouth opening. The outer pseudotrochal row is made up of 24–25 cirri of 10–12 adnate cilia each. The inner pseudotrochal row consists of about 75 cirri of 4–6 adnate cilia each (Figs 3 and 4). A bare surface dotted with numerous microvilli is visible between the cingulum and the pseudotrochus on each auricle (Figs 3 and 4).

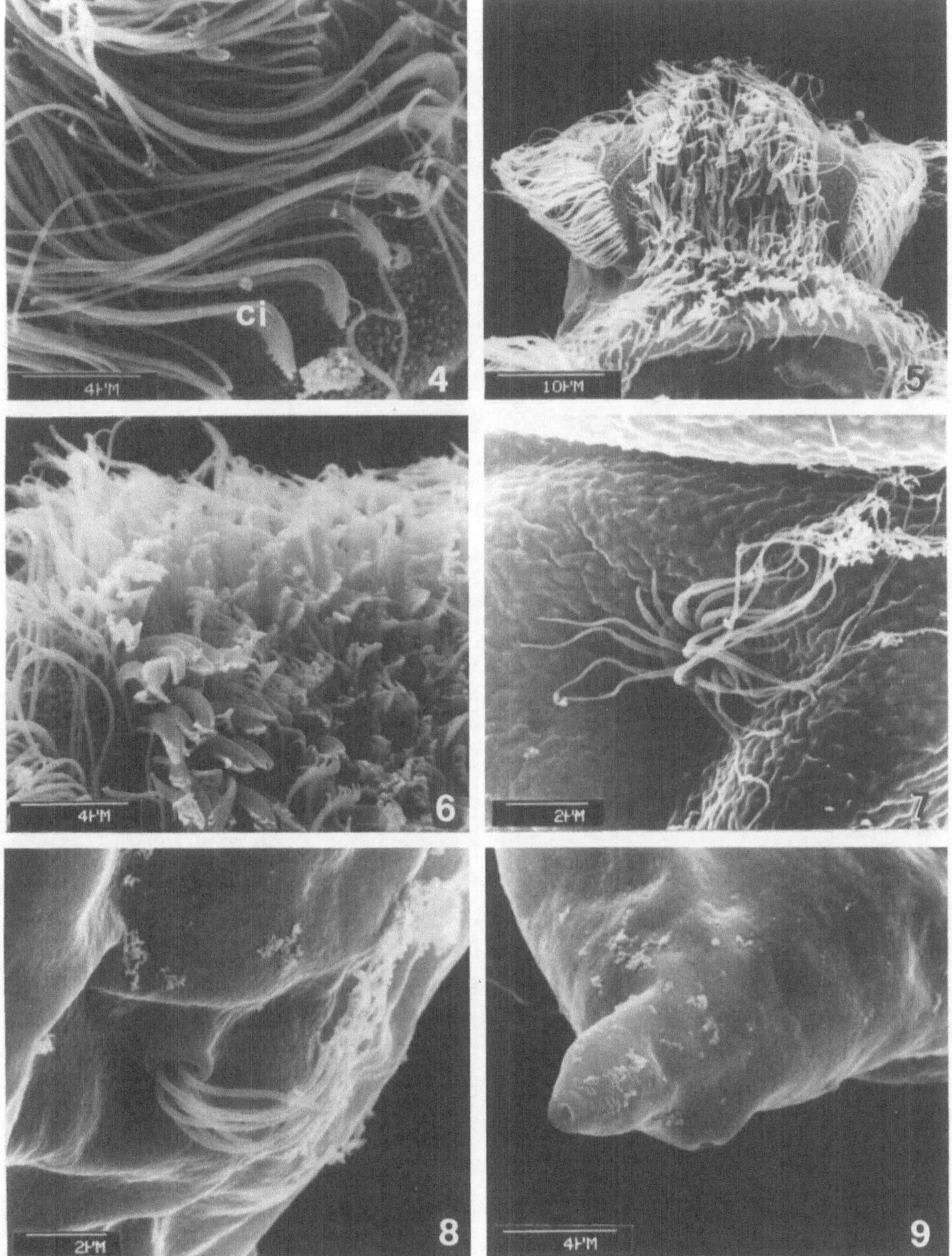


*Figures 4–9.* Female *Rhinoglena frontalis* by SEM. 4: Left pseudotrochus with rows of cirri; ci, cirrus. 5: Proboscis. 6: Cirri on the proboscis. 7: Dorsal antenna. 8: Lateral antenna. 9: Toes.

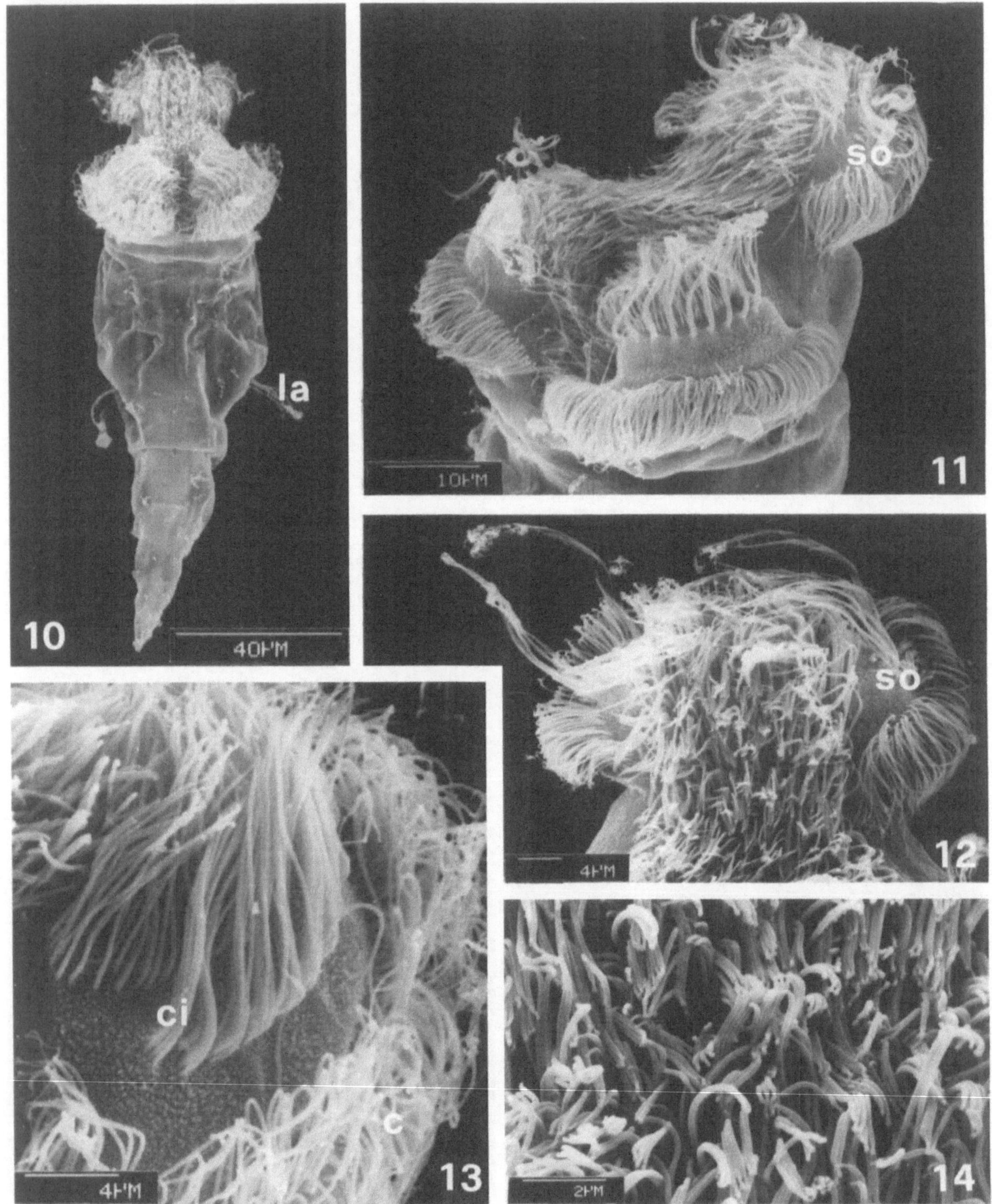


*Figures 10–14.* Male *Rhinoglena frontalis* by SEM. 10: General ventral view; la, lateral antenna. 11: Head with corona; so, sensorial organ. 12: Proboscis; so, sensorial organ. 13: Left lobe of corona; c, cingulum; ci, cirrus. 14: Cirri on the proboscis.

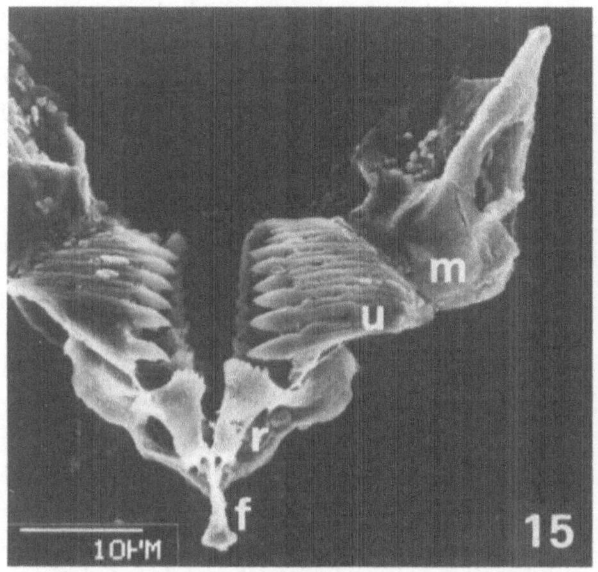


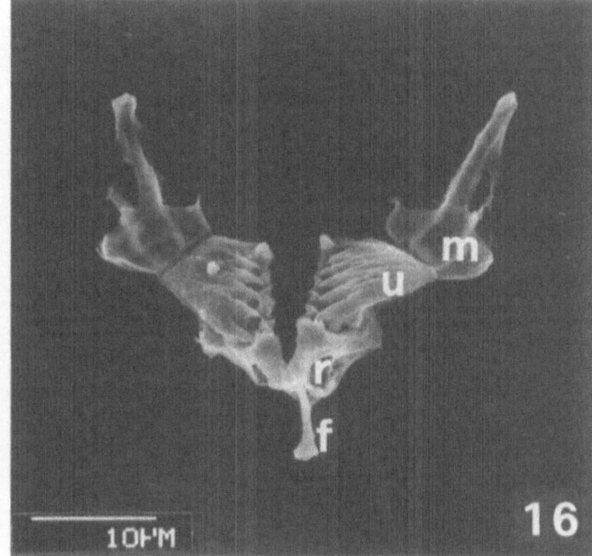


*Figures 15–16.* Trophi of *Rhinoglena frontalis* by SEM. 15: Female. 16: Male; f, fulcrum; m, manubrium; r, ramus; u, uncus.

The unpaired part of the corona is situated ventrally on the proboscis. An external wreath of cilia encircles an area covered by short cirri of 4–6 adnate cilia each, all bent toward the mouth opening (Figs 5 and 6). At the base of the proboscis, the field of short cirri extends laterally and reaches the auricles (Figs 2 and 5).

The dorsal antenna is a tuft of 24 cilia, emerging from a roundish opening on the dorsal epidermis of the proboscis (Fig. 7). The lateral antennae are similar to the dorsal one, have a tuft of 10 cilia and are located in the posterior third of the body (Fig. 8). The foot ends with two very small toes perforated by the openings of the pedal glands (Fig. 9). The trophi is malleate and has 6 major teeth plus 3–4 minor teeth on each uncus (Fig. 15).

*Male*

The overall shape of the male is similar to that of the female, but the male body is only about 150 $\mu$m long. Compared to the female, the male body is narrower in the lumbar region, the foot is proportionally longer, the proboscis larger and the lateral antennae are considerably more evident (Fig. 10).

The general organization of the male external morphology is similar to that of the female. The corona is made of cingulum and pseudotrochus and is located on the paired auricles and the proboscis (Fig. 11). Similarly to the female, the pseudotrochus is made of two concentric rows of cirri. In the male, the outer pseudotrochal row is made of about 18 cirri of 8–10 adnate cilia each. The inner pseudotrochal row is made of about 50 cirri of 4–6 adnate cilia each (Fig. 13).

On the proboscis, an external row of cilia encircles the central area with short cirri of 3–4 adnate cilia each, all bent toward the mouth (Figs 12 and 14). At the base of the proboscis the field of short cirri reaches the auricles (Fig. 11). At both sides of the male proboscis between the external row and the central area, there are two symmetrical sensorial structures. Each one consists of two tufts of 10–12 very long cilia and numerous very short cilia (Figs 11 and 12).

The dorsal antenna does not differ from that of the female. The lateral antennae are more evident because they are made of about 20 very long cilia.

The trophi is smaller but similar to that of the female. Each uncus plate has 5 major, plus 1–2 minor teeth.

*Remarks on biology*

Males of *R. frontalis* are able to swim very fast, while females are much slower swimmers. Both sexes feed on algae, as their guts contained green algae cells. In the mass culture, the mictic females produced a resting egg that remained inside the body till their death. As a consequence, each fertilized mictic female produces only one resting egg.

## Discussion

The coronae of female and male *R. frontalis* are organised in the same way: the external band of cilia on auricules and proboscis appears very similar to the cingulum of other rotifers (Clémont & Wurdak, 1991; Melone & Ricci, 1995; Melone, 1998). Possibly the cingulum is used to swim only. The pseudotrochus and the ventral area of the proboscis consist of cirri of different length and location, all bent toward the mouth. These parts of the corona seem specialized to collect and to direct food particles toward the mouth.

The similarity between female and male *R. frontalis* organization is remarkable and is to be related to the presence of a gut in the male. As far as I know, this species is the only monogonont rotifer known to have fully developed males (Wesenberg-Lund, 1923, 1930; Ruttner-Kolisko, 1974). Nogrady et al. (1993) considered the gut of male *R. frontalis* rudimentary. The present results demonstrate that the male is equipped with a fully functional digestive system, with corona, mouth, trophi similar to those of females, and ingested algae in the gut.

The size of the male is about 0.7 that of the female. This is likely due to the fact that males of monogonont rotifers are haploid. The male corona and trophi are smaller than those of the female. The size reduction of the male corona of *R. frontalis* is performed by a reduction of the number of the cilia present in the different parts of the apparatus, and not by miniaturizing the whole structure. Probably this is because the cilium (axoneme) has a fixed diameter. The trophi, too, is reduced to about 70% of the size of that of the female. It has fewer unci teeth than that of the female.

Major differences between the sexes concern sensorial structures, particularly the sensorial organs on the proboscis and the lateral antennae. These features can be related to a different swimming behavior of the male and to the need to locate females and mate.

From this first comparison between male and female *R. frontalis* follows that the smaller size of the male is accomplished by miniaturizing the structures (trophi) or, where this is not possible, by reducing the number of single elements (cirri). However, it should be born in mind that the male of *R. frontalis* represents an exception among monogononts, for being fully developed and able to feed. Information on other monogonont males is needed to determine a more general pattern of male organization.

## Acknowledgements

I am deeply indebted to Thomas Schröder for providing the resting eggs and for his advice.

## References

De Beauchamp, P. M., 1909. Recherches sur les Rotifères: les formations tégumentaires et l'appareil digestif. Arch. Zool. exp. gén. IV$^{e}$ Série, Tome X: 1–410.

De Beauchamp, P. M., 1965. Classe des Rotifères. In Grassé, P. P. (ed.), Traité de Zoologie. Masson et C. éd., Paris, Tome IV: 1225–1379.

Clément, P. & E. Wurdak, 1991. Rotifera. In Harrison, F. W. & E. E. Ruppert (eds.), Microscopic Anatomy of Invertebrates. Vol. 4. Aschelminthes, Wiley-Liss, Inc., New York: 219–297.

Hudson, C. T. & P. H. Gosse, 1886. The Rotifera; or Wheel-Animalcules. Longmans, Green & Co., London. 2 vols and Supplement.

Melone, G., 1998. The rotifer corona by SEM. Hydrobiologia 387/388: 131–134.

Melone, G. & C. Ricci, 1995. Rotatory apparatus in Bdelloids. Hydrobiologia 313/314: 91–98.

Nogrady, T., R. L. Wallace & T. W. Snell, 1993. Rotifera. Vol.I: Biology, Ecology and Systematics. Guides to the identification of the microinvertebrates of the continental waters of the world. 4. SPB Academic Publishers, The Hague: 142 pp.

Remane, A., 1929–33. Rotatoria. In: Bronn's Klassen und ordnungen des Tier-Reichs, 4, 2, 1: 1–576.

Ruttner-Kolisko, A., 1974. Planktonic rotifers: biology and taxonomy. Die Binnengewässer (Supplement) 26: 1–146.

Wesenberg-Lund, C., 1923. Contributions to the biology of the Rotifera. I. The males of the Rotifera. D. Kgl. Danske Vidensk. Selsk. Skrifter, Naturvidensk. og Mathem.Afd., 8, IV.3: 190–345.

Wesenberg-Lund, C., 1930. Contributions to the biology of the Rotifera. Part II. The periodicity and sexual periods. D. Kgl. Danske Vidensk. Selsk. Skrifter, Naturvidensk. og Mathem.Afd., 9, II.1: 1–230.

Hydrobiologia **446/447**: 297–304, 2001.
L. Sanoamuang, H. Segers, R.J. Shiel & R.D. Gulati (eds), Rotifera IX.




# The rotifer fauna of Lake Kud-Thing, a shallow lake in Nong Khai Province, northeast Thailand

La-orsri Sanoamuang & Sukonthip Savatenalinton
*Department of Biology, Faculty of Science, Khon Kaen University, Khon Kaen 40002, Thailand*
*E-mail: la_orsri@kku.ac.th*

*Key words:* Rotifera, Lake Kud-Thing, Nong Khai Province, northeast Thailand, *Lecane isanensis* n. sp., *Trichocerca orca* (Murray)

## Abstract

The species composition of the rotifer fauna of Lake Kud-Thing, Nong Khai Province, northeast Thailand was investigated in a monthly sampling program from January to December 1998. A remarkably rich rotifer community consisting of 183 taxa was recorded, including 32 new records to the Thai fauna. The most diverse genera were *Lecane* (23.5%), followed by *Trichocerca* (16.9%) and *Lepadella* (11.5%). *Lecane isanensis* n. sp. is described and figured. Thirteen species (*Aspelta circinator, Collotheca tenuilobata, Lecane nelsoni, Lepadella benjamini, Lepadella eurysterna, Sinantherina ariprepes, Testudinella* cf. *insinuata, Trichocerca abilioi, Trichocerca inermis, Trichocerca montana, Trichocerca orca, Trichocerca rosea* and *Trichocerca simonei*) are new to the Oriental region. Notably, the record of *Trichocerca orca* is the first since its discovery in New Zealand in 1913. The rotifers can be classified into three groups based on duration of appearance; (1) common, perennial; (2) uncommon, perennial; and (3) uncommon, sporadic species. The high species richness of rotifers, particularly members of the genera *Lecane* and *Trichocerca*, but low diversity of *Brachionus* together suggest that the lake is probably not eutrophicated.

## Introduction

The northeastern part of Thailand is known to have a large number of small ponds, swamps, rivers and reservoirs, but only few natural lakes. Although several investigations have been documenting the rotifer faunas of northeast Thailand (Segers & Sanoamuang, 1994; Sanoamuang et al., 1995; Sanoamuang, 1996; Sanoamuang & Segers, 1997; Sanoamuang & Savatenalinton, 1999), some habitats from the remote areas have not yet been investigated. Lake Kud-Thing is particularly interesting, because the lake is known to have a wide variety of freshwater fishes and birds, but no work on its zooplankton has been carried out. This natural lake is situated between 23° 13′ – 25° 7′ N and 59° 47′ – 62° 29′ E, in Nong Khai Province, northeast Thailand. It has an area of 4.80 $km^2$ and a maximum depth of 7 m. Margins of this shallow lake are covered with dense macrophytes stands, predominantly *Ceratophyllum demersum* Linn. and *Ottelia alismoides* Pers.

## Materials and methods

In order to obtain a comprehensive list of rotifer species in Lake Kud-Thing, qualitative samples were collected monthly between January and December 1998 from three stations using a 60 $\mu$m mesh plankton net. The rotifers were preserved in 4% formaldehyde. During the sampling period, water temperature ranged between 22.5° and 30 °C, pH 6.1–8.5, and conductivity 70–290 $\mu$S $cm^{-1}$. Specimens were searched under a dissecting microscope, and examined using an Olympus CHD microscope. Figures were drawn using a camera lucida. Scanning electron microscopy (SEM) was performed with a Hitachi S-3200N microscope.

## Results and discussion

Our investigation revealed that the rotifer community of Lake Kud-Thing consists of at least 183 taxa (Table 1). This number is not a final list, as some

*Table 1.* Rotifera recorded from Lake Kud-Thing, northeast Thailand, during this study; * new to Thailand, ** new to the Oriental region, *** new to science, + occurred in the samples of every month

**(1) Common, perennial species**

+*Anuraeopsis fissa* (Gosse)
*Ascomorpha ovalis* (Carlin)
*A. saltans* Bartsch
+*Brachionus quadridentatus* Hermann
*Cephalodella gibba* (Ehrenberg)
+*Colurella uncinata* (Müller)
*Conochilus coenobasis* (Skorikov)
*Dicranophorus caudatus* (Ehrenberg)
*Euchlanis dilatata* Ehrenberg
*E. incisa* Carlin
*Hexarthra intermedia* Wiszniewski
+*Keratella cochlearis* (Gosse)
*K. edmondsoni* (Ahlstrom)
+*Lecane bulla* (Gosse)
*L. closterocerca* (Schmarda)
*L. crepida* Harring
+*L. curvicornis* (Murray)
+*L. furcata* (Murray)
*L. haliclysta* Harring & Myers
+*L. hamata* (Stokes)
+*L. hornemanni* (Ehrenberg)
*L. inopinata* Harring & Myers
+*L. leontina* (Turner)
*L. ludwigii* (Eckstein)
*L. luna* (Müller)
+*L. lunaris* (Ehrenberg)
*L. pyriformis* (Daday)
+*L. signifera* (Jennings)
+*L. unguitata* (Fadeev)
*L. ungulata* (Gosse)
*Lepadella apsicora* (Myers)
*L. ehrenbergi* (Perty)
*L. quadricarinata* (Stenroos)
*Macrochaetus collinsi* (Gosse)
*Monommata* sp.
+*M. ventralis* (Ehrenberg)
*Notommata copeus* Ehrenberg
+*Plationus patulus* (Müller)
+*Polyarthra vulgaris* Carlin
**Ptygyra tacita* Edmondson
*Synchaeta stylata* Wierzejski
*Testudinella ahlstromi* Hauer
** *T.* cf. *insinuata* Hauer
+*T. parva* (Ternetz)
*T. patina* (Hermann)
*Trichocerca bidens* (Lucks)
+*T. braziliensis* (Murray)

*Table 1.* Continued

*T. capucina* Wierzejski & Zacharias
**T. collaris* (Rousselet)
*T. flagellata* Hauer
*T. hollaerti* De Smet
*T. insignis* (Herrick)
+*T. pusilla* (Lauterborn)
*T. similis* (Wierzejski)
*T. tenuior* (Gosse)
+*Trichotria tetractis* (Ehrenberg)

**(2) Uncommon, perennial species**

*Asplanchna brightwelli* (Gosse)
*A. priodonta* Gosse
*Brachionus dichotomus* Shephard f. *reductus* Koste & Shiel
*B. donneri* Brehm
*Collotheca* sp.
*Colurella adriatica* Ehrenberg
*C. obtusa* (Gosse)
* *C. sulcata* (Stenroos)
*Conochilus natans* (Seligo)
*C. unicornis* (Rousselet)
*Dicranophorus laviger* (Hauer)
*D. epicharis* Harring & Myers
*Dipleuchanis propatula* (Gosse)
*Euchlanis meneta* Myers
*Filinia camasecla* Myers
*F. opoliensis* (Zacharias)
*Floscularia conifera* (Hudson)
*Hexarthra mira* (Hudson)
*Keratella lenzi* Hauer
*K. tropica* (Apstein)
*Lecane aculeata* (Jakubski)
*L. arcula* Harring
*** *L. isanensis* new species
*L. lateralis* Sharma
*L. nana* (Murray)
*L. obtusa* (Murray)
*L. papuana* (Murray)
*L. pusilla* Harring
*L. quadridentata* (Ehrenberg)
*L. rhenana* Hauer
*L. shieli* Segers & Sanoamuang
*L. tenuiseta* Harring
*L. undulata* Hauer
****Lecane* sp.
*Lepadella acuminata* (Ehrenberg)
** *L. benjamini* Harring
*L. biloba* (Hauer)
*L. costatoides* Segers
**L. cristata* (Rousselet)
*L. lindaui* Koste

*Continued on p. 299*

*Table 1.* Continued

| |
|---|
| *L. ovalis* (Müller) |
| *L. patella* (Müller) |
| *L. quinquecostata* (Lucks) |
| *L. rhomboides* (Gosse) |
| *L. triptera* (Ehrenberg) |
| *L. vandenbrandei* Gillard |
| *Lophocharis salpina* (Ehrenberg) |
| *Macrochaetus longipes* Myers |
| *M. sericus* (Thorpe) |
| **M. subquadratus* (Perty) |
| *Mytilina bisulcata* (Lucks) |
| *M. unguipes* (Lucks) |
| *Notommata pachyura* (Gosse) |
| *Platyias quadricornis* (Ehrenberg) |
| **Ploesoma lenticulare* Herrick |
| *Scaridium elegans* Segers & De Meester |
| *S. grandis* Segers |
| *S. longicaudum* (Müller) |
| *Sinantherina spinosa* (Thorpe) |
| *Testudinella amphora Hauer |
| *T. brevicaudata* Yamamoto |
| *T. tridentata* Smirnov |
| *T. walkeri* Koste & Shiel |
| *Trichocerca bicristata* (Gosse) |
| *T. cylindrica* (Imhof) |
| *T. elongata* Gosse |
| ** *T. inermis* (Linder) |
| **T. longiseta* (Schrank) |
| ***T. montana* Hauer |
| *T. porcellus* (Goose) |
| ** *T. rosea* (Stenroos) |
| **T. scipio* Gosse |
| *T. siamensis* Segers & Pholpunthin |
| ***T. simonei* De Smet |
| *T. stylata* (Gosse) |
| *T. tigris* (Müller) |
| *T. tropis* Hauer |
| **T. vernalis* (Hauer) |
| **T. weberi* (Jennings) |
| **(3) Uncommon, sporadic species** |
| ** *Aspelta circinator* (Gosse) |
| *Brachionus angularis* Gosse |
| *B. bidentatus* Anderson |
| *B. calyciflorus* Pallas |
| *B. falcatus* Zacharias |
| *B. forficula* Wierzejski |
| *Cephalodella forficula* (Ehrenberg) |
| **C. mucronata* Myers |
| **C. tenuior* Gosse |

*Table 1.* Continued

| |
|---|
| ***Collotheca tenuilobata* (Anderson) |
| *Dicranophorus grandis* (Ehrenberg) |
| *Filinia novaezealandiae* Shiel & Sanoamuang |
| *Floscularia ringens* Linneus |
| *Lecane blachei* Berzinš |
| *L. braumi* Koste |
| *L. decipiens* (Murray) |
| *L. doryssa* Harring |
| *L.* cf. *doryssa* Harring |
| *L. elegans* Harring |
| *L. flexilis* (Gosse) |
| ** *L. nelsoni* Segers |
| *L. pertica* Harring & Myers |
| *L. ruttneri* Hauer |
| * *L. simmoneae* Segers |
| *L. thailandensis* Segers & Sanoamuang |
| *Lepadella apsida* Harring |
| *L. discoidea* Segers |
| *L.* cf. *elongata* Koste |
| ** *L. eurysterna* Myers |
| *L. latusinus* (Hilgendorf) |
| **L. monodactyla* Berzins |
| *L. triba* Myers |
| **L. triptera* Ehrenberg f. *alata* Myers |
| *Macrochaetus danneeli* Koste & Shiel |
| *Mytilina acanthophora* Hauer |
| *Ploesoma hudsoni* (Imhof) |
| *Ptygura elsteri* f. *thailandensis* Koste |
| *P. furcillata* (Kellicott) |
| *P. melicerta* Ehrenberg |
| ***Sinantherina ariprepes* Edmondson |
| *Squatinella lamellaris* (Müller) f. *mutica* (Ehrenberg) |
| *Testudinella greeni* Koste |
| ** *Trichocerca abilioi* Segers |
| *T. insulana* (Hauer) |
| * *T. jenningsi* Voigt |
| ***T. orca* (Murray) |
| *T. ruttneri* Donner |
| *Tripleuchlanis plicata* (Levander) |

contracted Bdelloids were encountered. The momentary species diversity ranged from 85 to 115 (mean 97) species. These numbers are comparable to the record from two localities in the floodplain of the River Nan in the north (Sanoamuang, 1998) and Thale-Noi Lake in Pattalung Province of the south of Thailand (Segers & Pholpunthin, 1997), where 73, 86 and 106 taxa were found, respectively. Since the present record is the res-

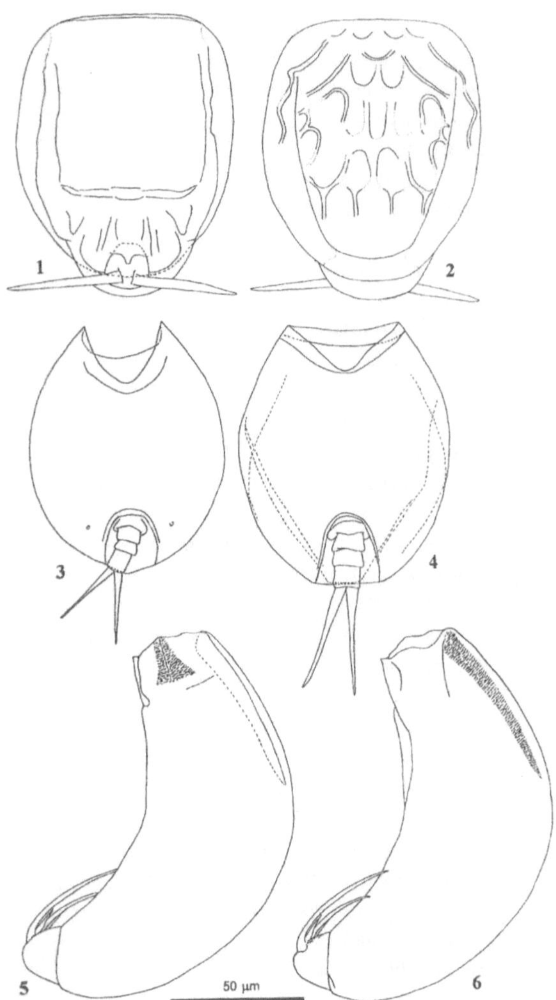


*Figures 1–6. Lecane nelsoni* Segers, (1) ventral view, (2) dorsal view. *Lepadella eurysterna* Myers, (3) ventral view. *Lepadella benjamini* Harring, (4) ventral view. *Trichocrca abilioi* Segers, (5–6) lateral views.

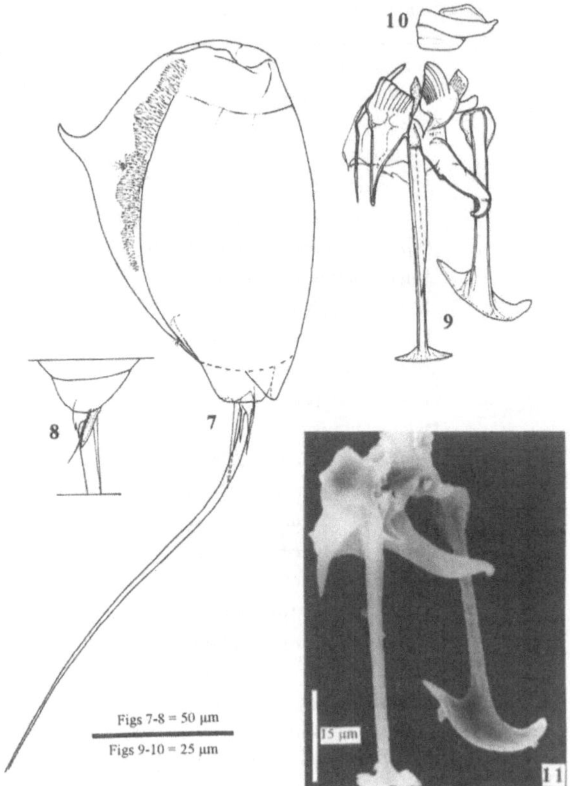


*Figures 7–11. Trichocerca orca* Murray, (7) right-lateral view, (8) posterior part, left-lateral view, (9) trophi; ventral view, (10) uncus, (11) trophi; scanning electron micrograph (Figs 7–10, drawn by H. Segers).

ult of examinations of monthly samples for a period of over 1 year, Lake Kud-Thing has the most diverse rotifer assemblage ever recorded in the country. The high species diversity can be compared with that of Lakes Oguta and Iyi-Efi in the floodplain of the River Niger, Nigeria, where 124 and 136 species were recorded (Segers et al., 1993). Segers & Dumont (1995) also recorded 102 rotifer species from 12 samples collected on a single date from Broa reservoir, Brazil. Moreover, the number of species recorded in Lake Kud-Thing is close to that estimate to be found (+210 spp.) from tropical habitats by Dumont & Segers (1996). This richness of the rotifer taxocoenosis supports the fact that the fauna of lakes is much more diverse than that of rivers or other habitats, particularly lakes with macrophyte-rich littoral as suggested by Segers et al. (1993).

Of the 183 species recorded from Lake Kud-Thing, 32 are new to Thailand (indicated by an asterisk in Table 1). Two of these, *Lecane isanensis* n. sp. and *Lecane* sp. are new to science. Additionally, 13 species; *Aspelta circinator, Collotheca tenuilobata, Lecane nelsoni* (Figs 1 and 2), *Lepadella eurysterna* (Fig. 3), *Lepadella benjamini* (Fig. 4), *Sinantherina ariprepes, Testudinella* cf. *insinuata, Trichocerca abilioi* (Figs 5 and 6), *Trichocerca inermis, Trichocerca montana, Trichocerca orca* (Figs 7–11), *Trichocerca rosea* and *Trichocerca simonei*, had not been recorded from the Oriental region before. This brings the Thai rotifer record to 310 valid species. Recently, *Lepadella eurysterna* is considered to be a synonym of *Lepadella arabica* by Baribwegure & Segers (2001).

The record of *Trichocerca orca* is particularly interesting, as it is the first since its discovery from New Zealand by Murray (1913). This very rare species can

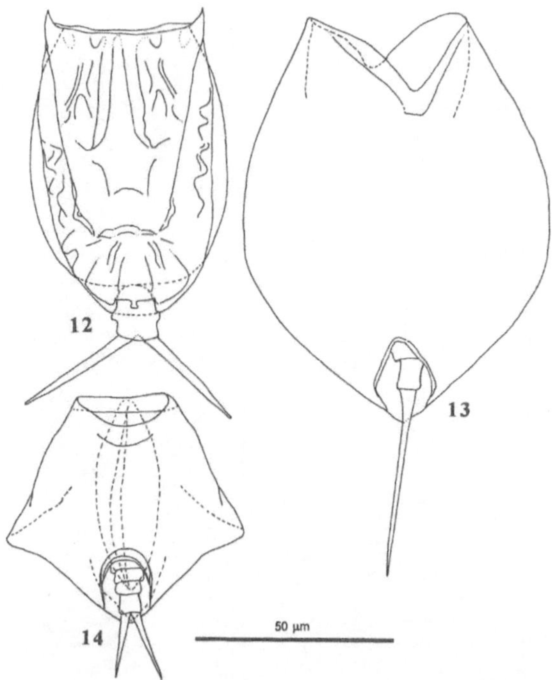


*Figures 12–14. Lecane simmoneae* Segers, (12) ventral view. *Lepadella monodactyla* Berzins, (13) ventral view. *Lepadella triptera* Ehrenberg f. *alata* Myers, (14) ventral view.

hardly be confused with any congener, by having a distinct forwardly directed spine (Fig. 7). It has to date been recorded only from Lake Kud-Thing in June and Lake Bung Khong Long (Nong Khai Province) in December (Sanoamuang, unpublished data).

It should be noted that the unidentified *Trichocerca* sp. in Sanoamuang et al. (1995, Figs 27–30) is in fact a variant specimen of *Trichocerca simonei*. This species was also found throughout the year in Lake Kud-Thing, and has to date been recorded from several localities in the northeast of Thailand. *T. simonei* was recorded previously from Zaïre (De Smet, 1989), the Galapagos island Santa Cruz (Segers, 1991) and Brazil (Turner & Da Silva, 1992). Representatives of the new records: *Lecane simmoneae, Lepadella monodactyla, Lepadella triptera* f. *alata,* and *Ploesoma lenticulare* are shown in Figures 12–18.

According to duration of appearance, the rotifer species of Lake Kud-Thing could be classified into three groups (Table 1): (1) common, perennial (56 spp.); (2) uncommon, perennial (79 spp.); and (3) uncommon, sporadic species (48 spp.). Common, perennial species were present throughout the year during a period of 9–12 months. Twenty taxa were found in every sample series, and indicated by the symbol + in Table 1. Uncommon, perennial species occurred for a period of 4–8 months, whereas uncommon, sporadic species were found during 1–3 months only.

Lake Kud-Thing contains a number of rotifer species typically found in Thai freshwaters (Sanoamuang et al., 1995; Segers & Pholpunthin, 1997; Sanoamuang, 1998; Chittapun et al., 1999). The most diverse genera were *Lecane* (23.5%, 43 species), followed by *Trichocerca* (16.9%, 31 species) and *Lepadella* (11.5%, 21 species). However, only eight taxa of *Brachionus* were recorded in this lake, compared with 21 from the list of the northeast (Sanoamuang et al., 1995). This is probably because most species of *Brachionus* are inhabitants of eutrophic waters while members of *Trichocerca* are indicative of oligotrophic waters (Sladecek, 1983). In addition, the lake's cladocerans are also diverse (53 species: Sanoamuang, unpublished data). Thus, the high species richness of rotifers and cladocerans, and the high diversity of *Lecane* and *Trichocerca*, together indicate that Lake Kud-Thing is probably not eutrophicated.

Although the majority of rotifers in Lake Kud-Thing consist of widely distributed tropical taxa, six species (3.3%) are restricted to the Oriental region. These are *Keratella edmondsoni, Lecane blachei, L. isanensis* n. sp., *L. shieli, L. thailandensis* and *Lecane* sp. Furthermore, four species (*Brachionus dichotomus* f. *reductus, Macrochaetus danneeli, Testudinella walkeri* and *Trichocerca orca*), or 2.2%, are Australasian taxa.

A description of *Lecane isanensis* n. sp. follows. Another unknown *Lecane* sp. belonging to the complicated *unguitata*-group will be published elsewhere.

*Lecane isanensis* n. sp. (Figs 19–24)

*Type locality:* Lake Kud-Thing, Bung Kan district, Nong Khai Province, northeast Thailand, 28 February 1998; water temperature 25 °C, pH 7.3, conductivity 80 $\mu$S cm$^{-1}$.

*Other localities:* 1. Hui Khee Hin, Phu Phan National Park, Sakon Nakhon Province, 24 April 1997; water temperature 26 °C, pH 6.8, conductivity 45 $\mu$S cm$^{-1}$. 2. Lake Nong-Han, Muang district, Sakon Nakhon Province, 6 December 1997; water temperature 25 °C, pH 6.9, conductivity 43 $\mu$S cm$^{-1}$.

*Material examined:* Female holotype and five paratypes deposited in the Science Museum, Khon Kaen University, Khon Kaen, Thailand. One and two female paratypes in the collections of the Royal Belgian Institute for Natural Sciences (K.B.I.N.), Brus-

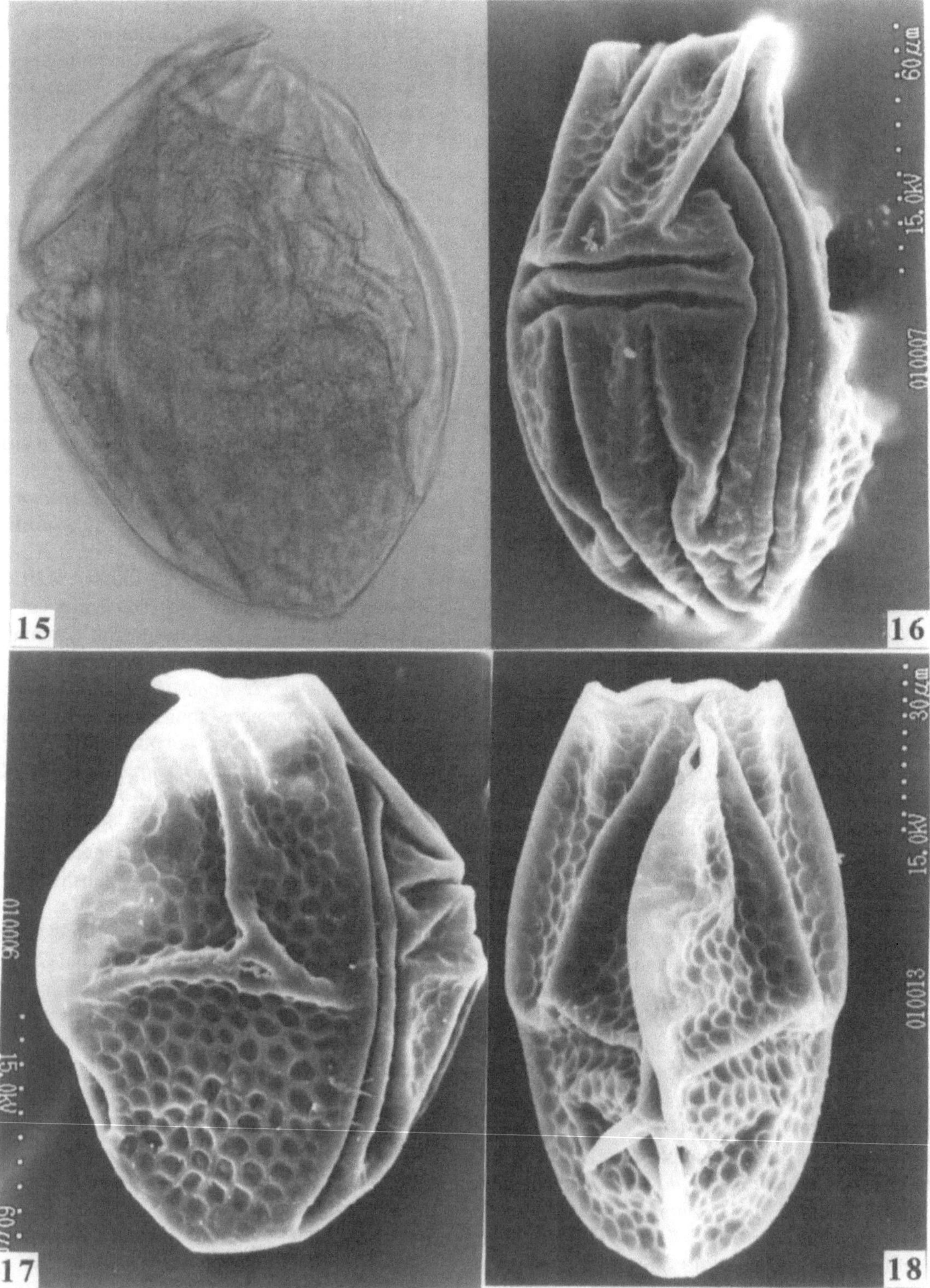


*Figures 15–18. Ploesoma lenticulare* Herrick, (15) photomicrograph, (16–18) scanning electron micrographs.

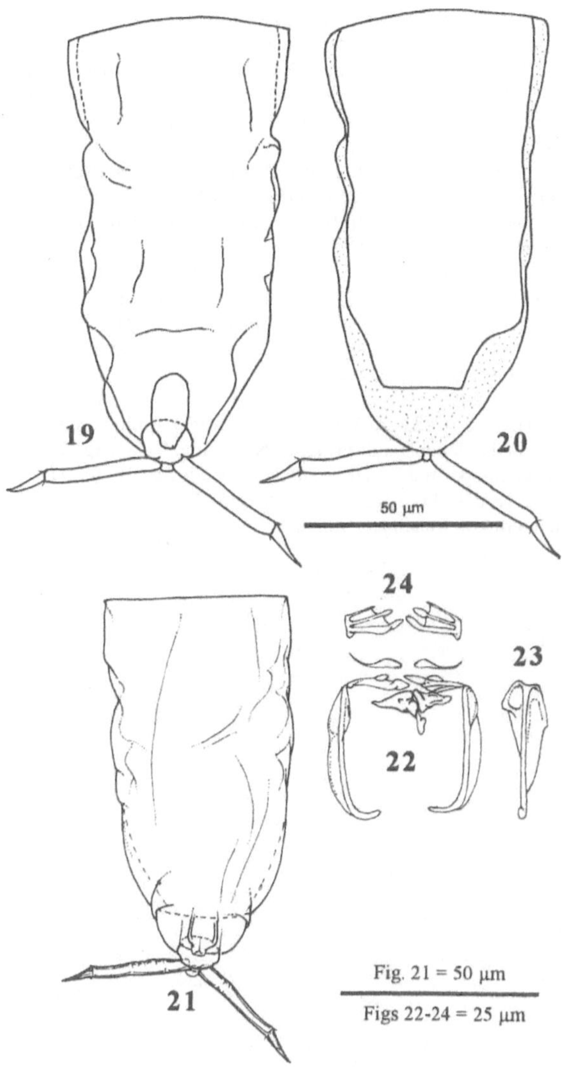


*Figures 19–24. Lecane isanensis* n. sp., (19) ventral view, (20) dorsal view, (21) another specimen; ventral view, (22) trophi, ventral view, (23) manubrium, (24) unci (Figs 21–24, drawn by H. Segers).

sels, and the Institute of Animal Ecology, University of Ghent (R.U.G.), Belgium, respectively.

*Differential diagnosis: Lecane isanensis* n. sp. can hardly be confused with any other Lecane, by having an elongate lorica, thick toes and peculiar claws. The species resembles *L. shieli* Segers & Sanoamaung, 1994, but is distinguished by the absence of antero-lateral spines.

The new species keys out to *L. elongata* Harring & Myers, 1926 following the key by Segers (1995). It is distinguished from *L. elongata* by its different head aperture, thick toes and peculiar claws.

*Description:* Parthenogenetic female: lorica relatively soft, elongate. Dorsal plate anteriorly as wide as, medially and posteriorly narrower than ventral plate, elongate. Head aperture margins straight, coincident, antero-lateral corners angulate. Ventral plate about twice as long as wide, generally parallel-sided, with indistinct transverse and some longitudinal folds. No lateral sulci. Foot plate with indistinct coxal plates. Prepedal fold narrow, elongate, distally with median projection. Foot pseudosegment short, not or slightly projecting. Toes thick, parallel-sided. Claws stout, weakly curved, inserted eccentrically. Several minute accessory claws present. Male unknown.

*Measurements (range and mean, in $\mu m$; n = 10):* Dorsal plate length 81–86 (83), width 41–49 (45), ventral plate length 90–95 (92), width 47–54 (50), toe length 35–37 (36), claw 9–11 (10).

*Etymology:* The species name is a toponym, refering to the area ('Isan', Thai for northeast Thailand) from where the species was first recognized.

*Distribution and ecology: Lecane isanensis* n. sp. is rare and usually occurs in small numbers. This species has to date been found in three localities in Sakon Nakhon and Nong Khai Provinces. In Lake Kud-Thing, it was present in February, May, June, September and November. It occurs over a temperature range of 24–30 °C, pH 6.5–8.1, and conductivity 80–310 $\mu S\ cm^{-1}$.

## Acknowledgements

This work was supported by the TRF/BIOTEC Special Program for Biodiversity Research and Training grant BRT 142011. We thank Dr H. Segers for the drawings of *Lecane isanensis* n. sp. and *Trichocerca orca*, and for providing copies of some rare publications, Mr Pipatphong Kanla for assistance with SEM; and Mrs Chutamas Sang-arun for collecting some samples. We greatly appreciate helpful comments on the manuscript from Drs R.J. Shiel, H. Segers and T.R. Rao.

## References

Baribwegure, D. & H. Segers, 2001. Rotifera from Burundi: the Lepadellidae (Rotifera: Monogononta). Hydrobiologia: 446/447 (Dev. Hydrobiol. 153): 247–254.

Chittapun, S., P. Pholpunthin & H. Segers, 1999. Rotifera from peat-swamps in Phuket Province, Thailand, with the description of a new *Colurella* Bory De St. Vincent. Int. Rev. Hydrobiol. 84: 587–593.

De Smet, W. H. 1989. Contribution to the rotifer fauna of the Bas-Zaïre. 2. Species composition and seasonal abundance of rotifers in a shallow pond. Biol. Jb. Dodonaea 57: 62–77.

Dumont, H. J. & H. Segers, 1996. Estimating lacustrine zooplankton species richness and complementarity. Hydrobiologia 341: 125–132.

Murray, J., 1913. Australasian Rotifera. J. R. Microsc. Soc.: 455–461.

Sladecek, V., 1983. Rotifers as indicators of water quality. Hydrobiologia 100: 169–201.

Sanoamuang, L., 1996. *Lecane segersi* n. sp. (Rotifera, Lecanidae) from Thailand. Hydrobiologia 329: 23–25.

Sanoamuang, L., 1998. Rotifera of some freshwater habitats in the floodplain of the River Nan, northern Thailand. Hydrobiologia 387/388: 27–33.

Sanoamuang, L. & S. Savatenalinton, 1999. New records of rotifers from Nakhon Ratchasima province, north-east Thailand, with a description of *Lecane baimaii* n. sp. Hydrobiologia 412: 95–101.

Sanoamuang, L. & H. Segers, 1997. Additions to the *Lecane* fauna (Rotifera: Monogononta) of Thailand. Int. Rev. ges. Hydrobiol. 82: 525–530.

Sanoamuang, L., H. Segers & H. J. Dumont, 1995. Additions to the rotifer fauna of south-east Asia: new and rare species from north-east Thailand. Hydrobiologia 313/314: 35–45.

Segers, H., 1991. Contribution to the knowledge of the rotifer fauna of the Galapagos Islands. Biol. Jb. Dodonaea 58: 113–119.

Segers, H., 1995. Rotifera. Vol 2: The Lecanidae (Monogononta). Guides to the Identification of the Microinvertebrates of the Continental Waters of the World 6. SPB Academic Publishing bv. The Hague, The Netherlands: 226 pp.

Segers, H. & H. J. Dumont, 1995. 102+ rotifer species (Rotifera: Monogononta) in Broa reservoir (SP., Brazil) on 26 August 1994, with the description of three new species. Hydrobiologia 316: 183–197.

Segers, H. & P. Pholpunthin, 1997. New and rare Rotifera from Thale-Noi, Pattalung Province, Thailand, with a note on the taxonomy of *Cephalodella*. Annls Limnol. 33: 13–21.

Segers, H. & L. Sanoamuang, 1994. Two more new species of *Lecane* (Rotifera, Monogononta) from Thailand. Belg. J. Zool. 124: 39–46.

Segers, H., C. S. Nwadiaro & H. J. Dumont, 1993. Rotifera of some lakes in the floodplain of the River Niger (Imo State, Nigeria) II. Faunal composition and diversity. Hydrobiologia 250: 63–71.

Turner, P. N. & C. Da Silva, 1992. Littoral rotifers from the State of Mato Grosso, Brazil. Environment 27: 227–241.

Hydrobiologia **446/447**: 305–313, 2001.
L. Sanoamuang, H. Segers, R.J. Shiel & R.D. Gulati (eds), Rotifera IX.




# Biodiversity of Rotifera in some tropical floodplain lakes of the Brahmaputra river basin, Assam (N.E. India)

B. K. Sharma & Sumita Sharma[1]
*Department of Zoology, North-Eastern Hill University, Umshing, Shillong - 793 022, Meghalaya, India*
*E-mail: profbksharma@hotmail.com*
[1]*Zoological Survey of India, Risa Colony, Shillong - 793 003, Meghalaya, India*

*Key words:* Rotifera, floodplain lakes, Brahmaputra basin, N.E. India, biodiversity, ecology

## Abstract

One hundred and sixteen species of Rotifera are recorded from seven floodplain lakes of the Brahmaputra basin (northeastern India), the highest rotifer biodiversity recorded from these biotopes in the Indian subcontinent to date. The Australasian *Brachionus dichotomus reductus* and *Lecane batillifer*; the Oriental *Keratella edmondsoni, Lecane blachei* and *L. acanthinula;* the Palaeotropical *Lecane braumi, L. lateralis, L. unguitata, Trichocerca tropis, Testudinella greeni* and *T. brevicaudata;* the Pantropical *Brachionus donneri* and a rather widely distributed *Horaella brehmi* represent taxa of biogeographical interest. Three species are new additions to the Indian rotifer fauna and eight are new to the N.E. region. Lecanidae > Brachionidae = Colurellidae > Trichocercidae > Testudinellidae comprise the largest fraction (68.0%) of the examined fauna. Comments are made on the general nature and composition of the rotifer taxocoenosis as well as on acidophilic elements, ecology of various taxa and on the species richness of different lakes.

## Introduction

Floodplain lakes comprise an important component (Sugunan, 1997) of inland aquatic resources of India (over 0.20 million ha) and its North-Eastern region (0.12 million ha). They cover about 93% of the total lentic fish-prone area of Assam state (Goswami, 1997). Very little is, however, known about their zooplankton diversity in general and that of Rotifera in particular (Sharma, 1996, 1998). Our knowledge of the qualitative richness of rotifers in these biotopes of Assam is to date confined to a number of unpublished works (Lahon, 1983; Goswami, 1985; Yadava, 1987; Goswami, 1997).

The present study deals with the biodiversity of Rotifera in seven floodplain lakes of the Brahmaputra basin, Assam (N.E. India) with special reference to the general nature and composition of the taxocoenosis, biogeography and ecology as well as to their species richness and community similarities between different lakes.

## Materials and methods

Seven floodplain lakes (locally called *beels)* of the Brahmaputra basin (Fig. 1A–C) Assam State (N.E. India) were studied. Of these, five, namely Puwa Saikia 1, Puwa Saikia 2, Kunwari, Butikor and Batua (94° 56′ E; 26° 75′ N) are located in the Dhemaji district of Upper Assam, Dhir beel (90° 50′ E; 26° 25′ N) is located in the Dhubri district and Dighali beel (91° 80′ E; 26° 28′ N) is located in the Kamrup district. Water and qualitative plankton samples were collected from the beels during summer (May), monsoon (July), post-monsoon (October) and winter (January) seasons during 1994–1995. Water samples were analyzed for temperature, specific conductivity, pH, dissolved oxygen and total alkalinity. Plankton samples were obtained by towing a nylobolt plankton net (No. 25; 50 $\mu$m) and preserved in 5% formalin. These samples were screened, various rotifer taxa isolated, and identified following Kutikova (1970), Koste (1978) and Segers (1995). Percentage similar-

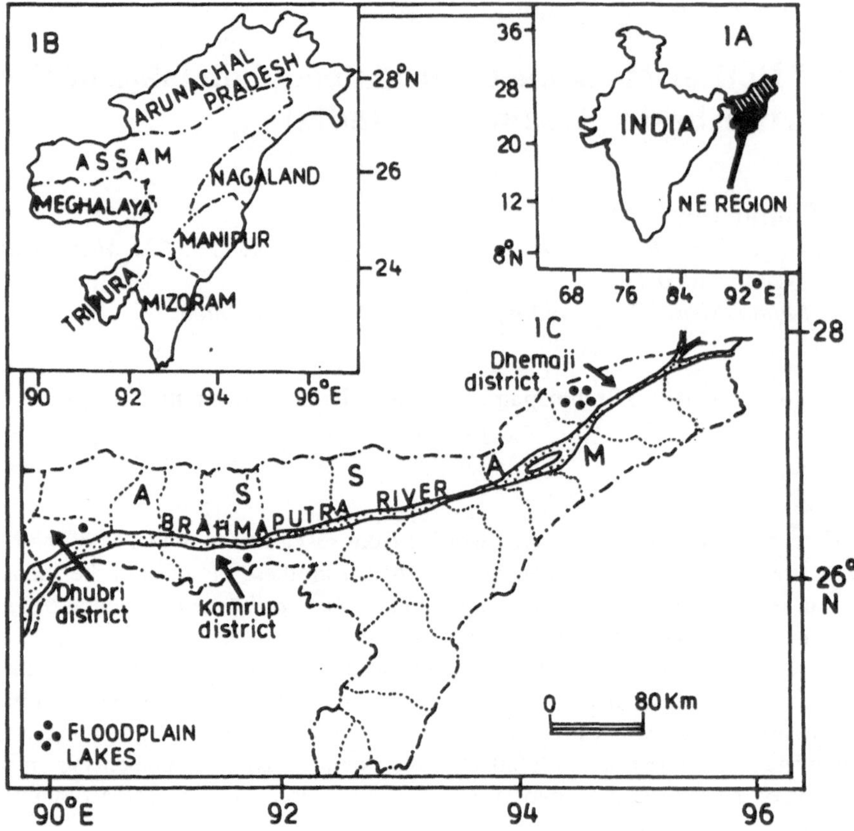


*Figure 1.* (A) Map of India showing North-Eastern Region. (B) Different states of North-Eastern India. (C) Map of Assam State showing various (seven) floodplain lakes.

ities between the rotifer communities of the different floodplain lakes were calculated *vide* Sorensen's index (Sorensen, 1948).

## Results and discussion

The present study reveals distinct variations in abiotic factors (Table 1) in the different beels; all these ecosystems, however, can be assigned to 'Class I' category *vide* Talling & Talling (1965). Puwa Saikia 1 and Puwa Saikia 2 indicate distinctly low ionic concentrations and low total alkalinity and, hence, depict 'very soft-acidic waters'. Dhir and Dighali are grouped by their 'soft-slightly acidic to alkaline waters'. Kunwari beel exhibits 'acidic hard water' character while Butikor and Batua beels show 'slightly acidic to slightly alkaline hard waters'. The specific conductivity and alkalinity values of all the beels are notably lower than those from the Kashmir valley (Khan, 1987), West Bengal (Vass, 1989) and Bihar (Singh & Roy, 1990). Water temperature varies within a range expected for water bodies in tropical regions and dissolved oxygen concentrations correspond with values recorded earlier from upper Assam (Sharma & Hussain, 1999).

Segers et al. (1993) hypothesize that (sub) tropical floodplains are the world's richest habitats for rotifers. The present report (116 species) from seven beels of the Brahmaputra basin (Table 2) supports this generalization as these biotopes contain the richest rotifer fauna ever recorded from the Indian subcontinent. The examined taxocoenosis is rich and diversified; the documented species comprise about 35% of the Indian Rotifera, about 80% of the fauna of North-Eastern India and raise the species record from this region to 145. Species richness is, however, relatively lower than in the floodplains of the river Niger, Nigeria (Segers et al., 1993) but nearly equals to the report of 118 species in the floodplains of the river Nan, northern Thailand (Sanoamuang, 1998). Interestingly, all 34 genera and 19 Eurotatorian families recorded to date from the N.E. region are represented in the present study and the generic and family diversity is,

*Table 1.* Variations in certain abiotic factors in different floodplain lakes

| Abiotic factors | Water temp. (°C) | Specific Conductivity ($\mu$S cm$^{-1}$) | pH | Dissolved oxygen (mg l$^{-1}$) | Alkalinity (mg l$^{-1}$) |
|---|---|---|---|---|---|
| **Floodplain lakes:** | | | | | |
| Puwa Saikia 1 | 17–29 | 21–32 | 5.0–6.6 | 4.0–6.4 | 16–22 |
| Puwa Saikia 2 | 18–29 | 22–33 | 5.0–6.7 | 4.0–6.4 | 14–21 |
| Dhir | 18–32 | 71–131 | 6.4–7.4 | 5.2–10.2 | 17–58 |
| Dighali | 18–30 | 56–94 | 6.5–7.2 | 4.3–9.2 | 28–45 |
| Kunwari | 20–33 | 110–123 | 5.0–6.8 | 4.8–9.6 | 50–76 |
| Butikor | 18–32 | 106–162 | 5.0–7.2 | 4.8–6.4 | 68–108 |
| Batua | 21- 32 | 136–184 | 6.5–7.5 | 3.2–10.4 | 86–110 |

therefore, rich in comparison with 60 genera and 25 families of Rotifera reported till now from India. In addition, *Lepadella lindaui, L. minoruoides* and *Filinia camasecla* are new records for this country while five other species namely *Keratella edmondsoni, Lecane blachei, Trichocerca tropis, Sinantherina spinosa* and *Filinia saltator* represent new records for the region.

The qualitative importance of two 'tropic-centred' genera namely *Lecane* and *Brachionus* imparts general tropical character to the rotifer taxocoenosis. In fact, the former genus alone accounts for 25.8% of the overall diversity as well as in individual beels (24.1–29.6%); such a feature compares well with its 26.9% and 28.5% contributions to the floodplain lakes of upper Assam (Sharma, unpublished data) and of the river Niger (Segers et al., 1993), respectively. The lecanid richness, however, corresponds with habitats in the floodplain of the Parana river, Brazil (Bonecker et al., 1994) and that of the river Nan, Thailand (Sanoamuang, 1998) but is in striking contrast to low *Lecane* richness in some beels of lower Assam (Lahon, 1983; Goswami, 1985, 1997). The tropical nature of the studied rotifer fauna is further supported by the low number of species of 'temperate-centred' *Keratella* (4 species) and *Synchaeta* (1 species), qualitative predominance of cosmopolitan species (64.7%) and high diversity of pantropical/cosmotropical elements (22.4%). All the stated aspects corroborate with salient features of many tropical rotifer faunas from different parts of the globe (Green, 1972; Pejler, 1977; Fernando, 1980; Dumont, 1983; Dussart et al., 1984; Sanoamuang et al., 1995; Sharma, 1996, 1998; Segers, 1996).

Biogeographically interesting elements constitute an important fraction (13%) of the rotifer fauna. Two Australasian elements, *Brachionus dichotomus reductus* and *Lecane batillifer* and the Palaeotropical *Testudinella greeni*, deserve special interest and are to date restricted to North-Eastern India. Comments on the distribution of *L. batillifer* were made by Sharma & Sharma (1997), while Sharma (1990) remarked on the occurrence of *T. greeni.* The present report comprises the second record of the last two species from India. In addition, this study includes three Oriental species - *Keratella edmondsoni, Lecane blachei* and *L. acanthinula.* Of these, the distributional ranges of *K. edmondsoni* and *L. blachei* are presently extended to the N.E. region. Furthermore, our material indicates five other Palaeotropical species i.e., *Lecane braumi, L. lateralis, L. unguitata, Trichocerca tropis* and *Testudinella brevicaudata.* Remarks on the distribution of the first three lecanids were made by Sharma & Sharma (1997) while the remaining species were commented on by Segers et al. (1993). The Pantropical *Brachionus donneri,* the Arctic-temperate *Dicranophorus luetkeni* as well as rather widely distributed *Trichocerca cylindrica* and *Horaella brehmi* also are interesting species; *T. cylindrica* occurs in Arctic-temperate, Oriental and Australian regions while *H. brehmi* is recorded from Palaearctic, Oriental, Australian and Neotropic regions.

Lecanidae (30 species) >Brachionidae (16 species) = Colurellidae (16 species) >Trichocercidae (10 species)>Testudinellidae (8 species), in the stated order, comprise the largest component (68.0%) of overall rotifer biodiversity and of the communities in the individual beels (64.8–72.2%). Such a pattern is in general confirmity with the composition of Indian Rotifera and also with the findings of Segers et al. (1993). The present report of 116 species is in distinct contrast to 64 species from five floodplain lakes of upper Assam (Sharma, unpublished data) but depicts about

*Table 2.* Species composition of Rotifera in different floodplain lakes

| Floodplain lakes | 1 | 2 | 3 | 4 | 5 | 6 | 7 |
|---|---|---|---|---|---|---|---|
| **Family: Brachionidae** | | | | | | | |
| *Anuraeopsis fissa* (Gosse) | – | – | + | + | + | – | – |
| *A. coelata* (De Beauchamp) | – | – | – | – | + | + | – |
| *Brachionus angularis* Gosse | – | – | – | – | + | + | – |
| *B. bidentatus* Anderson | + | + | – | + | – | – | + |
| *B. dichotomus reductus* Koste & Shiel | – | – | + | – | + | – | + |
| *B. diversicornis* (Daday) | – | – | + | + | + | – | – |
| *B. donneri* Brehm | + | + | – | + | – | – | – |
| *B. falcatus* Zacharias | – | + | + | + | + | + | + |
| *B. forficula* Wierzejski | – | – | + | – | + | – | + |
| *B. quadridentatus* (Hermann) | + | + | + | + | + | + | + |
| *Keratella cochlearis* Gosse | + | + | + | + | + | + | + |
| *K. edmondsoni* (Ahlstrom) | – | – | + | – | – | + | – |
| *K. tropica* (Apstein) | + | + | + | + | – | – | + |
| *K. lenzi* Hauer | + | + | – | – | – | + | + |
| *Platyias quadricornis* (Ehrenberg) | + | + | + | + | – | + | + |
| *Plationus patulus* (Müller) | + | + | + | + | + | + | + |
| *P. patulus macracanthus* (Daday) | + | + | – | – | + | – | + |
| **Family: Euchlanidae** | | | | | | | |
| *Euchlanis dilatata* Ehrenberg | + | + | + | + | + | + | + |
| *E. incisa* Carlin | + | – | + | + | – | – | – |
| *E. triquetra* Ehrenberg | – | + | – | + | – | – | – |
| *Dipleuchlanis propatula* (Gosse) | + | + | – | + | + | + | + |
| *Manfredium eudactylotum* Gosse | + | – | + | + | + | – | + |
| **Family: Mytilinidae** | | | | | | | |
| *Lophocharis salpina* (Ehrb.) | – | – | + | – | – | – | + |
| *Mytilina acanthophora* Hauer | – | – | – | + | – | – | – |
| *M. bisulcata* (Lucks) | + | + | – | – | – | + | – |
| *M. ventralis* (Ehrenberg) | + | + | + | + | + | + | + |
| **Family: Trichotriidae** | | | | | | | |
| *Macrochaetus collinsi* (Gosse) | + | – | – | + | + | + | – |
| *M. sericus* (Thorpe) | + | + | + | – | – | + | + |
| *Trichotria tetractis* (Ehrenberg) | + | + | + | + | + | + | + |
| **Family: Colurellidae** | | | | | | | |
| *Colurella obtusa* (Gosse) | – | – | – | + | + | – | + |
| *C. sulcata* (Stenroos) | + | + | + | – | – | – | – |
| *C. uncinata* (Müller) | + | + | + | + | – | + | – |
| *Lepadella acuminata* (Ehrenberg) | – | + | + | + | – | – | + |
| *L. apsida* Harring | + | – | – | – | – | + | – |
| *L. cristata* (Rousselet) | + | + | – | + | + | – | – |
| *L. lindaui* Koste | – | – | – | – | + | – | – |
| *L. minuta* (Montet) | – | – | – | – | – | + | – |
| *L. minoruoides* Koste & Robertson | – | – | – | – | – | – | + |
| *L. ovalis* (Müller) | + | + | + | + | + | + | + |
| *L. patella* (Müller) | + | + | + | + | + | + | + |

*Continued on p. 309*

*Table 2.* Continued

| Floodplain lakes | 1 | 2 | 3 | 4 | 5 | 6 | 7 |
|---|---|---|---|---|---|---|---|
| *L. rhomboides* (Gosse) | + | + | + | + | − | + | − |
| *L. triba* Myers | − | − | − | − | + | − | − |
| *L. triptera* Ehrenberg | − | − | + | + | − | − | − |
| *L. (Heterolepadella) ehrenbergi* (Perty) | + | + | + | + | + | − | + |
| *L. (H.) heterostyla* (Murray) | + | + | − | + | + | + | − |
| **Family: Lecanidae** | | | | | | | |
| *Lecane aculeata* (Jakubski) | − | − | + | + | + | + | − |
| *L. braumi* Koste | + | + | − | − | − | − | − |
| *L. curvicornis* (Murray) | + | + | + | + | + | + | + |
| *L. doryssa* Harring | − | − | − | + | + | − | − |
| *L. flexilis* (Gosse) | + | + | + | + | − | − | + |
| *L. hornemanni* (Ehrenberg) | − | − | + | + | + | + | + |
| *L. lateralis* Sharma | − | − | + | − | − | − | + |
| *L. leontina* (Turner) | + | + | + | + | + | + | + |
| *L. ludwigii* (Eckstein) | + | + | − | − | + | + | + |
| *L. luna* (Müller) | − | + | + | + | + | + | + |
| *L. nana* (Murray) | − | − | + | + | − | − | + |
| *L. ohioensis* (Herrick) | − | − | + | − | − | + | − |
| *L. papuana* (Murray) | − | − | + | + | + | + | + |
| *L. pertica* Harring & Myers | − | + | − | − | + | + | − |
| *L. signifera* (Jennings) | + | + | − | + | − | − | + |
| *L. ungulata* (Gosse) | + | − | + | + | + | + | − |
| *L. (Hemimonostyla) blachei* Berzins | − | − | − | − | − | + | − |
| *L. (H.) sympoda* Hauer | − | − | + | + | − | − | + |
| *L. (Monostyla) acanthinula* (Hauer) | − | + | − | − | − | − | − |
| *L. (M.) batillifer* (Murray) | + | − | − | − | − | − | − |
| *L. (M.) bifurca* (Bryce) | + | − | − | − | − | + | − |
| *L. (M.) bulla* (Gosse) | + | + | + | + | + | + | + |
| *L. (M.) closterocerca* (Schmarda) | + | + | + | + | + | + | + |
| *L. (M.) lunaris* (Ehrenberg) | + | + | + | + | − | − | + |
| *L. (M.) monostyla* (Daday) | − | − | − | − | + | − | − |
| *L. (M.) pyriformis* (Daday) | − | − | + | + | − | − | − |
| *L. (M.) quadridentata* (Ehrenberg) | + | + | − | − | + | + | + |
| *L. (M.) stenroosi* (Meissner) | − | − | + | − | − | − | + |
| *L. (M.) thienemanni* (Hauer) | − | − | − | − | + | − | − |
| *L. (M.) unguitata* (Fadeev) | − | + | − | + | − | + | − |
| **Family: Notommatidae** | | | | | | | |
| *Cephalodella forficula* (Ehrenberg) | + | + | − | + | − | − | + |
| *C. gibba* (Ehrenberg) | − | − | + | + | − | + | − |
| *C. mucronata* Harring & Myers | + | − | + | − | + | − | − |
| *Monommata longiseta* (Müller) | − | − | − | − | + | + | − |
| *Scaridium longicaudum* (Müller) | − | − | + | + | − | − | + |
| **Family: Gastropodidae** | | | | | | | |
| *Ascomorpha saltans* Bartsch | − | + | − | − | − | − | + |
| *A. ovalis* (Bergendal) | − | − | + | + | − | − | − |

*Continued on p. 310*

*Table 2.* Continued

| Floodplain lakes | 1 | 2 | 3 | 4 | 5 | 6 | 7 |
|---|---|---|---|---|---|---|---|
| **Family: Trichocercidae** | | | | | | | |
| *Trichocerca bicristata* (Gosse) | + | – | – | + | – | – | – |
| *T. braziliensis* Murray | – | – | + | – | – | + | – |
| *T. capucina* (Wierzejski & Zacharias) | – | + | – | + | – | – | + |
| *T. cylindrica* (Imhof) | + | + | + | – | + | + | – |
| *T. jenningsi* Voigt | – | – | + | + | – | – | + |
| *T. longiseta* (Schrank) | + | + | – | + | – | – | – |
| *T. rattus* (Müller) | – | – | + | – | – | – | + |
| *T. similis* (Wierzejski) | + | + | – | – | + | + | – |
| *T. tropis* (Hauer) | – | – | – | – | – | + | – |
| *T. weberi* (Jennings) | – | – | – | – | + | – | + |
| **Family: Asplanchnidae** | | | | | | | |
| *Asplanchna brightwelli* Gosse | + | – | + | + | – | + | – |
| *A. priodonta* Gosse | – | – | + | – | + | – | + |
| **Family: Synchaetidae** | | | | | | | |
| *Synchaeta oblonga* Ehrenberg | – | – | – | – | – | + | – |
| *Pleosoma lenticulare* Herrick | – | + | – | – | – | – | – |
| *Polyarthra vulgaris* Carlin | + | + | + | + | + | + | + |
| **Family: Dicranophoridae** | | | | | | | |
| *Dicranophorus forcipatus* (Müller) | + | + | – | – | + | – | – |
| *D. luetkeni* (Bergendal) | – | – | – | – | – | + | – |
| **Family: Floscularidae** | | | | | | | |
| *Sinantherina spinosa* (Thorpe) | + | + | – | – | – | – | – |
| **Family: Conochilidae** | | | | | | | |
| *Conochilus unicornis* Rousselet | – | – | + | + | + | – | + |
| **Family: Hexarthridae** | | | | | | | |
| *Hexarthra mira* (Hudson) | – | – | + | + | + | + | – |
| **Family: Filiniidae** | | | | | | | |
| *Filinia camasecla* Myers | + | + | – | – | – | – | + |
| *F. longiseta* (Ehrenberg) | – | + | + | + | + | + | – |
| *F. opoliensis* (Zacharias) | – | – | + | + | – | – | – |
| *F. saltator* (Gosse) | – | – | – | – | + | + | – |
| **Family: Testudinellidae** | | | | | | | |
| *Testudinella brevicaudata* Yamammoto | – | – | – | – | – | – | + |
| *T. emarginula* (Stenroos) | + | + | + | + | – | + | – |
| *T. greeni* Koste | – | – | – | – | + | – | – |
| *T. parva parva* (Ternetz) | + | + | – | + | – | + | – |
| *T. parva bidentata* (Ternetz) | + | + | – | + | – | – | – |
| *T. patina* (Hermann) | + | + | + | + | + | + | + |
| *T. tridentata* Smirnov | – | – | + | – | + | – | – |
| *Pompholyx sulcata* Hudson | – | – | + | + | – | – | + |

*Continued on p. 311*

*Table 2.* Continued

| Floodplain lakes | 1 | 2 | 3 | 4 | 5 | 6 | 7 |
|---|---|---|---|---|---|---|---|
| **Family: Trichosphaeridae** | | | | | | | |
| *Horaella brehmi* Donner | – | + | – | – | – | – | – |
| **Family: Philodinidae** | | | | | | | |
| *Philodina citrina* (Ehrenberg) | – | + | + | – | – | – | + |
| *Rotaria neptunia* (Ehrenberg) | – | – | + | + | – | + | – |
| *R. rotatoria* (Pallas) | + | – | – | + | + | – | – |
| Total No. of Species | 54 | 56 | 64 | 65 | 55 | 56 | 54 |

Abbreviations: 1 – Puwa Saikia 1; 2 – Puwa Saikia 2; 3 – Dhir; 4 – Dighali; 5 – Kunwari; 6 – Butikor; 7 – Batua; – = absent; + = present.

*Table 3.* Percentage similarities (*vide* Sorensen's index) between rotifer communities in different floodplain lakes

| Floodplain lakes | Puwa Saikia 1 | Puwa Saikia 2 | Dhir | Dighali | Kunwari | Butikor | Batua |
|---|---|---|---|---|---|---|---|
| Puwa Saikia 1 | – | 80.4 | 48.7 | 58.3 | 49.6 | 59.5 | 54.1 |
| Puwa Saikia 2 | | – | 49.6 | 60.7 | 48.7 | 56.6 | 63.7 |
| Dhir | | | – | 71.3 | 52.5 | 53.3 | 61.7 |
| Dighali | | | | – | 55.3 | 57.8 | 59.5 |
| Kunwari | | | | | – | 59.7 | 52.6 |
| Butikor | | | | | | – | 42.9 |
| Batua | | | | | | | – |

71% community similarity with the same. Total qualitative richness is, however, significantly higher than the reports of 37 species (Goswami, 1985) and 29 species (Goswami, 1997) from some beels of lower Assam and showed 30.0% and 27.6% community similarities with the mentioned works, respectively. Comparison with other Indian works is not feasible because of incomplete species inventories. Further, the qualitative diversity in individual lakes (54–65 species) is highest ever recorded in Indian floodplains but is yet significantly lower than reports of 136 species in Lake Iyi-Efi and 124 species in Lake Oguta in the Niger delta (Segers et al., 1993), 130 species in Lake Guarana, Brazil (Bonecker et al., 1994) and 104 species in Laguana Bufeos, Bolivia (Segers et al., 1998). The 'soft, slightly acidic to slightly alkaline waters of Dhir and Dighali beels indicate higher species richness (64–65 species),with maximum diversity in summer and post-monsoon seasons. Percentage similarity between the rotifer communities in the sampled beels (42.9–80.4%) is relatively higher (Table 3) than report (37.3–68.8%) by Sharma (unpublished data). Maximum percentage similarity is noticed between Puwa Saikia 1 and Puwa Saikia 2 which are identical in their abiotic factors. This is followed by 71.3% similarity between Dhir and Dighali beels which again are grouped together based on their abiotic parameters while similarity ranges between 42.9% and 63.9% in comparison between the rest of the beels.

The acidic to slightly alkaline waters (pH: 5.0–7.4) of the sampled beels are characterized by the presence of various acidophilic rotifers such as *Plationus patulus macracanthus, Dipleuchlanis propatula, Euchlanis triquetra, Mytilina bisulcata, Colurella sulcata, Lepadella cristata, Lecane pertica, Monommata longiseta, Testudinella parva* and *T. tridentata*. This is in conformity with remarks by Sharma (1991, 1996). The distribution of *Lepadella acuminata, Ascomorpha saltans, Trichocerca weberi* and *Testudinella emarginula* in the beels also reveals their acidic nature. Among about 22 species of the genus *Brachionus* known from the Indian waters (Sharma, 1996), only

8 species are included in this account. The relative paucity of brachionid species is attributed to the acidic nature of these biotopes and thus re-affirms the findings of Fernando & Zankai (1981) and Sharma (1996). *Anuraeopsis fissa, Brachionus bidentatus, B. forficula, Dipleuchlanis propatula, Manfredium eudactylotum, Macrochaetus collinsi, M. sericus, Colurella sulcata, Lepadella cristata, Lecane ludwigii* and *L. stenroosi* are observed primarily during warmer months and are, therefore, designated as warm-stenothermal species (*vide* Koste, 1978) in the present observations.

The rotifer fauna is characterized by a predominance of periphytic or littoral elements (86 species; 74.1%) and fewer planktonic rotifers (30 species; 25.9%). This important feature can be assigned to the lack of definite pelagic habitats (De Manuel, 1994) in the beels, their shallow nature and growth of aquatic macrophytes. Only 15 species i.e. *Lepadella lindaui, L. minoruoides, L. triba, Lecane acanthinula, L. batillifer, L. blachei, L. monostyla, L. thienemanni, Trichocerca tropis, Synchaeta oblonga, Pleosoma lenticulare, Dicranophorus luetkeni, Testudinella brevicaudata, T. greeni* and *Horaella brehmi* are noticed in only one of the beels. On the other hand, 12 species are observed in all these biotopes while 16 species occurred in at least two beels.

## Acknowledgements

Thanks are due to the Head, Department of Zoology, North-Eastern Hill University, Shillong for laboratory facilities and to Md. Hussain for help in field collections. One of the author's (BKS) is sincerely grateful to Dr La-orsri Sanoamuang, Organizing Committee Chairperson, Dr H. Segers and all other members of the Scientific Committee of the IXth International Rotifer Symposium, Khon Kaen, Thailand, as well as to Dr Elizabeth Wurdak, Chairperson, VIIIth International Rotifer Symposium, Minnesota (U.S.A.) for financial support for participation in the Khon Kaen Symposium. Finally, the senior author is personally thankful to Dr La-orsri Sanoamuang, Thailand and Dr H. Segers, Belgium for their help in various ways.

## References

Bonecker, C. C., F. A. Lansac-Toha & A. Staub, 1994. Qualitative study of Rotifers in different environments of the high Parana river floodplain (Ms). Brazil. Revista UNIMAR 16: 1–16.

De Manuel, J., 1994. Taxonomic and zoogeographic considerations on Lecanidae (Rotifera: Monogononta) of the Balearic archipelago, with description of a new species, *Lecane margalefi* n. sp. Hydrobiologia 288: 97–105.

Dumont, H. J., 1983. Biogeography of rotifers. Hydrobiologia 104: 19–30.

Dussart, B. H., C. H. Fernando, J. Matsumura-Tundisi & R. J. Shiel, 1984. A review of systematics, distribution and ecology of tropical freshwater zooplankton. Hydrobiologia 113: 77–91.

Fernando, C. H., 1980. The freshwater zooplankton of Sri Lanka, with a discussion of tropical freshwater zooplankton composition. Int. Rev. ges. Hydrobiol. 65: 411–426.

Fernando, C. H. & N. P. Zankai, 1981. The Rotifera of Malaysia and Singapore with remarks on some species. Hydrobiologia 78: 205–219.

Green, J., 1972. Latitudinal variation in associations of planktonic Rotifera. J. Zool. Lond. 167: 31–39.

Goswami, M. M., 1985. Limnological investigations of a tectonic lake of Assam, India and their bearing on fish production. Ph.D. thesis, Gauhati University, Assam. 395 pp.

Goswami, N., 1997. Studies on the productivity indicators in three different types of wetlands of Assam, India. Ph.D. thesis, Gauhati University, Assam: 217 pp.

Khan, M. A., 1987. Observations on zooplankton composition, abundance and periodicity in two flood-plain lakes of the Kashmir Himalayan valley. Acta hydrochem. hydrobiol. 15: 167–174.

Koste, W., 1978. *Rotatoria.* Die Rädertiere Mitteleuropas, begründet von Max Voigt. Überordnung Monogononta. Gebrüder Borntaeger, Berlin, Stuttgart. I. Text: 673 pp. U. II. Tafelbd. (T. 234).

Kutikova, L. A., 1970. The rotifer fauna of the USSR. Fauna SSSR 104, Academia Nauk: 744 pp (in Russian).

Lahon, B., 1983. Limnology and fisheries of some commercial beels of Assam, India. Ph.D. thesis, Gauhati University, Assam: 349 pp.

Pejler, B., 1977. On the global distribution of the family Brachionidae (Rotatoria). Arch. Hydrobiol. Suppl. 53: 255–306.

Sanoamuang, L., 1998. Rotifera of some freshwater habitats in the floodplain of the River Nan, northern Thailand. Hydrobiologia 387/388: 27–33.

Sanoamuang, L., H. Segers & H. J. Dumont, 1995. Additions to the rotifer fauna of south-east Asia: new and rare species from north-east Thailand. Hydrobiologia 313/314: 35–45.

Segers, H., 1995. Rotifera 2: Lecanidae. Dumont, H. J. & T. Nogrady (eds), In Guides to Identification of the Microinvertebrates of the Continental Waters of the World. SPB Academic Publishing, Amsterdam, The Netherlands: 6: 1–226.

Segers, H., 1996. The biogeography of littoral *Lecane* Rotifera. Hydrobiologia 323: 169–197.

Segers, H., N. L. Ferrufino & L. De Meester, 1998. Diversity and Zoogeography of Rotifera (Monogononta) in a flood plain lake of the Ichilo river, Bolivia, with notes on little known species. Int. Rev. Hydrobiol.83: 439–448.

Segers, H., C. S. Nwadiaro & H. J. Dumont, 1993. Rotifera of some lakes in the floodplain of the river Niger (Imo State, Nigeria). II. Faunal composition and diversity. Hydrobiologia 250: 63–71.

Sharma, B. K., 1990. The genus *Testudinella* (Eurotatoria: Gnesiotrocha: Testudinellidae) in North-Eastern India. Hydrobiologia 199: 29–33.

Sharma, B. K., 1991. Rotifera. In Animal Resources of India. Protozoa to Mammalia. State of the Art. Published by Zoological Survey of India, Calcutta: 69–88.

Sharma, B. K., 1996. Biodiversity of Freshwater Rotifera in India – a status report. Proc. zool. Soc. Calcutta 49: 73–85.

Sharma, B. K., 1998. Fauna diversity of India: Rotifera. In Alfred, J. R. B., A. K. Das & A. K. Sanyal (eds), Faunal Diversity of India. A Commemorative Volume in the 50th Year of India's Independence. ENVIS Centre, Zool. Surv. India, Calcutta: 57–70.

Sharma, B. K. & Md. Hussain, 1999. Temporal variations in abiotic factors of a tropical floodplain lake, Upper Assam (N. E. India). Rec. zool. Surv. India 97: 145–150.

Sharma, B. K. & Sumita Sharma, 1997. Lecanid rotifers (Rotifera: Monogononta: Lecanidae) from North-Eastern India. Hydrobiologia 356: 159–163.

Singh, J. P. & S. P. Roy, 1990. Investigations on the limnological profile of the Karwar lake (Begusarai, Bihar). Recent Trends Limnol. 457–467.

Sorensen, T., 1948. A method of establishing groups of equal amplitude in plant society based on similarity of species content and its application to analysis of the vegetation of Danish commons. Biol. Skr. 5: 1–34.

Sugunan, V. V., 1997. Fisheries management of small bodies in seven countries in Africa, Asia and Latin America. FAO Fisheries Circular No. 933. Rome: 149 pp.

Talling, J. F. & I. B. Talling, 1965. The chemical composition of African lake waters. Int. Rev. ges. Hydrobiol. 50: 421–463.

Vass, K. K., 1989. Beel fisheries resources in West Bengal. Bull. CICFRI Barrackpore 63: 29–35.

Yadava, Y. S., 1987. Studies on the limnology and productivity of an oxbow lake in Dhubri district of Assam (India). Ph.D. thesis, Gauhati University, Assam: 320 pp.

Hydrobiologia **446/447:** 315–322, 2001.
L. Sanoamuang, H. Segers, R.J. Shiel & R.D. Gulati (eds), Rotifera IX.




# Early contributions of molecular phylogenetics to understanding the evolution of Rotifera

David B. Mark Welch
*Department of Molecular and Cellular Biology, Harvard University, Cambridge, MA 02138, U.S.A.*
*E-mail: welch@fas.harvard.edu*

*Key words:* Eurotatoria, Gnathifera, Aschelminthes, 18S RNA, *hsp82*

## Abstract

The past decade has seen the application of DNA sequence data to phylogenetic investigations of Rotifera, both expanding and challenging our understanding of the evolution of the phylum. Evidence that Acanthocephala, long regarded as a separate but closely related phylum, is a highly derived class of Rotifera demonstrates the potential of molecular analyses to suggest relationships not obvious from morphological analysis. Phylogenies based on the sequence of the gene for the small ribosomal RNA suggest that rotifers and acanthocephalans are associated with Platyhelminthes and Gastrotricha, perhaps in a clade with Gnathostomula and Cycliophora; at present, this group lacks a clear morphological synapomorphy. A more complete resolution of the molecular phylogeny of Rotifera will require surveying multiple genes and several species from each clade under investigation.

## Introduction

The rapid growth of techniques facilitating DNA sequencing, coupled with the recent explosion of inexpensive computer processing power, has made the estimation of phylogenies based on DNA sequence data practical, even for phyla that are not the focus of major scientific investigation. There are distinct benefits of molecular phylogenetics that have encouraged its widespread use: the genes examined are generally present in all eukaryotes, allowing comparison of widely disparate taxa; characters (in the form of nucleotide positions) and character states (nucleotides) are unambiguous; and the data sets (sequence alignments) can be quite large, generally containing 1000 characters or more. An ever-growing number of algorithms are used to reconstruct phylogenies from nucleotide sequence data (for reviews see Felsenstein, 1988; Swofford et al., 1996; Li, 1997). Each method has merits, although none of them is guaranteed to produce the correct phylogeny (Cummings et al., 1995; Håstad & Björklund, 1998). The vast majority of analyses are performed using the gene which encodes the ribosomal small subunit RNA (18S). This is largely due to historical momentum and the number of sequences in databases to which new sequences can be compared, rather than advantageous characteristics of 18S sequence.

Unfortunately, the nature of nucleotide sequence data introduces complications that have been too often overlooked in the widespread acceptance of molecular phylogenetic results (but see Wägele & Wetzel, 1994; Maley & Marshall, 1998). Any evolutionary analysis must compare homologous characters, and thus the accuracy of molecular phylogenetic methods is dependent on aligning the sequences so that homologous nucleotide positions are compared between species. The correct alignment is not always obvious, and different alignment methods often produce different alignments that can result in different phylogenies, each well-supported by statistical tests (Winnepenninckx & Backeljau, 1996; Morrison & Ellis, 1997; Winnepenninckx et al., 1998a; Hickson et al., 2000). Also, phylogenetic algorithms presume that characters are evolving independently, when in fact there is often covariance among nucleotide positions (Tillier & Collins, 1995; Huelsenbeck & Nielsen, 1999; Parsch et al., 2000). Another critical handicap is that there are only four states in nucleotide sequence data, and the states are relatively unordered (each nucleotide is free to change to any of the other three, although the probabilities may vary). Thus, two taxa may easily have the same nucleotide at the same position for reasons other than common descent. This problem is most severe when highly divergent taxa are compared, and can result in the grouping of dissimilar taxa in phylogenetic trees (Felsenstein, 1978; Hendy & Penny, 1989; Philippe, 1997). Different rates of change between characters or between taxa can exacerbate the problem (Van De Peer et al., 1993; Rzhetsky, 1995; Aguinaldo et al., 1997; Giribet et al., 2000; Philippe & Germot, 2000). All of these complicating factors are particularly puissant in the case of 18S sequences, which can be difficult to align, clearly have covarying nucleotide positions (the function of the RNA transcript depends on its secondary and tertiary structure, which is conferred by interactions between nucleotides), and evolve at different rates along the length of the gene. In addition, recent studies suggest that aspects of the secondary structure of 18S are due to homeoplastic mutational processes rather than homology, calling into question the phylogenetic validity of structural models used for alignment (Schultes et al., 1999; Hancock & Vogler, 2000). There is a growing sense that genes that encode proteins may be less susceptible than 18S to some of these difficulties (Abouheif et al., 1998; Maley & Marshall, 1998; McHugh, 1998; Giribet & Wheeler, 1999).

Despite these complexities, phylogenetic analyses of 18S generally produce results consistent with morphological methods. They also may suggest new relationships that require a reassessment of morphological characters. For example, phylogenies using 18S sequences suggest two major clades of protostomes. One, referred to as the Ecdysozoa (Aguinaldo et al., 1997), includes priapulids, arthropods, tardigrades and onychophorans, and may also contain some phyla traditionally considered pseudocoelomates, such as nematodes, nematomorphs and kinorhynchs. The other clade is comprised of Ectoprocta, Phoronida and Brachiopoda (Lophophora), and of Mollusca, Annelida and allied phyla (Eutrochozoa), and has been termed the Lophotrochozoa (Halanych et al., 1995). These super-phylum groups are well supported in some molecular analyses, but may still prove to be ephemeral. As pointed out by McHugh (1998), support for the Ecdysozoa–Lophotrochozoa dichotomy tends to be best when only one or two species represent each phylum, but diminishes as more taxa are added. This demonstrates a property of molecular phylogenetics that is often overlooked: when only a single species is used to represent each group, it is impossible to know the degree to which the species is representative of the group. The choice of species can significantly affect the phylogeny (Lecointre et al., 1993; Aguinaldo et al., 1997; Philippe, 1997; Graybeal, 1998), in part because any given species may have random sequence similarity to another species to which it is not closely related, causing them to be drawn together in molecular phylogenetic analyses. Thus, the results of molecular analyses of deep phylogenies may be statistically precise but not accurate. This problem can be alleviated by including multiple species from each group of interest. These additional sequences can add resolution by reducing the chance that random similarities will outweigh homologies (Hillis, 1996). Furthermore, with multiple species representing each group, species can be successively removed from analyses to determine if the position or composition of a clade is dependent on a particular species which may not be representative of the group (Lanyon, 1985).

Before a convincing case can be made for a phylogeny based on sequence data, the results must be examined in a variety of ways. Particularly, critical nodes of a tree must be robust to a variety of tree-building methods, with several species representing key groups. The effects of any ambiguities in the alignment also must be assessed. Such investigations of the phylogeny of Rotifera have begun, and have provided fresh insights into the evolution of the phylum, although considerable work remains.

## Molecular evidence for the inclusion of Acanthocephala in Rotifera

The first evidence from molecular studies for a close relationship between rotifers and acanthocephalans, long suggested by morphological studies, was demonstrated in an 18S phylogeny of Metazoa in which the representative of Rotifera, the monogonont *Brachionus plicatilis*, and the representative of Acanthocephala, *Moniliformis moniliformis*, formed a clade (Winnepenninckx et al., 1995; see also Raff et al., 1994). This association has been supported in every ensuing 18S analysis of metazoans in which both species were included (Hanelt et al., 1996; Giribet & Ribera, 1998; Littlewood et al., 1998; Winnepenninckx et al., 1998a; Zrzavý et al., 1998; Wirz et al., 1999; Giribet et al., 2000). Garey and et al. (1996) extended the analysis of the rotifer-acanthocephalan relationship by including a bdelloid rotifer, *Philodina acuticornis*, and three additional acanthocephalan species in an analysis of 18S sequences (Garey et al., 1996). Surprisingly, they did not find that the bdelloid and the monogonont formed a clade (the super-class Eurotatoria). Rather, they found strong support for Acanthocephala + *P. acuticornis*, suggesting not only that acanthocephalans were rotifers, but that their closest relatives were bdelloids. This clade (Acanthocephala + Bdelloidea), which they named Lemniscea after a synapomorphy proposed by Lorenzen (1985), was well supported by the available data. However, the strength of their conclusion was tempered by the inclusion of only single representatives of Bdelloidea and of Monogononta, and from the lack of a representative of the third rotifer class, Seisonidea.

Additional support for Lemniscea was found by Garey's group in an analysis of the relationship between *B. plicatilis*, *P. acuticornis* and the acanthocephalan *M. moniliformis* based on the 16S gene, the mitochondrial homolog of 18S (Garey et al., 1998). Although this second analysis included fewer taxa than the 18S analysis, and relatively short sequences, the fact that two genes from two different genomes (nuclear and mitochondrial) of the taxa involved each yielded phylogenies in which the representatives of Acanthocephala and Bdelloidea formed a well-supported clade added considerable weight to the possibility that acanthocephalans are rotifers.

The Lemniscea hypothesis received further corroboration in preliminary analyses using *hsp82*, the gene encoding the 82 kD heat shock protein. In a data set that included four species of bdelloids, the monogonont *B. plicatilis*, and the acanthocephalan *M. moniliformis*, strong support was found for Lemniscea using codon third positions (pers. obs.; Fig. 1a). However, when three additional monogonont species and a representative of Seisonidea were added, the three new monogonont species formed a sister-clade with the bdelloid species, while *B. plicatilis* remained outside all other rotifers, including the seisonid (Fig. 1b). In some analyses, *B. plicatilis* competed with the outgroup as the taxon most dissimilar to other monogononts. This aberrant position of *B. plicatilis* was found in analyses using codon 3rd positions, which are relatively neutral to selection. When analyses included only the 1st and 2nd positions of codons (or the amino acid translation), *B. plicatilis* and *B. calyciflorus* were sister-taxa and both Monogononta and Eurotatoria were monophyletic (Mark Welch, 2000). These findings suggested that *B. plicatilis* is evolving in an unusual manner at nucleotide positions with relaxed selective constraints. Further examination revealed that *B. plicatilis* has a much higher percentage

*Table 1.* GC content (%GC) of *B. plicatilis* and other rotifers. For the two genes encoding proteins, %GC was determined for all codon positions (123), first and second positions only (12), and third positions only (3); for the two ribosomal small subunit genes, %GC was determined for all positions and for those positions where *B. plicatilis* differed from the other rotifer examined, the bdelloid *P. acuticornis*. Asterisks indicate significantly higher %GC in *B. plicatilis* (*, $p<0.05$; **, $p<0.01$). The 'other rotifers' were: for *hsp82*, three monogononts, four bdelloids, and a seisonid (see Fig. 1); for *tbp*, three bdelloids (Mark Welch & Meselson, this volume); for 18S and 16S, the bdelloid *P. acuticornis* (Garey et al., 1996, 1998)

| | *B. plicatilis* | Other rotifers |
|---|---|---|
| *hsp82* 123 | 51.2* | 35.0 |
| *hsp82* 12 | 39.7 | 37.8 |
| *hsp82* 3 | 74.3** | 30.0 |
| *tbp* 123 | 53.7* | 37.4 |
| *tbp* 12 | 45.4 | 47.0 |
| *tbp* 3 | 70.4** | 23.4 |
| 18S all pos | 48.5 | 43.0 |
| 18S var pos | 53.0* | 37.8 |
| 16S all pos | 30.1 | 28.5 |
| 16S var pos | 38.5** | 16.9 |

of guanines and cytosines (%GC) at codon 3rd positions than do other rotifers, a phenomenon much less apparent at codon 1st or 2nd positions (Table 1). This drastic deviation in %GC may well cause *B. plicatilis* to appear to be more similar to outgroup taxa with similar %GC (for discussion of this problem, see Steel et al., 1993; Hashimoto et al., 1994; Rzhetsky & Nei, 1995). Thus, the appearance of a clade of Bdelloidea and Acanthocephala may result more from *B. plicatilis* being drawn out of a clade of Bdelloidea + Monogononta than from the similarity of Bdelloidea and Acanthocephala.

If *B. plicatilis* is removed from the analysis of *hsp82* sequences, there is strong support for the super-class Eurotatoria. *B. plicatilis* was the sole representative of monogononts in the phylogenies of Garey et al. (1996, 1998) and it is possible that the unusual evolution of this species affected those analyses as well. The overall %GC of the 18S and 16S genes of *B. plicatilis* are similar to those of other rotifers, as is the %GC at codon 1st and 2nd positions of *hsp82* and other protein-coding genes (Table 1). Because 18S encodes a catalytic RNA rather than a protein, it does not have positions that are directly analogous to codon third positions. However, there are positions that are relatively free to change (and thus apparently relatively

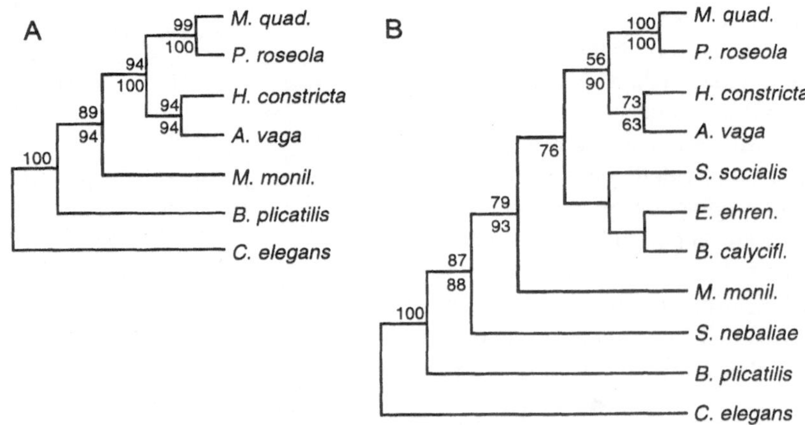


*Figure 1.* Relationship of acanthocephalan and rotifer species based on codon third positions of *hsp82*. When *B. plicatilis* is the sole representative of Monogononta, there is strong support for Lemniscea (a, left). However, adding more species demonstrates that *B. plicatilis* is not representative of other monogononts (b, right). Species examined were, Bdelloidea: *Philodina roseola* Ehrenberg 1832, *Macrotrachela quadricornifera* Milne 1886 [Philodinida, Philodinidae]; *Habrotrocha constricta* (Dujardin 1841) [Philodinida, Habrotrochidae]; *Adineta vaga* (Davis 1873) [Adinetida, Adinetidae]; Monogononta: *Brachionus plicatilis* Müller 1786, *Brachionus calyciflorus* Pallas 1766 [Ploima, Brachionidae]; *Eosphora ehrenbergi* Weber 1918 [Ploima, Notommatidae]; *Sinantherina socialis* (*Linnaeus* 1758) [Flosculariacea, Flosculariidae]; Seisonida: *Seison nebaliae* Grube 1859 [Seisonidea, Seisonididae]; Acanthocephala: *Moniliformis moniliformis* Bremser 1811 [Archiacanthocephala, Moniliformidae]. Numbers at nodes are percent of 1000 bootstrapped iterations of codon third positions supporting each clade by parsimony (above the line) and neighbor-joining of maximum likelihood distances (below the line). Analyses were performed as described (Mark Welch, 2000).

neutral to selection) and those that are much more constrained (Rzhetsky, 1995). At positions in 18S and 16S where *B. plicatilis* and *P. acuticornis* differ (an estimate of positions that are relatively neutral), the %GC of *B. plicatilis* is much higher, as with codon 3rd positions in coding genes (Table 1). Thus, it is possible that when *B. plicatilis* is the only representative of Monogononta in phylogenies, Bdelloidea and Acanthocephala may appear to be sister-taxa due to the disparity in %GC between *B. plicatilis* and other rotifers at phylogenetically important positions.

The indication from *hsp82* phylogeny that increased taxon sampling could affect the support for a monophyletic Eurotatoria has been bolstered by a recent analysis of 18S sequences that included several additional species of rotifers and acanthocephalans (García-Varela et al. 2000). This study employed successive searches for optimal trees to estimate parameters for a maximum likelihood model, an approach that is computationally intensive but may be very robust against rate heterogeneity along or between sequences. The authors re-examined the published 18S and 16S data (Garey et al., 1996, 1998) using this model, and found that trees containing Lemniscea or Eurotatoria received nearly identical likelihood scores (Table 2). When the authors examined their expanded group of sequences, which included three additional monogononts, a second bdelloid, and eight additional acanthocephalans, they found significantly superior support for Eurotatoria, using either their own alignment of the sequences or that of Garey et al. (1996).

*Table 2.* Maximum Likelihood Scores for Lemniscea and Eurotatoria. Numbers are -ln scores based on the likelihood model of García-Varela et al. (2000) using the datasets and alignments of Garey et al. (1998) (Garey 16S), Garey et al. (1996) (Garey 18S) or García-Varela et al. (2000) (García-Varela 18S). Asterisks (*) indicate cases where one clade received significantly better support than the other ($p<0.05$) by the test of Kishino & Hasegawa (1989). The data set of García-Varela et al. (2000) included the monogononts *B. plicatilis*, *B. platus*, *Asplanchna sieboldi* and *Lecane bulba*, the bdelloids *P. acuticornis* and *P. roseola*, and nine species of Acanthocephala representing the three major divisions

| Data set | Alignment | Eurotatoria | Lemniscea |
|---|---|---|---|
| Garey 16S | Garey 16S | 4384.4 | 4382.5 |
| Garey 18S | Garey 18S | 23499.1 | 23494.4 |
| García-Varela 18S | Garey 18S | 27004.1* | 27015.4 |
| García-Varela 18S | García-Varela 18s | 24231.7* | 24249.0 |

The different results obtained by Garey et al. (1996, 1998) and García-Varela et al. (2000) illustrate the ambiguities in molecular phylogenetics. A detailed examination of the effects of the alignments, methods, and species used by the two groups will be required to understand the reasons for these differences. Can

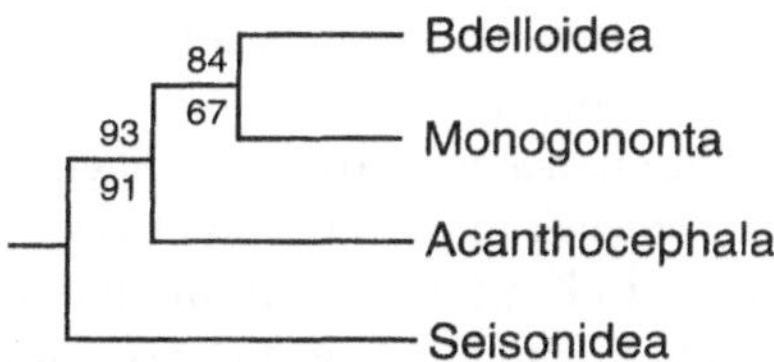


*Figure 2.* Phylogeny of Acanthocephala and the three traditional classes of Rotifera. The tree shown is obtained using *hsp82*; numbers at nodes are percent of 1000 bootstrapped iterations of codon third positions, not including *B. plicatilis*, supporting each clade by parsimony (above the line) and neighbor-joining of maximum likelihood distances (below the line). The outgroup (not shown) was the nematode *Caenorhabditis elegans*. Analyses were performed as described (Mark Welch, 2000).

the results of García-Varela et al. (2000) using their expanded dataset be replicated using maximum parsimony and distance methods? If so, how sensitive are the results to the species sampled?

If Eurotatoria is monophyletic, resolution of the relationship between Acanthocephala and Rotifera depends on the relationship of Acanthocephala to the third rotifer class, Seisonidea. At present, the only published molecular phylogeny that includes a seisonid is that of *hsp82*, for which there is strong bootstrap support for a clade of Acanthocephala + Eurotatoria, with a basal branch leading to Seisonidea (Fig 2; Mark Welch, 2000). This analysis included only a single acanthocephalan, *Moniliformis moniliformis*, and, while this species appears to be representative based on its 18S sequence (Garey et al., 1996; Near et al., 1998; García-Varela et al., 2000), it may not accurately represent Acanthocephala at the *hsp82* locus. The sequencing of *hsp82* from additional acanthocephalans is currently in progress.

## Molecular evidence for the inclusion of Rotifera in the Platyzoa

The position of Rotifera within Metazoa has not yet been the specific focus of molecular phylogenetic studies. When rotifers are included in molecular analyses, they have been found to be basal to other protostomes (Telford & Holland, 1993; Hanelt et al., 1996; Carranza et al., 1997; Campos et al., 1998; Winnepenninckx et al., 1998a) or to be a sister-group to Lophotrochozoa (Winnepenninckx et al., 1995; Garey et al., 1996; Aguinaldo et al., 1997; Giribet & Ribera, 1998; Halanych, 1998; Littlewood et al., 1998; Zrzavý et al., 1998; Ruiz-Trillo et al., 1999), generally in a clade with Platyhelminthes (Carranza et al., 1997; Halanych, 1998; Littlewood et al., 1998; Winnepenninckx et al., 1998a) or associated with a clade of Platyhelminthes and Gastrotricha (Winnepenninckx et al., 1995; Garey et al., 1996). The suggestion that the newly described phylum Cycliophora is a sister-taxon to Rotifera + Acanthocephala has been difficult to assess, as the study (Winnepenninckx et al., 1998b) did not include representatives of phyla that morphological studies indicate are the closest relatives to Rotifera: Gastrotricha, Gnathostomulida and Platyhelminthes (Wallace et al., 1996; Ahlrichs, 1997; Melone et al., 1998; Zrzavý et al., 1998).

Drawing in part on these studies, Cavalier-Smith (1988) has proposed grouping Rotifera (+Acanthocephala) with Gnathostomulida, Gastrotricha, and Platyhelminthes in an 'infrakingdom', Platyzoa, equal to other protostome super-phyla groups such as Ecdysozoa (Cavalier-Smith, 1998). The morphological synapomorphies defining this clade ("ciliated non-segmented acoelomates or pseudocoelomates lacking vascular system; gut (when present) straight") are rather nondescript, and the endurance of the clade may rely heavily on molecular analyses. However, the first report to include the 18S sequence of a gnathostomulid could not resolve the placement of this enigmatic phylum: a sister-group relationship with nematodes and chaetognaths received marginally better support than a similar relationship with rotifers (Littlewood et al., 1998). A recent examination of the 18S sequence of five gastrotrich species also failed to find statistical support for an association between Gastrotricha and Rotifera (Wirz et al., 1999). In a parsimony analysis of 18S sequences that included the gnathostomulid and two gastrotrich species, Zrzavý et al. (1998) did find that these two phyla formed a clade; however, Rotifera was not associated with this clade but with the Lophotrochozoa, and Platyhelminthes was not well-resolved (Zrzavý et al., 1998).

The appearance of a clade of Gnathostomulida, Nematoda and Chaetognatha found by Littlewood et al. (1998) probably resulted at least in part from the large number of substitutions that occurred in the 18S sequences of the species used to represent all three of these phyla relative to other metazoans. Relatively rapid evolutionary change results in long branches on phylogenetic trees, and these branches have been shown to cluster together regardless of true homology. This problem is particularly severe when assessing deep phylum-level relationships with single terminal taxa (Felsenstein, 1978; Hendy & Penny, 1989; Hillis et al., 1994; Aguinaldo et al., 1997). Similarly, Wirz et al. (1999) found that two of the five gastrotrich species they examined have very long branches, which may have affected the placement of the entire phylum. The study of platyhelminth molecular phylogeny also has been plagued by long branches, particularly of the acoel taxa (Katayama et al., 1995; Carranza et al., 1997; Balavoine, 1997; Campos et al., 1998; Ruiz-Trillo et al., 1999).

The strongest molecular phylogenetic support for Platyzoa comes from a recent analysis of metazoan phylogeny that included multiple species of Rotifera, Acanthocephala, Gnathostomulida, Gastrotricha and Platyhelminthes (Giribet et al., 2000). Using a novel parsimony-based approach that does not rely on sequence alignments, Giribet et al. (2000) explored a variety of character-change probabilities and the influence of highly variable regions of 18S. While they reported poor resolution (multiple paraphyletic groups) using the complete 18S sequence, they found that when highly variable regions of 18S were removed, the most parsimonious tree contained a clade of ((*B. plicatilis* + Acanthocephala) (Cycliophora + Gnathostomulida)) (Gastrotricha + some platyhelminthes) (Platyhelminthes + *P. acuticornis*). The presence of the cycliophoran in the platyzoan clade contrasts with most morphological analyses in which it is allied with entoprocts (cf Sørensen et al., 2000), but is consistent with earlier 18S phylogenies (Winnepenninckx, 1998b; note added in proof to Zrzavý et al., 1998).

The study by Giribet et al. (2000) represents a major advance in resolving the position of Rotifera and allied phyla, as it included several species from these phyla and specifically considered their relationship. However, the anomalous position of *P. acuticornis* and of several platyhelminthes indicate that further exploration of the Platyzoan hypothesis is necessary.

## Future directions

Rotifera remains one of the phylogenetically less well-resolved phyla, using either morphological or molecular approaches. Yet it contains the largest known group of obligately asexual species, wide morphological and ecological diversity, and, with the inclusion of Acanthocephala, an example of extreme adaptation to obligate parasitism. Clearly the phylum deserves additional study.

The potential of 18S to differentiate species-level relationships in Rotifera if demonstrated in the study of De La Riva & Walsh (pers. comm.). Their use of separate regions of the gene to resolve different levels of relationship and their treatment of alignment ambiguities should serve as an example for future explorations of rotifer relationships at the genus, species and sub-species levels.

The 18S sequence has now been determined from a number of species in each of the phyla likely to be closely related to Rotifera, and from multiple species within the phylum. Studies can now more fully address the relationships among Rotifera, Gnathostomulida, Gastrotricha, Cycliophora and Platyhelminthes, and support for Platyzoa. For example, it seems likely that the anomalous position of *P. acuticornis* in the tree of Giribet et al. (2000) could be due to the inclusion of an insufficient number of rotifer species in their analysis. It is also important to determine if the Platyzoa can be supported with phylogenetic reconstruction methods other than that used by Giribet et al. (2000). Attention should also be paid to the effect of particular species on tree topology.

There is only a single described species of Cycliophora, and a recently identified phylum, Micrognathozoa, with similarities to Gnathostomulida and Rotifera is also currently represented by a single species (Sørensen et al., 2000; P. Funch, pers. com.). It is unlikely that the 18S sequence of these single species will be sufficient to convincingly resolve the placement of these phyla. An alternative to the identification of additional species would be the examination of an additional region of the genome beyond 18S to increase the chance of recovering an accurate phylogeny (Cummings et al., 1995). If an analysis of the relationships among potential platyzoan phyla based on a protein-coding gene resulted in a tree similar to one obtained using 18S, this would dramatically increase the confidence in the molecular phylogeny, not only in resolving the position of Cycliophora and Micrognathozoa, but of platyzoan relationships in general (and, indeed, of 18S phylogenies in general). While the sequencing from various taxa of genes encoding such proteins as elongation factor 1$\alpha$, cytochrome oxidase, and the 82–90 kD heat shock protein is ongoing, it will be some time before phyla are as well represented by these genes as they are by 18S.

The difficulty resolving the relationship between acanthocephalan and eurotatorian 18S sequences implies a very close relationship between these groups. However, it cannot be assumed from this that Seisonidea is more distantly related to Eurotatoria than is Acanthocephala. Clearly, additional analyses that include Seisonidea are necessary. The partial 18S sequence of a seisonid has been made available by the Garey lab (GenBank Accession AF053612), which should soon lead to the sequencing of a longer region that can be used with the 18S sequences of Garey et al.

(1996) and García-Varela et al. (2000) to determine the phylogeny of Rotifera based on the 18S gene. As with the phylum-level relationships, the sensitivity of the resulting gene trees to alignment, species sampling, and methodology must be thoroughly explored. Unfortunately, there are only two described extant seisonid species; thus it may be difficult to determine if either (or both) of these has evolved in a manner representative of the history of the class. While it may be possible to alleviate this problem by examining several genes, this also illustrates that molecular phylogenetics must be conducted in concert with morphological cladistics using ultrastructural and biochemical characters for a unified interpretation of the evolution of the phylum.

## Acknowledgements

I thank Martin García-Varela, Gonzalo Giribet and Peter Funch for sharing results prior to publication, Jessica Mark Welch, Robert Wallace and Elizabeth Walsh for comments that greatly improved the manuscript, and Matthew Meselson for guidance and support. The work presented here was made possible by ongoing support from the Eukaryotic Genetics Program of the United States National Science Foundation.

## References

Abouheif, E., R. Zardoya & A. Meyer, 1998. Limitations of metazoan 18S rRNA sequence data: implications for reconstructing a phylogeny of the animal kingdom and inferring the reality of the Cambrian explosion. J. Mol. Evol. 47: 394–405.

Aguinaldo, A. M. A., J. M. Turbeville, L. S. Linford, M. C. Rivera, J. R. Garey, R. A. Raff & J. A. Lake, 1997. Evidence for a clade of nematodes, arthropods and other molting animals. Nature 387: 489–493.

Ahlrichs, W. H., 1997. Epidermal ultrastructure of *Seison nebaliae* and *Seison annulatus,* and a comparison of epidermal structures within the Gnathifera. Zoomorphology 117: 41–48.

Balavoine, G., 1997. The early emergence of platyhelminths is contradicted by the agreement between 18S rRNA and Hox genes data. Comp. Rendus III 320: 83–94.

Campos, A., M. P. Cummings, J. L. Reyes & J. P. Laclette, 1998. Phylogenetic relationships of platyhelminthes based on 18S ribosomal gene sequences. Mol. Phylogenet. Evol. 10: 1–10.

Carranza, S., J. Baguñà & M. Riutort, 1997. Are the platyhelminthes a monophyletic primitive group? An assessment using 18S rDNA sequences. Mol. Biol. Evol. 14: 485–497.

Cavalier-Smith, T., 1998. A revised six-kingdom system of life. Biol. Rev. 73: 203–266.

Cummings, M. P., S. P. Otto & J. Wakeley, 1995. Sampling properties of DNA sequence data in phylogenetic analysis. Mol. Biol. Evol. 12: 814–822.

Felsenstein, J., 1978. Cases in which parsimony and compatibility methods will be positively misleading. Syst. Zool. 27: 401–410.

Felsenstein, J., 1988. Phylogenies from molecular sequences: inferences and reliability. Ann. Rev. Genet. 22: 521–565.

García-Varela, M., G. Pérez-Ponce de León, P. de la Torre, M. P. Cummings, S. S. S. Sarma & J. P. Laclette, 2000. Phylogenetic relationships of Acanthocephala based on analysis of 18S ribosomal RNA gene sequences. J. Mol. Evol. 50: 532–540.

Garey, J. R., T. J. Near, M. R. Nonnemacher & S. A. Nadler, 1996. Molecular evidence for Acanthocephala as a subtaxon of Rotifera. J. Mol. Evol. 43: 287–292.

Garey, J. R., A. Schmidt-Rhaesa, T. J. Near & S. A. Nadler, 1998. The evolutionary relationship of rotifers and acanthocephalans. Hydrobiologia 387/388: 83–91.

Giribet, G. & C. Ribera, 1998. The position of arthropods in the animal kingdom: a search for a reliable outgroup for internal arthropod phylogeny. Mol. Phylogenet. Evol. 9: 481–488.

Giribet, G. & W. C. Wheeler, 1999. The position of arthropods in the animal kingdom: Ecdysozoa, islands, trees and the 'parsimony ratchet.' Mol. Phylogenet. Evol. 13: 619–623.

Giribet, G., D. L. Distel, M. Polz, W. Sterrer & W. C. Wheeler, 2000. Triploblastic relationships with emphasis on the acoelomates and the position of Gnathostomulida, Cycliophora, Platyhelminthes and Chaetognatha: a combined approach of 18S rDNA sequences and morphology. Syst. Zool. 49: 539–562.

Graybeal, A., 1998. Is it better to add taxa or characters to a difficult phylogenetic problem? Syst. Biol. 47: 9–17.

Halanych, K. M., 1998. Considerations for reconstructing metazoan history: signal, resolution and hypothesis testing. Am. Zool. 38: 929–941.

Halanych, K. M., J. D. Bacheller, A. M. A. Aguinaldo, S. M. Liva, D. M. Hillis & J. A. Lake, 1995. Evidence from 18S ribosomal DNA that lophophorates are protostome animals. Science 267: 1641–1643.

Hancock, J. M. & A. P. Vogler, 2000. How slippage-derived sequences are incorporated into rRNA variable-region secondary structure: implications for phylogeny reconstruction. Mol. Phylogenet. Evol. 14: 366–374.

Hanelt, B., D. Van Schyndel, C. M. Adema, L. A. Lewis & E. S. Loker, 1996. The phylogenetic position of *Rhopalura ophiocomae* (Orthonectida) based on 18S ribosomal DNA sequence analysis. Mol. Biol. Evol. 13: 1187–1191.

Hashimoto, T., Y. Nakamura, F. Nakamura, T. Shirakura, J. Adachi, N. Goto, K. I. Okamoto & M. Hasegawa, 1994. Protein phylogeny gives a robust estimation for early divergences of eukaryotes: phylogenetic place of a mitochondria-lacking protozoan, *Giardia lamblia*. Mol. Biol. Evol. 11: 65–71.

Håstad, O. & M. Björklund, 1998. Nucleotide substitution models and estimation of phylogeny. Mol. Biol. Evol. 15: 1381–1389.

Hendy, M. D. & D. Penny, 1989. A framework for the quantitative study of evolutionary trees. Syst. Zool. 38: 297-309.

Hickson, R. E., C. Simon & S. W. Perrey, 2000. The performance of several multiple-sequence alignment programs in relation to secondary-structure features for an rRNA sequence. Mol. Biol. Evol. 17: 530–539.

Hillis, D. M., J. P. Huelsenbeck & C. W. Cunningham, 1994. Application and accuracy of molecular phylogenies. Science 264: 671–677.

Hillis, D. M., 1996. Inferring complex phylogenies. Nature 383: 130–131.

Huelsenbeck, J. P. & R. Nielsen, 1999. Effect of nonindependent substitution on phylogenetic accuracy. Syst. Biol. 48: 317–328.

Katayama, T., H. Wada, H. Furuya, N. Satoh & M. Yamamoto, 1995. Phylogenetic position of the dicyemid mesozoa inferred from 18S rDNA sequences. Biol. Bull. 189: 81–90.

Kishino, H. & M. Hasegawa, 1989. Evaluation of the maximum likelihood estimate of the evolutionary tree topologies from DNA sequence data, and the branching order in Hominoidea. J. Mol. Evol. 29: 170–179.

Lanyon, S., 1985. Detecting internal inconsistencies in distance data. Syst. Zool. 34: 397–403.

Lecointre, G., H. Philippe, H. L.V. Lê & H. Le Guyader, 1993. Species sampling has a major impact on phylogenetic inference. Mol. Phylogenet. Evol. 2: 205–224.

Li, W-H., 1997. Molecular Evolution. Sunderland, MA, Sinauer Associates: 487 pp.

Littlewood, D. T. J., M. J. Telford, K. A. Clough & K. Rohde, 1998. Gnathostomulida – an enigmatic metazoan phylum from both morphological and molecular perspectives. Mol. Phylogenet. Evol. 9: 72–79.

Lorenzen, S., 1985. Phylogenetic aspects of pseudocoelomate evolution. In Morris, S. C., J. D. George, R. Gibson & H. M. Platt (eds), The Origins and Relationships of Lower Invertebrates. Clarendon Press, Oxford: 210–223.

Maley, L. E. & C. R. Marshall, 1998. The coming of age of molecular systematics. Science 279: 505–506.

Mark Welch, D. B., 2000. Evidence from a protein-coding gene that acanthocephalans are rotifers. Invert. Biol. 119: 17–26.

Mark Welch, D. B. & M. Meselson, 2000. A preliminary survey of intron conservation in bdelloid rotifers. Hydrobiologia, this volume.

McHugh, D., 1998. Deciphering metazoan phylogeny: the need for new molecular data. Amer. Zool. 38: 859–866.

Melone, G., C. Ricci, H. Segers & R. Wallace, 1998. Phylogenetic relationships of phylum Rotifera with emphasis on the families of Bdelloidea. Hydrobiologia 387/388: 101–107.

Morrison, D. A. & J. T. Ellis, 1997. Effects of nucleotide sequence alignment on phylogeny estimation: a case study of 18S rDNAs of Apicomplexa. Mol. Biol. Evol. 14: 428–441.

Near, T. J., J. R. Garey & S. A. Nadler, 1998. Phylogenetic relationships of the Acanthocephala inferred from 18S ribosomal DNA sequences. Mol. Phylogenet. Evol. 10: 287–298.

Parsch, J., J. M. Braverman & W. Stephan, 2000. Comparative sequence analysis and patterns of covariation in RNA secondary structure. Genetics 154: 909–921.

Philippe, H., 1997. Rodent monophyly: pitfalls of molecular phylogenies. J. Mol. Evol. 45: 712–715.

Philippe, H. & A. Germot, 2000. Phylogeny of eukaryotes based on ribosomal RNA: long-branch attraction and models of sequence evolution. Mol. Biol. Evol. 17: 830–834.

Raff, R. A., C. R. Marshall & J.M. Turbeville, 1994. Using DNA sequence to unravel the Cambrian radiation of the animal phyla. Ann. Rev. Ecol. Syst. 25: 351–375.

Ruiz-Trillo, I., M. Riutort, D. T. J. Littlewood, E. A. Herniou & J. Baguña, 1999. Acoel flatworms: earliest extant bilaterian metazoans, not members of Platyhelminthes. Science 283: 1919–1923.

Rzhetsky, A., 1995. Estimating substitution rates in ribosomal RNA genes. Genetics 141: 771–783.

Rzhetsky, A. & M. Nei, 1995. Tests of applicability of several substitution models for DNA sequence data. Mol. Biol. Evol. 12: 131–151.

Schultes, E. A., P. T. Hraber & T. H. LaBean, 1999. Estimating the contributions of selection and self-organization in RNA secondary structure. J. Mol. Evol. 49: 76–83.

Sørensen, M. V., P. Funch, E. Willerslev, A. J. Hansen & J. Olesen, 2000. On the phylogeny of the metazoa in the light of Cycliophora and Micrognathozoa. Zoologischer Anzeiger. 239: 297–318.

Steel, M. A., P. J. Lockhart & D. Penny, 1993. Confidence in evolutionary trees from biological sequence data. Nature 364: 440–442.

Swofford, D. L., G. J. Olsen, P. J. Waddell & D. M. Hillis, 1996. Phylogenetic inference. In Hillis, D. M., C. Moritz & B. K. Mable (eds), Molecular Systematics. 2nd edn. Sinauer Associates, Sunderland, MA: 407–514.

Telford, M. J. & P. W. H. Holland, 1993. The phylogenetic affinities of the chaetognaths: a molecular analysis. Mol. Biol. Evol. 10: 660–676.

Tillier, E. R. M. & R. A. Collins, 1995. Neighbor joining and maximum likelihood with RNA sequences – addressing the interdependence of sites. Mol. Biol. Evol. 12: 7–15.

Van de Peer, Y., J.-M. Neefs, P. De Rijk & R. De Wachter, 1993. Reconstructing evolution from eukaryotic small-ribosomal-subunit RNA sequences: calibration of the molecular clock. J. Mol. Evol. 37: 221–232.

Wägele, J. W. & R. Wetzel, 1994. Nucleic acid sequence data are not *per se* reliable for inference of phylogenies. J. Nat. Hist. 28: 749–761.

Wallace, R. L., C. Ricci & G. Melone, 1996. A cladistic analysis of pseudocoelomate (aschelminth) morphology. Inv. Biol. 115: 104–112.

Winnepenninckx, B., T. Backeljau, L. Y. Mackey, J. M. Brooks, R. De Wachter, S. Kumar & J. R. Garey, 1995. 18S rDNA data indicate that aschelminthes are polyphyletic in origin and consist of at least three distinct clades. Mol. Biol. Evol. 12: 1132–1137.

Winnepenninckx, B. & T. Backeljau, 1996. 18S rRNA alignments derived from different secondary structure models can produce alternative phylogenies. J. Zool. Syst. Evol. Res. 34: 135–143.

Winnepenninckx, B., Y. Van De Peer & T. Backeljau, 1998a. Metazoan relationships on the basis of 18S rRNA sequences: a few years later. Am. Zool. 38: 888–906.

Winnepenninckx, B. M. H., T. Backeljau & R. M. Kristensen, 1998b. Relations of the new phylum Cycliophora. Nature 393: 636–638.

Wirz, A., S. Pucciarelli, C. Miceli, P. Tongiorgi & M. Balsamo, 1999. Novelty in phylogeny of Gastrotricha: evidence from 18S rRNA gene. Mol. Phylogenet. Evol. 13: 314–318.

Zrzavý J., S. Mihulka, P. Kepka, A. Bezdek & D. Tietz, 1998. Phylogeny of the Metazoa based on morphological and 18S ribosomal DNA evidence. Cladistics 14: 249–285.

**Note added in proof:** A complete description of Micrognathozoa can be found in: Kristensen, R.M. S.P. Funch, 2000. Micrognathozoa: A new class with complicated jaws like those of Rotifera and Gnathostomulida. V. Morph. 246: 1–49.

Hydrobiologia **446/447:** 323–331, 2001.
L. Sanoamuang, H. Segers, R.J. Shiel & R.D. Gulati (eds), Rotifera IX.




# The approach to equilibrium of multilocus genotype diversity under clonal selection and cyclical parthenogenesis

Charles E. King[1] & Justin Schonfeld[2,3]
[1]*Department of Zoology, Oregon State University, Corvallis, OR 97331, U.S.A.*
*E-mail:kingc@ava.bcc.orst.edu*
[2]*Department of Computer Sciences, Oregon State University, Corvallis, OR 97331, U.S.A.*
[3]*Current address: Computational Biology & Bioinformatics, Iowa State University, Ames, IA 50011, U.S.A.*
*E-mail: schonfju@iastate.edu*

*Key words:* adaptive population structure, cyclical parthenogenesis, clonal selection, rotifers

## Abstract

Sexual generations in cyclical parthenogens are typically separated by multiple generations of clonal reproduction. In contrast to sexual reproduction, during parthenogenesis the genome of the parent is passed on to the offspring as a unit. The absence of recombination during parthenogenesis leads to differences in the action of natural selection in the two reproductive phases. In addition, since recombination is a sampling process, random genetic drift is potentially more important in sexual reproduction than in parthenogenesis. A recent development in the study of rotifer population genetics is the use of microsatellites to characterize natural populations. Microsatellites are selectively neutral, show patterns of Mendelian inheritance and tend to be much more variable than allozymes. An advantage over allozymes is that microsatellite DNA can be cloned with PCR and thus multiple loci can be assayed from a single individual. We use a new computer model in this paper to investigate the response of selectively active and selectively neutral genes to evolutionary forces during cyclical parthenogenesis. Selectively active alleles may respond differently to selection in the parthenogenetic and sexual phases of cyclical parthenogenesis. Even when strong clonal selection is acting on loci associated with adaptation, the view that emerges with microsatellites may be one of Hardy-Weinberg and linkage equilibrium. Thus studies using selectively neutral loci may fail to detect clonal selection even when it is an important feature of the rotifer population's adaptive structure.

## Introduction

In cyclical parthenogens such as monogonont rotifers, the life cycle is divided into two distinct phases. The sexual phase includes the classic genetic processes of segregation and recombination. Natural selection in this phase also conforms to the classical model; that is, alleles that make positive contributions to fitness at independent loci are expected to increase in frequency, and those which are deleterious are expected to become less common.

In contrast, during the parthenogenetic phase of the life cycle, the genome is effectively congealed so that the offspring of a single female are genetically identical to each other and their mother. Different parthenogenetic clones may have different relative fitnesses thereby creating a process of clonal selection. This clonal selection is based on a composite measure of fitness taken over all loci – those clones having high fitness increase in frequency while less well-adapted clones become relatively less common. Clonal selection may also be a major determinant of population structure in cyclical parthenogens.

Two methods have been used to characterize the genetic structure of rotifers. Allozyme studies have revealed the presence of significant amounts of genetic variation in a number of populations (reviewed by King & Serra, 1998). However, the identification of clones from field samples requires that each individual be genotyped at a number of loci so that multi-locus genotypes (MLGs) can be compared. This creates a major problem since rotifers are small, and even under ideal conditions single individuals can provide enough material to assay for only one or two allozyme loci.

The alternative is to isolate and clone field-captured individuals in the laboratory, a labor-intensive and expensive procedure. Consequently, most allozyme analyses of rotifer populations have both smaller numbers of individuals and fewer assayed loci than is desirable.

Accordingly, it is fair to ask whether the pattern of clonal selection inferred from allozyme studies is an accurate picture of the genetic structure of rotifer populations. That is, does our view of rotifer population structure reflect biological reality or simply the limitations of our analytical techniques? Might new analytical methods produce a different view?

A promising new method is the use of PCR-amplified DNA markers that require little material and generally have much higher levels of genetic variation than allozymes. Gómez et al. (1998), studying *Brachionus plicatilis*, scored seven polymorphic microsatellite loci having 34 alleles. (Microsatellites are tandemly repeated, short sequences of DNA found abundantly within the genomes of eukaryotes. They are generally less than six nucleotides in length and show a pattern of Mendelian inheritance.)

This system has recently been used by Gómez & Carvalho (2000) to characterize microsatellite variation in a population of *B. plicatilis* living in Poza Sur of Torreblanca Marsh on the Mediterranean coast of Spain.

Relative to allozyme studies of rotifers, but not to most microsatellite studies of other groups, an enormous amount of genetic variation was found in this population; of the 390 individuals assayed, 349 had unique MLGs. One interpretation of the high clonal diversities observed by Gómez & Carvalho (2000) is that clonal selection may be weak. In support of this, interpretation statistical analysis was unable to reject the null hypotheses of Hardy-Weinberg and linkage equilibrium in most of the samples, including a collection of resting eggs from the surface sediment of Poza Sur. However, the last sample from the parthenogenetic phase showed evidence of clonal selection and, as expected, it also displayed linkage disequilibrium.

A number of interesting questions arise from Gómez & Carvalho's (2000) study. First, how does the genetic structure of rotifers reflect differences in the selective régime between the sexual and parthenogenetic phases? That is, selection may produce quite different effects when it is accompanied by segregation and recombination than when it is directed at the entire genome as a unit. Second, how does the perceived genetic structure reflect the types of loci being investigated? Although microsatellites are thought to be selectively neutral (Jarne & Lagoda, 1996), selection may act on linked genes and thereby influence both the quantity of genetic variation and its distribution across the genome as a whole. Hitchhiking phenomena are particularly intense during the parthenogenetic phase of rotifer reproduction when all loci are effectively congealed. Finally, how does the genetic structure of cyclical parthenogens reflect levels of variation in the system being investigated? Microsatellite loci may be highly polymorphic and contain as many as 30–60 alleles per locus. While this high variability may seem ideal for studying the genetic structure of a population, it is accompanied by a price: as levels of variation increase, larger sample sizes are needed to accurately determine allele and genotype frequencies (Ruzzante, 1998). Gómez & Carvalho (2000) found much less variation in their study than the rates of polymorphism cited above; they detected 27 alleles in the seven loci they investigated for an average of about 4 alleles per locus. Still, this level of variation far exceeds that found in any study of rotifer populations using allozyme markers.

To explore these questions, and in particular to examine the role of cyclical parthenogenesis, we have constructed a computer model that compares the dynamics of alleles at loci subject to natural selection with alleles at selectively neutral loci. In this paper, we evaluate how the intensity of clonal selection affects both the genotype diversity and clonal diversity (number of MLGs) of cyclical parthenogens. Then we use our model to study the interactions of clonal selection, mutation, and random genetic drift as determinants of population structure. Finally, we explore potential effects of sexual recombination on clonal diversity in cyclical parthenogens.

## A model for natural selection, mutation and random genetic drift in cyclical parthenogens

### *Program structure*

Our model is written in Visual C++ and contains two loops, one for sexual reproduction and a second for parthenogenesis (Fig. 1). Input to the model includes information on the number of loci and the allele frequencies, selective values and mutation rates at each locus. A run is initiated by forming a pool of the input genotypes in Hardy-Weinberg equilibrium at each locus. Then one genotype is randomly selected from

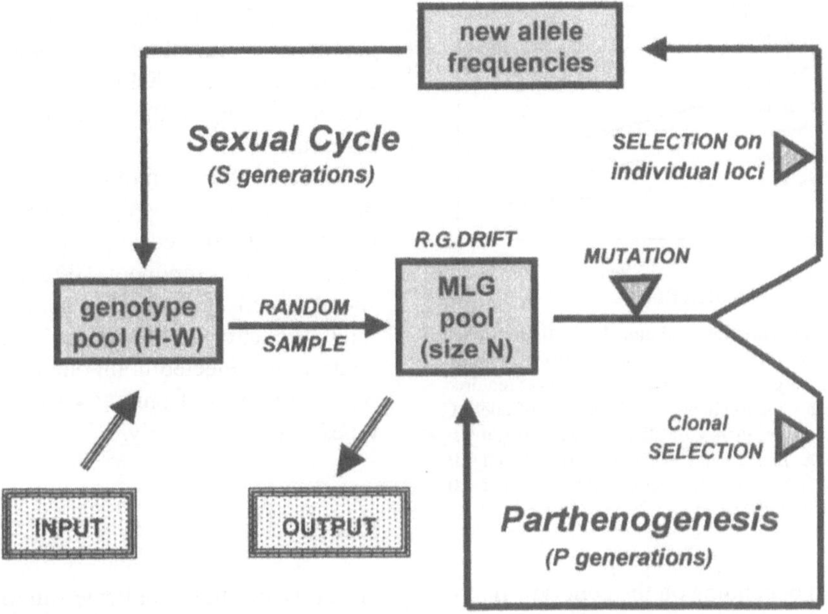


*Figure 1.* A model for analyzing the effects of mutation, selection, and random genetic drift on different types of genes in cyclical parthenogens. See text for further description.

each pool to constitute the first MLG. This process of random selection is repeated until the desired number of MLGs is obtained.

For both loops, first mutation and then selection is applied to the MLG pool. Mutation is handled by calculating the total number of mutations expected to occur at each locus and then randomly assigning the mutations to specific alleles at the locus. After all mutational events have been completed, MLG frequencies are adjusted to account for the losses and gains derived from mutation. At this point, the modeled population can either undergo sexual reproduction or go into the parthenogenetic loop for any desired number of generations.

Random numbers are used in two parts of the model, to form the MLG pool and to distribute mutations. The procedure used for both is nearly identical. In each sexual generation, a random number generator is seeded using the current time. To form the MLG pool, the Hardy-Weinberg frequencies of each of the genotypes present at the locus are placed in an array. A single genotype is randomly selected from the array for inclusion in the MLG being formed. Rare genotypes in the array will seldom be selected thus raising the possibility of loss of variation through random genetic drift. The process is then iterated for other loci until the complete MLG has been constructed. The same procedure is used for selecting the MLG alleles that will mutate except that MLG frequencies are used to create the array.

*Assumptions*

Loci are diploid, assort independently during sexual reproduction, and are inherited as a unit during parthenogenesis.

Alleles subject to natural selection are assumed to make additive contributions to the fitness of each genotype, so in this system there is no heterozygote advantage, frequency-dependent or fluctuating selection.

All alleles at a given locus are assumed to have the same mutation rate, but mutation rates may vary between any two loci in our model. Mutation rates of microsatellites ($10^{-4}$–$10^{-2}$) appear to be two to three orders of magnitude higher than those of allozymes (Weber & Wong, 1993; Jarne & Lagoda, 1996; Ruzzante, 1998). In the present paper, mutation rates of $10^{-6}$ have been assumed for loci under selection and $10^{-4}$ for selectively neutral loci.

In order to follow the effects of selection, the population is initiated with a single, MLG that is homozygous for the lowest fitness allele at each locus. This construction means that the first generation of individuals will produce only beneficial mutations. In subsequent generations, mutations can be deleterious,

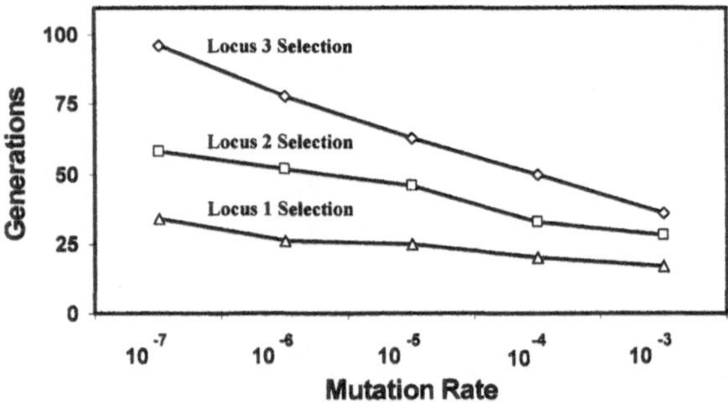


*Figure 2.* Interaction of mutation rate and clonal selection on three alleles (A, B and C) at three loci (treated independently). The ordinate indicates the number of generations of clonal selection required at the indicated mutation rates to change the frequency of allele C from 0 to 0.98. The initial frequency of allele A was 1.0. Fitness contributions of alleles A, B and C for Locus 1: 0.01, 0.5 and 1.0; Locus 2: 0.25, 0.625 and 1.0; and for Locus 3: 0.5, 0.75 and 1.0. Population size is $10^6$ individuals.

neutral or beneficial depending on the genotype that is mutating and the allele that is introduced.

All runs in the present paper were made by random sampling from the genotype pool for each locus to construct a population of 1 000 000 MLGs (individuals). We restrict our view in this paper to six loci. Loci 1–3 are under natural selection and each has three alleles (A–C). Loci 4–6 are neutral and each has six alleles (A–F). Fitnesses are assumed to be determined by additive effects. The fitness contributions of alleles A, B and C at locus 1 are 0.01, 0.5 and 1.0; for locus 2 they are 0.25, 0.625 and 1.0, and for locus 3 they are 0.5, 0.75 and 1.0. Thus, selection at each of the three loci will always favor allele B over A, and C over either A or B. Furthermore, selection will be more intense on 1 than on 2 or 3 and more intense on 2 than on 3. The response of a clonally reproducing population to selection at these levels over a wide range of mutation rates is presented in Figure 2. Note that the three curves do not overlap, and at a given level of mutation, selection intensity acts to scale the population response.

### *Output*

For each generation, output includes total genotype diversity (GDT), number and frequency of MLGs, and genotype and allele frequencies at each locus. Genotype diversity as given by Hoffman (1986) is:

$$\mathrm{GDT} = \frac{1}{\sum_{i=1}^{\mathrm{kT}} p_i^2},$$

in which KT is the total number of unique multilocus genotypes in the simulated population and $p_i$ is the frequency of the $i^{\text{th}}$ MLG. Stoddart and Taylor (1988) explored some of the statistical properties of GDT distributions.

Because we are interested in the role of selection on genotype diversity, we partition GDT into one component, GDN, based only on selectively neutral loci and a second component, GDS, based on loci that respond directly to natural selection. Similarly, KT, the total number of unique MLGs, can be partitioned into neutral and selected components, KN and KS, based on the numbers of unique MLGs considering the two types of loci separately.

## Results

A series of three replicate simulations of a model population with 100 generations of parthenogenesis is presented in Figure 3. These runs are included to illustrate population behavior under clonal selection alone; subsequent runs consider effects of sexual reproduction and variation in the relative intensity of clonal selection. Differences in the output of the three runs are solely determined by the stochastic timing and identity of the new variation produced by mutation.

All runs in this paper were initiated with a single MLG that was homozygous for the A alleles at each of its six loci. As mutation occurs, there is an increase in the number of unique MLGs at both selected and neutral loci as well as the total number of MLGs. Note that KT is not the sum of KN + KS because of the way the two types of loci are defined. For instance, if KS=1 and KN=4, KT=4. If a single new mutation occurs at a selected locus in one of the four MLG classes, KS=2 and KN=4 but KT=5. If the same new mutation occurs in all four MLG classes, KS=2, KN=4 but KT=8. In Figure 3 KN is larger than KS primarily because of the higher mutation rates at neutral loci and the larger number of neutral than selectively active alleles.

Genotype diversity is sensitive to variation in both MLG frequency and number. Accordingly, GDT patterns primarily reflect the effects of selection acting on new MLGs. For example, the first peak in the center panel of population 'a' is caused by the appearance and rapid increase of a new allele at locus 1 that produces MLGs having higher fitness that the original AA homozygote. A second peak is produced as allele 1C moves towards fixation. The more complex pattern in population 'b' is attributable to the added contribu-

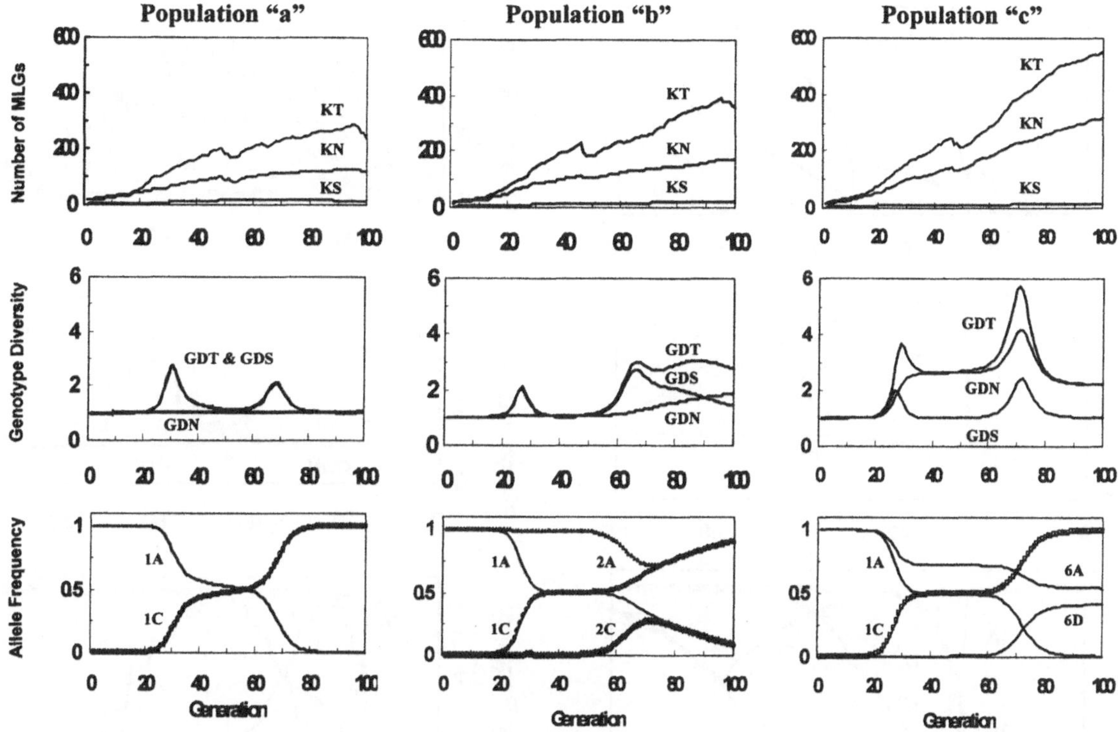


*Figure 3.* Three replicate runs of the model for 100 generations of parthenogenesis. Fitnesses of the three loci (1–3) under selection are given in the caption of Figure 2. In addition, the simulated population has three neutral loci (4 - 6) with six alleles each. Mutation rates of the three selected loci are $10^{-6}$ , and of the three neutral loci they are $10^{-4}$. Population size is $10^{6}$ individuals. KN, KS and KT are the clonal diversities (number of unique multilocus genotypes) of the neutral and selected loci considered separately and for the entire population taken over both neutral and selected loci. GDN, GDS, and GDT are measures of genotype diversity (see text). Gene identities are indicated by a locus number followed by an allele letter, e.g. 1A.

tions of a second locus that is also under selection. In general, the neutral alleles have a small initial impact on genotype diversity since their frequencies early in a run are determined by mutation and rare alleles make small contributions to GDT. However, under clonal selection rare neutral alleles may be associated with high fitness genotypes in MLGs and their frequencies can increase rapidly because of the hitchhiking effect (population 'c' in Fig. 3).

*Genotype diversity under strong clonal selection*

For a given set of allelelic contributions to fitness, the strength of clonal selection acting on a population is proportional to the number of parthenogenetic generations intervening between instances of sexual reproduction. Figure 4 shows a 100-generation simulation conducted with the 6-locus case described above. Again, the population was started with a single MLG that was homozygous for the 'A' alleles at each locus. Sexual reproduction occurs at intervals of 25 generations.

The most striking difference between the examples in Figures 2 and 3 relates to the change in population structure after sexual reproduction. Whenever sex occurs, there is a dramatic decline in the number of MLGs. Immediately prior to sexual reproduction most of the neutral alleles are present in low frequency. When the population is reconstituted by random sampling from the alleles at each locus, rare alleles are likely to be lost. This loss is due to random genetic drift, specifically a founder effect.

In the upper panel of Figure 4 there is a rapid increase in the number of MLGs following sexual reproduction. This increase is due to both new mutation and the replacement of low fitness alleles at the three loci subject to natural selection. The fitness of an MLG is taken over all loci that are subject to direct natural selection and expressed relative to that of other MLGs. In this system, the most fit MLG has a relative fitness of 1 and a genotype containing only *C* alleles at the

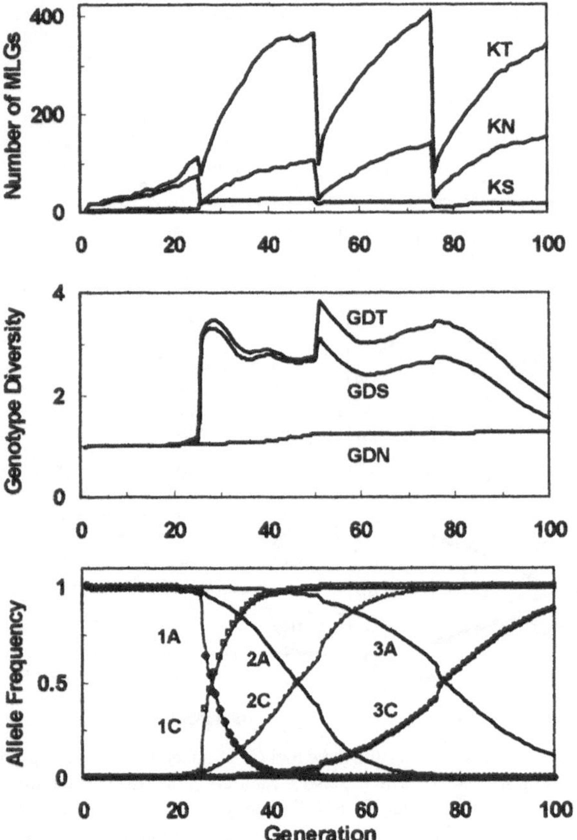


*Figure 4.* Strong clonal selection: Effects of mutation and clonal selection on a life history that has 24 generations of parthenogenesis between successive sexual generations. Other input parameters are the same as in Figure 3.

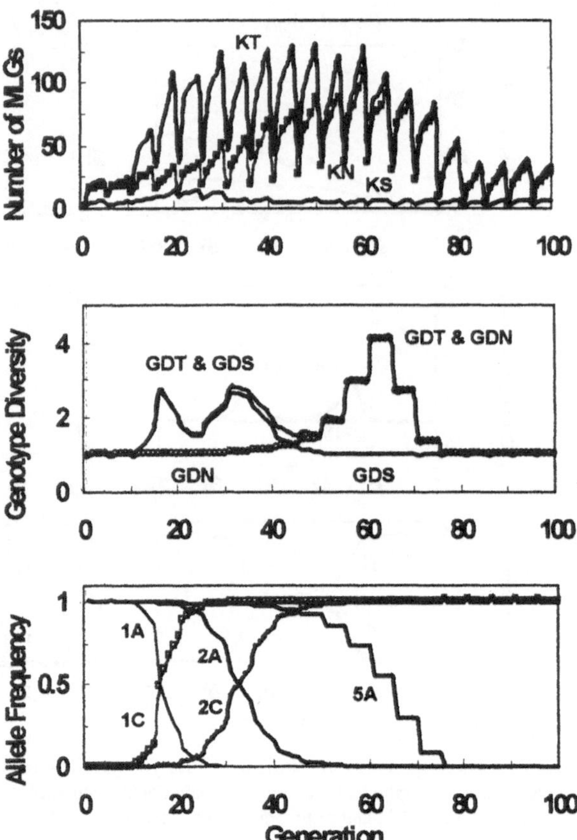


*Figure 5.* Weak clonal selection. Effects of mutation and clonal selection on a life history that has 4 generations of parthenogenesis between successive sexual generations. Other input parameters are the same as in Figure 3.

selection loci. As *B* and *C* alleles are generated by mutation they rapidly replace the A alleles and ultimately the *C* alleles at loci 1–3 will become fixed in the population. Although new *A* and *B* alleles continue to be generated by mutation, because of their low mutation rates and negative influence on fitness, they have little impact on the population's genetic structure. The frequencies of neutral alleles associated with high fitness MLGs increase as the selection occurs. A succession of selective changes is seen in the bottom panel of Figure 4. First, the *A* allele of locus 1 is replaced by the *C* allele and this is followed by the same change at the second locus and finally again at the third locus. The order of this change is determined by the relative contributions to fitness of the alleles at the three loci. Selection is strongest at selection locus 1 and weakest at selection locus 3.

In the center and bottom panels of Figure 4, GDT fluctuates as mutation creates and selection eliminates genetic variation. Not surprisingly, GDS mirrors the GDT variation. Clonal selection acting on the allele frequencies shown in the bottom panel is the major determinant of the patterns of genotype diversity.

*Genotype diversity of selected and neutral loci under weak clonal selection*

All else being equal, the intensity of clonal selection is directly proportional to the number of parthenogenetic generations intervening between sexual generations. In Figure 5, all input parameters are the same as in Figure 4 except that instead of 24 there are now 4 generations of parthenogenesis followed by a single generation of sexual reproduction. Again, a brief drop in the number of MLGs and the genotype diversity after each occurrence of sex is attributable to random genetic drift. Genotype diversity is buffered by the fact that the lost MLGs were present in low frequencies so

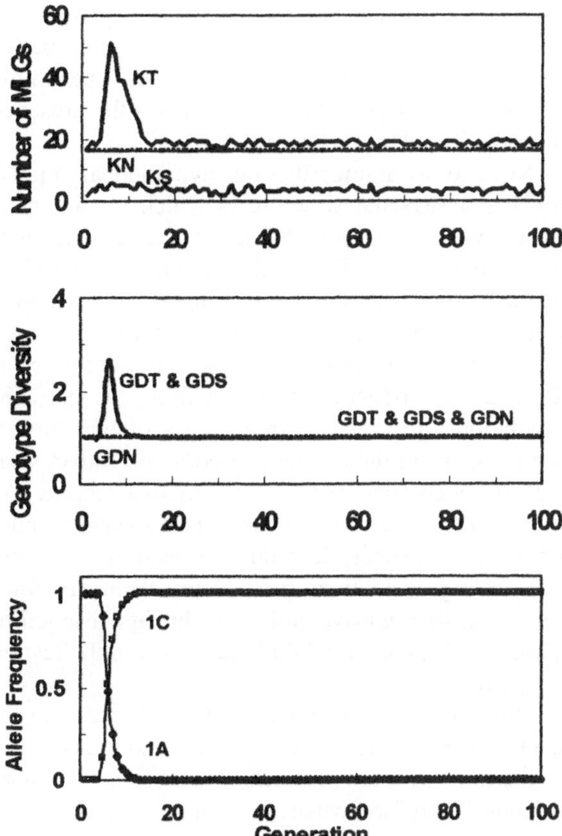


*Figure 6.* No clonal selection. Effects of mutation in concert with founder effects on a life history that has obligate sexual reproduction. Other input parameters are the same as in Figure 3.

the major changes in GDT reflect the effects of clonal selection displayed in the bottom panel. Notice that the frequency of allele 5A decreases as other alleles at this neutral locus are generated by mutation. Chance inclusions of neutral alleles may have important effects on genotype diversity even when sexual reproduction is frequent.

Although the qualitative outcomes of clonal selection are identical for the 4 and 24 parthenogenetic generation cases, there are fewer MLGs produced when sex is more frequent (Figures 4 vs 5). As explained above, this is because frequent sexual reproduction maximizes the effect of random genetic drift as the population is reformed.

### *Genotypic diversity of microsatellites and allozymes in the absence of clonal selection*

We next compare the above results with the case in which clonal selection is absent. In Figure 6, all input parameters are again the same except there is now obligate sexual reproduction.

In the absence of parthenogenesis, founder effects occur in every generation and new mutations are less likely to be incorporated into the population's gene pool when the next generation is formed. Although new multilocus genotypes are produced (Fig. 6, top panel), their numbers are small and almost exclusively dependent on the high mutation rates of the microsatellites. With the sole exception of the change from allele 1A to 1C due to natural selection, allele frequencies are not changed enough to show up in the bottom panel. Notice that the diversities of both selected and neutral multilocus genotypes are very close to their minimum possible levels, 1. Clearly, the predominant evolutionary force under these conditions is random genetic drift.

## Discussion

### *Equilibrium genotype and clonal diversities*

Expected equilibrium levels of both genotype diversity and clonal diversity in cyclical parthenogens are different for neutral and selected loci. In this paper we have used a model population with 3 selected and 3 neutral loci. Since there are 6 possible genotypes at each selected locus and 21 at each neutral locus, maximum clonal diversity at the selected loci (KS) is $(6)^3 = 216$ MLGs. At the neutral loci, maximum clonal diversity (KN) is $(21)^3 = 9261$ MLGs. Ignoring minor fluctuations due to recurrent mutation, under clonal selection the population in our model will always move towards KS = 1 at equilibrium. This MLG will be homozygous for the allele having highest fitness at each locus. Genotype diversity equilibrium at the selected loci will also occur at GDS= 1. Note, however, that attainment of these equilibrium values may take much longer than the 100 generations us in our simulations.

In striking contrast to loci under selection, mutation drives the population toward an equilibrium in which all possible neutral genotypes are present at each locus and in which all alleles at each locus have the same relative frequency. This state follows from our assumption of equivalence of mutation rates at a single locus. Genotype diversity of the neutral loci under these circumstances also moves towards a value of GDN=1 which requires that all neutral MLGs have the same frequencies. This condition would be possible only if the population had no sexual reproduction. At the average locus, recombination during sexual reproduction increases the frequencies of MLGs containing neutral heterozygotes relative to those containing individual classes of homozygotes. In sexual populations when the neutral genotypes at each locus are in Hardy-Weinberg equilibrium, an equilibrium GDN $>1$ will occur. Total genotype diversity (GDT) as the system approaches equilibrium under the combined action of mutation and selection reflects both of these forces. Maximum GDT is expected to occur when GDS = 1 and the single selected MLG is present on all possible neutral MLGs. Under these conditions, GDN equals GDT.

*Detection of clonal selection in natural population*

A frequency of 10% non-unique multilocus genotypes (41 of the 390 individuals scored by Gómez & Carvalho, 2000) might imply weak clonal selection unless the number of potential MLGs is considered. One of the loci studied by Gómez & Carvalho (2000) had 2 alleles, a second 3 alleles, four loci each had 4 alleles and the remaining locus had 5 alleles. None of the alleles in their samples was near fixation. This system potentially has $4.05 \times 10^6$ unique MLGs. In the absence of clonal selection, the probability of randomly drawing two individuals having identical genotypes at the seven loci with a sample size of 390 is very small. From this perspective, clonal selection in the Poza Sur population would appear to be quite strong.

What is the significance of the relative frequencies of unique and non-unique microsatellite genotypes in a population? Consider a highly simplified hypothetical population in which there is one locus, S, with three alleles (*A–C*) subject to selection favoring a homozygote, and one microsatellite locus, N, having 4 neutral alleles (*A–E*). As discussed above, when the genotype with highest fitness is a homozygote, clonal selection tends to eliminate genotypic variation in the locus under selection, while mutation tends to increase the number and uniformity of distribution of microsatellite alleles. Equilibrium is reached at KS=1 for locus S when all individuals in the population have the same homozygous genotype (ignoring new, unfavorable mutations). At this point clonal selection stops. New alleles continue to be produced by mutation at locus N until it reaches its maximum clonal diversity at KN=10. Presuming equality of mutation rates, mutational equilibrium is attained when the four alleles have the identical frequencies of 0.25. At this point, any sample of 10 + **n** individuals will contain at least **n** non-unique MLGs. However, since all selectively inferior genotypes have been eliminated, the relative frequency of non-unique genotypes in the sample does not provide information on the presence or absence of clonal selection.

Now, let us assume that our hypothetical population has a seasonal structure in which homozygous *AA* individuals at locus S have a selective advantage in the spring, BB in the summer, and CC in the fall. Clonal selection operating on locus S over a period of time would tend to produce a seasonal succession of the three genotypes. Furthermore, if all sexual reproduction occurs within seasonal subpopulations, the relative frequencies of the three S homozygotes in the resting egg pool will reflect variances in genotype- specific reproductive success and frequencies of the three environmental states. Under these circumstances, loci under selection are likely to have large deviations from Hardy-Weinberg equilibrium show extensive multilocus linkage disequilibrium, and display large Wahlund effects in the resting egg pool.

However, since the alleles at locus N are not subject to selection, they are driven toward identical states of maximum diversity in each of the seasonal subpopulations. Therefore, when we examine allele frequencies at locus N in the resting egg pool we expect to see Hardy-Weinberg equilibrium, linkage equilibrium and no Wahlund effects. These are the observations that were made by Gómez & Carvalho (2000). Based on this reasoning, we do not think that the relative frequency of non-unique genotypes or, by extension, whether the microsatellite allele frequencies are or are not in Hardy-Weinberg and linkage equilibrium, constitute informative indicators of clonal selection. This mechanism may also explain the failure of allozyme analyses to detect seasonal changes in population structure even though they have been quite successful in detecting frequency changes of sibling species (King & Serra, 1998).

In summary, selectively neutral loci are inappropriate genetic markers to use in the study of the adaptive dynamics of cyclical parthenogens. In the absence of genes that can respond directly and differentially to clonal selection as the environment changes, it is not possible to study such adaptive processes as seasonal selection. This requirement suggests that neither allozymes nor microsatellites are likely to shed much light on the adaptive structure of rotifer populations.

## Acknowledgements

We are deeply indebted to Dr Africa Gómez for valuable discussion and comments on the manuscript that have helped to resolve some of the differences of interpretation encountered by comparing natural and modeled populations.

## References

Gómez, A., C. Clabby & G. R. Carvalho, 1998. Isolation and characterization of microsatellite loci in a cyclically parthenogenetic rotifer, *Brachionus plicatilis*. Mol. Ecol. 7: 1613–1621.

Gómez, A., C. & G. R. Carvalho, 2000. Sex, parthenogenesis and genetic structure of rotifers: microsatellite analysis of contemporary and resting egg bank populations. Mol. Ecol. 203–214.

Hoffman, R. J., 1986. Variation in contributions of asexual reproduction to the genetic structure of populations of the sea anemone Metridium senile. Evolution 40: 357–365.

Jarne, P. & P. J. L. Lagoda, 1996. Microsatellites, from molecules to populations and back. Trends Ecol. Evol. 11: 424–429.

King, C. E. & M. Serra, 1998. Seasonal variation as a determinant of population structure in rotifers reproducing by cyclical parthenogenesis. Hydrobiologia 387/388: 361–372.

Ruzzante, D. E., 1998. A comparison of several measures of genetic distance and population structure with microsatellite data: bias and sampling variance. Can. J. Fish. aquat. Sci. 55: 1–14.

Stoddart, J. A. & J. F. Taylor, 1988. Genotypic diversity: estimation and prediction in samples. Genetics 118: 705–711.

Webber, J. L. & C. Wong, 1993. Mutation of human short tandem repeats. Hum. Mol. Genet. 2: 1123–1128.

*Hydrobiologia* **446/447:** 333–336, 2001.
*L. Sanoamuang, H. Segers, R.J. Shiel & R.D. Gulati (eds), Rotifera IX.*




# A survey of introns in three genes of rotifers

David B. Mark Welch & Matthew Meselson
*Department of Molecular and Cellular Biology, Harvard University, Cambridge MA 02138, U.S.A.*
*E-mail: welch@fas.harvard.edu*

*Key words:* Bdelloidea, *hsp82*, introns, Monogononta, *tbp*, *tpi*

## Abstract

Here we report on the occurrence and position of introns found in three genes of rotifers. A region of the gene for the TATA-box binding protein was examined in three species of Bdelloidea and one of Monogononta. There are two introns in both copies of this gene present in each of the three bdelloids examined – one at a position where introns occur in other eukaryotes and the other at a novel position; the monogonont has no introns in the region examined. A region of the gene encoding the 82 kD heat shock protein was examined in 10 species, with every rotifer class represented. Introns were found in only two species, both bdelloids: one of the species has an intron in all three copies of the gene; the other has an intron in only one of the three copies. Both introns occur at novel positions. The gene for triosephosphate isomerase was examined in one bdelloid. Both copies of the gene in this species contain introns, all at conserved positions: one copy contains five introns, the other copy three. These observations demonstrate the presence of introns in bdelloid rotifers, some in conserved positions, others apparently newly arisen during bdelloid evolution.

## Introduction

Introns, portions of a gene that are spliced out of its transcripts, have been found in all animals, plants and fungi that have been extensively investigated. The positions at which some introns occur are identical or very similar across evolutionarily distant taxa, consistent with the proposal that these introns originated in an ancestor common to all eukaryotes (Gilbert et al., 1986). However, evidence for this belief based on sequence phylogeny is lacking, owing to the rapid rate of intron sequence change. Moreover, it is clear that some nuclear introns have been acquired more recently (Logsdon et al., 1998). The hypothesis that all introns appeared after the origin of eukaryotes but tend to occur or persist at certain preferred positions is, therefore, not excluded.

Here, we present a preliminary survey of the occurrence and positions of introns in rotifers. Sequences were examined in the coding regions of three genes: the TATA-box binding protein (*tbp*), the 82 kD heat-shock protein (*hsp82*), and triosephosphate isomerase (*tpi*). Each of these genes exists in two or more divergent copies in individual bdelloid genomes; the other rotifer species examined appear to be diploids with only single loci of the genes investigated (Mark Welch & Meselson, 2000).

## Materials and methods

The species examined were, Bdelloidea:*Philodina roseola* Ehrenberg, 1832, *Macrotrachela quadricornifera* Milne, 1886 [Philodinida, Philodinidae]; *Habrotrocha constricta* (Dujardin, 1841) [Philodinida, Habrotrochidae]; *Adineta vaga* (Davis, 1873) [Adinetida, Adinetidae]. Monogononta: *Brachionus plicatilis* Müller, 1786 (Austrian strain), *Brachionus calyciflorus* Pallas, 1766 [Ploima, Brachionidae]; *Eosphora ehrenbergi* Weber, 1918 [Ploima, Notommatidae]; *Sinantherina socialis* (Linnaeus, 1758) [Flosculariacea, Flosculariidae]. Seisonida: *Seison nebaliae* Grube, 1859 [Seisonidea, Seisonididae]. Acanthocephala: *Moniliformis moniliformis* Bremser, 1811 [Archiacanthocephala, Moniliformidae]. Provenance and culture conditions are described elsewhere (Mark Welch, 2000).

A region of the coding sequence of each gene was amplified by PCR, cloned, and sequenced as described

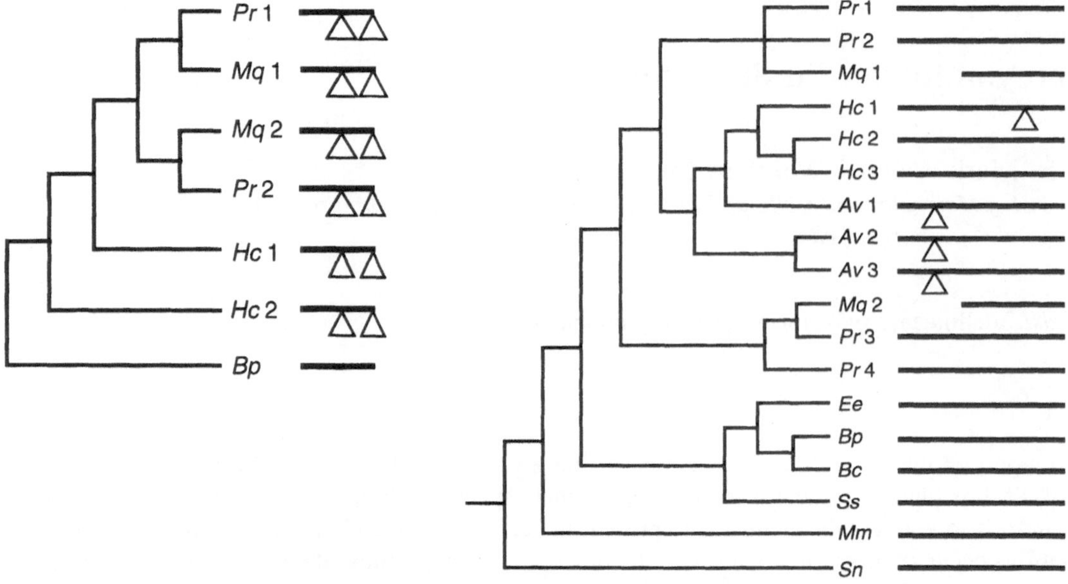


*Figure 1.* Occurrence and positions of introns in *tbp* (left) and *hsp82* (right) in Rotifera. Gene copies are identified by genus and species initials and, in bdelloids, also by copy number. Sequenced regions are designated by thick bars, intron locations by triangles. Phylogenetic relations are from Mark Welch (2000) and Mark Welch & Meselson (2000). The first intron in *tbp* occurs between two codons; the second interrupts a codon between the second and third positions. Both *hsp82* introns are located between codons.

(Mark Welch & Meselson, 2000). The region of *tbp* examined corresponds to *D. melanogaster* codons 187–295 (GenBank DROTATABF, Muhich et al., 1990); that of *hsp82* (also known as *hsp90*) to codons 13–302 (DROHSP82; Blackman & Meselson, 1986); and that of *tpi* to codons 71–242 (DMTPIG, Shaw-Lee et al., 1991). Sequences are available as GenBank AF142849–AF143850, AF249987–AF250004.

Sequences from other taxa were obtained from GenBank and were aligned with rotifer sequences using the Wisconsin Package (Genetics Computer Group). Introns were identified as interruptions in the consensus amino acid sequence.

## Results

### *tbp*

A 327 base pair (bp) region in the middle of the *tbp* coding sequence was examined in the bdelloids *H. constricta*, *P. roseola* and *M. quadricornifera*, and in the monogonont rotifer *B. plicatilis*. Each of the three bdelloids has two copies of *tbp*, and each copy contains introns, 46–70 bp in length, at the same two positions (Fig. 1). At both positions, the divergence between introns is similar to that at 4-fold degenerate codon sites in surrounding exons (Mark Welch & Meselson, 2000). No introns were found in *B. plicatilis*.

Introns have not been reported in other species at the position of the first bdelloid intron. If the structure of the bdelloid TATA-box binding protein is analogous to its structure in other species, this intron occurs in a region encoding a loop between two $\alpha$-helices (Nikolov et al., 1992). The second intron occurs at a conserved intron/exon boundary in a region encoding a $\beta$-sheet, position III in the notation of Goddemeier & Feix (1996).

### *hsp82*

An 865–888 bp region comprising the first third of the *hsp82* coding sequence was examined in the four bdelloid species and in the six species of other rotifers listed in 'Materials and methods' (Fig. 1). A 57–58 bp intron is present in all three copies of hsp82 in *A. vaga* and the divergence between copies is similar to that at 4-fold degenerate sites in the surrounding exons. A 58 bp intron downstream of the intron in *A. vaga* is present in only one of the three copies of hsp82 in *H. constricta*. There is no evident homology between the intron in *A. vaga* and that in *H. constricta*. No *hsp82* introns were found in any of the eight other rotifer species examined.

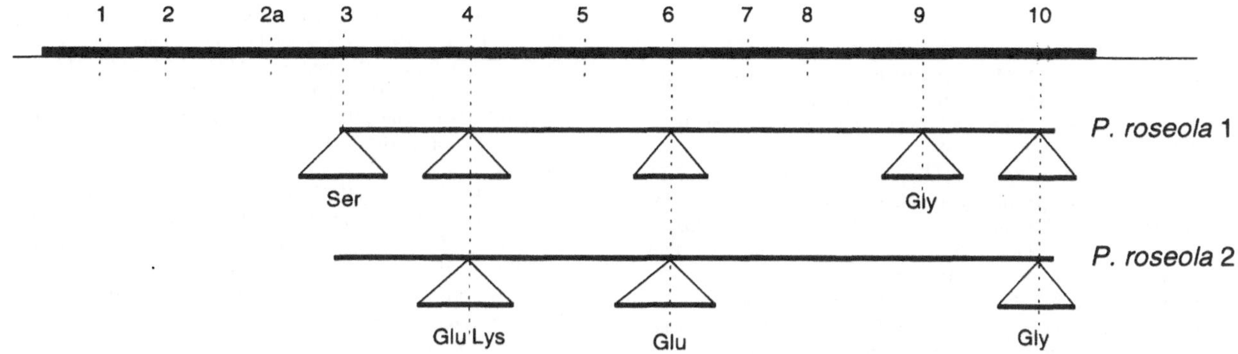


*Figure 2.* Occurrence and positions of *tpi* introns in the bdelloid *P. roseola*. The positions at which *tpi* introns have been found in other species are shown by dashed lines on a representation of the gene (solid bar) and are numbered after Gilbert et al. (1986) (position 10 is not precisely conserved). The lines labeled *P. roseola* 1 and *P. roseola* 2 represent the examined region of the two *P. roseola* gene copies, with intron locations indicated as triangles. Introns 3 and 9 in *P. roseola* 1 interrupt serine and glycine codons, respectively; introns 6 and 10 interrupt glutamine and glycine codons, respectively, in both copies.

The intron in *A. vaga* lies in a part of the gene that codes for a variable, charged region of asparagine, glutamine and lysine residues and has not been reported in other species. The intron in *H. constricta* lies 5 bp upstream of an intron/exon boundary found in the *hsp82* genes of several plants. If the structure of the hsp82 protein in *H. constricta* is like that in yeast, the intron occurs after the first codon of an $\alpha$-helix near the ATP binding domain (Prodromou et al., 1997).

*tpi*

A region spanning two-thirds of the *tpi* coding sequence was examined in *P. roseola*. The two copies of the gene in this bdelloid have three intron positions in common and there are introns at two additional positions in one of the copies. All five introns are at locations where introns are found in many other eukaryotes (Fig. 2) (Gilbert et al., 1986).

## Discussion

We found introns at nine positions in three genes of bdelloid rotifers. No introns were found in the examined regions of *tbp* in the monogonont *B. plicatilis* or of *hsp82* in four monogononts, an acanthocephalan and a seisonid. According to the phylogeny of *hsp82* in bdelloid rotifers shown in Figure 1b, the intron in *A. vaga* was lost in the lineage leading to *H. constricta*. Also, the different numbers of *tpi* introns in *P. roseola* require that these introns have been either lost or gained after the two copies of the gene diverged (Fig. 2). Both gain and loss of introns have been reported to occur in insects and in vertebrates (Tarrío et al., 1998; Waugh O'Neill et al., 1998; Venkatesh et al., 1999).

Three of the nine introns we found in bdelloids are at positions where introns have not been reported in other taxa. As is clear from their presence in only one of the 10 rotifer species examined and in no other reported sequence, the *hsp82* intron in *A. vaga* and that in *H. constricta* have almost certainly been gained since the origin of Bdelloidea. Also, it is likely that the first intron of *tbp* is novel. Each of the three novel bdelloid introns occurs between codons, and the *hsp82* intron in *A. vaga* and the first intron of *tbp* are in regions that may link protein structural modules, properties that have been cited as evidence for introns of pre-eukaryotic origin (Long et al., 1995; De Souza et al., 1996; Roy et al., 1999), which these introns are not.

## Acknowledgements

We thank J. Mark Welch, R. L. Wallace and E. Walsh for critically reading the manuscript, and the Eukaryotic Genetics Program of the U.S. National Science Foundation for consistent support. This work was originally funded by a U.S. N.S.F. Small Grant for Exploratory Research. D.M.W. was supported by a N.S.F. Graduate Research Grant.

## References


Blackman, R. K. & M. Meselson, 1986. Interspecific nucleotide sequence comparisons used to identify regulatory and structural features of the Drosophila hsp82 gene. J. Mol. Biol. 188: 499–515.

De Souza, S. J., M. Long, L. Schoenbach, S. W. Roy & W. Gilbert, 1996. Intron positions correlate with module boundaries in ancient proteins. Proc. nat. Acad. Sci. U.S.A. 93: 14632–14636.

Gilbert, W., M. Marchionni & N. G. McKnight, 1986. On the antiquity of introns. Cell 46: 151–154.

Goddemeier, M. L. & G. Feix, 1996. Genomic structure of the maize TATA-box binding protein 1 (TBP-1): conserved exon/intron structure in eukaryotic *TBP* genes. Gene 174: 111–114.

Logsdon, J. M. Jr., A. Stoltzfus & W. F. Doolittle, 1998. Molecular evolution: Recent cases of spliceosomal intron gain? Current Biology 8: R560–R563.

Long, M., C. Rosenberg & W. Gilbert, 1995. Intron phase correlations and the evolution of the intron/exon structure of genes. Proc. nat. Acad. Sci. U.S.A. 92: 12495–12499.

Mark Welch, D. B., 2000. Evidence from a protein-coding gene that acanthocephalans are rotifers. Invert. Biol. 19: 17–26.

Mark Welch, D. B. & M. Meselson, 2000. Evidence for the evolution of bdelloid rotifers without sexual recombination or genetic exchange. Science 288: 1211–1215.

Muhich, M., C. T. Iida, M. Horikoshi, R. G. Roeder, & C. S. Parker, 1990. cDNA clone encoding Drosophila transcription factor TFIID. Proc. natl. Acad. Sci. (U.S.A.) 87: 9148–9152.

Nikolov, D. B., S-H. Hu, J. Lin, A. Gasch, A. Hoffmann, M. Horikoshi, N-H. Chua, R.G. Roeder & S. K. Burley, 1992. Crystal structure of TFIID TATA-box binding protein. Nature 360: 40–46.

Prodromou, C., S. M. Roe, R. O'Brien, J. E. Ladbury, P. W. Piper & L. H. Pearl, 1997. Identification and structural characterization of the ATP/ADP-binding site in the Hsp90 molecular chaperone. Cell 90: 65–75.

Roy, S. W., M. Nosaka, S. J. de Souza & W. Gilbert, 1999. Centripetal modules and ancient introns. Gene 238: 86–91.

Shaw-Lee, R. L., J. L. Lissemore & D. T. Sullivan, 1991. Structure and expression of the triosephosphate isomerase (Tpi) gene of *Drosophila melanogaster.* Mol. Gen. Genet. 230: 225–229.

Tarrío, R., F. Rodríguez-Trelles & F. J. Ayala, 1998. New *Drosophila* introns originate by duplication. Proc. natl. Acad. Sci. USA 95: 1658–1662.

Venkatesh, B., Y. Ning & S. Brenner, 1999. Late changes in spliceosomal introns define clades in vertebrate evolution. Proc. natl. Acad. Sci. U.S.A. 96: 10267–10271.

Waugh O'Neill, R. J., F. E. Brennan, M. L. Delbridge, R. H. Crozier & J. A. Marshall Graves, 1998. *De novo* insertion of an intron into the mammalian sex determining gene, SRY. Proc. natl. Acad. Sci. U.S.A. 95: 1653–1657.

Hydrobiologia **446/447:** 337–353, 2001.
L. Sanoamuang, H. Segers, R.J. Shiel & R.D. Gulati (eds), Rotifera IX.




# Biotechnology and aquaculture of rotifers

Esther Lubzens[1], Odi Zmora[2] & Yoav Barr[2]
*Israel Oceanographic and Limnological Research, [1]National Institute of Oceanography, Haifa, Israel.*
*E-mail: esther@ocean.org.il;*
*[2]National Center for Mariculture, Eilat, Israel*
*E-mail: zmora@agri.huji.ac.il*

*Key words:* rotifers, *Brachionus plicatilis*, *Brachionus rotundiformis*, mass cultures

## Abstract

Biotechnology can be defined as any technology that involves living organisms or their derivatives. In applying this definition to rotifers, we focus on their contribution in culturing of early larval stages of marine fish. After almost four decades of marine fish culture in captivity, the success of this worldwide industry is still quite dependent on mass culture of the species *Brachionus plicatilis* and *B. rotundiformis*. In mass culture, the rotifers are continuously driven to reproduce at high rates, in relatively extreme environmental conditions of high population density and high loads of organic matter. Therefore, the success of mass cultures and future improvements in these systems relies on a close interaction between basic and applied studies of rotifers. In the present review, we will attempt to analyze why rotifers are suitable for early life stages of fish and to describe, in general, methodologies that have been devised for reliable supply of rotifers in large quantities. Problems associated with rotifer production, nutritional quality and effect on fish health and nutrition, will be discussed. Research on *B. plicatilis* and *B. rotundiformis* has increased enormously during the past three decades and these two species are the best-studied rotifers so far. While much of the research on these species is directed or devoted to the needs of aquaculture industry, they are also used as models for addressing basic biological questions, due to the relative ease of culture and their availability. Studies on feeding, pheromones, speciation in rotifers, the occurrence and putative hormones involved in sexual and asexual reproduction and production of resting eggs, are few examples of such studies. Rotifers will probably maintain their role as food organism for fish larvae, in spite of attempts to replace them with more accessible formulated food. Development of new culture methods that will improve the nutritional quality and production efficiency of rotifers may result in more diversified and flexible tasks for these organisms in aquaculture.

## Introduction

The flourishing cultures of marine finfish in various parts of the world, can be partly attributed to successful mass cultivation of the rotifers *Brachionus plicatilis* and *Brachionus rotundiformis*. These animals are provided as the first food to fish larvae, during the initial period of their life after they hatch from the eggs and close to the time of mouth opening. Aquaculture provided more than 23% of global fisheries production in 1997. About 11 million tonnes of marine fish at an estimated value of 2.9 billion U.S. $ (ex farm), were cultured in 1997, showing more than a two fold increase from 1990 (F.A.O., 1998). About 58 billion fry were produced in 1996, including freshwater species. Many of these hatchery produced fingerlings are released to the wild to replenish the dwindling natural populations. A large number of marine fish species are raised in captivity with rotifers, including the larvae of several major species such as yellowtail, red seabream, Barramundi, turbot, mullet, pufferfish, gilthead seabream and some European seabass. Seabream and seabass cultured in the Mediterranean basin can serve an excellent example for the fast development of mariculture that relies on mass produced rotifers. About 216 million of gilthead seabream fry were produced in 1998 in the Mediterranean basin and 247 million were expected to be produced in 1999 (Federation of European Aquaculture Producers, 2000).

Intensive marine fish culture imposes specific requirements that differ from those found in the natural environment (Table 1). The production of fish in captive systems today is more analogous to an industrial production line where broodstock are expected to continuously provide fertilized eggs, independently of the natural spawning season, larvae are raised to fingerlings in hatcheries and are later grown in sea pens to market size. In the natural environment, spawning is usually seasonal and the expected survival of larvae is extremely low due to lack of adequate nutrition and presence of predators. Larvae have at their disposal a wide range of food organisms in their natural habitats, but due to production cost considerations this choice is extremely limited in captive systems and usually consists of only two or three organisms; *Artemia, B. plicatilis* and/or *B. rotundiformis.* In nature, growth rate is slow and the prolific female that can produce several millions of fertilized eggs during her life time is expected to provide only two adult maturing fish, that will preserve the species (Blaxter, 1974; Kinne, 1977). In captive systems, however, economic feasibility requires high survival rate at early life stages, fast growth to market size and attractive external appearances of the fish. All cultures are, therefore, performed at significantly high densities, to save on space and energy that is required for pumping seawater and maintenance of optimal culture temperatures, thus exposing all developmental stages to stress and increasing their susceptibility to pathogens. Moreover, production must fit into a pre-conceived plan that will ensure the supply of fish to the consumers at the desired times.

Marine fish require live food at early life stages for survival and proper development, in contrast to most edible fish species that are cultured in fresh water and to salmonid juvenile stages cultured in seawater. The role of live exogenous food in early stages is controversial. It was suggested that marine fish larvae that have only a partially developed digestive tract after hatching, depend strongly on exogenous enzymes provided by the live food they consume for digestion of their prey. This means that rotifers or *Artemia* are partly digested by their own enzymes that are released when they reach the gut of the larvae (Dabrowski, 1979, 1984; Lauf & Hoefer, 1984; Munilla-Moran et al., 1990; Kolkovski et al., 1993; Walford & Lam, 1993). On the other hand, it has been shown that in some species the main enzymes already occur at the moment of first feeding, allowing the digestion of prey (Baragi & Lowell, 1986; Govoni et al., 1986; Cahu & Zambonino Infante, 1994). Diaz et al. (1997) have shown recently that alkaline proteases found in rotifers or *Artemia* differ from those found in fed or starved seabream larvae, suggesting that exogenous enzymes provided with live food only initiates protein digestion, but the food is fully digested by autogenous enzymes occurring within the larval gut.

Providing live food from natural sources is met with several difficulties; it is variable in species composition and nutritional quality, unpredictable and may expose the fish larvae to predators, natural toxins or pollutants. Most cultures today rely on providing rotifers for 10–30 days after mouth opening of hatched larvae, followed by *Artemia.* Rotifers were found as an adequate food source for the following reasons: their shape, size, and colour; their relatively slow motility; their chemical content that can be manipulated to meet the nutritional requirements of the raised fish larvae, and the ease with which they can be cultured at high densities to provide the large numbers required for raising larvae in captive systems. Moreover, rotifers can actually serve as a bio-capsule or vehicle, for transferring therapeutic agents to the fish larvae. (see Lubzens, 1987 and Lubzens et al., 1989 for detailed reviews on these topics; Verpraet et al., 1992). Recent experiments show that rotifers can be used for transferring probiotic bacteria to fish larvae (Markridis et al., 1999, 2000; Rombaut et al., 1999a,b – see below). Several reviews have been published on rotifer cultures (Hirata, 1979, 1980; Nagata & Hirata, 1986; Lubzens, 1987; Fukusho, 1989a,b; Fulks & Main, 1991). In the following review, we aim at concentrating on a few aspects including mass culture concepts, methods for increasing reliability of supplying rotifers and ways for improving their nutritional value to marine fish larvae. The present review attempts to concentrate on studies carried out in recent years on the euryhaline rotifer species *B. plicatilis* and *B. rotundiformis*.

## Mass culture methods

Following the first study by Ito (1960) on the biology and culture of rotifers, it was demonstrated that they can be used as food for marine fish larvae (Ito, 1963 cited in Hirata, 1980) and in 1964 the first methods for

*Table 1.* Comparison between fish growing in their natural environment and those in captive systems

| Natural Environment | Captive systems |
|---|---|
| Seasonal reproduction | Continuous reproduction regulated by photo-period or hormonally induced |
| Larvae feed on a variety of zooplankton species | Larvae feed on *Brachionus* and *Artemia* |
| Larvae are at low density | Larvae cultured at high density |
| Survival of larvae is extremely low (less than 0.1%) | Survival of larvae expected to be high (more than 20%) |
| Slow rate of growth to maturity | Fast growth to market size |
| High genetic variability | Low genetic variability |

mass production of rotifers were reported where rotifers were fed algae (Hirata, 1964). The production of rotifers in these cultures was dependent on the success in culturing algae and the rotifer density achieved in this system was limited (10–50 ind. $ml^{-1}$). A major breakthrough was achieved when baker's yeast was found to be appropriate for feeding rotifers (Hirata & Mori, 1967) and rotifer densities reached 150 ind. $ml^{-1}$ in most cases, but under specific culture conditions could reach 1500–2000 ind. $ml^{-1}$ (reviewed in Hirata, 1980). However, these cultures were not very reliable and often crashed (Hirayama, 1987). Scott (1981) and Hirayama & Funamoto (1983), found that rotifers require vitamin $B_{12}$ for proper development and adding it to mass cultures resulted in more predictable and stable cultures. While rotifers could be cultured on yeast as a single food source, these rotifers were found to be nutritionally deficient when fed to marine fish larvae (see below).

The number of rotifers that are required daily in a fish hatchery culturing gilthead seabream is illustrated in Table 2 and is based on methods and experience gained in the Ardag hatchery in Eilat, Israel, and is typical to most hatcheries around the Mediterranean basin. In general, about 1 million fertilized eggs are introduced every week into either 6 × 1700 l or one 8000 l tanks at a density of ~100 or ~120 hatched larvae $l^{-1}$, respectively. About 10 000–20 000 rotifers $l^{-1}$ are supplied daily for a period extending ~30 days, although *Artemia* nauplii are also introduced after day 14 into the larval tanks. Three or four batches are cultured simultaneously, with a weekly yield of ×200 000 post-larvae (~20% survival). This means that the daily need in the hatchery exceeds one billion rotifers in order to account also for losses during harvesting of rotifers from their culture tanks and the lipid enrichment process (see below). Additional mortality that occurs after the rotifer feeding stage, results in the production of about 6 million 1g fingerlings during 10 months every year. The low hunting skills of larvae that just hatch from eggs (Kinne, 1977; Theilacker & Dorsey, 1980), require that rotifers should be available at relatively high densities in order to assure an adequate number of successful encounters with the larvae. At this stage, most of the rotifers are not consumed but are washed out of the tank and replaced by newly fed rotifers, in order to assure the nutritional value of the rotifers supplied to the larvae (see below). At later stages, the hunting skills of the larvae are more efficient and the supplied rotifers are fully consumed. It had been calculated that 2.5 tons of rotifers were needed in addition to 3.5 tons of algae for culturing 10 million red sea bream larvae in 1974 (cited by Fukusho, 1989b). A more recent calculation shows that 1 ton (wet weight) of *B. plicatilis* rotifer (400 billion rotifers) and 150–240 kg of *Nannochloropsis* (dry weight) are needed for the production of 10 million ~1 g fry of gilthead seabream in the Mediterranean basin (Zmora et al., 1991). Clearly, a reliable daily supply of rotifers in the numbers required during each culture cycle, determines the success of culturing marine fish in captivity.

*Table 2.* The number of rotifers required daily for culturing guilthead seabream larvae in a hatchery

| | Stages of culture | |
|---|---|---|
| 1. | Number of fish eggs stocked every week (onebatch)- 6 × 1700 l tanks or 1× 8000 l tank | $10^6$ larvae |
| 2. | Stocking density | ~100 – 120 larvae/l |
| 3. | Rotifer density in fish tank | ~10 000–20 000/l/day |
| 4. | Total number of rotifers required every day per batch | ~200 × $10^6$ |
| 5. | Number of simultaneous batch cultures per week | 3–4 |
| 6. | Total number of rotifers required daily | 800–1200 × $10^6$ |
| 7. | Total number of days per cultured batch | ~30 |

In general, we can distinguish three main concepts for obtaining large quantities of rotifers; 1. Batch cultures, 2. Semi-continuous cultures and 3. Continuous (chemostat type) cultures.

## Batch cultures

This type of culture was initially developed in 1964 (for reviews see Hirata, 1980; Lubzens, 1987) and has been greatly modified in recent years by many fish hatcheries. Initially, rotifers were introduced at low density into '*green water*' that was produced in fertilized tanks or ponds. Rotifers were collected ('harvested') after all the algae were consumed, and used as food for fish larvae. A small number was used for inoculation of newly formed '*green water*' ponds. A recent adaptation (Lubzens et al., 1997) for batch culture relies on a series of tanks. The culture in each tank started with 1/5 of its volume with rotifers at a density of ~200–300 $ml^{-1}$. Rotifers were fed mainly with bakers' yeast and the supply of the daily ration was divided into 3–5 meals. The volume of culture medium was increased daily with seawater depending on the reproductive rate, thus assuring the relative high rotifer density in the culture volume and this may reach 300–500 rotifers $ml^{-1}$ at the end of the culture period. The final volume of the tank was reached after 4–5 culture days, rotifers were sieved, concentrated and the culture tank and its accessories (sieves, aeration tubes) were immersed in chlorine (10 ppm of commercial grade chlorine) for 24 h and washed thoroughly with seawater before re-use in culture. The concentrated rotifers were immersed briefly (~15 min) in de-chlorinated fresh water to remove ciliates and other undesired organisms. The rotifers were not affected by this brief immersion in fresh water. Most of the concentrated rotifers were enriched (see below) with essential fatty acids and proteins before being fed to fish larvae and a planned small fraction was used for inoculation of new culture tanks. This system is based on 6 tanks and every day one culture tank is used and one is inoculated. A different approach was based on inoculating rotifers into a culture tank, containing the final volume of culture medium, at an initial low rotifer density. The rotifers were fed with yeast, other dry food or concentrated algae and harvesting was performed after reaching high densities (>200 rotifers $ml^{-1}$).

The introduction of refrigerated and condensed freshwater *Chlorella* enriched with vitamin $B_{12}$ (Chlorella Industry Co., Japan) has facilitated the introduction of extremely high density rotifer mass cultures and changed dramatically the future potential for providing rotifers for fish hatcheries (Yoshimura et al., 1996, 1997; Balompapueng et al., 1997a). These high density culture systems consist of 1 $m^3$tank units and rotifers (*B. rotundiformis*) are batch cultured at 2–3 day intervals. Cultures are initiated at a density of 10 000 rotifers $ml^{-1}$ and after 2–3 days with condensed algae, the culture density reaches 20 000–30 000 rotifers $ml^{-1}$. The large number of rotifers that are needed for starting the first round of these cultures can originate from upscaling a small initial culture by traditional mass culture methods, or alternatively from hatched resting eggs, if available (see below for further discussion of this issue). Oxygen gas has to be supplied to these cultures in order to overcome the shortage of dissolved oxygen that results from the high amounts of food and the subsequent increase in the rotifer population. Also, at these high densities, ammonia excreted by the rotifers becomes a significant problem. The pH of the culture also increases presumably due to the liberation of carbon dioxide out of the water, by the oxygen gas supplied to the culture. This in turn, results in a higher proportion of toxic un-ionized ammonia of the total ammonia, which increases with the pH, salinity and the culture temperature. Regulation of the pH at 7.0 by hydrochloric acid minimizes these effects. A special nylon filtration mat is used for removal of large amounts of suspended organic material that may also include protozoans, fungi and bacterial flocs which may harm the fish larvae fed rotifers. The small space and reduced labor required, is the remarkable advantage of this ultra-high density culture system.

The relatively compact and closed system permits maintaining separate cultures in the hatcheries of several genetic rotifer strains or populations differering in size and will assist in providing more adequate food for larvae of different fish species, without enlarging the hatchery facilities. Fish larvae prefer smaller rotifers immediately after hatching and larger ones as they grow and providing them with these rotifers increases their survival (Lubzens et al., 1989). Moreover, small size rotifers are essential for culturing some fish species (e.g. grouper) that are equipped with extremely small mouth openings. Most hatcheries aim at culturing variety of fish species that may differ in their nutritional requirements at early life stages and these can be more easily provided by the small compact high density cultures. The relatively high culture temperatures (>25 °C) required by small size rotifers can be easily maintained in these compact systems. At the same time, fish cultured at low temperatures will benefit from this system by rotifers cultured at relatively low temperatures.

The convenience and accessibility of condensed algae also encourages its use for intensifying traditional culture systems.

## Semi-continuous cultures

This culture system relies on periodic (usually daily) harvesting of rotifers by removal of part of the culture medium and replacing it with new seawater (see Hirata, 1980). This method has been termed 'The Thinning Method' (Fukusho, 1989b). The removed volume depends on the reproductive rate of the rotifers and harvesting removes only the number of rotifers gained by reproduction from the previous harvesting period. In general, these systems often rely on large volume tanks ranging in volume from 3000–300 000 l (Hirata, 1980; Fukusho, 1989b; Lubzens et al., 1997). The cultures are characterized by the relatively lower rotifer density (100–300 rotifers $ml^{-1}$) and bakers' yeast is used for feeding the rotifers. Cultures continue for several days or even weeks, but eventually excretory products (solid wastes and nitrogenous products) that accumulate in the tank lead to their collapse. Continuous removal of the solid waste enhanced rotifer reproduction and resulted in densities reaching 400 ind. $ml^{-1}$ and cultures were stable for over 30 days with daily harvesting periods. In these cultures, termed 'feed back cultures' (Hirata 1980), the solid waste was transferred to decomposing tanks and after mineralization it was used as fertilizer for cultures of algae. However, these large volume tanks harbor many other organisms (Yu et al., 1990a,b; Colorni et al., 1991; Maeda & Hino, 1991; Comps et al., 1991a, b, 1993; Hagiwara et al., 1995a; Comps & Menu, 1997; Jung et al., 1997) that either compete for food with rotifers (e.g. ciliates, copepodes, cladocerans) or harm them, like viruses, bacteria, fungi or yeast. Since these and other pathogens could be transferred to the fish tanks (Gatesoupe, 1990; Blanch et al., 1991; Verdonck et al., 1997; Munro et al., 1999), it has been suggested to culture rotifers with selected bacterial strains (Gatesoupe, 1999; Markidis et al., 1999, 2000; Rombaut et al., 1999a,b) that will curtail the growth of undesired micro-organism and contribute to stabilization of the rotifer cultures (see below).

A high density semi-continuous culture, using a modified commercially available formulated rotifer diet with re-circulation of the culture media was described recently (Suantika et al., 2000). This system consists of a rotifer (*B. plicatilis*) 100 l culture tank, a settlement tank for suspended particulate matter, a protein skimmer and a bio-filter. The amount of food provided to the rotifers in this system was adjusted to the circulation flow rate and loss of food by the protein skimmer. Rotifer cultures were initiated at a density of ~250 rotifers $ml^{-1}$, the density increased after 8 days of culture to 3000 $ml^{-1}$ and was maintained at about this density for over a month. About 20% (30–60 $\times 10^6$ rotifers) of the standing stock culture was harvested every day. Adequate water quality was maintained by a daily 500% re-circulation rate. Although it was possible to obtain densities of 8000 rotifers $ml^{-1}$ after 8 days of culture from the day of inoculation, the authors recommended that rotifer production should be performed at a lower density, when the system is more stable. The total rotifer production in this system was $1.7 \times 10^9$ rotifers during 32 culture days.

An experimental 1 l semicontinuous culture system using dried *Nannochloropsis* powder system showed that *Brachionus* requires relatively high food concentrations to reach the maximum conversion efficiency; 47–64% of the dry weight in *B. plicatilis* and 77–136% in *B. rotundiformis.* There were significant differences in the system efficiency (percentage of dry microalgae transformed into dry rotifer biomass) between the two species with a maximal 67% efficiency for *B. plicatilis* at 0.3 $d^{-1}$ dilution rate and 24.5% at 0.2 $d^{-1}$, for *B. rotundiformis* (Navarro & Yufera, 1998a). In another study where dilution rate was kept at 10% per day, the efficiency of the system was ~45% and ~20% for *B. plicatilis* and *B. rotundiformis*, respectively (Navarro & Yufera, 1998b). The relatively low values for *B. rotundiformis* may be due to temperature (25 °C) used for this study, which is not in the optimal range for this species.

## Continuous cultures

These cultures are based on the chemostat or turbidostat models of micro-organisms and are fully controlled (temperature, pH, oxygen supply and density of cultured organisms) and highly dependable (Walz, 1993; Walz, et al., 1997). They offer easy manipulation of rotifer physiological and nutritional quality. Log phase produced rotifers can be harvested continuously and their nutritional quality is maintained by providing adequate food organisms (James et al., 1983, 1987; James & Abu-Rezeq, 1989a, b, 1990). A more recent adaptation of this method involves high density cultures (excess of 10 000 rotifers $ml^{-1}$) using the condensed freshwater *Chlorella.* (Fu et al., 1997). Part of the efficiency in rotifer production in these intensive systems may be attributed to bacterial growth and consumption by the cultured rotifers, which can reach ~20% of the particulate organic nitrogen ingested by the rotifers (Aoki & Hino, 1996; Hino et al., 1997). These systems are compact (1,000 l provide daily 1.7–3.5 billion *B. rotundiformis* and 500 l culture provide daily 0.13–0.27 billion *B. plicatilis*), their initial cost exceeds that of large installations and they depend on a constant supply of condensed algae. While more work is needed to optimize the culture system for *B. plicatilis,* it should be noted that their biomass is at least three times higher (expressed as dry weight – Yufera et al., 1997), depending on their size, than that of *B. rotundiformis.* This means that a similar biomass per culture volume will consist of a smaller number of individuals. Moreover, the lower number of rotifers produced per day for *B. plicatilis* may also be attributed to their lower metabolism and rate of reproduction at the optimal culture temperature (20–24 °C).

Cost of production, reliability and the practical experience of the staff are the main factors that dictate the choice of system. In Europe, North America and Japan, the cost of manpower is one of the main concerns in these calculations, while cost of equipment is the main concern in developing countries. Reliability of each system dictates the number of replicate culture tanks that should be installed for meeting the planned production, with fewer replicate cultures needed in more reliable systems.

## Factors affecting efficiency of rotifer production

The type (asexual or sexual reproduction) and rate of reproduction varies between *B. plicatilis* and *B. rotundiformis* and their respective zoogeographical strains (Hagiwara et al., 1988b; Rumengan & Hirayama, 1990; Hagiwara & Lee, 1991; Hirayama & Rumengan, 1993). Mass cultures benefit from rotifers that reproduce asexually, since sexual reproduction results in production of males and resting eggs. Males are nutritionally inferior to females due to their fast motility which reduces their chance of being consumed by the slower moving larvae, and lack of digestive system (Wesenberg-Lund, 1923) that prevents enriching them with nutrients required by fish larvae. The number of resting eggs produced by one female (1–8) is significantly smaller than the number of asexually (18–23) produced eggs (Snell, 1986; Lubzens & Minkoff, 1988; Hagiwara & Hino, 1990) and since sexually produced resting eggs do not hatch immediately, this means that sexual reproduction results in a smaller number of nutritionally adequate rotifers.

The rate of reproduction depends on salinity, temperature, food quantity and food quality and is strain specific (Snell, 1986; Lubzens, 1987; Miracle & Serra, 1989; Korstad et al., 1989a,b; Hirayama, 1990). Optimal salinity for asexual reproduction in *B. plicatilis* ranges between 10 and 25 ppt (Lubzens et al., 1985; Snell, 1986). Temperatures supporting maximal rates of asexual reproduction vary not only between species (Hirayama & Rumnegan, 1993) but also between zoogeographical populations (Lubzens et al., 1989). The direct effect of salinity on reproductive rates depends on the genotype, which is adapted to an optimal salinity range. A positive interaction was found between low temperature and low salinity or high temperature and high salinity in clones isolated in Spain (Miracle & Serra, 1989). Thus, the reproductive rate of a small size rotifer clone (clone FCA) was higher at relatively high salinities and high temperatures. Optimal reproductive rates for *B. plicatilis* rotifers ranges between 20 and 25 °C, while those for *B. rotundiformis* are higher and can reach 38 °C (Snell, 1986; Lubzens et al., 1989). Reproductive rates are also affected by the pH of the culture and a pH ranging from 7–8 is considered adequate for culturing rotifers (Furukawa & Hidaka, 1973; see discussion on this topic in Lubzens, 1987; Fulks & Main, 1991).

Rotifer reproductive rates are strongly affected by the food provided to cultures. Particle size and the species of algae determine the reproductive rates of *B. plicatilis* (reviewed in Lubzens 1987; Whyte & Nagata, 1990; Fulks & Main, 1991; Snell, 1991; Tamaru et al., 1991; Nagata & Whyte, 1992; Vadstein et al., 1993; Hansen et al., 1997. Most hatcheries culture *Nannochloropsis* (formerly known as 'marine *Chlorella*') for periodic feeding of rotifers and for providing *"green water"* to fish tanks. Reproductive rates of rotifers were higher when fed live or frozen *Nannochloropsis* than when fed bakes' yeast (Tamaru et al., 1991; Hamada et al., 1993; Lubzens et al., 1995a). Dried algae or formulated diets have also been used to feed rotifers and in several cases these sources of food are problematic for long term cultures (Hirayama & Nakamura, 1976: dried algae; Øie & Olsen 1997: Protein Selco or Super Selco, Inve, Belgium; Snell, 1991: dried algae, yeast; Yufera & Navarro, 1995; Navarro & Yufera, 1998a, b: dried *Nannochloropsis* and other species).

The effect of several other factors such as unionized ammonia levels, dissolved oxygen and light on efficiency of production in rotifer cultures have been discussed before (Yu & Hirayama, 1986; Fulks & Main, 1991).

## Early warning on state of cultures

Evaluating the physiological state of the rotifer culture is extremely important in hatcheries since larval production depends on a predictable and reliable daily supply of rotifers. Snell et al. (1987), Snell & Hoff (1988) and Korstad et al. (1995) have shown that swimming speed and egg ratio can serve as indicators for the state of the culture. Swimming velocity and egg ratio were reduced at high un-ionized ammonia concentration and at starvation. Extreme low or high values of temperature and pH had the same effect. The pH of the cultures plays an important role since the toxicity of un-ionized ammonia ($NH_3$–N) released from ammonium ($NH_4{}^+$–N) is a function of pH (Bower & Bidwell, 1978). While swimming activity can be a quick indicator for the current culture state, egg ratio can serve to predict the condition of the culture after 24 h. It has been suggested that rotifer cultures (Russian strain) showing an egg ratio of less than 0.13 will collapse due to the aging of the culture (Snell et al., 1987).

Hagiwara et al. (1998) suggested that measuring the viscosity of the rotifer culture can serve as an indicator for stress. They demonstrated that the relative viscosity of the culture medium increases with the age of the culture, resulting in reduced mean longevity and mean number of offspring of the rotifers. This phenomenon was attributed to the reduced swimming speed and reduced ingestion rate of the rotifers in the culture, and is of great importance in the high density continuous cultures in which the culture media contain large amounts of concentrated algae and excretory products.

Occurrence of diseases will eventually lead to the collapse of the culture and, therefore, early detection is important in taking appropriate measures. While there are no known remedies for rotifer diseases, it has been observed (Zmora, 1991) that providing *Nannochloropsis* to infected yeast-fed rotifer cultures will reduce infection rate. The individual rotifer is not cured by the algae but the accelerated reproduction of the cultured rotifers fed algae will subsequently result in new generations that show a low incidence of the disease (Zmora, unpublished results). These cultures remain susceptible and the diseases may recur if the cultures are exposed to additional stress. Daily inspection of swimming speed, egg ratio and health conditions must be performed on samples removed from each culture tank.

## Bacterial interactions in rotifer cultures

The bacteria prevalent in rotifer cultures, especially

those in mass cultures, are gaining increasing attention in recent years in two main aspects: 1. Bacteria that affect rotifer cultures and 2. Bacteria transferred from rotifer culture tanks to predator-larvae cultures.

It was mentioned before that rotifers use bacteria as food and ~20% of the nitrogen input in chemostat cultures originated from bacteria (Aoki & Hino, 1996; Hino et al., 1997). Bacteria are also important producers and sources of metabolites such as vitamin $B_{12}$, which is essential for rotifer growth (Scott, 1981; Hirayama & Funamoto, 1983; Yu et al., 1988, 1989). Rotifer asexual reproduction was improved by specific bacterial strains (Rombaut et al., 1999) and sexual reproduction and resting egg formation were increased after addition of bacterial extracts (Hagiwara et al., 1994). On the other hand, toxic bacteria in rotifer cultures have also been reported (Yu et al., 1990b).

The high organic load associated with intensive rotifer production selectively induces an increased proportion of opportunistic bacteria that may be pathogenic to fish larvae (Skjermo & Vadstein, 1999). Regulation of the type of bacteria inhabiting rotifer cultures has been offered as a solution, since it is extremely difficult to obtain bacteria-free cultures of rotifers (Rombaut et al., 1999a) and in most cases sterilization methods or use of antibiotics are not very effective (Maeda et al., 1997). Introduction of selected bacterial cultures not only serves for curtailing the proliferation of pathogens in the rotifer cultures, but they are also used as probiotics, for transmitting beneficial microbial fauna into the digestive system of fish larvae. Probiotics are defined as 'microbial cells that are administered in such a way as to enter the gastrointestinal tract and to be kept alive, with the aim of improving health' (Gatesoupe, 1999). Lactic acid bacteria and *Bacillus* sp. spores, that were introduced into the culture medium of rotifers increased rotifer production and growth of turbot larvae (Gatesoupe, 1991, 1993). Alternatively, rotifers can serve as vectors for probionts that will colonize the larvae gut (Fjelheim et al., 1999) and prevent disease outbreaks (Grisez et al., 1997). The bacterial microflora of rotifers from culture tanks can be replaced by 1 h incubation in bacterial suspensions consisting of one or more probiotic strains. The probiont bacteria persist as a dominant part of the bacterial flora for 4–24 h (Makridis et al., 2000) and the treated rotifers will transmit this flora to the fish larvae's gut, when supplied to the fish culture.

**Nutritional quality of rotifers**

Rotifers are used as the sole food of fish larvae during the early stages of development, therefore, they have to supply all the nutrients required for proper development of their predators during this period. These include lipids (Essential Fatty Acids – EFA), proteins, essential amino acids and vitamins. Most mass cultured rotifers may lack adequate amounts of these nutrients and must be enriched with them before being offered to the fish larvae. The nutritional value of rotifers depends on their dry weight, caloric value and chemical composition (reviewed in Lubzens et al., 1989). Dynamic physiological processes such as satiation, starvation and reproduction also affect the chemical composition of rotifers (Yufera & Pascual, 1989; Olsen et al., 1993; Øie & Olsen, 1997; Yufera et al., 1997; Makridis & Olsen, 1999). In evaluating the nutritional contribution of rotifers, it is necessary to consider that rotifer amictic eggs and lorica are not digested by early developmental larval stages (Lubzens et al., 1989).

The number of rotifers consumed determines the quantity of food reaching the gut of the larva. The number of rotifers consumed daily, increases with the size or age of the larva from 55–72 rotifers per 3.9 mm length larva to 4700 per 11.4 mm length larva of red seabream (Fukusho, 1989b). There is a large variation in sizes between *B. plicatilis* and *B. rotundiformis* and between various populations within each species (Fu et al., 1990, 1991a, b; Hagiwara et al., 1995b). This is reflected in their nutritional quality and consumption by fish larvae (Lubzens et al., 1989; Polo et al., 1992). Moreover, size changes occur during the life cycle of rotifers within each population or species, with rotifers growing in size from the time of hatching from the amictic egg until they reach reproductive maturity. Size also depends on culture conditions such as salinity, temperature or diet (Snell & Carillo, 1984). All these factors also contribute to the difficulties of comparing results from different publications.

The dry weight of rotifers depends on their size and nutritional state (Lubzens et al., 1989). Recently, Yufera et al. (1997) reported that the dry weight of *B. plicatilis* is 3–4 (~600–800 ng) times higher than that of *B. rotundiformis* (~200 ng) and this was affected by the rotifer reproductive rate. The caloric value was found to depend on the diet and ranged from 1.34 cal per $10^3$ rotifers fed on baker's yeast to 2.00 cal per $10^3$ rotifers after 6 h enrichment with a formulated enrichment food (Fernandez-Reiriz et al., 1993).

Rotifer protein content ranges from 28 to 63% and lipid content between 9 and 28% of the dry weight (Lubzens et al., 1989; Frolov et al., 1991; Frolov & Pankov, 1992; Nagata & Whyte, 1992; Fernandez-Reiriz et al., 1993; Reitan et al., 1993; Rainuzzo et al., 1994; Øie & Olsen, 1997; Makridis & Olsen, 1999). The carbohydrate content ranges from 10.5 to 27% of the dry weight (Whyte & Nagata, 1990; Frolov et al., 1991; Frolov & Pankov, 1992; Nagata & Whyte, 1992; Fernandez-Reiriz et al., 1993). It is composed of 61–80% glucose (which is present mainly as glycogen), 9–18% ribose and 0.8–7.0% of galactose, mannose, deoxyglucose, fucose and xylose (Nagata & Whyte, 1992).

Food ration affects the reproductive rate of rotifers and their protein, lipid and carbohydrate content. The protein content of individual rotifers increased by 60–80% with increasing food ration, but the proportion of protein in the dry weight (mg protein $g^{-1}$ dry weight) remained almost unaffected for a given diet. However, the proportion of proteins in the dry weight changes when rotifers are fed on different species of algae or on artificially prepared enrichment diets (Frolov et al., 1991; Frolov & Pankov, 1992; Nagata & Whyte, 1992; Fernandez-Reiriz et al., 1993; Øie & Olsen, 1997). The amino acid profiles of rotifers are unaffected by either food ration or type of food provided to rotifers (Lubzens et al., 1989; Frolov et al., 1991; Tamaru et al., 1991, 1993; Frolov & Pankov, 1992).

Lipids were found, unequivocally, to have the greatest influence on growth and survival of marine fish larvae. Eicosanpentaenoic (EPA) and Docosahexaenoic (DHA) acids (20:5 n-3 and 22:6 n-3, respectively) have been known for several decades to be essential fatty acids for survival of marine fish larvae (Owen et al., 1975; Fujita, 1979; Watanabe et al., 1983). Marine fish contain large amounts of EPA and DHA in the phospholipids of their cellular membranes and since they cannot synthesize them from linolenic acid (18:3 n-3), these acids are essential dietary constituents. More specifically, DHA is present in high concentrations in neural and visual membranes and, therefore, insufficiency in the larval diet may result in serious consequences for a wide range of physiological and behavioral processes. These include impaired pigmentation and vision at low light intensities leading to low hunting capabilities of the developing larvae and impaired development of the neuro-endocrine system (Sargent et al., 1997, 1999; Estevez et al., 1999). Similarly, fish have a limited capacity to convert linoleic acid (18:2 n-6) to n-6 highly unsaturated acids (n-6 HUFA), including arachidonic acid (AA, 20:4 n-6) which has gained increased attention in recent years (Sargent et al., 1997, 1999) and is now considered to be also an essential fatty acid. AA is the main precursor fatty acid of eicosanoids that are converted to biologically active compounds including prostaglandins, thromboxanes, leukotrienes or can function in response to hormone stimulation. Accumulating evidence points to the importance of supplying 'an optimal blend' of EPA, DHA and AA in the diet of fish larvae. Moreover, phospholipids rather than triacylglycerols are the preferred vehicle for delivery of these polyunsaturated fatty acids (PUFA). This probably relates to the limited ability of fish larvae to synthesize phospholipids *de novo*. It has been suggested that the optimal ratio for DHA/EPA/AA is 1.8:1:0.12 for turbot (Sargent et al., 1999). It was observed that EPA incorporation into the larval phospholipids was negatively affected by dietary AA content, suggesting that excess AA may have a deleterious affect (Bessonart et al., 1999).

About 34–43% of the lipids in rotifers are phospholipids and 20–55% are triacylglycerols, with small amounts of monoacylglycerols, diacylglycerols, sterols, sterol esters and free fatty acids (Teshima et al., 1987; Frolov et al, 1991; Nagata & Whyte, 1992; Fernandez-Reiriz et al., 1993; Rainuzzo et al., 1997). The phospholipids and triacyglycerols display similar profiles of fatty acids, but these are greatly affected by the lipids provided in the food of rotifers. Rotifers that are cultured on yeast lack adequate amounts of DHA, EPA or AA and should be enriched with these fatty acids before they can serve as an appropriate diet for fish larvae. Enrichment methods include feeding rotifers with algae, lipid emulsions, microparticulate or microcapsules containing lipids or lipids with protein and carbohydrates (reviewed in Lubzens et al., 1989; Rainuzzo et al., 1997; Sargent et al., 1997). The quantitative and qualitative lipid content of rotifers can be manipulated by long term and short term enrichment periods on various diets containing lipid emulsion. Rotifer phospholipids were less influenced by these diets than the triacylglycerol fraction (Rainuzzo et al., 1997). Commercial fish oils that consist mainly of triacylglycerols are the main source of DHA or EPA, and their content depends on the fish species but they usually are poor in AA, except for tuna orbital oil that may contain about 2% of AA. These oils are usually incorporated into most rotifers' artificial diets. Lipids can also be provided to rotifers by algae; e.g. *Nannochloropsis* being rich in EPA and

*Isochrysis* in DHA and both species contain AA. A mutant of *Nannochloropsis* that was found deficient in EPA (Schneider et al., 1995) contains abundant AA and can be used as an alternative source for this fatty acid. The algae can be supplied directly from cultures, as a live or frozen concentrated paste or after freeze-drying (Watanabe et al. 1983; discussed in Lubzens et al., 1989 and in Fulks & Main, 1991; Frolov et al., 1991; Lubzens et al., 1995a; Takeyama et al., 1996; Navarro & Yufera, 1998a). More recently, bacteria isolated from Antarctica and containing high levels of EPA were suggested as a potential alternative enrichment food for rotifers (Nichols et al., 1996). The lipid content of rotifers is usually lower than that of their food organism, indicating that lipids are utilized by the rotifers. Rotifers utilize relatively more DHA in highly reproducing cultures (Øie & Olsen, 1997) and lipid utilization is temperature dependent (Olsen et al., 1993). They accumulate about 3–5 times more total lipids when they are kept at 10 °C than at of 25 °C (Lubzens et al., 1995b). These results suggest that higher enrichment levels will be obtained if this procedure is performed at relatively low temperatures (depending on the rotifer strain), where reproductive rates and utilization rates are slowed down.

The importance of enriching rotifers with vitamins has not been extensively studied. In addition to vitamin $B_{12}$ that was mentioned before, fat soluble vitamins (A, D and E)were also found to promote rotifer reproduction (Hirayama, 1990). The content of water soluble vitamins in rotifers increased after changing their diet from baker's yeast and lipid emulsion to *Isochrysis*. The most significant increase was in the content of ascorbic acid and thiamin, and rotifers fed algae contained sufficient amounts of these vitamins for meeting the nutritional requirements of fish. However, lipid soluble vitamins that were found in rotifers raised in fish oil, were depleted during the transition to *Isochrysis*. Nevertheless, their final content in rotifers exceeded the recommended levels needed for proper growth of fish larvae (Lie et al., 1997). Vitamin C (ascorbic acid) not only stimulates rotifer growth (Satuito & Hirayama, 1991) but also contributes significantly to the survival of fish larvae (Dabrowski & Ciereszko, 1993; Dabrowski & Blom, 1994). Its content in rotifers was found to depend on the diet, as it is highly abundant in several species of algae. Maximal levels of enrichment were achieved in rotifers incubated with ascorbyl palmitate (Merchie et al., 1997).

One of the main problems in providing rotifers to larvae is the deterioration in their nutritional quality due to starvation that results from extended periods of residence in the fish tanks. About 40–50% of the rotifer body mass is lost during 4–5 days of starvation at 18–20 °C and the rate of decrease is positively related to temperature (Makridis & Olsen, 1999). During starvation, preferential degradation of lipids, carbohydrates and amino acid takes place leading to an increase in the proportion of proteins in the dry weight. The chemical composition of rotifers (e.g. protein content) can be presented in either mg/ dry weight or ng/ rotifer and while large differences may occur in the content per rotifer (especially during starvation), these may not be reflected in the ratio of mg per dry weight. During the first 8 h of starvation, free amino acids are used as the main energy source by rotifers, with lipids and carbohydrates being utilized later. Lipids serve as the main source of energy and different lipid classes are mobilized at different rates during starvation. An increase in the proportion of diacylglycerols, monoacylglycerols, sterols and free fatty acids results from mobilization of triacylglycerols and this also leads to an increase in the proportion of polar lipids. The mobilization of sterols and wax esters follows the hydrolysis of triacylglycerols and carbohydrates and coincides with the ensued mortality of rotifers (Frolov & Pankov, 1992). The loss of lipids depends on temperature and can reach 0.19 of total lipids per day at 18 °C. The content of n-3 fatty acids is reduced faster than other lipids (Olsen et al., 1993). The protein content of each rotifer is reduced during starvation but the amino acid composition is rather stable (Frolov & Pankov, 1992). The effect of starvation can be partially overcome by supplying algae to the fish tanks (Makridis & Olsen, 1999). These results suggest that newly fed and enriched rotifers should be supplied daily to cultured larvae.

## Providing adequate numbers of rotifers

Meeting the demands of fish larvae is a continuous effort from the day of first feeding up to the time that larvae are fed other food sources (e.g. *Artemia*). While the usual practice is to depend on daily harvesting of rotifers from live cultures, various ways of storing rotifers have been explored.

Frozen rotifers are usually not adequate as food due to leaching of nutrients after thawing, their lack of buoyancy and motility. Also, they may cause a deterioration of water quality if introduced into culture tanks. However, live rotifers can be stored at 4 °C at relatively high densities, for at least 1 month (Lubzens et al., 1990). Storage at low temperature can help in reducing the daily tasks of harvesting and enriching rotifers by performing them every few days. It can also be helpful in storing surplus rotifers produced on one day for later use, provide a safeguard against unpredicted crashes and also facilitate transport of rotifers between sites of production and culture facilities. While storage at 4 °C is feasible in most places, rotifers continue to reproduce at this temperature and require periodic feeding and exchange of culture media. Rotifers can be kept at –1 °C without feeding or water exchange for about 2 weeks (Lubzens et al., 1995b) but storage at this temperature requires more specific equipment. Storage at low temperature is only adequate for *B. plicatilis* and extending this method to *B. rotundiformis* is currently being examined (Assave-Aree et al., 2000).

Cryopreservation of rotifers is useful for long term preservation of genetically important strains. Amictic eggs (but not adults) are preserved in liquid nitrogen after they have been impregnated with cryoprotective agents such as DMSO or propanediol (Toledo & Kurokura, 1990; Toledo et al., 1991; Hadani et al., 1992). This method ensures full preservation of genetic traits of importance in aquaculture and is especially important for those strains that do not produce resting eggs. A small collection of cryopreserved *B. plicatilis* and *B. rotundiformis* strains is kept at the authors' laboratory and serves as a alternative source for live cultures. Since this is a relatively expensive method, it is not suitable for preservation of large numbers of rotifers for direct use as food after thawing.

Artificially produced rotifer resting eggs has been offered as an alternative route for supplying rotifers without depending on the daily production in marine hatcheries. Rotifer resting eggs are diploid, sexual eggs produced by mictic females after fertilization. Production is genetically determined (Hino & Hirano, 1976, 1977), with large variations between rotifers originating from eggs produced by one clone (Lubzens, 1989). The production of these eggs can be manipulated by environmental factors such as salinity, food quantity and quality, rotifer culture density, exchange of culture media, and temperature, and varies between *B. plicatilis* and *B. rotundiformis* (Hino & Hirano, 1976, 1977, 1984, 1985, 1988; Snell & Hoff, 1985; Lubzens et al., 1985; Snell, 1986; Lubzens, 1987; Lubzens & Minkoff, 1988; Hagiwara et al., 1988a, b; Haigiwara et al., 1989; Hagiwara & Lee, 1991; Hagiwara & Hirayama, 1993; Hamada et al., 1993; Kogane et al., 1997). Mass culture techniques have been developed (Hagiwara, et al., 1993a, 1997; Hagiwara & Hirayama, 1993; Hagiwara, 1994; Balompapueng et al., 1997a) and the resting eggs can be stored for extensive periods in large quantities and hatched at will under defined conditions (Minkoff et al., 1983; Pourriot & Snell, 1983; Hagiwara et al., 1985; Hagiwara & Hino, 1989, 1990; Hagiwara et al., 1995c; Hagiwara, 1996; Balompapueng et al., 1997a). These eggs can be used directly for feeding fish larvae (Hagiwara & Hirayama, 1993; Hagiwara et al., 1993b) or for initiating mass cultures. However, sexual or asexual reproduction of rotifers hatched from resting eggs differ from their parents (Lubzens, 1989), even though they may originate from a cloned culture. It will also require additional efforts to isolate a clone that does not produce resting eggs for more efficient mass cultures (Lubzens, 1989).

Based on the studies cited above, the cost of resting eggs exceeds by several fold the cost of rotifers produced in mass cultures due to the following reasons: 1. Only 1–8 resting eggs are formed by a mictic female vs. 16–24 eggs produced by an amictic female. 2. Resting egg production depends on presence of males and their ratio in the culture may exceed 50%. This adds to the cost of resting egg production. 3. Rotifers must be fed algae in order to produce resting eggs, thus increasing the cost of production. 4. Resting egg production depends on optimal, relatively low densities of rotifers, limiting the number of eggs produced per volume. 5. Eggs must be collected and cleaned from debris and stored. 6. Hatching rate of eggs seldom reaches more than 50%. If all these factors are taken into account, the estimated cost of one resting egg will be at least 20–30 fold that of an asexually produced rotifer (Lubzens, 1989). Resting eggs can be used to initiate mass cultures, thus saving the time needed for upscaling from small to large scale cultures (Snell & Hoff, 1988). However, asexual reproductive rates in rotifers hatched from resting eggs may show large fluctuations from those of the parent culture (Lubzens, 1989).

## Future directions

For almost four decades, biotechnologies have been developed to provide rotifers as food for raising marine fish larvae. Today, almost all fish species raised in captivity are fed rotifers. A continuous stable and reliable supply of adequate numbers of rotifers is essential for culturing a large variety of small mouthed marine fish larvae. There is an increased awareness that a variety of genetic strains within each of the cultured species (*B. plicatilis* and *B. rotundiformis*) are of importance in meeting the nutritional requirements of fish larvae. On one hand, small size rotifers (so called 'super small') are increasingly needed for culturing marine fish larvae with extremely small mouth size (e.g. some grouper species). On the other hand, large size rotifers may reduce the need for *Artemia* cysts, which are in short supply and are becoming extremely expensive. This means that several rotifer size varieties will be cultured in fish hatcheries and consequently used in research. Therefore, there is a need to identify as much as possible the strain or species used in each culture or study by size measurements (Hagiwara et al., 1995b) or dry weight. Genetic markers could be useful in further characterizations of clones, populations, strains or species (Gomez et al., 1998; Boehm et al., 2000; Gomez & Carvalho, 2000).

There is an increasing and pressing need for small size rotifers for new cultured marine species. Part of this need maybe fulfilled by small size genetic strains. Developing culture methods for synchronous production of a large number of gravid females that carry multiple eggs could serve as an alternative direction. The eggs that will hatch in the fish tanks will provide small size rotifers for at least 24 h.

The need for supplying live rotifers will continue until they are replaced by formulated foods. Major efforts are made to obtain these formulated foods (Tandler, 1984/1985; Kolkovski & Tandler, 1995; Sargent et al., 1997, 1999) for currently cultured fish species. However, even if they are found appropriate for some fish species, it will not assure their suitability for others. This means that for at least the near future, marine fish hatcheries will continue to rely on rotifers as the first food provided to fish larvae. The success in providing adequate numbers of rotifers relies on the extensive studies that were performed on the basic biology and physiology of *B. plicatilis* and *B. rotundiformis*. In several cases, the primary aim of these studies was directed towards elucidating ecological and evolutionary questions, including speciation, and successful long term survival of the species in an environment consisting of temporary water bodies with unpredictable food supply and variable temperatures. Continuous success in mass culturing of rotifers will rely on extending these studies to questions raised by aquaculturists but could also be relevant to the survival and successful propagation of rotifers in their natural environment.

Rotifers in mass cultures are exposed to high densities, high food input and high concentrations of excretory products. These conditions are far from those experienced by rotifers in their natural environment, suggesting that they are exposed to highly stressful conditions and, therefore, these mass cultures are inherently unstable. Studies into the relationship between stress and stability/instability of rotifer cultures could contribute at predicting the fate of these cultures. The methods available today for identifying newly synthesized stress proteins (Cochrane et al., 1991; Lubzens et al., 1995b) are cumbersome and long and are difficult to use for quick identification of stressful conditions evoked in rotifers. There is, therefore, a place for more accessible and quick methods. Furthermore, it would be useful to investigate whether it is possible to increase the tolerance of rotifers to stress and whether or not this can be regulated. Moreover, it is not possible to determine today how reversible is the effect of rotifers exposed to adverse conditions (e.g. high or low pH, high or low levels of ammonia, high levels of food, high rotifer densities), whether it depends on the duration of exposure and how rotifers can be treated to ensure better recovery.

The high cost of rotifer resting egg production prevents their routine use as a dependable source for rotifers in large scale cultures. Identification of hormones (Gallardo et al., 1997, 1999) and other factors (Hagiwara et al., 1994) that enhance resting egg formation could directly contribute to achieve this goal.

## Acknowledgements

We would like to thank the Samuel Lunenfeld Foundation for their continued support in maintaining the cultures of rotifer strains and species.

## References

Aoki, S. & A. Hino, 1996. Nitrogen flow in a chemostat culture of the rotifer *Brachionus plicatilis*. Fish. Sci. 62: 8–14.

Balompapueng, M. D., A. Hagiwara, A. Nishi, K. Imaizumi & K. Hirayama, 1997a. Resting egg formation of the rotifer *Brachionus plicatilis* using a semi-continuous culture method. Fish. Sci. 63: 236–241.

Balompapueng, M. D., A. Hagiwara, Y. Nozaki & K. Hirayama, 1997b. Preservation of resting eggs of the euryhaline rotifer *Brachionus plicatilis* O. F. Muller by canning. Hydrobiologia 358: 163–166.

Baragi, V. & R. T. Lowell, 1986. Digestive enzyme activity in striped bass from first feeding through larval development. Trans. am. Fish. Soc. 115: 478–484.

Bessonart, M., M. S. Izquierdo, M. Salhi, C. M. Hernandez-Cruz, M. M. Gonzalez & H. Fernandez-Palacios, 1999. Effect of dietary arachidonic acid levels on growth and survival of gilthead seabream (*Sparus aurata* L. ) larvae. Aquaculture 179: 265–275.

Blanch, A. R., M. Simo, J. T. Jofre & G. Minkoff, 1991. Bacteria associated with hatchery cultivated turbot: are they implicated in rearing success? In Lavens P, P. Sorgeloos, E. Jaspers & F. Ollevier (eds), Larvi '91 – Fish and Crustacean Larviculture Symposium, Eur. Aquacult. Soc. Spec. Publ. 15, Gent (Belgium): 392–394.

Blaxter, J. H. S., 1974. The Early life History of Fish. Springer-Verlag Berlin Heidleberg New York: 765pp.

Boehm, E. W., O. Gibson & E. Lubzens, 2000. Caharacterization of sattelite DNA sequences from commercially important marine rotifers *Brachionus plicatilis* and *Brachionus rotundiformis*. Mar. Biotechnol. 2: 38–48.

Bower, C. E. & J. P. Bidwell, 1978. Ionization of ammonia in seawater: effect of temperature, pH and salinity. J. Fish. Res. B. Can. 35: 1012–1016.

Cahu, C. L. & J. L. Zambonino Infante, 1994. Early weaning of sea bass (*Dicentrarchus labrax)* larvae with a compound diet: effect on digestive enzymes. Comp. Biochem. Physiol. 109A: 213–222.

Cochrane, B. J., R. B. Irby & T. W. Snell, 1991. Effects of copper and tributylin on stress protein abundance in the rotifer *Brachionus plicatilis.* Comp. Biochem. Physiol. 98C: 383–390.

Colorni, A., O. Zmora & E. S. Kutin, 1991. Systematic infection in the rotifer *Brachionus plicatilis* by an invasive yeast. Bull. Eur. Ass. Fish Path. 11: 116–117.

Comps, M. & B. Menu, 1997. Infectious diseases affecting mass production of the marine rotifer *Brachionus plicatilis.* Hydrobiologia 358: 179–183.

Comps, M., B. Menu & V. Moreau, 1993. Massive infections with fungus of the rotifer *Brachionus plicatilis.* Bull. eur. Ass. Fish Pathol. 13: 28–29.

Comps, M., J. Mari, F. Poisson & J. R. Bonami, 1991a. Biophysical and biochemical properties of an unusual birnavirus pathogenic for rotifers. J. Gen. Virol. 72: 1229–1236.

Comps, M., B. Menu, G. Breuil & J. R. Bonami, 1991b. Viral infection associated with rotifer mortalities in mass culture. Aquaculture 93: 1–7.

Dabrowski, K., 1979. The role of proteolytic enzymes in fish digestion. In Styczynska-Jurewicz, E., T. Backiel, E. Jaspers & J. Persoone (eds), Cultivation of Fish Fry and its Live Food. Eur. Maricult. Soc., Spec. Publ. 4, Bredene (Belgium): 107–126.

Dabrowski, K., 1984. The feeding of fish larvae: present 'state of the art' and perspectives. Reprod. Nutr. Develop. 24: 807–833.

Dabrowski, K. & J. H. Blom, 1994. Ascorbic acid deposition in rainbow trout (*Oncorhynchus* mykiss) eggs and survival of embryos. Comp. Biochem. Physiol. 108A: 129–135.

Dabrowski, K. & A. Ciereszko, 1993. Influence of fish size, origin, and stress on ascorbate concentration in vital tissues of hatchery rainbow trout. Prog. Fish Cult. 55: 109–135.

Diaz, M., F. J. Moyano, F. L. Garcia-Carreno, F. J. Alarcon & M. C. Sarasquete, 1997. Substrate-SDS–PAGE determination of protease activity through larval development in sea bream. Aquaculture Int. 5: 461–471.

Estevez, A., L. A. McEvoy, J. G. Bell & J. R. Sargent, 1999. Growth, survival, lipid composition and pigmentation of turbot *(Scophthalmus maximus)* larvae fed live-prey enriched in Arachidonic and Eicosapentaenoic acids. Aquaculture 180: 321–343.

F.A.O., 1998. The state of world fisheries and aquaculture. Food and Agricultural Organization of the United Nations, Rome: www. FAO.org.

Federation of European Aquaculture Producers, 2000. Mediterranean marine species juveniles: www. feap.org.

Fernandez-Reiriz, U. Labarta & M. J. Ferreiro, 1993. Effects of commercial enrichment diets on the nutritional value of the rotifer (*Brachionus plicatilis*). Aquaculture 112: 195–206.

Fjelheim, A. J., P. Markidis, J. Skjermo & O. Vadstein, 1999. Rotifers (*Brachionus plicatilis*) as vector for probiotic to turbot larvae *(Scophthalmus maximus).* In Towards predictable quality, Aquaculture Europe, EAS Special publication No. 27: 60–61.

Frolov, A. V. & S. L. Pankov, 1992. The effect of starvation on the biochemical composition of the rotifer *Brachionus plicatilis.* J. mar. biol. Ass. U. K. 72: 343–356.

Frolov, A. V., S. L. Pankov, K. N. Geradz, S. A. Pankova & L. V. Spektrova, 1991. Influence of the biochemical composition of food on the biochemical composition of the rotifer *Brachionus plicatilis*. Aquaculture 97: 181–202.

Fu, Y., K. Hirayama & Y. Natsukari, 1990. Strains of the rotifer *Brachionus plicatilis* having particular patterns of isozymes. In Hirano R. & I. Hanyu (eds), The Second Asian Fisheries Forum. Asian Fisheries Society, Manila, Philippines: 37–40.

Fu, Y., K. Hirayama & Y. Natsukari, 1991a. Morphological differences between the two types of the rotifer *Brachionus plicatilis* O. F. Muller. J. exp. mar. Biol. Ecol. 151: 29–41.

Fu, Y., K. Hirayama & Y. Natsukari, 1991b. Genetic divergence between S and L type stains of the rotifer *Brachionus plicatilis* O. F. Muller. J. exp. mar. Biol. Ecol. 151: 43–56.

Fu, Y., A. Hada, T. Yamashita, Y. Yoshida & A. Hino, 1997. Development of a continuous culture system for stable mass production of the marine rotifer *Brachionus*. Hydrobiologia 358: 145–151.

Fujita, S., 1979. Culture of red sea bream *Pagrus major,* and its food. In Styczynska-Jurewicz, E., T. Backiel, E. Jaspers & J. Persoone (eds), Cultivation of Fish Fry and its Live Food. Eur. Maricult. Soc., Spec. Publ. 4, Bredene (Belgium): 183–197.

Fukusho, K., 1989a. Biology and mass production of the rotifer *Brachionus plicatilis* (1). Int. J. Aquacult. Fish. Technol. 1: 232–240.

Fukusho, K., 1989b. Biology and mass production of the rotifer *Brachionus plicatilis* (2). Int. J. Aquacult. Fish. Technol. 1: 292–299.

Fulks, F. & K. L. Main, 1991. Rotifer and microalgae culture systems. Rotifer and Microalgae Culture Systems, Proc. U.S.– Asia Workshop. The Oceanic Insitute, Honolulu, HI: 364 pp.

Furukawa, I. & K. Hidaka, 1973. Technical problems encountered in mass culture of rotifer using marine yeast as food organisms. Bull. Plank. Soc. Jpn. 20: 61–71.

Gallardo, W. G. , A. Hagiwara, Y. Tomita & T. W. Snell, 1999. Effect of growth hormone and $\gamma$-aminobutyric acid on *Brachionus plicatilis* (Rotifera) reproduction at low food or high ammonia levels. J. exp. mar. Biol. Ecol. 240: 179–191.

Gallardo, W. G., A. Hagiwara, Y. Tomita, K. Soyano & T. W. Snell, 1997. Effect of some vertebrate and invertebrate hormones on the population growth, mictic female production and body size of the marine rotifer *Brachionus plicatilis* Muller. Hydrobiologia 358: 113–120.

Gatesoupe, F. J., 1990. The continuous feeding of turbot larvae, *Scophthalmus maximus,* and the control of the bacterial environment of rotifers. Aquaculture, 89: 139–148.

Gatesoupe, F. J., 1991. The effect of three strains of lactic bacteria on the production rate of rotifers, *Brachionus plicatilis,* and their dietary value for larval turbot, *Scophthalmus maximus*. Aquaculture 96: 335–342.

Gatesoupe, F. J., 1993. *Bacillus* sp. spores as food additive for the rotifer *Brachionus plicatilis*: improvement of their bacterial environment and their dietary value for larval turbot *Scophthalmus maximus* L. In Kaushik, S. J., P. Luquet (eds), Fish Nutrition in Practice. Institut National de la Recherche Agronomique, Paris, France, Les Colloques, Vol. 61: 561–568.

Gatesoupe, F. J., 1999. The use of probiotics in aquaculture. Aquaculture 180: 147–165.

Gomez, A., C. Clabby & G. R. Carvalho, 1998. Isolation and characterization of microsatellite loci in a cyclically parthenogenetic rotifer, *Brachionus plicatilis*. Mol. Ecol. 7: 1613–1621.

Gomez, A. & G. R. Carvalho, 2000. Sex, parthogenesis and genetic structure of rotifers: microsatellite analysis of contemporary and resting egg bank populations. Mol. Ecol. 9: 203–214.

Govoni, J. J., G. W. Boehlert & Y. Watanabe, 1986. The physiology of digestion in fish larvae. Envir. Bio. Fishes 16: 59–77.

Grisez, L., J. Reyniers, L. Verdonck, J. Swings & F. Ollevier, 1997. Dominant intestinal microflora of sea bream and sea bass larvae, from two hatcheries, during larval development. Aquaculture 155: 387–399.

Hadani, A, S. Beddig & E. Lubzens, 1992. Factors affecting survival of cryopreserved rotifers (*Brachionus plicatilis* O. F. Müller). In Moav, B., V. Hilge & H. Rosenthal (eds), Progress in Aquaculture Research. Eur. Aquacult. Soc. Spec. Publ. 17, Oostende (Belgium): 253–267.

Hagiwara, A., 1994. Practical use of rotifer cysts. Israel J. Aquaculture-Bamidgeh 46: 13–21.

Hagiwara, A., 1996. Appearance of floating resting eggs in the rotifers *Brachionus plicatilis* and *B. rotundiformis*. Bull. Fac. Fish. Nagasaki University 77: 111–115.

Hagiwara, A. & A. Hino, 1989. Effect of incubation and preservation on resting egg hatching and mixis in the derived clones of the rotifer *Brachionus plicatilis*. Hydrobiologia 186/187: 415–421.

Hagiwara, A. & A. Hino, 1990. Feeding history and hatching of resting eggs in the marine rotifer *Brachionus plicatilis*. Nippon Suisan Gakkaishi 56: 1965–1971.

Hagiwara, A. & K. Hirayama, 1993. Preservation of rotifers and its application in the finfish hatchery. In Lee, C.-S., M. S. Su & I. Liao (eds), Proc. Finfish Hatchery in Asia '91. TLM Conference Proceedings, Tungkang Marine Laboratory, Taiwan Fisheries research Institute, Tungkang, Taiwan 3: 61–71.

Hagiwara, A. & C.-S. Lee, 1991. Resting egg formation of the L- and S-type rotifer *Brachionus plicatilis* under different water temperature. Nippon Suisan Gakkaishi 57: 1645–1650.

Hagiwara, A., A. Hino & R. Hirano, 1985. Combined effects of environmental conditions on the hatching of fertilized eggs of the rotifer *Brachionus plicatilis* collected from an outdoor pond. Bull. Jap. Soc. Sci. Fich. 51: 755–758.

Hagiwara, A., A. Hino & R. Hirano, 1988a. Effects of temperature and chlorinity on resting egg formation in the rotifer *Brachionus plicatilis*. Nippon Suisan Gakkaishi. 54: 569–575.

Hagiwara, A., A. Hino & R. Hirano, 1988b. Comparison of resting egg formation among five Japanese stocks of the rotifer *Brachionus plicatilis*. Nippo Suisan Gakkaishi 54: 577–580.

Hagiwara, A., N. Yamamiya & A. Belem De Araujo, 1998. Effect of water viscosity on the population growth of the rotifer *Brachionus plicatilis* Muller. Hydrobiologia 387/388: 489–494.

Hagiwara, A., M. D. Balompapueng, N. Munuswamy & K. Hirayama, 1997. Mass production and preservation of the resting eggs of the euryhaline rotifer *Brachionus plicatilis* and *B. rotundiformis*. Aquaculture 155: 223–230.

Hagiwara, A., K. Hamada, S. Hori & K. Hirayama, 1994. Increased sexual reproduction in *Brachionus plicatilis* with the addition of bacteria and rotifer extracts. J. exp. mar. Biol. Ecol. 181: 1–8.

Hagiwara, A., M.-M. Jung, T. Sato & K. Hirayama, 1995a. Interspecific relations between marine rotifer *Brachionus rotundiformis* and zooplankton species contaminating in the rotifer mass culture tank. Fish. Sci. 61: 623–627.

Hagiwara, A., C.-S. Lee, G. Miyamoto & H. Hino, 1989 Resting egg formation and hatching of the S-type rotifer *Brachionus plicatilis* at varying salinities. Mar. Biol. 103: 327–332.

Hagiwara, A., K. Hamada, A. Nishi, K. Imaizumi & K. Hirayama, 1993a. Mass production of rotifer *Brachionus plicatilis* resting eggs in 50 $m^3$ tanks. Nippon Suisan Gakkaishi 59: 93–98.

Hagiwara, A. K. Hamada, A. Nishi, K. Imaizumi & K. Hirayama, 1993b. Dietary value of neonates from rotifer *Brachionus plicatilis* resting eggs for red sea bream larvae. Nippon Suisan Gakkaishi 59: 99–104.

Hagiwara A., T. Kotani, T. W. Snell, M. Assava-Aree & K. Hirayama, 1995b. Morphology, reproduction, genetics and mating behavior of small, tropical marine *Brachionus* strains (Rotifera). J. exp. mar. Biol. Ecol. 194: 25–37.

Hagiwara, A., N. Nishi, F. Kawahara, K. Tominaga & K. Hirayama, 1995c. Resting eggs of the marine rotifer *Brachionus plicatilis* Muller: development and effect of irradiation on hatching. Hydrobiologia 313/314: 223–229.

Hamada, K., A. Hagiwara & K. Hirayama, 1993. Use of preserved diets for rotifer *Brachionus plicatilis* resting egg formation. Nippon Suisan Gakkaishi 59: 85–91.

Hansen, B., T. Wernberg-Moller & L. Wittrup, 1997. Particle grazing efficiency and specific growth of the rotifer *Brachionus plicatilis* (Muller). J. exp. mar. Biol. Ecol. 215: 217–233.

Hino, A. & R. Hirano, 1976. Ecological studies on the mechanism of bisexual reproduction in the rotifer *Brachionus plicatilis*- I. General aspects of bisexual reproduction inducing factors. Bull. Jap. Soc. Sci. Fish. 42: 1093–1099.

Hino, A. & R. Hirano, 1977. Ecological studies on the mechanism of bisexual reproduction in the rotifer *Brachionus plicatilis*- II. Effect of cumulative parthogenetic generation on the frequency of bisexual reproduction. Bull. Jap. Soc. Sci. Fish. 43: 1147–1155.

Hino, A. & R. Hirano, 1984. Relationship between water temperature and bisexual reproduction in the rotifer *Bachionus plicatilis*. Nippon Suisan Gakkaishi 50: 1481–1485.

Hino, A. & R. Hirano, 1985. Relationship between the temperature given at the time of fertilized egg formation and bisexual reproduction pattern in the deriving strain of the rotifer *Brachionus plicatilis*. Nippon Suisan Gakkaishi 51: 511–514.

Hino, A. & R. Hirano, 1988. Relationship between water chlorinity and bisexual reproduction rate in the rotifer *Brachionus plicatilis*. Nippon Suisan Gakkaishi 54: 1139–1332.

Hino, A., S. Aoki & M. Ushiro, 1997. Nitroggen-flow in the rotifer *Brachionus rotundiformis* and its significance in mass cultures. Hydrobiologia 358: 77–82.

Hirata, H., 1964. Cultivation of live food organisms at the Yashima Station. Saibai-Gyogyo: 2–4, 4 (in Japanese).

Hirata, H., 1979. Rotifer culture in Japan. In Styczynska-Jurewicz, E., T. Backiel, E. Jaspers & J. Persoone (eds), Cultivation of Fish Fry and its Live Food. Eur. Maricult. Soc., Spec.l Publ. 4, Bredene (Belgium): 361–375.

Hirata, H., 1980. Culture methods of the marine rotifer *Brachionus plicatilis*. Min. Rev. Data File Res. 1: 27–46.

Hirata, H. & Y. Mori, 1967. Mass culture of the marine rotifer fed baker's yeast. Saibai Gyogyo 5: 36–40.

Hirayama, K., 1987. A consideration why mass culture of the rotifer *Brachionus plicatilis* with baker's yeast is unstable. Hydrobiologia 147: 269–270.

Hirayama, K., 1990. A physiological approach to problems of mass culture of the rotifer. NOAA Technical report No. NMFS 85. U.S. Dept. Commerce, U.S.A.: 73–79.

Hirayama, K. & H. Funamoto, 1983. Supplementary effect of several nutrients on nutritive deficiency of baker's yeast of population growth of the rotifer *Brachionus plicatilis*. Bull. Jpn. Soc. Sci. Fish. 49: 505–510.

Hirayama, K. & K. Nakamura, 1976. Fundamental studies on the physiology of rotifers in mass culture- V. Dry *Chlorella* powder as a food for rotifers. Aquaculture 8: 301–307.

Hirayama K. & I. F. M. Rumengan, 1993. The fecundity patterns of S and L type rotifers of *Brachionus plicatilis*. Hydrobiologia 255/256: 153–157.

Ito, T., 1960. On the culture of the mixohaline rotifer *Brachionus plicatilis* O. F. Muller, in sea water. Rep. Fac. Fish. Perfect. Univ. Mie 3: 708–740.

James, C. M., & T. Abu-Rezq, 1989a. Intensive rotifer cultures using chemostats. Hydrobiologia 186/187: 423–430.

James, C. M., & T. Abu-Rezq, 1989b. An intensive chemostats culture system for the production of rotifers for aquaculture. Aquaculture 81: 291–301.

James, C. M., & T. Abu-Rezq, 1990. Efficiency of rotifer chemostats in relation to salinity regimes for producing rotifers for aquaculture. J. Aqua. Trop. 5: 103–116.

James, C. M., P. A. Dias & A. E. Salman, 1987. The use of marine yeast (*Candida* sp.) and bakers yeast (*Saccharomyces cerevisiae*) in combination with *Chlorella* sp. for mass culture of the rotifer *Brachionus plicatilis*. Hydrobiologia 147: 263–268.

James, C. M., M. Bou-Abbas, A. M. Al-Khars, S. Al-Hinty & A. E. Salman, 1983. Production of the rotifer *Brachionus plicatilis* for aquaculture in Kuwait. Hydrobiologia 104: 77–84.

Jung, M.-M., A. Hagiwara & K. Hirayama, 1997. Interspecific interactions in the marine rotifer microcosm. 358: 121–126.

Kinne, O., 1977. Cultivation of animals. In Kinne, O., (ed.), Marine Ecology. John Wiley & Sons, Chichester, New York, Brisbane, Toronto, Vol. III. Part 2: 968–1004.

Kogane, T., A. Hagiwara & K. Imaizumi, 1997. Temperature conditions enhancing resting egg production of the euryhaline rotifer *Brachionus plicatilis* O. F. Muller (Kamiura strain). Hydrobiologia 358: 167–171.

Kolkovski, S. & A. Tandler, 1995. Why microdiets are still inadequate as aviable alternative to live zooplankters for developing marine fish larvae. Spec. Publ. Eur. Aquacult. Soc. 24: 265–266.

Kolkovski, S., A. Tandler, G. Wm. Kissil & A. Gertler. 1993. The effect of dietary exogenous enzymes on digestion, assimilation, growth and survival of gilthead seabream (*Sparus aurata,* Sparidae, Linnaeus) larvae. Fish Physiol. Biochem. 12: 203–209.

Korstad, J., Y. Olsen & O. Vadstein, 1989a. Life history characteristics of *Brachionus plicatilis* (Rotifera) fed different algae. Hydrobiologia 186/187: 43–50.

Korstad, J., Y. Olsen & O. Vadstein, 1989b. Feeding kinetics of *Brachionus plicatilis* fed *Isochrysis galbana*. Hydrobiologia 186/187: 51–57.

Korstad, J., A. Neyt, T. Danielsen, I. Overrein & Y. Olsen, 1995. Use of swimming speed and egg ratio as predictors of the status of rotifer cultures in aquaculture. Hydrobiologia 313/314: 395–398.

Lauf, M. & R. Hofer, 1984. Proteolytic enzymes in fish development and the importance of dietary enzymes. Aquaculture 37: 335–346.

Lie, O., H. Haaland, G.-I. Hemre, A. Maage, E. Lied, G. Rosenlund, K. Sandnes & Y. Olsen, 1997. Nutritional composition of rotifers following a change in diet from yeast emulsified oil to microalgae. Aquaculture Int. 5: 427–438.

Lubzens, E., 1987. Raising rotifers for use in aquaculture. Hydrobiologia 147: 245–255.

Lubzens, E., 1989. Possible use of rotifer resting eggs and preserved live rotifers (*Brachionus plicatilis)* in aquaculture and mariculture. In De Paw, N., E. Jaspers & H. Ackeford (eds), Aquaculture – A Biotechnology in Progress. Eur. Aquacult. Soc, Bredene, Belgium: 741–750.

Lubzens, E. & G. Minkoff, 1988. Influence of the age of algae fed to rotifers (*Brachionus plicatilis* O. F. Muller) on the expression of mixis in their progenies. Oecologia 76: 430–435.

Lubzens, E., G. Minkoff & S. Marom, 1985. Salinity dependence of sexual and asexual reproduction in the rotifer *Brachionus plicatils*. Mar. Biol. 85: 123–126.

Lubzens, E., A. Tandler & G. Minkoff, 1989. Rotifers as food in aquaculture. Hydrobiologia 186/187: 387–400.

Lubzens, E., O. Gibson, O. Zmora & A. Sukenik, 1995a. Potential advantages of frozen algae (*Nannochloropsis* sp.) for rotifer (*Brachionus plicatilis*) culture. Aquaculture 133: 295–309.

Lubzens, E., G. Minkoff, Y. Barr & O. Zmora, 1997. Mariculture in Israel-past achievements and future directions in raising rotifers as food for marine fish larvae. Hydrobiologia 358: 13–20.

Lubzens, E., G. Kolodny, B. Perry, N. Galai, R. Sheshinski & Y. Wax, 1990. Factors affecting survival of rotifers (*Brachionus plicatilis* O. F. Müller) at 4 ° C. Aquaculture 91: 23–47.

Lubzens, E., D. Rankevich, G. Kolodny, O. Gibson, A. Cohen & M. Khayat, 1995b. Physiological adaptations in the survival of rotifers (*Brachionus plicatilis* O. F. Muller) at low temperatures. Hydrobiologia 313/314: 175–183.

Maeda, M. & A. Hino, 1991. Environmental management for mass culture of rotifer, *Brachionus plicatilis*. In Fulks, W. & K. L. Main (eds), Rotifer and Microalgae Culture Systems. Proc. U.S.-Asia Workshop. The Oceanic Institute, Honolulu, HI: 125–133.

Maeda, M., K. Nogami, M. Kanematsu & K. Hirayama, 1997. The concept of biological control method in aquaculture. Hydrobiologia 358: 285–290.

Markridis, P. & Y. Olsen, 1999. Protein depletion of the rotifer *Brachionus plicatilis* during starvation. Aquaculture 174: 343–353.

Markridis, O., A. J. Fjelheim, J. Skjermo & O. Vadstein, 2000. Control of bacterial flora of *Brachionus plicatilis* and *Artemia franciscana* by incubation in bacterial suspensions. Aquaculture 185-207–218.

Markridis, P., O. Bergh, A. J. Fiellheim, J. Skjermo & O. Vadstein, 1999. Microbial control of live food cultures. In Laird, L. & H. Reinertsen (eds), Towards Predictable Quality. Aquaculture Europe 99. Eur. Aquacult. Soc. Spec. Publ. No. 27, Oostende (Belgium): 155–157.

Merchie, G., Pl. Lavens & P. Sorgeloos, 1997. Optimization of dietary vitamin C in fish and crustacean larvae: a review. Aquaculture 155: 165–181.

Minkoff, G., E. Lubzens & D. Kahan, 1983. Environmental facots affecting hatching of rotifer (*Brachionus plicatilis*) resting eggs. Hydrobiologia 104: 61–69.

Miracle, M. R. & M. Serra, 1989. Salinity and temperature influence on rotifer life history characteristics. Hydrobiologia 186/187: 81–103.

Munilla-Moran, R., J. R. Stark & A. Barbour, 1990. The role of exogenous enzymes in digestion in culture of turbot larvae (*Scophthalmus maximus* L.). Aquaculture 88: 337–350.

Munro, P. D., R. J. Henderson, A. Barbour & T. H. Birkbeck. 1999. Partial decontamination of rotifers with ultraviolet radiation: the effect of changes in the bacterial load and flora of rotifers on mortalities in start-feeding larval turbot. Aquaculture 170: 229–244.

Nagata, W. D. & H. Hirata, 1986. Mariculture in Japan: Past, present and future prespectives. Min. Rev. Data File Fish. Res. 4: 1–38.

Nagata, W. D. & J. N. C. Whyte, 1992. Effect of yeast and alagal diets on the growth and biochemical composition of the rotifer *Brachionus plicatilis* (Muller) in culture. Aquacult. Fish. Manage. 1992. 23: 13–21.

Navarro, N. & M. Yufera, 1998a. Influence of the food ration and individual density on production efficiency of semicontinuous cultures of *Brachionus*-fed mucroalgae dry powder. Hydrobiologia 387/388: 483–487.

Navarro, N. & M. Yufera, 1998b. Population dynamics of rotifers (*Brachionus plicatilis* and *Brachionus rotundiformis*) in semicontinuous culture fed freeze-dried microalgae: influence of dilution rate. Aquaculture 166: 297–309.

Nichols, D. S., P. Hart, P. D. Nichols & T. A. McMeekin, 1996. Enrichment of the rotifer *Brachionus plicatilis* fed an Antarctic bacterium containing polyunsaturated fatty acids. Aquaculture 147: 115–125.

Øie, G. & Y. Olsen, 1997. Protein and lipid content of the rotifer *Brachionus plicatilis* during variable growth and feeding conditions. Hydrobiologia 358: 251–258.

Øie, G., K. I. Reitan & Y. Olsen, 1994. Comparison of roifer culture quality with yeast plus oil and algal-based cultivation diets. Aquacult. Intern. 2: 225–238.

Olsen, Y., K. I. Reitan & O. Vadstein, 1993. Dependence of temperature on loss rates of rotifers, lipids and $\omega 3$ fatty acids in starved *Brachionus plicatilis* cultures. Hydrobiologia 255/256: 13–20.

Owen, J. M., J. W. Adron, C. Middleton & C. B. Cowey, 1975. Elongation and desaturation of dietary fatty acidsin turbot (*Scophthalmus maximus* L.) and rainbow trout (*Salmo gaidneri* Rich). Lipids 10: 528–531.

Polo, A., M. Yufera & E. Pascual, 1992. Feeding and growth of gilthead seabream ( *Sparus aurata* L.) larvae in relation to size of the rotifer strain used as food. Aquaculture 103: 45–54.

Pourriot, R. and T. W. Snell, 1983. Resting eggs in rotifers. Hydrobiologia 104: 213-224. Rainuzzo, J. R., K. I. Reitan & Y. Olsen, 1994. Effect of short -and long term enrichment on total lipids, lipid class and fatty acid composition. Aquacult. Int. 2: 19–32.

Rainuzzo, J. R., K. I. Reitan & Y. Olsen, 1997. The significance of lipids at early stages of marine fish: a review. Aquaculture 155: 103–115.

Reitan K. I., J. R. Rainuzzo, G. Øie & Y. Olsen, 1993. Nutritional effects of algal addition in first-feeding of turbot (*Scophthalmus maximus* L.) larvae. Aquaculture 118: 257–275.

Rombaut, G., Ph. Dhert, J. Vandenberghe, L. Verschuere, P. Sorgeloos & W. Verstraete, 1999a. Selection of bacteria enhancing the growth rate of axenically hatched rotifers (*Brachionus plicatilis*). Aquaculture 176: 195–207.

Rombaut, G., L. Vershuere, Ph. Dhert, P. Sorgeloos & W. Verstraete, 1999b. Multi-component probiotic for live feed (*Brachionus plicatilis*) cultures. In Laird, L. & H. Reinertsen (eds), Towards Predictable Quality. Aquaculture Oostende (Belgium): 201–202.

Rumengan, I. F. M. & K. Hirayama, 1990. Growth responses of genetically distict S and L type rotifer (*Brachionus plicatilis*) strains to different temperatures. In Hirano R. & I. Hanyu (eds), The Second Asian Fisheries Forum. Asian Fisheries Society, Manila, Philippines: 33–35.

Sargent, J. R., L. A. McEvoy & J. G. Bell, 1997. Requirements, presentation and sources of unsaturated fatty acids in marine fish larval feeds. Aquaculture 155: 117–127.

Sargent, J., L. McEvoy, A. Estevez, G. Bell, M. Bell, J. Henderson & D. Tocher, 1999. Lipid nutrition of marine fish during early development: current status and future directions. Aquaculture 179: 217–229.

Satuito, C. G. & K. Hirayama, 1991. Supplementary effect of vitamin C and squid oil on the nutritional value of baker's yeast for the population growth of the rotifer *Brachionus plicatilis*. Bull. Fac. Fish. Nagasaki Univ. 69: 7–11.

Schneider, J. C. A. Livne, A. Sukenik & P. Roussler, 1995. A mutant of *Nannochloropsis* deficient in eicosapentaenoic acid production. Phytochem. 40: 807–814.

Scott, J. M., 1981. The vitamin $B_{12}$ requirement of the marine rotifer *Brachionus plicatilis*. J. mar. biol. Ass. U. K. 61: 983–994.

Skejrmo, J. & O. Vadstein, 1999. Techniques for microbial control in the intensive rearing of marine larvae. Aquaculture 177: 333–343.

Snell, T. W., 1986. Effect of temperature, salinity and food level on sexual and asexual reproduction in *Brachionus plicatilis* (Rotifera). Mar. Biol. 92: 157–162.

Snell, T. W., 1991. Improving the design of mass culture systems for the rotifer *Brachionus plicatilis*. In Fulks W. & K. L. Main (eds), Rotifer amd Microalgae Culture Systems. Proc. U.S.– Asia Workshop. The Oceanic Insitute, Honolulu, HI: 61–71.

Snell, T. W. & K. Carrillo, 1984. Body size variation among strains of the rotifer *Brachionus plicatilis*. Aquaculture 37: 359–367.

Snell, T. W. & F. H. Hoff, 1985. The effect of environemental factors on resting egg production in the rotifer *Brachionus plicatilis*. J. World Maricult. Soc. 16: 484–497.

Snell, T. W. & F. H. Hoff, 1988. Recent advances in rotifer culture. Aquaculture Mag. 9/10: 41–45.

Snell, T. W., M. J. Childress, E. M. Boyer & F. H. Hoff, 1987. Assessing the status of rotifer mass cultures. J. World Aquacult. Soc. 18: 270–277.

Suantika, G., P. Dhert, N. Nurhudah & P. Sorgeloos, 2000. High-density production of rotifer *Brachionus plicatilis* in recirculated system: consideration of water quality, zootechnical and nutrient aspects. Aquacult. Engineer. 21: 201–214.

Takeyama, H., K. Iwamoto, S. Hara, H. Takano & T. Matsunaga, 1996. DHA enrichment of rotifers: a simple two-step culture using the unicellular algae *Chlorella reularis* and *Isochrysis galbana*. J. Mar. Biotechnol. 3: 244–247.

Tamaru, C. S., C.-S. Lee & H. Ako, 1991. Improving the larval rearing of stiped mullet (*Mugil cephalus*) by manipulating quantity and quality of the rotifer, *Brachionus plicatilis*. In Fulks W. & K. L. Main (eds), Rotifer and Microalgae Culture Systems. Proc. U.S.- Asia Workshop. The Oceanic Insitute, Honolulu, HI: 61–71.

Tamaru, C. S., R. Murashige, C.-S. Lee, H. Ako & V. Sato, 1993. Rotifers fed various diets of baker's yeast and/or *Nannochloropsis oculata* and their effect on the growth and survival of striped mullet (*Mugil cephalus*) and milkfish (*Chanos chanos*) larvae. Aquaculture 110: 361–372.

Tandler, A., 1985/1985. Overview: food for the larval stages of marine fish. Live or inert? Isr. J. Zool. 33: 161–166.

Teshima, S.-I., A. Kanazawa, K. Horinouchi, S. Yamasaki & H. Hirata, 1987. Phospholipids of the rotifer, prawn and larval fish. Nippon Suisan Gakkaishi 53: 609–615.

Theilacker, G. & K. Dorsey, 1980. Larval fish diversity. A summary of laboratory and field research. Workshop on the effects of environmental variation on the survival of larval pelagic fishes. Intergovernmental Oceanic Commision Workshop Rep. 28: 105–142.

Toledo, J. D. & H. Kurokura, 1990. Cryopreservation of the euryhaline rotifer *Brachionus plicatilis* embryos. Aquaculture 91: 385–394.

Toledo, J. D., H. Kurokura & H. Nakagawa, 1991. Cryopreservation of different strains of the euryhaline rotifer *Brachionus plicatilis*. Nippon Suisan Gakkaishi 57: 1347–1350.

Vadstein, O., G. Øie & O. Olsen, 1993. Particle size dependent feeding by the rotifer *Brachionus plicatilis*. Hydrobiologia 255/256: 261–267.

Verpraet, R., M. Chair, P. Leger, H. Nelis, P. Sorgeloos & A. De Leenheer, 1992. Live-Food mediated drug delivery as a tool for disease treatment in Larviculture. The enrichment of therapeutics in rotifers and *Artemia* Nauplii. Aquacult. Engineer. 11: 133–139.

Verdonck, L., L. Grisez, E. Sweetman, G. Minkoff, P. Sogeloos, O. Ollevier & J. Swings, 1997. *Vibrio* associated with routine production of *Brachionus plicatilis*. Aquaculture 149: 203–214.

Walford, J. & T. J. Lam, 1993. Development of digestive tract and proteolytic enzyme activity in seabass (*Lates calcarifer*) larvae and juveniles. Aquaculture 109: 187–205.

Walz, N., 1993. Plankton Regulation Dynamics. Experiments and Models in Rotifer Continuous Cultures. Springer Verlag, Berlin (Germany). Ecol. Stud. 98: 308 pp.

Walz, N., T. Hintze & R. Rusche, 1997. Algae and rotifer turbidostatas: studies on stability of live food cultures. Hydrobiologia 358: 127–132.

Watanabe, T., C. Kitajima & S. Fujita, 1983. Nutritional values of live organisms used in Japan for mass propagation of fish: a review. Aquaculture 34: 115–143.

Wesenberg-Lund, C., 1923. Contribution to the biology of the Rotifera, I. The males of Rotifera. Kgl. Dansk Vid. Selsk. Skr. Nat. math. Afd. Ser. 8, 4: 190–345.

Whyte, J. N. C. & W. D. Nagata, 1990. Carbohydrate and fatty acid composition of the rotifer , *Brachionus plicatilis*, fed monospecific diets of yeast and phytoplankton. Aquaculture: 89: 263–368.

Yoshimura, K., A. Hagiwara, T. Yoshimatsu & C. Kitajima, 1996. Culture technology of marine rotifers and implication for intensive culture of marine fish in Japan. Mar. freshwat. Res. 47: 217–222.

Yoshimura, K., K. Usuki, T. Yoshimatsu, C. Kitajima & A. Hagiwara, 1997. Recent developments of a high density mass culture system for the rotifer *Brachionus rotundiformis* Tschugunoff. Hydrobiologia 358: 139–144.

Yu, J. & K. Hirayama, 1986. The effect of un-ionized ammonia on the population growth of the rotifer in mass culture. Bull. Jap. Soc. Sci. Fish. 52: 1509–1513.

Yu, J.-P., A. Hino, R. Hirano & K. Hirayama, 1988. Vitamin $B_{12}$ producing bacteria as a nutritive complement for the culture of the rotifer *Brachionus plicatilis*. Nippon. Suisan Gakkaishi 54: 1873–1880.

Yu, J.-P., A. Hino, M. Ushiro & M. Maeda, 1989. Function of bacteria as vitamin $B_{12}$ producers during mass culture of the rotifer *Brachionus plicatilis*. Nippon. Suisan Gakkaishi 55: 1799–1806.

Yu, J.- P., A. Hino, R. Hirano & K. Hirayama, 1990a. The role of bacteria in mass culture of the rotifer *Brachionus plicatilis*. In Hirano, R. & I. Hanyu (eds), The Second Fisheries Society. Manila, Philippines: 29–32.

Yu, J.-P., A. Hino, T. Noguchi & H. Wakabayashi, 1990b. Toxicity of *Vibrio alginolyticus* on the survival of the rotifer *Brachionus plicatilis*. Nippon Suisan Gakkaishi 56: 1455–1460.

Yufera, M. & E. Pascual, 1989. Biomass and elemental composition (C.H.N.) of the rotifer *Brachionus plicatilis* cultured as larval food. Hydrobiologia 186/187: 371–374.

Yufera, M. & N. Navarro, 1995. Population growth dynamic of the rotifer *Brachionus plicatilis* cultured in non-limiting food condition. Hydrobiologia 313/314: 399–405.

Yufera, M., G. Parra & E. Pascual, 1997. Energy content of rotifers (*Brachionus plicatilis* and *Brachionus rotundiformis*) in relation to temperature. Hydrobiologia 358: 83–87.

Zmora, O., 1991. Management, production and disease interaction in rotifer culture. In Lavens P., P. Sorgeloos, E. Jaspers & F. Ollevier (eds), Larvi '91 – Fish and Crustacean Larviculture Symposium. Eur. Aquacult. Soc. Spec. Publ. 15, Ghent (Belgium): 104 pp.

Zmora, O., Y. Barr & A. Tandler, 1991. Report on a visit to several European commercial fish hatcheries (in Hebrew). Israel Oceanographic and Limnological Research Reports: 63 pp.

Hydrobiologia **446/447:** 355–361, 2001.
L. Sanoamuang, H. Segers, R.J. Shiel & R.D. Gulati (eds), Rotifera IX.




# Factors affecting low temperature preservation of the marine rotifer *Brachionus rotundiformis* Tschugunoff

Mavit Assavaaree[1], Atsushi Hagiwara[2] & Esther Lubzens[3]

[1]*Graduate School of Marine Science and Engineering, Nagasaki University, Bunkyo-machi 1-14, Nagasaki 852-8131, Japan*
*(Address in Thailand: National Institute of Coastal Aquaculture, Kao Saen Soi 1, Muang District, Songkhla 90000, Thailand).*
*E-mail: nicas@t-rex.hatyai.inet.co.th*
[2]*Faculty of Fisheries, Nagasaki University, Bunkyo-machi 1-14, Nagasaki 852-8521, Japan*
[3]*Israel Oceanographic & Limnological Research, National Institute of Oceanography, P.O. Box 8030, Haifa, 31080, Israel*

*Key words:* Rotifera, *Brachionus rotundiformis*, low temperature, preservation, salinity, feeding frequency

## Abstract

Experiments were performed to determine suitable conditions for low temperature preservation of small S (Fukuoka) and ultra-small SS (Thai) strains of *B. rotundiformis.* For this, single rotifers (an adult bearing one egg or a 4-h neonate) were incubated for 10 days in 1 ml seawater (22 ppt salinity). The highest survival was achieved at 10 and 12 °C for S-strain and 12 °C for SS-strain. The effect of salinity, change of culture medium and feeding regime were further tested on rotifers (300 ind. ml$^{-1}$) cultured in vials containing 10 ml seawater and microalgae at 12 °C. Survival of S-strain was highest (55.5±0.8%) at 35 ppt, while SS-strain survived best (43.1±2.6%) at 17 ppt. Survival was suppressed by changing the culture medium every 4 days. Feeding rotifers every 2 days yielded better survival (66.2±6.6%: S-strains, cultured at 35 ppt and 81.8±5.2%, SS-strains cultured at 17 ppt) than feeding them only at the beginning of the experiment or at 4-day intervals. An acclimation at 20 °C for 24 h before transferring them from their usual culture temperature (28 °C) to 12 °C resulted in higher survival of SS-strain. For S-strain, however, no significant improvement resulted from acclimation. SS-strain was more susceptible to lower temperature and higher salinity than S-strain.

## Introduction

The marine rotifers *Brachionus plicatilis* Müller and *B. rotundiformis* Tschugunoff (Segers, 1995) are important food organisms for feeding early stages of many marine fish species (Fukusho, 1983; Lubzens et al., 1989; Yoshimura et al., 1996; Fu et al., 1997). The major morphological and physiological differences between these two species have been summarized by several authors (Ito et al., 1981; Fukusho & Okauchi, 1982; Fu et al., 1991a, b; Fu et al., 1993; Hirayama & Rumengan, 1993). The species of interest to this study, *B. rotundiformis*, has a small and round lorica (adult lorica length = 212±20 $\mu$m (mean ± SD), *n*=740). Suitable temperatures for population growth of *B. rotundiformis* range from 28–35 °C. Among *B. rotundiformis*, ultra-minute strains with lorica length of 182.7±5.3 $\mu$m (*n*=3) have been reported from tropical regions (Hagiwara et al., 1995). The optimal temperature for population growth and mixis induction for these strains is 30–35 °C (Hagiwara et al., 1995; Rumengan et al., 1998).

Studies have been conducted to enhance rotifer population growth by manipulation of the environmental (reviewed by Maeda & Hino, 1991; Hino, 1993; Hirayama & Hagiwara, 1995), as well as by treatment with chemicals (Gallardo et al., 2000). Techniques have also been developed for assessing the status of rotifer cultures (Snell & Hoff, 1987; Araujo et al., 2000; Araujo et al., 2001). Despite these achievements, rotifer cultures occasionally collapse making them temporarily unavailable as food for larval fish.

The use of preserved rotifers could overcome this problem. Rotifer resting eggs can be mass-produced and preserved for long periods of time, and can be used directly as food source in smaller hatcheries, or to inoculate new rotifer cultures (reviewed by Lubzens, 1989; Hagiwara, 1994; Hagiwara et al., 1997). Aquaculturists generally produce an excess amount of rotifers to ensure sufficient rotifer supply to fish larvae. This can be stored at cold temperatures in live form (Lubzens, 1989). *B. plicatilis* can be preserved at high densities at 4 °C or below 0 °C for 3–4 weeks in laboratory conditions, however, its survival depends on quality and quantity of food, salinity and water exchange during preservation (Lubzens, 1989; Berghahn et al., 1990; Lubzens et al., 1990). Rotifers acclimated at 4 or 10 °C for 2–6 days before preserving at –1 °C had high survival up to 12–14 days without handling (Lubzens et al., 1995). In a previous paper, we determined the optimal conditions for keeping *B. plicatilis* at 4 °C (Assavaaree et al., 2001). The storage of *B. rotundiformis* strains at 4 °C, however, resulted in much lower survival. This lower survival may be explained by this smaller species having higher temperature optimum for population growth and preservation. Here we report on laboratory experiments aimed at determining the optimal temperature, salinity, food requirement and intermediate acclimation for low temperature preservation of *B. rotundiformis*.

## Materials and methods

### *Rotifers*

We used Fukuoka and Thai *B. rotundiformis* strains for this work. These two strains showed comparatively higher survival at 4 °C among 8 rotifer strains that were tested previously (Assavaaree et al., 2001). We designated Fukuoka and Thai strains as S- and SS-strains, respectively. The lorica size (mean ± SD of lorica length, $n$=64) of S-strain was 217.0±10.1 $\mu$m and that of SS-strain was 178.7±7.2 $\mu$m. Each strain was maintained in 250 ml culture media at 25 °C. Stock rotifers were fed fresh *Nannochloropsis oculata* (Droop) every 2–3 days. Rotifers were inoculated to new media every 4 weeks. Stock cultured rotifer strains (70–80% egg-bearing) were directly used for Experiment 1 (see below). For Experiments 2, 3 and 4, rotifers from stock cultures were further pre-cultured for a week in 500 ml beakers at 25 °C in darkness to acclimatize them at designated salinity levels. Unless otherwise stated, diluted (22 ppt salinity) and sterilized seawater was used for rotifer culture media throughout the experiments.

### *Algae*

Freshly condensed *N. oculata* was used as rotifer food. Algae were cultured in 2-l flasks with aeration, using the modified Erd-Schreiber medium (Hagiwara et al., 1994). Algal density was determined using a hemocytometer after harvest by centrifugation (1653 × ***g*** for 10 min). The condensed algae were re-suspended in experimental sterilized seawater and used as food for rotifers. For rotifer stock cultures and pre-cultures, feeding density of *N. oculata* was $7\times10^6$ cells/ml. At the start of experiments, the algal density was adjusted to $1\times10^6$ cells/ml for Experiments 1 and 3, and $7\times10^6$ cells/ml for all other experiments.

### *Experimental design*

Four experiments were conducted. The optimum temperature for cold-water preservation of rotifers was determined in Experiment 1; combined effects of salinity and medium renewal, and of feeding frequency and salinity were measured in Experiment 2 and 3, respectively. The effect of intermediate acclimation before preservation at 12 °C was established in Experiment 4. Experiment 1 was conducted by culturing rotifers individually. Experiments 2–4 started at 300 ind. ml$^{-1}$ in 30-ml screw-capped bottles containing 10 ml medium. Two 0.5 ml samples were taken every 2–4 days from each culture to count the rotifer density under a stereomicroscope. The observed rotifers were returned to the original cultures.

### *Experiment 1: Optimum low temperature for preservation*

Four temperatures (6, 8, 10 and 12 °C) were examined. For each rotifer strain, 72 neonate (4–6 h old) and 72 adult (24–26 h old) rotifers were prepared. These rotifers were inoculated at 1 ind. per well in 24-well plates. Each well contained 1 ml *N. oculata* suspension. We monitored rotifer survival daily under a stereomicroscope for 10 days to analyze percent survival of the neonate and adult rotifers at each temperature.

*Experiment 2: Combined effects of salinity and medium renewal*

Three salinity levels (17, 22 and 35 ppt) were examined. In Experiment 1, both two strains showed the highest survival at 12 °C; therefore, this temperature was used in further experiments. We prepared six rotifer cultures for each salinity level. In three of them, culture medium was exchanged every 4 days, and in the other three, the medium was not changed during preservation. The rotifers were fed *N. oculata* every 4 days, after monitoring their survival, for 16 days at 12 °C.

*Experiment 3: Combined effects of feeding frequency and salinity*

The optimal feeding frequency for rotifers at 12°C was tested at different salinities (17, 22 and 35 ppt) for 12 days. For each salinity level, nine cultures were prepared for three feeding intervals with three replicates. *N. oculata* was fed to rotifers at either 2 or 4 day interval. As a control, food was supplied to rotifers only on the first day of the experiment. Survival of rotifers was monitored every 2 days.

*Experiment 4: Effect of intermediate acclimation before preservation at 12 °C*

Rotifer cultures were first maintained at 28 °C for 7 days, and then transferred to either 15 or 20 °C for acclimatization for 24, 48 or 96 h. After these treatments, rotifers were cultured at 12 °C for 14 days. Control rotifers were transferred directly from 28 to 12 °C. Survival of rotifers was recorded every 3–4 days. *N. oculata* was fed to rotifers after these observations.

*Statistical analysis*

Two-way analysis of variance (ANOVA) was conducted to determine the effect of two factors in each experiment on rotifer preservation (Zar, 1999). Tukey HSD multiple comparison tests were conducted to determine which treatments were significantly different. Statistical analyses were conducted using SYSTAT ver. 5.0 (SAS Institute, Inc.).

## Results

*Optimum low temperature for preservation*

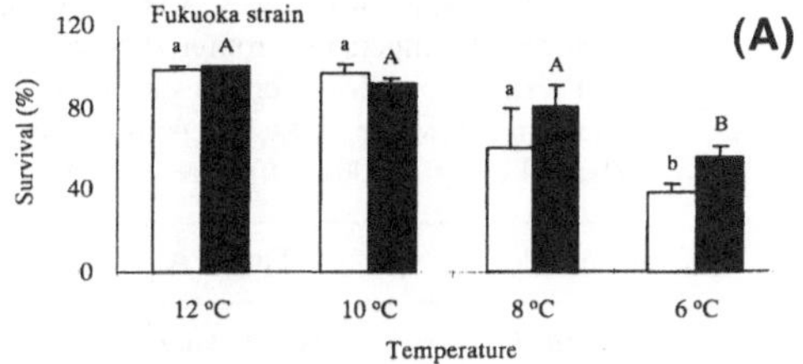


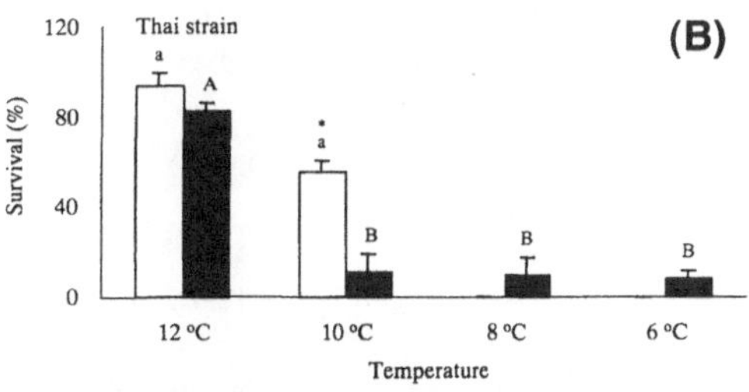


*Figure 1.* Survival (mean ± S.D.) of neonate (white column) and adult (dark column) of Fukuoka (A) and Thai (B) strain *Brachionus rotundiformis* on day 10 cultured at four different temperatures. Asterisks indicate significant differences between neonate and adult survival at the same temperature (* $P<0.05$). Among neonate (lower case) and adult (upper case) columns, values with the same letter are not significantly different. (Two way ANOVA, Tukey HSD multiple comparison, $P<0.05$, $n=3$).

For S-strain, rotifer age (see comparison between neonate vs adult) did not affect the survival at any temperature tested (Table 1, $P=0.069$), and no interaction between the age of rotifers and preservation temperature was detected. After 10 days, percent survival of rotifers preserved at 12, 10 and 8 °C was significantly higher than when preserved at 6 °C (Fig. 1). For SS-strain, a significant interaction between rotifer age and preservation temperature was found (Table 1, $P<0.001$). Stage of rotifers and preservation temperature both affected rotifer survival ($P=0.002$ and $P<0.001$, respectively). At 12 °C, rotifers survived better than at other temperatures, and this was more expressed with younger rotifers (Table 1, Fig. 1).

*Combined effects of salinity and medium renewal*

For both strains, salinity and change of culture medium affected survival after preservation at 12 °C for 16 days (Table 1, Fig. 2), but no interaction between these factors was observed (Table 1, S strain- $P=0.056$; SS strain-$P=0.200$). S-strain cultured in 35 ppt showed significantly higher survival than in 22 and 17 ppt seawater. At 22 and 35 ppt, not changing the culture medium resulted in higher rotifer survival than when the culture medium was changed every 4 days (Fig. 2). For SS-strain, rotifer survival among three salinity levels did not differ when the culture medium was not

*Table 1.* Results of two way ANOVA of low temperature preservation experiments to see the effect of treatments (temperature, rotifer age, salinity, water exchange, feeding interval, acclimation temperature and acclimation period) on the survival of Fukuoka (S-strain) and Thai (SS-strain) *B. rotundiformis*. Data shows the *F* value in each experiment at the end of preservation. Significant levels of $P>0.05$ (NS), $P<0.01$ (*), $P<0.001$ (**) were indicated

| *F*-ratio | Factor A | Factor B | Factor A × Factor B |
|---|---|---|---|
| *Exp. 1* | Temperature[a] | Rotifer age[b] | Temperature × Rotifer age |
| (Day 10) | | | |
| S-strain | 32.0** | 3.8$^{NS}$ | 2.2$^{NS}$ |
| SS-strain | 52.4** | 14.0* | 25.4** |
| *Exp. 2* | Salinity[c] | Medium exchange[d] | Salinity × Water exchange |
| (Day 16) | | | |
| S-strain | 56.0** | 35.7** | 3.2$^{NS}$ |
| SS-strain | 12.0** | 16.4** | 1.7$^{NS}$ |
| *Exp. 3* | Feeding interval[e] | Salinity[c] | Feeding interval × Salinity |
| (Day 12) | | | |
| S-strain | 107.4** | 262.2** | 4.8* |
| SS-strain | 394.6** | 217.6** | 22.6** |
| *Exp. 4* | Acclimation temperature[f] | Acclimation period[g] | Accl. temperature × Accl. period |
| (Day 14) | | | |
| S-strain | 12.8* | 235.3** | 9.5** |
| SS-strain | 30.0** | 5.1* | 5.6* |

[a] Preservation temperature at 6, 8, 10 and 12 °C.
[b] Rotifer age at 0–4 h and 24–28 h was analyzed.
[c] Salinity levels tested were 17, 22 and 35 ppt
[d] With or without medium exchange.
[e] Feeding frequency (feeding only on the first day of preservation, fed every 2 or 4 days).
[f] Acclimatization at 15 or 20 °C.
[g] Acclimatization for 0, 24, 48 and 96 h.

changed. But when the medium was changed, survival of preserved rotifers was significantly higher at lower salinities (Table 1, Fig. 2). Rotifers cultured without changing of the medium showed significantly higher survival than those cultured with medium changes in 22 ppt ($P$=0.016) and in 35 ppt ($P$=0.010), but no significant difference was detected in 17 ppt (Fig. 2).

*Combined effects of feeding frequency and salinity*

For both S and SS-strains, we found significant effects of feeding interval and culture salinity on rotifer survival, as well as a significant interaction between these two factors (each $P$=0.003 and $P<0.001$, respectively, Table 1). Fig. 3 shows the survival of S and SS-strains on day 12 after preservation at 12 °C. For both strains, rotifers survived best when they were fed every two days. The optimal salinity for rotifer survival was 35 ppt for S (mean±SD of percent survival = 66.2±6.6) but 17 ppt for SS strain (percent survival = 81.8±5.2).

*Effects of intermediate acclimation before preservation at 12 °C*

For the two strains, both temperature and period for acclimatization affected rotifer survival at 12 °C, and there was a significant interaction between these factors (Table 1, Fig. 4). These patterns were stronger for S-strain. For S-strain, survival of rotifers without acclimatization (62.0 ± 4.8%) was significantly higher than those acclimatized at 15 °C for 48 h (50.1±1.8%) and 96 h (32.0±1.3%) or at 20 °C for 24 h (50.6±2.7%), 48 h (53.1±2.9%) and 96 h (24.8±3.3%) (each $P<0.001$, Fig. 4). SS-strains acclimatized at 20 °C for 24 h (49.4±1.8%) showed significantly higher survival than those acclimatized at 15 °C for 24 h (43.0±3.8%, $P$=0.004), 48 h (41.3±2.4%,

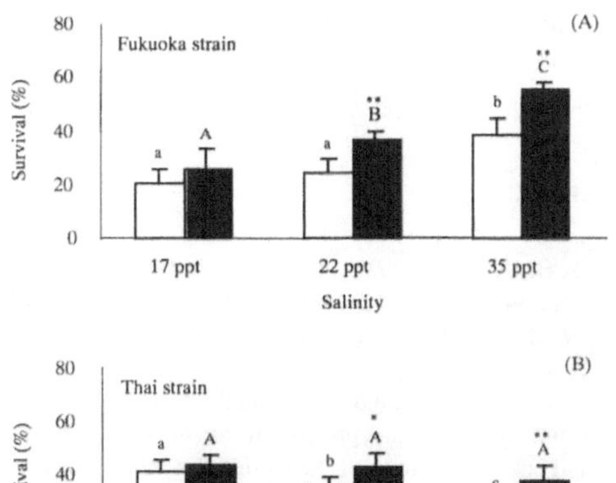


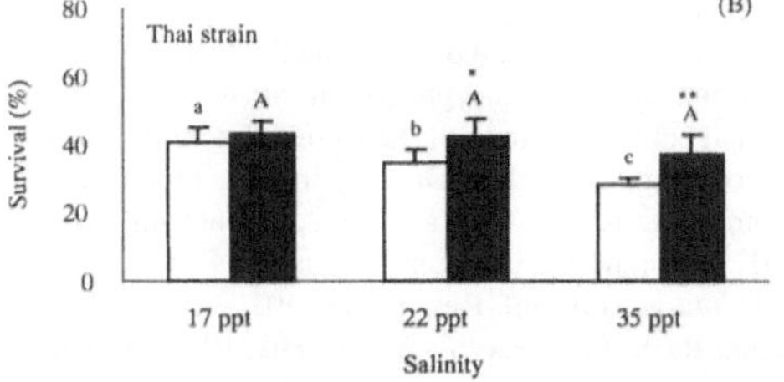


*Figure 2.* Combined effects of salinity and culture medium renewal on the survival of *B. rotundiformis* at 12 °C on day 16; (A) Fukuoka strain, (B) Thai strain (mean ± S.D.), for treatments with (white columns) and without (dark columns) culture medium renewal. Asterisks indicate significant differences between treatments at the same level of salinity (* $P<0.05$, ** $P<0.01$). Among treatments (lower case: medium renewal, upper case: no medium renewal), values with the same letter are not significantly different (Two way ANOVA, Tukey HSD multiple comparison, $P<0.05$, $n=3$).

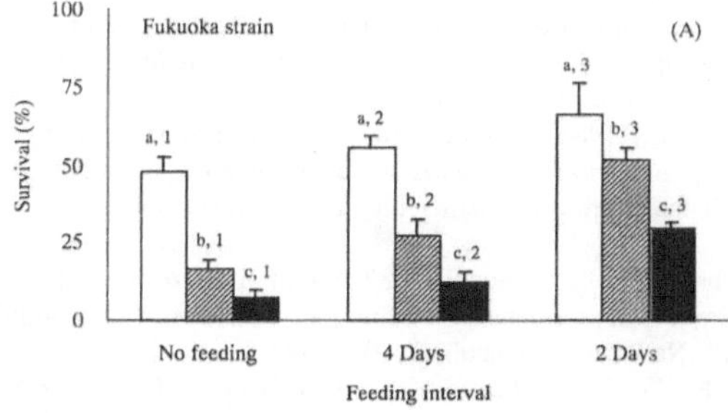


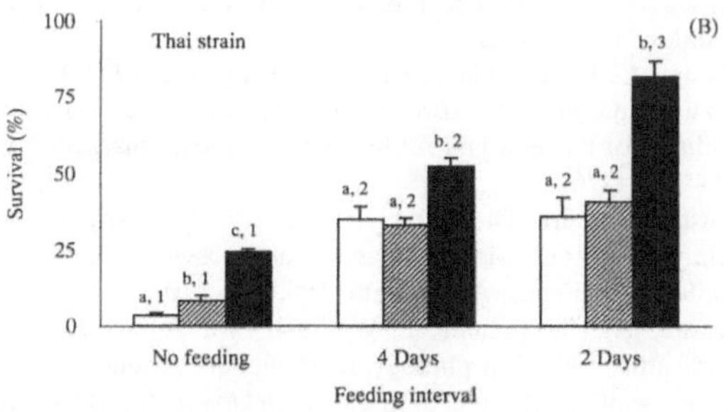


*Figure 3.* Combined effects of feeding frequency and salinity level on the survival of Fukuoka (A) and Thai (B) *B. rotundiformis* at 12 °C on day 12 (mean ± S.D.), for cultures in 35‰ (white columns), 22‰ (striped columns) and 17‰ (dark columns). Within feeding groups and salinity levels, values with different letters, resp. numbers on top of the columns are significantly different. (Two way ANOVA, Tukey HSD multiple comparison, $P<0.05$, $n=3$).

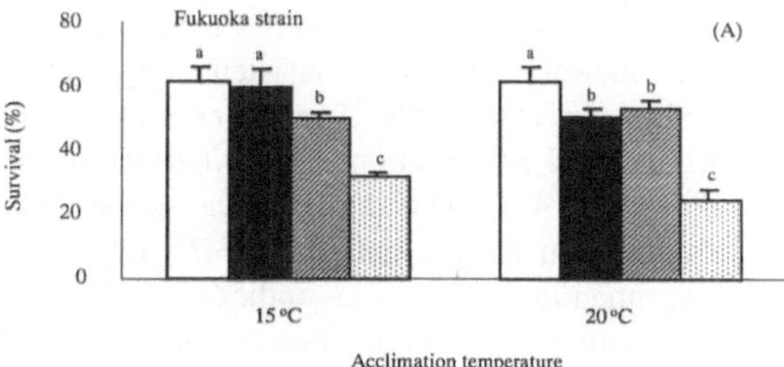


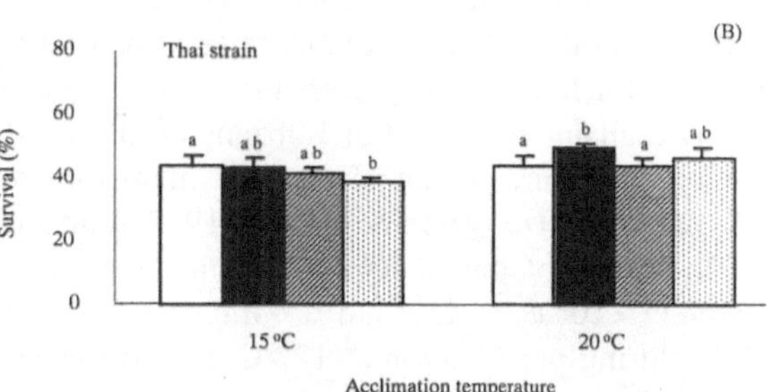


*Figure 4.* Effect of acclimation periods (0 h (control)- white columns, 24 h–black columns, 48 h–striped columns or 96 h–dotted columns) at 15 or 20 °C on survival of (A) Fukuoka strain, (B) Thai strain *B. rotundiformis*, kept later at 12 °C for 14 days. Mean (S.D.) with different letters on top of the columns are significantly different. (Two way ANOVA, Tukey HSD multiple comparison, $P<0.05$, $n=3$).

$P<0.001$) and 96 h (38.6±1.8%, $P<0.001$) or at 20 °C for 48 h (43.8±3.0%, $P=0.015$) and control (43.8±3.2%, $P=0.015$).

## Discussion

Results of this study indicated that both *B. rotundiformis* strains could be preserved at 12 °C for up to 10 days with more than 80% survival. Fukuoka strain (S-strain) can survive better at low temperature (6, 8, 10 and 12 °C) than Thai strain (SS-strain). The different tolerance to low temperature between the two strains may be attributed to differences in their habitats. S-strains are common in temperate region while SS-strains inhabit tropical regions (Hagiwara et al., 1995; Rumengan et al., 1998).

Lubzens et al. (1990, 1995) reported that salinity affects rotifer survival during low temperature preservation in *B. plicatilis*. Our study showed that salinity affects survival of the two tested *B. rotundiformis* strains as well, but in the opposite way: while the S-strain survived better at higher salinity, the tropical SS-strain did better at lower salinity. At 22 and 35 ppt, rotifers cultured without changing culture medium showed higher survival than those cultured with medium renewal. This is inconsistent with former studies (Lubzens et al., 1990; Assavaaree et al., 2001), which attempted preservation of *B. plicatilis* and *B. rotundiformis* at 4 °C. This may be because the initial rotifer density in the present study (300 ind. $ml^{-1}$) was lower than those in former studies (500–800 ind. $ml^{-1}$), perhaps causing less decline of water quality in the tested media. It is also possible that rotifers were stressed by the sieving treatment during water change.

The quantity and quality of phytoplankton diets are important for *B. plicatilis* preservation at low temperature (Berghahn et al., 1990; Lubzens et al., 1990). Salinity of culture media affects the ingestion rate of *B. plicatilis* (Hirayama & Ogawa, 1972; Lubzens, 1987). Results of this study showed that feeding *N. oculata* ($1\times10^6$ cells $ml^{-1}$) to *B. rotundiformis* every 2 days during preservation at 12 °C gave better survival for two strains of *B. rotundiformis*. We obtained highest survival in 35 ppt for S-strain and in 17 ppt for SS-strain.

The mechanism involved in these differences in salinity effect between strains is unknown. Rotifers preserved at 12 °C should be fed every 2–4 days during preservation.

Lubzens et al. (1995) succeeded in preserving *B. plicatilis* at –1 °C for 12–14 days by acclimatizing rotifers at 4 or 10 °C for 2–6 days before transfer to –1 °C. We obtained a similar effect of acclimatization before preserving *B. plicatilis* and *B. rotundiformis* at 4 °C (Assavaaree et al., in press). Such an effect, however, was not observed in the present study. Lubzens et al. (1995) isolated a 94 kD protein from rotifers exposed at 10 °C for 24 h. This protein may function to increase tolerance to low temperature environments. The final temperature for rotifer preservation was much higher in the present study (12 °C), which may explain the differences between the our and other studies.

We conclude that the preservation of *B. rotundiformis* is possible at 12 °C, without conducting temperature acclimatization. The use of adult rotifers as well as the manipulation of water salinity and feeding frequency at strain-specific levels improves survival after preservation.

## Acknowledgements

Part of this study was supported by the Sasakawa Scientific Research Foundation from the Japan Science Society.

## References


Araujo, A. B., T. W. Snell & A. Hagiwara, 2000. Effect of unionized ammonia, viscosity and protozoan contamination on the enzyme activity of the rotifer *Brachionus plicatilis*. Aquacult. Res. 31: 359–365.

Araujo, A. B., A. Hagiwara & T. W. Snell, 2001. Effect of unionized ammonia, viscosity and protozoan contamination on reproduction and enzyme activity of the rotifer *Brachionus rotundiformis*. Hydrobiologia, 446/447 (Dev. Hydrobiol. 153): 363–368.

Assavaaree, M., K. Ide, K. Maruyama., E. Lubzens & A. Hagiwara, 2001. Low temperature preservation (at 4 °C) of marine rotifer *Brachionus*. Aquacult. Res. 32: 29–40.

Berghahn, R., S. Euteneuer & E. Lubzens, 1990. High density storage of rotifers (*Brachionus plicatilis*) in cooled and undercooled water. Spec. Publ. Eur. Aquacult. Soc. 11: 267–274.

Fu, Y., A. Hada, T. Yamashita, Y. Yoshida & A. Hino, 1997. Development of a continuous culture system for stable mass production of marine rotifer *Brachionus*. Hydrobiologia 358: 145–151.

Fu, Y., A. Hagiwara & K. Hirayama, 1993. Crossing between seven strains of the rotifer *Brachionus plicatilis*. Nippon Suisan Gakkaishi 59: 2009–2016.

Fu, Y., K. Hirayama & Y. Natsukari, 1991a. Morphological difference between the two types of the rotifer *Brachionus picatilis*. J. exp. mar. Biol. Ecol. 151: 29–41.

Fu, Y., K. Hirayama & Y. Natsukari, 1991b. Genetic divergence between S and L type strains of the rotifer *Brachionus plicatilis*. J. exp. mar. Biol. Ecol. 151: 43–56.

Fukusho, K., 1983. Present status and problems in culture of the rotifer *Brachionus plicatilis* for fry production of marine fishes in Japan. Symp. Int. Aquacult., Coquimbo, Chile, Sept, 1983: 361–374.

Fukusho, K. & M. Okauchi, 1982. Strain and size of rotifer, *Brachionus plicatilis*, being cultured in Southeast Asia countries. Bull. Natl. Inst. Aquacult. 3: 107–109.

Gallardo, W. G., A. Hagiwara & T. W. Snell, 2000. GABA enhances reproduction of the rotifer *Brachionus plicatilis* Müller: Application to mass culture. Aquacult. Res. 31: 713–718.

Hagiwara, A., 1994. Practical use of rotifer cysts. Israel J. Aquacult-Bamidgeh. 46: 13–21.

Hagiwara, A., K. Hamada, S. Hori & K. Hirayama, 1994. Increased sexual reproduction in *Brachionus plicatilis* (Rotifera) with the addition of bacteria and rotifer extracts. J. exp. mar. Biol. Ecol. 154: 171–176.

Hagiwara. A., M. D. Balompapueng, N. Munuswamy & K. Hirayama, 1997. Mass production and preservation of marine rotifer resting eggs. Aquaculture 155: 223–230.

Hagiwara. A., T. Kotani, T. W. Snell, M. Assava-aree & K. Hirayama, 1995. Morphology, reproduction, genetic and mating behavior of small, tropical marine *Brachionus* strains (Rotifera). J. exp. mar. Biol. Ecol. 194: 25–37.

Hino, A., 1993. Present culture systems of rotifer (*Brachionus plicatilis*) and the function of microorganisms. In Lee C. S., M. S. Su & I. C. Liao (eds), Proceeding of Finfish Hatchery in Asia'91, Tungkang, Taiwan. TML Conference Proceedings 3: 51–59.

Hirayama, K. & A. Hagiwara, 1995. Recent advance in biological aspects of mass cultures of rotifers (*Brachionus plicatilis*) in Japan. *ICES* Marine Science Symposium 201: 153–158

Hirayama, K. & S. Ogawa, 1972. Fundamental studies on physiology of rotifer for its mass culture-I, Filter feeding of rotifer. Bull. Jpn. Soc. Sci. Fish. 38: 1207–1214.

Hirayama, K. & I. F. M. Rumengan, 1993. The fecundity patterns of S and L type rotifers of *Brachionus plicatilis*. Hydrobiologia 255/256: 153–157.

Ito, S., H. Sakamoto, M. Hori & K. Hirayama, 1981. Morphological characteristics and suitable temperature for the growth of several strains of rotifer, *Brachionus plicatilis*. Bull. Fac. Fish. Nagasaki Univ. 51: 9–16 (in Japanese with English abstract).

Lubzens, E., 1987. Raising rotifers for use in aquaculture. Hydrobiologia 147: 245–255.

Lubzens, E., 1989. Possible use of rotifer resting eggs and preserved live rotifers (*Brachionus plicatilis*) in aquaculture. In De Paul, N., E. Jaspers, H. Ackefords & N. Wilkins (eds): Aquaculture – A Biotechnology in Progress. Eur. Aquacult. Soc: 741–750.

Lubzens, E., A. Tandler & G. Minkoff, 1989. Rotifers as food in aquaculture. Hydrobiologia 186/187: 387–400

Lubzens, E., D. Rankevich, G. Kolodny, O. Gibson, A. Cohen & M. Khayat, 1995. Physiological adaptations in the survival of rotifers (*Brachionus plicatilis* O.F. Müller) at low temperatures. Hydrobiologia 313/314: 175–183.

Lubzens, E., G. Kolodny, B. Perry, N. Galai, R. Sheshinski & Y. Wax, 1990. Factors affecting survival of rotifers (*Brachionus plicatilis* O.F. Müller) at 4 °C. Aquaculture 91: 23–47.

Maeda, M & A. Hino, 1991. Environment management for mass culture of the rotifer, *Brachionus plicatilis*. In Fulks W. & K. L. Main (eds), Rotifer and Microalgae Culture System, Proceedings of a US–Asia Workshop. The Oceanic Institute, Honolulu, Hawaii: 125–133.

Rumengan, I. F. M., V. Warouw & A. Haigwara, 1998. Morphometry and resting egg production potential of the tropical ultraminute rotifer *Brachionus rotundiformis* (Manado strain) fed different algae. Bull. Fac. Fish. Nagasaki Univ. 79: 31–36.

SAS User's Manual, 1992. SYSTAT Statistic, version 5 Edition. Statistical Analysis System, Inc., Evanston, IL: 750 pp.

Segers, H., 1995. Nomenclatural consequences of some recent studies on *Brachionus plicatilis* (Rotifera, Brachionidae). Hydrobiologia 313/314: 121–122.

Snell, T. W. & F. H. Hoff, 1987. Fertilization and male fertility in the rotifer *Brachionus plicatilis*. Hydrobiologia 147: 329–334.

Yoshimura, K., A. Hagiwara, C. Yoshimatsu & C. Kitajima, 1996. Culture technology of marine rotifers and the implications for intensive culture of marine fish in Japan. Mar. Freshwat. Res. 47: 217–222.

Zar, J. H., 1999. Biostatistical Analysis (4th edn). Prentice-Hall, New Jersey: 663 pp.

Hydrobiologia **446/447:** 363–368, 2001.
L. Sanoamuang, H. Segers, R.J. Shiel & R.D. Gulati (eds), Rotifera IX.




# Effect of unionized ammonia, viscosity and protozoan contamination on reproduction and enzyme activity of the rotifer *Brachionus rotundiformis*

Adriana Belem de Araujo[1], Atsushi Hagiwara[2] & Terry W. Snell[3]
[1]*Graduate School of Marine Science and Engineering, Nagasaki University, Bunkyo 1-14, Nagasaki 852-8131, Japan*
[2]*Faculty of Fisheries, Nagasaki University, Bunkyo 1-14, Nagasaki 852-8521, Japan*
[3]*School of Biology, Georgia Institute of Technology, Atlanta, GA 30332-0230, U.S.A.*

*Key words:* Rotifera, *Brachionus rotundiformis*, culture diagnosis, environmental stress, enzyme activity

## Abstract

We determined the effect of environmental stressors on the physiological condition of *Brachionus rotundiformis*. For two morphologically distinct *B. rotundiformis* strains: Hawaii (average lorica length = 222 $\mu$m) and Langkawi strains (average lorica length 180 $\mu$m), neonates hatched from resting eggs were exposed to different levels of unionized ammonia (0.7–9.8 mg $l^{-1}$), viscosity (relative viscosity against natural seawater = 1–1.17) and *Euplotes* sp. (protozoan) contamination (1–40 cells $ml^{-1}$). Increasing stress decreased fecundity and lifespan of both rotifer strains. Glucosidase and phospholipase activities were correlated with reproductive responses of both the strains exposed to unionized ammonia. When culture water viscosity was changed, the activity of esterase and phospholipase was correlated with reproductive responses of the Hawaiian strain, and glucosidase activity was correlated with those of Langkawi strain.With the protozoan contamination, esterase and glucosidase activities were correlated only with reproductive responses of the Hawaiian strain, while activity of all three enzymes was correlated to those of the Langkawi strain. Glucosidase activity proved to be a reliable indicator of stress for cultured *B. rotundiformis.*

## Introduction

In ecotoxicology, techniques have been developed to use rotifers and cladocerans as test organisms to assess toxicity of water (Snell & Persoone, 1989b; Janssen & Persoone, 1993). These include measurement of swimming speed (Janssen et al., 1993, 1994), ingestion rate (Ferrando et al., 1993; Juchelka & Snell, 1994) and enzyme activity (Burbank & Snell, 1994; Moffat & Snell, 1995). We have been testing whether such techniques also are useful to assess the status of cultured rotifers. Araujo et al. (2000) applied an *in vivo* enzyme test to cultured *Brachionus plicatilis* (Müller) in order to evaluate the effect of unionized ammonia, culture water viscosity and protozoan contamination, which are common stressors in rotifers mass cultures (Yu & Hirayama, 1986; Maeda & Hino, 1991; Hung et al., 1997; Hagiwara et al., 1998). The correlation between enzyme activities and reproduction of *B. plicatilis* appeared to be sufficient for evaluating culture stressors. The activity of the endogenous enzymes esterase, glucosidase and phospholipase is a useful tool for rapid, early detection of environmental stress in mass cultures (Araujo et al., 2000). In marine larviculture, *Brachionus plicatilis* is mainly used for feeding fish larvae in colder regions, while *Brachionus rotundiformis* is common in regions with moderate temperatures. In sub-tropical and tropical regions, aquaculturists use ultra-small *Brachionus* strains (Doi et al., 1997; Rumengan et al., 1998). These strains are currently included in *Brachionus rotundiformis* (Hagiwara et al., 1995; Kotani et al., 1997), based on assays of the mate recognition pheromone (MRP) binding and anti-MRP antibody binding (Snell et al., 1995; Kotani et al., 2001). Reproductive features of these *Brachionus* spp are variable, and results obtained for *B. plicatilis* are not always applicable to *B. rotundiformis*. In the present study, *in vivo* enzyme activity of two morphologically distinct strains (Hawaii and Langkawi strains) of *B. rotundiformis* was tested. Our aim was to determine the response of fecundity, lifespan and enzyme activity of these strains of the *Brachionus* sp when exposed to environmental stressors. Correlation analysis will determine if the method for quantifying enzyme activity in *B. plicatilis* can also be applied to *B. rotundiformis* as an indicator of its physiological conditions.

## Materials and methods

### *Rotifers strains*

*Brachionus rotundiformis* strains from Langkawi (SS strain) and Hawaii (S strain) were used in this research. Based on morphological features, reproductive mode, allozyme patterns and mating behaviour, SS strains have similar characteristics to S strain (Hagiwara et al., 1995; Kotani et al., 1997). Hawaii strain has a small, round lorica, with average length of 222 $\mu$m (mature female) and pointed anterior spines (Fu et al., 1991). Langkawi strain has a shorter lorica length (180 $\mu$m) and sharply pointed anterior spines with a narrower spine base in relation to lorica width. Langkawi strain was sampled from a brackish water pond in Langkawi Island, Malaysia. Hawaii strain was collected from oxidation pond of the Oceanic Institute, Hawaii (Hagiwara et al., 1989). Neonates from resting eggs were used for individual culture experiments and enzyme activity measurements. The use of resting eggs is appropriate since it provides rotifers of similar age, size and physiological condition.

### *Individual culture*

The procedure for individual cultures and enzyme activity measurement followed the protocol of Araujo et al. (2000). Newly hatched neonates were individually introduced into wells of 96-well culture plates which contained 100 $\mu$l of *Nannochloropsis oculata* at $7\times10^6$ cells ml$^{-1}$ as food suspension. Every 24 h, the female was transferred to fresh food medium and fecundity and lifespan were monitored at 10$\times$ magnification. Except during feeding and observation, cultures were maintained at 24 °C in darkness.

The ciliate *Euplotes* sp. was added at densities of 0, 1, 3, 6, 10, 25 and 40 cells ml$^{-1}$ to investigate its effect on rotifer reproduction. Each treatment was replicated eight times. *Euplotes* sp. was collected from rotifer mass culture tanks in Kamiura Station, Japan Sea Farming Association, Japan in 1995 and maintained in laboratory at 25 °C (Jung et al., 1997). Unionized ammonia concentrations were prepared at 0, 0.7, 2.4, 3.1, 4.9, 7.4 and 9.8 mg l$^{-1}$, according to the method described by Yu & Hirayama (1986). Viscosity of experimental seawater was regulated by dissolving methyl cellulose (15cP - WAKO Pure Chemical Industries, LTD., Tokyo, Japan) at concentrations of 0.0125, 0.025, 0.05 and 0.1%. Compared with 22‰ control seawater without methyl cellulose, the relative viscosity of the treated seawater was 1.02, 1.03, 1.08 and 1.17, respectively (Hagiwara et al., 1998).

### *Enzyme activity measurement*

One hundred neonates hatched from resting eggs were inoculated into 1 ml in a well of a 24-well microplate and incubated for 2 h in the designated treatments. At the end of incubation, the neonates were transferred to fresh 22‰ seawater to rinse out the stressors. The rotifers were not fed during the test.

Neonates were then exposed to one of three fluorogenic substrates (Molecular Probes Inc). These substrates, which include cFDAam, PLA2 and FDGlu, were cleaved by endogenous esterase, phospholipase and glucosidase, respectively, to yield a fluorescein derivative that is highly fluorescent. Enzyme substrate solutions were prepared according to Burbank & Snell (1994) in 100% acetone. Each substrate solution was divided into several 50 $\mu$l aliquots and stored at –80 °C until use.

An aliquot of 1.3 $\mu$l of cFDAam, FDGlu and PLA2 solution was added to each of 3 replicate test wells and incubated at 24 °C in darkness for 15 min. The final test concentration of these substrates were 0.12, 1.9 and 15.2 mm. At the end of 15 min incubation, 20 $\mu$l of sodium dodecyl sulfate (SDS) was added to stop the enzyme reaction. Samples were then vortexed for 15 s and centrifuged at 9000 rpm for 5 min. About 450 $\mu$l of the supernatant was transferred to a 6 $\times$ 50 mm borosilicate microcuvette cylinder which was placed into a fluorometer (Turner TD-700) to measure the enzyme activity as fluorescent produced. Fluorescence emission was read at 515 nm with an excitation wavelength of 490 nm (Moffat & Snell, 1995). Enzyme activity was normalized to the mean fluorescence of the control animals (1.0).

Fecundity and lifespan LOECs (lowest observed effect concentration) were determined using Dunnett's

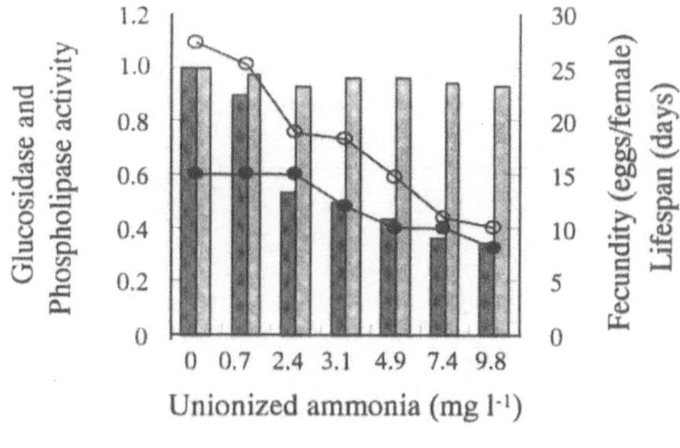


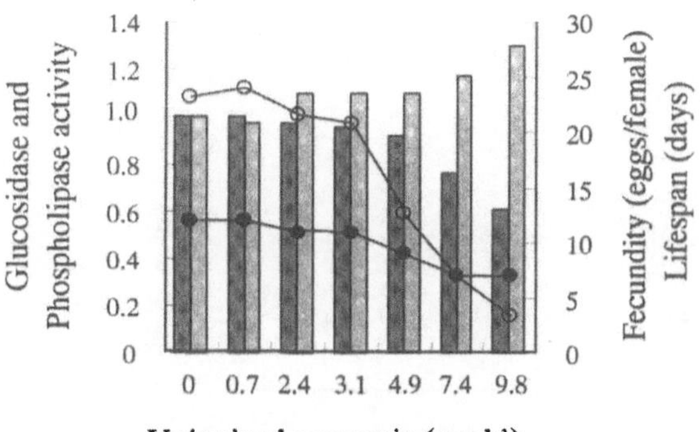


*Figure 1.* Fecundity (open circle), lifespan (closed circle) and enzyme activity of the Hawaiian (top) and Langkawi (bottom) *B. rotundiformis* strains under different unionized ammonia concentration. Columns illustrate activity of glucosidase (dark) and phospholipase (light).

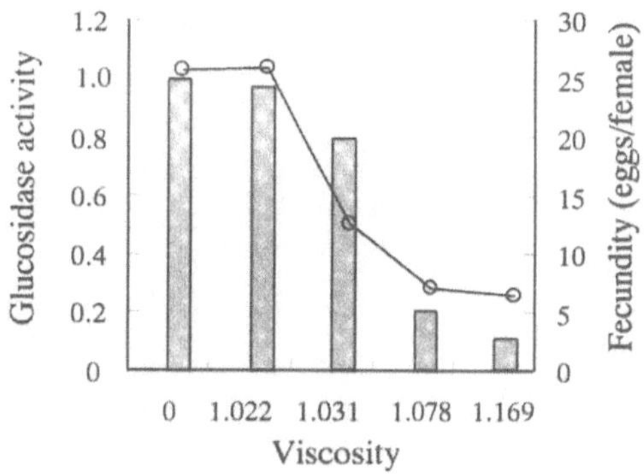


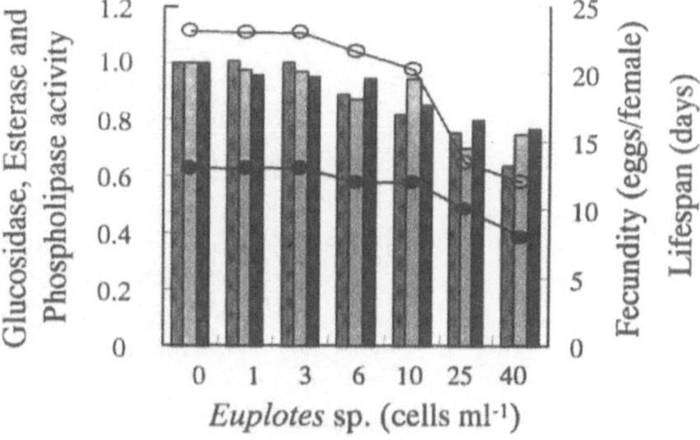


*Figure 2.* Fecundity (open circle) and glucosidase activity (columns) of Langkawi *B. rotundiformis* strain under different levels of viscosity (top); and fecundity and lifespan (closed circle) of Langkawi strain under *Euplotes* sp. presence (bottom). Columns (bottom) illustrate activity of glucosidase (dark), esterase (light) and phospholipase (black), respectively.

test for pairwise comparisons of data in each level of environmental stress relative to the control. Linear regression analysis was performed using enzyme activity as the dependent variable and reproductive characteristics of rotifers as independent variables (Systat version 7.0, SPSS Inc.).

## Results

For both rotifer strains, exposure to increasing unionized ammonia, viscosity and ciliate contamination caused rapid decline of fecundity, and comparatively slower decline of lifespan (Figs 1 and 2). Lifespan and fecundity LOEC of Hawaii and Langkawi strains of *B. rotundiformis* were compared for all treatments (Table 1). LOEC values ranged from 2.4 to 7.4 mg $l^{-1}$ for unionized ammonia, 1.08–1.17 for viscosity and 10–25 cells $ml^{-1}$ for *Euplotes* sp. contamination. Langkawi strain of *B. rotundiformis* showed higher LOEC levels for the treatments, except for viscosity.

Among enzymes tested, glucosidase activity also decreased with increasing environmental stress and showed similar changes to reproductive responses. Significant linear regressions were obtained between glucosidase activity and reproductive variables (e.g. fecundity of the Hawaiian strain; $y = 4.16x - 10.7$, $R^2=0.92$, $p<0.001$) with increasing unionized ammonia exposures (Fig. 1). Phospholipase activity of Hawaii strain was significantly correlated with its fecundity, and phospholipase activity of Langkawi strain was correlated with fecundity and lifespan (Table 2). For both strains, the esterase activity was not affected by unionized ammonia concentration.

A significant linear regression was found between fecundity and esterase activity and also between fecundity and phospholipase of the Hawaiian strain for increasing seawater viscosity (Table 2). For Langkawi strain, correlation between enzyme activity and reproductive variables was found only with glucosidase

*Table 1.* Lifespan and fecundity LOEC (lowest observed effect concentration) in *B. rotundiformis* and *B. plicatilis* under three environmental stressors

| | | Unionized-ammonia ($mg\ l^{-1}$) | Viscosity | *Euplotes* sp. ($cells\ ml^{-1}$) |
|---|---|---|---|---|
| *B. rotundiformis* | | | | |
| Hawaii strain | | | | |
| | Fecundity (eggs/female) | 2.4 | 1.169 | 10 |
| | Lifespan (days) | 3.1 | – | – |
| Langkawi strain | | | | |
| | Fecundity (eggs/female) | 4.9 | 1.078 | 10 |
| | Lifespan (days) | 7.4 | 1.078 | 25 |
| *B. plicatilis* [a] | | | | |
| NH1L strain | | | | |
| | Fecundity (eggs/female) | 2.4 | 1.022 | 3 |
| | Lifespan (days) | 4.9 | 1.169 | 10 |

[a] Data from Araujo et al. (2000).

and lifespan. Glucosidase activity was correlated with lifespan (Fig. 2; Table 2). Contamination of *Euplotes* sp. up to 40 cells $ml^{-1}$ reduced esterase activity of Hawaii, which was significantly correlated with fecundity and lifespan (Table 2). A decline was also observed in glucosidase activity, which was significantly correlated with fecundity. For Langkawi strain, correlation was detected between all three tested enzymes and rotifer reproductive variables under stress of *Euplotes* sp. contamination (Fig. 2; Table 2).

## Discussion

Studies on the effects of environmental stressors on reproduction of the euryhaline *Brachionus* have been exclusively conducted with *B. plicatilis*. These include the effects of unionized ammonia (Yu & Hirayama, 1986; Snell & Persoone 1989a), protozoan contamination (Maeda & Hino, 1991, Araujo et al., 2000) and viscosity (Hagiwara et al., 1998; Araujo et al., 2000). This study firstly reported how these stressors affect the reproductive characteristics of *B. rotundiformis* including two morphotypes. LOEC values indicate that test rotifer morphotypes (L and S-types) show different responses to the treatments (Table 1). The higher LOEC of fecundity and lifespan found for the Langkawi strain of *B. rotundiformis* in tests with unionized ammonia and protozoan contamination suggests that this strain was the least sensitive to the environmental stress than the other tested strains.

As in the case of *Brachionus plicatilis* (Araujo et al., 2000), it was confirmed that glucosidase, esterase and phospholipase activity can be quantified in *B. rotundiformis* using fluorescently labelled substrates. Effects of unionized ammonia, viscosity and protozoa on population growth of the two strains *B. rotundiformis* were similar to each other and to *B. plicatilis*. As reported earlier (Araujo et al., 2000), the sensitivity of enzyme activities of rotifers to environmental stress varied among enzymes. Furthermore, of the three enzymes, glucosidase activity in both rotifer strains showed the highest response to unionized ammonia and protozoa contamination. Only the reproductive variables of Hawaii strain responded significantly to increased viscosity. However, the glucosidase of both *B. rotundiformis* strains responded to unionized ammonia and viscosity. Similar results for esterase activity were obtained in *B. plicatilis*, as well as the response to protozoan contamination. Phosphol-

*Table 2.* ANOVA to the effects of unionized ammonia, viscosity and protozoa on *B. rotundiformis* lifespan, fecundity, and glucosidase, esterase and phospholipase activity. **A**- Langkawi strain, **B** - Hawaiian strain

| | df | *F* | SS | MS | $R^2$ | *P* |
|---|---|---|---|---|---|---|
| **A** Unionized ammonia | | | | | | |
| Glucosidase × lifespan | 5 | 26.674 | 118.448 | 23.060 | 0.842 | 0.004 |
| Glucosidase × fecundity | 5 | 45.176 | 74.773 | 14.955 | 0.900 | 0.001 |
| Phospholipase × lifespan | 5 | 17.380 | 8592.689 | 1718.538 | 0.777 | 0.009 |
| Phospholipase × fecundity | 5 | 20.769 | 7462.386 | 1492.477 | 0.806 | 0.006 |
| Viscosity | | | | | | |
| Glucosidase × fecundity | 3 | 15.621 | 1046.753 | 348.918 | 0.839 | 0.029 |
| Protozoa | | | | | | |
| Esterase × lifespan | 5 | 18.645 | 63.239 | 12.648 | 0.789 | 0.008 |
| Esterase × fecundity | 5 | 36.165 | 7.265 | 0.879 | 0.002 | |
| Glucosidase × lifespan | 5 | 55.153 | 132.537 | 26.507 | 0.917 | 0.001 |
| Glucosidase × fecundity | 5 | 39.138 | 180.628 | 36.126 | 0.887 | 0.002 |
| Phospholipase × lifespan | 5 | 27.038 | 4632.829 | 926.566 | 0.844 | 0.003 |
| Phospholipase × fecundity | 5 | 43.275 | 3074.607 | 614.921 | 0.896 | 0.001 |
| **B** Unionized ammonia | | | | | | |
| Glucosidase × lifespan | 5 | 9.961 | 1680.696 | 336.139 | 0.666 | 0.025 |
| Glucosidase × fecundity | 5 | 57.779 | 400.533 | 80.107 | 0.920 | 0.001 |
| Phospholipase × fecundity | 5 | 10.953 | 754.338 | 150.868 | 0.687 | 0.021 |
| Viscosity | | | | | | |
| Esterase × fecndity | 3 | 34.082 | 0.087 | 0.029 | 0.919 | 0.010 |
| Phospholipase × fecundity | 3 | 16.846 | 10209.157 | 34030.52 | 0.849 | 0.026 |
| Protozoa | | | | | | |
| Esterase × fecundity | 5 | 44.996 | 115.041 | 23.008 | 0.900 | 0.001 |
| Esterase × fecundity | 5 | 9.734 | 390.369 | 78.074 | 0.661 | 0.026 |
| Glucosidase × fecundity | 5 | 7.473 | 208.093 | 41.659 | 0.599 | 0.041 |

ipase activity of both the *B. rotundiformis* strains was sensitive to unionized ammonia and viscosity, whereas the *B. plicatilis* strain was unaffected. With exception of the Hawaiian strain, phospholipase activity responded to protozoan contamination. Linear regression for enzyme activity versus reproductive variables of the tested strains in high viscosity, ammonia and protozoa contaminated environments indicates that esterase, glucosidase and phospholipase respond differently to environmental stressors. Responses of glucosidase enzymes were the most sensitive in all of the three tested strains and appear to be the most suitable enzyme for applying in cultures of both *B. plicatilis* and *B. rotundiformis*. Because of the correlation between reproductive variables and enzyme activities of *B. rotundiformis,* enzyme activity appears to be a reliable indicator of culture condition. It suggests that the measurement of rotifer esterase, glucosidase and phospholipase activity as developed for ecotoxicology can be applied for rapid, early assessment of stress in rotifer *B. rotundiformis* mass cultures.

In order to apply this technique to rotifer mass culture practices, the preliminary treatment before enzyme activity test will be necessary; including rotifer culture water sampling, mesh filtration to remove mass cultured rotifers and exposure of neonates from resting eggs to the filtrate. For the direct enzyme activity measurement of mass cultured rotifers, further studies are necessary to test the physiological viability. These problems include the interference of the fluorescence of algal food in rotifer guts, standardization of rotifer enzyme activity that may be changed among rotifer age on size classes. Such research is currently under progress.

## Acknowledgements

The authors express thanks to the Ministry of Education, Science and Culture of Japan, which provided scholarship to A. B. Araujo. Langkawi strain was collected through Monbusho International Scientific Research Project between Malaysia Science University and Nagasaki University. A portion of this study was supported by a grant from Nagasaki Industrial Technology Foundation.

## References

Araujo, A. B., A. Hagiwara & T. W. Snell, 2000. Effect of unionized ammonia, viscosity and protozoan contamination on the enzyme activity of the rotifer *Brachionus plicatilis*. Aquacult. Res. 31: 359–365.

Burbank, S. E. & T. W. Snell, 1994. Rapid toxicity assessment using esterase biomarkers in *Brachionus calyciflorus* (Rotifera). Envir. Toxicol. Wat. Qual. 9: 171–178.

Doi, M., J. D. Toledo, M. S. N. Golez, M. De Los Santos & A. Ohno, 1997. Preliminary investigation of feeding performance of larvae of early red-spotted grouper, *Epinephelus coioides*, reared with mixed zooplankton. Hydrobiologia 358: 259–263.

Ferrando, M. D., C. R. Janssen & G. Persoone, 1993. Ecotoxicological studies with the freshwater rotifer *Brachionus calyciflorus*. III. The effects of chemicals on feeding behavior. Ecotoxicol. Envir. Safety 26: 1–9.

Fu, Y., K. Hirayama & Y. Natsukari, 1991. Morphological differences between two types of the rotifer *Brachionus plicatilis* O. F. Muller. J. exp. mar. Biol. Ecol. 151: 29–41.

Hagiwara, A., T. Kotani, T. W. Snell, M. Assava-Aree & K. Hirayama, 1995. Morphology, reproduction, genetics and mating behavior of small, tropical marine *Brachionus* strains (Rotifera). J. exp. mar. Biol. Ecol. 194: 25–37.

Hagiwara, A., C.-S. Lee, G. Miyamoto & A. Hino, 1989. Resting egg formation and hatching of the S-type rotifer *Brachionus plicatilis* at varying salinities. Mar. Biol. 103: 327–332.

Hagiwara, A., N. Yamamiya & A. B. Araujo, 1998. Effect of water viscosity on the population growth of the rotifer *Brachionus plicatilis* Müller. Hydrobiologia 386/387: 489–494.

Janssen, C. R., M. D. Ferrando & G. Persoone, 1993. Ecotoxicological studies with the freshwater rotifer *Brachionus calyciflorus*. IV. Conceptual framework and applications. Hydrobiologia 255/256 (Dev. Hydrobiol. 83): 21–32.

Janssen, C. R., M. D. Ferrando & G. Persoone, 1994. Ecotoxicological studies with the freshwater rotifer *Brachionus calyciflorus*. VI. Rotifer Behavior as a sensitive and rapid sublethal test criterion. Ecotoxicol. Envir. Safety 28: 244–255.

Janssen, C. R. & G. Persoone, 1993. Rapid toxicity screening tests for aquatic biota. 1. Methodology and experiments with *Daphnia magna*. Envir. Toxicol. Chem. 12: 711–717.

Juchelka, C. M. & T. W. Snell, 1994. Using rotifer ingestion rate rapid toxicity assessment. Arch. envir. Contam. Toxicol. 26: 549–554.

Jung, M. -M., A. Hagiwara & K. Hirayama, 1997. Interspecific interactions in the marine rotifer microcosm. Hydrobiologia 358: 121–126.

Kotani, T., A. Hagiwara & T. W. Snell, 1997. Genetic variation among marine *Brachionus* strains and function of mate recognition pheromone (MRP). Hydrobiologia 358: 105–112.

Kotani, T., M. Ozaki, K. Matsuoka, T. W. Snell & A. Hagiwara, 2001. Reproductive isolation among geographically and temporally isolated marine *Brachionus* strains. Hydrobiologia 446/447 (Dev. Hydrobiol. 153): 283–290.

Maeda, M. & A. Hino, 1991. Environmental management for mass culture of the rotifer *Brachionus plicatilis*. In: Rotifer and microalgae culture systems, Proceedings of the U.S.–Asia Workshop, Oceanic Institute, Honolulu: 125–133. The Oceanic Institute Honolulu, Hawaii, U.S.A.

Moffat, B. D. & T. W. Snell, 1995. Rapid toxicity assessment using an *in vivo* enzyme test for *Brachionus plicatilis* (Rotifera). Ecotoxicol. Envir. Safety 30: 47–53.

Rumengan, I. F. M., V. Warouw & A. Hagiwara, 1998. Morphometry and resting egg production potential of the tropical ultraminute rotifer *Brachionus rotundiformis* (Manado strain) fed different algae. Bull. Fac. Fish. Nagasaki Univ. 79: 31–36

Snell, T. W. & G. Persoone, 1989a. Acute toxicity bioassays using rotifers. I. A test for brackish and marine environments with *Brachionus plicatilis*. Aquat. Toxicol. 14: 65–80.

Snell, T. W. & G. Persoone, 1989b. Acute toxicity bioassays using rotifers. II. A freshwater test with *Brachionus rubens*. Aquat. Toxicol. 14: 81–92.

Snell, T. W., R. Rico-Martinez, L. S. Kelly & T. E. Battle, 1995. Identification of a sex pheromone from a rotifer. Mar. Biol. 123: 347–353.

SPSS inc., 1997. SYSTAT 7.0: New Statistics. SPSS inc., U.S.A.

Yu, J. -P. & K. Hirayama, 1986. The effect of un-ionized ammonia on the population growth of the rotifer in mass culture. Nippon Suisan Gakkaishi 52: 1509–1513.

*Hydrobiologia* **446/447:** 369–374, 2001.
*L. Sanoamuang, H. Segers, R.J. Shiel & R.D. Gulati (eds), Rotifera IX.*




# High density culture of the freshwater rotifer, *Brachionus calyciflorus*

Heum Gi Park[1], Kyun Woo Lee[1], Sung Hwoan Cho[2], Hyung Sun Kim[3],
Min-Min Jung[4] & Hyeung-Sin Kim[4]
[1]*Faculty of Marine Bioscience & Technology, Kangnung University, Kangnung 210-702, Korea*
[2]*Uljin Marine Hatchery, National Fisheries Research & Development Institute, Kyungbook 767-860, Korea*
[3]*Korean Ocean Research and Development Institute, Ansan, P.O. Box 29, Ansan 425-600, Korea*
[4]*National Fisheries Research and Development Institute (NFRDI), South Sea Fisheries Research Institute (SSFRI), 347, Anpo-ri Hwayang-myun Yosu-si Chullanam-do, 556-820, Korea*
*E-mail: jungminmin@nfrda.re.kr jungminmin@hanmail.net*

*Key words: Brachionus calyciflorus*, dissolved oxygen, growth rate, high density culture, pH, un-ionized ammonia

## Abstract

The freshwater rotifer, *Brachionus calyciflorus* is one of the live food organisms used for the mass production of larval fish. In this study possibility of obtaining high density cultures of the freshwater rotifer *B. calyciflorus* were investigated. The two culture systems used differed in their air and dissolved oxygen supplies using three temperatures in each case: 24, 28 and 32 °C. Rotifers were batch-cultured using 5 l-vessels and fed with the freshwater *Chlorella*. The growth rate of rotifers significantly increased with an increase in temperature. The maximum density of the rotifers with air-supply at 24 °C, 6500 ind. $ml^{-1}$, was significantly lower than those cultured at 28 and 32 °C, i.e. 8600 and 8100 ind. $ml^{-1}$, respectively. Dissolved oxygen levels decreased with time and ranged from 0.8 to 1.4 mg $l^{-1}$ when the density of freshwater rotifer was the highest at each temperature. The highest density (19 200 ind. $ml^{-1}$) of freshwater rotifer was obtained in cultures with a supply of oxygen at 28 °C. Densities of 13 500 and 17 200 ind. $ml^{-1}$ were found at 24 and 32 °C, respectively. Levels of $NH_3$-N increased with time and a dramatic increase of $NH_3$-N was observed at high temperatures. Levels of $NH_3$-N at 24, 28 and 32 °C were 13.2, 18.5 and 24.5 mg $l^{-1}$, respectively. These levels coincided with the highest rotifer density at each of the three temperatures. When rotifers were cultured with an oxygen-supply and pH was adjusted to 7, the maximum density of rotifer reached 33 500 ind. $ml^{-1}$ at 32 °C . These results suggested that high density culture of freshwater rotifer, *B. calyciflorus* could be achieved under optimal conditions with DO value of exceeding 5 mg $l^{-1}$ and $NH_3$-N values of lower than 12.0 mg $l^{-1}$.

## Introduction

Rotifers are usually mass-cultured as feed for the early stages of marine larvae because of their size, nutritional value and behavior (Snell & Carrillo, 1984; Hoff & Snell, 1989; Lubzens et al., 1989). The freshwater rotifer, *Brachionus calyciflorus* is a suitable organism for ornamental freshwater fish larvae and can serve as an adequate food source (Lim & Wong, 1997; Awaiss & Kestemont, 1998).

When freshwater rotifers were cultured in fertilized ponds supplied with human and animal excreta or artificial fertilizers (Groeneweg & Schlüter, 1981; Dahril, 1997; Lim & Wong, 1997), it is difficult to control the environmental factors that heavily affect growth of rotifers, such as temperature, pH or ammonia nitrogen concentration. In addition, the density reached was relatively low, probably due to competition with other zooplankters for same food (Gilbert, 1985; Jung et al., 1997) or due to predation by other rotifers, e.g. *Asplanchna* sp. on *B. calyciflorus* in ponds (Mitchell, 1986). Therefore, this culture method was found not to be effective for mass production of larval fish in terms of labor cost and space.

It was recently reported that the inhibiting factors for growth of marine rotifer, *B. rotundiformis* were low dissolved oxygen (DO) and high level of un-ionized ammonia ($NH_3$–N) in culture medium (Yoshimura et al., 1994, 1995, 1998; Park et al., 1999a, b). In order to resolve this problem, an oxygen-supply and controlled pH were used. In addition, high-density culture was achieved when the condensed freshwater *Chlorella* was fed to rotifers (Yoshimura et al., 1994). This culture method was quite effective for obtaining stable, high-density culture of rotifers, reducing labor cost, space and facilitated the supply of large quantities of food for larval fish.

We investigated the possibility of high-density culture of a freshwater rotifer species, *B. calyciflorus*. Two culture systems were tested: (1) with air and oxygen supply but without adjustment of pH at different culture temperatures (24, 28 and 32 °C); and (2) adjusted pH and at temperature of 32 °C.

## Materials and methods

The resting eggs of *B. calyciflorus* were isolated from water from a catfish farm (Okku, Korea) in the fall of 1995 (Hur & Park, 1996). They were stored in a refrigerator at 4 °C for 4 years and hatched after being incubated at 27 °C and 3000 lux light for 36 h. To isolate the fastest growing rotifer clone, 30 hatched females were individually cultured in 20-ml test tube for 12 days. The fastest growing rotifer clone was selected and cultured at 28 °C in a 5 l-culture vessel by feeding the concentrated freshwater *Chlorella* (Chlorella Ind. Co. Ltd., Japan; average cell density, $125\times10^9$ cells $ml^{-1}$; PCV, 500 ml $l^{-1}$). This culture was later used in other experiments. The lorica length of neonate and reproducing rotifers were 136±1.4 $\mu$m and 231±1.3 $\mu$m (mean±standard error, *n*=50), respectively.

The growth of rotifer was investigated at 3 different culture temperatures (24, 28 and 32 °C) and air and oxygen gas were supplied into 5 l-culture vessels, but the pH in these cultures was not controlled. The air and oxygen gas were supplied by an air blower and an electric oxygen concentrator (NIDEK Medical, Model Mark 5 plus), respectively, and the purity of oxygen was more than 90% at a rate of 1.5 l $min^{-1}$ (0.3 VVM).

In another experiment, the growth of rotifers was measured when dissolved oxygen (DO) was maintained at values over 5.0 mg $l^{-1}$ and pH was adjusted to 7 at 32 °C. Oxygen gas was supplied by the electric oxygen concentrator and pH was adjusted by an automatic pH controller followed by the method described by Yoshimura et al. (1995).

Rotifers were inoculated at the density of 1500 ind. $ml^{-1}$ into culture vessels after an acclimation period of one day at the each of the tested conditions. Culture vessels were placed in a water bath regulated by an electric heater at the defined temperature conditions. The condensed freshwater *Chlorella* was continuously supplied to the rotifer culture vessels through a peristalic pump (Model MP-N, Eyela, Japan) delivering 6 $\mu$l concentrated *Chlorella* suspension per 1000 rotifers per half day from 9 to 21 hrs every day. Concentrated *Chlorella* was added as required into the food supply vessel that was kept at 4 °C. A nylon filter net (Nippon Bilene Co. Ltd., Japan, $0.5\times10\times15$ cm) was placed in each culture vessel to remove the organic wastes produced by rotifers and the net was washed twice daily.

The number of rotifers was counted twice daily after the dilution of each culture that was maintained at 200–300 ind. $ml^{-1}$. The DO, pH and $NH_3$–N concentrations were monitored in culture media by an ion-selective electrode (Orion, Model 920A).

Population growth rate (*r*) of rotifer was calculated from:

$$r = 1/T \ln (N_T/N_0),$$

where, $T$ is culture days that rotifer density was the highest, and $N_0$ and $N_T$ are the initial and highest rotifer density, respectively.

The data of growth rate and maximum density of rotifer were subjected to one-way analysis of variance (AVOVA) and if significant ($P<0.05$) differences were found, Duncan's multiple test (Duncan, 1955) was used to rank the groups on the SPSS Version 7.5 (SPSS Inc., Michigan Avenue, Chicago, Illinois, U.S.A.). The data are presented as mean ± SEM of two replicate groups.

## Results

The growth rate of freshwater rotifer and environmental changes (pH, DO and $NH_3$–N) in the culture media at different temperatures and with an air-supply were given in Table 1 and Figure 1, respectively. The growth rate of rotifer significantly increased with an increase in temperature. The maximum density of rotifer at 24 °C was 6500 ind. $ml^{-1}$ and was significantly lower than those of rotifers at 28 and 32 °C, which reached 8600 and 8100 ind. $ml^{-1}$, respectively, but these values did not significantly differ from each other. DO levels decreased with time and ranged from 0.8 mg $l^{-1}$ to 1.4 mg $l^{-1}$ when the density of rotifer was highest at each temperature. This was followed

*Table 1.* Growth rate and maximum density of freshwater rotifer, *B. calyciflorus* at various culture conditions

| Aeration methods | pH control | Temperature (°C) | Growth rate ($r$) | Maximum density ($\times 10^3$ind.ml$^{-1}$) |
|---|---|---|---|---|
| | | 32 | 1.124±0.0000* | 8.1±0.00* |
| Air | No | 28 | 0.870±0.0030* | 8.6±0.05* |
| | | 24 | 0.488±0.0105* | 6.5±0.20* |
| | | 32 | 1.220±0.0030* | 17.2±0.10* |
| Oxygen | No | 28 | 1.020±0.0030* | 19.2±0.15* |
| | | 24 | 0.628±0.0080* | 13.5±0.38* |
| Oxygen | 7 | 32 | 1.036±0.0055* | 33.5±0.58* |

Values (Means ± SEM of duplication) in the same column not sharing a common superscript are significantly different, $P<0.05$).

by a gradual decline in density. The pH gradually increased with time. $NH_3$–N levels increased with time and showed a dramatic increase at high temperatures. $NH_3$–N levels at 32, 28 and 24 °C were 12.2, 9.2 and 8.9 mg l$^{-1}$, respectively, when the highest density of rotifer was observed.

The growth rate of freshwater rotifer and environmental changes of the culture media at different temperatures in cultures with an oxygen-supply are shown in Table 1 and Figure 2, respectively. The growth rate of rotifer significantly increased with an increase in temperature. The maximum densities of rotifer at 28 °C was 19 200 ind. ml$^{-1}$ and 17 200 and 13 500 ind. ml$^{-1}$ were reached at 32 and 24 °C, respectively. DO levels decreased with time. The lowest DO levels at 32, 28 and 24 °C were over 8.9, 5.9 and 24.3 mg l$^{-1}$, respectively. DO levels in the culture media with an oxygen-supply were relatively higher than an air-supply. pH gradually increased with time. $NH_3$–N levels increased with time and fast increased at high temperature. $NH_3$–N levels at 32, 28 and 24 °C were 24.5, 18.5 and 13.2 mg l$^{-1}$, respectively, when the highest density of rotifer was observed.

The growth rate of freshwater rotifer cultured at 32 °C with an oxygen-supply and pH adjusted at 7 and the changes in $NH_3$–N levels are presented in Table 1 and Figure 3, respectively. The highest density of rotifer, 33 500 ind. ml$^{-1}$, was reached in 3 days after inoculation, with the $NH_3$–N level at 12.9 mg l$^{-1}$.

## Discussion

The growth rate of freshwater rotifers, *B. calyciflorus* increased with the temperature as also reported by Rico-Martinez & Dodson (1992) and Park (1998). The growth rate of the *B. calyciflrous* strain used in the current study was faster than those of rotifers, *B. calyciflrous* strains used by Rico-Martinez & Dodson (1992).

The maximum density of rotifer cultured with an oxygen-supply, at all temperatures, was twice higher than those cultured with an air supply. The density of the animals rotifer gradually decreased in cultures supplied with air and DO level was especially, <2 mg l$^{-1}$. Thus, DO level is one of limiting factors for the growth of rotifers. Therefore, oxygen levels should be maintained at >2 mg l$^{-1}$. This observation is similar to that reported for the euryhaline rotifer, *B. rotundiformis* (Yoshimura et al., 1994; Park et al., 1999a, b).

$NH_3$–N level has also been reported as limiting factors for the growth of rotifer (Lincoln et al., 1983; Yu & Hirayama, 1986; Yoshimura et al., 1995). According to Yu & Hirayama (1986), the lethal concentration (24 h-$LC_{50}$) of $NH_3$–N to euryhaline rotifer, *B. plicatilis* was 17 mg l$^{-1}$ and acute toxicity suppressing the animals' growth was around 2.1 mg l$^{-1}$. As $NH_3$–N concentrations increased with time in high density culture of rotifer, the population growth of the rotifer sharply dropped when $NH_3$–N ranged from 16.6 to 22.6 mg l$^{-1}$ at 24–32 °C (Yoshimura et al., 1995; Park et al., 1999a). The value of 24 h-$LC_{100}$ of $NH_3$–N was 17 mg l$^{-1}$ for the freshwater rotifer, *B. rubens* (Lin-

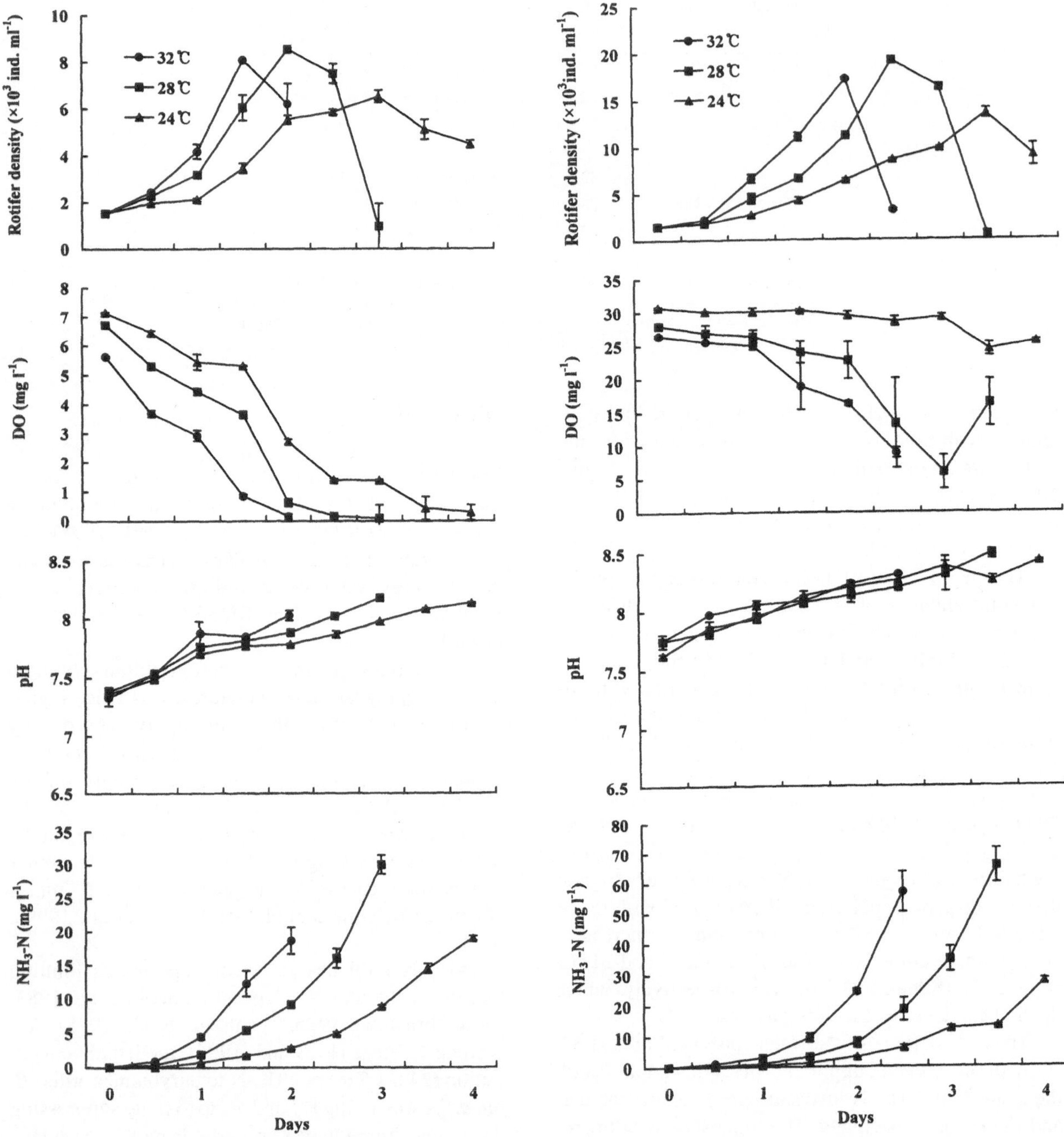


*Figure 1.* Population growth of freshwater rotifer, *Brachionus calyciflorus,* in cultures with a supply of air and at different temperatures.

*Figure 2.* Population growth of freshwater rotifer, *Brachionus calyciflorus,* in cultures with a supply of oxygen and at different temperatures.

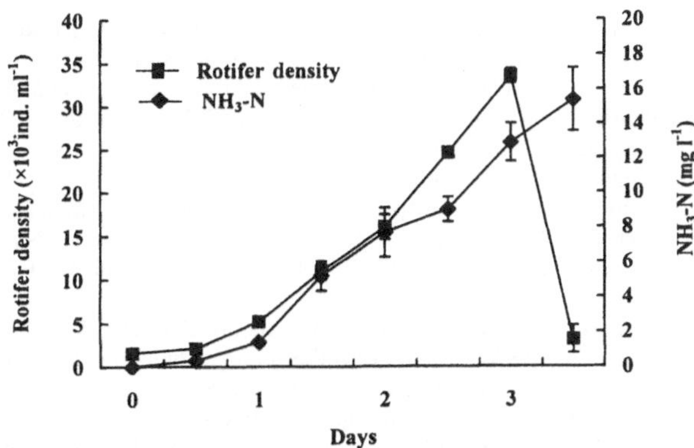


*Figure 3.* Population growth of freshwater rotifer, *Brachionus calyciflorus*, in cultures with an oxygen-supply and adjusted of pH to 7 at 32 °C.

coln et al., 1983). The population growth of rotifer *B. rubens* and *B. calyciflorus* was also suppressed when $NH_3$–N concentrations exceeded 3.0 and 5.0 mg $l^{-1}$, respectively (Schlüter & Groeneweg, 1985; Dahril, 1997). We observed that $NH_3$–N level suppressing the growth of rotifer to range from 12.4–24.5 mg $l^{-1}$ at 24 to 32 °C in the presence of oxygen supply. These ammonia values are in a similar range to that reported for the brackishwater rotifer, *B. rotundiformis* (Yoshimura et al., 1995; Park et al., 1999a). However, $NH_3$–N levels suppressing the growth of rotifer in the present study are higher than those reported for the freshwater rotifers, *B. rubens* or *B. calyciflorus* (Schlüter & Groeneweg, 1985; Dahril, 1997) and it was reduced with the decrease in culture temperature. These differences may have resulted from the differences in the length of exposure period to chronic toxicity.

Yoshimura et al. (1995) suggested that pH manipulation was effective in decreasing $NH_3$–N that inhibits the growth of rotifers and showed that stable growth of euryhaline rotifers, *B. rotundiformis* could be possible when pH was controlled at 7. Mitchell & Joubert (1986) and Mitchell (1992) reported that the optimal pH for rotifer culture ranged from 8.5 to 9.5 for *B. calyciflorus* without considering $NH_3$–N level. However, this pH range probably increases $NH_3$–N levels and it could result in the slower growth of the rotifer population. Also, population growth of rotifer, *B. calyciflorus* was suppressed at pH $<7$ (Mitchell, 1992). Yoshimura et al. (1995) demonstrated that pH of 7 was preferable for culturing euryhaline *B. rotundiformis* to pH 6 or 8. Therefore, when we adjusted the pH to 7, the rotifer reached a maximum density of 33 500 ind. $ml^{-1}$ mainly because of the slower increase in $NH_3$–N levels than in cultures without pH regulation.

Stable and cost effective cultures of rotifers are critical for healthy and effective larval fish production. High-density cultures of freshwater rotifer, *B. calycilforus* can be obtained by supplying oxygen gas and by adjustment of pH similar to the methods described for euryhaline rotifers. Therefore, our high-density-culture method of *B. calyciflorus* is more efficacious than the traditional culture method and provides a stable supply and cost-effective rotifers for larval production.

## Acknowledgements

This research was supported by the Special Grants Research of the Ministry of Maritime Affairs and Fisheries (MOMAF/SGR), Republic of Korea.

## References

Awaiss, A. & P. Kestemont, 1998. Feeding sequences (rotifer and dry diet), survival, growth and biochemical composition of African catfish, *Clarias gariepinus* Burchell (Pisces: Clariidae), larvae. Aquacult. Res. 29: 731–741.

Dahril, T., 1997. A study of the freshwater rotifer *Brachionus calyciflorus* in Pekanbaru, Riau, Indonesia. Hydrobiologia 358: 211–215.

Duncan, D. B., 1955. Mutilple-range and multiple F tests. Biometrics 11: 1–42.

Gilbert, J. J., 1985. Competition between rotifers and *Daphina*. Ecology 66: 1943–1950.

Groeneweg, J. & M. Schlüter, 1981. Mass production of freshwater rotifers on liquid wastes. II. Mass production of *Brachionus rubens* Ehrenberg 1838 in the effluent of high-rate algal ponds used for the treatment of piggery waste. Aquaculture 25: 25–33.

Hoff, F. H. & T. W. Snell, 1989. Plankton culture Manual. 2nd edn. Florida Aqua Farms, Florida: 126 pp.

Hur, S. B. & H. G. Park. 1996. Size and resting egg formation of Korean rotifer, *Brachionus plicatilis* and *B. calyciflorus*. J. Aquacult. 9: 187–194 (in Korean with English abstract).

Jung, M.-M., A. Hagiwara & K. Hirayama, 1997. Interspecific interactions in the marine rotifer microcosm. Hydrobiologia 358: 121–126.

Lincoln, E. P., T. W. Hall & B. Koopman, 1983. Zooplankton control in mass algal culture. Aquaculture 32: 331–337.

Lim, L. C. & C. C. Wong, 1997. Use of the rotifer, *Brachionus calyciflorus* Pallas, in freshwater ornamental fish larviculture. Hydrobiologia 358: 269–273.

Lubzens, E., A. Tandler & G. Minkoff, 1989. Rotifers as food in aquaculture. Hydrobiologia 186/187: 387–400.

Mitchell, S. A., 1992. The effect of pH on *Brachionus calyciflorus* Pallas (Rotifera), Hydrobiologia 245: 87–93.

Mitchell, S. A., 1986. Experiences with outdoor semi-continuous mass culture of *Brachionus calyciflorus* Pallas (Rotifera). Aquaculture 55: 289–297.

Mitchell, S. A. & J. H. B. Joubert, 1986. The effect of elevated pH on the survival and reproduction of *Brachionus calyciflorus*. Aquaculture 55: 215–220.

Park, H. G., 1998. Growth and production of resting eggs of freshwater rotifer, *Brachionus calyciflorus* Pallas at the different temperatures. J. Korean Fish. Soc. 31: 779–784 (in Korean with English abstract).

Park, H. G., K. W. Lee & S. K. Kim, 1999a. Growth of rotifer by the air, oxygen gas-supplied and the pH-adjusted and productivity of the high density culture. J. Korean Fish. Soc. 32: 753–757 (in Korean with English abstract).

Park, H. G., S. K. Kim, K. Y. Park & Y. J. Park, 1999b. High-density cultivation of rotifer, *Brachionus rotundiformis* in different diets. J. Korean Fish. Soc. 32: 280–283 (in Korean with English abstract).

Rico-Martinez, R. & S. I. Dodson, 1992. Culture of the rotifer, *Brachionus calyciflorus* Palls. Aquaculture 105: 191–199.

Schlüter, M. & J. Groeneweg, 1985. The inhibition by ammonia of population growth of the rotifer, *Brachionus rubens*, in continuous culture. Aquaculture 46: 215–220.

Snell, T. W. & K. Carrillo, 1984. Body size variation among strains of the rotifer *Brachionus plicatilis*. Aquaculture 37: 359–367.

Yoshimura, K., K. Usuki, T. Yoshimatsu, K. Tanaka, A. Ishizaki & H. Kamimura, 1998. Changes in the concentrations of ammonia and particulate organic matter and rotifer biomass in high density semi-continuous culture, Suisan Zoshoku 46: 183–192 (in Japanese with English abstract).

Yoshimura, K., C. Kitajima, Y. Miyamoto & G. Kishimoto, 1994. Factors inhibiting growth of the rotifer *Brachionus plicatilis* in high density cultivation by feeding condensed *Chlorella*. Nippon Suisan Gakkaishi 60: 207–213 (in Japanese with English abstract).

Yoshimura, K., T. Iwata, K. Tanaka, C. Kitajima & F. Ishizaki, 1995. A high-density cultivation of rotifer in an acidified medium for reducing undissociated ammonia. Nippon Suisan Gakkaishi 61: 602–607 (in Japanese with English abstract).

Yu, J. & K. Hirayama, 1986. The effect of un-ionized ammonia on the population growth the rotifer in mass culture. Bull. Jap. Soc. Sci. Fish. 52: 1509–1513.

Hydrobiologia **446/447:** 375–381, 2001.
L. Sanoamuang, H. Segers, R.J. Shiel & R.D. Gulati (eds), Rotifera IX.




# Acute toxicity tests on three species of the genus *Lecane* (Rotifera: Monogononta)

Ignacio Alejandro Pérez-Legaspi & Roberto Rico-Martínez*
*Universidad Autónoma de Aguascalientes, Centro de Ciencias Básicas, Departamento de Química, Avenida Universidad 940, C.P. 20100, Aguascalientes, Ags., México*
*(*Author for correspondence)*

*Key words:* aquatic toxicology, ecotoxicology, metal toxicity, sediment toxicity, rotifers


## Abstract

Three rotifer species, *Lecane hamata L. luna*, and *L. quadridentata*, were submitted to acute toxicity tests to compare their susceptibility to 11 toxicants. In acute tests with 48-h exposure of neonates of less than 24 h old, copper was most toxic with $LC_{50}$ values in the range of 0.06–0.33 mg $l^{-1}$, while acetone was the least toxic with $LC_{50}$ values in the range of 5000–7000 mg $l^{-1}$. Differences in $LC_{50}$ value of up to 22-fold were found in the susceptibility to lead between the three species. These data indicate large differences in toxicity among members of the same genus, and point out that it is necessary to submit several species to toxicity tests in order to assess the potential effects of toxicants to rotifers. The commonly used *Brachionus calyciflorus* cannot be considered representative of all freshwater rotifers in this respect.


## Introduction

Zooplankton composition and community structure can alter as a result of its high susceptibility to heavy metals, with consequences for higher trophic levels. Therefore, destruction of zooplankton can indirectly affect fish diversity of reservoirs. Knowledge of the toxic effects of heavy metals on zooplankton can be useful for the establishment of water quality criteria for the conservation of aquatic ecosystems and for the application of zooplanktonic as test organisms to assess effects of industrial waste (Baudouin & Scoppa, 1974).

Rotifers are preferred over other aquatic species for developing toxicity tests because of their size, ease of culture, high susceptibility to toxicants and the availability of cysts (Snell & Janssen, 1998). Snell et al. (1991) developed acute toxicity tests using the marine rotifer *Brachionus plicatilis,* and studied the susceptibility of different strains of this species to three toxicants. They found minimal differences in $LC_{50}$ values between strains, notwithstanding that one of the strains (HAW) was later found to belong to a different species, *Brachionus rotundiformis* (see Segers, 1997). Apart from this work, no data obtained by standard acute toxicity test is available on the susceptibility of closely related rotifer species to different toxicants. Recently, McDaniel & Snell (1999) obtained $LC_{50}$ values for 10 different rotifer species for cadmium and PCP. Ferrando & Andreu-Moliner (1992) investigated the effect of several petroleum derivatives, such as toluene and benzene on *Brachionus calyciflorus* and *B. plicatilis* using 24-h acute toxicity tests. They found that *B. calyciflorus* is more sensitive to these aromatic organic compounds than *B. plicatilis.* However, these two species are not among the most susceptible to these compounds (Rand & Petrocelli, 1985).

Most of the acute toxicity tests with rotifers measure mortality after an exposure period of 24 or 48-h. These tests have standardized protocols, approved by ASTM, and utilize *Brachionus calyciflorus* and *B. plicatilis* (Snell & Janssen, 1998). Perhaps the most accepted test worldwide is the 48-hour acute test using *Daphnia magna* Strauss (Hueck-Van Der Plas, 1978; Baudo, 1987). Mexico is one of the countries that embraced the *D. magna* acute test. However, since *D. magna* is a European cladoceran species which has never been found in any Mexican reservoir, there is some concern regarding the relevance of these tests to assess water quality in Mexican reservoirs (Rico-

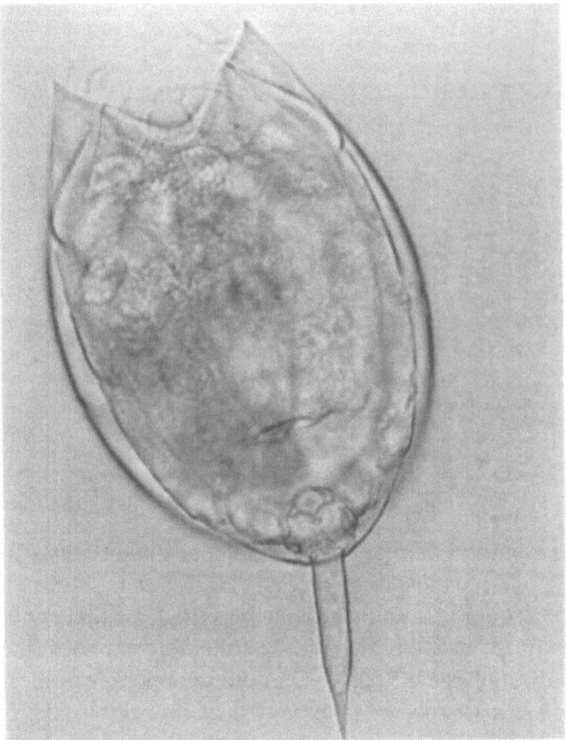

*Figure 1.* Photograph of *Lecane hamata* at 614 amplifications.

Martínez et al., 2000). In Mexico, the high influx of contaminants in to the major natural reservoirs is a constant preoccupation for local and federal authorities (see Limón et al., 1989; Flores-Tena & Silva-Briano, 1995; Rico-Martínez et al., 1997). Much of the contamination of reservoirs involves deposition of contaminants in sediments (Rand & Petrocelli, 1985; Snell et al., 1993). Yet, most recognized tests use planktonic species, which may not be the best model organisms to assess sediment toxicity. A 48-h acute toxicity tests similar to *Daphnia magna* acute test using animals better suited for living in sediments (benthic species) like members of the genus *Lecane* would circumvent these problems. Furthermore, there is little comparative information about the susceptibility of closely related species to contaminants. The aim of the our work is therefore to develop a 48-hour acute toxicity test using three littoral species of rotifers occurring in Mexico (*Lecane hamata*, *L. luna* and *L. quadridentata*), and to compare the results with data from other freshwater species.

## Materials and methods

We collected asexual females of *Lecane hamata* at Bordo Milpillas, Calvillo, Aguascalientes, Mexico, *L. luna* at Los Arquitos Dam and *L. quadridentata* at Lake Chapala (Pérez-Legaspi & Rico-Martínez, 1998). Figures 1, 2 and 3 show pictures of each of the species employed here. They belong to the genus *Lecane* (see Segers, 1995; Segers & Rico-Martínez, 2000), and were cultured continuously in our laboratory for more than 2 years. They were fed *Nannochloropsis oculata* (UTEX strain LB2194) grown in Bold's Medium (Nichols, 1973). Asexual eggs of the three species were collected from ovigerous asexual females grown at 25 °C in Petri dishes with EPA medium (U.S. EPA, 1985) 24 h prior to the start of the experiments. These were hatched by exposing the eggs to 25 °C and continuous light for a period of 24 h. The hatching percentage under these conditions is higher than 35% for all three species (Pérez-Legaspi & Rico-Martínez, 1998; Rico-Martínez unpublished observations). For the preparation of EPA we used deionized water obtained from a Water Pro PS deionization system (Labconco Co., U.S.A.). The conditions of the medium for the tests were as follows: 25 °C, pH 7.4–7.8, hardness 80–100 mg $CaCO_3$ $l^{-1}$, darkness. The tests were conducted in 24-well polystyrene plates (Costar Co, U.S.A.). Ten neonate females were placed in each well filled with a test volume of 0.5 ml. The test, being a 48-h static test, required no feeding or renewal of the medium.

We performed range-finding tests on *L. quadridentata* using five concentrations for each toxicant. Test concentrations are a logarithmic series of five concentrations and a control. These series ranged from $10^{-8}$ to $5 \times 10^4$ mg $l^{-1}$ for the 11 toxicants employed here. The highest concentration without mortality and the lowest concentration with 100% mortality were chosen as the lower and upper limits, respectively, for the definitive test. Three intermediate concentrations were included in the definitive test. Usually, the range for the definitive test found for *L. quadridentata* was appropriate to the other two species. Fifty test animals (10 per well, five replicates, at different times) were used for each control and toxicant concentration. Reference chemicals of the highest purity available were used. The toxicants tested were: acetone (J. T. Baker Co., U.S.A.), benzene, ethyl acetate, mercuric chloride, SDS and toluene (Sigma Co., U.S.A.) vinyl acetate (Supelco Co., U.S.A.) and atomic absorption standards (Sigma Co., U.S.A.) of the metals cadmium,

*Table 1.* Analysis of the acute toxicity tests performed on three different species of *Lecane*. Abbreviations correspond to the following; $LC_{50}$ = Lethal concentration where 50% of animals die. NOEC = No Observed Effect Concentration. $\chi_2$ = Results of the chi-square test. CV = Coefficients of variation. CL = 95% Confidence limits for the $LC_{50}$ values. $r^2$ = Coefficient of determination for those results not validated with the chi-square ($\chi^2$) test (see 'Materials and methods')

| Compound | | *L. hamata* | | | | | *L. luna* | | | | | *L. quadridentata* | | | | |
|---|---|---|---|---|---|---|---|---|---|---|---|---|---|---|---|---|
| | | Value $mg\ l^{-1}$ | CV % | CL $mg\ l^{-1}$ | $r^2$ | $\chi^2$ | Value $mg\ l^{-1}$ | CV % | CL $mg\ l^{-1}$ | $r^2$ | $\chi^2$ | Value $mg\ l^{-1}$ | CV % | CL $mg\ l^{-1}$ | $r^2$ | $\chi^2$ |
| Acetone | NOEC | 7000 | 19.2 | | | | 6000 | 22.2 | | | | 5000 | 26.6 | | | |
| | $LC_{50}$ | 7235 | 98.0 | 6872–7597 | 0.44 | * | 6833 | 101 | 6203–7527 | | ** | 5651 | 122 | 5147–6204 | | ** |
| Benzene | NOEC | 5000 | 17.0 | | | | 2000 | 38.3 | | | | 2000 | 56.5 | | | |
| | $LC_{50}$ | 6975 | 110 | 5828–8347 | | ** | 3762 | 47.0 | 2870–6230 | 0.97 | * | 2834 | 101 | 2415–3325 | | ** |
| Cadmium | NOEC | 0.05 | 36.6 | | | | 0.250 | 19.4 | | | | 0.05 | 34.2 | | | |
| | $LC_{50}$ | 0.23 | 60.0 | 0.090–0.300 | 0.99 | * | 0.350 | 63.0 | 0.27–0.45 | | ** | 0.28 | 53.0 | 0.02–0.81 | 0.87 | * |
| Chromium | NOEC | 3.00 | 19.2 | | | | 3.000 | 8.30 | | | | 3.00 | 47.9 | | | |
| | $LC_{50}$ | 4.41 | 89.0 | 1.730–8.720 | 0.91 | * | 3.260 | 83.0 | 2.91–3.64 | 0.71 | * | 4.50 | 82.0 | 1.80–10.5 | 0.91 | * |
| Copper | NOEC | 0.15 | 27.8 | | | | 0.020 | 45.1 | | | | 0.05 | 53.8 | | | |
| | $LC_{50}$ | 0.23 | 34.0 | 0.180–0.290 | | ** | 0.060 | 70.0 | 0.04–0.08 | | ** | 0.33 | 57.0 | 0.08–0.41 | 0.89 | * |
| Ethyl | NOEC | 50.0 | 57.6 | | | | 1000 | 25.6 | | | | 500.0 | 78.0 | | | |
| acetate | $LC_{50}$ | 1324 | 41.0 | 1258–1390 | 0.66 | * | 2606 | 118 | 1337–6011 | 0.91 | * | 1600 | 83.0 | 216–2207 | 0.78 | * |
| Lead | NOEC | <0.5 | 49.4 | | | | 0.100 | 37.9 | | | | 2.50 | 68.1 | | | |
| | $LC_{50}$ | 0.68 | 16.0 | 0.510–0.910 | | ** | 0.140 | 23.0 | 0.09–0.19 | | ** | 3.70 | 77.0 | 1.47–6.45 | 0.96 | * |
| Mercuric | NOEC | 0.50 | 31.7 | | | | 0.100 | 58.0 | | | | 0.05 | 67.7 | | | |
| chloride | $LC_{50}$ | 1.37 | 81.0 | 0.240–2.810 | 0.95 | * | 0.450 | 79.0 | 0.12–0.64 | 0.90 | * | 0.40 | 72.0 | 0.06–0.75 | 0.96 | * |
| Titanium | NOEC | 5.00 | 18.5 | | | | 5.000 | 18.5 | | | | <5.0 | 18.6 | | | |
| | $LC_{50}$ | 15.6 | 60.0 | 9.100–22.00 | 0.89 | * | 11.90 | 43.0 | 7.00–18.1 | 0.87 | * | 8.5.0 | 45.0 | 2.26–15.8 | 0.8 | * |
| Toluene | NOEC | 200.0 | 31.4 | | | | 200.0 | 17.0 | | | | 175.0 | 18.9 | | | |
| | $LC_{50}$ | 236.7 | 111 | 125.0–430.0 | 0.91 | * | 277.4 | 39.0 | 243–316 | | ** | 191.4 | 89.0 | 176–257 | 0.76 | * |
| Vinyl | NOEC | 200.0 | 3.40 | | | | 300.0 | 26.5 | | | | 200.0 | 63.9 | | | |
| acetate | $LC_{50}$ | 331.8 | 110 | 314.0–350.0 | | ** | 303.4 | 72.0 | 164–493 | 0.92 | * | 320.1 | 101 | 267–383 | | ** |

* $p>0.05$; ** $p<0.05$; The $\chi^2$ test with a value less than 0.05 means that the adjustment to linearity made by the $DL_{50}$ program is valid for the $LC_{50}$ value. For this reason a $r^2$ value is not presented in these cases. When the adjustment made by the program was not significant, we calculated the $LC_{50}$ value using the program Statistica and giving the correspondent coefficient of determination ($r^2$)

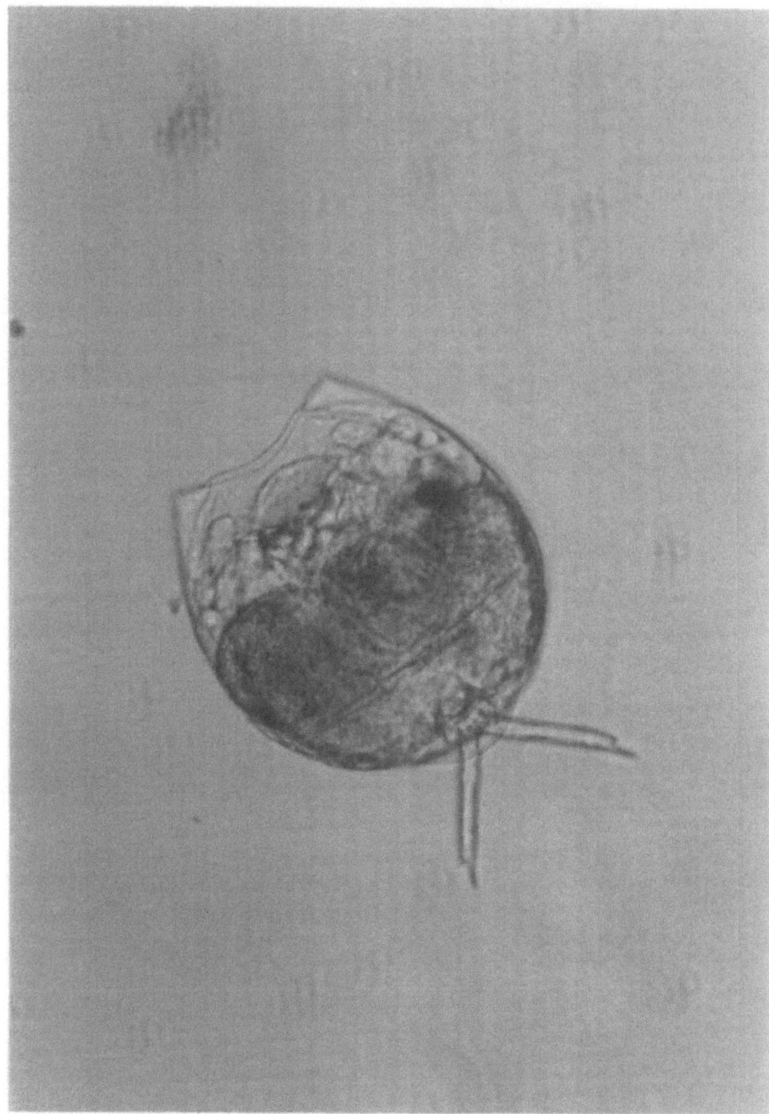

*Figure 2.* Photograph of *Lecane luna* at 338 amplifications.

chromium, copper, lead and titanium. No analytical chemistry was performed, so all toxicants were tested at their nominal concentrations. Mortality was scored after 48 h and the data were analyzed by means of the $DL_{50}$ software which calculates the $LC_{50}$ values and performs a chi-square to test the validity of the linearity of the regression ($p$-level $<0.05$) (Rico-Martínez et al., 2000). The software package Statistica (Statsoft Inc., 1993) was used to determine NOEC and $r^2$ values. Tests were considered valid only if mortality was less than 10% in the control. We performed solvent controls with DMSO (Sigma Co., U.S.A.) as the solvent, to monitor the toxicity due to solvents in the case of benzene.

## Results

We used sodium dodecyl sulfate (SDS) as reference chemical, as the Mexican Norm for Acute Toxicity with D. magna (NOM-074-ECOL-1994) suggests it as reference toxicant, because of its non-selective toxicity. We obtained a $LC_{50}$ value of 0.745 mg $l^{-1}$ using a range of concentrations of 0.3, 0.6, 0.9, 1.2 and 1.6 mg $l^{-1}$ for *Lecane quadridentata*. The 95% CL were 0.31 and 1.16, and the CV was of 41.4% for

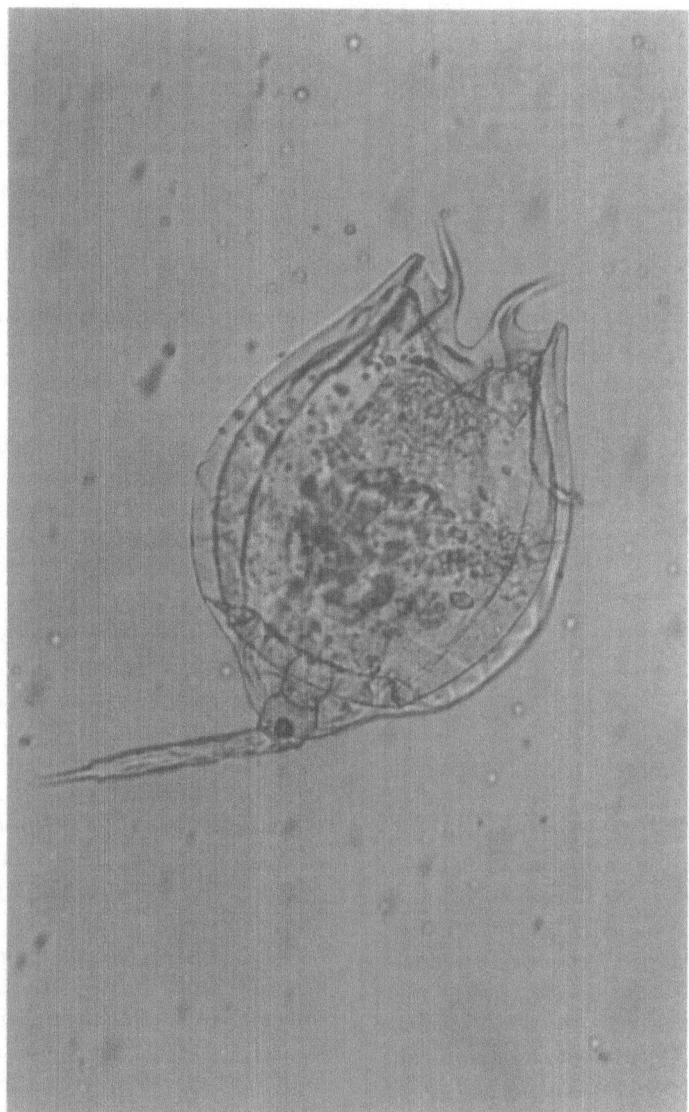

*Figure 3.* Photograph of *Lecane quadridentate* at 382 amplifications.

five replicates performed in two different months. The NOEC value for SDS was 0.3 mg $l^{-1}$.

Copper was the most toxic compound of the eleven chemicals investigated. The $LC_{50}$ values for copper are 0.61–0.33 mg $l^{-1}$ for the three species. On the other hand, acetone was the least toxic compound with $LC_{50}$ values of 5651 to 7235 mg $l^{-1}$ (Table 1). Lead was the toxicant with the highest difference among the three species. $LC_{50}$ values for lead ranged from 0.14 to 3.78 mg $l^{-1}$, which gives a ratio of 22.0 between the highest and the lowest $LC_{50}$values. Vinyl acetate was the toxicant whose $LC_{50}$ values were most stable among the three species, with values ranging from 303.4 to 331.8 mg $l^{-1}$ and a ratio of 1.09. Four other toxicants: acetone, cadmium, chromium and toluene have ratios of 1.28–1.50 between the highest and lowest $LC_{50}$ values. This result is understandable given the phylogenetic closeness of the three species (Segers & Rico-Martínez, 2000). Table 2 compares the susceptibility of each of the three *Lecane* species used in this work, and indicates the most susceptible species for each toxicant.

*Table 2.* Comparison of $LC_{50}$ values among the three *Lecane* species. Abbreviations correspond to the following; L.h. = *Lecane hamata*; L.I. = *Lecane luna*; L.q. = *Lecane quadridentata*. The columns represent the ratios of the $LC_{50}$ values

| Compound (mg $l^{-1}$ | L.h./L.q. | L.h./L.I. | L.I./L.q. | Most susceptible species |
|---|---|---|---|---|
| Acetone | 1.28 | 1.06 | 1.21 | L.q. |
| Benzene | 2.46 | 1.86 | 1.33 | L.q. |
| Cadmium | 0.83 | 0.66 | 1.26 | L.h. |
| Chromium | 0.97 | 1.56 | 0.62 | L.I. |
| Copper | 0.70 | 3.78 | 0.18 | L.I. |
| Ethyl acetate | 0.82 | 0.51 | 1.62 | L.h. |
| Lead | 0.18 | 4.87 | 0.04 | L.I. |
| Mercuric chloride | 3.42 | 3.04 | 1.12 | L.q. |
| Titanium | 1.84 | 1.31 | 1.40 | L.q. |
| Toluene | 1.24 | 0.85 | 1.45 | L.q. |
| Vinyl acetate | 1.04 | 1.09 | 0.95 | L.I. |

## Discussion

There is little information on standardized mortality tests using rotifers (Snell & Janssen, 1998). There are 24-h standard tests using neonates hatched from cysts for three species only (*Brachionus calyciflorus*, *B. plicatilis* and *B. rubens*). Furthermore, data on 48-h tests are rare (Buikema Jr. et al., 1982), and exist only for a non-standardized test using the bdelloid *Philodina acuticornis odiosa*. Moreover, these tests are without control of the age of test animals, and mostly use salts of different metals, instead of atomic absorption standards or chemicals of the highest purity.

Comparative studies on toxicant susceptibility among closely related species are equally rare. There are some data for *Daphnia* (Lewis & Weber, 1985) and *Brachionus* (Snell & Persoone, 1989a, b; Snell et al., 1991), for a small number of toxicants tested. Our results are generally similar to those of the above-mentioned authors. There is great similarity in the susceptibility of a particular toxicant among the three *Lecane* species tested. *L. quadridentata* is most susceptible to 5 of the 11 toxicants tested, *L. luna* is most susceptible to 4 toxicants, and *L. hamata* is most susceptible to cadmium and ethyl acetate.

Our coefficient of variation for $LC_{50}$s ranged from 3.4 to 78%, with an average of 33.6±8.4%. This value is higher than those reported for other rotifer species (Snell & Persoone, 1989a, b; Snell et al., 1991), but similar to those reported for *Daphnia magna* and *D. pulex* (see Lewis & Weber, 1985). *Lecane hamata* has an average CV of 28.3±14.7%, the CV for *L. luna* is 27.0±13.5% and for *L. quadridentata* 45.4±21.7%.

McDaniel & Snell (1999) found a $LC_{50}$value of 0.046 mg $l^{-1}$ for cadmium in a 24-h acute test for *Lecane quadridentata*. Our value of 0.28 mg $l^{-1}$ is more than six times higher, and was obtained in a 48-hour test. Both protocols use cadmium from an atomic absorption standard from Sigma. However, some differences exist between both protocols. Whereas McDaniel & Snell (1999) used small, non-ovigerous females, we used less than 24-h old neonates. However, this alone cannot explain the difference in test results obtained. In fact, neonates are thought to be more vulnerable to toxicants than adults. The Atlanta *Lecane quadridentata* strain could be more sensitive to cadmium than ours.

Our results of $LC_{50}$ values compare well with those of other freshwater rotifer species. For instance, the $LC_{50}$ values of the three *Lecane* species for cadmium and lead are lower than those of *Brachionus calyciflorus* (Snell et al., 1991). In contrast, our values for benzene and toluene are slightly higher for the three species than those of *B. calyciflorus* (Snell et al., 1991), and those of *B. plicatilis* (Ferrando & Andreu-Moliner, 1992). In the case of copper, *L. luna* is more sensitive than *B. calyciflorus*, but the other two *Lecane* species have $LC_{50}$ values higher than *B. calyciflorus*. The $LC_{50}$ values for cadmium and copper in the marine rotifer *Brachionus plicatilis* are in the range of our values (Snell & Persoone, 1989a). In the case of these two metals, *Daphnia magna* is more susceptible than rotifers (Lewis & Weber, 1985; Rico-Martínez et al., 2000). Our $LC_{50}$ values for acetone are much higher than those of *B. calyciflorus* (Snell et al., 1991). In a comparison of toxicant susceptibility between the three *Lecane* species used here and *Daphnia magna* EPA strain raised in our laboratory (Rico-Martínez, 1999), the *Lecane* species were more susceptible to lead than *D. magna*, and have similar susceptibility for cadmium, chromium, ethyl acetate and titanium. However, they were less sensitive to benzene, copper, mercuric chloride, toluene and vinyl acetate. The differences in all these 6 cases are within one order of magnitude or less. Similar results are reported by Snell et al. (1991) when they compared the susceptibilities of *Brachionus calyciflorus* to 12 toxicants with those of *Daphnia magna* and *Pimephales promelas*.

## Acknowledgement

We thank Martha Evelia Pérez Reyes for help in the laboratory.

## References

Baudo, R., 1987. Ecotoxicological testing with *Daphnia*. In Peters, R. H. & R. De Bernardi (eds), *Daphnia*. Mem. Ist. ital. Idrobiol. 45: 461–482.

Baudouin, M. F. & P. Scoppa, 1974. Acute toxicity of various metals to freshwater zooplankton. Bull. Envir. Cont. Toxic. 12: 745–751.

Buikema, A. L., Jr., B. R. Niederlehner & J. Cairns, 1982. Biological monitoring. Part IV – Toxicity Testing. Wat. Res. 16: 239–262.

Ferrando, M. D. & E. Andreu-Moliner, 1992. Acute toxicity of toluene, hexane, xylene and benzene to the rotifers *Brachionus calyciflorus* and *Brachionus plicatilis*. Bull. Envir. Cont. Toxic. 49: 266–271.

Flores-Tena, F. J. & M. Silva-Briano, 1995. A note on El Niàgara, a polluted reservoir in Aguascalientes, Mexico. Hydrobiologia 308: 235–241.

Hueck- Van Der Plas, E. H., 1978. Experiences with an inventory of ecological tests based on an enquiry by the OECD chemicals group. In Tests for the ecological effects of chemicals. Proc. Research Seminar, 7–9. Berlin. Erich Schmidt Verlag, Berlin: 63–73.

Lewis, P. A. & C. I. Weber, 1985. A study of the reliability of *Daphnia* acute toxicity tests. In Cardwell, R. D., R. Purdy & R. C. Bahner (eds), Aquatic Toxicology and Hazzard Assessment Seventh Symposium. ASTM STP 854, American Society for Testing and Materials. Philadelphia, U.S.A.: 73–86.

Limón, J. G. M., O. T. Lind, D. S. Vodopich, R. Doyle & B. G. Trotter, 1989. Long- and short-term variation in the physical and chemical limnology of a large, shallow, turbid tropical lake (Lake Chapala, Mexico). Arch. Hydrobiol. Suppl. 83(1): 57–81.

McDaniel, M. & T. W. Snell, 1999. Probability distributions of toxicant sensitivity for freshwater rotifer species. Envir. Toxicol. 14: 361–366.

Nichols, H. W., 1973. Growth media – freshwater. In Stein, J. R. (ed.), Handbook of Phycological Methods. Cambridge University Press. Cambridge, MA. U.S.A.: 7–24.

Pérez-Legaspi, I. A. & R. Rico-Martínez, 1998. Effect of temperature and food concentration in two species of littoral rotifers. Hydrobiologia 387/388: 341–348.

Rand, G. M. & S. R. Petrocelli, 1985. Fundamentals of Aquatic Toxicology. Washington Hemisphere Publishing Corporation. Washington D.C.: 670 pp.

Rico-Martínez, R., A. M. Jímenez-Rodríguez, C. A. Velázquez-Rojas & I. A. Pérez-Legaspi, 1997. Desarrollo de bioensayos toxicológicos y su aplicación en programas de monitoreo de la calidad de pozos y tomas de agua del Municipio de Aguascalientes. Memoria del Cuarto Simposio Estatal: La investigación y el desarrollo tecnológico en Aguascalientes, Aguascalientes, Mexico: 150–154.

Rico-Martínez R., 1999. Reporte final del proyecto RN-20/96 CONACyT SIHGO: Desarrollo de bioensayos toxicológicos y su aplicación en programas de monitoreo de la calidad de pozos y tomas de agua del Municipio de Aguascalientes. Consejo Nacional de Ciencia y Tecnología (CONACYT). Sistema Regional Hidalgo (SIHGO), México: 107 pp.

Rico-Martínez, R., C. A. Velázquez-Rojas & I. A. Pérez-Legaspi & G. E. Santos-Medrano, 2000. "The use of aquatic invertebrate toxicity tests and invertebrate enzyme biomarkers to assess toxicity in the states of Aguascalientes and Jalisco, Mexico." In (Butterworth, F. M., A. Gunatilake & M. E. Gonsebatt Bonaparte (eds), Biomonitors and Biomarkers as Indicators of Environmental Change, Volume 2. Plenum Press (in press).

Segers, H., 1995. Guides to the Identification of the Microinvertebrates of the Continental Waters of the World: Volume 2. Rotifera: The Lecanidae (Monogononta). SPB Academic Publishing: 191–193, 92–93, 144–145.

Segers, H., 1997. Nomenclatural consequences of some recent studies on *Brachionus plicatilis* (Rotifera: Brachionidae). Hydrobiologia 313/314: 121–125.

Segers, H. & R. Rico-Martínez, 2000. The male of *Lecane bulla* (Gosse, 1851): new support for the synonymy of *Lecane* Nitzsch, *Monostyla* Ehrenberg and *Hemimonostyla* Bartos. J. Nat. Hist. 34: 679–683.

Snell, T. W. & G. Persoone, 1989a. Acute toxicity bioassays using rotifers. I. A test for brackish and marine environments with *Brachionus plicatilis*. Aquat. Toxicol. 14: 65–80.

Snell, T. W. & G. Persoone, 1989b. Acute toxicity bioassays using rotifers. II. A freshwater test with *Brachionus rubens*. Aquat. Toxicol. 14: 81–92.

Snell, T. W., B. D. Moffat, C. Janssen & G. Persoone, 1991. Acute toxicity Tests Using Rotifers. IV. Effects of Cysts Age, Temperature and Salinity on the Sensitivity of *Brachionus calyciflorus*. Ecotoxicol. Envir. Safety 24: 308–317.

Snell, T. W., D. Dusenbery, L. Dunn & N. Walls, 1993. Biomarkers for managing water resources. Georgia Institute of Technology. Environmental Resources Center. ERC 02–93 Publication, Atlanta, Georgia, U.S.A.: 43 pp.

Snell, T.W. & C. R. Janssen, 1998. Microscale toxicity testing with rotifers. In Wells, P. G., K. Lee & Ch. Blaise (eds), Microscale Testing in Aquatic Toxicology, Advances, Techniques and Practice. CRC Press: 409–422

U.S. Environmental Protection Agency, 1985. Methods for measuring the acute toxicity of effluents to freshwater and marine organisms. EPA-600/4-85-013, U.S.A.: Environmental Protection Agency, Washington D.C., U.S.A. 159 pp.

Hydrobiologia **446/447:** 383–392, 2001.
L. Sanoamuang, H. Segers, R.J. Shiel & R.D. Gulati (eds), Rotifera IX.




# Studies on *Brachionus* (Rotifera): an example of interaction between fundamental and applied research

M. Yúfera
*Instituto de Ciencias Marinas de Andalucía (CSIC), Apartado Oficial, 11510 Puerto Real, Cádiz, Spain*
*E-mail: manuel.yufera@icman.csic.es*

*Key words: Brachionus*, bibliography, applied research, fundamental research

## Abstract

The genus *Brachionus* has been the main subject of studies reported in about 1000 papers published since 1950. About three-fourths of these deal with *Brachionus plicatilis* and *B. rotundiformis* and are mainly related to their use as prey for aquatic organisms. Also abundant, but to a lesser extent, are studies on *B. calyciflorus*, many of which are concerned with aquatic ecotoxicology. These studies constitute an interesting interaction between fundamental and applied research. For example, advances in fundamental biology have been applied to improve the production of rotifer biomass. Alternatively, new perspectives in fundamental research on rotifers have emerged while solving technical and biological problems related to the rearing of aquatic animal larvae. This review describes some aspects that have shown a significant advance due to such interaction between fundamental and applied research on rotifers, e.g. growth conditions, biochemical composition and morphotypes.

## Introduction

Species of the genus *Brachionus* (Brachionidae: Rotifera) are well represented in different water bodies worldwide (Pejler, 1977). They constitute a group of organisms that are quite important in aquatic ecosystems but also play a notable role in technological research for exploiting aquatic living resources. In the mid-1950s, Ito (1955) published the first paper of his series on the Mizukawari phenomenon, which culminated with the publication of the possibilities for using *B. plicatilis* as a controlled biomass source for the fledgling aquaculture industry (Ito, 1960). Twenty years later, studies of the mass production of this species had become universal. During the 1960s, other species began to be studied in laboratory experiments (Erman, 1962; Galkovskaya, 1963; Pourriot, 1965; Vasilieva, 1968).

The possibility of studying these organisms in the laboratory has promoted a strong increase in the quantity and variability of scientific papers on some species of *Brachionus*. The aim of this paper is to examine the scientific papers in the last 50 years, to analyse the research trends on the genus *Brachionus*, to assess whether applied research has benefited from fundamental biology and whether it has contributed to advances in the biology of the group.

## Scientific studies

The databases ASFA, AGRIS and BIOSIS were used to analyse the literature. These analyses were completed with help of Rotifer News (ISSN 1327-4007) and the author's own files. Single abstracts and unpublished material have not been included. Likewise, those papers with the word *Brachionus* in the title or in key words but in which the real information reported about this genus was not relevant were not considered.

Since 1950, more than 1000 papers have been published in which a species belonging to genus *Brachionus* is the main subject of the study (Fig. 1). During the 1970s, the number of publications reached 30 papers per year for the first time. The maximum, in 1993, was 81 published papers dealing with *Brachionus* species. Such high numbers of publications per year are due to the appearance of proceedings from rotifer and aquaculture symposia. About 75% of the papers deal with *Brachionus plicatilis* + *B. rotundiformis*. Also abundant, but to a lesser extent are studies

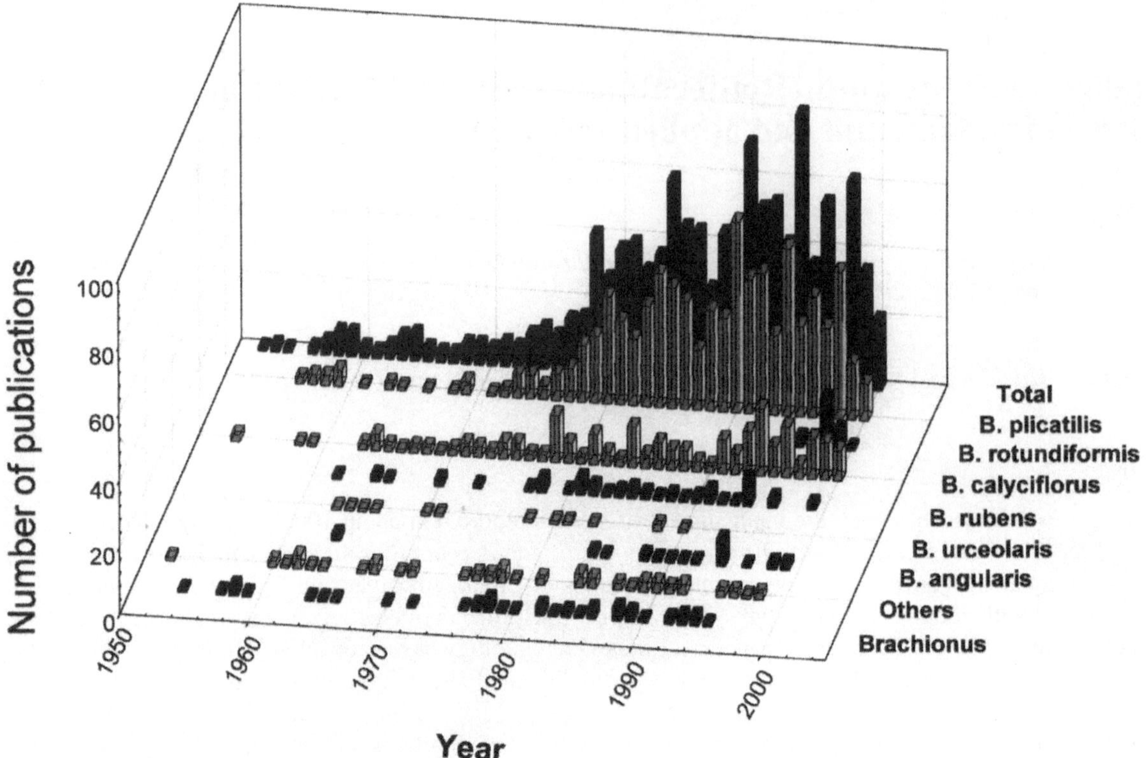


*Figure 1.* Publications on the different species of *Brachionus* from 1950 to 2000. Species below 10 publications are grouped in 'Other species'. Publications about the genus in general are grouped in '*Brachionus*'.

on *B. calyciflorus*, which account for about 20%. The third species with regard to number of publications is *B. rubens* (5%), followed by *B. angularis* (2.5%) and *B. urceolaris* (1%). Obviously, the main criterion for choice of the species studied has not been their abundance in nature but their usefulness in aquatic technology.

As for zooplanktonic organisms, the first step in acquiring information on the biology of *Brachionus* came from the studies on distribution and systematics. In many studies, several species of this genus appear in faunistic lists, but only a few field studies focused attention on the genus *Brachionus* spp. (e.g. Ito, 1955; King & Zhao, 1987; Arndt, 1988; Guisande & Toja, 1988; Arndt & Heerkloss, 1989; Roche, 1995). Thus, most of the information on the biology of these species has been obtained in laboratory experiments.

The main research fields in this genus can be classified as follows: (a) distribution and systematics; (b) population dynamics, physiology and energetics; (c) sexual reproduction and resting egg induction; (d) biochemical composition; and (e) ultrastructure, histology and histochemistry. These topics cannot be always considered in isolation and some of them are usually studied together. In addition, many of these studies are related to the technological applicability for aquaculture and ecotoxicology. In fact, most of studies on population dynamics and biochemistry are related to mass production and the use of rotifers as larval prey in the aquaculture industry. On the other hand, studies on toxicology also include population dynamics, and in general the influence of abiotic factors.

The relative attention that the different topics received is shown in Figure 2. The aspect that has been studied most is population dynamics. How much biomass is produced under given environmental conditions and how different populations are growing per time unit seem to be among the most interesting questions for *Brachionus* studies. Studies on sexual reproduction and induction of mixis also merit special attention, because this is one of the first topics that captured the interest of researchers. Many studies have used mainly *B. plicatilis* and *B. calyciflorus* (Pourriot, 1957; Gilbert, 1963; Ruttner-Kolisko, 1964;

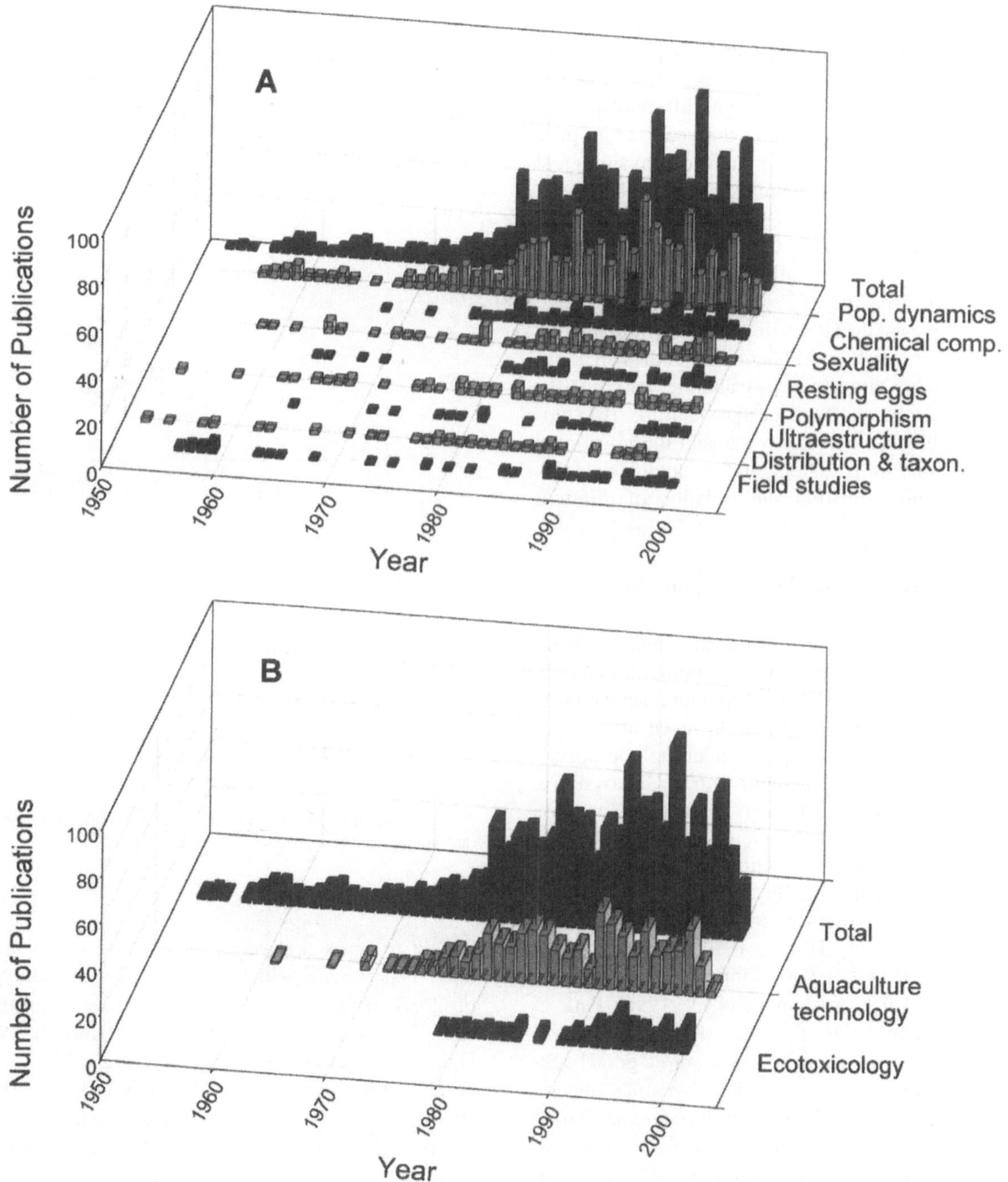


*Figure 2.* (A) Publications on *Brachionus* on the different fields of research from 1950 to 2000. (B) Publications related to aquaculture technology and ecotoxicology.

Pozuelo, 1977; Ben-Amotz & Fishler, 1982; Snell, 1986; Rico-Martínez. & Snell, 1995).

Another classic field of research in rotifers, including *Brachionus*, is morphometry. The intraspecific variability in the body shape and size induced by environmental factors or as an expression of genotype differences has been studied in different species. Methodology has changed over the years, from pure morphometric analysis to the study of physiological responses and genetic variation. Studies on biochemistry (Ogino & Watanabe, 1978; Scott & Baynes, 1978) and ultrastructure (Clément, 1977; Clément et al., 1980; Kleinow & Wratil, 1995; Yu & Cui, 1997) were started during the 1970s. Some of these topics are good examples of the interaction between fundamental and applied research: (a) fundamental studies on population dynamics and physiology and the searching for the best growth conditions in mass culture; (b) studies on the biochemistry of rotifer and the nutritional quality as food for predators; and (c) studies on polymorphism and speciation and isolation of different sized strains to be used as prey.

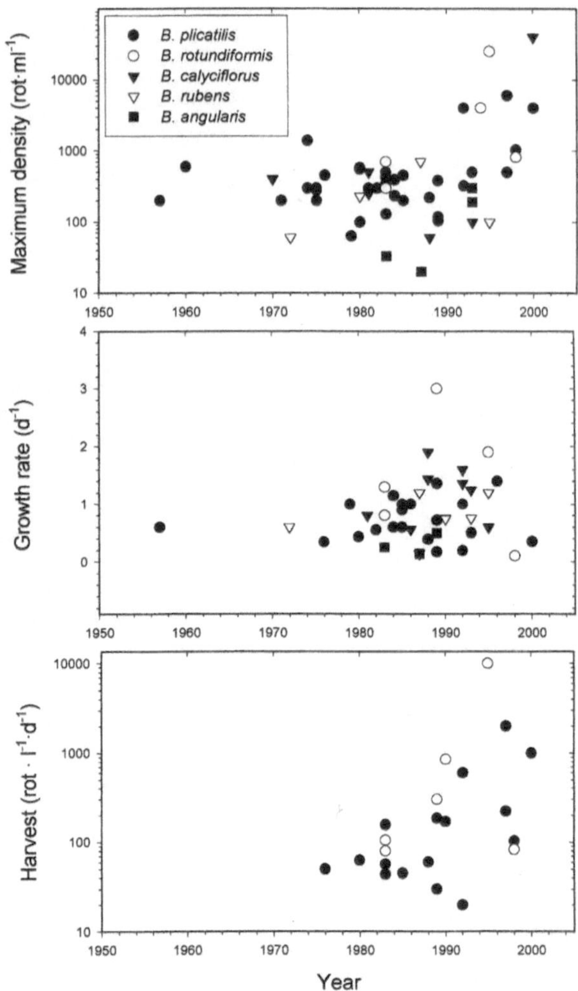


*Figure 3.* Maximum values reported in population density, growth rate and mass culture production in different species of *Brachionus*.

## Population dynamics and growth conditions

The main aim of the studies on *Brachionus* has been to examine the best conditions for population growth. The first studies in laboratory population showed that different *Brachionus* species could reach high population density and growth rates under the appropriate conditions (Ito, 1960; Erman, 1962; Galkovskaya, 1963). During the last 30 years, techniques of mass culture have allowed a continuous increase in the efficiency of mass production, with new culture systems, which yielded progressively higher rotifer densities and productions (Hirata & Mori, 1967; Theilacker & McMaster, 1971; Hirata, 1974; Yúfera & Pascual, 1980; Gatesoupe & Robin, 1981; James & Abu-Rezeq, 1989; Yoshimura et al, 1996; Park et al., 2001; Suantika et al., 2000) (Fig. 3). Improvements in mass culture technology required good background knowledge on population dynamics and on the microalgae-rotifer system of *Brachionus*. These studies have at the same time contributed strongly to advances in the knowledge of factors affecting population growth while searching for the best growth conditions (Hirayama & Kusano, 1972; Hirayama et al., 1973; Yúfera et al., 1983; Nagata, 1985a; Yúfera, 1987).

Data on population dynamics and physiology also have been obtained in studies on species that are not related to mass production technology, from laboratory experiments which contributed significantly to progress in this field (Halbach, 1970a; Pourriot & Rougier, 1975; Pilarska, 1977a, b; Boraas, 1983; Galkovskaya et al., 1987; Schlüter at al., 1987; Walz et al., 1989; Charoy, 1995; Galkovskaya, 1995; Walz, 1995). Many of these studies deal with feeding behaviour and regulation of the feeding activity in these suspension feeders (Hirayama & Okawa, 1972; Pilarska, 1977a; Gilbert & Starkweather, 1977; Chotiyaputta & Hirayama, 1978; Starkweather et al., 1979; Schlosser & Anger, 1982; Nagata, 1985b; Yúfera & Pascual, 1985; Walz & Gschloessl, 1988; Rothhaupt, 1990; Lebedeva & Orlenko, 1995; Hansen et al., 1997).

The other important application of population dynamics and physiology research is in ecotoxicology studies. The sensitivity of changes in demographic and physiological characteristics to toxicant compounds, the high fecundity and short life cycle and their universal distribution make the *Brachionus* species a useful tool for testing natural or artificial toxicants in aquatic ecosystems (Capuzzo, 1979; Halbach et al., 1983; Jansen et al., 1993; Ferrando et al., 1993; Charoy et al., 1995; Snell & Jansen, 1995; Moreno-Garrido et al., 1999). The two most commonly used *Brachionus* species in toxicological studies are *B. calyciflorus* and *B. plicatilis* (Snell & Jansen, 1995), although other species have also been used (Halbach et al., 1983; Rao & Sarma, 1986).

On the other hand, although mainly related with research in sexual reproduction, a particular field in population dynamics, due to the applicability of the results, is the study on the resting egg production (Gilbert, 1974; Lubzens, 1981; Pourriot & Snell, 1983; Hagiwara, 1994; Hagiwara et al., 1997) (Fig. 2).

## Biochemical composition

Biochemical composition and energy content in *Brachionus* have been analysed with two objectives in mind: Firstly, for examining the growth conditions and health status in relation to biotic and abiotic factors, i.e. considering the global process of transfer of matter and energy to higher trophic levels. Secondly, because of the importance of the nutritional quality of rotifers as food for larval fish.

The calorimetric value of *Brachionus* has been determined in several studies (Doohan, 1973; Pilarska, 1977b; Scott, 1980; Theilacker & Kimball, 1984; Szyper, 1989; Yúfera & Pascual, 1989; Yúfera et al., 1997). The biochemical composition of rotifers has been studied almost exclusively in *B. plicatilis* and *B. rotundiformis* though there are also some studies on *B. calyciflorus* (Mityanina, 1985; Galkovskaya et al., 1987; Guisande & Serrano, 1989; Awais et al., 1992). Figure 4 shows the frequency in which a given value of protein, lipids, carbohydrates and ash, have been reported in the literature. The source of variation in these data is not only growth and feeding conditions of the population but methodology, too. The latter mostly affects the protein determination due to the different methods used to determine nitrogen, nitrogenous material including free amino acids or crude proteins (Scott & Baynes, 1978; Caric et al., 1993; White, et al., 1994; Yúfera et al., 1997; Øie & Olsen, 1997; Makridis & Olsen, 1999).

Studies on the lipid content and composition of fatty acids are the most numerous among the papers dealing with chemical composition. Fatty acid composition has generally been determined from the point of view of larval nutrition and of manipulating rotifer feeding. The main conclusion of this long series of studies is that in *B. plicatilis* and *B. rotundiformis*, the composition of polyunsaturated fatty acids depends on the composition of these acids in food supplied to rotifers (Watanabe et al., 1983; Lubzens et al., 1985; Ben-Amotz et al., 1987; Dendrinos & Thorpe, 1987; Whyte & Nagata, 1990; Frolov et al., 1991; Olsen et al., 1993; Reitan, 1993; Rainuzzo et al., 1994). There are only a few studies on *B. calyciflorus* (Awais et al., 1992; Isik et al., 1999), but according to this last study (Isik et al., 1999), the fatty acid composition remains unaltered regardless of food composition. There is virtually no information on other rotifer species.

Contrary to the numerous studies on fatty acids, the determination of amino acid composition, also of primary nutritional importance, has received less attention although there are several studies on their composition (Ogino & Watanabe, 1978; Dendrinos & Thorpe, 1987; Frolov et al., 1991; Awais et al., 1992; Øie & Olsen, 1997). Likewise, there are few studies on the properties of enzymes of rotifers (Hara et al., 1984; Kühle & Kleinow, 1985; Kleinow, 1993; Díaz et al., 1997; Hara et al., 1997), another interesting aspect from the viewpoint of food for fish larvae.

## Morphotypes and speciation

The morphological variability in *Brachionus* species enables different subspecies to be described (Alhstrom, 1940; Koste, 1978; Suzuki, 1987). In addition, changes in appearance and length of spines induced by environmental conditions have been investigated in different *Brachionus* species (Gilbert 1967; Halbach 1970b). In the case of *B. plicatilis*, the changes in size and shape of the rotifers were investigated after they were detected in routine mass cultures as well as while searching for different sized prey for fish larva. Two varieties in relation to lorica length, large (L-type) and small (S-type), were reported in Japan during the 1970's (Oogami & Maeda, 1977). Fukusho (1980) reported cyclomorphosis in *B. plicatilis* mainly related to changes in temperature. Description of strains with different sizes and shapes suggested genetic dif-

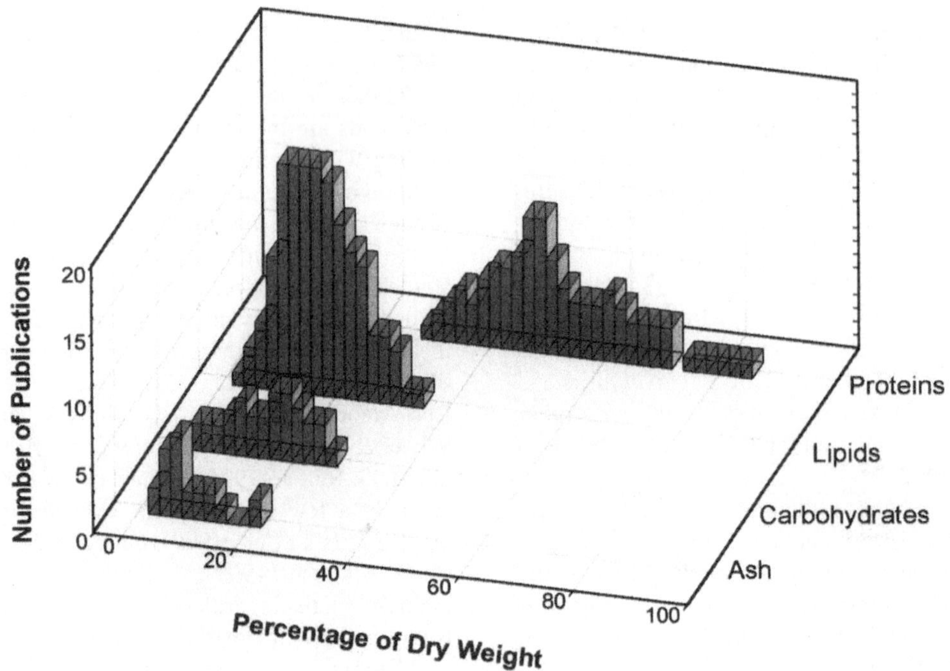


*Figure 4.* Frequency in which given values of protein, lipids, carbohydrates and ash has been reported in *B. plicatilis.*

ferences attributable to different organisms by Ito et al. (1981) and Yúfera (1982) working in laboratory strains, and by Fukusho & Okaushi (1983, 1984) and Serra & Miracle (1983) working on both laboratory and wild populations. Snell & Carrillo (1984) and Fu et al. (1991a) working with several strains observed clear biometric differences between the two morphotypes. Throughout the studies during the 1980s on production characteristics evidenced the physiological differences between L and S types, (Yúfera et al., 1983; Yúfera & Pascual, 1985; Yamasaki et al., 1987; Yúfera, 1987). Fu et al. (1991a, b) reported differences in chromosomes. Finally, Segers (1997) positively identified S-type as *B. rotundiformis,* and L-type as *B. plicatilis* according to the historical taxonomical descriptions in the literature. There are recent new records on *B. plicatilis* complex (Gómez & Serra, 1995; Hagiwara et al., 1995; Levedeva & Orlenko, 1997; Serra et al., 1998) and the number and range of size and morphotypes will probably continue to increase.

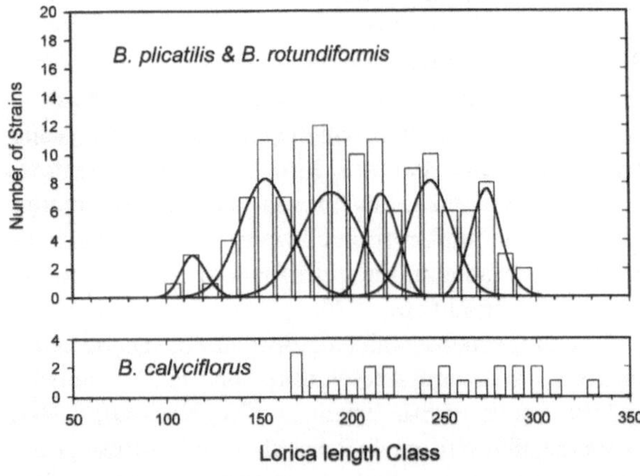


*Figure 5.* Size of the different *B. plicatilis* and *B. calyciflorus* strains and morphotypes reported in the literature. Lines indicate the separation of hypothetical size-morphotypes using the Bhattacharya method. Three peaks on the left correspond to *B. rotundiformis* and three peaks of the right to *B. plicatilis*.

A compilation of the information on lorica length of the different laboratory strains reported in the last 30 years shows that the former division by size into small and large *B. plicatilis* is unclear. In fact, multiple size-forms can be detected when plotting the number of strains whose adult-size average falls into each size class (Fig. 5). The lorica length without spines of the different strains in culture ranges from 100 $\mu$m to 300 $\mu$m approximately, although more than 400 $\mu$m has been reported in nature (Koste & Shiel, 1980; Levedeva & Orlenko, 1995). Using Bhattacharya's method for separating populations (Bhattacharya, 1967), it is possible to identify several groups distanced by approximately 20 $\mu$m in lorica length that could reflect the dominant morphotype-sizes in nature. Whether such a pattern is a consequence of polyploidy, multi-alellic expression or because of an ecological selection of more competitive body sizes remains an open question. Nevertheless, it is not possible to find a similar pattern in other species due to the scarcity of data. Only in *B. calyciflorus* has the lorica length been reported in a series of studies. These data show a similar body size variability for the different strains in culture and suggest that a similar situation could occur as in *B. plicatilis* complex.

## Concluding remarks

The three cases explained above show that fundamental research on the biology in the genus *Brachionus* has progressed alongside applied research. Without such technological interest, the level of our knowledge on this genus would be noticeably less, as in other rotifer genera. As field studies are relatively scarce, most of the information has been obtained in laboratory conditions, from only three species (*B. plicatilis*, *B. rotundiformis* and *B. calyciflorus*), and to a lesser extent from other *Brachionus* species (*B. rubens*, *B. variabilis*, *B. urceolaris*, *B. patulus*). This is because of the commercial applicability of the three first species and the relatively high number of researchers working in aquatic technology. Efforts to maintain other species under laboratory cultures will contribute in obtaining a more comprehensive knowledge on the biology of the genus *Brachionus*, although it is necessary to bear in mind the limitations of such studies.

Laboratory studies are simplified in that the culture deals with only one food species (plus an unknown bacteria load), the population is cloned from a single female, and natural predators and competitors are absent. Important gaps in our knowledge of the genus exist, because most studies have been carried out in species inhabiting brackish and saline waters. In addition, the phenotypic and genetic polymorphism observed in *B. plicatilis* and *B. rotundiformis* probably also occurs in other species. The size variability among strains as observed in *B. plicatilis* and *B. calyciflorus*, could be a general rule in the species belonging to *Brachionus*. This fact has some implications for the food availability threshold concept in relation to body biomass and its ecological and physiological significance for each species (Stemberger & Gilbert, 1985; Walz, 1995). Competition for optimal exploitation of resources in relation to food level in aquatic ecosystems would occur among different sized strains of the same species and among strains with similar size of different species. The use, however, of *Brachionus* as a research model in ecophysiological and genetic studies is on the increase and such studies will play an important role in rotifer research in the future.

## Acknowledgements

I would like to thank to Dr P. Drake for her helpful suggestions and the reviewers for the linguistic revision. The Comisión Interministerial de Ciencia y Tecnología, Spain (CICYT Project MAR97-0924-C02-01) supported this work.

## References

Ahlstrom, E. H., 1940. A revision of the rotatorian genera *Brachionus* and *Platyias* with descriptions of one new species and two new varieties. Bull. am. Mus. Nat. Hist. 77: 143–184.

Arndt, H., 1988. Dynamics and production of a natural population of *Brachionus plicatilis* (Rotatoria, Monogononta) in a eutrophicated inner coastal water of the Baltic. Kieler Meereforsch 6: 147–153.

Arndt, H. & R. Heerkloss, 1989. Diurnal variation in feeding and assimilation rates of planktonic rotifers and its possible ecological significance. Int. Rev. ges. Hydrobiol. 74: 261–272.

Awais, A., P. Kestemon & J. C. Micha, 1992. Nutritional suitability of the rotifer, *Brachionus calyciflorus* Pallas for rearing freshwater fish larvae. J. Appl. Ichthyol. 8: 263–270.

Ben-Amotz, A. & R. Fishler, 1982. Induction of sexual reproduction and resting egg production in *Brachionus plicatilis* by a diet of salt-grown *Nannochloris oculata*. Mar. Biol. 67: 289–298.

Ben-Amotz, A., R. Fishler & A. Schneller, 1987. Chemical composition of dietary species of marine unicellular algae and rotifers with emphasis on fatty acids. Mar. Biol. 95: 31–36.

Bhattacharya, C. G., 1967. A simple method of resolution of distribution into Gaussian components. Biometrics 97: 115–135

Boraas, M. E., 1983. Population dynamics of food-limited rotifers in two-stage chemostat culture. Limnol. Oceanogr. 28: 546–563.

Capuzzo, J. M., 1979. The effects of halogen toxicants on survival, feeding and egg production of the rotifer *Brachionus plicatilis*. Estuar. coast. mar. Sci. 8: 307–316.

Caric, M., Sannko-Njire & B. Skaramuca, 1993. Dietary effects of different feeds on the biochemical composition of the rotifer (*Brachionus plicatilis* Müller). Aquaculture 110: 141–150.

Charoy, C., 1995. Modification of the swimming behaviour of *Brachionus calyciflorus* (Pallas) according to food environment and individual nutritive state. Hydrobiologia 313/314: 197–204.

Charoy, C., C. R. Janssen, G. Persoone & P. Clément, 1995. The swimming behaviour of *Brachionus calyciflorus* (rotifer) under toxicity stress. I. The use of automated trajectometry for determining sublethal effects of chemicals. Aquat. Toxicol. 32: 271–282.

Chotiyaputta, C. & K. Hirayama, 1978. Food selectivity of the rotifer *Brachionus plicatilis* feeding on phytoplankton. Mar. Biol. 45: 105–111.

Clément, P., 1977. Ultrastructural research on rotifers. Arch. Hydrobiol. Beih. Ergebn. Limnol. 8: 270–297.

Clément, P., J. Amsellem, A.-M. Cornillac, A. Luciani & C. Ricci, 1980. An ultrastructural approach to feeding behaviour in *Philodina roseola* and *Brachionus calyciflorus* (rotifers) I. The buccal velum. Hydrobiologia 73: 127–131.

Dendrinos, P. & J. P. Thorpe, 1987. Experiments on the artificial regulation of the amino acid and fatty acid contents of food organisms to meet the assessed nutritional requirements of larval, post-larval and juvenile Dover sole (*Solea solea* (L.)). Aquaculture 61: 121–154.

Díaz, M., F. J. Moyano, F. L. García-Carreño, F. J. Alarcón & M. C. Sarasquete, 1997. Substrate-SDS–PAGE determination of protease activity through larval development in sea bream. Aquacult. Int. 5: 461–471.

Doohan, M., 1973. An energy budget for adults *Brachionus plicatilis* Müller (Rotatoria). Oecologia 13: 351–362.

Erman, L. A., 1962. Nutrition and multiplication of planktonic rotifer *Br. calyciflorus* Pall. in mass cultures. Dokl. Akad. Nauk USSR 144: 926–929.

Ferrando, M. D., C. R. Jansen, E. Andreu & G. Persoone, 1993. Ecotoxicological studies with the freshwater rotifer *Brachionus calyciflorus*. II. An assessment of the chronic toxicity of lindane and 3,4-dichloroaniline using life tables. Hydrobiologia 255/256: 33–40.

Frolov, A. V., S. L. Pankov, K. N. Geradze, S. A. Pankova & L. V. Spectorova, 1991. Influence of the biochemical composition of food on the biochemical composition of the rotifer *Brachionus plicatilis*. Aquaculture 97: 181–202.

Fu, Y., K. Hirayama & Y. Natsukari, 1991a. Morphological differences between two types of the rotifer *Brachionus plicatilis* O.F. Müller. J. exp. mar. Biol. Ecol. 151: 29–41.

Fu, Y., K. Hirayama & Y. Natsukari, 1991b. Genetic divergence between S and L type strains of the rotifer *Brachionus plicatilis* O.F. Müller. J. exp. mar. Biol. Ecol. 151: 43–56

Fukusho, K. & H. Iwamoto, 1980. Cyclomorphosis in size of the cultured rotifer *Brachionus plicatilis*. Bull. natl. Res. Inst. Aquacult. 1: 29–37.

Fukusho, K. & M. Okaushi, 1983. Sympatry in natural distribution of the two strains of a rotifer, *Brachionus plicatilis*. Bull. natl. Res. Inst. Aquacult. 4: 135–138.

Fukusho, K. & M. Okaushi, 1984. Seasonal isolation between two strains of rotifer *Brachionus plicatilis* in an eel-culture pond. Bull. Jap. Soc. Sci . Fish. 50: 909.

Galkovskaya, G. A., 1963. On the utilization of food for growth and conditions of highest yield in rotatoria *Brachionus calyciflorus*. Zool. Zh. 42: 506–512.

Galkovskaya, G. A., 1995. Oxygen consumption rate in rotifers. Hydrobiologia 313/314: 147–156.

Galkovskaya, G. A., J. Ejsmont-Karabin & V. N. Evdokimov, 1987. Relative protein metabolism in rotifer *Brachionus calyciflorus* Pallas, in relation to temperature. Int. Rev. ges. Hydrobiol. 72: 559–69.

Gatesoupe, F. J. & J. H. Robin, 1981. Commercial single-cell proteins either as sole source or in formulated diets for intensive and continuous production of rotifer *Brachionus plicatilis*. Aquaculture 25: 1–15.

Gilbert, J. J., 1963. Mictic female production in the rotifer *Brachionus calyciflorus*. J. exp. Zool. 153: 113–124.

Gilbert, J. J., 1967. *Asplachna* and posterolateral spine production in *Brachionus calyciflorus*. Arch. Hydrobiol. 64: 1–62.

Gilbert, J. J., 1974. Dormancy in rotifers. Trans. am. Microsc. Soc. 93: 490–513.

Gilbert, J. J. & P. L. Starkweather, 1977. Feeding in the rotifer *Brachionus calyciflorus*. I. Regulatury mechanisms. Oecologia 28: 125–131.

Gómez, A. & M. Serra, 1995. Behavioural reproductive isolation among sympatric strains of *Brachionus plicatilis* Müller, 1786: insights into status of this taxonomic species. Hydrobiologia 313/314: 111–119

Guisande, C. & J. Toja, 1988. The dynamic of various species of the genus *Brachionus* (Rotatoria) in the Guadalquivir river. Arch. Hydrobiol. 112: 579–595.

Guisande, C. & L. Serrano, 1989. Analysis of protein, carbohydrate and lipid in rotifer. Hydrobiologia 186/187: 339–346.

Hagiwara, A., 1994. Practical use of rotifer cysts. The Israeli Journal of Aquaculture-Bamidgeh 46: 136–21.

Hagiwara, A., M. D. Balompapueng, N. Munuswamy & K. Hirayama, 1997. Mass production and preservation of the resting eggs of the euryhaline rotifer *Brachionus plicatilis* and *B. rotundiformis*. Aquaculture 155: 223–230.

Hagiwara, A., T. Kotani, T. W. Snell, M. Assava-Aree & K. Hirayama, 1995. Morphology, reproduction, genetics, and mating behaviour of small, tropical marine *Brachionus* strains (Rotifera). J. exp. mar. Biol. Ecol. 194: 25–37.

Halbach, U., 1970a. Einfluss der Temperatur auf die Population-dynamik des planktischen Rädertieres *Brachionus calyciflorus* Pallas (Rotatoria). Oecologia 4: 176–207.

Halbach, U., 1970b. Die Ursachen der Temporalvariation von *Brachionus calyciflorus* Pallas (Rotatoria). Oecologia 4: 262–318.

Halbach, U., M. Sievert, M. Westermayer & C. Wiessel, 1983. Population ecology of rotifers as a bioassay tool for ecotoxicological test in aquatic environments. Ecotoxicol. envir. Safety 1: 484–513.

Hansen, B., T. Wernberg-Moller & L. Wittrup, 1997. Particle grazing efficiency and specific growth efficiency of the rotifer *Brachionus plicatilis*. J. exp. mar. Biol. Ecol. 215: 217–233.

Hara, K., H. Harano & T. Ishihara, 1984. Some enzymatic properties of alkaline proteases of the rotifer Brachionus plicatilis. Bull. Japan Soc. Sci. Fish. 50: 1611–1616.

Hara, K., H. Pangkey, K. Osatomi, K. Yatsuda, A, Hagiwara, K. Tachibana & T. Ishihara, 1997. Some properties of $\beta$-1,3-glucan hydrolyzing enzymes from *Brachionus plicatilis*. Hydrobiologia 358: 89–94.

Hirata, H., 1974. An attempt to apply an experimental microcosm for the mass culture of marine rotifer, *Brachionus plicatilis* Müller. Mem. Fac. Fish., Kagoshima Univ. 32: 163–172.

Hirata, H. & Y. Mori, 1967. Culture of the rotifer *Brachionus plicatilis* fed on baker's yeast. Saibai-Gyogyo 5: 36–40.

Hirayama, K. & S. Okawa, 1972. Fundamental studies on physiology of the rotifer for its mass culture. I. Filter feeding of rotifer. Bull. Japan Soc. Sci. Fish. 38: 1207–1214.

Hirayama, K. & T. Kusano, 1972. Fundamental studies on physiology of the rotifer for its mass culture. II. Influence of water temperature on population growth of rotifer. Bull. Japan Soc. Sci. Fish. 38: 1357–1363.

Hirayama, K., K. Watanabe & T. Kusano, 1973. Fundamental studies on physiology of the rotifer for its mass culture. III. Influence of phytoplankton density on population growth. Bull. Japan Soc. Sci. Fish. 39: 1123–1127.

Isik, O., E. Sarihan, E. Kusvuran, O. Gül & O. Erbatur, 1999. Comparison of the fatty acid composition of the freshwater fish larvae *Tilapia zillii*, the rotifer *Brachionus calyciflorus*, and the microalgae *Scenedesmus abundans*, *Monoraphidium minutum* and *Chlorella vulgaris* in the algae-rotifer-fish larvae chains. Aquaculture 174: 299–311.

Ito, T., 1955. Studies on the 'Mizukawari' in eel-culture ponds. 1. The feeding activity of *Brachionus plicatilis* on phytonanoplankton (as a cause of 'Mizukawari'). Rep. Fac. Fish. Prefect. Univ. Mie 2: 162–167.

Ito, T., 1960. On the culture of the mixohaline rotifer *Brachionus plicatilis* O.F. Müller in the sea water. Rep. Fac. Fish. Prefect. Univ. Mie 3: 708–740.

Ito, S., H. Sakamoto, M. Hori & K. Hirayama, 1981. Morphological characteristic and suitable temperature for growth of several strain of the rotifer *Brachionus plicatilis*. Bull. Fac. Fish. Nagasaki Univ. 51: 9–16.

James, C. M. & T. Abu Rezeq, 1989. Intensive rotifer cultures using chemostats. Hydrobiologia 186/187: 423–430.

Jansen, C. R., M. D. Ferrando & G. Persoone, 1993. Ecotoxicological studies with the freshwater rotifer *Brachionus calyciflorus*. I. Conceptual framework and applications. Hydrobiologia 255/256: 21–32.

King, C. E. & Y. Zhao, 1987. Coexistence of rotifer (*Brachionus plicatilis*) clones in Soda Lake, Nevada. Hydrobiologia 147: 57–64.

Kleinow, W., 1993. Biochemical studies on *Brachionus plicatilis*: hydrolytic enzymes, integument proteins and composition of trophi. Hydrobiologia 255/256: 1–12.

Kleinow, W. & H. Wratil, 1995. SEM of internal structures of *Brachionus plicatilis* (Rotifera). Hydrobiologia 313: 129–132.

Koste, W., 1978. Rotatoria. Borntraeger, Berlin, 2 vols: 673 pp., 234 plates.

Koste, W. & R. J. Shiel, 1980. Preliminary remarks on the characteristics of the rotifer fauna of Australlia (Notogaea). Hydrobiologia 73: 221–227.

Kühle, K. & W. Kleinow, 1985. Measurements of hydrolytic enzymes in homogenates from *Brachionus plicatilis* (Rotifera). Comp. Biochem. Physiol. 81B: 437–442.

Lebedeva, L. I. & O. N. Orlenko, 1995. Feeding rate of *Brachionus plicatilis* O.F. Müller on two types of food depending on ambient temperature and salinity. Int. Rev. Ges. Hydrobiol. 80: 77–87.

Lebedeva, L. I. & O. N. Orlenko, 1997. Morphometry of *Brachionus plicatilis* (Rotifera, Brachionidae) from the Caspian Sea region. Zool. Zh. 76: 771–776.

Lubzens, E., 1981. Rotifer resting eggs and their application to marine aquaculture. Europ. Maricult. Soc., Spec. Publ. 6: 163–179.

Lubzens, E., A. Marko & A. Tietz, 1985. De novo syntesis of fatty acids in the rotifer *Brachionus plicatilis*. Aquaculture 47: 27–37.

Lubzens, E., Y. Wax, G. Minkoff & F. Adler, 1993. A model evaluating the contribution of environmental factors to the production of resting eggs in the rotifer *Brachionus plicatilis*. Hydrobiologia 255/256: 127–138.

Makridis, P. & Y. Olsen, 1999. Protein depletion of the rotifer *Brachionus plicatilis* during starvation, Aquaculture 174: 343–353.

Mityanina, I. F., 1985. Glycogen content in the rotifer *Brachionus calyciflorus* Pallas. Vesti AN BSSR, ser. Bijal. Navuk 1: 109–111.

Moreno-Garrido, I., L. M. Lubián & A. M. V. M. Soares, 1999. Growth differences in cultured populations of *Brachionus plicatilis* Müller caused by heavy metal stress as function of microalgal diet. Bull. Environ. Cont. Toxicol. 63: 392–398.

Nagata, W., 1985a. Long-term acclimation of a parthenogenetic strain of *Brachionus plicatilis* Müller to subnormal temperatures I. Influence on size, growth and reproduction. Bull. mar. Sci. 37: 716–725.

Nagata, W., 1985b. Long-term acclimation of a partenogenetic strain of *Brachionus plicatilis* Müller to subnormal temperatures II. Effect on clearance and ingestion rates. Bull. Fac. Fish. Hokkaido Univ. 36: 1–11.

Ogino, C. & T. Watanabe, 1978. Nutritive value of proteins contained in activated sludge, photosynthetic bacteria and marine rotifer for fish. J. Tokio Univ. Fish. 64: 101–108.

Øie, G. & Y. Olsen, 1997. Protein and lipid content of the rotifer *Brachionus plicatilis* during variable growth and feeding conditions. Hydrobiologia 358: 251–258.

Olsen, Y., J. R. Rainuzzo, K. I. Reitan, & O. Vadstein, 1993. Manipulation of lipids and n-3 fatty acids in *Brachionus plicatilis*. In: Reinertsen, H., L. A. Dahle, L. Jorgensen & K. Tvinnereim (eds), Proceedings of the First International Conferenceon Fish Farming Technology. Trondheim, Norway, 9–12 August 1993, A.A. Balkema, Rotterdam: 101–108.

Oogami, H. & Y. Maeda, 1977. Studies on variation of the rotifer *Brachionus plicatilis*-I. Size and shape. Proc. Japan Soc. Sci. Fish. 1977: 47 pp.

Park, H. G., K. W. Lee, S. H. Cho, H. S. Kim & M.-M. Jung, 2001. High density culture of the freshwater rotifer, *Brachionus calyciflorus*. Hydrobiologia, 446/447 (Dev. Hydrobiol. 153): 369–374.

Pejler, B., 1977. On the global distribution of the family Brachionidae (Rotifera). Arch. Hydrobiol. Suppl. 53: 255–307.

Pilarska, J., 1977a. Eco-physiological studies on *Brachionus rubens* Ehrbg (Rotatoria). I. Food selectivity and feeding rate. Pol. Arch. Hydrobiol. 24: 319–328.

Pilarska, J., 1977a. Eco-physiological studies on *Brachionus rubens* Ehrbg (Rotatoria). II. Production and respiration. Pol. Arch. Hydrobiol. 24: 329–341.

Pourriot, R., 1957. Influence de la nourriture sur l'apparition des femelles mictiques, chez deux espèces et une variété de *Brachionus* (Rotifères). Hydrobiologia 9: 60–65.

Pourriot, R., 1965. Recherches sur l'écologie des rotifères. Vie et Milie, Suppl. 21: 1–224.

Pourriot, R. & C. Rougier, 1975. Dynamic d'une population experimentale de *Brachionus dimidiatus* (Bryce) (Rotifere) en fonction de la nourriture et de la temperature. Ann. Limnol. 11: 125–143.

Pourriot, R. & T. W. Snell, 1983. Resting eggs in rotifers. Hydrobiologia 104: 213–224.

Pozuelo, M., 1977. Male production in seawater cultured *Brachionus plicatilis*. Arch. Hydrobiol. Beih. 8: 172–173.

Rainuzzo, J. R., K. I. Reitan & Y. Olsen, 1994. Effect of short- and long-term lipid enrichment on total lipids, lipid class and fatty acid composition in rotifers. Aquacult. Int. 2: 19–32.

Rao, T. R. & S. S. S. Sarma, 1986. Demographic parameters of *Brachionus patulus*, Rotifera, exposed to sublethal DDT concentrations at low and high food levels. Hydrobiologia 139: 193–200.

Rico-Martínez, R. & T. W. Snell, 1995. Male discrimination of female *Brachionus plicatilis* Müller and *Brachionus rotundiformis* Tschugunoff (Rotifera). J. exp. mar. Biol. Ecol. 190: 39–49.

Roche, K. F., 1995. Growth of the rotifer *Brachionus calyciflorus* Pallas in dairy waste stabilization ponds. Wat. Res. 29: 2255–2260.

Rothhaupt, K. O., 1990. Changes of the functional responses of the rotifers *Brachionus rubens* and *Brachionus calyciflorus* with particle size. Limnol. Oceanogr. 35: 24–32.

Ruttner-Kolisko, A., 1964. Über die labile Periode im Fortpflnzungszyklus der Rädertiere. Int. Rev. ges. Hydrobiol. 49: 473–482.

Schlosser, H. J. & K. Anger, 1982. The significance of some methodological effects on filtration and ingestion rates of the rotifer *Brachionus plicatilis*. Helgoländer Meeresunters. 35: 215–225.

Schlüter, M., C. J. Soeder & J. Groeneweg, 1987. Growth and food conversion of *Brachionus rubens* in continuous culture. J. Plankton Res. 9: 761–783.

Scott, A. P. & S. M. Baynes, 1978. Effect of algal diet and temperature on the biochemical composition of the rotifer *Brachionus plicatilis*. Aquaculture 14: 247–260.

Scott, J. M., 1980. Effect of growth rate of the food alga on the growth/ingestion efficiency of a marine herbivore. J. mar. biol. Ass. U.K. 60: 681–702.

Segers, H., 1997. Nomenclatural consequences of some recent studies on *Brachionus plicatilis* (Rotifera, Brachionidae). Hydrobiologia 313/314: 121–122.

Serra, M. & M. R. Miracle, 1983. Biometric analysis of *Brachionus plicatilis* ecotypes from Spanish lagoons. Hydrobiologia 104: 279–291.

Serra, M., A. Gómez & M. J. Carmona, 1998. Ecological genetics of *Brachionus* sympatric species. Hydrobiologia 387/388: 373–384.

Snell, T. W., 1986. Effect of temperature, salinity and food level on sexual and asexual reproduction in *Brachionus plicatilis* (Rotifera). Mar. Biol. 92: 157–162.

Snell, T. W. & K. Carrillo, 1984. Body size variation among strains of the rotifer *Brachionus plicatilis*. Aquaculture 37: 359–367.

Snell, T. W. & C. R. Jansen, 1995. Rotifers on ecotoxiclogy. A review. Hydrobiologia 313/314: 231–247.

Starkweather, P. L., J. J. Gilbert & T. M. Frost, 1979. Bacterial feeding by the rotifer *Brachionus calyciflorus*: clearance and ingestion rates, behaviour and population dynamics. Oecologia 44: 26–30.

Stemberger, R. S. & J. J. Gilbert, 1985. Body size, food concentration and population growth in planktonic rotifers. Ecology 66: 1151–1159.

Suantika, G., P. Dhert, M. Nurhudah & P. Sorgeloos, 2000. High-density production of the rotifer *Brachionus plicatilis* in a recirculation system: consideration of water quality, zootechnical and nutritional aspects. Aquacult. Eng. 21: 201–214.

Sudzuki, M., 1987. Intraspecific variation of *Brachionus plicatilis*. Hydrobiologia 147: 45–47.

Szyper, J. P., 1989. Nutritional depletion of the aquaculture feed organisms *Euterpina acutifrons*, *Artemia* sp. and *Brachionus plicatilis* during starvation. J. Word Aquacult. Soc. 20: 162–169.

Theilacker, G. H. & M. F. McMaster, 1971. Mass culture of the rotifer *Brachionus plicatilis* and its evaluation as food for larval anchovies. Mar. Biol. 10: 183–188.

Theilacker, G. H. & A. S. Kimball, 1984. Comparative quality of rotifers and copepods as food for larval fishes. Calif. Coop. Oceanic Fish. Invest. Rep. 15: 80–86.

Vasilieva, G. L., 1968. Rearing of *Brachionus rubens* Ehrbg. as feed for fish larvae. Some data on the species biology. Hidrobiol. Zh. 4: 39–45.

Walz, N., 1995. Rotifer populations in plankton communities: energetics and life history strategies. Experientia 51: 437–453.

Walz, N. & T. Gschloessl, 1988. Functional response of ingestion and filtration rate of the rotifer *Brachionus angularis* to the food concentration. Verh. Internat. Verein. Limnol. 23: 1993–2000.

Walz, N., T. Gschloessl & U. Hartmann, 1989. Temperature aspects of ecological bioenergetics in *Brachionus angularis* (Rotatoria). Hydrobiologia 186/187: 363–369.

Watanabe, T., C. Kitajima & S. Fujita, 1983. Nutritional values of live organisms used in Japan for mass propagation of fish. Aquaculture 34: 115–143.

White, J. M. C. & W. D. Nagata, 1990. Carbohydrate and fatty acid composition of rotifer, *Brachionus plicatilis*, fed monospecific diets of yeast or phytoplankton. Aquaculture 89: 263–272.

White, J. M. C., W. C. Clarke, N. G. Ginther, J. O. T. Jensen & L. D. Townsend, 1994. Influence of composition of *Brachionus plicatilis* and *Artemia* on growth of larval sable fish (*Anoplopoma fimbria* Pallas). Aquaculture 119: 47–61.

Yamasaki, S., D. H. Secor & H. Hirata, 1987. Population growth of two types of rotifer (L and S) *Brachionus plicatilis* at different dissolved oxygen levels. Nippon Suisan Gakkaishi 53: 1303.

Yoshimura, K., A. Hagiwara, T. Yoshimatsu & T. Kitajima, 1996. Culture technology marine rotifers and the implication for intensive culture of marine fish in Japan. Mar. Freshwat. Res. 47: 217–222.

Yu, J. P. & S.-J. Cui, 1997. Ultrastructure of the rotifer *Brachionus plicatilis*. Hydrobiologia 358: 95–103.

Yúfera, M., 1982. Morphometric characterisation of a small-sized strain of *Brachionus plicatilis* in culture. Aquaculture 27: 55–61.

Yúfera, M., 1987. Effect of algal diet and temperature on the embryonic development time of the rotifer *Brachionus plicatilis* in culture. Hydrobiologia 147: 319–322.

Yúfera, M. & E. Pascual, 1980. Estudio del rendimiento de cultivos masivos del rotífero *Brachionus plicatilis* alimentados con levadura de panificación. Inv. Pesq. 44: 55–61.

Yúfera, M. & E. Pascual, 1985. Effect of algal food concentration on feeding and ingestion rates of *Brachionus plicatilis* in mass culture. Hydrobiologia 122: 181–187.

Yúfera, M. & E. Pascual, 1989. Biomass and elemental composition (C.N.H.) of the rotifer *Brachionus plicatilis* cultured as larval food . Hydrobiologia 186/187: 371–374.

Yúfera, M., L. M. Lubián & E. Pascual, 1983. Efecto de cuatro algas marinas sobre el crecimiento poblacional de dos cepas de *Brachionus plicatilis* (Rotifera: Brachionidae) en cultivo. Inv. Pesq. 47: 325–337.

Yúfera, M., G. Parra & E. Pascual, 1997. Energy content of the rotifers *Brachionus plicatilis* and *Brachionus rotundiformis* in relation to temperature. Hydrobiologia 358: 83–87.

*Hydrobiologia* **446/447:** 393–395, 2001.
*L. Sanoamuang, H. Segers, R.J. Shiel & R.D. Gulati (eds), Rotifera IX.*


# Subject index

Page numbers refer to the first page of a paper in which the entry is employed